Laser in der Technik
Laser in Engineering

Vorträge des 11. Internationalen Kongresses
Proceedings of the 11th International Congress

Laser 93

Herausgegeben von/Edited by
Wilhelm Waidelich

Mit 776 Abbildungen/With 776 Figures

Springer-Verlag Berlin Heidelberg GmbH

Dr. rer. nat. Wilhelm Waidelich

Universitätsprofessor, em. Vorstand des Instituts
für Medizinische Optik der Universität München, em. Direktor
des Instituts für Angewandte Optik der Gesellschaft
für Strahlen- und Umweltforschung , Neuherberg

ISBN 978-3-540-57444-6

Die Deutsche Bibliothek - CIP-Einheitsaufnahme
Laser in der Technik: Vorträge des 11.Internationalen Kongresses Laser 93 =

ISBN 978-3-540-57444-6   ISBN 978-3-662-08251-5 (eBook)
DOI 10.1007/978-3-662-08251-5

Vorwort

Im Juni 1993 fand in München die 11. Internationale Fachmesse mit Kongressen LASER 93 statt. Der fortschreitenden Spezialisierung der Lasertechnologie trägt die Messe Müchen mit einer Aufteilung in nach Anwendungsschwerpunkten strukturierte Fachkongresse Rechnung.

Ziele dieser Kongresse sind Informationstransfer für potentielle Anwender der neuen Lasertechnologien, kritische Diskussion neuer Forschungs- und Entwicklungsergebnisse sowie eine Vorausschau auf zukünftige Trends.

Den Kongressen Laser in der Umwelttechnik und Laser in der Medizin ist je ein eigener Band gewidmet.Im vorliegenden Band Laser in der Technik sind die Vorträge folgender Kongresse zusammengestellt:

MODERNE FESTKÖRPERLASER: Grundlagen und technisch-industrielle Anwendungen - Congress-Chairman: B.Wallenstein

OPTISCHE MESSTECHNIK - Congress-Chairmen: H.-J.Tiziani, K. Biedermann

LASER IN DER FERTIGUNG - Congress-Chairmen: G.Herziger, E.Kreutz, H.Weber

OPTOELEKTRONISCHE KOMPONENTEN UND SYSTEME - Congress-Chairman: F.Lanzl

MIKROSENSORIK UND FASEROPTIK - Congress-Chairman: E.Wagner

OPTISCHE KOMMUNIKATIONSTECHNIK - Congress-Chairman: J.Franz

LASER IN DER FORSCHUNG - Congress-Chairmen: F.P. Schäfer, M. Stuke

POST DEADLINE PAPERS: Laser Physik - Congress-Chairman: H. Seidlitz

Der reichhaltige Inhalt der in diesem Band zusammengestellten Kongresse ist eine Fundgrube für die vielseitigen Möglichkeiten der neuen Lasertechnologien. Dieser aktuelle Schatz wird mit der Publiikation der Vorträge allen Interessenten zugänglich gemacht.

Für das Erscheinen dieses Kongressbandes gilt allen Autoren, den Congress-Chairmen, der Messe München und dem Springer-Verlag besonderer Dank.

München, Oktober 1993   Wilhelm Waidelich

Preface

LASER 93, the 11th International Trade Fair and International Congress, was held at the Munich Trade Fair Center in June 1993.

To reflect the ongoing specialization in the development and application of laser technolgy, the Munich Trade Fair Corporation organized a series of technical congresses according to major application sectors. The objective of these congress events is to transfer information to potential users of new laser technologies, promote critical discussions of the latest research and development findings and examine future development trends.

The congresses Laser in Environmental Technology and Lasers in Medicine are covered in separate volumes. This volume, Lasers in Engineering, features summaries of presentations made during the following congress events:

MODERN SOLID-STATE LASERS: Basic Principles and Technical/Industrial Applications
Congress-Chairman: B. Wallenstein

OPTICAL MEASURING - Congress-Chairmen: H.J. Tiziani, K. Biedermann

LASER IN PRODUCTION - Congress-Chairmen: G. Herziger, E.W. Kreutz, H. Weber

OPTOELECTRONIC COMPONENTS AND SYSTEMS - Congress-Chairman: F. Lanzl

MICRO-SENSORS AND FIBER OPTICS - Congress-Chairman: E. Wagner

OPTICAL COMMUNICATIONS TECHNOLOGY - Congress-Chairman: J. Franz

LASERS IN RESEARCH - Congress-Chairmen: F.P. Schäfer, M. Stuke

POST DEADLINE PAPERS: Laser Physics - Congress-Chairman: H. Seidlitz

The wealth of information contained in the lectures and summarized in this volume is a treasury of the diverse range of potential applications of state-of the-art laser technologies.Publication of these lectures is intended to make this wealth of up-to-date information available to all interested parties.

We would like to thank the authors, the Congress-Chairmen, the Munich Trade Fair Cooperation and the Springer-Verlag for making this publication possible.

Munich, October 1993   Wilhelm Waidelich

Inhaltsverzeichnis - Contents

MODERNE FESTKÖRPERLASER:
Grundlagen und technisch-industrielle Anwendungen
MODERN SOLID-STATE LASERS:
Basic Principles and Technical/Industrial Applications

Chairman: R. Wallenstein, Universität Kaiserslautern

X

OPTISCHE MESSTECHNIK
OPTICAL MEASURING

Chairmen:
H.-J. Tiziani, Institut für Technische Optik, Univ. Stuttgart
K. Biedermann, K.T. Högskolan, Stockholm/S

LASER IN DER FERTIGUNG
LASER IN PRODUCTION

Chairmen:
G. Herziger, Fraunhofer Institut für Lasertechnik, Aachen
H. Weber, Festkörper-Laser-Institut, Berlin
E.W. Kreutz, Fraunhofer Institut für Lasertechnik, Aachen

PLENARVORTRÄGE I / PLENARY SESSION I

ANWENDERFORUM: QUALITÄTSSICHERUNG IN DER LASERMATERIALBEARBEITUNG
USER'S FORUM: QUALITY ASSURANCE IN MATERIAL PROCESSING

VARIA / VARIOUS

OPTOELEKTRONISCHE KOMPONENTEN UND SYSTEME
OPTOELECTRONIC COMPONENTS AND SYSTEMS

Chairman: F. Lanzl
DLR, Institut für Optoelektronik, Oberpfaffenhofen

MICROSENSORIK UND FASEROPTIK
MICRO-SENSORS AND FIBER OPTICS

Chairman:
E. Wagner, Fraunhofer-Institut für Physikalische Messtechnik, Freiburg

OPTISCHE KOMMUNIKATIONSTECHNIK
OPTICAL COMMUNICATIONS TECHNOLOGY

Chairman: J. Franz, FH Düsseldorf, Fachgebiet opt. Nachrichtentechnik

LASER IN DER FORSCHUNG
LASERS IN RESEARCH

Chairmen:
F.P. Schäfer und M. Stuke
Max-Planck-Institut für biophysikalische Chemie, Göttingen

Experimental Study of 248nm and 308 nm Ablation in Dependence on Optical
Illumination Parameters
B. Burghardt, U. Sarbach, B. Klimt, H.-J. Kahlert/D 809

Change of the Ablation Rates with Ablation Structure Size
B. Wolff-Rottke, J. Ihlemann, H. Schmidt, A. Scholl/D 816

Laser Induced Etching of Silicon with Fluorine and Chlorine and with
Mixtures of Both Gases
U. Köhler, A. Guber, W. Bier/D 820

PLD of Thin Films for Applications
A. Voss, W. Pfleging, M. Alunovic, E.W. Kreutz/D 823

Tunable Subpicosecond Light Pulses in the Mid Infrared Produced by
Difference Frequency Generation
C. Lauterwasser, P. Hamm, M. Zurek, W. Zinth/D 828

Self-Phase Modulation of Ultrashort Light Pulses in the Second-Harmonic
Nonlinear Mirror
K.A. Stankov/D, V.P. Tzolov/BG 832

Generation and Amplification of Subpicosecond Pulses in the VUV
- Towards Soft X-Rays with Nonlinear Optics
B. Wellegehausen/D 838

Parametric Picosecond Laser System with Very Broad Tunability for
Nonlinear Optical Interface Spectroscopy
H.-J. Krause, U. Reichel, W. Daum/D 841

Ultrashort Laser Pulses in Surface Mass Spectrometry
M. Schütze, C. Trappe, M. Raff, H. Kurz/D 845

Second Harmonic Generation on Chemically Treated Surfaces of Vicinal
Si<111>-Wafers
U. Emmerichs, C. Meyer, K. Leo, H. Kurz/D, C.H. Bjorkman,
C.E. Shearon Jr., Y. Ma, T. Yasuda, G. Lucovsky/USA 849

Application of an Intracavity Laser Spectroscopy Based on Ar$^+$ Laser
for the Detection of Unstable Molecules in Liquid Phase
S.A. Mulenko, A.I. Chaus/UKRAINE 853

Deposition of Diamond-Like-Carbon Films by Laser PVD Technique Using
Nanosecond and Femtosecond Excimer Lasers
K. Mann, F. Müller/D 857

Ein Modell für die Optimierung der PLD / A Model for Optimizing Laser PVD
G. Granse, H. Mai, B. Schultrich, S. Völlmar/D 861

POST DEADLINE PAPERS
LASERPHYSIK / LASER PHYSICS

Chairman:
H. Seidlitz, Forschungszentrum für Umwelt und Gesundheit, Neuherberg

Sitzungsleiter - Session Chairmen

Biedermann, K.	Meß- und Prüftechnik / Measuring and Testing
Dumbs, A.	Sensorik u. Faseroptik / Sensors and Fiber Optics
Fogarassy, E.	Forschung / Research
Franz, J.	Kommunikationstechnik / Communications Technology
Geiger, M.	Fertigung / Production
Herziger, G.	Fertigung / Production
Jüptner, H.	Meß- und Prüftechnik / Measuring and Testing
Kompa, K.	Forschung / Research
Kreutz, E.W.	Fertigung / Production
Lanzl, F.	Componenten u. Systeme / Components and Systems
Marowsky, G.	Forschung / Research
Milberg, J.	Fertigung / Production
Schäfer, F.P.	Forschung / Research
Seidlitz, H.	Laserphysik / Laser Physics
Siegbahn, K.	Forschung / Research
Stuke, M.	Forschung / Research
Tiziani, H.-J.	Meß- und Prüftechnik / Measuring and Testing
Treusch, H.G.	Fertigung / Production
Tschudi, T.	Meß- und Prüftechnik / Measuring and Testing
Wagner, E.	Sensorik u. Faseroptik / Sensors and Fiber Optics
Waidelich, W.	Kongresskoordinator / General Chairman
Wallenstein, R.	Festkörperlaser / Solid State Lasers
Weber, H.	Fertigung / Production

Referenten - Contributors

Congress A

Congress-Chairman: B. Wallenstein

Moderne Festkörperlaser: Grundlagen und technisch-
industrielle Anwendungen
Modern Solid-State Lasers: Basic Principles and Technical/
Industrial Applications

Properties and Applications of High Power Diode Laser Oscillator-Amplifier Systems

K.-J. Boller

Universität Kaiserslautern, Fachbereich Physik
Erwin-Schrödinger-Str. W-6750 Kaiserslautern, Germany

Abstract

We discuss the spatial and spectral properties of high power diode laser arrays by injection locking. Compared to free running diode lasers, the injection locking increases the brightness of the output beam by orders of magnitudes. Due to this substantial improvement, these systems are attractive laser sources for many applications. This is demonstrated by the use as a pump source for a monolithic Nd:YAG ring laser and for a laser made of the non-linear material NYAB.

The spectral and spatial beam quality of high power diode lasers can be substantially improved by injection locking or comparative designs, as master oscillator power amplifier systems.[1,2] This approach, which is a promising alternative to external cavity designs [3,4] and, so far, also to resonant optical antiguide systems[5], was first demonstrated in the mid eighties[6] and has since been under investigation[7]. During the past few years master-oscillator-power-amplifier systems (MOPA) were developed which emit a cw output power in the range of several Watts, and of ~20 Watts of peak power in pulsed operation[8]. A first version (~1W cw) with a narrowband master laser and a power amplifier on a single chip (MMOPA) has already become commercially available this year at 980nm[9].

The working principle of MOPAs is the same as with injection locking of arrays and broad area diodes, only, that an amplifier is AR-coated to a higher degree. Such amplifiers, on the other hand, require an input power of several hundreds of mW for successful depletion of the gain. Although injection locking in a discrete element setup may be vieved the precedetor of MOPAs and MMOPAs, it is an extremely flexible technique and a powerful method for exploring numerous possible applications of high brightness beams from diode lasers.

In this paper we demonstrate the superiour capabilities of high brightness laser diodes when applied as a pump source for solid state lasers, at the example of two specially designed Nd-doped lasers.

For injection locking, the master laser, a low power single-stripe diode laser of high spectral and spatial purity, is focused into the emitter facet of a high power diode laser, termed the slave laser, a multiple-stripe diode laser array or a broad area diode with reduced reflectivity of its facets. When the master laser is tuned to proper super-modes of the slave laser array, locking generates a single, near diffraction limited beam which contains a major fraction of the total power of the free running array.

An injection locked diode array, operated with a stabilized drive current, was used to pump a monolithic Nd:YAG ring laser[10]. Injection locking, as compared to the free running array, led to an improvement of the slope efficiency from 23% to 44% (see fig.1), with a reduction of the threshold pump power from 150mW to 80mW.

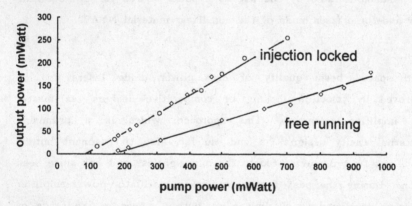

Fig.1: Improvement of a MISERs output power by injection locked pumping

The explanation of these effects is simple: a diffraction limited pump beam can be focused to a complete spatial overlap with the MISER's mode volume, even when thermal lensing may lead to a small beam waist of the mode; the narrow bandwidth of the pump radiation facilitates a spectrally complete overlap with the Nd absorption lines.

We further observed that an injection locked pump beam provides the maximum slope efficiency that can be obtained with a specific MISER design. The efficiency with a narrowband TEM_{00} Ti:Sapphire laser which can be perfectly matched to the MISER mode, is the same as with injection locked diode pumping.

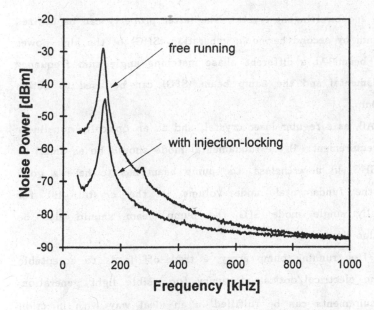

Fig.2: Reduction of a MISERs intensity noise by injection locked diode pumping

Injection locking can also improve the noise characteristics of the MISER. Fig.2 (upper trace) shows the intensity noise spectrum, when pumping with a free running array. The noise is substantially reduced when the pump array is injection locked (lower trace).

The explanation for the above observation is that the pump radiation from a free running array is spectrally distributed over several nanometers and subject to mode competition noise. Since the pump absorption line of the Nd-ions is much narrower ($\sim$1cm^{-1}), the pump rate of the Nd-laser fluctuates and causes intensity noise at the MISERs relaxation oscillation frequency, although the arrays total pump power is stable. An injection locked pump beam, on the other hand, carries the total power in a narrow frequency band, well within the width of the pump absorption line, so that pump fluctuations are much reduced.

Thus, the method of injection locking of the pump array may be an efficient tool to first passively reduce the noise in diode array pumped solid state lasers, before an active intensity or frequency stabilization may be applied.

In a second application, the injection locked source was used to pump a laser made of 3%-doped Nd:Yttrium-Aluminium-Borate (NYAB)[11]. In this material the Nd-ions can be diode pumped at 803nm in order to produce gain at 1062nm, while the host material is a non-linear and birefringent crystal. The latter property can be utilized to produce light at 531nm by second harmonic generation (SHG) of the high power intracavity fundamental beam. At a different phase matching angle sum frequency generation of the fundamental and the pump beam (SFG) can be used to obtain tunable emission at 460nm.

The twofold use of NYAB, as a regular laser crystal, and as an optically non-linear crystal, imposes some requirements: i) the fundamental mode should be as small as possible for efficient SHG; ii) nevertheless, the pump beam has to have a good spatial overlap with the fundamental mode volume in the crystal; iii) for narrowband and spatially single mode SFG the pump beam should also be narrowband and diffraction limited.

We show that, with a free running pump array, a trade-off leads to acceptable results in terms of the electrical/optical efficiency for visible light generation. It is clear that all requirements can be fulfilled in an ideal way with injection locked pumping. So far, we obtained, with 500mW pump power, 40mW at 531nm and ~0.2mW at 460nm, while a free running array shows a much lower efficiency, decreasing with the length of its emitter. We verified the optimum operation of the diode pumped NYAB laser at 531nm vs. the output generated with the "perfect" pump source, a narrowband TEM_{00} Ti:Saphhire: the performance of the investigated systems is identical, for injection locked diode pumping, and for Ti:Sapphire pumping.

The performance of the Nd:YAG laser and NYAB lasers, that is to be expected with future high brightness diodes at several Watts, was explored experimentally. The maximum output of the monolithic Nd:YAG ring laser was determined to be 1.4 Watts, before transverse mode operation sets in. The maximum visible output from the NYAB laser was 450mW at 531nm and 6.4mW at 456nm, limited only by the pump power of the Ti:Sapphire laser, and tunable from 452nm to 466nm.

The above presented two applications demonstrate clearly that high brightness beams from high power diode lasers will lead to a number of highly interesting applications, as are, i.e., efficient frequency coversion, fiber laser pumping, single mode fiber transmission, or direct use for material processing.

References

1. M.K.Chun, L. Goldberg, and J.F.Weller, Opt.Lett.**14**, 272 (89)

2. L. Goldberg, J.F.Weller, D.Mehuys, D.F.Welch, D.F.Scifres, Electron.Lett**27**, 929 (91)

3. R.Waarts, D.Mehuys, D.Nam, D.Welch, W.Streifer, and D.Scifres, Cleo'91, CWE7

4. J.E.Epler, N.Holonyak Jr.,R.D.Burnham, T.L.Paoli, and W.Streifer, J.Appl.Phys.**57**, 1489 (85)

5. L.J.Mawst, D.Botez, M.Jansen, T.J.Roth, C.Tu, and C.Zmudzinski, Electron.Lett.**27**, 1586 (91); L.J.Mawst, D.Botez, M.Jansen, T.J.Roth, and J.Rozenbergs, Electron. Lett.**27**, 369 (91)

6. L. Goldberg, H.F.Taylor, and J.F.Weller, Appl. Phys. Lett.**46**, 236 (85)

7. J.-M.Verdiell, R.Frey, and J.-P.Huignard, IEEE J.Quantum Electron.**27**, 296 (91)

8. L.Goldberg, D.Mehuys, CLEO'91, CTuI1

9. R.Parke, D.F.Welch, D.Mehuys, and S.O'Brien, LEOS'92, DLTA11.4; available from Spectra Diode Labs., 80 Rose Orchad Way, San Jose, CA 95134, USA

10. T.J.Kane, R.L.Byer, Opt.Lett.**10**, 65 (85)

11. I.Schütz, I.Freitag, and R.Wallenstein, Opt. Commun.**27**, 221 (90)

Abstimmbare Mikrokristall-Laser

N. P. Schmitt, S. Heinemann, A. Mehnert, P. Peuser

Deutsche Aerospace - Technologieforschung, Postfach 80 11 09, D-81663 München

Abstract:

Microcrystal lasers are characterized by a resonator mode separation that is of the order of the laser gain bandwidth of the active medium, which is responsible for several interesting laser physical phenomena that can be observed very pronouncedly. Especially by thermally tuning the resonator frequency with respect to the centre-of-gain not only broadly tunable single frequency lasers can be built: the measured data allow the simulation of the microcrystal laser behaviour by means of which optimisation of such lasers can be performed as well as laser performance can be understood better. Microcrystal lasers therefore are a good means to study relevant physical parameters by measuring the phenomena of stimulated emission themselves. This has been already suggested very early by Kamiskii [1] as 'stimulated emission spectroscopy', which up to now has been applied to macroscopic laser systems only.

Experiments are described for which a 300 μm cavity length microcrystal resonator was built that was tuneable in single frequency operation over a range of more than 130 GHz at an output power of 23 mW. The characterization of the laser emission includes relaxation oscillation measurements and observation of birefringence effects in Nd:YAG.

Einführung:

Bei Mikrokristall-Lasern können aufgrund der gleichen Größenordnung von Verstärkungsbandbreite und Resonator-Modenabstand eine Reihe von laserphysikalischen Phänomenen in besonders reiner Form beobachtet werden. Speziell durch thermisches Durchstimmen der Resonatorfrequenz gegen den Schwerpunkt der Verstärkungslinie können nicht nur über einen relativ breiten Bereich abstimmbare single-frequency-Laser realisiert werden: die hierbei gewonnenen laserphysikalischen Daten erlauben zudem eine gute Simulation des Verhaltens von Mikrokristall-Lasern und tragen außer zur Optimierung des Systemes auch zu einem besseren Verständnis der Vorgänge in Mikrokristal-Lasern bei. Mikrokristall-Laser bieten sich somit als ideales Medium zur Messung laserrelevanter physikalischer Größen durch die Messung von Phänomenen der stimulierte Emission selbst dar, wie sie von Kaminskii z.B. in [1] bereits 1971 als "stimultated emission spectroscopy" bezeichnet, an allerdings noch makroskopischen Lasern durchgeführt worden ist.

Experimente:

Zur Messung einiger wichtiger laserphysikalischer Konstanten ist es zunächst zweckmäßig, Mikrokristall-Laser mit einem etwas kleineren longitudinalen Modenabstand als die Verstärkungslinienbreite zu verwenden. So können mehrere longitudinale Moden innerhalb eines Laserüberganges verstärkt und emittiert werden. Die hierbei auftretenden Modensprünge erleichtern die Messung von Verstärkungsbandbreite und thermischer Verschiebung der am Laserübergang beteiligten Stark-Niveaus innerhalb eines moderaten Temperaturbereiches. Abb. 1 zeigt den Ausschnitt der A und A'-Linie des Emissionsspektrum eines 900 μm langen, monolithisch mit Laserspiegeln versehenen Nd:YAG-Kristalles. Die Laser-Emissionswellenlänge wurde von T= -10°C bis 90°C in Schritten von 5°C gemessen, was im Diagramm als ausgefüllter Punkt dargestellt ist. Wie bereits berichtet [5], emittieren Mikrokristall-Laser typisch auf mehreren Laserübergängen eines Stark-aufgespaltenen Niveaupaares gleichzeitig. Im vorliegenden Falle sind die Übergänge $R_1 \rightarrow Y_1$, $R_2 \rightarrow Y_3$ und $R_1 \rightarrow Y_2$ des Überganges $^4F_{3/2}$ nach $^4I_{11/2}$, auch als B, A und A'-Linie bezeichnet, im Emissionsspektrum zu beobachten. Gemäß einer Besonderheit sind bei Nd:YAG die beiden Linien A und A' nicht getrennt sondern überlappen vielmehr [1,3]. Aus der Differenz der maximalen und minimalen

9

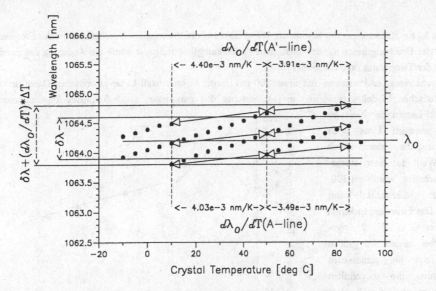

Abb. 1: Laseremissions-spektroskopische Daten, gemessen an einem 900 μm
Nd:YAG Mikrokristall-Laser

auftretenden Emissionswellenlänge der Laserübergänge ergibt sich eine virtuell verbreiterte Linienbreite $\delta\lambda + (\partial\lambda_0/\partial T)*\Delta T$ der (A+A´)-Linie sowie der B-Linie. Virtuell verbreitert deshalb, da durch thermische Verschiebung der Starkniveaus eine Verschiebung $(\partial\lambda_0/\partial T)*\Delta T$ der Zentralwellenlänge λ_0 der Übergänge überlagert ist. Diese Verschiebung läßt sich aus dem Emissionsspektrum durch Bestimmung der Differenz der Emissionswellenlängen an den Punkten des Modensprunges der jeweiligen Laserübergänge wie eingezeichnet ermitteln. Mit den so gewonnenen Werten kann dann die virtuell verbreiterten Linienbreiten jeweils korrigiert werden.

Es zeigt sich aus der Messung auch, daß $\partial\lambda_0/\partial T$, hergeleitet aus dem Phononenspektrum des Kristalles [4], strenggenommen eine nichtlineare Funktion der Temperatur ist, ebenso wie die Linienbreite selbst. Für den hier betrachteten

Bereich einer thermischen Durchstimmung über einen Bereich von etwa 100°C ist jedoch die Annahme einer temperaturinvarianten Linienbreite $\delta\lambda$ sowie Linienverschiebung $\delta\lambda_0/\delta T$ hinreichend genau (vergl. [4]).

Ausgehend von der thermischen Verschiebung der Zentralwellenlänge und der gemessenen virtuell verbreiterten Linienbreite läßt sich die Zentralwellenlänge des

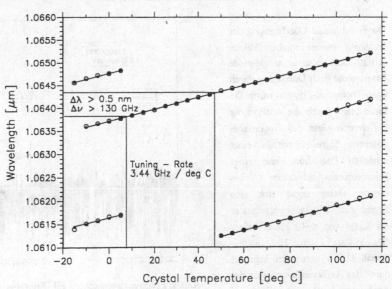

Abb.2: Thermische Abstimmung eines 300 μm Nd:YAG Mikrokristall-Lasers

Überganges λ_0 bei Raumtemperatur bestimmen. Weiter sind aus der thermischen Durchstimmrate der Frequenzänderung $\partial f/\partial T$ der Brechungsindex n_0, der thermische Ausdehnungskoeffizient α sowie die Änderung des Brechungsindexes mit der Temperatur $\partial n/\partial T$ ableitbar [5]

Ausgehend von diesen und weiteren, mit einem 800 µm langen Mikrokristall-Laser gewonnenen Daten und anhand eines theoretischen Modelles (publiziert in [5]), welches den maximalen single-frequency-Abstimmbereich von

Mikrokristall-Lasern bei Emission nur einer einzigen Linie unter Unterdrückung der anderen durch geeignete Wahl der Resonatorparameter beschreibt, wurde ein 300 µm langer Mikrokristall-Laser untersucht. Das Emissionsspektrum dieses Lasers ist in Abb. 2 dargestellt. Hierbei zeigen die großen offenen Kreise die gemessenen Emissionslinien, die ausgefüllten Punkte stellen die gemäß dem Modell nach [5] berechneten theoretisch zu erwartenden Linien dar.

Es zeigt sich hier eine ausgezeichnete Übereinstimmung zwischen der theoretisch zu erwartenden und der tatsächlichen spektralen Emis-

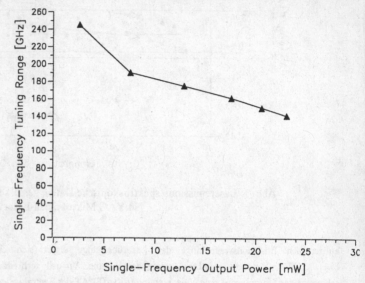

Abb. 3: Single-Frequency-Abstimmbereich bei reduzierter Pumpleistung

sion des Lasers. Weiterhin konnte nachgewiesen werden, daß ein optimierter Mikrokristall-Laser bei einer Ausgangsleistung von 23 mW und einer Pumpleistung zwanzigfach über der Schwelle über einen Bereich größer 0.5 nm ($\cong$ 130 GHz) monoton ohne Modensprünge, monofrequent und ohne Auftreten von simultaner Emission auf anderen Laserübergängen, thermisch durchgestimmt werden kann. Bei einer Reduktion der Pumpleistung ergaben sich noch höhere monofrequente Durchstimmbereiche gemäß Abb. 3.

Durch kohärente Überlagerung der Strahlung zweier solcher Mikrokristall-Laser in einer single-mode Faserweiche nach Durchgang durch Glan-Thompson-Prismen wurde die Linienbreite durch die Analyse der Differenzfrequenz des frequenzgemischten Signales mittels einer schnellen Photodiode und eines Hochfrequenz-Analysators bestimmt. Dabei ergab sich eine Breite des Überlagerungssignales der Laser von < 40 Hz bei einer "Sweep-Rate" von 100 ms/div gemäß Abb. 4, vermutlich begrenzt durch das Auflösungsvermögen des Analysators. (Die Auflösungsbandbreite für die Trennung zweier

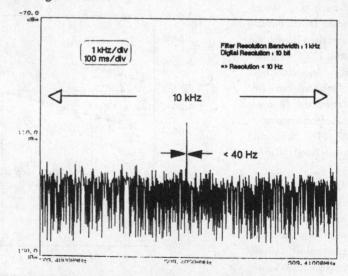

Abb.4: Frequenzgemischtes HF-Signal aus der kohärenten Überlagerung zweier 300 µm Nd:YAG Mikrokristall-Laser

Peaks betrug 1 kHz, die Auflösung der Digitalisierung war 10 Bit, was bei einem Gesamtbereich von 10 kHz einer Auflösung von etwa 10 Hz für das letzte Digit entspricht). Die Glan-Thompson-Prismen waren notwendig, da sich anhand von Fabry-Perot-Messungen gezeigt hat, daß Nd:YAG-Mikrokristalle durchweg doppelbrechend sind. Zwar läßt ein isotropes Gitter wie das YAG dies zunächst nicht vermuten, jedoch ist bekannt [6], daß jeder Nd:YAG thermisch doppelbrechend ist.

Der Grad der Doppelbrechung ist hierbei von der Kristallzuchtrichtung wie auch in bestimmten Fällen vom Kristallschnitt abhängig.

Die hier verwendeten 300 μm und 900 μm langen Kristalle waren in der Orientierung [111] gezüchtet und geschnitten, der 800 μm Kristall in der [100]-Richtung. Der 300 μm Kristall war zudem nach der Grobbearbeitung getempert worden. Bei allen Kristallen wurde, unabhängig von der Zuchtrichtung, eine Doppelbrechung festgestellt. Die Dif-

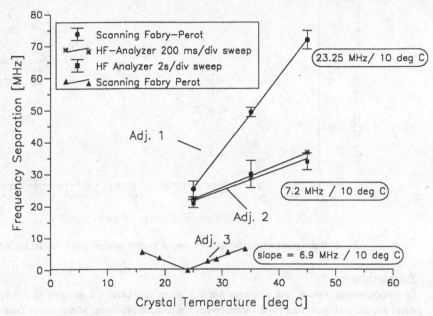

Abb. 5: Doppelbrechung eines 300 μm Nd:YAG Mikrokristall-Lasers bei drei unterschiedlichen Justagen Adj.1-3

ferenzfrequenz der beiden senkrecht zueinander stehenden Laserpolarisationen ließ sich hierbei thermisch durchstimmen, wobei die Temperatur auch so gewählt werden konnte, daß beide Polarisationslinien in ihrer Frequenz genau übereinstimmten (Differenzfrequenz null). Die thermische Änderung der Frequenzaufspaltung der beiden Polarisationen wurde sowohl mit einem Fabry-Perot-Interferometer als auch durch eine HF-Analyse des kohärent überlagerten Signales in oben beschriebenem Aufbau, aber ohne Glan-Thompson-Prismen, für 300 μm Mikrokristall-Lasern vermessen (s. Abb. 5). Es trat hierbei eine Abhängigkeit der Abstimmrate der Doppelbrechung versus Temperatur vom Ort der Laseranregung im Kristall auf. Dieser Effekt könnte unter anderem auch für das bei manchen Kristallen stärker oder schwächer ausgeprägte örtliche inhomogene Laserverhalten (vgl. [7]) mit verantwortlich sein.

Letztlich wurden noch die Relaxationsoszillationen von Mikrokristall-Lasern untersucht. Auch wenn das Relaxationsrauschen von Mikrokristall-Lasern einen Signal-Rauschabstand von etwa 65 dBm aufweist, muß es als der dominante Rauschterm doch für manche Anwendungen in Betracht gezogen werden. Daher wurde die Abhängigkeit der Frequenz des Relaxationsrauschpeaks von der Länge des Mikrokristalles und der Pumpleistung untersucht. Aus Abb. 6 ergibt sich eine relativ gute Übereinstimmung der gemessenen Relaxationsoszillations-Frequenzen mit den theoretisch zu erwartenden Werten. Die Relaxationsoszillations-Frequenz liegt für Mikrokristall-Laser relativ hoch zwischen 0.6 und 2.7 MHz für Kristall-Längen zwischen 300 und 900 μm und Pumpleistungen zwischen 50 und 250 mW. . Die resonatorinternen Verluste lagen bei 2.9×10^{-3} bis 3.6×10^{-3} und somit in der Größenordnung des Auskoppelgrades (0.3%). Die sich ergebenden Resonator-Lebensdauern lagen bei dieser geringen Auskopplung zwischen 290 und 930 ps.

12

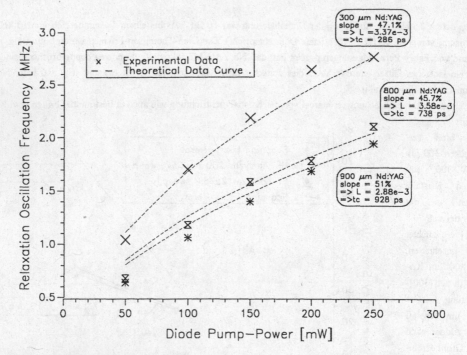

Abb. 6: Relaxationsoszillations-Frequenzen verschiedener Nd:YAG Mikrokristall-Laser

Zusammenfassung:

Es konnte gezeigt werden, daß diodengepumpte Mikrokristall-Laser gut geeignet sind, wichtige Laserparameter direkt aus der spektralen Analyse der stimulierten Emission zu gewinnen. Mittels dieser Daten und einer geeigneten theoretischen Modellierung konnte das Emissionsverhalten von Mikrokristall-Laser gut simuliert werden. Im Resultat der sich aus der Simulation ergebenden Optimierung konnte ein Mikrokristall-Laser realisiert werden, der bei einer Ausgangsleistung von 23 mW über einen Bereich von größer 130 GHz monofrequent, modensprungfrei und ohne simultane Emission weiterer Laserübergänge durchgestimmt werden konnte.

Danksagung:

Besonderer Dank gilt Professor Dr. W. Waidelich, Ludwig-Maximilians-Universität, München, für die Begleitung all dieser Arbeiten über nunmehr mehrere Jahre hinweg. Dank gebührt auch Dr. L. Wetenkamp, Deutsche Aerospace AG, Dr. K. D. Salewski, Ernst-Moritz-Arndt-Universität, Greifswald, und Dipl. Phys. K. H. Bechstein, Carl Zeiss Jena GmbH, für viele hilfreiche Anregungen und Diskussionen.

Referenzen:

[1]   A.A. KAMINSKII, D.N. VYLEGZHANIN, *IEEE J. QE* 7(1971) 329.

[2]   N.P. SCHMITT, S. HEINEMANN, A. MEHNERT, P. PEUSER, in: *Laser in Engineering*, edited by W. Waidelich, (Springer-Verlag, Berlin, 1991) S.599.

[3]   T. KUSHIDA, H.M. MARCOS, J.E. GEUSIC, *Phys. Rev.* **167**(1968) 289

[4]   S.Z. XING, J.C. BERGQUIST, *IEEE J. QE* **24**(1988) 1829

[5]   N.P. SCHMITT, P. PEUSER, S. HEINEMANN, A. MEHNERT, *Opt. and Quant. Electr.*, im Druck; erscheint vor. Sept. 1993.

[6]   W. KOECHNER, D.K. RICE, *J. Opt. Soc. Am.*, **61**(1971) 758

[7]   S. HEINEMANN, A. MEHNERT, N.P. SCHMITT, P.PEUSER, *Laser und Optoelektronik* **5** (1992) 48

MIKROLAS

Th. Halldórsson[1], W. Kroy[1], H. Schmidt-Bischoffshausen[1] und P. Thoren[2]

[1]Deutsche Aerospace AG, Technologieforschung, D-81663 München,

[2]VDI/VDE Technologiezentrum Informationstechnik GmbH,

Hanseatenhof 8 W-2800 Bremen 1

Aufgabenstellung

Im Rahmen des Forschungsschwerpunktes "Mikrosystemtechnik" des Bundesministeriums für Forschung und Technologie wird in dem Zeitraum 1992-1994 das Verbundprojekt "Miniaturisierter diodengepumpter Festkörperlaser (μ-Las) durchgeführt. Die Projektpartner sind 3 Industrieunternehmen und 4 Hochschulinstitute, die auf den verschiedenen Gebieten der Optik und Lasertechnologie tätig sind.

Das Projekt hat das Ziel, diodengepumpte Festkörperlaser mit Fertigungstechniken der Mikrosystemtechnik zu erstellen. Hierzu werden Demonstratoren von Mikro-Lasern verschiedener Betriebsart im mittleren Leistungsbereich von einigen zehn milliwatt bis einige Watt bis zum Funktionsnachweis entwickelt und damit die Eignung der Mikrosystemtechnik für die Herstellung von Festkörperlasern im mittleren bis höheren Leistungsbereich gezeigt.

Außer der Miniaturisierung mit der typischen Größe der Laser im Bereich 1 cm^3 bis 10 cm^3 stehen insbesondere die Kostenvorteile bei der Serienherstellung im Vordergrund, so daß insgesamt neue Anwendungs- und Marktfelder in Massenanwendungen eröffnet werden können

Die Anwendungen umfassen verschiedene Gebiete der Materialfeinbearbeitung, Meß- und Prüftechnik, Biologie, Medizin, Analytik und Chemie. Trotz des sehr hohen Innovationspotentials von Lasern in diesen Gebieten ist ihr Einsatz bisher noch durch großen apparativen Aufwand und hohe Herstellungs- und Betriebskosten begrenzt.

Zur Zeit bieten nur die Halbleiterlaser (Laserdioden) Vorteile wie Kompaktheit, hohe Effizienz, Robustheit, lange Lebensdauer und geringe Herstellkosten in der Massenfertigung. Sie konnten die Massenmärkte der Telekommunikation, CD-Schallplattenabtastung, Laserdrucker und -scanner erobern.

Seitdem es in den letzten Jahren gelungen ist Festkörperlaser aus seltenen Erden dotierten Gläsern und Kristallen (z.B. Nd:YAG) mit Laserdioden zu pumpen, konnten die vorgenannten Vorteile der Laserdioden auch auf diese Art von

Lasern übertragen werden. Grundsätzlich bietet der Festkörperlaser mehrere Vorteile und zusätzliche Möglichkeiten, die dem Laser Zugang zu neuen Anwendungsbereichen eröffnet und damit den erhöhten Herstellungsaufwand gegenüber dem Halbleiterlaser rechtfertigt.

Wo die Gestalt des Halbleiter-Laserresonators durch den geometrischen Aufbau des Verstärkungsmediums bestimmt ist, kann die Geometrie des Laserresonators des Festkörperlaser und damit seine Strahleigenschaften variabel gestaltet werden. Es ist möglich, außer dem Verstärkungsmedium andere Funktionselemente wie Polarisatoren, Güteschalter und Modulatoren in dem Resonator zu integrieren. Die lange Lebensdauer der oberen Laserniveaus ermöglicht eine einfache Speicherung der Laserenergie und Pulsbetrieb mit hoher Spitzenleistung. Auf Grund der geringeren Störungen durch Strom- und Temperaturschwankungen kann der Festkörperlaser leichter in einen stabilen Einfrequenzbetrieb mit sehr geringer Frequenzbandbreite versetzt werden.

Hier sind die folgenden weiterführenden Eigenschaften des diodengepumpten Festkörperlasers (DFKL) zu nennen:

- Erweiterung der verfügbaren Spektralbereiche von Dauerstrichlasern und gepulsten Lasern in den kurzwelligen sichtbaren (0,2 μm-0,7 μm) und dem nahen IR-Bereich 1,5 μm-3,5 μm.

- Verfügbarkeit von abstimmbaren Lasern über einen weiten Wellenlängenbereich

  (Titan-Saphir-Laser und optisch parametrische Oszillatoren)

- Geringer Aufwand bei der Erzeugung von hoher Frequenzstabilität in Einmodenbetreib (<10 kHz)

- Schnelle reproduzierbare Durchstimmbarkeit über einen weiten Frequenzbereich (<100 GHz).

- Geringes Intensitäts- und Frequenzrauschen der Emission (<-160dB/Hz)

- Hohe Strahlqualität (5-10 millirad Strahldivergenz, rotationssymmetrisches Gauß-Profil)

Technologieeinsatz

Im Aufbau des μ-Lasers sind eine Reihe wichtiger Mikrosystem-Komponenten integriert. So sind der Festkörperlaser und die zu seinem Betrieb nötigen Komponenten, wie eine Laserdiode als Pumplichtquelle, Optik, Kühlung und Regelung auf einem Siliziumchip vereinigt. Auf einem Siliziumwafer können mit Hilfe von Photolithographie und anisotropem Ätzen Gräben zur Montage von optischen Bauteilen und Mikrokühlkanäle zur Flüssigkeitskühlung von Laserdiode, Laserkristall und elektronischen Bauteilen strukturiert werden. Weiterhin kann die Versorgungs- und Regelungselektronik direkt auf den Wafer aufgebracht werden. Mit mikromechanischen Aktoren in Silizium können Schalter und Spiegelverstellelemente realisiert werden. Gleichzeitig mit der Miniaturisierung der mechanische und elektrischen Funktionsementen ist die Verwendung von mikro-optischen Bauteilen notwendig. Hier können sowohl lichtbrechende Optiken, Gradienten-Optiken als auch diffraktive und holographische Elemente eine Verwendung finden.

Die Arbeiten an dem Vorhaben sind gegliedert in fünf Hauptkategorien:

- Simulation
- Prozesstechnologien
- Aufbau- und Verbindungstechniken
- Intelligente Systemsteuerung
- Experimente und Tests

Die Rechnersimulation ist notwendig um die experimentellen Ergebnisse systematisch interpretieren zu können, eine Vorhersage über das Systemverhalten zu ermöglichen und nicht zuletzt um ein Werkzeug für den Design verschiedener Laserstrukturen zu schaffen. Die Rechnersimulation umfaßt:

- Wellenberechnung des Aufbaus einer stabilen Lasermode im Resonator unter Berücksichtigung der Resonatorparameter und der thermisch optischen Beeinflussung des Lasermaterials
- Auslegung der Kühlung und der Temperaturregelung
- Auslegung der Regelung und Steuerung

Die Siliziumtechnologie stellt die umfangreichste Prozesstechnologie des Vorhabens dar. Weitere Technologien, die zum Einsatz kommen, sind Laserkristallziehverfahren, Aufdampf- und Ion-Plating Techniken für optische Funktionsschichten, Photoresisttechnik für Hologrammme, Diamantbeschichtung für gezielte Wärmeverteilung, Löttechniken für Laserdioden-Chips und Bondtechniken und Herstelltechniken für SMD und ASICS .

Zielsetzungen und erste Resultate

Das Technologiepotential der Mikrosystemtechnik soll anhand von einigen beispielhaften Funktionseinheiten demonstriert werden:

- Leistungslaser über 1W bei 1,06 μm

- Laser im sichtbaren Bereich 0,53 μm mit einigen hundert mW Leistung

- Laser im augensicheren Bereich (gepulst oder Dauerstrich) > 1,5 μm

- Gepulster Laser (evt. mit aktivem Spiegel als Q-Schalter)

- Einfrequenzlaser durchstimmbar

Bisher lag das Schwergewicht auf der Entwicklung des Simulationswerkzeugs für die thermischen und optischen Funktionen, der Silizium Prozesstechnik, dem ziehen neuer Lasermaterial und der Ion-Plating Aufdampftechnik.

Es konnten die ersten μ-Laser auf einer optischen Bank in Silizium mit aufgelöteten Laserdioden und integrierten μ-Kühlern in Betrieb genommen werden.

Die ersten mit Piezoelementen durchstimmbaren Laser und Laser bei 2,9 μm wurden untersucht.

Es konnte die Funktion eines Mikro-Spiegels im Betrieb eines 1,06 μm Lasers demonstriert werden.

Diodengepumptes Single-Frequency Nd:YAG Lasersystem hoher Leistung

I. Freitag, H. Zellmer, W. Schöne, D. Golla und H. Welling

Laser Zentrum Hannover e.V.

Hollerithallee 8, D-30419 Hannover

Abstract

Vorgestellt wird ein diodengepumptes 15W cw Nd:YAG Lasersystem. Single-Frequency Betrieb dieses Ringlasers wird durch Injection Locking mit einem diodengepumpten monolithischen Miniatur Ringlaser als frequenzstabilen Single-Frequency Master-Oszillator erreicht.

I.) Einleitung

Die nächste Generation von großen Michelson-Interferometern zur Detektion von Gravitationswellen soll ein Auflösungsvermögen von 10^{-21} erreichen. Grundvoraussetzung hierfür ist eine extrem stabile und rauscharme Lichtquelle. Für ein hohes Signal zu Rausch-Verhältnis wird zusätzlich eine große optische Leistung der Laserlichtquelle im Single-Frequency Betrieb benötigt. Darüber hinaus sollte der Laser eine hohe Effizienz besitzen, um die Betriebskosten in vernünftigen Grenzen zu halten. Festkörperlaser haben das Potential diese Anforderungen zu erfüllen. Insbesondere diodengepumpte Nd:YAG Laser werden in Zukunft den bisher als Strahlquelle verwendeten Argon-Ionen Lasern weit überlegen sein. Diodengepumpte Festkörperlaser sind skalierbar zu großen Ausgangsleistungen, und die Anregung mit Diodenlasern verursacht weniger technisches Rauschen als die konventionelle Anregung mit Bogenlampen. Schließlich ist die Lebensdauer von diodengepumpten Systemen hoch und der Gesamtwirkungsgrad typisch um einen Faktor 100 größer als bei Argon-Ionen Lasern.

Ein diodengepumpter Nd:YAG Miniatur Ringlaser [KANE] erfüllt die geforderten Eigenschaften bezüglich Single-Frequency Betrieb und Frequenzstabilität. Der direkte Einsatz dieses Systems in einem Gravitationswellendetektor ist allerdings nicht möglich, da die maximale Leistung im Single-Frequency

Betrieb auf Werte unterhalb von etwa 2W begrenzt ist. Mit konventionellen Verstärkern könnte zwar die Leistung verstärkt, aber nicht das Signal zu Rauschverhältnis verbessert werden. Deshalb muß die Technik des Injection Locking in Master/Slave-Anordnung verwendet werden, um die benötigten hohen Leistungen zu erzeugen, bei denen das störende Rauschen nur durch das Quantenrauschen bestimmt ist.

Verschiedene Injection Locking Experimente mit lampengepumpten Nd:YAG Lasern als Verstärkerstufe sind bisher beschrieben worden. Aufgrund des geringen technischen Rauschens von Diodenlasern ist es allerdings wünschenswert, auch den Hochleistungslaser mit Diodenlasern anzuregen. Dieser Beitrag berichtet über die Realisierung eines diodengepumpten Lasers, in dem durch Injection Locking eine maximale Single-Frequency Leistung von 15W erreicht wird.

II.) Diodengepumpter Nd:YAG Hochleistungslaser

Die rasanten Fortschritte im Bereich von Hochleistungsdiodenlasern [FAN] in den letzten Jahren haben zu einer großen Aktivität im Bereich der Entwicklung diodengepumpter Festkörperlaser geführt. Hohe Gesamtwirkungsgrade von diodengepumpten Festkörperlasern im cw-Betrieb sind in verschiedenen endgepumpten Systemen erzielt worden. Die Skalierung der Ausgangsleistung dieser Systeme ist allerdings begrenzt, da nur das Licht von wenigen Diodenlasern in den Stab eingekoppelt werden kann.

Für Ausgangsleistungen im Bereich von 100W haben wir uns daher für eine transversale Pumpgeometrie entschieden. Erste Versuche mit kleinen Slablasern zeigten, daß das stark astigmatische Strahlprofil für das Injection Locking ungünstig ist. Daher entwickelten wir einen seitengepumpten Nd:YAG Stablaser. Der Laserstab ist von einer Flowtube zur direkten Wasserkühlung umgeben. Die Pumpquelle für den Nd:YAG Stab besteht aus 28 linearen Diodenarrays (SDL-3490 S) mit einer nominellen Ausgangsleistung von jeweils 10W cw und einer Linienbreite von etwa 2nm. Die Wellenlänge der Diodenlaser wird über die Temperatur auf die stärkste Absorptionslinie in Nd:YAG bei 808nm abgestimmt. Sechs Pumpmodule sind symmetrisch um den Stab angeordnet, um eine hohe Verstärkung auf der Stabachse zu erzielen. Jedes Pumpmodule besteht aus bis zu 5 Diodenlasern, die auf einer wasserdurchflossenen Wärmesenke aus Kupfer befestigt sind. Über den Wasserdurchfluß wird die Temperatur der Wärmesenke geregelt. Die Strahlung von jedem Diodenlaser wird mit einer für die Pumpwellenlänge entspiegelten Zylinderlinse in den Stab eingekoppelt. Die Pumpleistung hinter der Optik beträgt insgesamt 260W.

Zur Charakterisierung des Lasers wurde ein linearer plan-konkaver Resonator mit einem hochreflektierenden Spiegel mit Krümmungsradius 1500mm aufgebaut. Die optimale Auskopplung des planen Spiegels beträgt 5%. Bei einer Resonatorlänge von 200mm kann eine maximale Multimodeleistung von über 50W cw erzielt werden. Der differentielle Wirkungsgrad beträgt hierbei 23% bei einer Schwellpumpleistung von 37W.

Die Strahlqualität kann verbessert werden, wenn nur plane Spiegel verwendet werden. Der Resonator wird dann durch die thermisch induzierte Linse im Nd:YAG Stab stabil, die eine Brennweite von etwa 40cm besitzt. Durch Vergrößerung des Spiegelabstandes wird der Modenradius des Grundmodes im Stab vergrößert. TEM$_{00}$-Mode Betrieb des Lasers wird durch die Unterdrückung höherer Moden mit einer resonatorinternen Blende erzwungen. Optimaler Grundmodebetrieb mit einer Leistung von 17W wird in einem symmetrischen Resonator bei einem Spiegelabstand von 1.4m erreicht. Aufgrund des Spatial Hole Burning Effektes oszilliert der Laser auf mehreren axialen Moden mit starken Frequenzschwankungen.

III.) Injection Locking

Modenstruktur und Rauschverhalten eines Hochleistungslasers (Slave) können durch Injection Locking mit einem stabilen Oszillator (Master) verbessert werden. Als Master-Laser für Nd:YAG Hochleistungslaser sind diodengepumpte Miniatur Ringlaser aus Nd:YAG am besten geeignet. Wir haben verschiedene dieser longitudinal gepumpten Systeme entwickelt [FREITAG]. Für das Injection Locking wurde ein System verwendet, das mit einem Diodenlaser mit einer Apertur von 1x200µm (Siemens SFH 487401) und einer Ausgangsleistung von 1W cw gepumpt wird. Die maximale Ausgangsleistung dieses Masterlasers beträgt 400mW im Single-Frequency Betrieb mit einer Linienbreite kleiner 3kHz/100ms.

Zur Ankopplung an den Miniatur Ringlaser, wurde für den Hochleistungslaser ein Ringresonator aufgebaut. Im Einrichtungsbetrieb wird so eine Rückkopplung in den Master-Oszillator und die Ausbildung von stehenden Wellen im Hochleistungslaser verhindert. Der Ringresonator mit einer Länge von 255cm besteht aus zwei planen Spiegeln und zwei gekrümmten Spiegel mit einem Krümmungsradius von 500mm. Ein planer Auskoppelspiegel besitzt eine Transmission von 5% für die Laserwellenlänge 1064nm. Im freilaufenden Betrieb oszilliert der Ringlaser in beiden Umlaufrichtungen mit einer maximalen Ausgangsleistung im transversalen Grundmode von 7.5W pro Richtung.

Um Injection Locking zu realisieren, muß die Frequenz der vom Master-Oszillator eingekoppelten Strahlung in einem begrenzten Bereich um eine Resonanzfrequenz des Slave-Lasers liegen. Da die Resonanzfrequenz des Hochleistungslasers aufgrund thermischer Drift und akustischer Störungen große Schwankungen aufweist, wird sie mit einer aktiven Regelung über die Pound-Drever Seitenbandtechnik auf die Frequenz der eingekoppelten Strahlung abgestimmt [GOLLA]. Die Rückkoppelelektronik enthält einen PID-Regler und kontrolliert die Resonatorlänge über Spiegel, die sich auf Piezoelementen befinden. Ein langsames Piezoelement mit einem großen dynamischen Bereich kompensiert die thermische Drift, während ein schnelles Piezoelement die hochfrequenten Störungen ausregelt.

Die Modenstruktur des Hochleistungslasers wird mit einem Fabry-Perot Interferometer beobachtet. Ohne Injection Locking oszilliert der Slave-Laser in beiden Umlaufrichtungen und auf mehreren axialen Moden. Beim Auftreten von Injection Locking zeigt das Modenspektrum des Hochleistungslasers nur eine einzige Linie mit einer maximalen Leistung von 15W und oszilliert nur in die Richtung der eingekoppelten Strahlung. Bei etwa 50mW eingekoppelter Strahlung kann eine maximale Verstärkung von mehr als 250 erzielt werden.

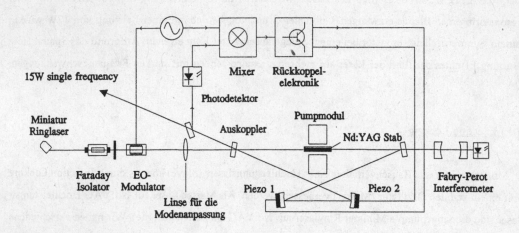

Abbildung: Diodengepumptes Single-Frequency Nd:YAG Lasersystem hoher Leistung in Master/Slave-Anordnung

IV.) Zusammenfassung

Wir haben über ein diodengepumptes Nd:YAG Hochleistungslasersystem berichtet. Durch Ankopplung an einen diodengepumpten Miniatur Ringlaser, konnten dessen exzellente spektrale Eigenschaften auf den Hochleistunglaser übertragen werden. Eine maximale Ausgangsleistung im Single-Frequency Betrieb von 15W wurde bisher erreicht. Durch Erhöhung der Pumpleistung wird dieses System in naher Zukunft den Leistungsbereich von 30W im Single-Frequency Betrieb erreichen.

V.) Literatur

T. J. KANE, and R. L. Byer, Opt. Lett. 10 (1985), 65

T. Y. FAN, and D. F. Welch, IEEE J. Quantum Electron. QE-28 (1992), 997

I. FREITAG, I. Kröpke, A. Tünnermann, and H. Welling, Optics Comm., accepted for publication

D. GOLLA, I. Freitag, H. Zellmer, W. Schöne, I. Kröpke, and H. Welling, Optics Comm. 98 (1993), 86

Diode-Array Side-Pumped Multiwatt Nd:YAG Laser

H.H. Klingenberg, U. Greiner
Institut für Technische Physik der Deutschen Forschungsanstalt für Luft und Raumfahrt (DLR)
Pfaffenwaldring 38-40, 70503 Stuttgart, FR Germany

The availability of ever more powerful laser diodes operating in the infrared or near visible spectral range led to an increased application of the diodes for pumping solid-state lasers. Diode pumping of solid-state lasers is performed either by end or side-pumping geometries. As well known, end or longitudinal pumping of Nd:YAG lasers is the method for obtaining higher conversion efficiencies of pump-to-TEM_{00}-laser outputs compared to side-pumped techniques /1/. As reported, in the more efficient end-pumped scheme for generating diffraction-limited output power, the pump light and the TEM_{00} laser mode spatially overlap. However, end-pumped systems yielded lower output powers, while side-pumping seems easier scalable towards higher powers due to the larger lateral dimensions of the present diode laser arrays, and the simpler pump-light focusing optics. Optical conversion efficiencies of pump-to-Nd:YAG-laser output powers were reported to be approximately 30 % when end-pumped, and 10 % for side-pumped geometries /2/. It depends on what the diode-pumped laser system is finally used for, whether a high multimode output power or a fairly Gaussian mode profile is needed for a specific application. Therefore, considerable efforts are being made to improve both pumping schemes by using sophisticated imaging optics for focusing the pump light into the laser material. For example, a high power Nd:YAG laser achieved 1.9 W of TEM_{00} mode output at 1.064 μm wavelength when end-pumped with a 10 W laser diode array /3/. Various geometries were studied using quasi longitudinal pumping techniques /4/. The authors have scaled the angular multiplex pump geometry to focus the power of eight 15 W laser arrays into the rod ends obtaining an optical-to-optical efficiency of 26 %. The expected thermal distortion and stress-induced birefringence at these high pump power levels could be minimized.

In this paper, a pump scheme is investigated based on three highly reflective elliptically shaped focusing mirrors arranged around the laser rod. A cross sectional view of the geometry is depicted in Fig. 1. In a first configuration three cw operating 10 W diode arrays were used to pump a 3 mm x 40 mm Nd:YAG rod. This tripel set of diodes was also applied for pumping a 2 mm x 35 mm Nd:YAG rod. Since the rod was not cooled, the system was pumped with a duty cycle of 10 %, pulse duration of 500 μs, to reduce the heating up of the laser rod. The obtained output powers in the TEM_{00} mode were 1.5 W for the 3 mm rod, and 2.7 W for the 2 mm laser rod. In both cases, the cavity length was 30 cm, and the output coupler had a transmission of 7.5 % with a radius of curvature of 5 m. The second cavity mirror was a plane high reflector. A maximum multimode power of 5 W was produced from the 2 mm rod.

In a second approach the laser head was improved for cw operation by putting a flow-through tube around the rod. Instead of placing one diode array in the sidewise pump configuration, *a pair* of 10 W laser diode arrays was installed for scaling the pump power. By this means a total of six 10 W arrays pumped the 3 mm Nd:YAG rod.

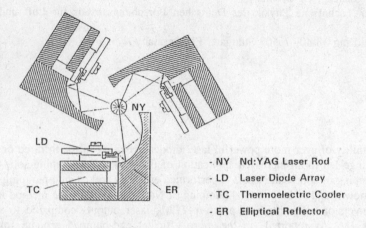

- NY   Nd:YAG Laser Rod
- LD   Laser Diode Array
- TC   Thermoelectric Cooler
- ER   Elliptical Reflector

Fig.1: Cross sectional view of the pump geometry

The cavity length was 75 cm. Both cavity mirrors were plane, and the output coupler had a transmission of 7.5 %. For the experiment the six 10 W diode arrays provided only 45 W of pump power. At the rod site itself approximately 39 W was available, due to reflexion losses from the surfaces of the elliptial mirrors and the surfaces from the flow-through tube. Also absorption losses account for the lower pump power. The distribution of the pump light luminescence in the laser crystal provided by the tripel set of diode arrays facilitated by the elliptically shaped focusing mirrors is shown in Fig. 2. The coupled-in pump power is higher at the rod center and along the rod axis, according to the length of the lasing region of the individual array, thus encouraging TEM$_{00}$ laser mode of operation. The thermal lens of the rod at this pump power level was interferometrically measured to have a focal length

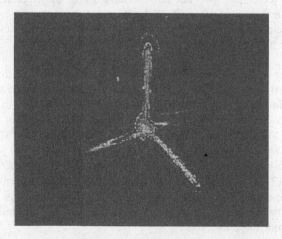

Fig.2: Distribution of the pump light luminescence into the Nd:YAG rod produced by three 10 W laser diode arrays

of 15 cm. A TEM$_{00}$ mode output power of 3 W was obtained. A 3-D plot of the transverse mode profile is depicted in Fig. 3.

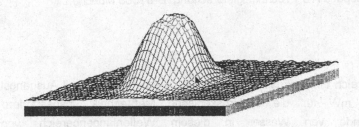

Fig.3: 3-D plot of the transverse mode profile

A reduction of the cavity length to 30 cm, and by applying a diode pump power of 45 W, yielded a maximum multimode output power of 5.5 W for the plane-plane, and 6.4 W for the plane-concave resonator, respectively. In the latter case, the output coupler had a radius of curvature of 5 m. Both results were achieved at 7.5 % transmission of the output coupler which resulted in a slope efficiency of 18 % for the plane-concave resonator.

In a third step the 3 mm diameter rod was replaced by a selected high quality Nd:YAG laser rod with 2mm diameter and 40 mm length. The smaller diameter is expected to give a better match of the pumped volume realized by the line focused pump light of the diode arrays and the laser mode volume. Furthermore, the new design is using a common cooling system for all assembled arrays, rather than the previously employed individual Peltier element cooling of the arrays. With a new set of diodes a pump power of 58 W was provided at the flow tube site. As a first result a TEM$_{00}$ mode laser output power of 4.1 W was obtained. The 30 cm long laser resonator consisted of plane mirrors. The output coupler had 7.5 % transmission.

This research project was supported by the Ministerium für Wirtschaft, Mittelstand und Technologie, Baden-Württemberg.

References

/1/ W. KOECHNER, Solid-State Laser Engineering, 3rd edition 1992. Springer Series in Optical Sciences, Vol. 1
/2/ D.C. SHANNON and R.W. WALLACE, Opt. Lett. Vol. 16, 318(1991), and further literature therein
/3/ T.S.ROSE, J.S.SWENSON and R.A.FIELDS, in Digest of Lasers and Electro-Optics Society Conference Proceedings (LEOS '90), paper TuC4-1, p. 186
/4/ S.C. TIDWELL, J.F.SEAMANS,and M.S. BOWERS, Opt. Lett. Vol. 18, 116(1993)

Laserdioden-gepumpte Er^{3+}-Mikrokristall-Laser bei 3 μm

L. Wetenkamp, D. Kromm, P. Steinbach

Deutsche Aerospace AG - Technologieforschung, D-81663 München

Im Spektralbereich um 3 μm sind miniaturisierte Festkörperlaser mit Ausgangsleistungen von bis zu 100 mW für die spektroskopische Meßtechnik, insbesondere wegen der Absorptionsbande von Wasser in diesem Wellenlängenbereich, von Bedeutung. Dauerstrichlaserbetrieb um 3 μm in den Er^{3+}-dotierten Kristallen YLF [1], GSGG [2] sowie YSGG, GGG und YAG [3] wurde bei den für die Laserdioden(LD)-Anregung interessanten Pumpwellenlängen um 790 nm (AlGaAs-LD) und 960 nm (InGaAs-LD) realisiert, obwohl normalerweise sich der Laserbetrieb für den Übergang $^4I_{11/2} \rightarrow\ ^4I_{13/2}$ in Er^{3+} aufgrund der längeren Lebensdauer des unteren Niveaus gegenüber dem oberen Niveau selbst beenden und damit nur Pulsbetrieb möglich sein sollte. Eine hohe Er^{3+}-Konzentration in den Wirtskristallen ermöglicht jedoch den CW-Betrieb infolge einer Entleerung der Niveaus $^4I_{13/2}$ durch Energietransfer (ET) zwischen benachbarten Er^{3+}-Ionen (Bild 1).

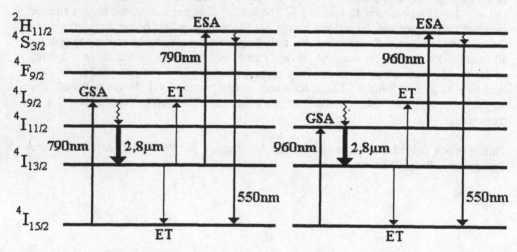

Bild 1: Pump- und Relaxationsschemata von Er^{3+}-Ionen für 790 und 960 nm

Im Vergleich zur Pumpwellenlänge bei 790 nm kann man bei 960 nm eine höhere Effizienz beim CW-Laserbetrieb erwarten, da bei dieser direkt ins obere Laserniveau (Bild 1) gepumpt wird. In Bild 1 ist auch das prinzipielle Zustandekommen der grünen Fluoreszenz dargestellt, die man im Laserbetrieb infolge der Anregung höher gelegener Niveaus durch weitere Absorptionsprozesse der Pumpstrahlung sowie durch Energietransferprozesse beobachten kann.

Im Rahmen dieser Arbeit wurden mit Ti:Saphir-Laser (TSL) bzw. Laserdioden (LD) gepumpte, monolithisch aufgebaute Mikrokristall-Laser aus 38% bzw. 50% Er^{3+},2% Cr^{3+}:YSGG, 15% Er^{3+}, 0,5% Cr^{3+}:GGG und 38% bzw. 50% Er^{3+}:YAG untersucht. Die Monolithen hatten eine Dicke von 2 mm und waren HR von 2,8 - 2,95 µm und T = 80 % bei 790 bis 960 nm auf der Eingangsseite beschichtet. Der Auskoppelgrad betrug 99,5 % von 2,8 bis 2,95 µm. Die Monolithen wurden als plan/plan-Resonator und als plan/konkav(r = 20 mm)-Resonator hergestellt. Die Strahlung des TSL wurde mit einer Linse (f = 25 mm) auf den Kristall fokussiert. Für die Fokussierung der 1W-AlGaAs-LD bzw. der 1W-InGaAs-LD wurde eine GRIN-Linse eingesetzt. Beide Optiken hatten einen Transmissionsgrad von 90 %. Die angegebenen Pumpleistugen sind um die Transmissionsverluste der Einkoppeloptik und -beschichtung korrigiert.

Bis auf die Anregung mit der 790 nm-LD konnte mit allen Pumpquellen und für alle Wirstkristall- sowie Resonatorkonfigurationen CW-Betrieb um 2,8 µm erzielt werden. Die Tabellen 1 und 2 enthalten die Zusammenfassung der Lasereigenschaften für einige Wirtskristalle für die Pumpwellenlänge um 790 nm bzw. um 960 nm. Bild 2 zeigt den Einfluß der Pumpquellen und -wellenlängen auf die Effizienz des Laserbetriebs am Beispiel des Er^{3+}-YSGG Materials. Aufgrund der besseren Pumpstrahlqualität des TSL sind jeweils bei beiden Pumpwellenlängen die Laserschwellen niedriger bzw. deren Kennliniensteigungen größer als bei Anregung mit den Laserdioden. Dabei werden auch die Vorteile hinsichtlich eines effizienteren CW-Betriebs durch Pumpanregung bei 960 nm gegenüber 790 nm deutlich. Diese Verbesserung wird neben der direkten Anregung des oberen Laserniveaus durch die größere Breite der drei Absorptionslinien bei 960 nm von ca. 6 nm gegenüber ca. 2 nm bei 790 nm und der damit verbundenen besseren Absorption der LD-Strahlung erreicht.

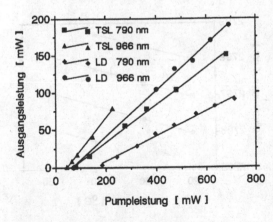

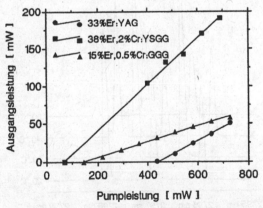

Bild 2: Laserkennlinien von 38 % Er:YSGG (pl/pl) für verschiedene Pumpquellen

Bild 3: Laserkennlinien für die verschiedenen Wirtskristalle bei Pumpanregung mit der 1W-InGaAs-Laserdiode

Beim Vergleich der Laserdaten der drei Wirtskristalle (Bild 3) ergaben die Untersuchungen für die verwendeten Monolithen, daß Er^{3+}-Dotierungskonzentrationen von 30 - 50% für das Pumpen mit LD den optimalen CW-Betrieb bewirken. Insgesamt ergeben sich die besten Leistungsmerkmale des 2,8 μm-Lasers für das Wirtsmaterial Er^{3+}-YSGG. Bei LD-Anregung bei 966 nm liegen die Laserschwellen unter 100 mW, und differentielle Wirkungsgrade von mehr als 30 % sind erreichbar. Der Wert von ca. 39 % bei Anregung mit dem TSL bei 966 nm liegt höher als der theoretisch erreichbare Wirkungsgrad von ca. 34 %. Durch den schon erwähnten Energietransfer (Bild 1) und die damit zusätzliche Bevölkerung der $^4I_{11/2}$-Niveaus wird ein Quantenwirkungsgrad von größer als 1 verursacht. Die Ergebnisse für das Er^{3+}-GGG liegen aufgrund der geringen Er^{3+}-Dotierung und der damit unvollständigen Absorption der Pumpstrahlung unter denen von Er^{3+}-YSGG. Eine Dotierung von GGG mit mehr als 25 % Er^{3+} sollte den Laserbetrieb deutlich verbessern. Die schlechteren Laserdaten des Er^{3+}-YAG finden ihre Ursache in dem ungünstigen Lebensdauerverhältnis der beteiligten Laserniveaus $^4I_{11/2}$ und $^4I_{13/2}$ von 0,2 ms zu 2 ms. Die um einen Faktor 5 geringere Lebensdauer des Niveaus $^4I_{11/2}$ im YAG gegenüber denen im GGG und YSGG bewirkt die merkliche Erhöhung der Laserschwelle.

Die Laser mit Er^{3+}:YSGG emittierten um 2,797 μm und für höhere Pumpleistungen auch um 2,823 μm. Die Emissionslinie von Er^{3+}:GGG lag bei 2,822 μm, während Lasertätigkeit für Er^{3+}:YAG nur bei 2,83 μm und nicht wie sonst berichtet [1] bei 2,94 μm beobachtet werden konnte. Bild 4 zeigt die Intensitätsspektren des Er^{3+}-YSGG-Lasers um 2,797 μm für unterschiedliche Temperaturen des Monolithen. Erkennbar sind drei longitudinale Moden mit einem Abstand von $\Delta\lambda = 1,0$ nm zueinander, deren Lagen sich mit steigender Temperatur zu längeren Wellenlängen verschieben. Der maximale Durchstimmbereich beträgt 3 nm.

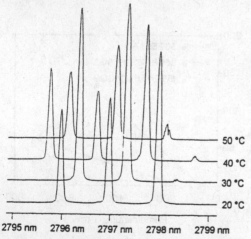 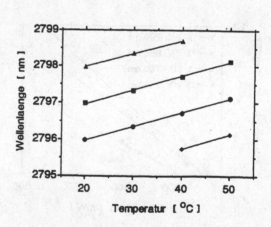

Bild 4: Emissionsspektrum von Er^{3+}-YSGG für verschiedene Kristalltemperaturen, LD-gepumpt bei 966 nm

Bild 5: Temperaturabstimmung der Zentralwellenlängen der longitudinalen Moden des Er^{3+}-YSGG-Lasers (Bild 4)

Aus Bild 5 läßt sich für diesen Kristall eine Abstimmrate von ca. 0,04 nm/°C ermitteln. Unter Verwendung der Gleichung für den longitudinalen Modenabstand $\Delta\lambda = \lambda^2/2\,n\,L$ und der Brechzahl für YSGG von n = 1,897 bei 2,8 µm ergibt sich eine berechnete Kristalldicke von L = 2,06 mm in guter Übereinstimmung mit der hergestellten Kristalldicke. Um Laserbetrieb auf nur einer longitudinalen Mode zu erreichen, dürfte der Er^{3+}YSGG Kristall unter Berücksichtigung der Durchstimmbreite maximal nur 0,75 mm dick sein.

Wirtsmaterialien	38 % Er:YSGG		15 % Er:GGG		33 % Er:YAG	
Pumpwellenlänge	966 nm		966 nm		964 nm	
Laserwellenlänge	2,797 µm		2,822 µm		2,830 µm	
Laserschwelle	158 mW	(72 mW)	324 mW	(90 mW)	-	(389 mW)
max. Laserleistung bei [...]-Pumpleistung	90 mW [720 mW]	(72 mW) [677 mW]	18 mW [720 mW]	(82 mW) [720 mW]	-	(60 mW) [691 mW]
diff. Wirkungsgrad	16,0 %	(25,2 %)	4,5 %	(13,0 %)	-	(19,8 %)
Polarisation	nein		ja		ja	

Tabelle 1: Lasereigenschaften der 2,8 µm-Laser in YSGG, GGG, YAG gepumpt mit der 1W-AlGaAs-Laserdiode. Geklammerte Daten für Pumpanregung mit dem TSL

Wirtsmaterialien	38 % Er:YSGG		15 % Er:GGG		33 % Er:YAG	
Pumpwellenlänge	966 nm		966 nm		964 nm	
Laserwellenlänge	2,797 µm, 2,823 µm		2,822 µm		2,830 µm	
Laserschwelle	72 mW	(43 mW)	144 mW	(< 5 mW)	418 mW	(133 mW)
max. Laserleistung bei [...]-Pumpleistung	201 mW [720 mW]	(74 mW) [234 mW]	58 mW [720 mW]	(41 mW) [216 mW]	51 mW [720 mW]	(33 mW) [248 mW]
diff. Wirkungsgrad	31,1 %	(38,8 %)	10,1 %	(19,5 %)	16,9 %	(28,4 %)
Polarisation	nein		ja		ja	

Tabelle 2: Lasereigenschaften der 2,8 µm-Laser in YSGG, GGG, YAG gepumpt mit der 1W-InGaAs-Laserdiode. Geklammerte Daten für Pumpanregung mit dem TSL

Im Rahmen dieser Arbeit wurden mit Ti:Saphir-Laser bzw. Laserdioden gepumpte, monolithisch aufgebaute Mikrokristall-Laser aus Er^{3+}:YSGG, Er^{3+}:GGG und Er^{3+}:YAG mit unterschiedlichen Dotierungskonzentrationen untersucht. Durch optisches Pumpen sowohl mit AlGaAs-LD um 790 nm als auch mit InGaAs-LD um 965 nm konnte effizienter Dauerstrichbetrieb der Mikrokristall-Laser in den drei Wirtsmaterialien realisiert werden, wobei die Pumpanregung bei 960 nm erwartungsgemäß einen effizienteren Laserbetrieb mit niedrigeren Laserschwellen, höheren Ausgangsleistungen und Wirkungsgraden ermöglicht.

Literatur

[1]   B.J. Dinerman, P.F. Moulton, OSA Proceedings on Advanced Solid-State Lasers, Santa FE (USA), 1992, Vol. 13, S. 152-155.

[2]   G.J. Kintz, R. Allen, L. Esterowitz, Appl. Phys. Lett., Vol. 50, 1987, S. 1553 -1555.

[3]   R.C. Stoneman, J.G. Lynn, L. Esterowitz, Technical Digest CLEO, Baltimore (USA), 1991, Vol. 10, S. 134-135.

ES-Laser

Ch. Kolmeder, L. Langhans
Carl Baasel Lasertechnik GmbH,
Petersbrunner Str. 1b, D-82319 Starnberg

Konzept

ES-Laser ist ein neues Konzept für Festkörperlaser. Es stellt eine Weiterentwicklung des slab-Konzeptes dar.

Beim slab-Laser wird im Prinzip die Linsenwirkung der thermischen Gradienten im Laserstab dadurch beseitigt, daß der Strahl im zick-zack die thermischen Gradienten abwechselnd in Gegenrichtung durchläuft, so daß sich ihre Wirkung aufhebt [1].

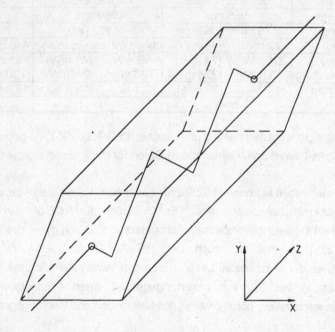

Durch den rechteckigen Querschnitt des slabs und durch den Strahlverlauf in der y-z-Ebene entsteht eine Asymmetrie in der x- bzw. y-Komponente des Strahls. Strahlabmessung, thermische Linse, Fresnel-Zahl und Strahlparameterprodukt haben stark unterschiedliche Komponenten in x- und y- Richtung [2].

Beim ES-Prinzip wird die Kompensation der thermischen Effekte durch einen zick-zack-förmigen Strahlverlauf beibehalten. Statt einer ebenen Dreieckskurve wie beim slab durchläuft der Strahl jedoch eine räumliche Dreieckskurve, die wie eine eckige Schraubenlinie aussieht (daher auch die Bezeichnung ES-Laser von Eckige Schraube).

Man kann sich diese eckige Schraubenlinie als eine Überlagerung von einer ebenen, senkrechten Dreieckskurve mit einer ebenen, waagerechten Dreieckskurve vorstellen mit jeweils gleicher Periodenlänge, gleicher Amplitude, aber phasenverschoben um eine viertel Periodenlänge.

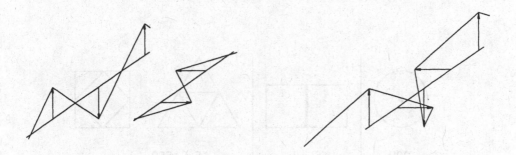

An dieser Darstellung als Überlagerung zweier ebener Dreieckskurven erkennt man deutlich, daß bei der räumlichen Dreieckskurve des ES-Lasers die x- und y-Komponenten völlig gleichwertig sind, d.h. die entsprechende Asymmetrie des slabs ist aufgehoben.

Der ES-Laserstab, der diesen Strahlweg erzeugt, ist dann ein Stab mit quadratischem Querschnitt, dessen Längsseiten optisch poliert sind. Der Strahl durchläuft diesen Stab in einer eckigen Schraubenlinie, wobei er an den polierten Längsseiten jeweils totalreflektiert wird. Die Abmessungen des Stabes werden so gewählt, daß der Strahl eine ganze Zahl von Umläufen macht. Im Querschnitt gesehen durchläuft der Strahl ein Quadrat, das in dem quadratischen Umriß des Stabes liegt.

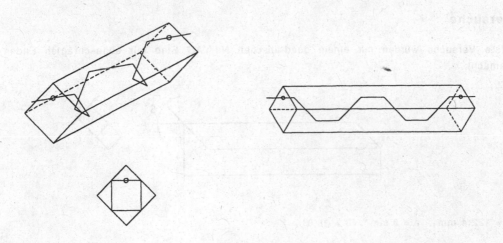

Für den Strahleintritt und -austritt kann man die gleichen Anordnungen wählen wie beim Slab (d.h. z.B. gerader Strahleintritt, Stabendfläche unter Brewsterwinkel).

Während für den slab spezielle, aufwendige Pumpkammern notwendig sind, kann man für den ES-Laser alle gängigen Pumpanordnungen verwenden, die vom rod-Laser bekannt sind, d.h. abbildende Ellipse oder close coupled, diffuse cavity, für eine oder mehrere Lampen.

Einen ES-Laser kann man nicht nur mit quadratischem Querschnitt realisieren. Alle gleichseitigen Polygone sind als Querschnitt möglich. Technisch sinnvoll sind wohl nur das gleichseitige Dreieck und das Quadrat [3]. Zusammen mit rod und slab lassen sie sich in einem systematischen Schema anordnen.

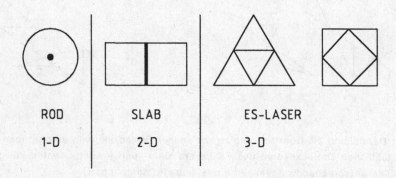

In dieser Systematik erkennt man den prinzipiellen Zusammenhang der drei Bauformen:

rod   - Laser: Strahlverlauf eindimensional, Gerade
slab   - Laser: Strahlverlauf zweidimensional, ebene Dreieckskurve
ES   - Laser: Strahlverlauf dreidimensional, räumliche Dreieckskurve

Versuche

Erste Versuche wurden mit einem quadratischen Nd-YAG Stab mit abgeschrägten Enden gemacht.

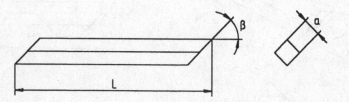

L = 122,4 mm   a = 6 mm   ß = 28,8°

Der Stab wurde in einer elliptischen Goldcavity mit einer Pumplänge von 80 mm getestet.

Dabei war bis 3,5 kW Pumpleistung eine thermische Linse nicht meßbar. Die Leistung im cw-Betrieb betrug wenige Watt. Weitere Versuche wurden mit einer Blitzlampe gemacht. Dabei wurden im Grundmode Pulsenergien von 80 mJ erreicht bei einer Pulsdauer von 0,3 ms. Das ist etwa doppelt so viel, wie bei einem vergleichbaren rod-Laser.

Weitere Versuche wurden mit einem dreieckigen ND-YAG Stab gemacht.

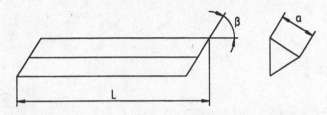

L = 112,8 mm   a = 7 mm   ß = 40°

Die thermische Linse hatte eine Brennweite von 6 m bei 3,5 kW Pumpleistung, d. h. 0,05 Dioptrien/kW. Das ist etwa 1/10 des Wertes, den man bei einem vergleichbaren rod-Laser mißt. Die maximale Ausgangsleistung cw betrug 16 W.

Beim Pumpen mit einer Blitzlampe wurden im Grundmode 170 mJ bei 0,3 ms erreicht. Das ist etwa viermal so viel, wie bei einem vergleichbaren rod-Laser.

Zusammenfassung

Das ES-Laserkonzept ist eine Weiterentwicklung des slab-Konzepts. Es ermöglicht eine weitgehende Kompensation der thermischen Linse in beiden Strahlachsen bei Verwendung von konventionellen Pumpanordnungen, d. h. es vereint die Vorteile von rod- und slab- Konzept. Erste Versuche bestätigen die konzeptionellen Überlegungen.

Diese Arbeit wurde gefördert vom Bayerischen Staatsministerium für Wirtschaft und Verkehr.

Literaturverzeichnis

[1]   W.S. Martin, J.P. Chernock, Multiple internal reflection facepumped laser,
       US Patent 3633 126, 1972

[2]   N. Hodgson, Q. Lü, S. Dong, B. Eppich, U. Wittrock, Hochleistungs-Festkörper-Laser in Stab-, Slab- und Rohrgeometrie, Laser und Optoelektronik 23 (3), 82, 1991

[3]   L. Langhans, Festkörperlaser-Stab, Europäische Patentschrift 0301 526 B1, 1992

Small Fiber Bundle Laser Structures Delivering 100 W Average Power

U.Griebner, R.Grunwald, R.Koch

Max-Born-Institut für nichtlineare Optik und Kurzzeitspektroskopie

Rudower Chaussee 6, D-12489 Berlin

Introduction

Due to their excellent thermal properties, flashlamp-pumped fiber bundle lasers (FBL) are a prospective approach to realize high average power solid state lasers /1,2/. The lasing medium is composed of a bundle of thin fibers doped with active ions. The fibers are waterproof bunched at the ends and spread out in the middle for water cooling. The extraordinary high thermal loading capability and small relaxation time allow extremely high pump energy and pulse repetition frequencies (PRF) delivering high gain, output energy and average power.

Furthermore, the high structural flexibility enables numerous variations. An example is the simultaneous generation of multiple wavelengths by combining different active fiber materials in the FBL in a parallel arrangement.

Experimental results

Experiments concerning high average power FBL have been done with highly Nd^{3+}-doped phosphate glass fibers. The fibers were designed for multimode operation with 9.0 wt% Nd_2O_3 in the core (core diameter d_{core} = 100 µm) and without doping in the cladding (cladding diameter d_{clad} = 140 µm). The fibers with a numerical aperture of 0.15 and an attenuation of 4 dB/m at the lasing wavelength of 1.053 µm were drawn by Fiberware GmbH Berlin (Kigre Q-100 preforms). An active length of 130 mm has been side-pumped by two Kr-flashlamps within a diffuse reflecting cavity.

The small-signal gain has been measured time-resolved by a probe-beam technique (cw-Nd:YLF-laser at λ = 1.053 µm). For this purpose, a selected fiber in the bundle was antireflection coated on both sides. A maximum gain of 0.30 cm^{-1} has been demonstrated for a pump energy of 160 J (Fig.1).

The increase of the fiber number corresponds to two opposite tendencies, i.e. the addition of the single fiber energies in the bundle on the one hand and the reduction of the pumping field

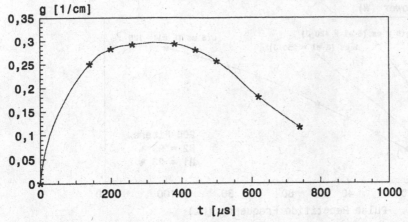

Fig.1: Time resolved small signal gain (E_{pump} = 160 J, τ_{pump} = 0.5 ms)

caused by absorption and other losses on the other hand. Thus it is necessary to optimize the fiber number. Since the lasing strongly depends on the outcoupling losses, the optimum number of fibers also depends on the total reflectivity R_{total} of the bundle endfaces which has to be optimized simultaneously. Raytracing calculations deliver reasonable results only for small n. We present the experimental results concerning this 2D-optimization problem. For a bundle with highly reflecting back mirror (R_1 = 99%, optical coating directly deposited on the polished endface), R_2 has been varied systematically by changing the outcoupling mirror in contact with the opposite polished bundle endface. Fig.2 clearly shows an optimum for n = 200 and R_2 = 23% (R_{total} = 47%) delivering an pulse energy in the free running regime of 900 mJ for a pump pulse energy of E = 250 J (pump pulse length τ_{pump} = 1 ms). For this optimum bundle configuration the average power has been optimized by a variation of the pump pulse energy and the PRF (Fig.3).

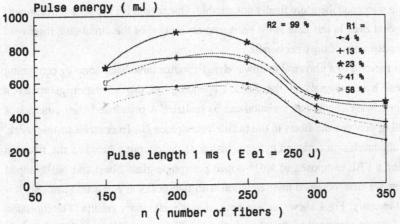

Fig.2: FBL energy in dependence on n for different reflectivities R

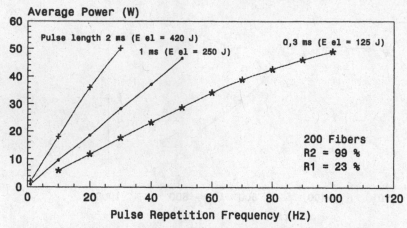

Fig.3: Average power of the FBL in dependence on the PRF for different τ_{pump}

For a reliable bundle consisting of fibers with an additional acrylate coating for water protection, a maximum average power of approximately 50 W has been obtained. It should be noted that the pulse energy doesn't show significant decrease up to a PRF of 100 Hz . This indicates that almost no thermal accumulation appears up to this PRF in good correspondence with the expected thermal relaxation time of 10 ms. The average power is limited by the power supply and the damage threshold of the acrylate. The average power corresponds to an efficiency of $\eta = 0.4\%$ and an extracted power per volume of 250 W/cm^3 (active volume 0.2 cm^3). The thermal loading tolerated within the laser glass can be estimated to be above 1 kW/cm^3 what is several orders of magnitude more than corresponding values for other glass laser designs. For a bundle consisting of fibers without organic coating, an average power almost two times higher up to 100 W (n = 200, R_2 = 99%, R_1 = 13 %) has been obtained. The absence of the coating further improves the cooling properties and the pumping efficiency. But such bundles, produced from uncoated fibers, are very fragile due to the solubility of phosphate glass in water and therefore finally not reliable. One reason for the poor efficiency of less than one percent even in this case is the enhanced absorption of the circulating laser field in the uncooled endsections of the fiber bundle.

First experiments forward to a universal multiwavelength source have been done by combining fibers of different host materials in the FBL. Depending on the resonator geometry a broadband or a multiwavelength laser emission can be realized. A broadband laser emission is expected if coupling between the fibers in the bundle takes place /3/. In contrast to this work, we realized a multiwavelength emission by suppression of the coupling between the fibers in the FBL. By using a FBL composed of Nd^{3+}-doped phosphate-glass fibers and Nd^{3+}-doped silicate-glass fibers we have obtained laser emission with two peaks at λ = 1.053 μm and λ = 1.061 μm simultaneously. Fig.4 shows an example of our preliminary results. The emission around λ = 1.053 μm is generated by the phosphate-glass fibers (core Kigre Q-100, 2.0 wt% Nd_2O_3, core diameter 100 μm, acrylate cladding). The Nd^{3+}-doped silicate-glass fibers

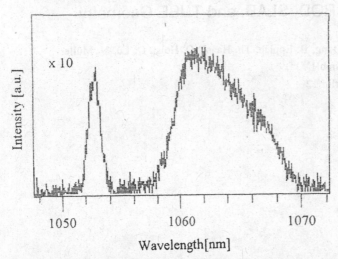

Fig.4: Output spectrum of the FBL composed of two kinds of Nd^{3+}-doped active fibers

(GLS-1, about 1.0 wt% Nd_2O_3, core diameter 180 μm, silicon rubber cladding) deliver the superimposed emission with a centre wavelength of λ = 1.061 μm. Depending on the unfavourable ratio of the doping concentration caused by the available glasses, the peak intensity of the Nd^{3+}-doped phosphate-glass component was about 10 times higher compared to that of the Nd^{3+}-doped silicate-glass.

Conclusions

It has been verified that the fiber bundle laser is a prospective concept for generating high average power (up to 100 W), high pulse energy (in excess of 1J) extractable of small active volumes (0.2 cm^3). These results confirm the predictions for the high thermal load and the short thermal relaxation time for the FBL. Furthermore, we demonstrated the potential of multiwavelength emission of a FBL consisting of different active materials. An optimization of different parameters, especially the fiber designs, to increase the efficiency has to be the subject of further investigations.

This work has been supported by the Bundesministerium für Forschung und Technologie (BMFT) under contract no. 13 N 5895.

References

1. L. E. Zapata, "Continuous-wave 25-W Nd^{3+}:glass fiber bundle laser," *J. Appl. Phys.* **62** (1987), 3110-3115.
2. U. Griebner, R. Grunwald, R. Koch, "1J Nd:Glass Fiber Array Laser," Int. Conf. on LASERS '92, Houston, Texas, 1992, paper ThD.6, Technical Digest p.26.
3. G. P. Banfi, P. G. Gobbi, L. Mussone and G. C. Reali, "A wide bandwidth, high power laser source for plasma interaction studies," *Opt. Comm.* **44** (1983), 192-195.

High Power Lasers in ROD, SLAB, and TUBE Geometry

U. Wittrock, G. Bostanjoglo, S. Dong, B. Eppich, Th. Haase, O. Holst, Q. Lü, N. Müller
Festkörper-Laser-Institut Berlin GmbH
Str. des 17. Juni 135, D-1000 Berlin 12

INTRODUCTION

This talk focuses on research and development of high power solid state lasers for industrial materials processing. For these lasers one is generally aiming at *compact* laser heads with *high output power*, *good beam quality*, and *high electrical-to-optical efficiency*. However, these qualitative goals depend on the specific application and can be quite different. For example, very often the beam quality is "good enough", if the beam can be coupled to a 600-μm quartz fiber with a numerical aperture of 0.2. This is an important benchmark for the beam quality because many industrial laser systems use such fibers of 600 μm or 1000 μm core diameter.

Some other applications require much better beam quality, i. e. beam parameter products close to the diffraction limit. To date, commercially available lasers with a beam parameter product below 5 mm·mrad (86%-definition, beam radius times far field half-angle) deliver no more than 350 W. However, it is very likely that many new applications will emerge once we have kW- and multi-kW-solid state lasers with this beam quality. At the moment, the only viable concept for this combination of output power and beam quality is clearly the slab laser.

High output power and very good efficiency at low beam quality is available with the concept of the internally-pumped tube laser. Efficiency is an important issue because power supplies and cooling units present about 2/3 of the total system costs. Improved efficiency thus not only means reduced operating costs of the laser but also less investment costs because smaller or less power supplies and cooling units would suffice.

MULTI-ROD LASERS

The traditional concept for kW-solid state lasers is the multi-rod laser. There are, however, several limitations to this concept.[1] First of all, as the number of rods is increased it becomes exceedingly difficult to align the laser. Also, the whole laser becomes quite large; a 2-kW multi-rod laser head with 4 to 6 rods typically has a footprint of about 0.4 m by 2 m. The maximum electical-to-optical efficiency is about 4% and can't be raised significantly for lamp-pumped systems. The stress-induced birefringence makes it very difficult to obtain linearly polarized output and makes the thermal lens bifocal. In addition, the thermal lens has severe aberrations due to inhomogeneities of the heat source distribution in the crystal and due to the temperature dependence of the thermal conductivity.

Stable Resonators

In order to obtain a low-order mode one needs a resonator with a large fundamental mode diameter. These resonators can be found in the g-diagram at the limits of stability. If the thermal lens would produce a purely parabolic index profile, free of aberrations, we could easily design a resonator that reaches the limit of stability at the maximum pumping power. The laser would thus have a nearly diffraction-limited beam, the TEM_{00} -intensity profile being only slightly distorted by the gain-profile. In practice this has never been achieved at output powers above a few tens of Watts. The reasons for this are the bifocal nature of the thermal lens and the spherical aberrations.

The stress-induced birefringence in the rod has rotational symmetry. It makes the thermal lens bifocal with different refractive powers for the radial and the azimuthal polarization component of the field. Radially or azimuthally polarized modes have a singularity with zero intensity at the optical axis. In general, the laser beam will contain a mixture of both polarization eigenstates. At the limits of stability in the g-diagram the r-polarized component of the field becomes unstable while the resonator is still stable for the φ-polarization that sees a lens of less power. With a proper layout for the resonator it should thus be possible to generate the TEM_{01}. hybrid mode, which has an almost diffraction limited beam parameter product, at maximum average power. However, just as for the pure TEM_{00} mode this has been only achieved experimentally at low output powers.

The g-diagram with its stable and unstable zones reflects the theory of spherical resonators. The boundaries between these zones are less sharp when spherical aberrations are introduced into the resonator. Different annular regions of the rod have different refractive powers and formally reach the limit of stability at slightly different pumping powers. The transition between stable and unstable operation smears out. Therefore it is impossible to obtain the azimuthally polarized eigenstate with an Nd:YAG rod at high average power.

Unstable Resonators

Pulsed Nd:YAG lasers have sufficient gain so that unstable resonators can be used. Unstable resonators have the advantage over stable resonators that they produce a nearly diffraction limited beam parameter product in combination with a large mode diameter and low misalignment sensitivity. Best beam quality is obtained with graded reflectivity mirrors which minimize diffraction at the reflectivity profile. In addition, the risk of damage to the active medium is reduced as compared with hard-edged unstable resonators. Hard-edged unstable resonators have a top-hat reflectivity profile with 100% reflectivity. This can lead to a stable resonator with two 100% mirrors if the thermal lens is not carefully taken into account during the resonator design.

The beam emanating from unstable resonators has a phase front curvature that varies with the focal power of the thermal lens. Since the thermal lens is bifocal, the radial and the azimuthal polarization components acquire slightly different phase front curvatures. Therefore, after a lens they have their waists at slightly different locations. This effectively increases the beam parameter product. (Even though, in principle, the two polarization components could be made to have the same phase front curvature by means of another radially/azimuthally birefringent element.) Of course, the spherical aberration also increases the beam parameter product of unstable resonators. At high average power this can become more than 15 times diffraction limited. Unlike in the case of a multi-rod laser with a stable resonator, the bifocal thermal lens and the spherical aberration cause the beam parameter product of multi-rod lasers with unstable resonators to increase with the number of rods; the diffraction losses are also increasing. At the moment it appears to be impractical to use more than 2 or 3 rods with an unstable resonator.

SLAB-LASERS

The slab principle was invented to overcome the thermal problems of rods.[2] Thermal lensing is compensated by the zig-zag path of the beam in the slab. The polarization eigenstates are the two orthogonal polarizations parallel to the slab surfaces, it is thus possible to obtain linearly polarized output. As brilliant as the slab principle is, it is hard to build real units that emitt several hundred Watts and actually do have better beam quality than rod lasers. It took about 18 years of research at General Electric from the time when the patent was filed[2] to the first kW-laser and the first 500-W laser with a few times diffraction limited beam quality.[3] Several papers have been written and many patents have been filed about slab lasers. The key issue in making full use of the slab principle is to obtain the right heat source distribution in the crystal and the right thermal boundary conditions at the slab surfaces. This can only be accomplished with the aid of ray-tracing programs for the design of the pumping chamber and finite element programs for analyzing the thermal effects in the crystal. It is still not clear how the beam propagation through a slab with spatially varying birefringence can be modelled numerically with sufficient accuracy.

Since the slab technology is too complex to be described in a short article like this one, I will merely present some of our experimental results.[4] Fig. 1 shows the output power and the beam parameter product of a multi-slab system with three heads. One head operates as an oscillator and the other two as amplifiers. The laser heads were designed before we had the full modelling capability for the thermal effects in the crystal. The slab cooling was thus not appropriate. At high pumping loads this caused the output power to decrease with time and settle at a lower value than the maximum that is reached shortly after switch-on. The thermal effects also lead to depolarization at high average power. Since the slab tips are cut under the Brewster-angle this creates considerable losses for the beam. Fig. 2 shows how much the depolarization losses can be reduced with a recently tested new setup as compared to the old thermal boundary conditions. With this cooling, the output power should also be stable.

We believe that the current slab technology still has ample room for further development. The main issue to be addressed is the thermo-optical aberrations that lead to reduced beam quality at high average

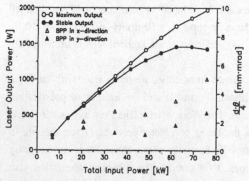

Figure 1. Output power and beam parameter product of the multi-slab laser with unstable resonator.

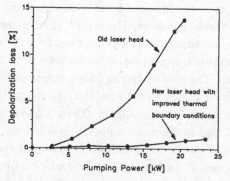

Figure 2. Depolarization losses (at 632 nm) of the slab with improved thermal boundary conditions.

power. With unstable resonators it also leads to pump-power dependent phase front curvature and astigmatism. Also, cw-operation of slab lasers at high power has so far only be achieved at GE. In view of the great progress that is being made in diode-pumping of solid state lasers the slab technology will become even more important in the near future.

TUBE-LASER

The first internally-pumped Nd:YAG tube laser has been built at the Festkörper-Laser Institut Berlin.[5] The concept proved very successful in terms of output power and electrical-to-optical efficiency. 1860 W of output power at over 9% efficiency has been achieved. At up to 1.2 kW output power the efficiency is 10%.[6] However, the beam parameter product of tube lasers with stable resonators is several times larger than that of rod lasers. The laser employs an Nd:YAG tube of 35 mm inner, 53 mm outer diameter and 130 mm length. It is enclosed by a cylindrical reflector made of Spectralon and is being pumped from the inside by four Kr-flashlamps. Both the inside and the outside tube surfaces are water-cooled. One end surface has a high-reflecting coating, forming one resonator mirror, the other end surface is AR-coated.

The annular beam of the tube laser has some peculiar properties not known from rod or slab lasers.[7] If a beam is focused with an ordinary lens and has a waist at the position of the lens, the spot size of the focus is determined by the far field divergence of the beam, while the Rayleigh range of the caustic is determined by the beam waist diameter at the lens. In contrast to the case of rod or slab lasers, for the tube laser the beam parameter product defined by these two quantities, spot size and Rayleigh range, is not the phase space volume of the beam. The reason is that the beam is not compact, i. e. it has a hole in the near field. If the beam had an annular Gaussian intensity profile and a perfectly flat phase front (the fundamental tube-laser mode) it would still have a beam parameter product of $2 \cdot (\lambda/\pi) \cdot R_o/(R_o - R_i)$, where R_o and R_i are the outer and inner beam radii, respectively. The beam parameter product of a mode of higher radial order is also larger by this factor than that of a conventional Gauss-Laguerre mode of the same order. In principle, the beam parameter product could be reduced to its phase space value by optics outside of the resonator. If the beam has no azimuthal structure this means that it can be reduced by a factor of $R_o/(R_o - R_i)$. This could be accomplished with a pair of axicons that compact the beam. However, in practice this doesn't work because of the high azimuthal mode order.

A new type of unstable resonator for annular gain media has been proposed recently.[8] We call it *azimuthally unstable resonator*. The beam is guided in the radial direction by a toric mirror surface or, as in the case of our tube laser, by the toric thermal lens in the tube wall. A slight tilt of the mirror causes the beam to wander under multiple bounces along the azimuthal direction to the coupling hole (Fig. 3). Using this resonator we obtained a beam parameter product of less than 50 mm·mrad at up to 700 W of output power.[6]

Research on the tube laser continues to focus on this and other suitable resonators for better beam quality than that of stable toric resonators.

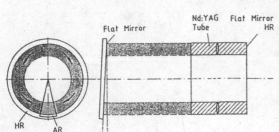

Figure 3. Azimuthally unstable resonator for the tube laser.

40

LITERATURE

1. M. Kumkar, B. Wedel, K. Richter, *Beam quality and efficiency of high-average-power multirod lasers*, Opt. and Laser Tech., **24**, 67 (1992).
2. W. S. Martin and J. P. Chernoch, *Multiple internal reflection face pumped laser*, U. S. Patent 3 633 126 (1972).
3. J. Chernoch, paper presented at ICALEO 1990 '90, 4-9 November 1990, Boston, MA.
4. N. Hodgson, R. Dommaschk, N. Müller, Q. Lü, S. Dong, Multi-kW Nd:YAG-Slab-Laser hoher Effizienz und Strahlqualität (Multi-kW Nd:YAG slab laser with high efficiency and high beam quality) Laser und Optoelektronik **24** (6) (1992).
5. U. Wittrock, B. Eppich, and H. Weber, Opt. Lett. **16**, 1092 (1991).
6. U. Wittrock, B. Eppich, and O. Holst, *Internally pumped Nd:YAG tube laser with 10% efficiency and 1.8-kW output power*, in Conference on Lasers and Electro-Optics, 1993 (Optical Society of America, Washington, D.C., 1993).
7. U. Wittrock, *High Power Rod, Slab, and Tube Lasers*, to be published in the NATO ASI SERIES "Solid State Lasers: New Developments and Applications", Plenum Press, 1993.
8. German Patent Application DE 41 23 024 A1 (1992).

Skalierung longitudinal gepumpter Festkörperlaser

J. Plorin*, K. Altmann, P. Durkin, A. Mehnert, N. Schmitt, P. Peuser
DASA Technologieforschung, D-81663 München
*Festkörperlaserinstitut Berlin

Die longitudinale Anregung von Festkörperlasern mit Laserdioden zeichnet sich durch einen beinahe idealen Überlappungsbereich von optisch gepumptem Volumen und der Ausdehnung der Lasermode im Festkörper-Laserresonator aus. Daher wurden sehr hohe Wirkungsgrade von 61,8 % [1] optisch zu optisch und 19% [2] elektrisch zu optisch erzielt.

Weitere Vorteile sind die relativ geringe Wellenlängenabhängigkeit der Pumplichtabsorption bei längeren Kristallen und die Möglichkeit durch "härteres" Pumpen auch Quasi-drei-Niveau-Übergänge zum Anschwingen zu bringen [3].

Mit zunehmender Ausgangsleistung tritt aber auch immer mehr die Schwierigkeit auf, das Pumplicht in der Festkörper-Lasermode zu konzentrieren. Die beschränkte Leistungsfähigkeit von Laserdioden und deren emittierende Flächen mit relativ großer lateraler Ausdehnung machen besondere Techniken notwendig, um longitudinal gepumpte Lasersysteme auf höhere Leistung zu skalieren. Für die erforderliche Transferoptik bieten sich in erster Linie drei Ausführungsarten an:

Das Pumplicht mehrerer Laserdioden läßt sich mit Linsen in den Laserkristall abbilden; diese Übertragung kann auch über ein Glasfaserbündel erfolgen, oder man verwendet eine Art Lichtkonzentrator, um das Licht eines breiten Lasediodenarrays zusammenzuführen.

Im Kristall entstehen hohe Temperaturen und große Temperaturdifferenzen, die nur für wenige Spezialfälle analytisch berechnet werden können. Eine umfangreichere Rechnung liefert eine Temperaturverteilung $\vartheta(r,z)$ in einem zylindrischen Laserkristall mit Radius r und Länge z als Lösung der partiellen Differentialgleichung zweiter Ordnung, mit der pro Zeiteinheit zuströmenden Wärmemenge $Q(r,z)$ und der Wärmeleitfähigkeit K:

$$\nabla^2 \vartheta + \frac{Q}{K} = 0 \quad \Leftrightarrow \quad \frac{d^2\vartheta}{dr^2} + \frac{1}{r}\frac{d\vartheta}{dr} + \frac{d^2\vartheta}{dz^2} + \frac{Q(r,z)}{K} = 0$$

Die Lösung dieser Differentialgleichung ist in Abbildung 1 für folgende Eckwerte dargestellt: Nd:YAG (1 at-%), Kristallänge von 8 mm, longitudinal angeregt mit einem rotationssymmetrischen und gaußförmigen Pumpstrahl mit Strahlradius ω_p = 1,4 mm, der nicht fokussiert wird.

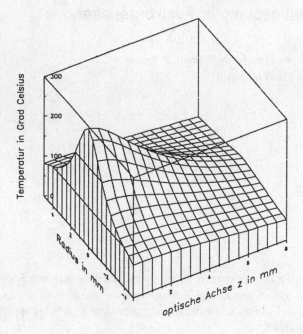

Abbildung 1: analytische Lösung der Temperaturverteilung in einem Laserkristall (siehe Text).

In Folge der beträchtlichen Aufheizung des Laserkristalls in der Nähe der optischen Achse entstehen radiale und axiale Spannungen, die zu einer thermischen Linsenbildung und zur optischen Doppelbrechung mit den damit verbundenen internen Resonatorverlusten führen. Darum kann man feststellen, daß die Laserkennlinie (optische Pumpleistung über Laserausgangsleistung) bei Pumpleistungen von mehreren Watt nicht mehr eine Gerade ist, sondern sich bei hohen Pumpleistungen zunehmend abflacht [4]. Wegen der mit der Linsenbildung einhergehenden Phasenfrontverzerrungen steigt auch der M^2-Faktor und die damit zusammenhängende Strahldivergenz. Im Extremfall kann es bei etwa 115 Watt pro cm absorbierter Pumpleistung zu einer Zerstörung des Laserkristalls kommen [5].

Der Schlüssel zur Lösung dieser Probleme ist die Verringerung der Pumpleistungsdichte im Laserkristall. Dies kann man auf folgende Arten erreichen:

a. Vergrößern des Modenvolumens im Festkörper-Laserresonator, so daß der Pumpstrahl nicht so stark fokusiert werden muß.

b. Verwendung von Laserkristallen mit niedriger Dotierung, um die Absorptionslänge im Laserkristall zu erhöhen.

c. Verwendung von Laserdioden mit breitem Emissionsspektrum oder Verstimmen der Laserdiode neben ein Hauptabsorptionsmaximum, um die Pumplichtabsorption auf einen größeren Bereich im Laserkristall zu verteilen.

d. Verwendung von mehreren Laserkristallen, um die Pumpleistung auf mehrere räumliche Bereiche zu verteilen.

Durch die Methode der finiten Elemente läßt sich das Temperaturprofil genauer berechnen, da Effekte wie die Strahldivergenz des Pumpstrahls und dessen Intensitätsprofil sowie verschiedene Arten der Pumplichteinkopplung berücksichtigt werden können. Außerdem kann man unterschiedliche Kristallgeometrien und Modenvolumina miteinberechnen. Als Beispiel sei hier die Temperaturverteilung in einem rotationssymmetrischen Kristall von 4 mm ∅ und 8 mm Länge angeführt, der bei einem Modenradius von 1,4 mm und Absorptionskoeffizienten von $\alpha = 0,4$ mm^{-1} (entspricht 1 at-% Nd:YAG) longitudinal mit 30 W Pumpleistung angeregt wird. Im wesentlichen entspricht die Temperaturverteilung derjenigen der analytischen Lösung.

Abbildung 2: Ergebnis einer Finite-Elemente-Rechnung für die Temperaturverteilung bei longitudinalem Pumpen (siehe Text).

Eine wie unter a vorgeschlagene Vergrößerung des Modenvolumens auf das Doppelte kann die Maximaltemperatur in der Kristallmitte nahezu halbieren. Diese Vergrößerung des Modenvolumens im Resonator führt aber zu einem signifikanten Anstieg des M^2-Faktors und damit zu einer starken Verringerung der Strahlqualität, da die Phasenfrontverzerrungen bei einer ausgedehnten Mode sich stärker auswirken.

Ein Verringern des Absorptionskoeffizienten durch Verwendung von Laserkristallen niedriger Dotierung oder absichtliches Pumpen mit nicht idealer spektraler Überdeckung von Pumplicht und Absorptionsband, wie unter b und c erwähnt, verringert die thermische Kristallbelastung nur unwesentlich.

Das Pumpen von zwei Seiten des Laserkristalls mit je 30 W läßt bei einem 8 mm langen Kristall die Maximaltemperatur nicht wesentlich über die Werte gegenüber einer Anregung mit nur einmal 30 W ansteigen.

Verringert man jetzt aber zusätzlich den Absorptionskoeffizienten von $\alpha = 0,4$ mm^{-1} auf $\alpha = 0,3$ mm^{-1}, so zeigt sich, daß die axialen Temperaturgradienten fast völlig verschwinden (siehe Abbildung 3). Außerdem bleibt man bei diesen Pumpleistungen noch deutlich unterhalb der vermutlichen Zerstörungsgrenze von ca. 45 W pro Kristallfläche ausgekoppelter Leistung [6].

	0
	12.863
	25.727
	38.59
	51.453
	64.316
	77.18
	90.043
	102.906
	115.769

Abbildung 3: Temperaturverteilung eines Laserkristalls mit gleichen Abmessungen wie in Abb. 2, von zwei Seiten mit jeweils 30 W optisch angeregt.

Mit derart optimierten Pumpgeometrien kann man verschiedene Laserkristalle in einem gemeinsamen Resonator einsetzen. Hierfür wurden Modellversuche unternommen, zunächst nur mit 3 W longitudinal gepumpt, womit eine Ausgansleistung von 1,1 W erzielt werden konnte.

Mit 6 Watt Gesamtpumpleistung ergaben sich 1,8 Watt, mit 8,5 Watt Pumpleistung 2,2 Watt TEM_{oo} Ausgangsleistung. Die schon bei diesen relativ geringen Pumpleistungsdichten entstehenden Temperatur-differenzen bewirken die zuvor beschriebenen Phasenfrontfehler, die zusammen mit zusätzlichen Beugungseffekten an den Kristallflächen den Gesamtwirkungsgrad des Lasersystems deutlich reduzierten.

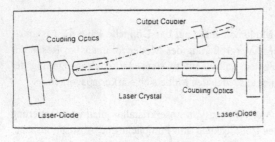

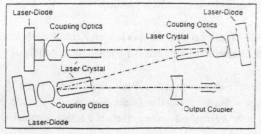

Abbildung 4: V- und Z-förmiger Festkörperlaser, gepumpt mit zwei bzw. drei Laserdioden.

Für eine Leistungsskalierung mit mehreren Laserkristallen sind weitere longitudinal gepumpte Aufbauten in Erprobung.

In der Literatur wurden auch verschiedene andere, longitudinal gepumpte Aufbauten beschrieben [8] [9], mit denen Ausgangsleistungen von inzwischen bis zu 60 Watt TEM_{00} bei 235 Watt Pumpleistung erzielt werden konnten [10].

Alternativ dazu bietet sich die Möglichkeit, jeden Laserkristall in einem eigenen Resonator zu betreiben und deren Strahlung anschließend kohärent zu addieren. Dies soll jedoch ein Thema von folgenden Arbeiten sein.

[1]   N. Schmitt, P. Peuser, W. Waidelich: Optisches Pumpen von Nd:YAG-Lasern mit Laserdioden
      Proceedings of the 9th International Congress Laser '89, pp. 166, Springer-Verlag 1990

[2]   R. Fields, T. Rose, M. Innocenzi, H. Yura, C. Fincher: Diode Laser End-Pumped Neodymium Lasers: The Road to
      Higher Powers; OSA Proceedings on Tunable Solid State Lasers, May 1989, North Falmouth, Massachusetts, pp. 301

[3]   T. Fan, R. Byer: Continous-wave operation of a room-temparature, diode-laser-pumped, 946 nm Nd:YAG laser
      Optics Letters, Vol. 12, No. 10, pp. 809 - 811, October 1987

[4]   Y. Kenada, M. Oka, H. Masuda, S. Kubota: 7.6 W of continous wave radiation in a TEM_{00} mode from a laser-diode
      end-pumped Nd:YAG laser; Optics Letters, Vol. 17, No. 14, pp. 1003 - 1004, July 15, 1992

[5]   D: L. Sipes, Jr., TDA Progress report 42 - 80, Oct. - Dec. 1984

[6]   S.Tidwell, J. Seamans, M. Bowers, A. Cousins: Scaling cw diode-end-pumped Nd:YAG lasers to high average powers;
      IEEE Journal of Quantum Electronics, Vol. 28, No. 4, pp. 997 -1009, April 1992

[7]   K.Driedger, R. Iffländer, H. Weber: Multirod resonators for high-power solid-state lasers with improved beam quality;
      IEEE Journal of Quantum Electronics, Vol. 24, No. 4, pp. 665 - 674, April 1988

[8]   C.Pfisterer, P. Albers, H.P.Weber: Efficient Nd:YAG slab longitudinally pumped by diode lasers
      IEEE Journal of Quantum Electronics, Vol. 26, No. 5, pp. 827 - 829, May 1990

[9]   S.Tidwell, J. Seamans, C. Hamilton, C. Muller, D. Lowenthal: Efficient, 15-W output power, diode-end-pumped
      Nd:YAG laser; Optics Letters, Vol. 16, No. 8, pp. 584 - 586, April 15, 1991

[10] S.Tidwell, J. Seamans, M. Bowers: Highly efficient 60-W TEM_{00} cw diode-end-pumped Nd:YAG laser
      Optics Letters, Vol. 18, No. 2, pp. 116 - 118, January 15, 1993

Aufbau und Beurteilung eines Nd:YAG Grundmodelasers

Ch. Kolmeder, Carl Baasel Lasertechnik GmbH
Petersbrunner Str. 1b, D-8130 Starnberg

Die Motivation für diesen Vortrag kam aus der Erfahrung, daß der Anwender einen Laser meist nur nach dessen Leistung beurteilt und andere Kriterien, wie etwa das Strahlparameterprodukt, außer acht läßt. Deshalb richtet sich dieser Vortrag hauptsächlich an den Laseranwender.

Aufbau eines güte-geschalteten, cw-gepumpten Nd:YAG-Grundmodelasers

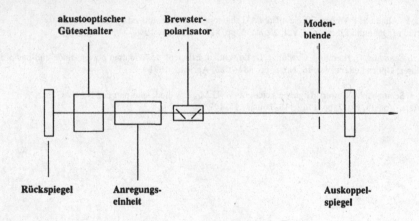

Abbildung 1

Der akusto-optische Güteschalter erzeugt Laserpulse von ca. 100 - 200 ns, abhängig von der Repetitionsrate, die im Bereich 1 - 30 kHz liegt. TEM_{00}-Strahlung wird erreicht mit Hilfe der Kombination aus Modenblende und Brewsterpolarisator.

Beurteilungskriterien, Meßmethoden

Die wichtigsten Kriterien zur Beurteilung eines Lasers sind:

○ Leistung
○ Pulsstabilität
○ Strahlparameterprodukt $w_0 \cdot \theta$

w_0 ist der Bündelradius in der Strahltaille; w ist der Radius, bei dem die Intensität eines Gauß-Bündels auf $1/e^2$ abgefallen ist. θ ist der volle Divergenzwinkel.

Da die Methoden zur Messung von Laserleistung und Pulsstabilität als bekannt vorausgesetzt werden können, soll hier nur auf Meßmethoden des Strahlparameterprodukts eingegangen werden.

Eine einfache und genaue Methode besteht darin, w_0 und θ mit kreisförmigen Lochblenden zu ermitteln. Voraussetzung dafür ist eine zirkulare Modenstruktur. Die Blendengröße wird an den entsprechenden Orten z so eingestellt, daß jeweils 86.5 % der Laserleistung gemessen werden, was dem $1/e^2$ Abfall der Intensität entspricht.

Gleichungen: (1) $w(z) = w_0 \left[1 + \left(\dfrac{\lambda z}{\pi w_0}\right)\right]^{1/2}$ Hyperbelgleichung

(2) $\theta \cdot 2w_0 = 1.27 \cdot \lambda$ Gleichung des Strahlparameterprodukts für einen beugungsbegrenzten Strahl

(3) $M^2 = \dfrac{(\theta 2w_0) \, real}{(\theta 2w_0) \, theor}$

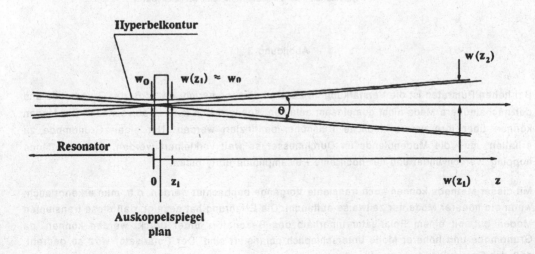

Abbildung 2

Bei einem planen Auskoppelspiegel liegt die Strahltaille genau an der Spiegeloberfläche. Außerhalb des Resonators unmittelbar am Spiegel kann mit einer Blende w_0 bestimmt werden. θ wird ermittelt, in dem $w(z)$ in ausreichend großer Entfernung gemessen wird, d.h. dort wo die Hyperbel in die Asymptote übergeht.

Aus den gemessenen Größen θ und w_0 kann das Produkt $\theta 2 w_0$ gebildet und mit Gleichung (2) überprüft werden, inwieweit der Laserstrahl beugungsbegrenzt ist, d.h. ob Grundmode vorliegt. Mit M^2 in Gleichung (3) wird der Faktor der Beugungsbegrenzung angegeben.

Eine andere Methode, den Laser auf Grundmode zu überprüfen, stützt sich auf das Phänomen des "Mode hopping" bei hohen Pulsraten (>10 kHz). Diese Meßmethode ist für den Produktions- und Serviceeinsatz geeignet, da zur Messung nur ein Oszilloskop benötigt wird. Eine schnelle Photodiode zur Erfassung der Lichtpulse ist in kommerziellen Lasern meistens vorhanden. In Abb. 3 ist das "Mode hopping" dargestellt.

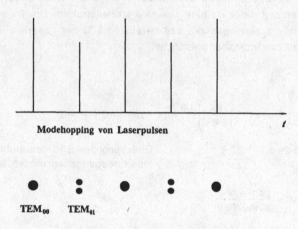

Modehopping von Laserpulsen

Korrelation von Pulsamplitude und Modenstruktur

Abbildung 3

Bei hohen Pulsraten ist die Verstärkung im Lasermedium so gering, daß z.B. der Grundmode und der nächsthöhere Mode nicht gemeinsam anlaufen, sondern sich zeitlich abwechseln. Die Moden können über ihre unterschiedliche Pulshöhe identifiziert werden. Um einen Grundmode zu erhalten, muß die Modenblende im Durchmesser so weit verkleinert werden, daß das "Mode hopping" verschwindet und nur noch eine Pulsamplitude übrig bleibt.

Mit dieser Methode können auch transiente Vorgänge beobachtet werden, d.h. man erkennt auch, wenn ein höherer Mode nur zeitweise auftaucht. Die Erfahrung hat gezeigt, daß diese transienten Moden gut mit einem Polarisator innerhalb des Resonators unter drückt werden können, da Grundmode und höherer Mode unterschiedlich polarisiert sind. Der Polarisator wird so gedreht, daß der Grundmode bevorzugt wird.

Einfluß der Lasereigenschaften auf die Applikation

Wichtige Größen bei Applikationen mit Grundmodelasern sind die erreichbare Intensität und der minimale Fokusdurchmesser bei ausreichender Tiefenschärfe. In Abb. 4 ist ein typischer optischer Aufbau dargestellt.

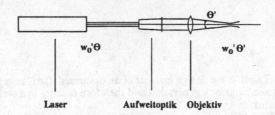

Abbildung 4

Ein Laserstrahl mit einem Strahlparameterprodukt $w_0\theta$ wird aufgeweitet und anschließend mit einem Objektiv fokussiert. Falls das optische System ausreichend aberrationsfrei ist, bleibt das Strahlparameterprodukt erhalten ($w_0\theta = w_0'\theta'$).

Die Intensität im Fokus ist definiert:

$$I = \frac{P}{A} \quad \text{mit } A = \pi w_0'^2, \text{ P ist die Laserleistung.}$$

mit Gleichung $w_0\theta = w_0'\theta'$ bzw. $w_0' = \frac{w_0\theta}{\theta}$

folgt:
$$I = \frac{\theta'^2}{\pi} \frac{P}{(w_0\theta)^2}$$

Die Intensität ist vom Strahlparameterprodukt quadratisch, von der Laserleistung nur linear abhängig.

In der folgenden Tabelle sind Daten von kommerziell erhältlichen Grundmodelasern angegeben:

Hersteller:	A	B	C
P [w]:	20	25	30
M^2	1,1	1,5	2,0
I_{norm}:	1	0,67	0,45

Die Beispiele zeigen, daß das Strahlparameterprodukt wesentlich stärkeren Einfluß (quadratisch) auf die Intensität hat als die Leistung, die üblicherweise immer in den Vordergrund gestellt wird.

Optimal Design of a Flashlamp Pumped free Running Nd:YAG Slab Laser for Harmonic Frequency Generation

T.Sidler, LOA/EPFL, CH-1015 Lausanne; P.Verboven, LASAG AG, CH-3600 Thun

1. Objective

The prime objective of this work is the design and test of an optimized flashlamp pumped free-running Nd-YAG slab laser for harmonic frequency generation and therefore running in a well controlled fundamental mode. Explicitely, this means:
– lowest possible wave front deformations
– no or very low pump dependent thermal lensing
– high pump efficiency (lamp to slab light transfer efficiency) in conjunction with most homogeneous pump light distribution
– highest possible fundamental mode volume inside the active material, compatible with sufficient stability of the resonator and the maximum pump induced thermal lensing.

2. Basic procedure

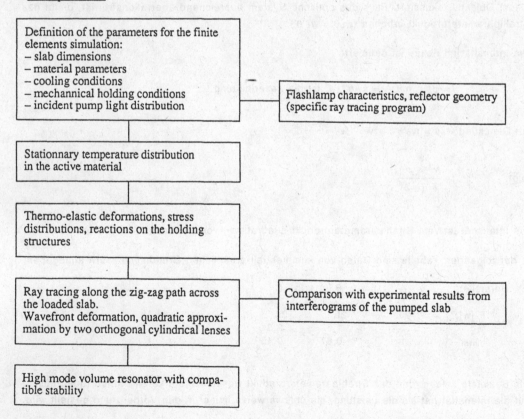

3. Simulation results

Slab length optimization: The principal factor for the wave front deformation is the ratio of the total slab length to the pumped slab length.The longer the unpumped end regions and the smoother the transitions, the smaller the resulting wave front deformations. The principal effect comes from a bending of the Brewster end faces which produce a divergent cylindrical lens over the slab thickness (y-direction), as a result of the refraction over the curved Brewster face and the total internal reflection at the surface just opposed (R1 of the figure 1).The curvature of the Brewster face in the orthogonal direction (x-direction) is practically negligible. The curvature in the x-direction of the adjacent total reflection is taken into account in the next point.

Slab width optimization: The thickness to width ratio of the slab determines the transverse profile (x-direction at the surfaces of total reflection) and therefore principally the wave front deformation in the x-direction.The absolute deformation at the center of the slab (x=0, $-z_{max}/2 < z < z_{max}/2$) is rather independent of the width of the slab, therefore the transverse curvatures of the surfaces decrease with increasing slab width. The transverse curvatures change notably at the ends of the slab as shown in figure 1.The absolute slab thickness and width are of some concern in conjunction with the pump reflector design and trade-offs between slab thickness, (absorption efficiency) and thermal deformations, and slab height, (pump reflector efficiency) and aspect ratio of the laser beam have to be solved.

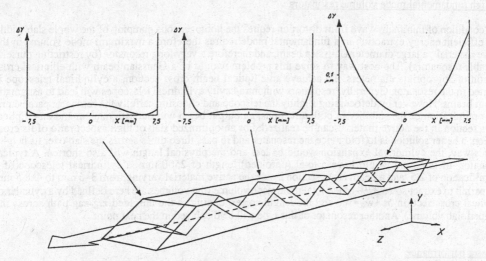

Figure 1   Three dimentionnal view of a slab with zig-zag beam. Wave front deformation results principally from bending of the totally reflecting surfaces. Three typical transverse profiles are given.

Table 1 gives the global results of the simulations in a quadratic approximation as represented by two orthogonal cylindrical lenses of specific focussing power ϕ_x/P, and ϕ_y/P. The transverse effect is mostly dependent on the slab width; in the other direction, the slab length is the principal factor.

	Standard 6x15x110 mm	long 6x15x130 mm	long+wide 6x20x130 mm	long+very wide 6x22.5x130 mm
ϕ_x/P	$2.2 \; 10^{-3}$	$4.0 \; 10^{-3}$	$0.41 \; 10^{-3}$	$0.20 \; 10^{-3}$ m^{-1}/kW
	Standard 6x15x110 mm	long 6x15x130 mm	very long 6x15x150 mm	
ϕ_y/P	$-1.5 \; 10^{-2}$	$-0.85 \; 10^{-2}$	$-0.14 \; 10^{-2}$	

Table 1 Pump induced specific focussing power in m^{-1}/kW, for different slab dimensions.

4. Interferometric wave front deformation

The interferograms of the pumped zig-zag slabs have been measured in a Mach-Zehnder interferometer. For both, the standard slab [1] and the optimized slab the two orthogonal transverse directions have been measured, the corresponding focussing power (quadratic approximation of the measured wave-front over the central part of the slab) is given in table 2.

	Standard 6x15x110 mm	Optimized 6.8x24x125 mm
$\emptyset_x/P$	$2.7\ 10^{-2}$	$1.6\ 10^{-2}$
$\emptyset_y/P$	$-3.0\ 10^{-2}$	$1\ 10^{-2}$

Table 2 Interferometrically measured specific focussing power for the standard and the optimized slab, in m^{-1}/kW.

5. High fundamental mode volume resonators

The condition of minimum wave front distortion reqires the homogeneous pumping of the whole slab width. The efficient energy extraction in a fundamental mode reqires therefore a maximum mode volume in the active material , a large diameter Gaussian beam, and therefore a very long resonator (by restricting ourselfs to stable resonators). The best way to solve this problem would be a Gaussian beam with elliptical cross section of appropriate diameters. To achieve an elliptical beam cross section, a cylindrical telescope is inserted in the resonator. Generally, resonators with intra-cavity cylindrical telescopes will lead to astigmatic output beams. However by defocussing slightly the telscope and choosing carfully the resonator parameters, the optimization of the stability in both directions, an output beam without astigmatism and an elliptical cross section in the active material can be realized. For an optimized slab of high aspect ratio of its cross section, a better solution is to fold twice the resonator and to pass three times across the slab over its height. In this way, the cylindrical expansion would be less and the physical length will also shorten. A typical resonator of this kind would have a total (unfolded) length of 2500 mm, a cylindrical telescope of a magnification of 2.5 and a beam cross section inside the active material varying from 3*6 mm to 4*8.5 mm and permit to extract some 37% of the total multimode volume (the volumes are here defined by a cylinder of elliptical cross section $\pi * w_x * w_y$ and the length corresponding to the unfolded zig-zag path across the pumped slab volume). Another resonator of this type will be displayed in the next point.

6. Laser performance

The optimized laser head has been tested in different resonator configurations and compared to the standard slab head. Table 3 gives the principal results of these tests. The high multimode power diagrams show clearly the superiority of the optimized design. For the quasi TEM 00 resonator, with the standard slab head the laser beam becomes instable for input powers above approx. 5 kW. The optimized slab head shows no efficiency decrease at high input power but the beam quality decreases continually with increasing pump power. Figure 2 shows the <Pin> - <Pout> diagrams for the results of table 3. For the quasi TEM00 resonator, the beam quality q* dependence is also given.

	$<P_{max}>$	efficiency	q*
High multimode	268 W	3.84 %	---
Low order mode	187 W	2.71 %	2.5 x 3.2 mm*mrad
Quasi TEM 00	51.5 W	0.77 %	0.50 - 1.2 mm*mrad
Stand.slab, high m.m.	130 W	2.39 %	---

Table 3 Laser performance for standard and optimized slab head with different resonators (* max. value, power supply limited).

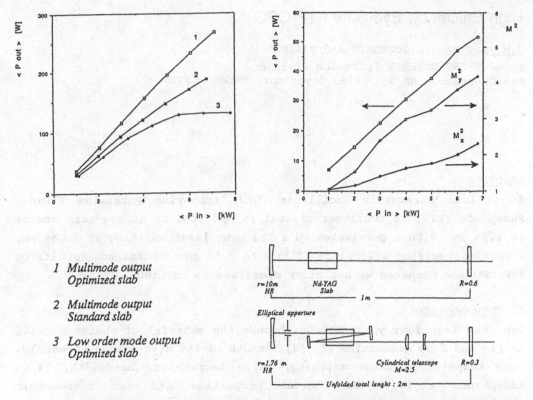

1 Multimode output
   Optimized slab

2 Multimode output
   Standard slab

3 Low order mode output
   Optimized slab

Figure 2   <Pin>-<Pout> diagrams for different resonators. a) Multimode resonator A, with standard and optimized slab head, and low order mode resonator B (without aperture) with optimized slab. b) Optimized slab in the resonator B, with elliptical fundamental mode selecting aperture.

7. Conclusions

The simulation of the thermo-elastic and optic behavior of a thermally loaded Nd-YAG slab by a numerical finite elements method allows to separate the influence of the different parameters, slab length, slab width and thickness, and also pump light distribution. Based on the simulation results, an optimized slab geometry has been designed and realized. The improvement in waveform deformation, measured in a Mach-Zehnder interferometer, is roughly a factor of 2. The multimode laser efficiency has improved from 2.4% for the standard slab laser [1] to 3.8% for the optimized slab head (measured in the same resonator). For quasi fundamental mode operation of the new slab head, a doubly folded resonator with an intra cavity cylindrical telescope has been implemented and yields an efficiency of 0.77% at a beam quality of 0.5 to 1.2 mm∗mrad (M^2 from 1.2 to 4) , depending on the mean input power. A rough extrapolation for the harmonic frequency generation shows that with the following beam parameters: $E_p = 1J$, t = 0.25 ms, $M^2 = 1.2$, for an intensity of 50 MW/cm^2, the Rayleigh range z_R would be of 6.3 mm and the beam divergence in this range of 4 mrad. This result compares favorably with the requirements for NCPM in LBO [2] for instance, with acceptance angles of $\Delta O \ast l^{1/2}$ of 72 mrad∗cm$^{1/2}$ and $\Delta \phi \ast l^{1/2}$ of 99 mrad∗cm$^{1/2}$.

[1]   H.P.von Arb,C.Lüchinger, F.Studer, U.Dürr, A.Gressli, T.Sidler, J.Steffen, J.C.Poli, High average power slab geometry solid state lasers, SPIE Proceedings Vol 1021, 1988, pp 565-568.

[2]   T.Ukachi, R.J.Lane, W.R.Bosenberg, C.L.Tang, Measurements of noncritically phase-matched second harmonic generation in a LBO-crystal, Appl.Phys.Lett.57 (10), 1990 pp 980-982.

High Efficiency, Eye-Safe KTP OPO

J.M. BRETEAU, C.JOURDAIN and F.SIMON
THOMSON-TRT-DEFENSE Optronics Division
Rue Guynemer - BP 55 78283 Guyancourt Cedex - France

ABSTRACT
An Optical Parametric Oscillator (OPO) employing Potassium Titanyl Phosphate (KTP) as nonlinear crystal is operated as an eye-safe source at 1.58 μm. With a Q-switched Nd : YAG pump laser emitting at 1.064 μm, a total conversion efficiency of 35 % to 1.58 μm is obtained. Results on KTP OPO are compared with similar measurements on $LiNbO_3$ OPO.

1. INTRODUCTION
Over the last four years KTP has become the material of choice for 1.6 μm [1] and 2 μm generation [2] [3] because of its high damage threshold, room temperature phase matching, large temperature bandwidth, large acceptance angle, good thermal properties and non hygroscopic characteristics. All these properties make it very attractive to use KTP OPO as eye-safe sources with an output wavelength beyond 1.5 μm. While such eye-safe outputs can be obtained from erbium lasers [4] or by Raman shifting in high pressure methane cells [5], Optical Parametric Oscillators can provide a more efficient and relatively simpler means of frequency shifting at higher Pulse Repetition Frequency (PRF).
We report the performance of a 100Hz (PRF) Nd : YAG pumped OPO operating at 1.58 μm and producing pulse energies of 15 to 20 mJ for both signal and idler emissions. The energy conversion of pump to eye-safe energy is 35 %.

2. EXPERIMENTAL
The experimental arrangement is shown in Figure 1. The OPO cavity was composed of one KTP crystal. The crystal was 4 x 4 mm² in cross-section and 17 mm in length.

Since noncritical phase matching maximizes the effective non linear coefficient and eliminates walk-off, the crystal was cut to achieve the optical fields propagation along the crystal's X axis, with the pump and 1.58 μm signal polarized along the Y axis and the idler polarized along the Z axis.

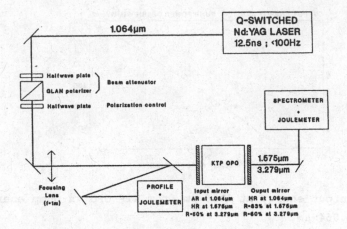

FIGURE 1 : SCHEMATIC DIAGRAM OF THE EXPERIMENT

The crystal was centred inside an optical cavity formed by a pair of plane parallel mirrors. The input mirror, through which the pump is launched, was anti-reflection coated at the pump wavelength (1.064 μm) and highly reflecting at the signal wavelength (1.58 μm); whilst the output mirror was highly reflecting at the pump wavelength and 17 % transmitting at the signal wavelength.

The pump laser was a Q-switched injection seeded Nd : YAG oscillator operating in a quasi gaussian transverse mode with a 18 ns pulse width and up to 100 Hz pulse repetition rate. The collimated pump beam had a cross-sectional area of 0.02cm^2 at OPO crystal. The OPO cavity and KTP crystal were misaligned by 4 mrad in φ from the pump beam to prevent feedback to the pump laser. Pump energy was varied by an attenuator consisting of a half wave plate and a Glan prism polarizer. Input pump energy was monitored by measuring the rejected component from a glass slide pickoff before the cavity. The OPO output wavelengths were recorded on a ¼ meter spectrometer capable of 0.3 nm resolution and calibrated to within 1nm accuracy.

3- OPO RESULTS

The output energy at 1.58 μm is plotted as a function of 1.064 μm pump energy in Figure 2. Threshold for parametric oscillation is reached for 5.7 mJ pump energy corresponding to a pump intensity around 35 MW/cm^2. The 1.58 μm energy increases with pump energy, with a slope efficiency of 44 %. The energy conversion efficiency to 1.58 μm was 34 %.

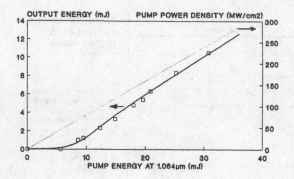

FIGURE 2 : output energy at 1.58 μm from KTP OPO vs pump energy at
1.064 μm.

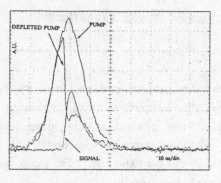

FIGURE 3 : Oscilloscope trace showing input and depleted pump at
1.064 μm, and signal output at 1.58 μm.

An oscillogram showing the input and backward depleted pump together with
the generated signal pulse, is presented in Figure 3. From the difference
between input and depleted pump pulses, about 50 % energy conversion of
pump to both signal and idler outputs can be estimated. At 1.58 μm this
pump depletion figure gives 33 % energy conversion to the signal alone.
The estimate is in good agreement with the efficiency determined directly
from 1.58 μm energy measurements in Figure 2.

The wavelength measurements revealed a signal wavelength at 1.575 μm and
an idler wavelength at 3.279 μm. Using the Sellmeier's equations
published by KATO [6] the wavelength of the signal was calculated to be
1.579 μm for an X cut crystal and type II non critical phase matching.
This value is in good agreement with the measured wavelength. The
effective non-linear coefficient value could be determined earlier
[7] and was found to be 1.25 pm/V which is a factor of two lower than
reported in [8].

4- COMPARISON WITH LiNbO₃ OPO

In order to compare OPO performances in terms of total efficiency and thermal sensitivity, we have replaced the KTP crystal with a LiNbO₃ crystal cut at θ = 47° and φ = 90° for type I phase matching. This crystal was 7 x 7 mm² in cross-section and 50mm in length.

Under similar experimental conditions, the output energy at the defined eye-safe wavelength 1.54 μm is plotted as a function of 1.064 μm pump energy in Figure 4. Threshold for parametric oscillation is reached for 7 mJ pump energy corresponding to a pump intensity of 45 MW/cm².The 1.54μm energy increases with pump energy, with a slope efficiency of 41 %. The energy conversion efficiency to 1.54 μm was limited at 28 % due to optical damage in LiNbO₃ at pump intensities beyond 250 MW/cm².

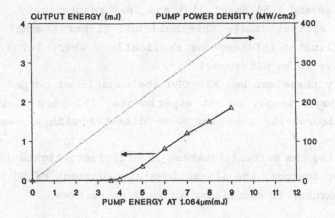

FIGURE 4 : Output energy at 1.54 μm from LiNbO₃ OPO vs.pump energy at 1.064 μm.

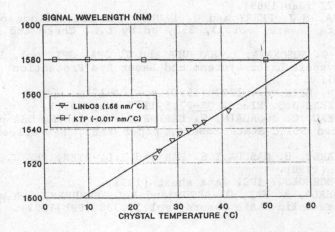

FIGURE 5 : Thermal dependence of the signal wavelength for KTP and LiNbO₃ OPO's.

In spite of a higher effective non-linear coefficient for $LiNbO_3$ ($d_{eff} \approx 5$ pm/V [9]), the conversion efficiency is also limited by a large walk-off effect ($\rho = 35$ mrad) resulting in a rather poor overlap between the optical fields interacting inside the crystal.

The thermal dependence of the signal wavelength for KTP and $LiNbO_3$ OPO's is shown in Figure 5. High sensitivity to thermal fluctuation is observed for $LiNbO_3$ OPO with a wavelength shifting rate of 1.58 nm/°C as compared to -0.017nm/°C in KTP OPO.

5- CONCLUSIONS

We have operated KTP as well as $LiNbO_3$ OPO's in the eye-safe range 1.5-1.6 μm and have demonstrated energy conversion efficiencies from 1.064 μm pump to 1.58 μm and 1.54 μm of 34 % and 28 % respectively.

Due to lower optical damage threshold and higher thermal sensitivity $LiNbO_3$ has a limited interest for applications where large temperature variations have to be withstood.

Non-critically phase matched KTP OPO are capable of output energies in the 10 millijoule range. Recent experiments [10] show that scaling is possible by increasing the pump beam diameter with a beam expanding telescope.

When considering the current interest in efficient, lightweight, rugged, eye-safe laser sources, the all-solid-state approach of Nd : YAG pumped KTP optical parametric oscillator is very attractive.

[1]   L.R. MARSHALL, A. KAZ and R.L. BURNHAM, SPIE Vol. 1627, Solid-State Lasers III (1992) p. 262-272.
[2]   R. BURNHAM, R.A. STOLZENBERGER and A. PINTO, IEEE Phot. Tech. Lett.1,1,27 (Jan.1989).
[3]   S. CHANDRA, M.J. FERRY and G. DAUNT, OSA proceedings on Advanced Solid-State Lasers, vol.13, 353, ed.by L.L. Chase and A.A. Pinto (Feb.1992).
[4]   K. ASABA, T. HOSOKAWA, Y. HATSUDA and J. OTA, SPIE vol. 1207, Laser Safety, Eyesafe Laser Systems and Laser Eye Protection (1990) p. 164-171.
[5]   J.F. RUGER, SPIE vol. 1207 (1990) p. 164-171.
[6]   K. KATO, IEEE J.Q. Elect. QE-27,5,1137 (May 1991).
[7]   J.M. BRETEAU, C. JOURDAIN, T. LEPINE and F. SIMON, OSA proceedings on Advanced Solid-State Lasers, paper AME9, New-Orleans Feb.1-3 1993.
[8]   R.C. ECKHARDT, H. MASUDA,Y.S. FAN and R.L. BYER,IEEE J.Q. Elect. QE-26,922 (1990).
[9]   CRYSTAL TECHNOLOGY INC. data sheet (1991).
[10] L.C. MARSHALL, A. KAZ, O. AYTUR and R.L. BURNHAM, OSA proceedings on Advanced Solid-State Lasers, vol.13,338 (Feb.1992).

Analyzing Astigmatic Beam Propagation by Means of the Complex-Ray Concept

Baida Lü   Guoying Feng   Bangwei Cai
(Department of Opto-Electronic Science & Technology,
Sichuan University, Chengdu 610064 China)

1. Introduction

Experiments[1]-[3] have shown, that beams emerging from solid-state slab lasers, semiconductor lasers exhibit astigmatic characteristics. In addition, the application of axis-asymmetric optical elements and/or resonators to solid-state slab lasers leads to the generation of generally astigmatic beams, hybrid beams (i.e. Gaussian beam-spherical wave) and astigmatic spherical waves[4]-[7]. As yet, some useful methods such as the generalized ABCD law and the eigenray-vector method were presented to analyze the characteristics of generally astigmatic beams and resonators, details are found in Refs[5]-[7]. The subject of the present paper is to extend the complex ray concept, originally proposed by Arnaud[8], to the general case, and then to study the astigmatic beam propagation by using this method.

2. Theoretical analysis

The basic idea of representing Gaussian beams by complex rays[8] is that in the paraxial approximation a circular Gaussian beam can be simulated by a complex ray with the position of $r(z) = r_r(z) + i r_i(z)$ and the direction of $p(z) = p_r(z) + i p_i(z)$, or equally, by two real rays $\begin{pmatrix} r_r(z) \\ p_r(z) \end{pmatrix}$, $\begin{pmatrix} r_i(z) \\ p_i(z) \end{pmatrix}$. The complex parameter $q(z)$ of the Gaussian beam is related with $r(z)$, $p(z)$ by

$$q^{-1}(z) = \frac{p(z)}{r(z)} = \frac{p_r(z) + i p_i(z)}{r_r(z) + i r_i(z)} \qquad (1)$$

where the subscripts r, i denote the real and imaginary parts, respectively. It is well-known that an arbitrary ray in the space can be represented by coordinates r_x, r_y in the transverse plane (x,y) and directional derivatives p_x, p_y, i.e., $\begin{pmatrix} r_x \\ r_y \\ p_x \\ p_y \end{pmatrix}$. In the paraxial approximation we have

$$p_x = \frac{dr_x}{dz}, \quad p_y = \frac{dr_y}{dz} \qquad (2)$$

In the extension of Arnaud's method to the astigmatic beams, which can be Gaussian beams with simple or general astigmatism, hybrid beams or astigmatic spherical waves in general, it is necessary to represent the beam by two complex rays $\begin{pmatrix} r_{1x} \\ r_{1y} \\ p_{1x} \\ p_{1y} \end{pmatrix}$, $\begin{pmatrix} r_{2x} \\ r_{2y} \\ p_{2x} \\ p_{2y} \end{pmatrix}$ (3)

written in terms of the matrix form

$$\mathbf{r} = \begin{pmatrix} r_{1x} & r_{2x} \\ r_{1y} & r_{2y} \end{pmatrix} , \quad \mathbf{p} = \begin{pmatrix} p_{1x} & p_{2x} \\ p_{1y} & p_{2y} \end{pmatrix} \tag{4}$$

or equally, by four real rays, which are the real and imaginary parts of (3) respectively

$$\begin{pmatrix} r_{jx} \\ r_{jy} \\ p_{jx} \\ p_{jy} \end{pmatrix} = \begin{pmatrix} r_{jrx} \\ r_{jry} \\ p_{jrx} \\ p_{jry} \end{pmatrix} + \iota \begin{pmatrix} r_{jix} \\ r_{jiy} \\ p_{jix} \\ p_{jiy} \end{pmatrix} \quad (j = 1, 2) \tag{5}$$

The transformation of $\mathbf{r}$, $\mathbf{p}$ through an axis-asymmetric optical system with the transfer matrix $\begin{pmatrix} \mathbb{A} & \mathbb{B} \\ \mathbb{C} & \mathbb{D} \end{pmatrix}$ obeys the geometrical law

$$\begin{pmatrix} \mathbf{r'} \\ \mathbf{p'} \end{pmatrix} = \begin{pmatrix} \mathbb{A} & \mathbb{B} \\ \mathbb{C} & \mathbb{D} \end{pmatrix} \begin{pmatrix} \mathbf{r} \\ \mathbf{p} \end{pmatrix} \tag{6}$$

or written as

$$\begin{cases} \mathbf{r'} = \mathbb{A}\mathbf{r} + \mathbb{B}\mathbf{p} \\ \mathbf{p'} = \mathbb{C}\mathbf{r} + \mathbb{D}\mathbf{p} \end{cases} \tag{7}$$

where $\mathbb{A}$, $\mathbb{B}$, $\mathbb{C}$, $\mathbb{D}$ are all 2×2 matrices, the quantities without/with prime are those in the input / output reference planes.

Assume that the complex wavefront matrices of an astigmatic beam are $\mathbb{Q}^{-1}$ and $\mathbb{Q}^{-1'}$ in the input and output reference planes, it follows from the generalized ABCD law[5][6] that

$$\mathbb{Q}^{-1'} = (\mathbb{C} + \mathbb{D}\mathbb{Q}^{-1})(\mathbb{A} + \mathbb{B}\mathbb{Q}^{-1})^{-1} \tag{8}$$

which can be reformed as

$$\mathbb{Q}^{-1'} = (\mathbb{C} + \mathbb{D}\mathbb{Q}^{-1}) \mathbf{r}\mathbf{r}^{-1} (\mathbb{A} + \mathbb{B}\mathbb{Q}^{-1})^{-1} = (\mathbb{C}\mathbf{r} + \mathbb{D}\mathbb{Q}^{-1}\mathbf{r})(\mathbb{A}\mathbf{r} + \mathbb{B}\mathbb{Q}^{-1}\mathbf{r})^{-1} \tag{9}$$

Recalling the relation[5]

$$\mathbf{p} = \mathbb{Q}^{-1}\mathbf{r} \tag{10}$$

and combing with Eq.(7) lead to

$$\mathbb{Q}^{-1'} = (\mathbb{C}\mathbf{r} + \mathbb{D}\mathbf{p})(\mathbb{A}\mathbf{r} + \mathbb{B}\mathbf{p})^{-1} = \mathbf{p'} \mathbf{r}^{-1'} \tag{11}$$

or rewritten as

$$\mathbf{p'} = \mathbb{Q}^{-1'} \mathbf{r'} \tag{12}$$

which has the same form of Eq.(10). Consequently, Eq.(10) is a basic relation between the beam and ray characteristic parameters while representing astigmatic beams by complex rays. It can be easily shown, that in the case discussed by Arnaud[8], Eq.(10) is reduced to Eq.(1).

From the above discussion the procedures modeling the astigmatic beam propagation by complex rays can be summarized as follows
1) Starting from $\mathbb{Q}^{-1}$ of astigmatic beams (its physical significance and calculation methods are found in [5],[6] and omitted here), we find out the complex ray parameters $\mathbf{r}$, $\mathbf{p}$ corresponding to $\hat{\mathbb{Q}}^{-1}$, which satisfy Eq.(10). In the paraxial approximation $\mathbf{r}$,$\mathbf{p}$ are not unique, but lead to the same result of $\mathbb{Q}^{-1'}$, as long as Eq.(10) is satisfied. e.g.,(1) For simply astigmatic Gaussian beams it is available to use Arnaud's waist and divergence rays with a small modification, i.e.,

$$\mathbf{r}=\begin{pmatrix} w_{0x}-i\dfrac{\lambda L_{0x}}{\pi w_{0x}} & 0 \\ 0 & w_{0y}-i\dfrac{\lambda L_{0y}}{\pi w_{0y}} \end{pmatrix} \tag{13}$$

$$\mathbf{p}=\begin{pmatrix} -i\dfrac{\lambda}{\pi w_{0x}} & 0 \\ 0 & -i\dfrac{\lambda}{\pi w_{0y}} \end{pmatrix} \tag{14}$$

where L_{0x} ,L_{0y} are the distances between the waists w_{0x} , w_{0y} and the reference planes, λ is the wave length. (2) For astigmatic resonators the eigenrays[7] can be used. (3) Assume

$$\mathbf{r}=\begin{pmatrix} 1 & 0 \\ 0 & 1 \end{pmatrix} (mm) \tag{15}$$

from Eq.(10), we get $\qquad\qquad \mathbf{p}=\mathbb{Q}^{-1} \tag{16}$
It is worth noting that an additional restriction for choosing $\mathbf{r}$, $\mathbf{p}$ is the existence of the inverse matrices of $\mathbf{p}$, $\mathbf{r}$, it means that their determinant should not be zero.
2) The propagation of $\mathbf{r}$, $\mathbf{p}$ through axis-asymmetric optical systems is described by Eq.(6). In accordance with Eqs.(5)-(7), it is possible to draw the trace of the four real rays, and $\mathbb{Q}^{-1'}$ is calculated by means of Eq.(11).
3) Finally, the beam parameters are obtained from $\mathbb{Q}^{-1'}$ by using the well-known matrix diaganalized procedures[6].

3.Application examples
The widely applicable characteristics of the complex ray and the equivalence with the methods discussed in [6,7] have been confirmed by our numerical calculation results. Here the propagation of astigmatic hybrid beams and spherical waves through axis-asymmetric optical systems are chosen as typical application examples. Assume that the astigmatic beam is generated from a crossed cylindrical resonator shown in Fig.1, which is formed by two cylindrical mirrors M_1 , M_2 with principal curvature radii R_1,

R_2, and their generatrices are oriented at angles θ_1, θ_2 with respect to the y axis, respectively, and the resonator length is L. The laser beam reflected on M_1(RP) is chosen as the initial one for our calculation, which passes then through three cylindrical lenses F_1, F_2, F_3 of focal lengths f_1, f_2, f_3 successively, and finally arrives at the output reference plane RP'. The generatrices of

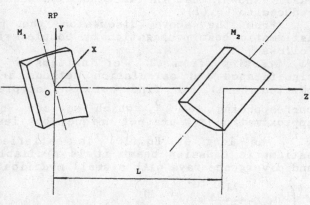

Fig.1.A crossed cylindrical mirror resonator.

the three lenses make angles ϕ_1, ϕ_2, ϕ_3 with respect to the y axis, and the distances between RP and F_1, F_1 and F_2, F_2 and F_3, F_3 and RP' are L_1, L_2, L_3 and L_4, respectively. In the following, two cases are of consideration.

1) Astigmatic Gaussian beam-spherical wave

Let R_1=6m, R_2=-4m, L=1m, θ=-40°, θ=90°. It turns out from the astigmatic resonator theory[6], that the two cylindrical mirrors constitute a stable-unstable resonator with general astigmatism. By means of the matrix method, the complex wave front Q^{-1} of the reflected astigmatic Gaussian beam-spherical wave on the mirror M_1 can be calculated, which is given by

$$Q^{-1} = \begin{pmatrix} -5.436\times10^{-5}-\iota3.026\times10^{-4} & -2.052\times10^{-4}-\iota8.540\times10^{-5} \\ -2.052\times10^{-4}-\iota8.540\times10^{-5} & 2.803\times10^{-4}-\iota2.410\times10^{-5} \end{pmatrix} \text{ (mm)}$$

(17)

and the corresponding eigenray parameters are

$$r = \begin{pmatrix} 0.5 & 2.0 \\ -1.772 & 0.7055 \end{pmatrix} \text{ (mm)}$$

(18)

$$p = \begin{pmatrix} 3.364\times10^{-4} & -2.535\times10^{-4}-\iota6.654\times10^{-4} \\ -5.991\times10^{-4} & -2.126\times10^{-4}-\iota1.878\times10^{-4} \end{pmatrix}$$

(19)

Assume that f_1=1.5m, f_2=2.5m, f_3=-2.5m, ϕ_1=0°, ϕ_2=-65°, ϕ_3=-30°, L_1=1.5m, L_2=0.5m, L_3=1m, L_4=1m, from which the 4×4 transfer matrix of the optical system is obtained[6][9]. Then, according to the calculation procedure discussed in Sec.2, r', p', $Q^{-1'}$ can be computed. The numerical calculations and ray-tracing are all performed on a super-386 computer, and the results are shown in Figs.2,3.

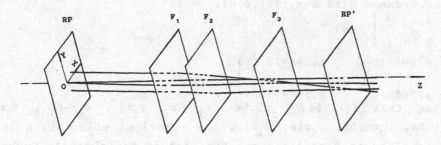

Fig.2. Ray tracing through a system of three cylindrical lenses. The initial values of **r** , **p** are given by Eqs.(18)(19). Parameters are seen in the text.

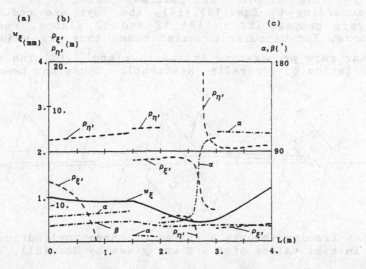

Fig.3. (a) Spot size w_ξ (———) ; (b) wavefront curvature radii ρ_ξ, ,ρ_η, (– – – –) of the astigmatic hybrid beam and (c) diagonalized angles α ,β (–·–·–·–) of the real and imaginary parts of Q^{-1} vs propagation distance z. Parameters for the calculation are the same as in Fig.2.

2).Astigmatic spherical waves

Let R_1=-6m, R_2=-4m, L=1m, θ_1=-30°, θ_2=90°, the resonator becomes unstable-unstable, and the calculations similar to the above yields Q^{-1} on the mirror M_1

$$\mathbb{Q}^{-1} = \begin{pmatrix} 4.904 \times 10^{-4} & 1.765 \times 10^{-4} \\ 1.765 \times 10^{-4} & 5.387 \times 10^{-4} \end{pmatrix} (mm^{-1}) \tag{20}$$

In accordance with Eqs.(15),(16), we get

$$\mathbb{r} = \begin{pmatrix} 1 & 0 \\ 0 & 1 \end{pmatrix} (mm) \tag{21}$$

$$\mathbb{p} = \begin{pmatrix} 4.904 \times 10^{-4} & 1.765 \times 10^{-4} \\ 1.765 \times 10^{-4} & 5.387 \times 10^{-4} \end{pmatrix} \tag{22}$$

Assume that $f_1 = 1.5m$, $f_2 = 2.5m$, $f_3 = -1m$, $\phi_1 = 0°$, $\phi_2 = -65°$, $\phi_3 = -30°$, $L_1 = 1.5m$, $L_2 = 0.5m$, $L_3 = 1m$, $L_4 = 1m$, the numerical calculation results are complied in Figs.4,5. From Figs.2-5 it follows: (1) Generally, we have to use four real skew rays to simulate the propagation of astigmatic beams (see Fig.2), which are reduced to tree ones in some cases exemplarily shown in Fig.4 , where one of the three rays is the ray passing through the optical axis z. In the simple astigmatism case , if $\mathbb{r}$, $\mathbb{p}$ are chosen according to Eqs.(13),(14), the rays are reduced to the planar rays propagating in the XZ and YZ planes, respectively. Furthermore, for circular Gaussian beams they are reduced to the two planar rays propagating in the rz plane[5].(2) The propagation characteristics of generally astigmatic Gaussian beam-spherical

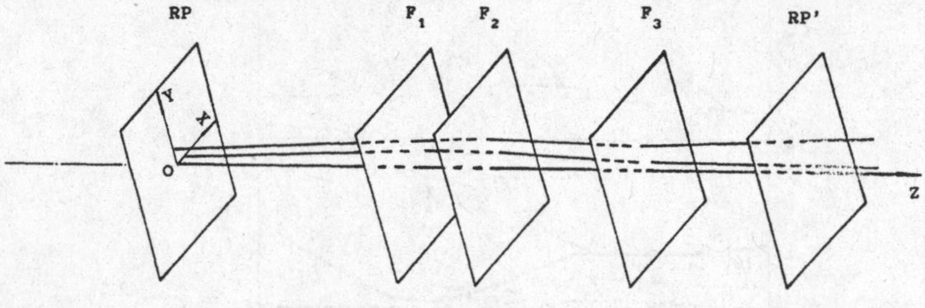

Fig.4. Ray tracing through a system of three cylindrical lenses. The initial values of $\mathbb{r}$, $\mathbb{p}$ are given by Eqs.(21), (22).

wave and the corresponding ray tracing are described by Figs.2.3, i.e., $\rho_\xi \neq \rho_\eta$, , $\alpha \neq \beta$, and α, β are variable with the propagation distance z . (3) Fig.5 points out that the diaganalized angle α of the real part of $\mathbb{Q}^{-1}$ is a constant while the astigmatic spherical waves propagate in the free space between the cylindrical lenses, demonstrating simply astigmatic characteristics, which can also be shown more strictly by means of matrix algebras[10].

4.Conclusion

Arnaud's method, i.e., the representation of Gaussian beams by complex rays, has been extended to modeling the propagation of astigmatic beams through axis-asymmetric optical systems, showing generally applicable characteristics. The equivalent complex rays satisfying Eq.(10) carry the important informations of the corresponding astigmatic beam parameters, such as beam sizes, curvature radii and diagonalized angles of the complex wavefront

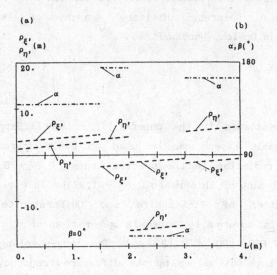

Fig.5. (a) Wavefront curvature radii ρ_ξ, ,ρ_η, (- - -)of the astigmatic spherical waves and (b) diagonalized angle α (—·—·—) of the real part of Q^{-1} as a function of propagation distance z. Parameters for the calculation are the same as in Fig.4.

matrix, which can be obtained simply from the ray parameters. Finally, we would like to point out that there are some invariants of complex rays related to the astigmatic beam properties[11], and the further applications of the complex-ray concept to the design of optical systems and resonators are possible, the relevant results will be published elsewhere.

This work was supported by the State Education Commission of China and the National Science and Technology Foundation 410-01-9.

References

1. E.Acosta et al., Opt.Lett., 16(1991)627.
2. N.Hodgson et al., Laser & Optoelektronik, 23(1991) 82.
3. N.Hodgson et al., Laser & Optoelektronik 24(1992) 54.
4. M.K.Chun, USP 4(1985) 559, 627.
5. J.A.Arnaud, BSTJ, 49(1970)2310.
6. B.Lü et al., Opt. & Quant. Electron., 24(1992)619.
7. B.Lü et al., Optik 90(1992)158.
8. J.A.Arnaud, Appl. Opt., 24(1985)538.
9. B.Lü, "Laser Optics", 2nd edition, Sichuan University Press, 1992.
10.G.Feng et al., "Propagation Characteristics of astigmatic beams in free space", J. Sichuan University (to be published)
11.B.Lü et al., "Invariant relations of complex rays in propagation and transformation" (to be published)

Subharmonic Generation of Tunable Ti:Sapphire Laser

G. A. Skripko, I. G. Tarazewicz, Z. Jankiewicz[*], R. Wodnicki[**]
FOTEK, Minsk/rep. of Belarus, [*]Military Academy of Technology, Warsaw/PL; [**]Solaris Optics, Warsaw/PL

As it was mentioned in the papers [1,2] the Ti:Sapphire and Cr:Forsterite tunable lasers can be used as basic sources for getting tunable coherent radiation in the spectral range 0,2 - 5,5 µm and even wider. The central part of this range 0,68 - 1,37 µm is covered by the main frequency generation of Ti:Sapphire and Cr:Forsterite lasers. The short-wavelength is covered by harmonic generation of the new nonlinear crystals, such as BBO, LBO, KTP [3 - 5]. The long-wavelength part of the above mentioned range was gained by the differece frequency generation of tunable lasers (in the dual-wavelength regime) or tunable and pump lasers. In this case the main problem is the synchronization of pulses to be mixed. Different methods were used for this problem solution [6,7]. Besides the synchronization difficulty, this method has a principal deficiency - energetic efficiency limit at a given wavelength. In this proces the long-wavelength is loosing and the energetic efficiency is less than one. The middle and far IR generation efficiency is especially noneffective. In this case the long-wavelength quantum energy ($h\nu_1$) is comparable with short-wavelength quantum $h\nu_s$.

This work deals with the investigation of tunable lasers conversion in the IR region with the theoretical energetic efficiency one. The process of subharmonic generation $\omega/2$, $\omega/4$ etc. gives such an ability. Subharmonic generation is realized in the degenerated OPO. In this case the output has two frequency-equal quantums and the theoretical energetic efficiency can reach value one. The OPO tuning is carried out by pump wavelength and nonlinear crystal simultaneously tuning.

It is possible to get the spectral range 1,32 - 2,74 µm, with $\omega/2$ generation and 2,64 - 5,48 µm for $\omega/4$, using for pumping Ti:Sapphire and Cr:Forsterite lasers.

The experiments were carried out with the use of the tunable Ti:Sapphire laser which has the characteristics presented in the Table 1.

TABLE 1

The output characteristics of Ti:Sapphire laser.

Tuning range, λ, μm	Linewidth, nm	Output energy at λ=780 nm	Pulsewidth, t, ns	Divergence θ, mrad
0,69 - 0,98	0,1	10	8	1

The nonlinear KTP crystal was used for the OPO (KTP crystal was grown in the laboratory, headed by V. N. Semenenko, Novosibirsk). The crystal was 14 mm long and cut at an angle θ = 57°, φ = 90° (type II interaction). There was no antireflected coating.

OPO cavity (Fig. 1) was formed by two flat mirrors: M1 (R = 99,5%, λ = 1,45 - 1,9 μm) and M2 (R = 90%, λ = 1,48 - 1,9 μm). The cavity length was 18 mm.

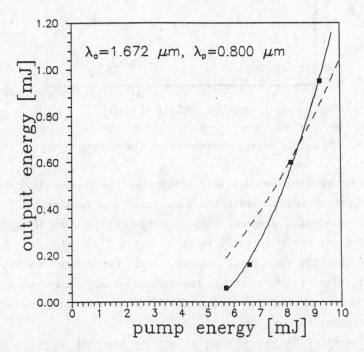

Fig. 1. OPO output energy dependence on pump energy for degenerated regime and the OPO optical set-up.

λ_p - pump wavelength, λ_o - output wavelength.

Figure 1 shows the experimental set-up and the generation energy dependence on pump energy for λ_p = 756 nm. It is evident that almost 31% of slop efficiency is achieved.

In Fig.2 the OPO output energy dependence on the cavity length L is shown.

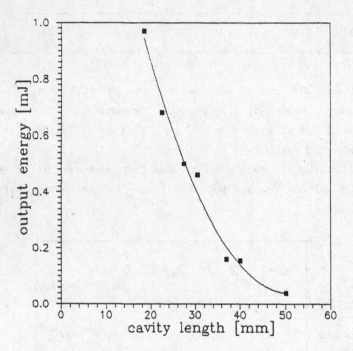

Fig. 2. OPO output energy dependence on the cavity length.

There is one inconvenience while realizing the degenerative OPO. It is necessary to tune synchronously the pump laser and nonlinear OPO crystal. OPO output wavelength changes, while tuning the pump wavelength without nonlinear crystal rotation, as it is shown in Fig.3.
It is seen that the orthogonal component (OC) (in comparison with a pump beam polarization) in OPO output is tuned considerably faster than collinear one (CC). It may be used for realization of widely tunable source, without nonlinear crystal tuning.

Experimentally it can be carried out by the OPO output mirror (OM) coating in such a way, that it was high-reflective for CC and transparent for OC. In our experiment the cavity was formed by two flat mirrors with the following characteristics: M1 (R ~ 99,5%, λ = 1,45-1,9 μm) and M2 (τ ~

90-95%, λ = 1,63-1,8 μm; τ ~ 99,5%, λ = 1,45-1,58 μm). The OPO output energy dependence on pump energy for this regime is shown in Fig.4. It is seen that the slope efficiency in this case is higher than in the degenerated OPO (dotted line) and it is about 36% . It is stipulated, to our mind, by the absence of subharmonic transformation into the pump wave. The higher threshold is due to the only one wave , taking part in the generation development.

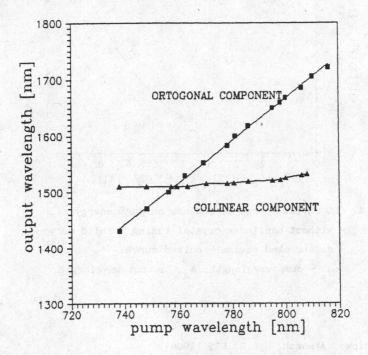

Fig. 3. Generation wavelength dependence on pump wavelength.

As in this case there is no need in tuning of OPO crystal, it gives the possibility to coat the cavity mirrors directly to the crystal surfaces, to minimize the cavity length (see Fig.2).

Suuming up the results presented above we can conclude:
- the capability of subharmonic generation of tunable lasers with more than 30% slope efficiency was demonstrated;
- the OPO without nonlinear crystal tuning was realized. This permits to achieve the higher efficiency and wider OPO tuning range.

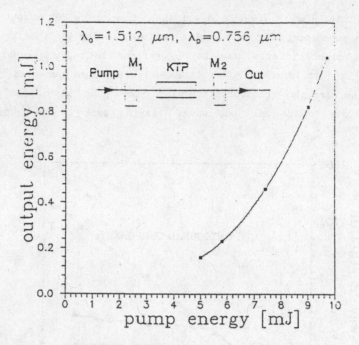

Fig. 4. OPO output energy dependence on pump energy:
- without nonlinear crystal tuning - solid curve;
- degenerated regime - dotted curve.

λ_p - pump wavelength, λ_o - output wavelength.

REFERENCES

1. G.A.Skripko, Atmosph. Opt.2, 675 (1989).

2. G.A. Skripko, Laser and Optoelectronic Engineering. Minsk: Universitetskoye, 14 (1989).

3. A.Borutzky, R.Brunger, Ch.Huang and R.Wallenstein, Appl. Phys. B.52, 55 (1991).

4. G.A.Skripko, S.G.Bartoshevich, I.V.Mikhnyuk and I.G.Tarazevich, Opt.Lett.16, 1 1991).

5. J.O.Bierlein and H.Vanherzeele, J.Opt.Spc.Am.B.6, 622 (1989).

6. S.G.Bartoshevich, I.V.Mikhnyuk, G.A.Skripko and I.G.Tarazevich, IEEE QE, 27, 2234 (1991).

7. H.H.Zenzie and P.E. Perkins, in Advanced Solid-State lasers. Technical Digest Series, 147 (1989).

Forsterite Laser Generation Characteristics Dependence on Upper Laser Level Kinetics

S. G. Bartoszewicz, R. Wodnicki, B. V. Minkow, G. A. Skripko, I. G. Racewicz

Solaris Optics, Warsaw/PL

The forsterite laser $Mg_2SiO_4:Cr^{4+}$ is an effective tunable source in the spectral range 1,15-1,35 μm [1] and it is a subject of interest for a number of researches. The laser in this wavelength range is expected to be useful in medical, diagnostic and laboratory applications.

This work deals with the research of the Forsterite laser generation characteristics dependent on wavelength, pulsewidth and pump energy density.

In our experiments we used Forsterite crystals (6x6x24 mm, C + 0,3% wt) with the Brewster - angle - cut faces. We used the Nd:YAG lasers (λ = 1,064 μm) working in different regimes and the Ti:Sapphire laser (λ = 0,78 μm) for pumping.

The dependence of the generation efficiency of the Forsterite laser from pump energy density of Nd:YAG and Ti:Sapphire lasers (pulsewidth near 10 ns) is shown in Fig.1. As we can see the Forsterite laser has the greatest efficiency with the Ti:Sapphire laser pumping in equal conditions.

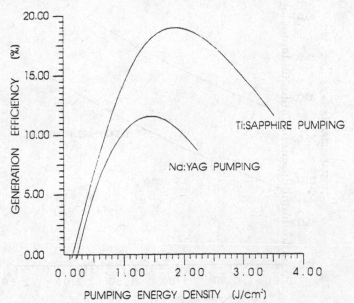

Fig. 1. GENERATION EFFICIENCY OF THE FORSTERITE LASER
AS A FUNCTION OF PUMPING ENERGY DENSITY
(pulsed lasers, τ_i 10ns).

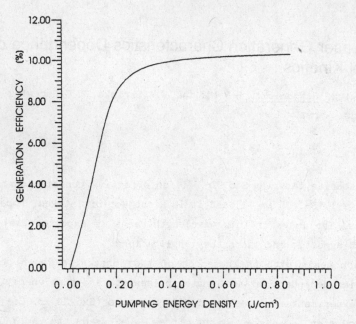

Fig. 2. GENERATION EFFICIENCY OF THE FORSTERITE LASER
AS A FUNCTION OF PUMPING ENERGY DENSITY
(FREE-RUNNING Nd:YAG, τ 180ms).

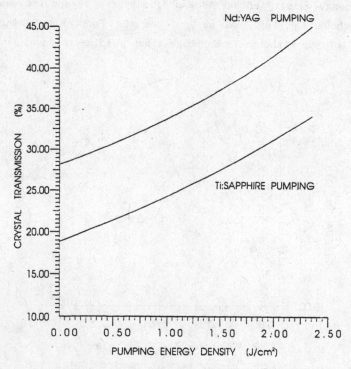

Fig. 3. ABSORPTION OF THE FORSTERITE CRYSTAL AS A FUNCTION
OF PUMPING ENERGY DENSITY.

The Forsterite laser pulse is delayed in relation to the pump pulse. The delay near the threshold is 1 ms and it is decreasing when the pump energy density increases. When it exceeds the threshold in 5-7 times, it reaches its minimum (near 75 ns and 60ns) in case of pumping Nd:YAG laser or Ti:Sapphire laser, respectively. Minimum of the delay corresponds to the maximum of generation efficiency. The delay dependence of the pump energy density is oposite to the generation efficiency dependence.

The Forsterite laser generation efficiency in dependence on the pump energy density for free-running Nd:YAG laser (pulsewidth near 160 ms) is shown in Fig.2. In this case the threshold of the Forsterite laser generation is 20-30 kW/cm^2 and the maximum efficiency is reached at 150-200 kW/cm^2. The temporary characteristics of the Forsterite laser are determined by the pump pulse. If the pump pulse consists of a regular series of pulses, then a series of pulses with the characteristic pulsewidth of 1,5 ms and with the period of 2 ms was generated. The general efficiency in this case is a little higher than for 10 ns - pulses pumping.

All the above mentioned give the possibility to suppose the availability of nonlinear excited - state absorption. Such a process is characteristic for the incomplete 3d - shell ions and it has been described earlier in our work for Cr - activated crystals [2]. In particular, it is quite enough to suppose two - photon process where the probability of particles traveling from the upper laser level to the highest levels is quadric function of the pump intensity.

In this case too intensive pumping can be as unacceptable as too weak one. All the particles from the upper laser level will travel to the highest levels, from which they relaxate quickly (foremost non-radiatevely, warming up the host). The upper laser level population can be lower then the threshold one. There exists pumping speed range where lasing process is possible for all the active mediums with the excited state absorption. We call this process "throat" effect.

Except the excited-state absorption, the absorption of the Cr:Forsterite crystal decreases when the pump energy density increases (Fig.3). This is due to the relatively low saturation of the pump energy density ($\sim$ 1,7 J/cm^2).

We have also investigated the Cr:Forsterite laser action with CW and Q-CW pumping. Fig.4 shows the generation efficiency dependence on the pump energy for output mirrors with different reflection coefficients.

It is evident, that the efficiency in this case is comparable with the efficiency in pulsed regime.

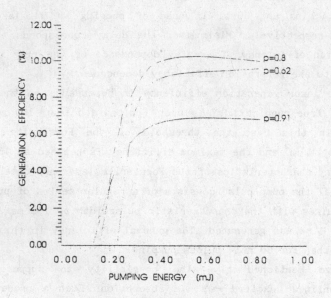

Fig. 4. GENERATION EFFICIENCY OF THE FORSTERITE LASER WITH Q-CW PUMPING FOR DIFFERENT OUTPUT MIRRORS.

In the CW regime we have achieved the 5% conversion efficiency. Based on the above results we have developed two commercial Cr:Forsterite laser models.

REFERENCES

1. V.Petricevic, A.Seas and R.R.Alfano, Laser Focus 11, 109 (1990).

2. G.S.Kruglik, G.A.Skripko and A.P.Shkadarevich, Eds., Tunable Lasers on Activated Crystals. Minsk: BPI, 1984, p.33.

Tunable Laser Ti:Al$_2$O$_3$ with the SBS Cell

Z. JANKIEWICZ, W. ŻENDZIAN
Institute of Optoelectronics, Military Academy of Technology
Kaliski St. 2, 01 489 Warsaw-49, POLAND

The progress in a field of a technology of crystals doped by metal ions, which occured in 70s and 80s years caused the appearance of the series of new crystalic, active materials which are characterized by the wide luminescence bands and thus suitable for the tunable lasers. Such crystals as an alexandrite (BeAl$_2$O$_4$:Cr^{3+}), sapphire doped titanium (Al$_2$O$_3$:Ti^{3+}), forsterite (Mg$_2$SiO$_4$:Cr^{4+}) and the so called black YAG (Y$_3$Al$_5$O$_{12}$:Cr^{4+}) enable the construction of tunable lasers operating at room temperature in the range from 700 nm to 1580 nm [3].

The very important parameter of tunable lasers is a width of a generation line. The narrowing of the lasers generation spectra which are expected to generate the mono-pulses with the high energy faces with the following difficulties: losses in the laser resonator caused by the elements with the high dispersion result in the decrasing of a generation efficiency, a low threshold of a damage of diffraction gratings surfaces and interferometers F-P makes it impossible to apply them in the lasers operating with the high threshold exceeding.

Moreover, the energetic pulse parameters (energy, peak power and pulse duration) get worse together with the offset of generation wavelegth from the center of a luminescence line.

Self-injection locking obviates many of these problems [1]. A self-injection laser operates sequentially in two laser resonators. In the first resonator, nascent pulse travels through the wavelength selective elements reeatedly, being further narrowed on each pass. However, before significant energy extraction occurs, the laser is switched to operate on the second nondispersive resonator. A self-injection laser can be passively or actively switched.

In the presented paper we suggest to use the SBS phenomenon in the dispersive resonator of the mono-pulse tunable laser Ti:Al$_2$O$_3$ for the automatic swiching over of the resonator from the dispersive branch into the non-dispersive one.

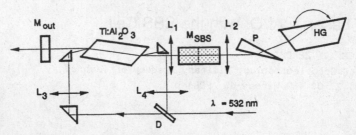

Fig.1. Optical scheme of the titanium laser with the SBS cell
in a configuration of the resonator with the diffractive grating HG
(P-prismatic telescope, D- energy divider R=0.5,T=0.5 for λ=0.53 μm.

Experimental set-up.

The scheme of an optical system of the tunable laser with the SBS cell is shown in Fig.1.The dispersive branch of the presented tunable laser consists of the holographic volumetric diffractive grating (3600 lines/mm, efficiency of 80% in the range of λ=700÷900 nm). The measured width of a generation line of the laser there was $\delta\lambda$=17pm. The remain optic elements of the laser there are:

1) output mirror M_{out} - double-plate selector with the energetic reflection coefficient R = 45%;

2) active medium – Ti:Al_2O_3 crystal of the l_a = 20 mm length with the cut faces at the Brewster angle, concentration N_a = 2.5 10^{19} cm^{-3} (0.1% of weight), absorption coefficient for λ = 532 nm, α = 1.75 cm^{-1}, FOM = 35, energetic treshold of a damage about 400 MW/cm^2.

3) lenses telecope L_1 - L_2 - lenses with the 50 mm focal length without the anti-reflexive layers, applied in order to increase the intensity of radiation in a cuvette with the acetone, and the same with the increase of the SBS efficiency;

4) the cuvette M_{SBS} filled with acetone having on the both sides closed by the plane-parallel plates without the antireflexive layers.

The titanium laser have been pumped by the second harmonic of Nd:YAG laser with the Q modulation by means of the Pockels cell (puls duration τ_p = 8 ns, pulse energy E_{in} = 45 mJ). In order to obtain homogenity of the cross-section of the pumping channel there was applied the double-side pumping of the Ti:Al_2O_3. Moreover, in this way there was obtained a double decrease of the intensity of pumping radiation at the Ti:Al_2O_3 crystal surface.

Theoretical description

The SBS mirror divides the resonator of the titanium laser (Fig.1) into two parts: non-dispersive /output mirror (M_{out}), active medium, SBS mirror (M_{SBS})/ and the dispersive part /SBS mirror, dispersive elements (HG, P)/. The energetic reflection coefficient SBS (R_{SBS}) depends on the intensity of radiation inciding the cuvette with the dispersive medium:

for $\quad E_s \left[-\dfrac{W_o}{W_1}-\right]^2 I^+(z_o,t) < I_p,$ $\qquad R_{SBS} = 0$ $\hfill$ /1a/

for $\quad E_s \left[-\dfrac{W_o}{W_1}-\right]^2 I^+(z_o,t) \geq I_p,$ $\qquad R_{SBS} = 1 - \dfrac{I_p}{E_s \left[-\dfrac{W_o}{W_1}-\right]^2 I^+(z_o,t)}$ $\hfill$ /1b/

where: $\quad 2W_o$ - diameter of the cross-section of a laser beam in a resonator at the L_1 lens of the (L_1 - L_2) telescope,

$2W_1$ - diameter of the cross-section of a laser beam in a focus of the (L_1-L_2) telescope,

E_s - $h\nu/2\sigma(\nu)$ - saturation energy,

h - Planck constant,

ν - frequency of generated radiation,

σ (ν) - emission cross section,

$I^+(z_o,t)$ - value of a photon stream inciding the M_{SBS} mirror,

I_p - value of the threshold intensity SBS,

z_o - coordinate of the telescope (L_1- L_2) focus.

The $(W_o/W_1)^2$ ratio describes the multiplication factor of the increase of a power density of the radiation inciding the SBS cuvette in the focus of the (L_1-L_2) telescope. From the relationship /1/ it results that the generation process of the tunable laser with the SBS cell can be divided in two stages:

1. The linear generation development which lasts untill the moment when the density of power radiation in the resonator exceeds the threshold density SBS - (I_p). At this stage the generation spectrum is created;

2. The avalanche development of generation - exceeding of the threshold power density causes the reverse reflection of radiation from the SBS mirror with high efficiency and thus immediate swiching off of the losses introduced to the resonator by the dispersive elements (switching off of the dispersive branch) and the shortening of the resonator. At this stage there are created the energetic - time characteristics of a generated mono-pulse. The SBS mirror protects the dispersive elements against the damage by laser radiation.

For the analysis of the process of mono-pulse generation of Ti:Al_2O_3 laser with the SBS cell there were used the equations of energy transport, which take into account the changes of amplification coefficient and the changes of density of photons stream in a time and space. These equations are as follows:

$$\frac{\partial I_\lambda^+(z,t)}{\partial z} + \frac{1}{V_\lambda}\, \frac{\partial I_\lambda^+(z,t)}{\partial z} = [k_\lambda(z,t) - \rho m]\, I_\lambda^+(z,t) \qquad \text{/2a/}$$

$$-\frac{\partial I_\lambda^-(z,t)}{\partial z} + \frac{1}{V_\lambda}\, \frac{\partial I_\lambda^-(z,t)}{\partial z} = [k_\lambda(z,t) - \rho m]\, I_\lambda^-(z,t) \qquad \text{/2b/}$$

$$\frac{\partial k_\lambda(z,t)}{\partial z} = - [I_\lambda^+(z,t) - I_\lambda^-(z,t)]\, k_\lambda(z,t) - \frac{k_\lambda(z,t)}{\tau} \qquad \text{/2c/}$$

where: I_λ^+, I_λ^- - photon streams propagating in a resonator within the solid angle (Ω) of a generation, accordingly - in the "+" direction and "-" direction of the optical axe of the resonator, $[cm^{-2}s^{-1}]$,

ρm - material losses of the active medium $[cm^{-1}]$,

k_λ - gain coefficient $[cm^{-1}]$, τ - emission lifetime,

$V_\lambda = c/n_\lambda$ - velocity of a propagation in the active medium, (c - in air),

n_λ - refraction coefficient of the active medium.

Index "λ", in the above equations denotes that these equations are valid for the generation wavelength λ, therefore λ - is a parameter.

It results from this fact that in order to obtain the pulse characteristic as a function of wavelength of tunable laser generation, there is necessary, in case of λ changes, to solve the system of equations each time. These equations are useful for the analysis of dynamics of phenomena occuring in lasers.

Equations /2/ have been solved using the initial conditions for the photon streams I^+, I^- and for gain coefficient $k(\lambda)$ as well as using the initial conditions at the output SBS mirror (taking into consideration the relationship /1/) and at the substitute tunable mirror which reflectivity regarded the losses of dispersive branch of a resonator. The initial and boundary conditions regarded the real parameter values of the resonator what is presented in Fig.1.

Experimental investigation

In the following figures there are shown the characteristics of an energy generation and a pulse duration of generation as a function of wavelength λ. The continuous curves were obtained on the basis of numerical analysis of equations /2/, whereas the experimental values are denoted as squares for the Ti:Al_2O_3 laser with the SBS cell and as triangles for the Ti:Al_2O_3 laser with the classical resonator (without the SBS cell).

The relationship of a pulse duration as a function of generation wavelength is shown in Fig.2. This figure univocally shows that the application of SBS cell enables the shortening of a generation pulse duration. One of the reasons of these

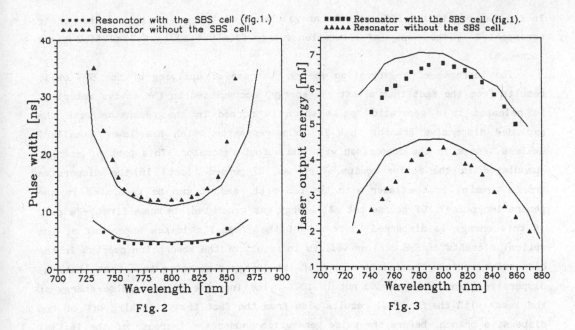

••••• Resonator with the SBS cell (fig.1.)
▲▲▲▲▲ Resonator without the SBS cell.

Fig.2

Fig.3

changes is an effective shortening of resonator length occuring at the moment of switching on of the SBS mirror. Simultanously there follows the decreasing of resonator losess for losses of the switched off branch.

Since the total resonator losses of titanium laser with the diffraction grating amount $\rho=0.491$ cm^{-1}, that is why the switching off of rather great losses of the dispersive branch $\rho_d = 0.127$ cm^{-1} as well as the half of the SBS mirror losses and the telescope ($L_1 - L_2$) losses $\rho_{SBS}=0.124$ cm^{-1} before the start of the avalanche development of generation should significantly influence on the increase of energy of pulse generation. This fact is confirmed by the energetic characteristics of a generation of the lasers with SBS cell as well as with the classical resonator, as a function of a generation wavelength. These characteristics are presented in Fig.3. They univocally show that of using the SBS cell in resonator of the tunable laser, the energies of generated mono-pulses can increase. Application of the SBS cell in the titanium laser with the diffractive grating has caused 2.5 times increase of generation energy for extreme wavelengths of tunable characteristic and 1.7 times increase of energy in the maximum of this characteristic. Due to this fact one can see that SBS mirror additionaly stabilizes the generation energies. For a laser with the classical resonator the process of tunning the generation wavelength from the short-wave limit of tunable band to the maximum of the tunable characteristic is accompanied by the change of generation energy from 2 mJ to 4.3 mJ (2.1 times increase).

In the laser with the SBS cell the above change of the generation wavelength is accompanied by the change of generation energy from 5.4 mJ to 6.7 mJ (1,2 times increase).

The increase of a generation energy, in case of applying of the SBS cell, results from the fact that almost all energy, accumulated in the active material, is released in a generation pulse which is formed in the resonator with the excluded dispersive branch, that is in a resonator which has lower (smaller) useless losses in the comparison with the output resonator. This part of energy, cummulated in the active medium which was dispersed (lost) in the dispersive branch remains in the laser with the SBS cell, and it can be generated in the generation pulse?. Of course not all energy was generated, because first - a part of this energy is dispersed in result of the Fresnel's losses appearing at the optical elements of SBS cell as well as in result of the absorption process in the dispersive medium. The second reason of the energy dispersion is a fact that the dispersive branch is excluded not in 100% . The increase of mono-pulse energy of the laser with the SBS cell results also from the fact that switching off of the dispersive branch, before the pulse generation causes the increase of the initial threshold exceeding and thus more effective use of the energy which is cummulated in the active medium, because the final amplification coefficient is the smaller the higher an initial threshold exceeding is.

Moreover, it was found that there is insignificant increase of the divergency angle of radiation which is generated by the titanium laser with the SBS cell ($\theta \approx$ 1 mrad) in comparison with the divergence angle of the radiation of a laser with the classical resonator ($\theta \approx 0.8$ mrad). It can be caused by an occurrence of the competitive non-linear phenomena in relation to the SBS phenomena, in the dispersive medium what make it worse the coupling of the wave front of the forcing wave e.g. self-focusing of a radiation. The increase of a divergence angle of a generated radiation is not accompanied by a disturbance of the distribution of a radiation intensity in a beam cross-section.

Both for the titanium laser with the classical resonator and as for the laser with the SBS cell, the distributions of the intensity of radiation at the cross-section of the beam, there were Gaussian distributions.

Conclusions

On the basis of the results obtained from the theoretical analysis as well as from the experimental investigations one can give the following results:
1. The SBS cell placed in the resonator of the dispersive laser in an effective way switched off the losses of the dispersive branch of a resonator before the start of the avalanche stage of a generation development. Switcging off of losses

and a shortening of a resonator, during the generation, causes next, that the generated mono-pulses are characterized, in comparison with the mono-pulses generated by the laser with the classical resonator, by:

a) higher energy in all the tunable range,

b) shorter duration in all the tunable range,

c) higher peak power in all the tunable range.

2. The SBS cell stabilizes a duration of pulse generation and the energies of generated mono-pulses. That means that offset the generation wavelength from the maximum of tunable charakteristic $\lambda = 800$ nm does not cause such great changes of the above parameters as it is for the tunable laser with the classical resonator.

3. An application of the SBS cell is especially useful in the tunable lasers with the resonator of high dispersion. The experimental investigations carried out on the titanium laser with the branch with significantly lower dispersion ($\delta\lambda \simeq 0.3$ nm) gave worse results.

4. Insignificant increase of the divergence is not essential and moreover it seems to be possible to eliminate.

REFERENCES

1. N.P. Barnes, J.A. Williams, J.C. Barnes, G.E. Lockard, "A self-injection locked, Q-switched, line-narrowed Ti:Al$_2$O$_3$ laser", IEEE J. Quantum Electronics, vol. 24, no. 6, pp. 1021-1028, 1988.

2. P.F. Moulton, "Spectroscopic and laser characteristics of Ti:Al$_2$O$_3$", J. Opt. Soc. Amer. B, vol. 3, pp. 125-133, 1986.

3. J. Hecht, "Tunability makes vibronic lasers versatile tools", Laser Focus World, october 1992, pp. 93-103.

PS-Pulse Generation Using Flashlamp Pumped 9wt % Nd^{3+}-Doped Multimode Phosphate Glass Fibres

P.Glas, M.Naumann, A.Schirrmacher, H.Schönnagel

Max-Born-Institut für Nichtlineare Optik und Kurzzeitspektroskopie

12489 Berlin / Germany

Rudower Chaussee 6

Pulses of 15 ps duration at λ=1054 nm , an energy of ~ 1µJ characterized by a time-bandwidth product of 1.5 were obtained without dispersion compensation using a single active fibre within a resonator applying active/passive loss modulation.

In approximately the same experimental configuration a fibre array laser (FAL) consisting of 150 gain elements has been mode locked to give pulses as short as 15 ps with an energy of ~ 3 µJ and a spectral bandwidth of ~ 0.3 nm.

Introduction

Highly Nd^{3+} - doped Phosphate glass fibres are well suited for realizing a flashlamp pumped, solid state, high power, short pulse laser.

There are some papers on short pulse generation in cw diode pumped rare earth doped silica fibres /1/ - /4/. Our main goal is a considerable enhancement of the energy as well as of the mean power of the generated ultrashort pulses.

Phase locking of an FAL using highly doped multimode Phosphate glass fibres provides a good starting point for designing a high power short pulse laser /5/.

A high doping level requires a transversal pumping regime due to the short absorption length, being realized in a transversal flashlamp pump configuration.

On that account we report in the first part on mode locked ps-pulse generation from a heavily Nd^{3+}- doped multimode Phosphate glass single fibre laser and in a second part about first results obtained from FAL investigations.

Experimental

A multimode optical fibre has been placed together with an acousto-optical modulator and a dye cell (saturable absorber) contacted to one of the resonator mirrors within a Fabry-Perot resonator, cf. Fig.1.

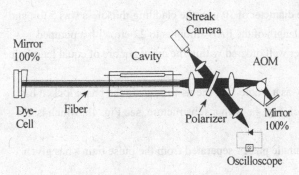

Fig.1 Experimental setup of the single fibre laser resonator

The fibres are made of Phosphate glass with a core diameter of 100 μm possessing a numerical aperture of 0.15. The Nd^{3+} doping level was chosen to 9 wt%, the fibre length was 50 cm with a flashlamp pumped section of 8 cm.

Fig.2 shows the result of a pulse duration measurement, carried out with a streak camera (maximum available time resolution ~ 15 ps) after frequency conversion (2ω). Typical pulse widths are of the order of 50 ps, the shortest measured pulse width was ~ 15 ps.

The single fibre laser provided radiation at λ=1054 nm with a spectral bandwidth of ~ 0.15 nm. The detected energy/pulse was ~ 1 μJ giving an intrafibre intensity of ~ 1 $GWcm^{-2}$ /6/.

The number of transverse modes was reduced from about 1000 to 1-2 in going from the passive fibre to the lasing one.

In a second step a FLA consisting of 150 gain elements (fibres) was placed in the same cavity with similar resonator configuration instead of a single fibre , cf. Fig.3.

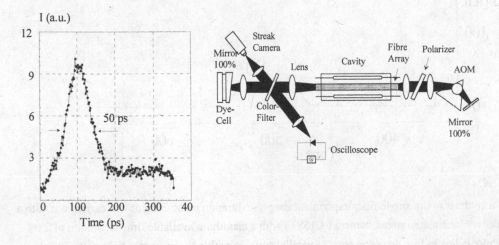

Fig. 2 Streak camera record of a 50 ps-pulse   Fig. 3 Resonator configuration with a fibre array laser

The single fibre of the FAL had a core diameter of 30 μm, the cladding thickness was 5 μm and the numerical aperture 0.12. The total length of the FAL amounts to 22 cm with a pumped section of ~ 8 cm. The fibres are neither well ordered within the array nor are of equal length due to technological imperfections.

We have locked the modes of the array as a whole. On the ns-time scale the phase locked FAL has provided a single Q-switch pulse, cf. Fig. 4a, with a substructure, see Fig. 4b, which is a typical mode locked pulse train.

The pulse duration measurement of a single pulse - separated from the pulse train - has given $\tau_p < 15$ ps, cf. Fig. 4c.

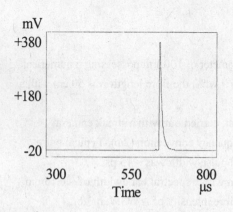

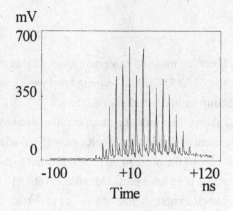

Fig. 4a Pulse train envelope (Q-switch)

Fig. 4b Pulse train under the Q-switch envelope

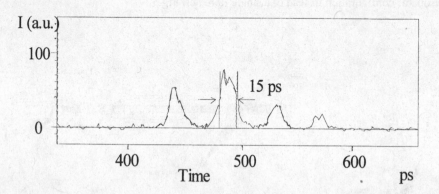

In contrast to the single fiber experiment the pulse duration measurement was carried out with a faster Hamamatsu streak camera (C1587) with a maximum available time resolution of 2 ps. Pulse echos in Fig. 4c correspond to parallel surfaces within the resonator light path (Fabry-Perot).

Fig. 5b shows the near field pattern of the lasing array. The fact that there is only a limited number of lasing fibres - in comparision to the passively illuminated FAL, see Fig. 5a - is caused by angle misalignment with respect to the optical axis of the array as well as to internal defects due to mechanical stress.

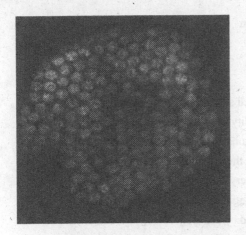

Fig.5a Passively illuminated FAL endface   Fig. 5b FAL endface above threshold

We have obtained single pulse energies around 3 μJ and a spectral bandwidth of ~ 3 nm at λ=1054 nm.

Summary

It was shown, that highly Nd^{3+}-doped Phosphate glass fibres are well suited for realizing a flashlamp pumped, high power, short pulse laser. Active/passive mode locking of a single multimode fibre under flashlamp pumping conditions yields ps-pulses and synchronized emission from a fibre laser array placed in an external resonator containing active/passive loss modulation yields ps-pulses too. It was demonstrated, that mode locked emission from different members of a fibre laser array as well as spatial coupling of these elements is possible. Further detailed investigations will be reported.

This work was supported by the Bundesministerium für Forschung und Technologie under contract No. 13 N 5896.

References
/1/   F. Krauß et. al., Appl.Phys.Lett. 55, (1989) p. 2386
/2/   C. Spielmann et.al., Appl.Phys.Lett. 58, (1991) p.2470
/3/   M. E. Fermann et. al., Opt.Lett. 16, (1991) p.244
/4/   M. H. Ober et. al., Appl.Phys.Lett. 60, (18), (1992) p. 2177
/5/   U. Griebner et.al., '1 J Nd:Glass Fibre Array Laser',
       Proceedings of the Int. Conf. LASERS '92, Houston, Texas, p.319
/6/   M. W. Phillips et.al., SPIE Vol. 1171, (1989) p.280

Miniaturised Diode Pumped Solid State Lasers

S.Heinemann[1], A.Mehnert[2], P.Peuser[2], N.P.Schmitt[2]

[1]Technische Universität Berlin, D - 1000 Berlin 10

[2]DASA Technologieforschung, D - 81663 München

Miniaturised diode-pumped solid state lasers are well suited for measurement and probing techniques because of their inherent capability of single frequency emission, narrow linewidth, excellent frequency stability and tunability.

To build a single frequency laser requires first of all the selection of a certain wavelength by means of the mirror coating and the laser crystal. Second the operation in the fundamental Gaussian mode is easily achieved by longitudinally pumping the laser crystal. Longitudinal mode selection can be achieved by using short resonator lenghts. As the longitudinal mode spacing $\Delta\upsilon$ is inversely proportional to the resonator length it is possible to have only one longitudinal mode within the gain bandwidth of the active medium, if the cavity length is short enough [1]. Several monolithic lasers with a resonator length of 0.7 mm made of Nd:YAG and Nd:GGG were investigated for which the efficiency curves are shown in fig.1.

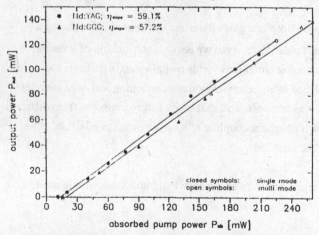

Fig.1: Slope efficiencies of 0.7 mm long monolithic Nd:YAG and Nd:GGG microcrystal-lasers ($\lambda = 1064$ nm).

Fast frequency modulation of single-frequency microcrystal-lasers with a resonator length of less than 1 mm can be achieved in a quasi-monolithic setup. A piezoelectric element with a hole in the center, e.g. a 25 µm thick PVDF foil, serves as the tuning element, which is glued between the laser crystal and an external mirror as shown in fig.2.

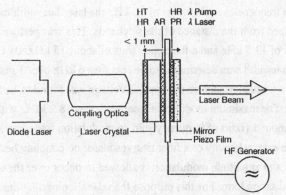

Fig.2: Setup of a quasi-monolithic micro-crystal-laser for fast frequency modulation.

The emission spectrum reveals laser activity at 1319 nm on one single longitudinal mode. Though Nd:YAG is an isotropic laser crystal laser emission takes place on two orthogonal polarisations most likely induced by thermal birefringence. The frequency difference between these two polarisations is independent of the crystal temperature but can be influenced by external mechanical stress. The laser emission can be slowly tuned over an interval of 30 GHz with an amplitude stability of better than 5 % by varying the resonator temperature.

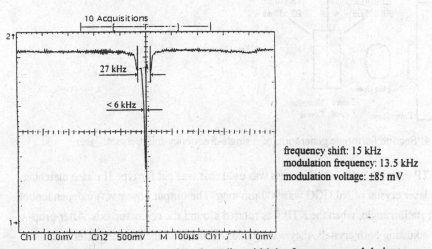

Fig.3: "Intrinsic" determination of the laser linewidth by frequency modulation.

The laser was coupled to a fiber ring resonator with a variable optical length. This optical spectrum analyzer has a free spectral range of 66 MHz and a resolution band-width of 140 kHz. When a voltage of about ±100 mV was applied to the piezoelectric element, sidebands were observed for modulation frequencies from 10 kHz up to 1

MHz. For modulation frequencies in the range of 10 kHz the laser linewidth can be "intrinsically" determined from the distance of the sidebands. This was performed for a modulation frequency of 13.5 kHz and a frequency shift of about 15 kHz. By this experiment the laser linewidth was determined to be less than 6 kHz over 1 ms (fig.3). The dependence of the frequency shift on the applied voltage was found to be 5 MHz per 1 V peak to peak. The maximum modulation frequency was 18.2 MHz with a modulation index of about 1 (ratio of frequency shift to modulation frequency). If the laser is stabilised to an external cavity or a fiber ring resonator no coupling between frequency modulation and amplitude modulation is allowed to occur over the entire bandwidth of the stabilisation loop. For this purpose the relaxation oscillations and their harmonics must be eliminated, which is achieved by an intensitiy modulation of the pump diode. Therefore part of the laser output is directed on a photodiode, which is connected to an electronic feedback loop that controls the current of the pump diode. First experiments yielded a reduction of the noise level by 25 dB at the main frequency, and the higher harmonics of the relaxation oscillations were completely eliminated when the electronic control loop was on.

Microcrystal-lasers are also well suited for intracavity frequency doubling as longitudinal mode competition is a priori excluded. A setup is shown in fig.4.

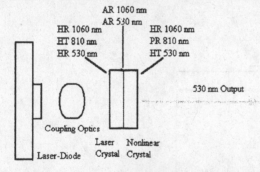

Fig.4: Second harmonic generation of a single-frequency microcrystal-laser.

A KTP crystal of 400 μm thickness was used that was cut for type II phase matching. The laser crystal of Nd:GGG was 900 μm long. The output power was independent of the rotation angle, when the KTP was rotated around the resonator axis. After properly adjusting both crystals, they were glued together and the resulting laser resonator was pumped with a laser diode of 250 mW optical ouput power. This resulted in a green output power of 200 μW. Phase matching was achieved by varying the temperature of the laser resonator. The spectrum reveals single frequency emission on one longitudinal mode. The RIN spectrum of the green microcrystal-laser shows only weak relaxation oscillations. No signal occurs in the low frequency range, which would correspond to the "green problem".

Further investigations on miniaturised diode pumped solid state lasers aim at the exploitation of the higher power range. The main problem that has to be solved is the highly efficient cooling and the special mounting of the laser diodes dissipating heat densities of up to 700 W/cm². Silicon microchannel-coolers are well suited for this. Small channels of about 25 µm thickness and about 200 µm height are formed in silicon by anisotropic etching and are typically closed with glass by anodic bonding. If water is forced through the microchannels with a pressure drop of about 2 bar, heat densities of 1 kW/cm² can be dissipated [2]. By this way a specially designed micro-channel-cooler was developed with microchannels of 200 µm depth and 10 µm width. For simulating the heat generation of a laser diode high power transistors were bonded on the upper side of this microchannel-cooler.

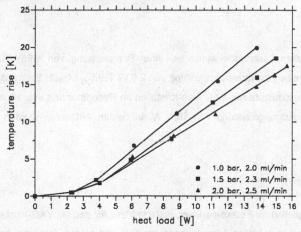

Fig.5: Performance of a silicon micro-channel-cooler with a cooled area of 3.1∗3.5 mm².

A current controlled by the emitter/base voltage was imprinted to the emitter/collector circuit. The emitter/collector voltage was varied to simulate different heat loads. The temperature on the upper side of the micochannel-cooler was measured by NTC-resistors for different heat loads and different pressure drops of the cooling fluid. This device was tested for heat loads up to 16 W resulting in a temperature rise of about 18 K. The pressure drop was 2 bar and the flow rate 2.5 ml/min (fig.5). The resulting thermal resistance is 1.13 K/W. Since the cooled area is only 3.1∗3.5 mm² a thermal resistance of 0.1 K/W would result for a cooled area of 1 cm².

The authors thank Dr.Raab and Mr.Quast from the company BGT, Überlingen, for providing the fiber ring and the control loops as well as testing the fast frequency modulated microcrystal-lasers.

References:
[1]   A.E.Siegman, Lasers, University Science Books, Mill Valley, California 1986
[2]   D.B.Tuckerman and R.F.W.Pease, IEEE Electron. Device Letters 5 (1981), 126

Diodengepumpte Festkörperlaser: Praktische Lösungsansätze für das longitudinale und transversale Pumpen im Dauerstrichbetrieb

P. Zeller, K. Altmann, Th. Halldórsson, S. Heinemann, A. Mehnert, G. Reithmeier,

P. Steinbach, P. Peuser

Deutsche Aerospace AG, Technologieforschung, D-81663 München

Kurzfassung

Mit einer fasergekoppelten Laserdiode wurde bei einer Pumpleistung von 4,95W und longitudinaler Anregung von Nd:YAG eine Laserausgangsleistung von 2,6 W TEM_{00} erzielt. Ein transversal gepumpter Aufbau lieferte in der Ausbaustufe mit zwei Slab-Kristallen im Resonator und vier 15 W-Laserdioden als Pumpquelle eine Laserausgangsleistung von 16,1 W. In beiden Aufbauten wurden Mikrokanalkühler eingesetzt.

1. Einleitung

Laserdioden als Pumpquellen für Festkörperlaser, insbesondere für den Nd:YAG-Kristall als laseraktives Medium, haben sich bereits in vielen Anwendungen bewährt und zeigen dabei die bekannten Vorteile. Durch immer höhere optische Leistungen, die von einer einzigen Laserdiode geliefert werden, ist ein effizienter Einsatz in Lasergeräten möglich. Doch die hohen Leistungen der Laserdioden sind nur bei entsprechend vergrößerter emittierender Fläche möglich, da sonst die zulässige Leistungsdichte der optischen Flächen überschritten würde und eine effiziente Ableitung der Verlustwärme nicht mehr realisiert werden kann. Kommt man bei einer kontinuierlich emittierenden 1-Watt-Laserdiode noch mit einem Array von 100 bis 200 µm Breite aus, so vergrößert sich dieses Array bei einer 20-Watt-Diode auf 10mm. Das ungünstige Aspektverhältnis der emittierenden Fläche (sie ist nur jeweils 1 µm hoch) und die großen Abstrahlwinkel erfordern eine sorgfältige Auslegung der Transferoptik, wenn eine optimale Ankopplung an den Nd:YAG-Kristall erreicht werden soll. Zwei Pumpanordnungen mit jeweils spezifischen Vorteilen werden nachfolgend dargestellt.

2. Longitudinales Pumpen über Lichtwellenleiter

Das longitudinale Pumpen eignet sich besonders zur Erzeugung von Laserstrahlung im Grundmode (TEM_{00}) bei hohem Wirkungsgrad. Im vorliegenden Aufbau wurde in der Transferoptik ein Lichtwellenleiter verwendet, wodurch ein sehr kompakter Laserkopf realisiert werden konnte. Das Aufbauprinzip ist in Bild 1 dargestellt. Als Pumpquelle diente zunächst eine Laserdiode mit Gehäuse und Fenster und interner thermoelektrischer Kühlung mit einer optischen Leistung von 3 W (2,2 W am Faserende). Außerdem wurde eine bereits fasergekoppelte Diode mit externer Kühlung und einer Leistung von 4,95 Watt am Faserende eingesetzt.

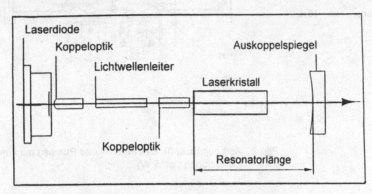

Bild 1: Aufbauprinzip für das longitudinale Pumpen mit Faserkopplung

Die Laserleistung bei einer Wellenlänge von 1064 nm betrug 1,05 Watt für die 3 Watt-Pumpquelle und 2,6 Watt für die 5 Watt-Pumpquelle. Weitere Daten sind in der nachfolgenden Tabelle 1 zusammengestellt:

Optische Leistung der Laserdiode	3 W	10 W
Leistung am Faserende	2,2 W	4,95 W
Laserleistung (1064 nm)	1,05 W	2,6 W
Wirkungsgrad (opt./opt.)	48%	53%
Wirkungsgrad (elektr./opt.)	12,3%	10,2%
Diff. Wirkungsgrad (opt./opt.)	55%	55%
Strahlqualität	TEM_{00}	TEM_{00}

Tabelle 1: Ergebnisse beim longitudinalen Pumpen mit Faserkopplung

3. Transversales Pumpen für mittlere und höhere Leistungen im Dauerstrichbetrieb

Für eine effiziente Ankopplung von Hochleistungslaserdioden mit breiter emittierender Fläche wurde eine transversale Pumpanordnung verwendet (Bild 2). Insgesamt standen vier 15 Watt-Laserdioden zur Verfügung. In dieser Konfiguration erwies sich die Verwendung der Slab-Geometrie für den Kristall als vorteilhaft, weil dadurch die Ankopplung der Laserdioden erleichtert wird. Als Koppeloptik wurden Zylinderlinsen (in diesem speziellen Fall runde Stäbe aus SF57, undodiertem GGG und Zirkonia) verwendet. Im ebenen Aufbau und bei Verwendung von zwei Slab-Kristallen wurde eine maximale

92

Ausgangsleistung von 16,1 Watt bei 1064 nm im Dauerstrichbetrieb erzielt. Weitere Daten sind in der Tabelle 2 zusammengestellt. Die Intensitätsverteilung bei einer Laserausgangsleistung von 16,1 Watt zeigt Bild 3.

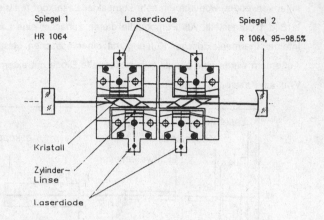

Bild 2: Aufbauprinzip für das transversale Pumpen mit vier Laserdioden (je 15 W)

Pumpleistung	4 x 15 W
Laserleistung (1064 nm)	16,1 W
Strahldurchmesser (x/y)	3,6 / 1,8 mm
Divergenz (Vollwinkel)	10 / 5 mrad
Diff. Wirkungsgrad (opt./opt.)	35%
Wirkungsgrad (el./opt.)	8,5%
Polarisation	750:1

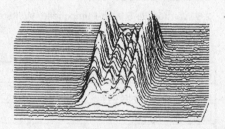

Tabelle 2: Ergebnisse beim transversalen Pumpen mit 4 x 15 W

Bild 3: Intensitätsverteilung bei P_{out} =16,1 W

Die oben beschriebene Anordnung wurde auch mit nur einem Kristall im Resonator betrieben, wobei in einem Fall die Dioden schräg (Drehung um die Strahlachse z) angeordnet waren. Das ermöglicht später den Einsatz von mindestens vier Laserdioden pro Kristall. Bild 4 zeigt die Laserausgangsleistungen in Abhängigkeit der jeweiligen elektrischen Pumpleistung für die verschiedenen Konfigurationen im Vergleich.

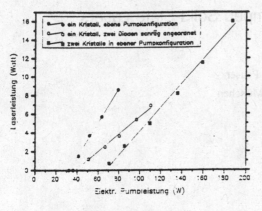

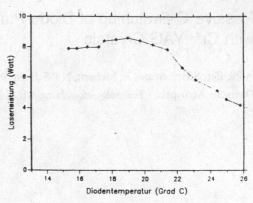

Bild 4: Laserausgangsleistungen für verschiedene Konfigurationen

Bild 5: Abhängigkeit der Laserausgangsleistung von der Diodentemperatur

Die Laserdioden wurden zum Betrieb mit einer einfachen Stromversorgung in Reihe geschaltet. Die unterschiedlichen Betriebsströme der einzelnen Laserdioden wurden mit Hilfe eines Parallelwiderstandes angepaßt. Im Kühlkreislauf wurde ein Umlaufkühler mit Kompressor und Pumpe verwendet. Zur effizienten Wärmeabfuhr auf kleinem Volumen dienten Mikrokanalkühler. Es zeigte sich, daß sich bei einer Temperaturdifferenz der Laserdioden von 6°C die Leistung um weniger als 10% änderte (Bild 5). Diese Werte gelten für ein Modul mit zwei Laserdioden annähernd gleicher zentraler Wellenlänge. Wird eine größere Anzahl von Laserdioden verwendet, kann über eine entsprechende Selektion der Laserdioden die Temperaturabhängigkeit der Laserleistung definiert werden.

Literatur

D. Golla, A. Berndt, W. Schöne, I. Kröpke, H. Schmidt: Mit Diodenlasern transversal angeregte Slab-Laser, Laser und Optoelektronik 25(1)/1993

S.C. Tidwell, J.F. Seamans, M.S. Bowers: Highly efficient 60-W TEM_{00} diode-end-pumped Nd:YAG laser, OPT. LETT./Vol. 18, No. 2/1993

Passive Q-Switching of Diode-Pumped Solid-State Lasers with Cr^{4+}:YAG Crystals

A.Pfeiffer, S.Heinemann, A. Mehnert, N.P.Schmitt, P.Peuser

Deutsche Aerospace - Technologieforschung, 81663 München

Tests of Cr^{4+}-doped garnets concerning their behavior as saturable absorbers have already been reported by several authors [1]. These materials show several well known good properties (high thermal conductivity, hardness, etc.) and especially for their high photostability they seem to be good candidates for passive Q-switching.

1. Theory

The rate equations describing the dynamics of passive Q-switching are [2,3]:

$$\text{I)} \qquad \frac{d\Phi}{dt} = (c\sigma_a n_a + c\sigma_s n_s - \frac{1}{\tau_c}) \cdot \Phi$$

$$\text{II)} \qquad \frac{dn_a}{dt} = -\gamma c \, \sigma_a n_a \Phi$$

$$\text{III)} \qquad \frac{dn_s}{dt} = -2c \, \sigma_s n_s \Phi$$

Φ : photon density in laser cavity

n_a : inversion population density of active material $\qquad n_a = n_{a2} - n_{a1} \cdot \dfrac{g_2}{g_1}$

n_s : inversion population density of saturable absorber

$\qquad n_s = n_{s2} - n_{s1}$

$\qquad$ (equal level degeneracies assumed)

σ_a : stimulated emission cross section of active material

σ_s : absorption cross section of saturable absorber

τ_c : cavity photon lifetime

g_1 , g_2 : level degeneracies of active material ;

$\qquad \gamma = 1 + g_2/g_1$

$\qquad$ (Index 1 : ground level ; Index 2 : excited level)

The fluorescence decay times of the active material and the saturable absorber (several μs) are assumed to be very long compared to the expected pulse width (some ns).

Once the first threshold has been reached (i.e. gain equals the sum of initial, saturable and constant, losses ⇔ $d\Phi/dt = 0$) the photon density has to increase rapidly, i.e. the second derivation of Φ with time has to be at least positive: $d^2\Phi/dt^2 > 0$.

For good Q-switching behavior (i.e. short pulses with high peak power) $d^2\Phi/dt^2$ should be rather large. This condition is sometimes called "*second threshold*".

From Eqs. I) to III) we get :

$$d^2\Phi/dt^2 > 0 \iff \frac{\sigma_s^2}{\sigma_a^2} > \frac{\gamma\, n_{ai}}{2\, n_{s,tot}}$$

n_{ai} : initial value of n_a

$n_{s,tot}$: total doping concentration of saturable absorber centers ; initial value of $|n_s|$

So for efficient passive Q-switching σ_s should be rather large compared to σ_a.

2. Experiments

At the DASA laser labs we have verified the performance of Cr^{4+}:YAG as Q-switch in three different very compact diode pumped Nd:YAG (respectively Nd:GGG) laser systems, two longitudinally and one side-pumped configurations.

In the transversal configuration one side of the Cr^{4+}:YAG crystal was coated as high reflecting flat end mirror. In one of the two longitudinal configurations the Cr^{4+}:YAG was designed with one side as curved, partially reflecting output mirror. So it was possible to further reduce the resonator length, which is an important point to produce very short laser pulses.

2.1. Longitudinally pumped lasers

The experimental setup for the longitudinally pumped Nd:YAG laser is shown in fig.1.

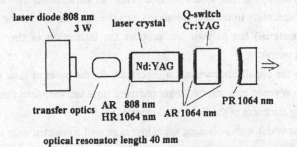

Fig.1: Longitudinally pumped Nd:YAG laser

The results for three Cr^{4+}:YAG crystals of different lengths but equal doping concentration are plotted in figs.2-4.

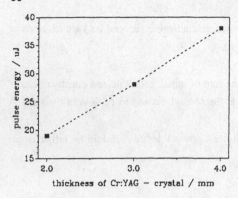

Fig.2: Pulse energy vs thickness of saturable absorber, quasi cw pumping - rep.rate 1 kHz

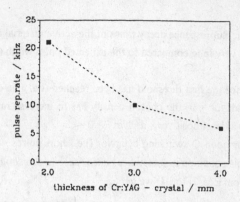

Fig.3: Pulse rep.rate vs thickness of saturable absorber, cw pumping

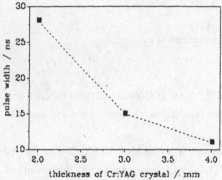

Fig.4: Pulse width vs thickness of saturable absorber

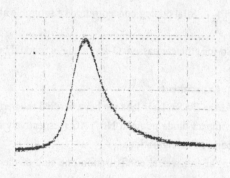

Fig.5: Oscilloscope trace of 10.000 shots, timebase 10 ns/div

The increase in pulse energy with increasing absorber thickness is due to the higher initial population inversion density (i.e. higher stored energy) of the laser material that is reached for higher initial saturable losses (fig.2). In an optimised passively Q-switched laser system the maximum attainable population inversion density of the active material should be used. This can be achieved by installing the maximum possible saturable loss (according to the first threshold) into the laser cavity. So the optimisation process (for a given absorber material) has to take into account the interaction of the initial losses and the reflectivity of the output mirror.

In the cw pumping case the population inversion n_{ai} necessary for the onset of laser action corresponding to a higher initial loss is reached only after a longer pumping time and therefore the pulse repetition rate decreases with increasing thickness of the Cr4+:YAG crystal (fig.3).

The decrease of the pulse width with increasing initial loss is as well connected with the higher population inversion n_{ai}. Higher n_{ai} means higher gain (after the saturable absorber has been bleached) which results in stronger respectively faster amplification of the laser pulse; the higher the gain the faster the population inversion gets depleted and the smaller is the pulse width (fig.4).

The oscilloscope trace of 10.000 laser pulses demonstrates an excellent pulse stability (fig.5).

For further reducing the size of the laser device a short Nd:GGG laser crystal (1.5 mm length) was used which was pumped by a 1W diode laser, and the Cr^{4+}:YAG crystal was partially reflecting coated and used as output coupling mirror. The optical resonator length was 5 mm. This laser configuration exhibited also excellent pulse stability with a pulse energy of 7 μJ and a pulse width of 8 ns at TEM_{00} operation.

2.2. Side-pumped laser

In fig.6 the setup of the side pumped Nd:YAG laser is depicted. The slab crystal was cut at brewsters angle and the optical resonator length was 32mm. The laser crystal was pumped by a high power quasi cw laser diode array and the repetition rate was adjusted to 10 Hz but could be enhanced to about 100 Hz (according to the maximum duty cycle of 2 % of the laser diode). We attained 1.6 mJ pulse energy in a higher order mode and 0.9 mJ in TEM_{00} operation. The pulse width was 10 ns and the energy efficiency E_{pulse}/E_{pump} reached a value of about 5%.

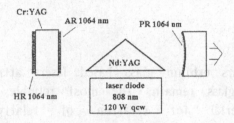

Fig.6:   Side pumped Nd:YAG laser

3. Conclusions

The experiments have shown that Cr^{4+}:YAG can be used as reliable passive Q-switch especially for long-term use. The energy efficiency (E_{pulse}/E_{pump}) is low compared to active Q-switching methods; this slight drawback is due to the rather small absorption cross section of Cr^{4+}:YAG ($\sigma_s \approx 1...10 \cdot \sigma_a$) compared to other materials like $LiF:F^{2-}$ ($\sigma_s \approx 60 \cdot \sigma_a$) or organic dyes ($\sigma_s \approx 1000 \cdot \sigma_a$) and to the appearance of excited state absorption. On the other hand the last two mentioned materials suffer from rather bad photostability. The doping concentration and therefore the cross section of Cr^{4+} in YAG is not known with sufficient accuracy to allow precise theoretical calculations. Cr^{4+}:YAG as passive Q-switch allows a very compact and easy laser design. For the future a saturable absorber material with higher absorption cross section would be desirable in order to increase energy efficiency and to achieve even shorter pulses.

Acknowledgements

The authors would like to thank Dr. M.Kokta, Union Carbide Corp., USA and U.Linnekuhle, Roditi GmbH, Hamburg for supplying the Cr^{4+}:YAG and for valuable discussions.

A.Pfeiffer wants to thank Prof. Dr. H.Schrötter, Universität München, for accompanying his diploma thesis and T.Halldórsson, Deutsche Aerospace, for supporting this work.

References

[1] D.M.Andrauskas, C.Kennedy, OSA Proc. on Advanced Solid-State Lasers 1991, Vol. 10, G.Dubé and L.Chase (eds.), p. 393-397

[2] A.Szabo, R.A.Stein, "Theory of laser giant pulsing by a saturable absorber", J. Appl. Phys., 36 (1965), p. 1562-1566

[3] Siegman, Lasers, University Science Books, Mill Valley, California 1986, ch 26.3

A Compact FTIR Q-Switched 1,54 Micron Erbium Glass Laser

B. I. Denker[1], A. P. Fefelov[2], I. Kertesz[3], S. I. Khomenko[2], N. Kroo[3], A. L. Luk'yanov[2], V. V. Osiko[1], S. E. Sverchkov[1], Yu. E. Sverchkov[1]

[1]General Physics Institute, 117942, Moscow, Vavilov str, 38, Russia, Fax (095)1350270, [2]State Technical University, Moscow, Russia
[3]Research Institute for Solid State Physics, Budapest, Hungary

In the last years erbium glass based lasers attract more and more attention. Erbium glass remains the most reliable, available and easy-to-use laser material for creation of relatively eye-safe laser radiation in the spectral band near 1,54 μm. Besides eye-safety, this wavelength has some additional advantages: presence of highly sensitive room temperature photodetectors, low adsorption losses in fused silica waveguides and in earth atmosphere, relatively high absorption in biological tissues.

3-level laser scheme of Er ions forces to low its concentration in glass. Narrow and weak absorption lines of Er ions in glass is a drawback that can be overcomed by codoping the glass by sensitizer ions such as Yb, Cr, Nd. There are few effective erbium laser glasses in present time. A set of such glasses was designed by General Physics Institute [1]. High doping concentration of sensitizer ions (up to 4×10^{21} cm^{-3} of Yb^{3+}) make it possible to miniaturize the laser rods Figure 1 illustrates some possibilities of these glasses in the free-run mode under lamp pumping. Small elements demonstrate lasing thresholds as low as 6 J; the larger ones allow to obtain output energy up to 30 J with overall efficiency up to 2,8%. Thus it is demonstrated that compact and effective lasers can be designed on the base of these glasses.

Erbium glass laser can be Q-switched by rotating prism or by lithium niobate Pockels cell. Both methods suffer from definite drawbacks. For example, most lithium niobate cells have low optical damage threshold and loose the depolarized component of laser radiation.

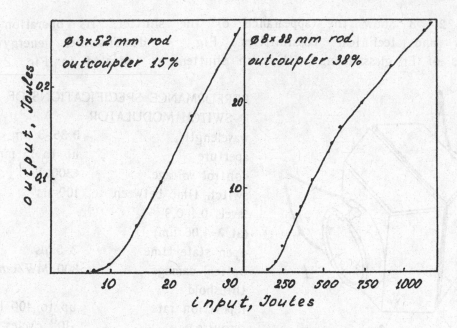

Ø 3×52 mm rod
outcoupler 15%

Ø 8×88 mm rod
outcoupler 38%

Fig.1. Free-run operation of erbium laser glasses

For our laser we have chosen Frustrated Total Internal Reflection (FTIR) Q-switcher. This type of shutters was previously tested with various solid-state lasers (see for example [2-4]).

Low stimulated emission cross-section σ and high energy density $E_0 = \frac{\hbar\omega}{\sigma}$ (approx 18 J/cm^2) make erbium glass laser very sensitive to all kinds of optical losses and impede efficient energy extraction due to danger of optical damage. We consider FTIR shutters to be the best for low-gain lasers such as Er glass ones due to very low losses and high optical damage threshold of the materials they are made from. Low driving voltage (<300 V) and switching of both polarizations are additional useful features of such shutters. As for their limited Q-switching rate, it was found to be quite enough for low gain Er glass lasers.

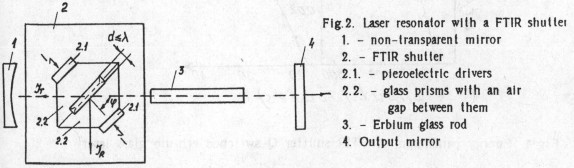

Fig.2. Laser resonator with a FTIR shutter
1. - non-transparent mirror
2. - FTIR shutter
2.1. - piezoelectric drivers
2.2. - glass prisms with an air
       gap between them
3. - Erbium glass rod
4. Output mirror

Fig. 2, 3 show the appearance of the shutter, its operation principles and technical possibilities. Fig. 4 demonstrates energy parameters of Er glass lasers with such shutters as indcated in Fig. 2

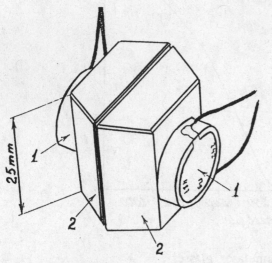

PERFORMANCE SPECIFICATIONS OF Q-SWITCH MODULATOR

wavelength	0,35–3 μm
aperture	up to 15 mm
control voltage	<300 V
switch time between levels 0,1–0,9 (at λ=1,06 μm)	100 ns
open state time	3–5 μs
optical damage threshold	>800 MW/cm^2
repetition rate	up to 400 Hz
resourse	>10^6 cycles

Fig.3. Overall view of FTIR shutter.

1. – piezoelectric drivers
2. – glass prisms

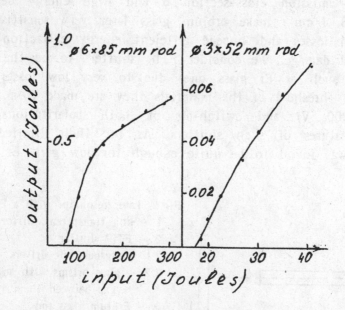

Fig.4. Energy parameters of FTIR shutter Q-switched erbium glass laser

optical scheme. In all these cases pulse duration was about 30÷50 ns. Another optical schemes makes it possible to generate pulse trains at frequencies of tens of kHz, to produce single giant pulses with variable duration in microsecond range, or to get free-running pulses when the shutter is off [4].

On the base ofthese experiments a compact laser was designed. Its technical parameters are listed in Table 1 . FTIR shutter has the size

Table 1.
Specification for MPQ-Er FTIR Q-swithched Erbium glass minilaser

active media	Cr-Yb-Er phosphate glass (type LGE-C)	beam diameter	3 mm
active element	3 mm diameter	output mirror reflectivity	80%
size	55 mm long AR coated rod	repetition rate	up to 0,25 Hz
		active rod cooling	by water circuit and air heat exchanger
wavelength	1,54 micrometers		
operation regime	Q-swithched		
Q-swithcer	frustrated total internal reflection shutter	power supply size (including cooling system)	15x22x40 cm
output energy	up to 25 mJ	laser head size	20x6,5x4,5 cm
pulse duration	50 ns	total weight	approx. 6 kg
beam structure	multimode	power consumption	220 V 40

25x25x25 mm and working aperture ⌀7 mm. The air gap between its two prisms is 0,6 μm. The outer facets of the shutter were AR coated at 1,54 μm.

The laser may find its application in surgery, science and technology, for example it can be used for plasma production in spectral analysis.

REFERENCES
1. Denker B.I., Maximova G.V., Osiko V.V., Sverchkov S.E, Sverchkov Yu.E., Lasing parameters of new erbium glasses. SPI vol 1627, 1992, p. 39-41.
2. K.Asaba et al Development of Near Infrared Q-switched laser. NEC Res. and Dev. N 93, April 1989, Japan, p. 21-27
3. Fefelov A.P., Khomenko S.I., Mikhailov V.A., et al. Application of optomechanical modulators for IR solid state   laser schemes, SPIE vol. 1625, 1992, p 113-119
4. Denker B.I., Osiko V.V., Sverchkov S.E., Sverchkov Yu. E. ,Fefelov A.P., Khomenko S.I. Effective eyesafe frustrated total internal reflection Q-switched erbium glass lasers SPIE vol. 1627, 1992, p p 42-45, Russian J. of Quantum Electron., 1992, N 6 p. 544-547.

Multiion and Multiphoton Effects in YAG:Tm³⁺ Laser Crystals

W. Woliński, M. Malinowski, P. Szczepański, R. Wolski, and Z. Frukacz[*]

Institute of Microelectronics and Optoelectronics, Warsaw University of Technology ul.Koszykowa 75, 00-662 Warsaw, Poland
[*]Institute of Electronic Materials Technology, ul.Wólczynska 133, 01-919 Warsaw, Poland

Yttrium aluminum garnet, $Y_3Al_5O_{12}$ (YAG), crystal doped with Tm^{3+} ions has demonstrated useful performance as a solid state laser in 2 μm range [1]. Recently, the emission properties of YAG:Tm^{3+} have been studied in the near UV and visible ranges [2] and an upconversion pumped blue laser on the $^1G_4 \rightarrow {}^3H_6$ transition has been demonstrated at cryogenic temperatures [3]. The possibility of simultaneous emission at various wavelengths after infrared (IR) excitation, where high power GaAlAs laser diodes are available, has renowed our interest in the study of excitation mechanisms in YAG:Tm^{3+}.

Excited state absorption (ESA) and cooperative energy transfer processes contribute significantly to losses for 2 μm thulium laser [1,4,5]. On the other hand these processes may result in up-conversion pumping and short wavelength anti-Stokes Tm^{3+} emission.

In this work various ESA processes in Tm doped YAG crystals are presented and discussed.

The basic spectroscopic data of investigated crystals are listed below.

Table 1 Room temperature spectroscopic data on YAG:Tm^{3+} crystals.

Sample at %	Tm concent. x 10^{20}cm^{-3}	$\alpha_{GS}(^3H_4)$ 1/cm 780 nm	$\alpha_{GS}(^3F_3)$ 1/cm 681 nm	$\alpha_{GS}(^3F_2)$ 1/cm 658 nm	$\tau_{fl}(^3F_4)$ ms	$\tau_{fl}(^3H_4)$ μs
5.0	6.925	4.70	17.8	0.90	8.3	25
0.5	0.692	0.50	1.8	0.10	12.0	350
0.1	0.138	0.15	0.4	≈0.02	15.0	827

α_{GS} - ground state abs. coeff.(trans. from the $^3H_6$, to $^3H_4$, $^3F_3$,$^3F_2$ states)
τ_{fl} - fluorescence lifetime of the considered excited states.

The ESA cross sections (σ_{ESA}) for these crystals were recorded using IR pump beam and tunable dye laser beam as a probe. A simple energy level diagram in Fig.1A [6] explain the ESA measurements. In this experiments the transmission of a cw probe beam was compared in the presence and in the absence of a pump pulse. As the pump was used laser diode 780 nm (output power of a few tens of mW), corresponding to the $^3H_6 \rightarrow ^3H_4$ transition. The highest ESA peak at 632 nm (vary closed to He-Ne laser radiation) results from the $^3F_4(1) \rightarrow ^1G_4(3)$ transition and the corresponding σ_{ESA} was measured to be 0.18 x $10^{-21} cm^2$, Fig.2.   This upconversion process results in the blue $^1G_4 \rightarrow ^3H_6$ emission Fig.3.

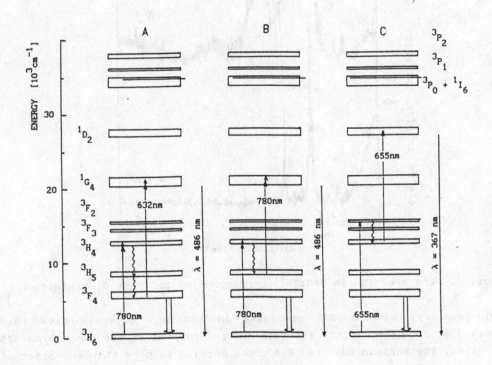

Figure 1. Energy level diagram YAG:Tm^{3+}, 5% at Tu.

Room temperature anti-Stokes $^1G_4$ emission was also observed with only one laser resonantly exciting the Stark components of the $^3H_4$ manifold. This can be explained by the quasi resonant upconversion as shown in Fig. 1B. In this process, after absorption of the first photon and rapid nonradiative relaxation to the lowest Stark level in the $^3H_5$ manifold, there could be ESA from the $^3H_5$ level to a phonon band associated with the $^1G_4$ manifold.

Another room temperature one colour upconversion process was observed in YAG:Tm^{3+} system after red excitation at 655 nm. This could be explained by the double resonant ESA from the $^3H_4$ to $^1D_2$ state according to scheme presented in Fig.1C. This process results in anti-Stokes UV emission, Fig.4

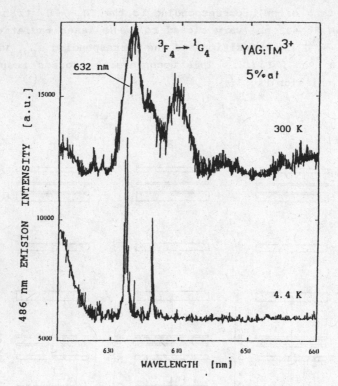

Figure 2. ESA spectrum in YAG:Tm^{3+} corresponding to $^3F_4 \rightarrow {}^1G_4$ transition.

In summary, several ESA processes in YAG:Tm^{3+} crystals have been observed at room temperature after IR and red pumping in the range from 620 to 785 nm. The maximum measured ESA cross section is more than one order of magnitude smaller than the effective emission cross section for the 2.01 μm laser determined by Kintz [1], which is $\sigma_e = (3 \mp 1) \times 10^{-21}$ cm^2. However, our measurements showed that several ESA processes are active in the investigated system, they could create the significant losses for 2 μm laser especially at high excitation densities. It is also demonstrated that YAG: Tm^{3+} is also attractive as a potential UV and blue upconversion laser material. There exists also possibility of several upconversion transitions by energy transfer between Tm ions in YAG. Decay curves from all the investigated metastable states, namely $^1I_6$, $^1D_2$, $^1G_4$ and $^3H_4$ are non exponential, indicating a strong ion-ion interaction.

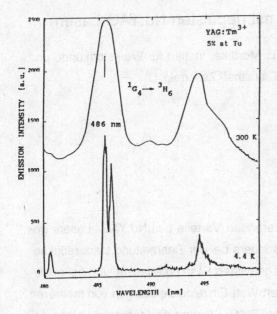
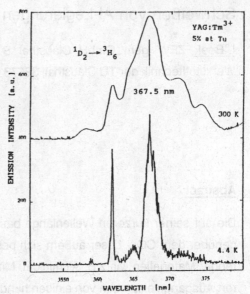

Figure 3.
Upconverted emission from the
$^1G_4$ manifold after 780 nm
excitation.

Figure 4.
Upconverted emission from the
$^1D_2$ manifold after 655 nm
excitation.

This work was supported by KBN grants 3 0479 9101 and 3 3706 9102.

References

1. G.J. Kintz, LEOS'90 Conference Proceedings, November 4-9, 1990, Boston, MA, USA, paper SSL1.2/ThL2

2. J.A. Mares, H. Landova, and M. Nikl, Phys.Stat.Sol (a) **133** (1992) 515

3. B.P. Scott, F. Zhao, R.S.F. Chang, and N. Djeu, Optics Letters **18** (1993) 113

4. S.R. Bowman, G.J. Quarles, and B.J. Feldman, Advanced Solid-State Lasers, Technical Digest, February 17-19, 1992, Santa Fe, NM, paper 100/ME8-1

5. G. Armagan, A.M. Buoncristiani, A.T. Inge, and B.Di Bartolo, OSA Proc. Advanced Solid-State Lasers, 1991, p. 222

6. J.B. Gruber, M.E. Hills, R.M. Macfarlane, C.A. Morrison, G.A. Turner, G.J. Quarles, G.J. Kintz, and L. Esterowitz, Phys.Rev.B **40** (1989) 946

Schweißen von Al-Legierungen mit gepulsten Nd:YAG-Lasern

I. Beek, ZFW gem. GmbH, Clausthal; B.L. Mordike, Institut für Werkstoffkunde und Werkstofftechnik der TU Clausthal, 38678 Clausthal- Zellerfeld, D.

Abstract

Die auf seiner kürzeren Wellenlänge basierenden Vorteile des Nd:YAG- Lasers gegenüber dem CO_2- Laser äußern sich besonders bei der Bearbeitung stark reflektierender Materialien wie z.B. Aluminium. Mit gepulsten Festkörperlasern sind bei mittleren Ausgangsleistungen von einigen hundert Watt Einzelpulsleistungen von mehreren Kilowatt zu erreichen. Zudem bietet das sog. Pulse Shaping die Möglichkeit einer modifizierten Energieeinbringung in den zu bearbeitenden Werkstoff. Daraus ergeben sich zahlreiche neue Perspektiven für das Schweißen von Werkstoffen, die mit konventionellen Verfahren nicht oder nur sehr schwer schweißbar sind, z.B. sehr dünne Al- Bleche oder ausgehärtete Al- Legierungen.

Zielsetzungen

Die Arbeiten zum Thema " Schweißen mit Festkörperlasern " sind Bestandteile eines BMFT- Verbundprojektes, das in Zusammenarbeit mit dem VDI- Technologiezentrum durchgeführt wird, wobei in Clausthal das Schweißen von Werkstoffen mit geringer Schweißeignung im Vordergrund steht. Diese Werkstoffe lassen sich in drei Gruppen unterteilen:

1. Werkstoffe mit hoher Wärmeleitfähigkeit

Beim Schweißen dieser Materialien treten häufig Poren und Risse auf, die zu Festigkeitsverlusten führen. Zudem bewirken die meist sehr großen Schmelzbäder, die bei anderen Verfahren auftreten, eine große Wärmeeinflußzone, die sich in ihren Eigenschaften gravierend vom Grundmaterial unterscheidet. Durch Verwendung des Nd:YAG- Lasers soll hier Abhilfe geschaffen werden.

2. ausgehärtete Werkstoffe

In diesem Bereich soll untersucht werden, inwieweit es möglich ist, ausgehärtete Materialien zu schweißen, ohne die durch Wärmebehandlungen eingestellten Eigenschaften zu zerstören. Gepulste Nd:YAG- Laser bieten durch die Möglichkeit der Pulsformung interessante Perspektiven.

3. artfremde Werkstoffe (z.B. Ti-Fe, Al-Fe u.a.)

Beim Fügen unterschiedlicher Werkstoffe treten oft harte und spröde intermetallische Phasen auf, die eine gute Schweißung unmöglich machen. Durch Einsatz geeigneter Zusatzwerkstoffe soll die Bildung dieser Phasen unterbunden werden, wobei der Nd:YAG-Laser aufgrund seiner Strahleigenschaften eventuell große Vorteile gegenüber anderen Verfahren hat.

Ergebnisse

Für die Bearbeitung des Projektes stehen in Clausthal zwei Nd:Yag- Laser mit maximalen Ausgangsleistungen von 400 bzw. 1200 W zur Verfügung. Beide Laser bieten die Möglichkeit des Pulse Shaping. Dadurch ist es z.B. möglich, Leistungsüberhöhungen, die am Pulsanfang auftreten, zu beseitigen oder Pulse mit abfallender Flanke zu formen, um ein abruptes Abfallen der Leistung und damit eine zu hohe Abkühlgeschwindigkeit zu vermeiden. Prinzipiell läßt sich nahezu jede denkbare Pulsform programmieren.

Am Beispiel des Schweißens von Aluminiumlegierungen erwies sich eine Leistungsspitze am Pulsbeginn als vorteilhaft im Hinblick auf die zu erzielende Einschweißtiefe. Diese Überhöhung läßt sich durch das Voransetzen eines kurzen aber hohen Vorimpulses noch verstärken. Dadurch wird zunächst die hochschmelzende Oxidhaut aufgebrochen, um dann mit dem schwächeren Hauptsektor die eigentliche Schweißung durchzuführen. Vergleichsweise sehr große Einschweißtiefen sind mit sog. LD- Resonatoren (LD = Low Divergence) zu realisieren. LD- Resonatoren sind abgestimmte Resonatoren, d.h. ihre maximale Leistung wird bei einer ganz bestimmten Lampenleistung erreicht. Bei dem in Clausthal zur Verfügung stehenden 400 W- Laser liegt die maximale Ausgangsleistung für den LD- Resonator bei 130 W. Dennoch können aufgrund des kleinen Strahldurchmessers und der damit verbundenen hohen Leistungsdichte auf dem Werkstück Einschweißtiefen von 1,8 mm erreicht werden (Keyhole-Effekt).

Um Poren in der Schweißnaht zu vermeiden, sind längere Pulse nötig, da Gasblasen im Schmelzbad eine gewisse Zeit benötigen, um aufsteigen zu können. Programmiert man nun einen Puls mit abfallender Flanke, so wird die Erstarrung verlangsamt, was die Spannungen und somit die Rißneigung verringert.

Literatur

1. Laserstrahltechnologien in der Schweißtechnik,
   DVS Düsseldorf 1989

2. L. Dorn u.a.: Fügen von Aluminiumwerkstoffen,
   Grafenau 1983

3. The Industrial Laser Annual Handbook,
   Tulsa, Oklahoma 1987

4. J. Ruge : Handbuch der Schweißtechnik,
   Berlin 1980

5. VDI- Handbuch : Schweißen mit CO_2- Hochleistungslasern,
   Düsseldorf 1987

6. Materials Processing, theory and practices, Vol. 3,
   North- Holland Publishing Company 1983

Nonisothermal Electron Behaviour in Injection Laser

V. K. Batovrin, N. N. Evtikhiev, L. A. Rivlin
Moscow Institute of Radioengineering,
Electronics and Automation
N 78 Vernadsky Prospect, Moscow 117454, Russia

Most of the current investigations of the dynamics of semiconductor injection lasers allow only for three characteristic time constants: the spontaneous recombination time of electrons and holes $t(sp)$ amounting to ~1000 psec, photon lifetime $t(ph)$ and cavity round-trip transit time $t(c)$, both amounting to ~10 psec. This approach is used, for instance, in [1] and it implies that intraband relaxation processes are instanteneous. It is then possible to use quasiequilibrium distribution functions of carriers with constant temperature T and to assume that the gain profile of the semiconductor is quasihomogeneously broadened.

The search out the very fast bleaching mechanism in a semiconductors (needed in particular for passive laser mode-locking) leads up to the investigation of the non-isothermal electron behaviour, in other words, to the investigation of the time variation of the carrier temperature T, which may have significant influence on the injection laser dynamics, as the change in the carrier temperature affects the absorption coefficient or the gain g even when the electron density remains constant [2,3].

This analysis should take into account at least two factors: the temperature of the quasi-Fermi distribution may not be constant and the carrier distribution function may not be quasiequilibrium at all. Time constants representing the rates of the processes of thermal equilibrium establishment with the lattice $t(l)$ and the electron relaxation inside the band $t(b)$ both lie in the picosecond and subpicosecond range for typical laser semiconductor materials.

Therefore these factors are important only in the case of sufficiently short and/or intence excitation of a semiconductor, for example, by laser's own ultrashort pulse.

If the intraband relaxation time $t(b)$ is short compared with the reciprocal line width $t(lw)$ of spontaneous recombination in a given semiconductor, the broadening is essentially homogeneous; that is any local perturbation of distribution function affects this function all through the energy spectrum of carriers due to uncertainty principle. The change in the distribution function in time is clearly described by the time dependence of quasi-Fermi level F or of the electron density, exactly as in the quasiequilibrium approach [1]. In a typical case this is true if the intraband relaxation time $t(b)$ is shorter than 100 femtoseconds.

The second effect is the time variation of the carrier temperature T. It is convenient to begin the analysis of nonequilibrium kinetics of the carrier distribution function in the operating semiconductor laser by considering the reaction of a semiconductor characterized by $t(b)<t(lw)$ to the light pulse of duration $t(p)$ shorter than the carrier-lattice relaxation time $t(l)$ [2]. It is this reaction in the form of, for example, the time derivative dg/dt that determines (together with other factors) the saturation rate of the laser sections [1] and is given by the sum of two derivatives dF/dt and dT/dt.

The sign of the time derivative of temperature changes with photon energy $E(ph)$:

1) $dT/dt > 0$ if $\qquad\qquad E(ph) < F \qquad\qquad$ and $g > 0$

2) $dT/dt < 0$ if $\qquad F < E(ph) < F+kTa(1)/a(0) \qquad$ and $g < 0$

3) $dT/dt > 0$ if $\quad F+kTa(1)/a(0) < E(ph) \qquad\qquad$ and $g < 0$

(the values of ratio $a(1)/a(0)$ are calculated in [2], k is the Boltzmann constant).

Both heating and cooling processes are caused by the stimulating action of the laser light. In the first region the carrier heating is due to the extraction of the "cold" electrons from the low energy wing of the carrier distribution function. In the second region the cooling is due to the creation of the surplus "cold" electrons. And in the third region the heating is due to the creation of surplus "hot" electrons in the high energy wing of the distribution function.

Electrons react to the ultrashort light pulse by the fast heating-cooling or cooling-heating pulses so that electron temperature becomes time dependent. Such "febril" or feverish reaction creates equally fast pulsations of the laser gain which are comparable in magnitude with contribution due to variation of electron density.

It is this process that is responsible for multisided pico- and subpicosecond transient effects in semiconductor laser, especially in passive mode-locking experiments.

REFERENCES

1. L. A. Rivlin, A. T. Semenov, S. D. Yakubovich
   Dynamics and Emission Spectra of Semiconductor Lasers.
   Ed. by L. A. Rivlin.
   J. Sov. Laser Res. v. 7, #2, 1986. Plenum Publ. Co. N. Y.
2. L. A. Rivlin - Sov. J. Quant. Electr. v. 15(4), 1985. Publ. by AIP.
3. L. A. Rivlin - Sov. J. Quant. Electr. v. 19(10), 1989. Publ. by AIP.

Lasing Characteristics of a 1.644 µm Er3+:YAG Monolith Pumped at 647 nm, 787 nm and 964nm

S. Nikolov[1] and L. Wetenkamp[2]
1) DLR, Inst. of Optoelectronics, 82230 Oberpfaffenhofen
2) DASA, Technological Research, 81663 München

For several applications, for example in the optical communication the eyesafety of lasers is important. The lasing wavelength of the Er:YAG at 1.644 µm meets the eyesafe region beyond 1.45 µm.

Experimental Setup

An Er:YAG laser monolith has been pumped with a Krypton Ion laser at 647 nm, Ti:Sapphire laser at 787 nm and a laserdiode at 964 nm (Fig.1).

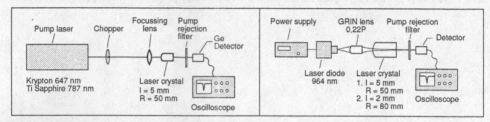

Fig.1: Experimental Setup with various pump sources

The 5 mm monolith has a curved input coupler with a radius of 50 mm and a plane output coupler with 98.7% reflectivity at 1.644 µm. The geometrical dimension corresponds to a beam waist of 65 µm. The input coupler is not specially AR coated for all three pump wavelengths. The coupling losses at the entrance surface of the crystal (approx. 20% for 647 nm, 10% for 787 nm and 964 nm) must be taken into account for the estimation of the input power P_o.

Calculation of the small signal gain g_0 [1,2]

For a better comparison of the pumping configurations, the small signal gain was calculated using the three pump wavelengths.

The Expression for small signal gain is given by:

$$g_0(l) = -\alpha_a l + (\gamma_s + 1) \frac{\sigma_e \tau_2}{h \nu_p} \frac{P_{abs}}{A} \xi(l) \qquad \text{with} \quad P_{abs} = P_0(1 - p(l)) \qquad [2]$$

The first term represents the loss due to ground state signal absorption (GSA) and the second term is the positive gain term with the additional loss factor $\xi(l)$. This factor includes the losses due to ESA (Excitet State Absorption) and ground state depletion.

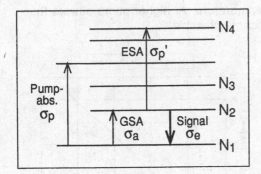

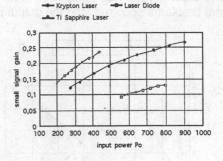

Fig.2: Simplified energy diagram for 3-level
system with ESA at the pump wavelength

Fig.3: Small signal gain g_0 with
respect to the input power P_0

The theoretical model outlined in Fig.2 does not include the upconversion and cross relaxation processes, because these only become significant for an Er concentration of over 2 atomic %. The model can also be applied to Quasi-3-level systems with and without ESA in addition to 4-level systems. In the population inversion of 3-level lasers more than one half of the laser ions need to be excited to the upper laser level. This causes, at high pump intensities, a saturation of the pump transition due to ground state depletion. In this case, Lambert´s absorption law cannot be applied. For the calculation of the transmission factor p(l) [2] a saturation term needs to be added to the Lambert formula. In 4-level systems and in 3-level systems at low pump intensity (e.g. spectrometer measurements) this saturation term can be neglected.

Measurements of the reflected power from the front and rear surfaces and the transmitted power lead to an estimation of the absorbed power. These results gave good agreement with the calculated power, e.g. transmission factor p(l) for input power Po = 300 mW at 647 nm: p(l) = 0.52 (measured) and p(l) = 0.48 (calculated). Due to the saturation of the pump transition the absorbed power in the crystal depends nonlinearly on the pump power. For a higher pump power a greater percentage of the pump light will be transmitted, then for low pump power. This needs to be taken into account for the calculation of the absorbed pump power and the small signal gain. Fig.3 shows the plots of the small signal gain g_0 with respect to the measured input power P_0.

Results of the Laser Experiments

For a better comparison of the laser characteristics, the output power P_{out} is indicated with respect to the absorbed pump power in the crystal P_{abs} and to the input power P_0 on the front surface.

Krypton Ion laser pumping:
With the focussing lens (f´=182 mm) a pump mode radius of approx. 50 µm was obtained. The slope efficiency of the function P_{out}/P_{abs} is 9.5% and that for the In-Out-Characteristic P_{out}/P_0 2.1%, with the respective thresholds of 109 mW and 198 mW.

The laser crystal was pumped at room temperature. At higher pump powers thermal problems occurred, and the output became unstable.

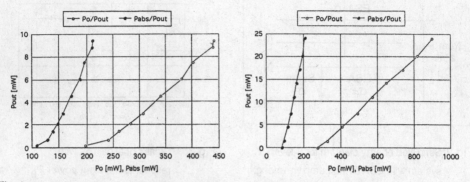

Fig.4: Er:YAG Monolith (l=5mm) pumped with Krypton Ion laser (left) at 647nm and with Ti:Sapphire laser (right) at 787 nm

Ti:Sapphire pumping:

The pump mode within the rod has approx. the same radius as with Krypton Ion laser. The laser operates at room temperature with good stability and without any thermal problems. The slope efficiency for P_{out}/P_{abs} is 20.4% and that for the In-Out-Characteristic is 3.9%, with the respective thresholds of 81 mW and 275 mW . P_{out}=24 mW was the highest output power obtained by any of the pump configurations.

Diode pumping:

To our knowledge laser action with diode pumping of Er:YAG at 964 nm was demonstrated first time. With laser diode pumping one has to take into account the large divergence and the spectral width of the pump beam. For the calculation of P_{abs} it should be noticed that only approx. a third of the spectral width of the diode will overlap the absorption line. This explains the high threshold (P_0 =560 mW) and low small signal gain g_0 in comparison to Kr:- and Ti:Sapphire pumping. The laser rod needs to be cooled to 10 °C. Even then thermal problems still occured.The slope efficiency for P_{out}/P_{abs} is approx. 16 % and for the In-Out-Characteristic 2.3%.

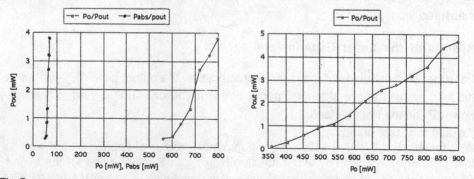

Fig.5: Er:YAG Monoliths pumped with laser diode - l=5mm (left) and l=2mm (right)

For laser diode pumping, shorter crystals with HR coating at the pump wavelength are a better solution. First experiments with a 2 mm crystal (pl/80 mm, reflectivity 99.8% at 1.644µm) have been completed. The laser operates with good stability at 18 °C.

For a 2 mm resonator 5 longitudinal modes oscillate. The FSR is 41.4 GHz, this corresponds to 0.37 nm mode separation. The gain bandwidth of Er:YAG is approx. 210 GHz as in Nd:YAG.

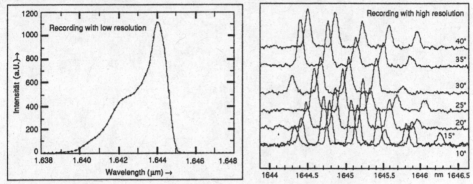

Fig 6: Lasing spectrum of the 5mm (left) and the 2mm (right) Er:YAG Monolith

Conclusion

The pumping with the Ti:Sapphire laser at 787 nm gave the best results due to the best spatial and spectral overlap of signal and pump mode in comparison with diode pumping at 964 nm. For comparison with Krypton Ion laser pumping one has also to take into account the lower absorption at 787 nm (approx. 30%) and the higher absorption at 647 nm (approx. 50%). The laser threshold with respect to P_0 therefore is higher than with pumping at 647nm, but the efficiency for pumping at 647 nm with respect to the absorbed power is lower (ref. also Fig.3).

Further experiments with crystals of various lengths and coupling radii are currently in progress.

The authors would like to thank Dr. M. Fickenscher, DLR, Institute of Optoelectronics for his helpful suggestions and discussions.
Prof. Wiederhold, Friedrich-Schiller-University Jena, We´d like to thank for his encouragement during the experiments with the Krypton laser.

Ref.:

1) M. Digonnet / C. Gaeta: "Theoretical analysis of optical fiber laser amplifiers and oscillators"; Applied Optics Vol. 24 No. 3, S. 333 - 342; 01.02.1985

2) M. Digonnet: "Closed form expressions for the gain in Three- and Four- level-laser fibers"; IEEE Quantum Electronics Vo. 26, No. 10, S. 1788 - 1796,Oct. 1990

Congress B

Congress-Chairmen: H.-J.Tiziani and K. Biedermann

Optische Meßtechnik
Optical Measuring

Phasenschiebe-Verfahren in der interferometrischen Meßtechnik: Ein Vergleich

Th. Kreis, J. Geldmacher, W. Jüptner
BIAS - Bremer Institut für angewandte Strahltechnik
Klagenfurter Str. 2, D-2800 Bremen 33

Kurzfassung

Das Gleichungssystem für die Phasenschiebe-Verfahren der interferometrischen Meßtechnik wird aufgestellt sowie eine Reihe von Lösungen angegeben. Hierbei werden Verfahren mit bekannten und unbekannten, sowie mit konstanten und nicht konstanten Phasenschiebungen behandelt. Ein Vergleich der Verfahren im Hinblick auf die erreichbare Genauigkeit gibt Hinweise zur Wahl des optimalen Verfahrens für ein gegebenes Meßproblem.

Abstract

Based on the system of equations for phase-shifting in interferometry several solutions are derived. These solutions consider methods with known and unknown as well as with constant and non-constant phase-shifts. A comparison of the methods with regard to the achievable accuracy gives recommendations how to choose the optimum method for a given measurement task.

Einleitung

Interferometrische Verfahren erlauben die hochgenaue Messung von Änderungen optischer Weglängen und damit von Verschiebungen, Verformungen, Brechzahlvariationen etc. Zu den Verfahren gehören nicht nur die klassischen interferometrischen Methoden zur Messung von Abstandsänderungen oder Dichteverteilungen, sondern auch die holografische Interferometrie, die Speckleinterferometrie, die Scherografie oder die Moire-Meßtechnik.

Im Idealfall liefert jede Zweistrahl-Interferometrie eine Intensitätsverteilung, die über den Cosinus von der Interferenzphase, welche durch die Meßgröße und die Geometrie des Meßaufbaus definiert ist, abhängt. In der Praxis treten jedoch verschiedene Störungen auf, so daß die resultierende Intensität $I(P)$ als

$$I(P) = I_0(P)[1 + V(P)\cos\phi(P)]R_S(P) + R_E(P) \tag{1}$$

mit der Hintergrundausleuchtung $I_0(P)$, der Kontrastvariation $V(P)$, dem Specklerauschen $R_S(P)$ und dem elektronischen Rauschen $R_E(P)$ geschrieben werden kann [1]. $\phi(P)$ ist die aus der gemessenen Intensitätsverteilung zu bestimmende Interferenzphasenverteilung, P gibt die Bildpunktkoordinaten an. Die Störgrößen lassen sich zusammenfassen zu den additiven Störungen $a(P) = I_0(P)R_S(P) + R_E(P)$ und den multiplikativen Störungen $b(P) = I_0(P)R_S(P)V(P)$, so daß die gemessene Intensitätsverteilung als

$$I(P) = a(P) + b(P)\cos\phi(P) \tag{2}$$

geschrieben werden kann. Weiter ist zu beachten, daß die Interferenzphase aufgrund der Mehrdeutigkeit des Cosinus

$$\cos\phi = \cos(s\phi + 2\pi m) \qquad s \in \{-1, 1\}, \quad m \in \mathcal{Z} \tag{3}$$

nur bis auf ein ganzzahliges additives Vielfaches von 2π und bis auf das Vorzeichen aus einer

Intensitätsverteilung bestimmt werden kann. So ist aus nur einer einzigen Intensitätsverteilung nicht zu erkennen, ob die Interferenzphase steigt oder fällt oder ihre Richtung wechselt.

Ein allgemeines Prinzip zur Unterdrückung von Störungen und zur Auflösung von Mehrdeutigkeiten ist die gleichzeitige Aufnahme mehrerer Meßinformationen mit gezielt geänderten Parametern und anschließender Lösung der entstehenden Gleichungssysteme. Eine Art, in der Interferometrie derartige Redundanz einzuführen, ist die Aufnahme mehrerer Muster mit relativer zusätzlicher Verschiebung der Interferenzphase. Technisch läßt sich die Phasenschiebung auf vielerlei Weise realisieren, z. B. polarisationsoptisch in zirkular polarisiertem Licht durch Drehung einer $\lambda/2$-Platte, in der $+1$. oder -1. Beugungsordnung bei Verschiebung eines Beugungsgitters, durch Verkippung einer Planparallelplatte mit von der Umgebung unterschiedlicher Brechzahl, durch Vakuumkammern, Lichtleitfasern oder durch Wellenlängenänderungen z. B. mit Diodenlasern. Die am häufigsten eingesetzte Methode zur Phasenschiebung ist die piezoelektrische Verschiebung eines Spiegels in einem der Interferometerarme [2].

Die einzelnen Interferenzmuster können zwischen den Phasenschiebungen aufgenommen werden, wobei die Phase während der Aufnahme stillsteht. Bei diesem, auch Phasenstufen-Verfahren genannten Vorgehen erhält man die Intensitätsverteilungen

$$I_i(P) = a(P) + b(P)\ \cos(\phi(P) + \alpha_i) \qquad i = 1, 2, \ldots, N \tag{4}$$

Wird dagegen die Phase kontinuierlich linear mit der Zeit verschoben und die Intensität über gleiche Zeiträume Δt integriert, während derer sich die Phase um $\Delta\alpha$ ändert, so erhält man

$$I_i(P) = \frac{1}{\Delta\alpha} \int\limits_{\alpha_i-\Delta\alpha/2}^{\alpha_i+\Delta\alpha/2} a(P) + b(P)\ \cos(\phi(P) + \alpha)\ d\alpha(t)$$

$$= a(P) + b(P)\ \frac{\sin(\Delta\alpha/2)}{\Delta\alpha/2}\ \cos(\phi(P) + \alpha_i) \qquad i = 1, 2, \ldots, N \tag{5}$$

Der einzige Unterschied dieses Phasenschiebe-Verfahrens zum Phasenstufen-Verfahren besteht in dem sinc-Faktor, der mit $b(P)$ zusammengefaßt werden kann, so daß im folgenden ohne Beschränkung der Allgemeinheit von Gl. (4) ausgegangen werden kann.

Das nichtlineare Gleichungssystem Gl. (4) ist für jeden Bildpunkt P nach $\phi(P)$ zu lösen. Sind die α_i durch das Experiment vorgegeben, sind nur $a(P)$, $b(P)$ und $\phi(P)$ unbekannt, so daß zur Lösung mindestens 3 Intensitätsverteilungen benötigt werden. Unterscheiden sich die α_i dagegen um einen konstanten Wert und dieser Wert ist unbekannt, so müssen mindestens 4 Intensitäten aufgenommen werden.

Lösung des Gleichungssystems bei bekannten Phasenschiebungen

Bei bekannten Phasenschiebungen α_i wird das Gleichungssystem Gl. (4) nach der Methode der kleinsten Quadrate gelöst. Hierzu werden u und v wie folgt eingeführt, die Koordinaten P der Übersichtlichkeit halber weggelassen.

$$\begin{aligned} I_i &= a + b\ \cos(\phi + \alpha_i) \\ &= a + b\ \cos\phi\ \cos\alpha_i - b\ \sin\phi\ \sin\alpha_i \\ &= a + u\ \cos\alpha_i + v\ \sin\alpha_i \end{aligned} \tag{6}$$

Die Summe der quadratischen Fehler

$$\sum_{i=1}^{N}(a + u\ \cos\alpha_i + v\ \sin\alpha_i - I_i)^2 \tag{7}$$

ist zu minimieren. Differenzieren nach a, u und v sowie Nullsetzen der Ableitungen ergibt das Gleichungssystem

$$\begin{pmatrix} N & \sum \cos \alpha_i & \sum \sin \alpha_i \\ \sum \cos \alpha_i & \sum \cos^2 \alpha_i & \sum \sin \alpha_i \cos \alpha_i \\ \sum \sin \alpha_i & \sum \sin \alpha_i \cos \alpha_i & \sum \sin^2 \alpha_i \end{pmatrix} \begin{pmatrix} a \\ u \\ v \end{pmatrix} = \begin{pmatrix} \sum I_i \\ \sum I_i \cos \alpha_i \\ \sum I_i \sin \alpha_i \end{pmatrix} \tag{8}$$

Dieses System ist nach a, u und v zu lösen, dann wird aus

$$\arctan \frac{-v}{u} = \arctan \frac{b \sin \phi}{b \cos \phi} = \phi \tag{9}$$

die Interferenzphase modulo 2π bestimmt.

Besonders einfach werden die Lösungen für α_i, die Vielfache von 30°, 45°, 60° oder 90° sind [3]. Die Auswertegleichung für $N = 4$ Intensitäten mit Phasendifferenzen 90° und Startwert $\alpha_1 = 0$° erhält man nach

i	α_i	$\sin \alpha_i$	$\cos \alpha_i$	$\sin^2 \alpha_i$	$\cos^2 \alpha_i$	$\sin \alpha_i \cos \alpha_i$
1	0°	0.	1.	0.	1.	0.
2	90°	1.	0.	1.	0.	0.
3	180°	0.	-1.	0.	1.	0.
4	270°	-1.	0.	1.	0.	0.
Σ		0.	0.	2.	2.	0.

Gleichungssystem (8) ist damit für diesen Fall

$$\begin{pmatrix} 4 & 0 & 0 \\ 0 & 2 & 0 \\ 0 & 0 & 2 \end{pmatrix} \begin{pmatrix} a \\ u \\ v \end{pmatrix} = \begin{pmatrix} I_1 + I_2 + I_3 + I_4 \\ I_1 - I_3 \\ I_2 - I_4 \end{pmatrix} \tag{10}$$

Dessen Lösung $u = (I_1 - I_3)/2$, $v = (I_2 - I_4)/2$ führt zu der Interferenzphase

$$\phi = \arctan \frac{I_4 - I_2}{I_1 - I_3} \tag{11}$$

Eine Reihe weiterer Lösungen, die nach dem gleichen Schema gewonnen wurden, sind in Tabelle 1 angegegben.

Lösung für unbekannte Phasenschiebungen

Im folgenden sollen Lösungen für unbekannte Phasenschiebungen α angegeben werden, wobei nur vorausgesetzt wird, daß die Differenz aufeinanderfolgender Phasenwerte konstant bleibt und diese Differenz hinreichend verschieden von ganzzahligen Vielfachen von π ist. Als erster Fall wird $N = 4$ betrachtet und ohne Beschränkung der Allgemeinheit der Startwert $\alpha_1 = -3\alpha/2$ angenommen. Bei konstanter Phasenschiebung ist dann das Gleichungssystem

$$\begin{aligned} I_1 &= a + b \cos(\phi - \frac{3\alpha}{2}) = a + b \cos \phi \cos \frac{3\alpha}{2} + b \sin \phi \sin \frac{3\alpha}{2} \\ I_2 &= a + b \cos(\phi - \frac{\alpha}{2}) = a + b \cos \phi \cos \frac{\alpha}{2} + b \sin \phi \sin \frac{\alpha}{2} \\ I_3 &= a + b \cos(\phi + \frac{\alpha}{2}) = a + b \cos \phi \cos \frac{\alpha}{2} - b \sin \phi \sin \frac{\alpha}{2} \\ I_4 &= a + b \cos(\phi + \frac{3\alpha}{2}) = a + b \cos \phi \cos \frac{3\alpha}{2} - b \sin \phi \sin \frac{3\alpha}{2} \end{aligned} \tag{12}$$

Phasensch.	Startwert	N	Auswertegleichung
$30°$	$0°$	3	$\phi = \arctan \dfrac{(3\sqrt{3}-5)I_1 + (\sqrt{3}-2)I_2 + (7-4\sqrt{3})I_3}{(5-3\sqrt{3})I_1 + (2\sqrt{3}-3)I_2 + (\sqrt{3}-2)I_3}$
$30°$	$-30°$	3	$\phi = \arctan \dfrac{I_1 - I_3}{(2+\sqrt{3})(-I_1 + 2I_2 - I_3)}$
$45°$	$0°$	3	$\phi = \arctan \dfrac{(2+\sqrt{2})I_1 - (2+2\sqrt{2})I_2 + \sqrt{2}I_3}{-\sqrt{2}I_1 + (2+2\sqrt{2})I_2 - (2+\sqrt{2})I_3}$
$45°$	$-45°$	3	$\phi = \arctan \dfrac{\sqrt{2}(I_3 - I_1)}{(2+\sqrt{2})(I_1 - 2I_2 + I_3)}$
$60°$	$0°$	3	$\phi = \arctan \dfrac{2I_1 - 3I_2 + I_3}{\sqrt{3}(I_2 - I_3)}$
$60°$	$0°$	4	$\phi = \arctan \dfrac{5(I_1 - I_2 - I_3 + I_4)}{\sqrt{3}(2I_1 + I_2 - I_3 - 2I_4)}$
$60°$	$0°$	5	$\phi = \arctan \dfrac{\sqrt{3}(2I_1 - 3I_2 - 4I_3 + 5I_5)}{8I_1 + 3I_2 - 4I_3 - 6I_4 - I_5}$
$90°$	$0°$	3	$\phi = \arctan \dfrac{I_1 - 2I_2 + I_3}{I_1 - I_3}$
$90°$	$45°$	3	$\phi = \arctan \dfrac{I_3 - I_2}{I_1 - I_2}$
$90°$	$0°$	4	$\phi = \arctan \dfrac{I_4 - I_2}{I_1 - I_3}$
$90°$	$0°$	5	$\phi = \arctan \dfrac{7(I_4 - I_2)}{4I_1 - I_2 - 6I_3 - I_4 + 4I_5}$
$90°$	$-180°$	5	$\phi = \arctan \dfrac{2(I_2 - I_4)}{-I_1 + 2I_3 - I_5}$
$120°$	$0°$	3	$\phi = \arctan \dfrac{\sqrt{3}(I_3 - I_2)}{2I_1 - I_2 - I_3}$
α	$-\alpha$	3	$\phi = \arctan \dfrac{(1-\cos\alpha)(I_3 - I_1)}{\sin\alpha(I_1 - 2I_2 + I_3)}$
$\frac{2\pi m}{N}$ $N, m \in \mathcal{N}$	0	N	$\phi = \arctan \dfrac{-\sum_{i=1}^{N} I_i \sin \frac{2\pi m}{N}(i-1)}{\sum_{i=1}^{N} I_i \cos \frac{2\pi m}{N}(i-1)}$

Tabelle 1: Phasenschiebeverfahren

Mit $S_1 = I_1 + I_4$, $S_2 = I_2 + I_3$, $S_3 = I_1 - I_4$, $S_4 = I_2 - I_3$, $u = 2b\cos\phi$, $v = 2b\sin\phi$, $w = \cos(\alpha/2)$ erhält man das Gleichungssystem

$$
\begin{aligned}
S_1 &= 2a + u(4w^3 - 3w) \\
S_2 &= 2a + uw \\
S_3 &= v\sqrt{1 - w^2}(4w^2 - 1) \\
S_4 &= v\sqrt{1 - w^2}
\end{aligned}
\tag{13}
$$

Eine Lösung dieses Gleichungssystems ist die wohlbekannte Carre-Formel [4]

$$
\phi = \arctan\frac{v}{u} = \arctan\frac{\sqrt{I_1 + I_2 - I_3 - I_4}\sqrt{3I_2 - 3I_3 - I_1 + I_4}}{I_2 + I_3 - I_1 - I_4}
\tag{14}
$$

Nach Gl. (14) wird ϕ direkt aus den 4 aufgenommenen Intensitäten berechnet. Es hat sich jedoch als vorteilhaft herausgestellt, zuerst punktweise die unbekannte Phasenschiebung α zu berechnen [5]. Nach obigen Definitionen ist $\cos\alpha = 2w^2 - 1$ und man erhält aus der Lösung des Gleichungssystems nach w für jeden Bildpunkt P

$$
\alpha(P) = \arccos\frac{I_1 - I_2 + I_3 - I_4}{2(I_2 - I_3)}
\tag{15}
$$

Da vorausgesetzt werden kann, daß α auch über alle P konstant sein muß, kann $\alpha(P)$ über alle P gemittelt werden, um die Fluktuationen, die durch Störungen wie Specklerauschen verursacht wurden, zu vermindern. Ausreißer, die besonders dort auftreten, wo der Nenner von (15) Null oder nahe Null wird, werden vor der Mittelung eliminiert.

Nach der Berechnung der gemittelten Phasenschiebung α kann die Interferenzphase ϕ dann entweder nach

$$
\phi_1 = \arctan\frac{I_3 - I_2 + (I_1 - I_3)\cos\alpha + (I_2 - I_1)\cos 2\alpha}{(I_1 - I_3)\sin\alpha + (I_2 - I_1)\sin 2\alpha} + \frac{3\alpha}{2}
\tag{16}
$$

aus den ersten 3 gemessenen Intensitäten oder aus den nächsten 3 aufeinanderfolgenden Intensitäten nach

$$
\phi_2 = \arctan\frac{I_4 - I_3 + (I_2 - I_4)\cos\alpha + (I_3 - I_2)\cos 2\alpha}{(I_2 - I_4)\sin\alpha + (I_3 - I_2)\sin 2\alpha} + \frac{\alpha}{2}
\tag{17}
$$

berechnet werden. Die 2π-Sprünge liegen bei ϕ_1 und ϕ_2 an verschiedenen Stellen, dies kann zur sichereren Demodulation benutzt werden.

Als nächster Fall werde $N = 5$ mit ebenfalls konstanten Phasenschritten behandelt. Ohne Beschränkung der Allgemeinheit sei der Startwert $\alpha_1 = -2\alpha$. Dann sind die aufgenommenen Intensitäten

$$
\begin{aligned}
I_1 &= a + b\cos(\phi - 2\alpha) = a + b\cos\phi\cos 2\alpha + b\sin\phi\sin 2\alpha \\
I_2 &= a + b\cos(\phi - \alpha) = a + b\cos\phi\cos\alpha + b\sin\phi\sin\alpha \\
I_3 &= a + b\cos\phi \\
I_4 &= a + b\cos(\phi + \alpha) = a + b\cos\phi\cos\alpha - b\sin\phi\sin\alpha \\
I_5 &= a + b\cos(\phi + 2\alpha) = a + b\cos\phi\cos 2\alpha - b\sin\phi\sin 2\alpha
\end{aligned}
\tag{18}
$$

Wieder werden $u = 2b\cos\phi$, $v = 2b\sin\phi$ und diesmal $w = \cos\alpha$ definiert. Damit erhält man das Gleichungssystem

$$
\begin{aligned}
I_1 + I_5 &= 2a + u(2w^2 - 1) \\
I_2 + I_4 &= 2a + uw \\
2I_3 &= 2a + u \\
I_1 - I_5 &= 2vw\sqrt{1 - w^2} \\
I_2 - I_4 &= v\sqrt{1 - w^2}
\end{aligned}
\tag{19}
$$

124

Nach Lösung dieses nichtlinearen System ergibt sich

$$\phi = \arctan \frac{v}{u} = \arctan \frac{\sqrt{4(I_2 - I_4)^2 - (I_1 - I_5)^2}}{2I_3 - I_1 - I_5} \qquad (20)$$

Auch für $N = 5$ empfiehlt sich jedoch der Zwischenschritt der Berechnung der Phasenschiebung α und deren Mittelung. Für die Berechnung von α gibt es hier verschiedene Formeln. Am einfachsten ist

$$\alpha(P) = \arccos \frac{I_1 - I_5}{2(I_2 - I_4)} \qquad (21)$$

Weitere Lösungen erhält man durch konvexe Kombination zweier Lösungen für den Fall $N = 4$, z. B. bei gleicher Gewichtung beider Lösungen

$$\begin{aligned} \alpha(P) &= \arccos \frac{1}{2} \left[\frac{I_1 - I_2 + I_3 - I_4}{2(I_2 - I_3)} + \frac{I_2 - I_3 + I_4 - I_5}{2(I_3 - I_4)} \right] \\ &= \arccos \frac{I_1(I_3 - I_4) + (I_2 - I_3 + I_4)(I_2 - 2I_3 + I_4) - I_5(I_2 - I_3)}{4(I_2 - I_3)(I_3 - I_4)} \end{aligned} \qquad (22)$$

Eine weitere Lösung, welche ebenfalls alle 5 gemessenen Intensitäten berücksichtigt, ist

$$\alpha(P) = \arccos \frac{(I_2 - I_4)(I_1 - 2I_2 + 2I_3 - 2I_4 + I_5)}{(I_2 - I_4)(I_1 - 2I_3 + I_5) - (I_1 - I_5)(I_2 - 2I_3 + I_4)} \qquad (23)$$

Nach Mittelung über alle α kann ϕ nach jeder der Gleichungen

$$\phi = \arctan \frac{I_{i+2} - I_{i+1} + (I_i - I_{i+2})\cos\alpha + (I_{i-1} - I_i)\cos 2\alpha}{(I_i - I_{i+2})\sin\alpha + (I_{i+1} - I_i)\sin 2\alpha} - (i+1)\alpha \qquad i = 1, 2, 3 \qquad (24)$$

berechnet werden. Die mehrfache Berechnung von ϕ erlaubt eine weitere Störunterdrückung sowie eine einfachere Demodulation.

Berechnung der Phasenschiebung durch Fourier-Transformation

Falls nicht von konstanten Phasenschiebungen zwischen aufeinanderfolgenden Intensitätspaaren ausgegangen werden kann, sind die Gleichungssysteme nicht mehr geschlossen lösbar. Dann wird die Phasenschiebung zwischen aufeinanderfolgenden Intensitäten über das Fourier-Transformations-Verfahren berechnet [6], sofern wenigstens davon ausgegangen werden kann, daß die Phasenschiebung über alle Bildpunkte eines Musters konstant bleibt.

Die zwei Intensitätsverteilungen seien

$$\begin{aligned} I_1(P) &= a(P) + b(P)\cos\phi(P) \\ I_2(P) &= a(P) + b(P)\cos(\phi(P) + \alpha) \end{aligned} \qquad (25)$$

Mit

$$\begin{aligned} c_1(P) &= \frac{1}{2}b(P)\exp(j\phi(P)) \\ c_2(P) &= \frac{1}{2}b(P)\exp(j\phi(P) + j\alpha) \end{aligned} \qquad (26)$$

werden

$$\begin{aligned} I_1(P) &= a(P) + c_1(P) + c_1^*(P) \\ I_2(P) &= a(P) + c_2(P) + c_2^*(P) \end{aligned} \qquad (27)$$

Die Fouriertransformation der aufgenommenen Intensitätsverteilungen ergibt somit

$$\mathcal{I}_1(u) = \mathcal{A}(u) + \mathcal{C}_1(u) + \mathcal{C}_1^*(u)$$
$$\mathcal{I}_2(u) = \mathcal{A}(u) + \mathcal{C}_2(u) + \mathcal{C}_2^*(u) \tag{28}$$

Durch Bandpassfilterung wird der verbreiterte Gleichanteil $\mathcal{A}(u)$ und einer der beiden symmetrisch zum Ursprung liegenden Anteile $\mathcal{C}_i(u)$ oder $\mathcal{C}_i^*(u)$ jeweils eliminiert [7]. Dieser sei $\mathcal{C}_i^*(u)$, so daß nach der Filterung nur die $\mathcal{C}_i(u)$ übrig bleiben. Die inverse Fouriertransformation, angewandt auf $\mathcal{C}_1(u)$ und $\mathcal{C}_2(u)$ ergibt die komplexen $c_1(P)$ und $c_2(P)$ aus denen nach

$$\alpha(P) = \arctan \frac{\text{Re } c_1(P)\text{Im } c_2(P) - \text{Im } c_1(P)\text{Re } c_2(P)}{\text{Re } c_1(P)\text{Re } c_2(P) + \text{Im } c_1(P)\text{Im } c_2(P)} \tag{29}$$

punktweise $\alpha(P)$ berechnet wird. Nach Voraussetzung sollte α zwischen 0 und π liegen, deshalb wird α nach Eliminieren von Ausreißern durch Mittelung der Absolutbeträge der $\alpha(P)$ ermittelt

$$\alpha = \overline{|\alpha(P)|} \tag{30}$$

Ist die Vorausetzung $0 < \alpha < \pi$ nicht erfüllt, so kann man α mit umgekehrtem Vorzeichen erhalten.

Mit den so bestimmten Phasenschiebungen läßt sich nun das Gleichungsystem (8) auch für nicht konstante Phasendifferenzen entsprechend dem angegebenen Schema aufstellen und nach der Interferenzphase ϕ lösen.

Vergleich der Verfahren

Die hier vorgestellten Phasenschiebe-Verfahren mit den verschiedenen Auswertealgorithmen wurden in umfangreichen Rechnersimulationen im Hinblick auf die erreichbare Meßgenauigkeit verglichen [8]. Es zeigte sich, daß additive und multiplikative Störungen, sofern sie die gemeinsam ausgewerteten Muster gleichartig beeinflussen, inhärent kompensiert werden. Eine Herabsetzung der Genauigkeit rührt hauptsächlich von einer nicht exakten Phasenschiebung her, wenn der gewählte Auswertealgorithmus eine solche vorschreibt, oder von einer fehlenden Konstanz der Phasenschiebung. Die Algorithmen mit unbekannter Phasenschiebung sind unempfindlich gegenüber einem linearen Phasenschiebefehler. Es empfiehlt sich, die phasengeschobenen Interferenzmuster so schnell wie möglich, - Video-Bildfolge -, aufeinanderfolgend aufzunehmen, da dann Störungen aus der Meßumgebung, wie Stöße oder Schwingungen, nicht oder nur linear als zusätzliche Phasenschiebungen eingehen.

Die Berechnung der vorher unbekannten Phasenschiebungen aus den aufgenommenen Intensitätsverteilungen bietet eine Reihe von Vorteilen: Aus der Form der Verteilung der Phasenschiebung lassen sich experimentelle Fehler wie Nicht-Konstanz der Phasenschiebung frühzeitig erkennen und können entweder berücksichtigt werden, - z. B. bei linearer Variation -, oder die Messung wird wiederholt. Ausreißer können eliminiert und durch sicherere Nachbarwerte ersetzt werden. Eine Mittelung steigert die Genauigkeit und Auflösung erheblich. Die mehrfache Berechnung der Interferenzphase läßt eine Genauigkeitssteigerung durch Mittelung sowie eine einfachere und sicherere Demodulation zu.

Bei Rechnersimulationen und praktischen Experimenten mit weitgehend identischen Meßketten und gleichen statistischen Störeinflüssen war die Varianz der Meßergebnisse für die Phasenschiebe-Verfahren mitkonstanter aber vorher unbekannter Phasenschiebung und Mittelung der punktweise berechneten Phasenschiebung geringer als beim Heterodyn-Verfahren. Als geeignete Phasenschiebungen haben sich alle Winkel zwischen etwa 30° und 150° erwiesen.

Die Einflüsse von Detektornichtlinearitäten werden stärker, je weniger Intensitätsverteilungen der jeweilige Algorithmus zu verarbeiten hat und je kleiner die Phasenschiebung wird. Insbesondere Sättigungseffekte des A/D-Wandlers sind zu vermeiden.

Quantisierungsfehler spielen bei der üblichen Quantisierung in 256 Graustufen (8 bit) keine signifikante Rolle.

Die Algorithmen mit bekannter Phasenschiebung sind dort ungenau, wo die Nenner klein werden. Jedoch haben die Algorithmen für unterschiedliche Startwerte bei gleicher Anzahl von Intensitäten und gleicher Phasenschiebung unterschiedliche Kombinationen von Intensitäten im Nenner, so daß hier bei entsprechender Berücksichtigung des Startwerts zwischen den jeweils optimalen Algorithmen gewechselt werden kann.

Beim Vergleich aller Kriterien haben sich die Algorithmen mit unbekannter aber konstanter Phasenschiebung und Mittelung der punktweise berechneten Phasenschiebung als überlegen erwiesen. Sollte sich bei der Beurteilung der berechneten Phasenschiebung herausstellen, daß die Phasenschiebung nicht konstant war, und die Messung kann nicht wiederholt werden, dann müssen die Phasenschiebungen zwischen jeweils aufeinanderfolgenden Intensitätsverteilungen nach dem Fourier-Transformations-Verfahren bestimmt werden.

Zusammenfassung

Der vorliegende Artikel gibt eine systematische Herleitung der vielfältigen Algorithmen zur Auswertung von Interferenzmustern mit den Phasenschiebe-Verfahren. Im Vergleich zeigten die Algorithmen mit unbekannter konstanter Phasenschiebung die höchste erreichbare Genauigkeit. Auch für nicht konstante Phasenschiebungen ist ein Lösungsweg angegeben. Die Forschungsarbeiten, die zu den dargestellten Ergebnissen führten, wurden von der DFG im Vorhaben Kr 953/2-1 gefördert, wofür ausdrücklich gedankt sei.

Literatur

[1] N. Eichhorn, W. Osten. An algorithm for the fast derivation of line structures from interferograms. Journal of Modern Optics, 35(10), 1717–1725, 1988.

[2] K. Creath. Phase-measurement interferometry. In E. Wolf, editor, Progress in Optics, vol. 25, 349–393, 1988.

[3] Th. Kreis. Automatic evaluation of interference patterns. Proc. of Soc. Photo-Opt. Instr. Eng., vol. 1026, 80–89, 1988.

[4] P. Carre. Installation et utilisation du comparateur photoelectrique et interferentiel du Bureau International des Poids et Mesures. Metrologia, 2(1), 13–23, 1966.

[5] W. Jüptner, Th. Kreis, H. Kreitlow. Automatic evaluation of holographic interferograms by reference beam phase shifting. Proc. of Soc. Photo-Opt. Instr. Eng., vol. 398, 22–29, 1983.

[6] Th. Kreis. Digital holographic interference-phase measurement using the Fourier-transform method. Journ. Opt. Soc. Amer. A, 3(6), 847–855, 1986.

[7] Th. Kreis, W. Jüptner. Fourier-transform evaluation of interference patterns: the role of filtering in the spatial-frequency domain. Proc. of Soc. Photo-Opt. Instr. Eng., vol. 1162, 116–125, 1989.

[8] Th. Kreis, J. Geldmacher, R. Biedermann. Vergleichende theoretische und experimentelle Untersuchungen zur Genauigkeit der verschiedenen Verfahren zur Auswertung von Interferenzmustern. Bericht zum DFG-Vorhaben Kr 953/2-1, 1993.

Bestimmung einer absoluten Verformungskomponente eines belasteten Bauteils aus zwei Interferenzmustern, die bei unterschiedlichen Belastungen aufgenommen wurden

Th. Bischof, W. Jüptner

BIAS – Bremer Institut für angewandte Strahltechnik (FRG)

1. Problemstellung

Soll eine Verformung eines Bauteils holografisch erfaßt werden, so treten bei der quantitativen Auswertung der Interferenzmuster unter anderem folgende Problemstellungen auf:

- Die Standard-Auswerteverfahren [1, 2] liefern allesamt eine Berechnung des relativen Phasenverlaufs aus den Interferenzstreifen. Zusammen mit dem Empfindlichkeitsvektor, der durch die Geometrie des holografischen Aufbaus gegeben ist, erhält man eine relative Verformungskomponenten des Bauteil unter der anliegenden Last. Um den vollständigen Verformungsvektor des Bauteils bestimmen zu können, werden drei Aufnahmen aus linear unabhängigen Richtungen benötigt, die zudem noch die absolute Verformungskomponenten liefern müssen.

- Bei der Verwendung des holografischen Real-Time-Verfahren ist einerseits eine genaue Repositionierung der Hologrammplatte nach dem photochemischen Prozeß nicht möglich. Andererseits kann es zwischenzeitlich durch thermische Ausdehnungen zu Weglängenänderungen im optischen Aufbau gekommen sein. Diese Effekte führen zu einem Interferenzstreifen-Grundmuster, obwohl noch keine Belastung stattgefunden hat.

Im folgenden soll nun ein Verfahren vorgestellt werden, das einerseits die Bestimmung einer absoluten Verformungkomponente erlaubt und andererseits unempfindlich gegen Repositionierungsfehler und thermische Effekte ist.

2. Theoretische Betrachtungen

2.1. Spannungszustand im Kontinuum

Ein Körper befindet sich im Gleichgewicht, wenn die Summe der inneren Volumenkraft und der äußeren Flächenkraft, die auf das Volumen wirken, gleich Null ist. Für ein infinitesimales Volumenelement mit der Volumenkraft $\vec{V} = (X, Y, Z)^T$ gilt: (Normalspannungen σ, Schubspannungen τ)

$$\frac{\partial \sigma_x}{\partial x} + \frac{\partial \tau_{yz}}{\partial y} + \frac{\partial \tau_{zx}}{\partial z} + X = \frac{\partial \tau_{xy}}{\partial x} + \frac{\partial \sigma_y}{\partial y} + \frac{\partial \tau_{zy}}{\partial z} + Y = \frac{\partial \tau_{xz}}{\partial x} + \frac{\partial \tau_{yz}}{\partial y} + \frac{\partial \sigma_z}{\partial z} + Z = 0 \tag{1}$$

Das Materialgesetz liefert uns den Zusammenhang zwischen Dehnungen und Spannungen. Das verallgemeinerte Hook'sche Gesetz lautet:

$$\sigma_x = \lambda e + 2\nu \epsilon_x, \; \sigma_y = \lambda e + 2\nu \epsilon_y, \; \sigma_z = \lambda e + 2\nu \epsilon_z, \; \tau_{xy} = \nu \gamma_{xy} \; \tau_{yz} = \nu \gamma_{yz} \; \tau_{zx} = \nu \gamma_{zx} \tag{2}$$

mit λ, ν Lamésche Materialkonstanten. u, v, w sind die Verformungskomponenten des Punktes in x, y, z-Richtung und ferner gilt $e = \epsilon_x + \epsilon_y + \epsilon_z$. Ein Auseinanderklaffen oder ein Überlappung des Werkstoffes wird ausgeschlossen:

$$\epsilon_x = \frac{\partial u}{\partial x}, \; \epsilon_y = \frac{\partial v}{\partial y}, \; \epsilon_z = \frac{\partial w}{\partial z}, \; \gamma_{xy} = \frac{\partial u}{\partial y} \frac{\partial v}{\partial x}, \; \gamma_{yz} = \frac{\partial v}{\partial z} \frac{\partial w}{\partial y}, \; \gamma_{zx} = \frac{\partial w}{\partial x} \frac{\partial u}{\partial z} \tag{3}$$

Sind die an der gesamten Oberfläche der Körpers angreifenden Lasten bekannt, so ist das System eindeutig lösbar [3].

2.2. Superpositionsprinzip

Die Gleichungen (1), (2) und (3) sind linear und homogen in den: Verschiebungen, Spannungen und in den Komponenten der Volumenkräfte. Es gilt also das Superpositionsprinzip [4].

Hat man eine Lösung $\mathcal{L}_1$ für einen isotropen linear elastischen Körper mit den Randbedingungen $\mathcal{R}_1$ und eine zweite Lösung $\mathcal{L}_2$ für den selben Körper unter den Randbedingungen $\mathcal{R}_2$, so ist die Lösung für die Randbedingung $\mathcal{R} = \mathcal{R}_1 + \mathcal{R}_2$ gegeben durch $\mathcal{L} = \mathcal{L}_1 + \mathcal{L}_2$

Betrachten wir nun zwei Lastfälle $\mathcal{R}_1$ und $\mathcal{R}_2$, wobei sich die beiden Randbedingungen nur um einen konstanten Faktor k unterscheiden, $\mathcal{R}_2 = k \cdot \mathcal{R}_1$, so gilt für die Lösungen:

$$\mathcal{L}_2 = k \cdot \mathcal{L}_1 \tag{4}$$

3. Verfahren zur Bestimmung einer absoluten Verformungskomponenten

Wird ein Bauteil, das linear elastisches Materialverhalten zeigt und keine geometrischen Nicht-Linearitäten aufweist, wie z.B. Kontaktflächen, mit einer äußeren Last σ_0 beaufschlagt, so ergeben sich ensprechend den Differentialgleichungen (1), (2) und (3) Verschiebungen $\vec{d}_0$ auf der Oberfläche des Bauteils. Wird nun die anliegende Last um den Faktor k verändert, also $\sigma_1 = k \cdot \sigma_0$ so verändern sich die Verformungen·an der Oberfläche proportional zu der zusätzliche Last (Gleichung (4)):

$$\vec{d}_1 = k \cdot \vec{d}_0 \tag{5}$$

In der HNDT wird pro Aufnahme eine Verformungskomponente in Richtung des Empfindlichkeitsvektor $\vec{e}$ in Form der relativen Phase φ modulo 2π aufgezeichnet.

$$\varphi = \mathrm{mod}_{2\pi}\left(\frac{2\pi}{\lambda}\,\vec{d}\cdot\vec{e}\right) \tag{6}$$

Angenommen wir hätten zwei Interferenzmuster einer belasteten Probe, wobei die Art der Belastung identisch, die Höhe der Belastungen jedoch unterschiedlich ist. Bei der ersten Aufnahme sei die Belastung σ_0; bei der zweiten Aufnahme sei die Belastung auf $k \cdot \sigma_0$ verändert worden. Weiterhin sei vorausgesetzt, daß für mindestens einen Punkt im Interferenzmuster die Änderung der relative Phase $(\varphi_0 \to \varphi_1)$ bei der Belastungssteigerung $(\sigma_0 \to k \cdot \sigma_0)$ bekannt ist. Der Einfachheit halber kann die Zusatzbelastung so gewählt werden, daß die Änderung der relativen Phase im gesamten auszuwertenden Bild kleiner als $0,5\,\pi$ ist $(\varphi_1 - \varphi_0 < 0,5\pi)$, damit für jeden Punkt eine direkte Zuordnung der Streifenmuster vor und nach der Zusatzbelastung gegeben ist und die absolute Änderung der Phase aufgrund der Belastungssteigerung direkt berechnet werden kann:

$$\varphi_1 - \varphi_0 = \frac{2\pi}{\lambda}\,(k-1)\cdot\vec{d}_0\cdot\vec{e}$$

Die Größen φ_1 und φ_0 können unter der Bedingung $(\varphi_1 - \varphi_0 < 0,5)$ eindeutig mit dem Auswerteverfahren bestimmt werden, so daß gilt:

$$\vec{d}_0\cdot\vec{e} = \frac{\lambda}{2\pi}\,\frac{(\varphi_1-\varphi_0)}{k-1} \tag{7}$$

Für die Bestimmung von k ist zusätzlicher Meßaufwand notwendig, wohingegen die Bestimmung der Streifenordnung $(\varphi_1 - \varphi_0)$ bei dem verfügbaren computergestützten Auswerteverfahren keine weiteren Probleme bereiten dürfte, wenn die Belastung genügend gleichförmig aufgebracht werden kann.

4. Experiment

Um die theoretischen Ableitungen zu bestätigen wurden ein Biegebalken, wie in Bild 1 dargestellt, mit einer Einzelkraft belastet und die resultierende Biegelinie holografisch ausgewertet.

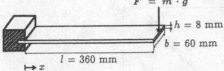

Bild 1: Geometrie des Biegebalkens

Für Durchbiegung $w(x)$ eines Biegebalkens gilt die Differenzialgleichung:

$$\frac{w''(x)}{\left(1 + w'(x)^2\right)^{\frac{3}{2}}} = -\frac{M}{E\,J} \qquad (8)$$

Für kleine Auslenkungen, also insbesondere in der HNDT, gilt $w' \ll 1$.

Mit den Randbedingungen $w(x=0) = 0$ und $w'(x=0) = 0$ sowie Balken unter Einzellast, ergibt sich für die Biegelinie:

$$w(x) = \frac{F}{6\,E\,J}\left(3 \cdot l\,x^2 - x^3\right) \qquad (9)$$

Die Untersuchungen werden mit einem Argon-Laser durchgeführt (Wellenlänge: $\lambda = 512$ nm). Die Ausmessung der Aufbaugeometrie ergab in erster Näherung einen normal zur Oberfläche des Balkens stehenden Empfindlichkitsvektor mit der Einheitslänge 1.8, so daß für die absolute Streifenordnung im Punkt x ergibt:

$$n \cdot \lambda = 1.8 \cdot w(x)$$

$$n = \frac{1.8}{\lambda} \cdot \frac{m\,g}{6\,E\,J}\left(3\,l\,x^2 - x^3\right) \qquad (10)$$

Die Grundmasse getrug 30,55 g, die Zusatzmasse 1,895 g. Bild 2 zeigt das resultierende Interferenzstreifenmuster. Oben: unter der Grundlast; unter: unter Grundlast plus Zusatzlast.

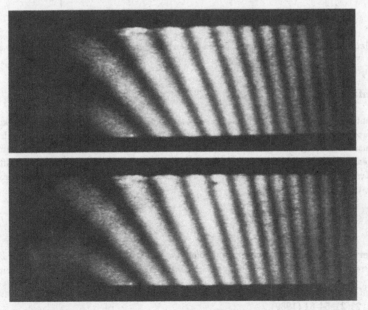

Bild 2: Biegebalken: Grundlast oben, Grundlast + Zusatzlast unten

Zur Ermittlung der Phasenverteilung wurde das Fourier-Transformationsverfahren verwendet. Bild 3 zeigt die resultierende Phasenverläufe. Die Differenz der beiden Phasenverläufe $(\varphi_1 - \varphi_0)$ eingesetzt in Gleichung 7 zusammen mit der analytischen Lösung des Biegebalkenproblems ist in Bild 3 dargestellt.

Sowohl qualitativ, wie auch quantitativ ist ein gute Übereinstimmung zu beobachten.

Wie zu erwarten war, hatte der leichte Repositionierungsfehler, der durch die schräge Lage der Interenzferenzstreifen deutlich wird, keinen Einfluß auf das Verformungsergebnis.

130

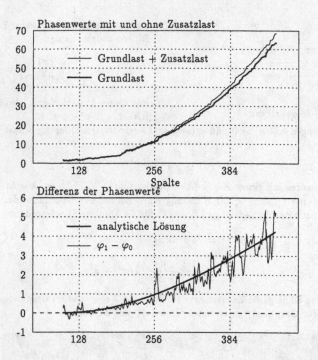

Bild 3: Phasenverlauf der beiden Belastungszustände (oben), Differenz der Phasenverläufe mit analytischer Lösung (unten)

5. Zusammenfassung

Unter den Voraussetzungen, daß das Bauteil linear elastisches Verhalten unter Belastung zeigt, konnte durch die Nutzung des Superpositionsprinzips der linearen Elastizitätstheorie gezeigt werden, daß durch die gezielte Aufnahme eines zusätzlichen Belastungszustandes eine absolute Verformungskomponente bestimmt werden kann. Das Verfahren bietet weiterhin den Vorteil, daß Repositionierungsfehler und thermische Effekte keinen Einfluß auf das Messergebnis haben, da diese Fehler bei der Differenzbildung der Phasenverläufe aus den beiden Belastungsstufen elminiert werden.

Literatur

[1] W. Jüptner: *Automatisierte Auswertung holografischer Interferogramme mit dem Zeilen-Scan-Verfahren*, Proc. Frühjahrsschule 78, Holografische Interferometrie in Technik und Medizin (1978)

[2] Th. Kreis:*Methoden zur Auswertung holografischer Interferenzmuster : ein Vergleich*, Laser und Optoelektronik, 21(2), 54-61 (1989)

[3] N. I. Muskhelishvili: *Some Basic Problems of the Mathematical Theory of Elasticity*, Noordhoff International Publishing, Moskau 1954

[4] S. P. Timoshenko, J. N. Goodier: *Theory of Elasticity*, McGraw-Hill Book Company, Third Edition 1970

3-Dimensionale Verformungsmessung mittels 3D-Phasenmessung

Junli Sun
STEINBICHLER OPTOTECHNIK GmbH
Am Bauhof 4, W-8201 (83115) Neubeuern

1. Einleitung

Seit einem Jahrzehnt wird eine 3-dimensionale Verformung bekannterweise mit der holografischen Interferometrie in 3 verschiedenen Empfindlichkeitsrichtungen gemessen, ausgewertet und dann in einem Koordinatensystem berechnet. Einer der Nachteile ist, daß solche Methoden aufwendig und unzuverlässig sind. Die größte Schwierigkeit liegt darin, daß die absoluten Null-Ordnungen (Offset) vor der Auswertung und Berechnung bekannt sein müssen, doch wird diese wichtige Voraussetzung in der Praxis selten erfüllt.

Nach der expansiven Entwicklung der Speckle-Technik in der letzten Zeit wurde dieser Nachteil mit einem sogenannten 2-Objektstrahl-Verfahren vermieden. Mit diesem Verfahren kann man die in-plane und out-of-plane Verformung getrennt und direkt messen. Nachteil aber ist, daß für eine 3-dimensionale Verformungsmessung mindestens 5 Beleuchtungen benötigt werden. Der optische Aufbau ist kompliziert und unflexibel.

Um oben genannte Nachteile zu vermeiden, wurde die 3D-Phasenmessung im LABOR DR. STEINBICHLER entwickelt und angewandt.

2. Allgemeines

Im Allgemeinen kann die Intensität eines Interferogramms durch folgende Gleichung beschrieben werden:

$$I(\vec{P}) = H(\vec{P})\{1 + K(\vec{P}) cos\, [\, \Delta\Phi(\vec{P})\,]\} \qquad (1)$$

$\vec{P}$ ist der Ortsvektor, der die Objektpunkte bezeichnet, $H(\vec{P})$ und $K(\vec{P})$ sind die Hintergrundhelligkeit und der Kontrast. $\Delta\Phi(\vec{P})$ ist die mit dem bekannten Phasenshift-Verfahren gesuchte lokale Phasenänderung, die durch eine vorliegende Verschiebung $\vec{V}$ (V_x, V_y, V_z) des Punktes $\vec{P}$ erzeugt wurde,

$$\Delta\Phi = \frac{2\pi}{\lambda}\, \vec{E}\cdot\vec{V} \qquad (2)$$

wobei $\vec{E}$ der Empfindlichkeitsvektor ist.

Bei einer 3-dimensionalen Verformungsmessung werden die drei Interferogramme bei unterschiedlicher Ausrichtung der Empfindlichkeitsvektoren $\vec{E}_1$, $\vec{E}_2$ und $\vec{E}_3$ aufgezeichnet. So

132

treten drei entsprechende unterschiedliche Phasenänderungen $\Delta\Phi_1$, $\Delta\Phi_2$ und $\Delta\Phi_3$ auf. Wie in Gleichung (2) für eine Empfindlichkeitsrichtung dargestellt, ergibt sich nun für die drei Phasenänderungen folgendes lösbare Gleichungssystem:

$$\Delta\Phi_1 = \frac{2\pi}{\lambda} (E_{1x}V_x + E_{1y}V_y + E_{1z}V_z)$$

$$\Delta\Phi_2 = \frac{2\pi}{\lambda} (E_{2x}V_x + E_{2y}V_y + E_{2z}V_z) \tag{3}$$

$$\Delta\Phi_3 = \frac{2\pi}{\lambda} (E_{3x}V_x + E_{3y}V_y + E_{3z}V_z)$$

Die Lösung $\vec{V}$ (V_x, V_y, V_z) des Gleichungssystems (3) ist die Verschiebung des Punktes $\vec{P}$. Für eine komplette Verformungsmessung gibt es ca. 250.000 - 1 Mio. Bildpunkte und die gleiche Anzahl linear unabhängiger Gleichungssysteme zu lösen. Bei den herkömmlichen Methoden müssen die Phasenbilder der Interferogramme noch vorher demoduliert (ausgewertet) werden und die Phasenänderungen $\Delta\Phi_1$, $\Delta\Phi_2$ und $\Delta\Phi_3$ "absolut" sein, aber das ist oft sehr schwierig, da die holografische Interferometrie ein relatives Meßverfahren ist.

3. 3D-Phasenmessung

Das Ziel der 3D-Phasenmessung ist der industrielle Einsatz, d.h. daß der optische Aufbau kompakt und handlich, die Berechnung schnell und zuverlässig sein soll. Die Philosophie der 3D-Phasenmessung ist, durch die symmetrisch optische Anordnung und mathematisch-physikalische Näherung das Ziel zu erreichen. Die systematischen Meßfehler, die durch die Näherung entstehen, sollten nach den kompletten Messungen je nach Bedarf und Anspruch korrigiert werden. Danach erhält man einen Überblick und mehrere Informationen über eine räumliche Verformung in einem gewöhnlich rechtwinkligen kartesischen Koordinatensystem, und kann damit wesentlich leichter die Verformungen zuordnen.

Bild 1 zeigt eine optische Anordnung der 3D-Phasenmessung. $\vec{K}_r$, $\vec{K}_l$ und $\vec{K}_o$ sind die drei Beleuchtungseinheitsvektoren, die sich jeweils in den Ebenen XZ und YZ befinden und einen gleichen Winkel Θ zur Beobachtungsrichtung (Z-Achse) haben. $\vec{K}_B$ ist der Beobachtungseinheitsvektor.

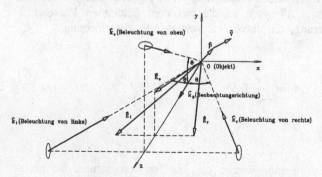

Bild 1:   Optische Anordnung der 3D-Phasenmessung

Normalerweise sind die Empfindlichkeitsvektoren $\vec{E}_r$, $\vec{E}_1$ und $\vec{E}_o$ ortsabhängig, nicht überall identisch. Nur wenn die Geometriegröße des Objektes, verglichen mit den Entfernungen der Beleuchtungslichtquellen und des Beobachtungspunktes, so klein und damit vernachlässigbar ist, kann man die Empfindlichkeitsvektoren am Punkt O auch für andere Objektpunkte $\overline{P}$ verwenden. Bei der 3D-Phasenmessung wird diese Näherung von vornherein schon angenommen, und aus den ursprünglich erfassten Phasenänderungen $\Delta\Phi_r$, $\Delta\Phi_1$, $\Delta\Phi_o$ werden nicht direkt die Verformungen V_x, V_y, V_z berechnet, sondern die Phasenänderungen $\Delta\Phi_x$, $\Delta\Phi_y$, $\Delta\Phi_z$ mit der einfachen Bildverarbeitung ins neue rechtwinklige Koordinatensystem transformiert. In diesem kann man die Null-Ordnungen der Verformungen wesentlich leichter als in einem ursprünglich schiefwinkligen Koordinatensystem ($\vec{E}_1$, $\vec{E}_2$, $\vec{E}_3$ oder $\vec{E}_r$, $\vec{E}_1$, $\vec{E}_o$) physikalisch zuordnen. Die Ergebnisse, die Verformungen V_x, V_y und V_z erhält man durch Demodulation:

$$V_x = \frac{\lambda}{2\pi E_x}\, \Delta\Phi_x$$

$$V_y = \frac{\lambda}{2\pi E_y}\, \Delta\Phi_y \tag{4}$$

$$V_z = \frac{\lambda}{2\pi E_z}\, \Delta\Phi_z$$

4. Anwendungsbeispiele

Die 3D-Phasenmessung kann ohne weiteres sowohl in der holografischen und elektronischen Speckle- Interferometrie bei einer Verformungsmessung als auch in der Phasen-Shearografie bei einer Messung der ersten Ableitungen der Verformung angewandt werden.

Im Folgenden sind einige beispielhafte Ergebnisse der Untersuchungen angeführt, bei denen die Verformung bzw. die ersten Ableitungen der Verformung separat mit einem 3D-ESPI und einem 3D-Phasen-Shearing-Interferometer (3D-PSI) an einer ebenen, gekerbten Metall-Probe (eine Aluminiumplatte 210 x 70 x 2 mm^3) gemessen wurden. Die Probe wurde an den beiden Seiten mit Schrauben auf einer Vorrichtung festgeschraubt und in der Mitte mit einer zunehmenden Kraft auf Biegung beansprucht (Bild 2).

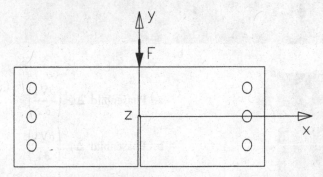

Bild 2:   Biegebelastung an einer gekerbten Probe

Bild 3 zeigt die drei transformierten Phasenbilder $\Delta\Phi(V_x)$, $\Delta\Phi(V_y)$ und $\Delta\Phi(V_z)$, deren ursprüngliche Phasenbilder mit einem 3D-ESPI aufgenommen wurden.

134

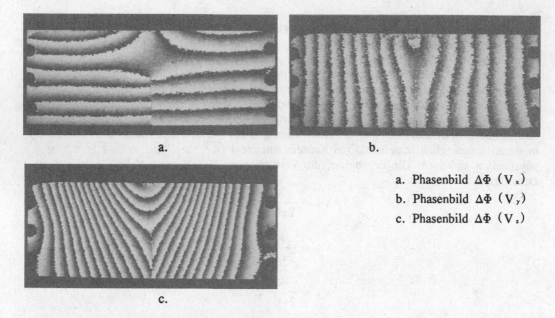

a.

b.

a. Phasenbild $\Delta\Phi\,(V_x)$

b. Phasenbild $\Delta\Phi\,(V_y)$

c. Phasenbild $\Delta\Phi\,(V_z)$

c.

Bild 3:   3-dimensionale Verformungsmessung mittels eines 3D-ESPI

In Bild 4 und 5 sind die Phasenbilder der ersten Ableitungen der Verformung dargestellt. Die Messungen wurden mittels eines 3D-Phasen-Shearing-Interferometers mit jeweils einer Shearweite von $\Delta x = 6$ mm in x-Richtung und $\Delta y = 6$ mm in y-Richtung durchgeführt. Am auffälligsten weist die Dehnung $\partial V_x/\partial x$ im Bild 4 (a) ein Maximum an der Spitze der Kerbe auf.

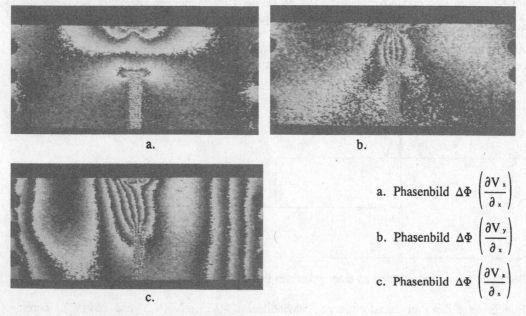

a.

b.

a. Phasenbild $\Delta\Phi\left(\dfrac{\partial V_x}{\partial x}\right)$

b. Phasenbild $\Delta\Phi\left(\dfrac{\partial V_y}{\partial x}\right)$

c. Phasenbild $\Delta\Phi\left(\dfrac{\partial V_z}{\partial x}\right)$

c.

Bild 4:   Messung der ersten Ableitungen einer Verformung mittels eines 3D-PSI mit einer Shearweite von $\Delta x = 6$ mm

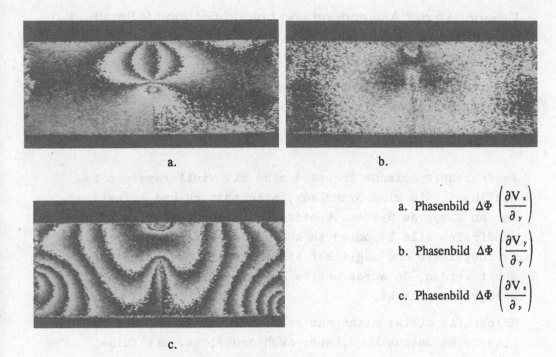

a.

b.

c.

a. Phasenbild $\Delta\Phi\left(\dfrac{\partial V_x}{\partial y}\right)$

b. Phasenbild $\Delta\Phi\left(\dfrac{\partial V_y}{\partial y}\right)$

c. Phasenbild $\Delta\Phi\left(\dfrac{\partial V_z}{\partial y}\right)$

Bild 5: Messung der ersten Ableitungen einer Verformung mittels eines 3D-PSI mit einer Shearweite von $\Delta y = 6$ mm

5. Diskussion

Die bisherigen Untersuchungen weisen nach, daß die Meßfehler der 3D-Phasenmessung, die systematischen Fehler durch die Näherung oft deutlich kleiner als die Fehler bei den herkömmlichen Methoden durch falsche Zuordnung sind, besonders, wenn das Geometrieverhältnis zwischen dem Objekt und der optischen Anordnung beachtet und optimiert wird. Andererseits kann man bei Bedarf oder hohem Anspruch immer durch eine absolute Koordinatenmessung des Objekts diese systematischen Meßfehler korrigieren.

Gegenüber dem 2-Objektstrahl-Verfahren hat die 3D-Phasenmessung die folgenden 3 Vorteile: der optische Aufbau ist einfacher und kompakter, die Ausgangsleistung des Lasers wird besser ausgenutzt und die ersten Ableitungen einer räumlichen Verformung können mit der Phasen-Shearografie optisch direkt gemessen werden.

6. Anerkennung

Ein Teil der Arbeiten wird durch das Bundesministerium für Forschung und Technologie in einem Verbundprojekt ("Holografische Meßtechnik") gefördert.

Ergebnisse der Anwendung von holografischen Meßmethoden in der Verfahrenstechnik

Judit J. TIMKÓ
Ungarische Akademie der Wissenschaften,
Budapest, Ungarn

Verfahrenstechnische Prozesse sind mit vielParametern beeinflusst. Sie sind dynamisch, stochastisch und meisstens werden disperse Systeme behandelt. Um von dem Ablauf einwandfreien Bild bekommen zu können, müssen informationsreiche,exakte und möglichst störungsfreie Messmethode gewählt werden. So wurde unsere Aufmerksamkeit auf die Puls-Holografie gelenkt.

Holografie bietet nicht nur mehr an Information, als klassische Messmethode, aber auch neuartige. Mit Pulslaser können schnell ablaufende Prozesse berührungsfrei abgebildet werden, von dem Hologramm in beliebiger Zeit wieder hergestellt und räumlich ausgewertet werden.
In der chemischen Verfahrenstechnik ist es noch zusetzlich grosser Vorteil, dass mikro-Anomalien -- die fast unmöglich wieder hervorrufbar sind, aber für die Beurteilung der Arbeitsweise einer Konstruktion /z.B. einem Zerstäuber/ enorm wichtig sind -- einwandfrei im ganzem Versuchsvolumen zu beurteilen werden.

Viele Jahre verwenden wir Puls-Holografie erfolgreich bei Mehrphasenströmungen der chemischen Verfahrenstechnik, im Gebiet des Umweltschutzes, der Energieversorgung oder der Landwirtschaft. Unser mehrfach patentiertes Messverfahren wurde auch bei Versuchen in Industriegrössen verwendet.

Unser Messverfahren hat eine Anordnung bei den Aufnahmen, wo gleichzeitig zwei Hologramme entstehen können. Als Beleuchtungsquelle dient ein Impulsrubinlaser, mit 20 ns Blitzdauer, die Hologramme werden auf AGFA-Gevaert "Holotest" Filme oder Platten aufgenommen. Dem abbildenden

Prozess gemäss, werden die Aufnahmen entweder mit einer
Streulinse aufgeweitetem Laserstrahl, oder mit parallelem
Laserbündel durchgeführt. Dabei kann die Geradeaus-Methode,
oder optischer Aufbau mit getrenntem Referenzbündel verwendet werden.

Die Wiedergabe erfolgt durch einem kontinuierlichen HeNe-
Gaslaser und geschlossenem TV System. Die nacheinander-
folgende scharfgestellte Ebenen können automatisch ausgewertet werden. So wird der ganze Raum durchsucht und die
überlagerten Objekte getrennt werden. Photographieren ist
inzwischen auch möglich. Die Methode gibt qualitative und
quantitative Information. Mit dieser Methode wurden Tropfen
in Luft,Feststoff-Partikeln in Luft und in Flüssigkeit und
Blasen in Flüssigkeit abgebildet. Also verfahrenstechnisch
ausgedrückt – wir haben Versuche mit Zerstäubern fast
aller Art gemacht, kalte und brennende Nebel, newtonsche
und nicht-newtonsche Tropfenbildung untersucht. Brennende
Kohlenpartikeln und Wirbelschichtanomalien abgebildet,
Luftfilter getestet; sich bildende Kristallen und ihre
Wachstum untersucht. Sedimentationsgeschwindigkeit durch
Hologramme festgestellt und Flockenbildner getestet.
Energiewirtschaft und Umweltschutz gehörten auch zu den
bekempfenden Terrainen. Die gewonnenen Ergebnisse haben
bisher fast in allen Verwendungsgebieten etwas neues gebracht, die sowohl in der Theorie, als in der Praxis ver-
wendbar wurden.

Holografie ist eine konzentrationsabhängige Messung, die
es möglich macht Grösse, Grössenverteilung und räumliche
Lage von allen Partikeln oder Tropfen in einem gewissen
Volumen in dem Zeitpunkt der Aufname des Hologrammes zu
erfassen. Die Ergebnisse unseren vielseitigen Versuchen
zeigen, wie der gewonnene Informationszusatz zur besseren
Lösung der Problematik von strömenden dispersen Systemen
der Verfahrenstechnik führt.

138

Veröffentlichungen

TIMKÓ, J.J. : Bild und Ton 32 269 p. /1979/
TIMKÓ, J.J. BLICKLE,T. : Journal of Power Bulk
Solids Techn. 3 22 p. /1979/
TIMKÓ, J.J. : "METROP" Spie Vol. 210-140 /1979/
TIMKÓ, J.J. : "FLOW VISUALIZATION II." 535 p.
Hemispere Publ. Corp. /1982/
TIMKÓ, J.J. : "IUTAM Symp." Liblice 29 p. Springer Verlag
/1984/
TIMKÓ, J.J. : "ECOOSA'84 Amsterdam, SPIE Vol. 482-81 /1984/
TIMKÓ, J.J. : "LASER 89" Optoelektronik in der Technik 309 p.
München, Springer Verlag /1989/
TIMKÓ, J.J. : "LASER 91", Laser in der Technik 178 p.,
München, Springer Verlag /1991/

Patents   Hung. Patent 175.498
               UK Patent GB 2.042.754 B
               U.S.A. Patent 4.278.319

Weitere Information : Dr. Ing. J.J. TIMKÓ
                      1026 Bufapest Ervin u. 7.
                      Ungarn

Dreidimensionale holografische Interferometrie und Modalanalyse

M. Sellhorst, H. Ostendarp *, C.R. Haas, R. Noll

Fraunhofer-Institut für Lasertechnik, Steinbachstraße 15, D-5100 Aachen

*Fraunhofer-Institut für Produktionstechnologie, Steinbachstraße 17, D-5100 Aachen

1 Einleitung

Bei Neuentwicklungen im Werkzeugmaschinen- und Automobilbau ist die Kenntnis und entsprechende Berücksichtigung des dynamischen Verhaltens komplexer mechanischer Baugruppen häufig ausschlaggebend für die Funktion und Qualität des neuen Produktes. Die Analyse an Hand von theoretischen Modellen ist in vielen Fällen unzureichend. Eine direkte Messung des dynamischen Verhaltens ist erforderlich.

Das in der Praxis dafür bewährte Meßverfahren ist die Experimentelle Modalanalyse (EMA) [1]. Dabei wird der Zeitverlauf der räumlichen Bewegung mehrerer ausgewählter Strukturpunkte in Abhängigkeit von einem eingeleiteten Kraftsignal bestimmt. Die Meßdaten liefern die interessierenden dynamischen Eigenschaften des Meßobjekts. Dieses Meßverfahren ist sehr zeitaufwendig, und beeinflußt durch das Aufbringen von elektromechanischen Beschleunigungssensoren die dynamischen Eigenschaften des Meßobjekts. Außerdem gibt die EMA nur an einigen diskreten Punkten Aufschluß über die räumliche Bewegung der untersuchten Struktur.

Die holografische Interferometrie ist ein bekanntes Verfahren, um die Verlagerung mechanischer Strukturen zwischen zwei diskreten Zeitpunkten zu bestimmen [2]. Bei den herkömmlichen Meßsystemen wird die Verlagerung lediglich in einer Raumrichtung erfaßt. Bei harmonischer Anregung des Meßobjekts bietet sich die holografische Interferometrie zur flächenhaften Analyse des dynamischen Verhaltens an. Durch die berührungslose Messung der Objektverlagerungen bleibt das dynamische Verhalten des Objekts unbeeinflußt.

Die Kombination der holografischen Interferometrie mit dem Verfahren der EMA erlaubt die Ausnutzung der Vorteile beider Meßverfahren. Ein solches kombiniertes Meßsystem erfaßt das dynamische Verhalten mechanischer Strukturen vollständig, trägheits- und berührungslos über der gesamten Bauteiloberfläche.

2 Experimentelle Modalanalyse

Bei der experimentellen Modalanalyse werden die komplexen Nachgiebigkeiten zwischen diskreten Punkten bestimmt. Die Nachgiebigkeit ist als das Verhältnis zwischen einer Komponente der Verlagerung und einer Komponente der eingeleiteten Kraft definiert. Die Nachgiebigkeiten aller Komponenten ergeben den Nachgiebigkeitstensor. Ist dieser für zwei Strukturpunkte bekannt, so ist es möglich, den zeitlichen Verlauf der Verlagerung an dem einen Strukturpunkt für ein an dem anderen Punkt eingeleitetes Kraftsignal zu bestimmen.

Die Nachgiebigkeit ergibt sich aus dem Kraftsignal, das ein Schwingerreger in die Struktur einleitet, und dem zeitlichen Verlauf der Verlagerung an dem betrachteten Strukturpunkt. Beschleunigungssensoren, die auf der Struktur aufgebracht sind, messen sowohl das eingeleitete Kraftsignal, als auch die Verlagerung des Bauteils am interessierenden Strukturpunkt.

140

Die Digitalisierung des Kraft- und Wegsignals und die anschließende schnelle Fourieranalyse liefern den Frequenz- und Phasengang der Nachgiebigkeit.

Bild 1 zeigt einen typischen Frequenzgang des Absolutbetrages der Nachgiebigkeit für zwei Strukturpunkte wie sie in Bild 2a gekennzeichnet sind.

Mit Hilfe der Nachgiebigkeitsfrequenzgänge für hinreichend viele Strukturpunkte wird die Schwingungsform der untersuchten Struktur bei der Einleitung eines bestimmten Kraftsignals bestimmt.

Die Modellierung der untersuchten Struktur durch ein Gittermodell wie es in Bild 2a gezeigt ist, dient als Ausgangspunkt für die Schwingungsformanalyse. Die einzelnen Gitterpunkte repräsentieren die Meßorte auf der untersuchten Struktur, wo der zeitliche Verlauf der Objektverlagerung bekannt ist. Die phasenrichtige Darstellung der Objektbewegung in einer dreidimensionalen grafischen Animation stellt die räumliche Bewegung des Bauteils bei einer bestimmten Anregungsfrequenz dar.

Eine Anregung des untersuchten Bauteils durch ein Rauschsignal liefert mit einer einzigen Messung die Information über das dynamische Verhalten im gesamten interessierenden Frequenzbereich.

Die Ergebnisse der EMA bzw. das Gittermodell der Struktur werden zur Simulation des dynamischen Verhaltens bei Krafteinleitungen, wie sie in der Realität auftreten, herangezogen.

Solche Simulationen geben beispielsweise Aufschluß über die dynamischen Ursachen, die für das Auftreten von Bearbeitungsungenauigkeiten beim Betrieb hochpräziser Werkzeugmaschinen verantwortlich sind.

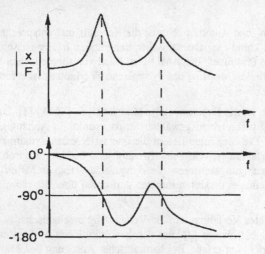

Bild 1: Frequenz- und Phasengang des Absolutbetrages der Nachgiebigkeit

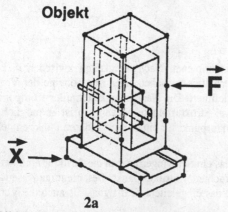
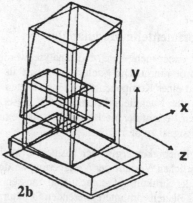

2a   **2b**

Bild 2a: Strukturpunkte für die Krafteinleitung und die Verlagerungsmessung
b: Analyse der Schwingungsform aus den Nachgiebigkeitsfrequenzgängen

3 Holografische Interferometrie und Modalanalyse

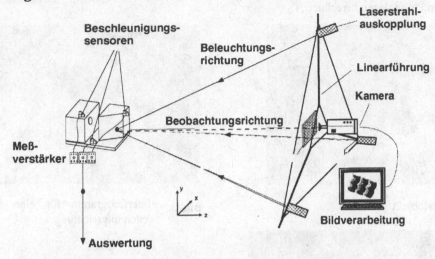

Bild 3: Skizze des Aufbaus zur dreidimensionalen Holografie

Um mit der holografischen Interferometrie die Verlagerung über der gesamten Objektoberfläche vollständig dreidimensional zu erfassen, ist eine Erweiterung des herkömmlichen Aufbaus erforderlich. Bild 3 zeigt die Skizze einer solchen Anordnung.

Das Objekt wird aus 3 nicht koplanaren Richtungen beleuchtet und aus einer Richtung beobachtet, so daß sich 3 Empfindlichkeitsrichtungen ergeben.

Die von Objekt- und Referenzstrahl durchlaufenen Wegstrecken sind für jeden Kanal unterschiedlich. Dieser Wegunterschied ist größer als die Kohärenzlänge des verwendeten Lasers, so daß der Objektstrahl des einen Kanals mit dem Referenzstrahl des anderen Kanals nicht interferenzfähig ist. Auf diese Weise entstehen mit einer einzigen Aufnahme drei Hologramme.

Eine einzige Messung liefert die Komponenten des Vektorfeldes der Verlagerungen in den 3 Empfindlichkeitsrichtungen für die gesamte Bauteiloberfläche. Gleichzeitig erfassen Beschleunigungssensoren, die an ausgewählten Referenzpunkten des Objekts angebracht sind, zeitkontinuierlich die räumliche Bauteilverlagerung.

Bei der Anregung des Objekts mit einem harmonischem Testkraftsignal, ist es möglich aus den Referenzsignalen und den Interferogrammen für die Objektverlagerungen die räumliche Schwingungsform des Objekts herzuleiten.

Das in Bild 4 dargestellte Testobjekt weist bei bestimmten Frequenzen Schwingungsformen auf, die mit dem herkömmlichen EMA-Verfahren nur mit unverhältnismäßig großem Aufwand erkannt werden können. Das kombinierte Meßsystem erfaßt diese charakteristischen Eigenschwingungsformen jedoch ohne weiteren Aufwand.

In Bild 5 ist ein holografisches Interferogramm dargestellt, das die Objektverlagerung in einer Raumrichtung charakterisiert. Aus den Interferogrammen für drei verschiedene Raumrichtungen werden die Komponenten der Verlagerung im kartesischen Objektkoordinatensystem ermittelt. Bild 6 zeigt die grauwertcodierte Darstellung der Verlagerung der Bauteiloberfläche in einer Richtung des Objektkoordinatensystems.

Durch die Verknüpfung der dadurch gewonnenen Informationen mit den von den Referenzsensoren gelieferten Daten wird die räumliche Schwingungsform des Objekts rekonstruiert. Bild 7 zeigt

142

die Darstellung der Verformung für eine Schwingungsendlage. Die Verformung wird für die interessierenden Phasenlagen berechnet.

Bild 4: Testobjekt

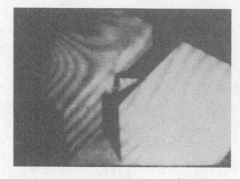

Bild 5: Interferogramm für eine Beleuchtungsrichtung

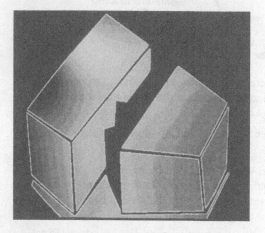

Bild 6: Verformung in einer Richtung des Objektkoordinatensystems

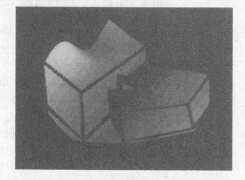

Bild 7: Darstellung der Verformung in einer Schwingungsendlage

4 Zusammenfassung

Durch die Verknüpfung des Meßverfahrens der holografischen Interferometrie mit den Methoden der experimentellen Modalanalyse (EMA) ist ein Meßsystem realisiert worden, mit dem die räumliche Schwingbewegung komplexer mechanischer Strukturen flächenhaft erfaßt wird. Der zeitliche Aufwand für die Vorbereitung und Durchführung der Messung ist gegenüber dem herkömmlichen Verfahren der EMA erheblich geringer, da für jede stetige Teilfläche der Objektoberfläche nur ein Referenzsignal zur Auswertung erforderlich ist. Der Meßaufwand hierfür ist wesentlich geringer als bei der herkömmlichen EMA. Die dynamischen Eigenschaften der untersuchten Struktur werden bei diesem Verfahren nur minimal verändert.

Die vorgestellten Arbeiten wurden von der Volkswagenstiftung gefördert.

Digitale Scherografie mit Hilfe von Phaseschiebeverfahren

J. Geldmacher, Th. Kreis, W. Jüptner
BIAS – Bremer Institut für angewandte Strahltechnik (FRG)

1. Problemstellung

Die Scherografie ist ein Verfahren zur Bestimmung von Verschiebungsänderungen technischer Oberflächenstrukturen. Eine effektive Anwendung dieses Verfahrens ist aber nur gewährleistet, wenn die scherografischen Streifenmuster automatisiert quantitativ ausgewertet werden können [1]. Bei der Auswertung aller Arten von Interferenzmustern hat sich das Phaseschiebe-Verfahren als robustes und genaues Verfahren erwiesen. Die Aufnahme- und Speicherzeiten für die dazu benötigten zeitlich nacheinander phasengeschobenen Bilder lagen bisher aufgrund der verfahrensbedingten Zeitverzögerung im Millisekundenbereich. Daher war eine Anwendung der scherografischen Meßtechnik mit Auswertung nach dem Phaseschiebeverfahren nur auf quasistationäre Dehnungsvorgänge begrenzt.

Im folgenden soll nun eine Methode vorgestellt werden, mit der durch räumliche Phasenschiebung drei oder mehr phasengeschobene Bilder zeitgleich aufgenommen und mit Hilfe des Phasenschiebeverfahrens dann quantitativ ausgewertet werden können.

2. Theorie

Die Speckle-Meßmethoden werden benutzt, um das Verschiebungsvektorfeld der Punkte einer rauhen Objektoberfläche bei Belastung zu messen. Hierzu werden das von der Oberfläche reflektierte Wellenfeld und eine Referenzwelle auf dem Target einer Videokamera zur Interferenz gebracht und das entstehende Specklefeld aufgenommen. In der digitalen Scherografie werden zwei gegeneinander lateral verschobene Specklebilder der rauhen Oberfläche kohärent überlagert und das durch Interferenz entstehende Specklemuster digital aufgenommen und gespeichert [2, 3]. Hier wird keine informationsfreie Referenzwelle gewählt, sondern ein verschobenes Bild derselben Objektoberfläche.

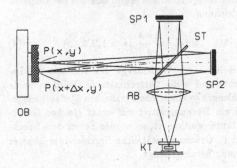

Bild 1: Scherografie mit Oberflächenspiegeln

Zur Erzeugung der Scherung werden in den letzten Jahren zunehmend zwei gegenseitig verkippte Oberflächenspiegel in einem Interferometeraufbau verwendet. Dabei werden die von einen beliebigen Objektpunkt reflektierten Lichtwellen durch einen optischen Strahlteiler ST in zwei gleichgroße Intensitätsanteile räumlich zerlegt. Die beiden Lichtwellen werden über zwei justierbare Oberflächenspiegel SP1 und SP2 auf den Strahlteiler ST zurückgeworfen. Beide überlagern sich und werden über eine Optik AB auf das Kameratarget in der Bildebene abgebildet, Bild 1.

Die beiden Oberflächenspiegel können so justiert werden, daß die in der Bildebene abgebildeten Bilder wechselseitig verschoben sind. Ist die Ausrichtung der Spiegel so, daß eine Verschiebung der Bilder in $x-$ Richtung auftritt, so interferieren in der Bildebene die von $P(x, y)$ kommenden mit den von $P(x + \Delta x, y)$ kommenden Wellen.

Die Vorteile dieser optischen Anordnung mit zwei Spiegeln sind, daß jede der beiden Lichtwellen zwischen Strahlteiler und Oberflächenspiegel separat durch Phasenschieber, Polarisationsfilter oder λ/Viertelplättchen beeinflußt werden kann und daß der Scherwinkel durch ein elektronisch gesteuertes Verkippen der

144

Oberflächenspiegel stufenlos eingestellt werden kann. Bei einer Veränderung $\vec{d} = (d_x, d_y, d_z)$ des Objekts tritt eine relative Verschiebung zwischen den beiden interferierenden Punkten auf, die zu der optischen Phasendifferenz $\Delta\phi$ führt [4]:

$$\Delta\phi = k\left[(1 + \cos\theta)(d_z(x + \Delta x, y) - d_z(x, y)) + \sin\theta(d_x(x + \Delta x, y) - d_x(x, y))\right] \tag{1}$$

Dabei ist θ der Winkel zwischen Beleuchtungs- und Beobachtungsrichtung. Für hinreichend kleine Scherungen Δx können die relativen Verschiebungen durch die Ableitungen ersetzt werden:

$$\Delta\phi = k\left[(1 + \cos\theta)\frac{\partial d_z}{\partial x} + \sin\theta\frac{\partial d_x}{\partial x}\right]\Delta x \tag{2}$$

Die Subtraktion oder Addition der Specklebilder vor und nach der Veränderung des Objekts durch Belastungsänderung erzeugt ein Interferenzstreifenmuster entsprechend der Phasenverteilung $\Delta\phi$. Es sind in der Bildebene dunkle Streifen dort zu sehen, wo:

$$\Delta\phi = (2n + 1)\pi \tag{3}$$

Bisher konnte die quantitative Auswertung von Scherogrammen nach dem Phasenschiebe-Verfahren bei schnellen Dehnungsabläufen (im Mikrosekundenbereich) nicht durchgeführt werden. Grund hierfür ist, daß die Zeit, die das Bildverarbeitungssystem für die Aufnahme der zur quantitativen Auswertung erforderlichen drei oder mehr nacheinander phasengeschobenen Scherogrammbilder benötigt, sehr viel größer ist als die Zeit, in der die dynamischen Oberflächenverformungsänderungen ablaufen. Daher wurde die Methode der räumlichen Phasenschiebung in Verbindung mit dem scherografischen Verfahren entwickelt und eingesetzt. Sie erlaubt die gleichzeitige Aufnahme und Speicherung von drei oder mehr phasengeschobenen Specklebildern.

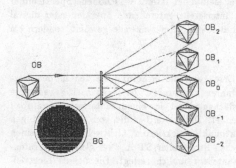

Bild 2: Beugung am Gitter

Wenn ein von einem Objekt OB kommendes Lichtfeld ein optisches Beugungsgitter BG durchläuft, entstehen neben dem Objektbild 0. Ordnung weitere gleichartige Objektbilder höherer Ordnung als reelle Bilder [5]. Der Beugungswinkel ergibt sich aus der Dispersionsgleichung

$$\sin\alpha = \frac{k * \lambda}{s} \quad (k = 1, 2, 3, \ldots) \tag{4}$$

Dabei ist λ die Wellenlänge des verwendeten Laserlichts und s die Gitterkonstante. Wird in beide Strahlengänge eines scherografischen Meßaufbaus jeweils ein Beugungsgitter mit exakt gleicher Gitterkonstante positioniert, so werden neben dem Specklebild 0. Ordnung zusätzliche Specklebilder höherer Ordnung herausgebeugt.

Die Phasenschiebung zwischen den einzelnen Beugungsbildern wird erzeugt, indem eines der Beugungsgitter senkrecht zu den Gitterlinien verschoben wird. Es stellt sich eine wellenlängenunabhängige und beliebig einstellbare Phasenschiebung in der n-ten Beugungsordnung ein.

Die Intensitätverteilung zwischen den Beugungsordnungen wird durch die Gitterform bestimmt. Ein Sinusgitter wird die Intensitäten in den Specklebildern der niedrigen Ordnungen verstärken, ein Rechteckgitter die Intensität in den Bildern höherer Ordnung. Für die Anwendung in der Scherografie ist eine Gitterform anzustreben, die eine gleichgroße Intensität in allen herausgebeugten und zur Auswertung herangezogenen Specklebildern gewährleistet. Durch exakte Positionierung der Beugungsgitter in beide Strahlengänge des scherografischen Aufbaus wird sichergestellt, daß die herausgebeugten Specklebilder der Größe und Lage nach deckungsgleich sind.

3. Experiment

3.1. Versuchsaufbau

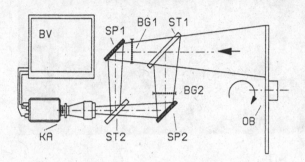

Der experimentelle Teil der Untersuchungen wurde an dem in Bild 3 schematisch dargestellten Meßaufbau durchgeführt. Das Objekt war der Ausschnitt einer mittig eingespannten rotierenden Kreisscheibe aus einer Aluminiumlegierung, deren Durchmesser 250 mm und deren Dicke 2 mm betrugen. Das System war auf der Welle eines regelbaren Gleichstrommotors befestigt. Zusätzlich war auf der Motorwelle ein Winkelcodierer mit 500 Schritten pro Umdrehung angebracht, der die Winkellage der Motorwelle und des Objektes in Form von elektrischen Signalen zur Weiterverarbeitung abgab. Bei den bisher erreichten Drehzahlen von ≈ 1000 1/min wurde die rotierende Kreisscheibe durch magnetische Fremdbelastung verändert.

Bild 3: Scherografischer Meßaufbau

Zur Beleuchtung des Meßobjekts wurde ein temperaturstabilisierter Rubin-Riesenimpulslaser, der Licht mit der Wellenlänge 694.3 Nanometern emittierte, eingesetzt. Die Ausgangsleistung des Impulslasers ist im Einzelpulsbetrieb 30 mJoule bei einer Pulsbreite von etwa 28 Nanosekunden. Mit dem Laser können im Doppelpulsbetrieb zeitliche Abstände zwischen den Belichtungen von wahlweise 200 Mikrosekunden bis herunter bis auf wenige Mikrosekunden erreicht werden. Diese Zeitabstände sind kurz genug, um die Untersuchung dynamischer Objektveränderungen zu ermöglichen.

Die diffus vom beleuchteten Objekt reflektierten Lichtwellen werden durch einen optischen Strahlteiler ST1 in zwei Anteile gleicher Intensität räumlich zerlegt. Diese Lichtanteile werden jeweils über die Beugungsgitter BG1 und BG2 auf justierbare Oberflächenspiegel SP1 und SP2, die das für die Scherografie notwendige wechselseitige Verschieben der in der Bildebene abgebildeten Bilder ermöglichen, gelenkt.

Die eigens für diesen Zweck hergestellten Beugungsgitter sind in der Gitterkonstanten und Gitterform identisch und wurden durch Interferenz zweier paralleler Lichtwellen, die sich in der Gitterebene unter einem spitzen Winkel überlagerten, auf hochauflösenden Fotoplatten erzeugt. Die Gitterkonstante betrug bei dem hier beschriebenen Experiment 20 Mikrometer. Neben dem Specklebild 0. Ordnung werden die Bilder der 1. Ordnung und -1. Ordnung herausgebeugt. Gespeichert wurden diese Daten in einem digitalen Bildverarbeitungssystem (BV).

3.2. Ablauf der Messungen

Die Ausrichtung des Strahlenganges wurde so gewählt, daß eine Scherung von 20 mm in $x-$ Richtung auftritt. Bild 4 zeigt das Signalablaufschema, aus dem das Zusammenspiel zwischem dem auf der Motorachse fixierten Winkelcodierer, dem Rubin-Riesenimpulslaser und der Bildverarbeitung zu ersehen ist. Zu einem ausgewählten Impuls des Winkelcodierers werden der Rubin-Riesenimpulslaser und das Bildverarbeitungssystem für die Bildaufnahme und -speicherung getriggert.

146

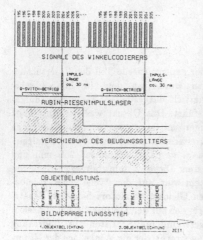

Bild 4: Signalablaufschema

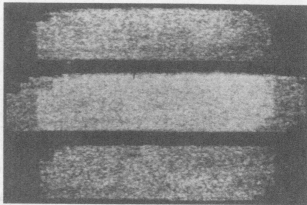

Bild 5: Specklebild

Bild 5 stellt das gespeicherte Specklebild der -1., 0. und 1. Ordnung der rotierenden unbelasteten Kreisscheibe bei etwa 320 1/min dar. Nach der Speicherung des 1. Bildes wurde das rotierende Objekt durch eine elektromagnetische Belastungseinrichtung um wenige Mikrometer verformt. Zusätzlich wurde mit Hilfe eines Piezokristalls die Lage eines der Phasengitter im Vergleich zum anderen stationären Phasengitter in Richtung senkrecht zu den Gitterlinien um exakt ein Viertel der verwendeten Gitterkonstanten, $s/4 = 20/4$ Mikrometer, verstellt. Dadurch tritt zwischen der -1., 0. und 1. Ordnung eine Phasenschiebung von jeweils 90 Grad auf. Dieses durch Belastung und Gitterverschiebung veränderte Specklebild wurde als 2. Bild aufgenommen und gespeichert.

3.3. Auswertung und Ergebnisse

Durch Subtraktion der Intensitäten in den Specklebildern ergeben sich die drei phasengeschobenen Scherogramme, Bild 6. Die Interferenzstreifen entsprechen der Objektveränderung; die Interferenzphasendifferenz zwischen den einzelnen Scherogrammen beträgt 90 Grad. Die in drei Einzelbilder aufgesplitteten phasengeschobenen Scherogramme werden nach dem Phasenschiebeverfahren ausgewertet [6]. Eine systematische Darstellung der Phasenschiebeverfahren, insbesondere eine Auswertung von 5 Mustern mit unbekannter konstanter Phasenschiebung gibt [7]. Die aus der Auswertung resultierende stetige Interferenzphasenverteilung ist in der Grauwertdarstellung in Bild 7 zu sehen. Die perspektivische Darstellung der richtungsabhängigen Objektveränderungen zeigt Bild 8.

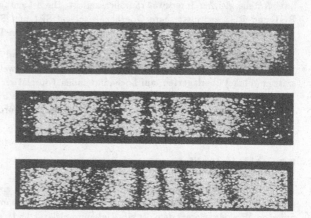

Bild 6: Scherogramme

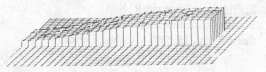

Bild 7: Stetige Interferenzphasendarstellung   Bild 8: Pseudo-3D-Darstellung

Es sind kleinere Unstetigkeiten im Verlauf der Oberflächenänderung zu erkennen, die auf Intensitätsstörungen durch den Einfluß der Speckles zurückzuführen sind. Durch Anpassung der Specklegröße, vorzugweise durch Änderung der Aufweitungsoptik bei der Objektbeleuchtung oder durch Anpassung der Blende des Aufnahmeobjektivs, können diese Störungen minimiert werden. Eine Verbesserung der hier gezeigten Auswerteergebnisse kann erreicht werden, wenn die Beugungsgitterstruktur optimiert wird. Sie ist für die Intensitätsverteilung zwischen den einzelnen herausgebeugten Specklebildern ausschlaggebend.

4. Zusammenfassung

Die in den vorliegenden Untersuchungen erzielten Ergebnisse zeigen, daß das scherografischen Meßverfahren in Verbindung mit einer quantitativen Auswertung nach dem Phasenschiebeverfahren auch bei Messungen von schnellen dynamischen Oberflächenveränderungen eingesetzt werden kann. Dazu sind drei oder mehr phasengeschobene Scherogramme erforderlich, die durch räumliche Phasenschiebung mittels optischer Beugungsgitter gleichzeitig erzeugt wurden.

Fragen in bezug auf Specklegröße und Beugungsgitterstruktur müssen in zukünftigen Arbeiten noch eingehender untersucht werden. Erst dann kann das scherografische Verfahren mit räumlicher Phasenschiebung und automatisierter quantitativer Auswertung aus dem Laborstadium heraustreten und in der Praxis eingesetzt werden.

Literatur

[1] **Ettemeyer, A.:** *Shearographie - ein optisches Verfahren zur zerstörungsfreien Werkstoffprüfung,* Technisches Messen, Heft 6, 247-252, 58. Jahrgang 1991

[2] **Hung, Y.Y.:** *Shearography versus Holography in Nondestructive Evaluation,* Proc. SPIE 704, 18, 1986

[3] **Hung, Y.Y.:***A speckle shearing interferometer: A tool for measurement derivatives of surface displacement,* Optics Communications, Vol. 11, 132-135, 1974

[4] **Gasvik, K.J.:***Optical Metrology,* John Wiley & Sons Ltd., Cichester, 1987

[5] **Kujawinska, M., Wojciak, J.:** *Spatial phase-shifting techniques of fringe pattern analysis in photomechanics,* Proc. SPIE 1554, 503-513, 1991

[6] **Jüptner, W., Kreis, Th. and Kreitlow, H.:** *Automatic evaluation of hologaphic interferogramm by reference beam phase shifting,* Proc. SPIE 398, 22-29, 1983

[7] **Kreis, Th., Geldmacher, J., Jüptner, W.:** *Phasenschiebe-Verfahren in der interferometrischen Meßtechnik: ein Vergleich,* Proc. Laser 93, 1993

Doppelpuls-Interferometrie mit Lichtwellenleitern

S. Pflüger, F. Legewie, R. Noll
Fraunhofer-Institut für Lasertechnik
Steinbachstr. 15, 52074 Aachen, Germany

Doppelpuls-Interferometrie ist ein Meßverfahren, mit dem Schwingungen und Verformungen mechanischer Baugruppen flächenhaft und berührungslos erfaßt werden können. Die bisher komplexen optischen Aufbauten lassen sich für den Anwender vereinfachen, indem das Laserlicht mit Lichtwellenleitern bis zum Meßobjekt geführt wird. Durch Modifikation eines Rubin-Lasers wird die über Lichtwellenleiter übertragbare Pulsenergie gesteigert und damit eine Voraussetzung für flexible und anwenderfreundliche Doppelpuls-Interferometer geschaffen.

Double-exposure-interferometry allows to measure contact-free vibrations and deformations of the whole surface of mechanical structures. The construction of the test equipment becomes considerably more user-friendly if the laser light is carried flexibly to the measuring object using fiber optics instead of mirror systems. Modification of a ruby laser increases the energy transmittable through fibers and leads to an improved interferometer.

1 Einleitung

Die Speckle-Interferometrie und die holografische Interferometrie mit Pulslasern sind Präzisionsmeßverfahren, mit denen Verformungen bzw. Verlagerungen von Bauteiloberflächen berührungslos mit einer Genauigkeit im Sub-Mikrometerbereich vermessen werden können. Diese optischen Meßverfahren stellen eine Erweiterung der Untersuchungsmethoden der Akustik und der Schwingungsmeßtechnik wie Modalanalyse und Schallintensitätsmessung dar. Die interferometrischen Doppelpuls-Verfahren werden bisher überwiegend für spezielle Anwendungen in der Bauteilentwicklung eingesetzt, siehe z.B. /1/. Das Anwendungsfeld soll erweitert werden, indem der Aufbau flexibler gestaltet wird und z.B. auch Messungen an Orten erlaubt, die mit Spiegeloptiken nicht erreichbar sind.

2 Vereinfachung durch Lichtwellenleiter

In der Lasertechnik wachsen die Anwendungsgebiete von Lichtwellenleitern (LWL) ständig, da der Einsatz des Lasers in vielen Bereichen der Technik erst durch die Kombination mit LWL ermöglicht wird. In der Interferometrie werden durch LWL die Aufbauten kompakter und einfacher handhabbar. Bild 1 zeigt schematisch einen konventionellen Holografieaufbau, bei dem Beleuchtungs- und Referenzstrahlen über Spiegel geführt werden.

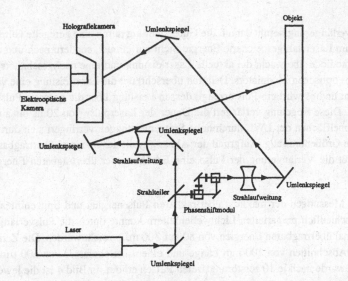

Bild 1   Konventioneller Aufbau eines Holografiesystems mit Strahlführung und -formung mit diskreten optischen Bauteilen

Der Aufbau besteht aus einer großen Anzahl optischer Komponenten zur Strahlführung und -formung. Viele dieser Komponenten werden bei Verwendung von LWL überflüssig oder können, wie das Phasenshiftmodul, in eine direkt an den Laser angeflanschte Einheit integriert werden, die der Anwender nicht mehr justieren muß. Bei der Holografie und der Speckle-Interferometrie mit kontinuierlicher Laserstrahlung gehört der Einsatz von Lichtwellenleitern zum Stand der Technik /2/. Die langen Belichtungszeiten erfordern jedoch einen schwingungsisolierten Aufbau von LWL, Meßobjekt und Meßkopf. Würde ein solches Meßsystem in der Produktion verwendet, müßten die zu prüfenden Bauteile aus der Produktionslinie genommen und schwingungsisoliert vermessen werden.

Dahingegen sind die bei der Interferometrie mit Doppelpulsverfahren auftretenden Zeitabstände von 1 - 800 µs zwischen den Belichtungen klein gegenüber der Periodendauer von Störschwingungen zwischen Meßsystem und Meßobjekt, so daß Untersuchungen unter praxisnahen Bedingungen möglich werden. Im technischen Bereich wird bei den Doppelpuls-Verfahren meist mit gütegeschalteten Rubinlasern gearbeitet. Diese eignen sich aufgrund ihrer Strahleigenschaften und der hohen Pulsenergien von bis zu 10 Joule in verfügbaren Systemen besonders für die Holografie. Der Versuch, diese Pulsenergien bei üblichen Pulsdauern von ca. 20 ns über LWL zu übertragen, führt aufgrund der hohen Intensitäten zur Zerstörung der LWL.

3   Erhöhung der Zerstörschwellen von LWL durch Modifikation des Lasers

Die durch LWL übertragbare Energie kann nur durch Herabsetzen der Spitzenintensitäten des Lasers gesteigert werden. Dies läßt sich durch eine Modifikation des Lasers, die zu einer zeitlichen Streckung der Laserpulse führt, erreichen. Das verwendete Lasersystem besteht aus einem elektro-optisch gütegeschalteten Oszillator, der Laserpulse von bis zu 40 mJ erzeugt, sowie zwei Verstärkerstufen, die die Pulse auf bis zu 3 J nachverstärken. Bei der Güteschaltung des Oszillators wird die gesamte im Laser gespeicherte Energie in einem Puls mit einer Länge von 20-30 ns abgerufen.

Das Prinzip der Pulsverlängerung beruht darauf, die Güteschaltung durch eine geregelte Gütemodulation zu ersetzen, d.h. die im Laserstab gespeicherte Energie nicht auf einmal, sondern nach und nach zu extrahieren /3/. Eine Fotodiode überwacht die aktuelle Laserleistung und eine nachgeschaltete Elektronik regelt die Öffnung des optischen Modulators. Dadurch überschreitet die Laserleistung eine vorgegebene Maximalleistung nicht und es wird ein Puls erzeugt, dessen Leistung über die gesamte Pulslänge möglichst konstant bleibt. Diese Regelung verlängert die Dauer der Laserpulse von 20 ns bis auf 1 μs. Die Belastung der Einkoppelflächen der LWL durch hohe Spitzenleistungen verringert sich durch die Pulsverlängerung um eine Größenordnung. Aufgrund der Abhängigkeit der maximal übertragbaren Energie von der Pulslänge läßt die Verlängerung der Pulse eine Erhöhung der übertragbaren Energie um den Faktor 3-4 erwarten.

Dies bestätigen auch Messungen der maximal übertragbaren Pulsenergien und Spitzenintensitäten: An verschiedenen, unterschiedlich präparierten Lichtwellenleitern konnte durch die Pulsverlängerung eine Steigerung der maximal übertragbaren Energien von 80 auf 300 mJ erreicht werden. Die Zerstörschwellen wurden an LWL-Abschnitten von 100 mm Länge und einem Durchmesser von 600 μm gemessen. Die Laserpulsenergie wurde nach je 10 zerstörungsfreien Pulsen erhöht. In Bild 4 ist die jeweils höchste von einem LWL-Abschnitt vor der Zerstörung transmittierte Pulsenergie aufgetragen. Durch eine weitere Verbesserung der Pulsverlängerungsschaltung soll eine Übertragung von einem Joule pro Puls über Lichtwellenleiter möglich werden.

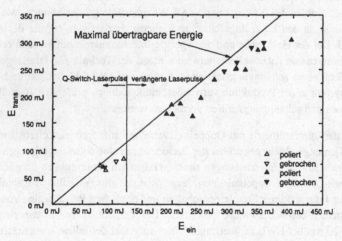

Bild 4   Zerstörschwellen verschieden präparierter Quarz-Quarz-Lichtwellenleiter (d = 600 μm) bei der Übertragung von Q-Switch-Pulsen (t = 20 ns) und verlängerten Pulsen (t = 200 ns)

4   Gesamtsystem

Die Erhöhung der über LWL übertragbaren Energie ermöglicht es, auch bei der Beleuchtung größerer Objekte, den Beleuchtungsstrahl über Lichtwellenleiter zu führen. Für die Referenzstrahlen ist nur ein Bruchteil der Energie des Beleuchtungsstrahls erforderlich. Diese Energie kann problemlos von Lichtwellenleitern übertragen werden. Der Durchmesser dieser Lichtwellenleiter sollte unter 50 μm liegen, da

151

beim Referenzstrahl im Gegensatz zum Beleuchtungsstrahl die räumliche Kohärenz des Lichtes erhalten bleiben muß. Um den Aufbau für die holografische Interferometrie bedienerfreundlich zu gestalten, wird die gesamte erforderliche Optik in einem Strahlteilermodul, das unmittelbar an den Laser angeflanscht wird, integriert. In das Strahlteilermodul ist auch die Phasenschiebeeinheit integriert. Ein Aufbau für holografische Doppelpulsinterferometrie mit Lichtwellenleitern ist schematisch in Bild 3 dargestellt.

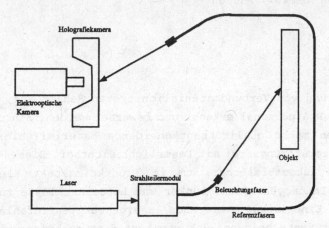

Bild 3   Aufbau eines Holografiesystems mit Strahlführung und -formung über Lichtwellenleiter

5   Zusammenfassung

Konventionelle holografische oder Speckle-Doppelpulsinterferometer lassen sich durch eine einfache Modifikation des Rubinlasers derart umrüsten, daß eine Strahlführung und -formung von Beleuchtungs- und Referenzstrahlengang auch bei größere Objekten mit Lichtwellenleitern möglich ist. Damit steht die Doppelpulsinterferometrie als Werkzeug nicht mehr nur für spezielle Anwendungen in optischen Labors zur Verfügung, sondern ist als einfaches und flexibles Präzisionsmeßsystem auch in propuktionstypischen Umgebungen einsetzbar.

6   Danksagung

Die Arbeiten, die zu den hier dargestellten Ergebnissen führten, werden im Rahmen eines Verbundprojektes vom BMFT und dem Labor Dr. Steinbichler unterstützt, wofür ausdrücklich gedankt sei.

7   Literatur

/1/ Ch. Wanders, C.R. Haas, S. Lampe, R. Noll, *Bestimmung der absoluten Verfomung einer schwingenden Motorhaube mit holografischer Doppelpulsinterferometrie,* Proc. Laser 91, 1992, p. 22-27

/2/ T.D. Dudderar et al., *Potential Applications of Fiber Optics and Image Processing in Industrial Holo-Interferometry,* Proc. SPIE Vol. 746, 1987, p. 20-28

/3/ R.V. Lovberg et al., *Pulse Stretching and Shape Control by Compound Feedback in a Q-switched Ruby Laser,* IEEE Journal of Quantum Electronics, QE-11, No. 1, 1975, p. 17-21

Neues Lasermeßverfahren zur Fehlererkennung in Verbundmaterialien

B. Hellenthal, M. Krauhausen, H. Wiesel
NoKra Optische Prüftechnik und Automation GmbH
Steinbachstr. 15, D-52074 Aachen

Zusammenfassung

Bei der Herstellung von Verbundmaterialien treten Fehler auf, die heute
meist noch durch Prüfpersonal erkannt und bewertet werden. Ein neuartiges
Lasermeßverfahren macht qualitätsentscheidende Materialfehler in Verbundmaterialien zerstörungsfrei mit Laserlicht sichtbar. Dieses Verfahren
ist vollständig automatisierbar, um im Produktionstakt Klebefehler,
Delaminationen, Inhomogenitäten, Lunker und Lufteinschlüsse zu erkennen
und zu bewerten. Eine darauf basierende lasergestützte Prüfanlage trifft
eine objektive Prüfentscheidung, dokumentiert die Ergebnisse und liefert
Signale für die Produktionssteuerung.

Einleitung

Für die Qualitätsprüfung von Verbundmaterialien werden derzeit noch
manuelle, visuelle und zerstörende Prüftechniken eingesetzt. Lasergestützte zerstörungsfreie Verfahren wie holografische Interferometrie,
Speckle-Interferometrie und Speckle-Shearing-Interferometrie sind bereits
ausgereifte Techniken, um Fehlstellen zu detektieren. Die Meßergebnisse
liegen als Interferenz-Streifenbilder vor, die es subjektiv zu beurteilen
gilt oder die durch aufwendige Streifenanalysesysteme bewertet werden
müssen. Ein neuartiges Verfahren, das auf den angesprochenen Lasermeß-
techniken aufbaut, lokalisiert direkt Fehler im Material. Ferner ist es
mit diesem Verfahren möglich, Fehler nach ihrer Größe einzustufen und die
Tiefenlage des Fehlers im Material zu bestimmen. Ein Modell zur Fehler-
analyse klassifiziert die Fehler und es läßt sich eine Nachweisgrenze für
jeweilige Materialien aufzeigen.

Prüfaufgabe

Die wenigsten Produktionsverfahren sind frei von Mängeln. Eine qualitätssichernde Fertigungssteuerung und -prüfung soll verhindern, daß bereits
bei der Herstellung Mängel und Fehler entstehen. Typische Fehler in
Verbundmaterialien sind Klebefehler, Delaminationen, Inhomogenitäten,
Lunker, und Lufteinschlüsse.

Tabelle 1 Verbundmaterialien, die in Großserienprodukten eingesetzt werden und deren typische herstellungsbedingte Defekte

Prüfling	Schaumformteil	Gummi-Metallteil	Schichtstoffplatte
Materialien	Polyurethan, Kunst-stoffolie, Metall- oder Kunststoffträger	Elastomer, Metall- oder Textilgewebe, Metalleinsatz	Spanplatte, Beschichtungs-material
Defekt	Lunker	Verbindungsfehler	Klebefehler
Auswirkung des Defekts	Blasenbildung unter Wärmeeinwirkung an der Oberfläche	Funktionsbeeinträch-tigung, Herabsetzung der Belastungs-grenze	Ablösung des Beschichtungs-materials, Blasen-bildung
typische Produk-tionsrate [Teile/min]	0,5 - 1	0,2 - 60	2 - 6
gegenwärtige Prüfart	manuelle Abtastung jedes Teils an kriti-schen Stellen	zerstörende Stich-probenprüfung, subjektive interfero-metrische- oder Röntgenprüfung	visuelle Stichpro-beninspektion
Beispiel	Armaturentafel	Autoventil, Fahr-zeugreifen	HPL-beschichtete Schichtstoffplatte

In Tabelle 1 sind drei verschiedene Verbundmaterialien, die in Großse-rienprodukten eingesetzt werden, und deren herstellungsbedingte Defekte zusammengestellt.

Meßverfahren
Das Meßverfahren beruht auf der Laserinterferometrie.

In Bild 1 ist der Meßablauf dargestellt. Die zu prüfenden Objekte werden in die Meßposition geführt, wo sie durch eine Belastungsvorrichtung angeregt werden. Durch diese Anregung wird bewirkt, daß eine eventuelle Fehlstelle im Matreial eine Deformation an der Oberfläche hervorruft, die durch das Meßsystem erfaßt wird. Das Objekt wird mit Laserlicht beleuch-tet und mit einem Sensor beobachtet. Die Auswertung der Sensorsignale liefert ein Bild des Objektes in dem die Fehler lokalisiert sind. Mit

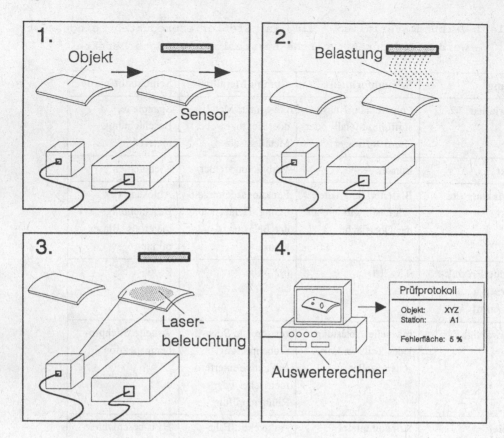

Bild 1 Beschreibung des Prüfablaufs

diesem Verfahren ist somit eine direkte Fehlerzuordnug im Bauteil möglich. Die aufwendige Auswertung von Interferenz-Streifenmustern der
konventionellen Lasermeßverfahren entfällt hier gänzlich.

Fehlerklassifizierung
Durch das neuartige Meßverfahren können genauere Aussagen zu Fehlern in
Verbundmaterialien gemacht werden. So werden beispielsweise durch eine
entsprechende Kalibrierung Daten über die Fehlergröße und die Lage eines
Fehlers im Material generiert. Bild 2 zeigt, wie das Meßsystem mittels
Master-Teilen mit bekannten Fehlern kalibriert werden kann, so daß die
Fehlerfläche im Meßergebnisbild mit der tatsächlichen Fehlergröße korre-
spondiert. In Bild 3 ist dargestellt, wie durch Betrachtung der
Belastungs- und Zeitparameter Informationen über die Tiefenlage eines
Fehlers im Material gewonnen werden. Ein tiefer im Material liegender

Fehler 2 bewirkt zu einem anderen Zeitpunkt und bei einem anderen Belastungszustand, beispielsweise bei einem größeren Unterdruck, eine Oberflächendeformation als ein Fehler 1, der dicht unter der Oberfläche des Materials liegt.

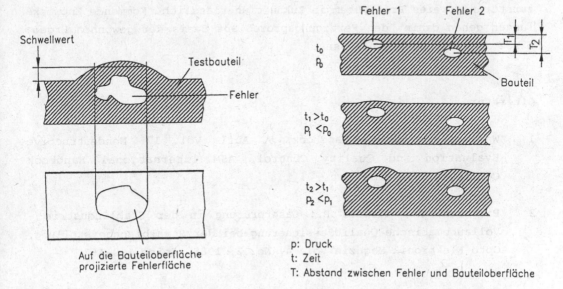

Bild 2 Bestimmung der Fehlergröße   **Bild 3** Bestimmung der Fehlerlage

Mit diesen Informationen über Lage und Größe von Fehlern lassen sich detektierte Fehler klassifizieren und für jeweilige Materialien Nachweisgrenzen aufzeigen.

Anwendungen

Die Praxistauglichkeit des neuen Laserprüfsystems konnte bereits in vielen Testapplikationen nachgewiesen werden. Anwendungsbeispiele sind die Fehlererkennung in PKW- und LKW-Reifen. Das Prüfsystem erkennt Verbindungsfehler zwischen dem Gummi und den Stahl- bzw. Gewebeeinlagen sowie Risse im Gewebe. Bei der Prüfung von Schichtstoffplatten erkennt das System Klebefehler zwischen der Grundplatte und der Kunststoffbeschichtung. In Gummi-Metall-Verbindungen werden Verbindungsfehler zwischen Metallkörpern und der Gummiummantelung detektiert. In Schaumformteilen, wie beispielsweise Auto-Armaturentafeln, werden Lufteinschlüsse lokalisiert.

Ausblick

Das Verfahren hat einen Entwicklungsstand erreicht, der in naher Zukunft einen breitgefächerten Einsatz in der Fertigung von Verbundmaterialien erwarten läßt. Hohe Anforderungen an die Qualitätssicherung in Hinblick auf die DIN ISO 9000 machen ein solches vollautomatisches, objektives und zerstörungsfreies Prüfsystem in Zukunft unerlässlich. Kommende Entwicklungen gehen dahin, den Fertigungsprozeß auf Basis der gewonnen Ergebnisse gezielt zu beeinflussen.

Literatur

1   Wagner, W.; Metals Handbook, 9. Aufl. Vol. 17: Nondestructive Evaluation and Quality Control, ASM International Handbook Committee, USA 1989

2   Pietzsch, K.; Müller, P.: Laserprüfung in der Stahlindustrie - Vollautomatische Qualitätssicherung bei der Stahlblechherstellung, Opto Elektronik Magazin Vol. 4, No. 2, 1988, Seite 160-167

Holographic Vibration Analysis for Noise Reduction

H. Steinbichler, H. Klingele, T. Franz, S. Leidenbach, R. Huber
Am Bauhof 4
D-83115 Neubeuern

1. Holographic Vibration Measurement

1.1 Principles

Two different holographic methods are currently used to measure vibrations:

In double pulse holography, a ruby laser is triggered such that the interesting object vibration is recorded on a thermoplast film by two consecutive laser pulses. The interference fringes can be evaluated at reconstruction time by the phase shift technique, if two reference beams are used.

The double pulse method is very useful for the investigation of operational vibration modes, or of transient processes like shock waves.

In speckle interferometry ("TV-Holography"), two images of the speckle field at different times are directly recorded with a CCD camera, and then subtracted from each other by an image processing system. The resulting fringes of interference are such available in real-time.

If the first image was taken with no vibration, the modulated laser beam is triggered to the vibration maximum of a stationary resonance. By using a phase shift method again, the amplitude distribution of the mode pattern can be calculated and displayed very fast and with a high resolution.

1.2 Measurement of the 3D Deformation Vector

For a plane object vibrating in eigenmodes, the deformation vector is assumed to be perpendicular to the plane, so that a 1D-holographic technique gives the full vibration information.

In the case of curved surfaces however, the deformation vectors may have different orientations. Because for sound calculation the surface normal component must later be extracted, all three components of the deformation vector have to be measured.

This can be done by introducing three illumination directions in the holographic setup, going along with three sensitivity directions.

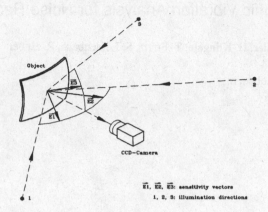

Object

CCD-Camera

$\overline{E1}, \overline{E2}, \overline{E3}$: sensitivity vectors
1, 2, 3: illumination directions

The resulting deformation data represent projections of the real deformation vector onto the three sensitivity directions

$$\delta_i = \vec{\epsilon}_i \cdot \vec{d} \qquad i = 1,2,3$$

which can be written using a sensitivity matrix, and then be solved for the deformation vector:

$$\vec{\delta} = \underline{E} \cdot \vec{d}$$

$$\vec{d} = \underline{E}^{-1} \cdot \vec{\delta}$$

For speckle interferometry with stationary vibration excitation, the laser beam may be switched to the three illumination directions one after the other, because recording can be triggered to the same phase angle of the vibration signal.

In double pulse holography however, the three illumination sources must be active at the same time if there is no stationary vibration. The 3 object and reference beams are then separated by delay lines. The delay of each beam is defined by the coherence length of the laser.

An example for 3-dimensional vibration measurement (tube) is given on the penultimate side of this text.

1.3 Multi Pulse Options

The analysis of operational vibration modes is significantly simplified if more laser pulses in a short time are available ("train pulse"), and the corresponding interferograms can be recorded.

Already existing technology for the decomposition of operational modes is based on the knowledge of eigenmodes and 2 consecutive interferograms.

2. Contour Measurement

Sound calculation for curved surfaces requires contour data for the generation of boundary elements. There are two possibilities:

a) Adaption of CAD or FEM data, if the interesting object is already in a data base. The computer model must be matched to the holographic view.

b) Optical measurement of the contour with fringe projection methods. The object orientation can then be adjusted to the holographic view.

There are optical methods of fringe projection, which offer as output data the laboratory coordinates (x, y, z) of all object points appearing in the image.

3. Sound Calculation
3.1 Theory

The complete treatment of the acoustic radiation problem for arbitrary 3D object geometry invokes the use of the Helmholtz integral representation for harmonic vibrations. Hereby the sound pressure in an exterior point is expressed by an integral over the surface normal deformation and the surface sound pressure on the object.

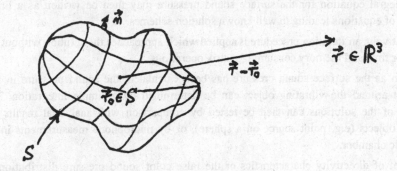

Geometry of the exterior radiation problem

$$p(\vec{r}) = \int_s \{p(\vec{r}_0) \frac{\partial}{\partial n} G(\vec{r},\vec{r}_0) - G(\vec{r},\vec{r}_0) \frac{\partial}{\partial n} p(\vec{r}_0)\} d\vec{r}_0(S)$$

$$G(\vec{r},\vec{r}_0) = \frac{e^{ik|\vec{r} - \vec{r}_0|}}{4\pi|\vec{r} - \vec{r}_0|}$$

The surface sound pressure is not known by measurement and must be calculated by solving the integral equation on the surface, where the factor ½ holds for sufficiently smooth surfaces.

$$\frac{1}{2} \, p(\vec{r}_0) = \int_S \{p(\vec{r}_0) \, \frac{\partial}{\partial n} \, G(\vec{r},\vec{r}_0) \, - \, G(\vec{r},\vec{r}_0) \, \frac{\partial}{\partial n} \, p(\vec{r}_0)\} d\vec{r}_0(S)$$

3.2 Implementation

For the numerical evaluation of the Helmholtz integrals the object surface must first of all be separated into boundary elements by a mesh generation on the contour data. The deformation vectors at the nodes of the mesh are taken from the evaluated 3D holograms for the selected frequency.

The deformation component normal to the surface is then defined for each element and represents the vibration boundary condition for the radiation problem,

$$\frac{\partial p}{\partial n} = i\omega\rho v_n \, .$$

The integral equation for the surface sound pressure may then be written as a linear system of equations leading to well known solution schemes.

Alternatively, an iteration procedure is applied which approaches the solution without the need for time and memory consuming matrix operations.

As soon as the surface sound pressure has been calculated, the sound pressure in any location around the vibrating object can be determined by a simple integration. The quality of the solutions can then be tested by comparison with analytical results for simple objects (e.g. point source on a sphere), or by microphone measurements in an anechoic chamber.

The plot of directivity characteristics or the false color sound pressure distribution in planes or on spheres, and intensity or power calculations only depend on the choice of points where to evaluate the exterior Helmholtz integral.

An example for a sound calculation (tube) is given on the last side of this text.

Literature:

- Praxis der Holografie. Grundlagen, Standard- und Spezialverfahren. expert Verlag 1990
- P. M. MORSE, K.U. INGARD: Theoretical Acoustics. Princeton University Press
- C. JUNGER, FEIT: Sound, Structures, and Their Interaction. MIT Press
- R. CISKOWSKI, C.A. BREBBIA: Boundary Element Methods in Acoustics. Computational Mechanics Publ. Southhampton 1991
- M. P. NORTON: Fundamentals of noise and vibration analysis for engineers. Cambridge University Press 1989

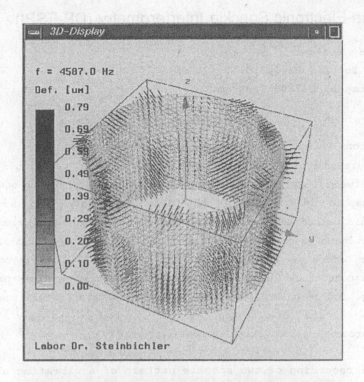

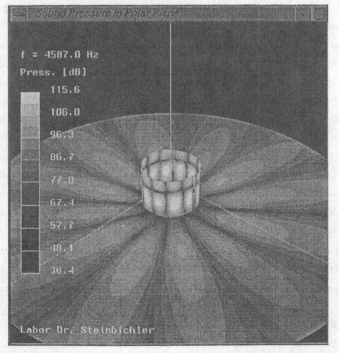

Double Pulse-Electronic Speckle Interferometry (DP-ESPI)

G. Pedrini, H. Tiziani
Institut für Technische Optik, Universität Stuttgart
Pfaffenwaldring 9, D-7000 Stuttgart 80

1. Introduction

For the measurement of vibrations using double-pulse techniques, pulses separations between 1 and 1000 microseconds are necessary. Until now double-pulsed holographic interferometry has been extensively used to measure vibrations. This method has the disadvantage that it needs the recording and the reconstruction of a hologram. For the recording a photographic plate or a thermo-plastic camera is usually used. The hologram is then reconstructed usually with a continuous laser and viewed with a CCD camera. This process is time consuming. Double pulsed ESPI [1] enables to obtain correlation fringes corresponding to the displacement without recourse to any form of photographic processing and plate relocation.

2. Electronic recording of two speckle pattern of a vibrating object

The system used is shown in Fig.1. The beam coming from the ruby laser is splitted into two beams, the object beam and the reference beam. The object beam is enlarged by a diverging lens and it illuminates the object(O). The object is imaged on the CCD camera by the lens L. With the aperture (AP) in front of the lens L it is possible to choose the mean dimension of the speckle in the sensor plane. The CCD-camera records the interference between the light coming from the object and a reference. When the object vibrates the interference pattern changes. We record the first image with the first pulse and the second image with the second pulse. The two images are then subtracted one from the other and correlation fringes corresponding to the object deformation appear. For our experiments we used a ruby laser (wavelength 694 nm), which can emit two high energy pulses separated be few microseconds. The problem is to record two images corresponding to the two pulses by using a CCD-camera. To perform this task we used an interline transfer CCD-camera. This camera consists of an array of photosensors each connected to a tap on a vertical shift register. When illuminated, the photo sensors generate charges that after a period of time are transfered in the shift register which is covered to prevent generation of new charges. The time necessary to transfer the charges from the photosensors to the shift register is short (2 or 3 microseconds for the camera used in our experiment) since it involves only a parallel transfer from each photosensor to the adjacent one. After the charge transfer the photosensors of the camera are ready to capture a new image. For our particular case we recorded the first pulse and we transfered the charges to the shift register, after this transfer we recorded the second pulse. The two images (first image in the shift register and second image in the photosensors) can be read in two normal readout cycles, digitalized and stored into the frame memory. Since the two laser pulses usually do not have the same energy, a normalization of the two recorded speckle images is necessary. The images are the substracted one from the other and the absolute value is taken and stored into the frame grabber. Figure 2 shows the result for a vibrating plate after the subtraction between the two speckle pattern. The puls separation was 100 microseconds. It was even possible to record two separated images using puls separation of 5 microseconds. Since the camera used in our experiment is an interlaced, it is not possible to transfer the charges of all the elements at the same time but only the odd or the even lines, therefore we can use only half of the vertical resolution, but still good results can be obtained.

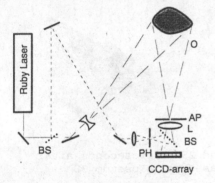

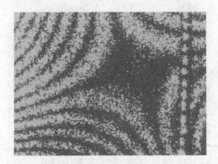

Fig.1. Optical set-up

Fig.2. Speckle interferogram of a vibrating plate recorded with puls separation of 100 μsec.

3. Quantitative analysis of the fringes

The spatial-carrier phase-shifting method [1], [2], [3], is particularly well suited to be used for a quantitative analysis in the case of a pulsed laser, since all the information necessary to reduce an interferogram to a phase-map is recorded simultaneously. The reference beam is tilted by an angle θ with respect to the optical axis. In the image plane (where the CCD sensor is located) the speckle image of the object to be tested is then modulated with a carrier frequency having a period $p_M = \lambda / \sin\theta$. The angle θ is chosen so that the phase difference between the reference and object beam changes by a

164

constant α from one pixel of the CCD camera to the other. In order to apply
this method it is necessary that the speckle are still correlated after the
image-shift of one pixel, this involves that the pixel size should be greater
than the period p_M. The first speckle pattern SP1 with the object in
position O1 and the second SP2 with the object in position O2 are recorded
and stored in the frame grabber. Three phase-shifted fringes-patterns can
then be obtained. For the first fringe pattern (phase-shift -α) the speckle
pattern SP2 is shifted one pixel (Δx) left (digital shift in the frame
grabber) with respect to SP1, a subtraction between the pattern is then
performed (SP1(x,y)-SP2(x-Δx,y)). The second fringe pattern (phase-shift 0)
is obtained by subtracting the speckle image SP2 from SP1 (SP1(x,y)-
SP2(x,y)). The third interferogram is obtained by shifting to the right by
one pixel SP2 with respect to SP1, by subtracting (SP1(x,y)-SP2(x+Δx,y)). The
three fringe-pattern obtained are then filtered and the phase is calculated.
We used the spatial-carrier shift-method to study the vibration of a plate
few microseconds after a choc. One result is represented in Figure 3.

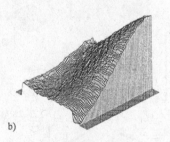
a) b)

Fig.3. Deformation of a plate between 150 and 250 microseconds after
the impact of a pendulum on the plate. a) Phase-map, b) pseudo 3D
representation of the deformation

4. Two and three dimensional measurements

The results presented in sections 2 and 3 are only onedimensional, this means
that they give us only the deformation of the object along one sensitivity
vector. In some cases a two or three dimensional analysis of the deformation
is necessary. More sensitivity vectors can be generated by observing the
object from different direction or by illuminating the object from different
directions (thre directions of illumination and one direction of
observation). We choose the second possibility because it has the advantage
that it does not needs rectifications due to the distortion by different
observation directions. Figure 4 show the arrangement used for the
measurement of two dimensional deformations. The sensitivity vectors are
given by the half-angle between illumination and observation directions. The

camera 1 records the interference between the reference 1 and the illumination 1, it give also the information of the deformation along the sensitivity vector **e1**, and analogously the camera 2 measure the deformation along the vector **e2**. In order to avoid unwanted interference, the second reference/illumination beam pair is delayed by 5 m (coherence lenght of the ruby laser). For the three dimensional case we use the same principle but with 3 cameras and three illumination directions. Figure 5 shows the results of the measurement of a vibrating cognac-glas. The two dimensional arrangement was used and the figure shows the projection (on a horizontal plane) of the objects deformation along a line at a constant high.

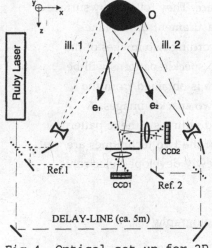

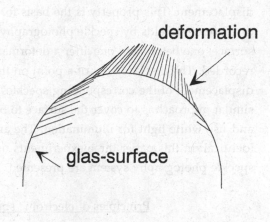

Fig.4. Optical set-up for 2D Speckleinterferometry

Fig.5. 2D measurement on a cognac-glas. Deformation along a horizontal line.

5.Conclusions

The double-pulsed speckle·interferometry method is very much simpler than the double pulse holographic interferometry and allows a quick analysis of the interferograms without the development of films and hologram reconstructions. It is thus well suited to be used in an industrial environment. Using 3 cameras and three illumination directions (in order to have three sensitivity vectors) it is possible to measure 3-D deformations.

References
[1] G.Pedrini, B. Pfister, H. Tiziani, "Double pulse-electronic speckle interferometry", J. of modern Optics, Vol. 40, 89-96 (1993).
[2] B. Pfister, M. Beck and H. J. Tiziani, "Speckleinterferometrie mit alternativen Phasenshiebe-methode an Beispielen aus der Defektanalyse", in Proc. **Laser 91**, (München 1991).
[3] S.Leidenbach, "Die direkte Phasenmessung-ein neues Verfahren zur Berechnung von Phasenbildern aus nur einem Intensitätsbild", in Proc. **Laser 91**, (München 1991).

Electronic Speckle Photography: Some Applications

Mikael Sjödahl

Division of Experimental Mechanics, Luleå University of Technology,
S-971 87 Luleå, Sweden.

Introduction

A diffusely reflecting surface illuminated by coherent (laser) light appears grainy when viewed by eye or by a camera. These so called laser speckles form a pattern which behaves as being attached to the surface. They follow any surface displacement. This property is the basis for the measurement of in-plane displacement fields by speckle photography[1]. By recording two images of the surface, one before and one after a deformation, two speckle patterns will be recorded. The displacement of a point on the surface is obtained from the displacement of the corresponding speckles in the two speckle images. A similar approach is to cover the surface to be studied with a synthetic pattern and use white light for illumination. The analysis of the two techniques are identical. In this paper three experiments using a newly developed electronic speckle photography system are presented.

Principles of electronic speckle photography

An experimental set-up for electronic speckle photography is shown in figure 1. A measurement is started by recording two speckled images of the object surface, one before and one after deformation of the object, using a CCD-camera, frame-grabber and a computer. A speckle pattern has a random distribution of light intensity between zero and a maximum value[2]. The typical speckle size is determined by the imaging geometry ($\sigma=1.22\lambda f_\#$ where σ is the speckle size, λ the wave length of the laser and $f_\#$ is the effective f-number of the imaging system). Since the speckle patterns are digitized, the speckle size is essential. It is shown[3] that recording the patterns using a CCD-array with high fill factor, undersampling at 70% of the Nyquist frequency is tolerable before systematic errors are introduced in the analysis. The analysis of the speckle patterns are made on a finite area basis. To measure the displacement at a specific point on the object, one chooses the corresponding subimage (usually 32x32 pixels) from the displaced and undisplaced speckle patterns and performs a two-dimensional cross correlation between the subimages. The position of the peak coincides with the displacement at that

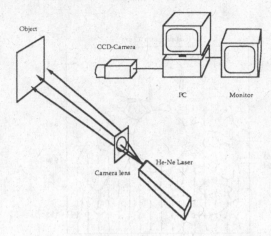

Figure 1. Schematic of the experimental set-up used in the experiments.

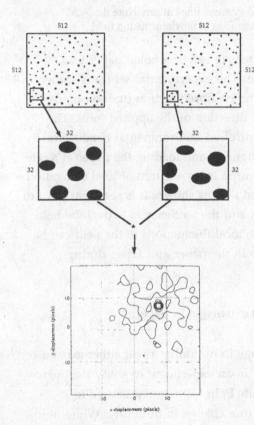

Figure 2. Principle of the analysis. ∗ means cross correlation.

point, see figure 2. To obtain subpixel accuracy in the location of the peak position, a Fourier series expansion of the peak is performed[4]. The accuracy in the described procedure is about 0.05 pixels[3,4] which for a 1:2 magnification gives an accuracy of about 1µm. An accuracy suitable in many engineering applications.

Using laser speckles instead of randomly attached speckles is preferable in most applications. First object preparation is minimized since the speckles are formed in the imaging procedure. Secondly the size of the speckles are magnification invariant, i.e. they can be optimized to provide as accurate results as possible. Last the random nature of the speckles provides an equally distributed pattern, which is crucial for the reliability of the system. There are, however situations where the white light technique is superior; When the micro-structure of the surface changes with time and/or deformation the two laser speckle patterns might become decorrelated. This change in micro-structure will not affect the white light technique.

Application 1: Tensile test of paper

With ordinary contact measuring techniques such as strain gauges, it is in general difficult to measure strength properties of soft and nonisotropic materials such as paper

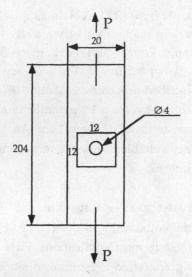

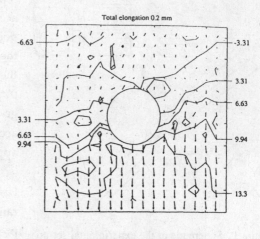

Figure 4. The deformation field at the hole.
Iso-contour lines in μm. Note the local
variations in the deformation field.

Figure 3. Geometry of the paper sample.
Imaged area 12x12 mm².

without influencing the strain field. An tensile test of a holed paper was
performed in an Instron testing machine. The experimental set-up is shown in
figure 3, where the statistically most common fibredirection (the so called
machine direction) of the paper is in the direction of the applied force. The
experiment was performed under uncontrolled environmental conditions. By
using an experimental set-up similar to figure 1 and loading the paper at a rate
of 0.2 mm/minute, images could be captured every 0.1 mm of total elongation.
In figure 4 an example of the deformation field is shown. It is seen that most of
the strain appear at the sides of the hole, and the tendency is to oval the hole.
By careful study of the deformation field, local fluctuations in the field can be
seen. This is due to local inhomogenities in the paper and arises during
manufacturing.

Application 2: Hygro expansion in paper

When a paper sheet is exposed to a change in humidity it will either expand or
shrink differently in different directions. In an experiment to study the hygro
properties of paper (see figure 5), the white light technique was used to
measure the in-plane deformations due to a change in humidity. White light
was used since the micro-structure of the paper is changing with changing
humidity, which will cause the speckles to decorrelate. The paper samples are

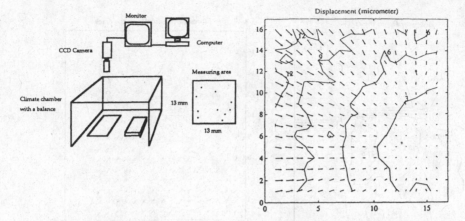

Figure 5. Experimental set-up for measuring hygroexpansion in white light. The measuring area has synthetic speckles.

Figure 6. The hygroexpansion in a sheet of paper obtained by decreasing the humidity from 30 % to 10 %.

put into a climate chamber in which the humidity can be controlled. Coal powder is put onto the surface to give a random pattern. Images at different humidities are captured in white light and processed using the electronic speckle photography system. A reference paper is put onto a balance to keep a record of the relative amount of water in the paper sheet. The results of one such experiment is shown in figure 6. It is seen that the deformation forms ellipses of constant deformation, which proves the anisotropic properties of paper for a change in humidity.

Application 3: In-plane deformations of composites around a cracktip

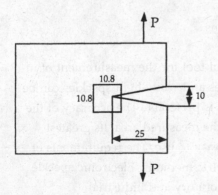

It is of interest to study the deformation field around the cracktip of composite materials when the crack is loaded perpendicular to the crack. Therefore two test objects, having the same geometries (figure 7), made of uni-directional glass fibre/epoxy laminate were fabricated. One of the objects had the fibres oriented parallel to the crack and the other perpendicular to the crack.

Figure 7. Geometry of the prefabricated crack. The imaged area was 10.8x10.8 mm^2.

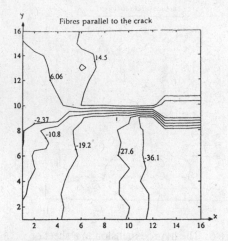

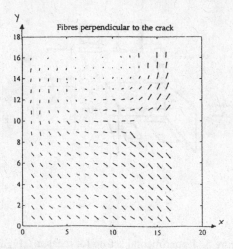

Figure 8. The y-component of the opened crack in the specimen having the fibres oriented perpendicular to the loading.

Figure 9. Displacement field of the specimen having the fibres oriented along the loading.

An experimental set-up like figure 1 was used. An Instron testing machine was used to apply the load and images were captured every 0.1 mm of elongation at a stroke rate of 0.4 mm/min. An example of the result can be seen in figure 8 and 9. For the composite that had the fibres oriented parallel to the crack, a clearly visible crack starts to grow at the tip of the prefabricated crack. The crack width can be followed throughout the loading cycle, while the rest of the composite mainly acts as a hinge. For the composite that had the fibres perpendicular to the prefabricated crack, no crack starts to grow. The deformation field appears as coming from pure bending.

Conclusions

Electronic speckle photography is a powerful tool for the measurement of in-plane displacement fields. Both laser speckles and white-light speckles can be used, but in most applications laser speckles is preferable. The accuracy of the present system is about $1/10^4$ of the side of the measured area. Its greatest advantage is perhaps the fact that it is very easy to use. The requirements of stability are less severe than for interferometric methods. Electronic speckle photography is easily performed both in laboratory and industrial environments, which increases its usefulness.

171

References

1. A. E. Ennos, "Speckle Interferometry," in Laser Speckle and Related Phenomena, J. C. Dainty, Ed. (Springer-Verlag, Berlin, 1975), pp. 203-253.
2. J. W. Goodman, "Statistical Properties of Laser Speckle Patterns," in Laser Speckle and Related Phenomena, J. C. Dainty, Ed. (Springer-Verlag, Berlin, 1975), pp. 9-75.
3. M. Sjödahl and L. R. Benckert, "Systematic and random errors in electronic speckle photography," submitted to Applied Optics, (April 1993).
4. M. Sjödahl and L. R. Benckert, "Electronic speckle photography: Analysis of an algorithm giving the displacement with subpixel accuracy," accepted for publication in Applied Optics, (December 1992).

Interferometrische Dehnungsmessung - Aufbau und Anwendung eines DSPI-Meßsystems

Petra Aswendt, Roland Höfling

Fraunhofer Einrichtung für Umformtechnik und Wekzeugmaschinen IUW

Reichenhainer Str. 88, 09126 Chemnitz

Einleitung

Die Speckleinterferometrie hat in den letzten Jahren als eine vielseitig einsetzbare Meßtechnik zunehmende Akzeptanz gefunden, was sich unter anderem darin ausdrückt, daß eine Reihe kommerzieller Geräte angeboten wird. Die meisten Systeme eignen sich für Schwingungsanalysen, zerstörungsfreies Prüfen und out-of-plane Verformungsmessungen. Dabei kommt die Phasenschiebetechnik zum Einsatz, die eine vollautomatische Berechnung der Verschiebungsfelder erlaubt. Gegenstand dieser Arbeit ist es, ein bisher nicht verfügbares, mobiles und miniaturisiertes DSPI-Meßsystem zu entwickeln, das geeignet ist, in-plane Dehnungen feldweise, mit einer Genauigkeit in der Größenordnung herkömmlicher Dehnmeßstreifen zu bestimmen. Das heißt, das optische Dehnmeßsystem soll qualifiziert werden, den ebenen Verzerrungstensor von Proben- und Bauteiloberflächen zu erfassen.

Aufbau des Meßkopfes

Ausgangspunkt bildet ein am IUW entwickeltes und patentiertes faseroptisches Interferometer. Es handelt sich um eine geschlossene Lösung, bei der Strahlteilung und Phasenschieben in die Faseroptik integriert sind. Das Interferometer kommt ohne jegliche mechanische Komponenten aus und ist damit sehr robust und anwenderfreundlich. Der prinzipielle Aufbau für eine Empfindlichkeitsrichtung ist im Bild 1 (links) angegeben. Als Lichtquelle dienen zwei Laserdioden mit stabilem, longitudinalen Single-Mode und einer Ausgangsleistung von 50 mW bei einer Wellenlänge von 820 nm. Die Ankopplung an das Interferometer erfolgt über ein Faserpigtail, dessen Qualität entscheidend ist für die Lichtausbeute. Bidirektionale Koppler teilen das Licht im Verhältnis 1:1 und eine geeignete Piezokeramik realisiert das Phasenschieben in einer der beiden Fasern. Die Faserenden sind so fixiert, daß das Objekt symmetrisch aus zwei Richtungen beleuchtet wird. Um eine weitere Empfindlichkeit senkrecht dazu zu erreichen, ist diese Anordnung im Meßkopf zweifach enthalten. Die über eine serielle Schnittstelle mit dem Rechner verbundene Laserdiodensteuerung regelt den Diodenstrom und gewährleistet gleichzeitig das Umschalten zwischen beiden Empfindlichkeitsrichtungen. Eine Miniatur-CCD-Kamera komplettiert den Meßkopf, wie er im Bild 1 (rechts) zu sehen ist.

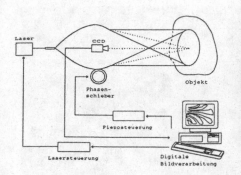 

Bild 1. Faseroptisches DSPI-Meßsystem zur 2D-Dehnungsmessung

Digitale Verarbeitung der Specklemuster

An die Aufnahme der Specklemuster schließt sich deren digitale Verarbeitung mit dem modularen Hochleistungssystem Imaging Technology-151 an. Dieses erlaubt die Verknüpfung von Bildern des Formats 512x512 im Videotakt und damit die Echtzeitdarstellung von Korrelationsstreifen. Die um $\pi/2$ phasenversetzten Streifenmuster werden unter Ausnutzung der Möglichkeiten des Pipeline-Prozessors soweit vorverarbeitet, daß die Verschiebungsgradienten $d_{x,x}$, $d_{x,y}$, $d_{y,x}$ und $d_{y,y}$ berechnet werden können. Die Komponenten des Verzerrungstensors ergeben sich dann zu

$$\varepsilon_{xx} = d_{x,x}$$
$$\varepsilon_{yy} = d_{y,y}$$
$$\varepsilon_{xy} = 1/2(d_{x,y} + d_{y,x}).$$

Auf diese Weise läßt sich die Scherung ε_{xy} separieren und Rotationsanteile können das Meßergebnis nicht verfälschen. Diese Berechnung wird für 511x511 Bildpunkte ausgeführt und liefert zunächst eine qualitative Darstellung des Dehnungsfeldes. Darüberhinaus ist das Auswerteprogramm in der Lage, Dehnungen zuverlässig quantitativ zu bestimmen. Die Auswertezeit beträgt 3 min von der Bildaufnahme bis zum Endergebnis. Bei einer Objektgröße von 10x10 mm^2 liegt der Meßbereich zwischen 10 $\mu\varepsilon$ und 800 $\mu\varepsilon$ und wird begrenzt durch die minimal bzw. maximal auflösbare Streifenzahl. Das quantitative Dehnungsfeld wird für ein Raster von 9x12 Elementen berechnet und entspricht damit 100 DMS-Rosetten mit je 1 mm Meßlänge. Außerdem ist zu beachten, daß die Empfindlichkeit der Dehnungsmessung mit zunehmender Objektgröße (bisher max. 100x100 mm^2) steigt. Durch einen fortlaufenden, lückenlosen Meßprozeß können andererseits beliebig hohe Dehnungen schrittweise analysiert werden, wenn die Belastung in einzelne Intervalle unterteilbar ist.

Bild 2 demonstriert die Auswertung am Beispiel eines synthetischen Streifenmusters, bei dem die Dehnung ε_{xx} in y-Richtung stetig wächst.

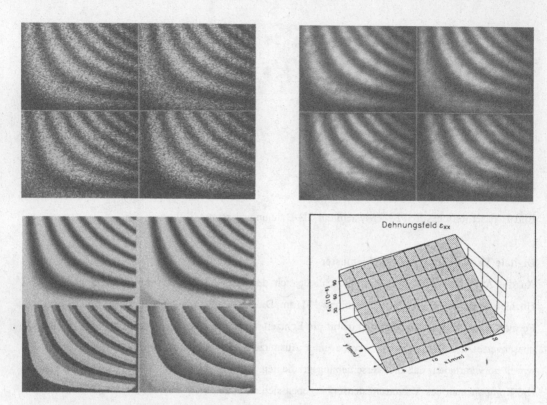

Bild 2. Verarbeitungsschritte zur automatischen Berechnung des Dehnungsfeldes

Anwendungsbeispiele

Anhand von drei Beispielen sollen die Einsatzmöglichkeiten des DSPI-Meßsystems demonstriert werden. Im ersten Fall handelt es sich um die Bestimmung der Wärmedehnung an einer Schweißverbindung eines plattierten Stahles. Für die im Bild 3 (links oben) dargestellte Probe wird aufgrund der Werkstoffkombination ein komplexes Dehnungsverhalten erwartet, dessen Analyse mit herkömmlichen Verfahren einen hohen Aufwand erfordern würde. Aus den DSPI-Streifen wurde das inhomogene Wärmedehnungsfeld berechnet (Bild 3, rechts) in dem sich Werkstoffgrenze und Schweißnaht genau separieren lassen.

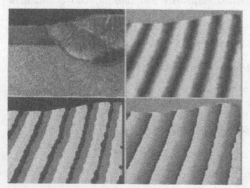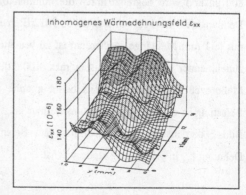

Bild 3. Thermisch belastete, plattierte und geschweißte Stahlprobe

Da das Meßsystem in der Lage ist, gleichzeitig die Dehnung in x- und y-Richtung zu erfassen, eignet es sich auch sehr gut zur Kennwertermittlung an anisotropen Verbundmaterialien. Untersuchungen an kohlenstofffaserverstärktem Aluminium lieferten in einem Schritt die Wärmeausdehnungskoeffizienten längs (α_x) und quer (α_y) zur Faser. Ein Beispiel dafür ist im Bild 4 wiedergegeben, es zeigt neben den Streifen in beiden Empfindlichkeitsrichtungen (x-oben, y-unten) das quantitative Resultat.

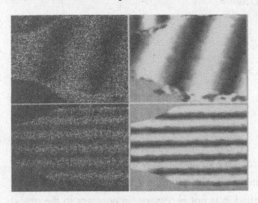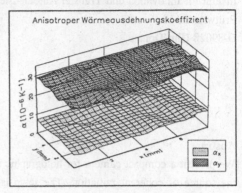

Bild 4. Thermisch belastete C/Al-Zugprobe

Der komplette Verzerrungstensor ist von Interesse, wenn mehrachsige Beanspruchungen vorliegen, wie es z.B. bei einem zugbelasteten Motorpleuel der Fall ist. Das Meßsystem erlaubt die feldweise Erfassung der drei Komponenten ε_{xx}, ε_{yy} und ε_{xy}, woraus direkt Vergleichsspannungen abgeleitet wurden. Im Bild 5 (links) sind wiederum zwei der insgesamt acht Streifenmuster angegeben, die in die Berechnung eingehen. Die Grafik (Bild 5 rechts) verdeutlicht die Spannungsüberhöhung an den Flanken des Pleuels.

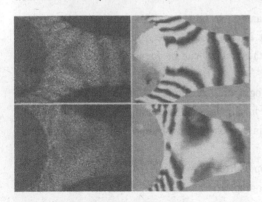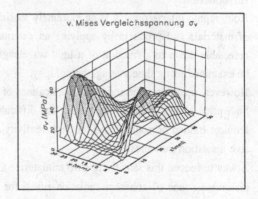

Bild 5. Zugbelastetes Motorpleuel

Zusammenfassung

Das hier beschriebene, miniaturisierte DSPI-Meßsystem befindet sich nach umfangreichen Erprobungen im Vergleich zum DMS bereits beim Auftraggeber in Anwendung. Aufgrund der genannten Eigenschaften und dem Vorzug, am Ende ein für den Ingenieur direkt verwertbares Ergebnis zu liefern, werden künftig vielfältige Ensatzmöglichkeiten für diese Technik gesehen.

Compact Camera for Holography with a 670 nm Semiconductor Laser Without Vibration Insulation

B. Lau, E. Mattes

Fachhochschule Ulm

Institut für Innovation und Transfer Automatisierungssysteme

Prittwitzstr. 10

D-89028 Ulm/Donau

Abstract

We present a compact camera for holographic interferometry equipped with a semiconductor laser as the source of coherent radiation. The setup is extremely simple, alignment and operation are very easy. Due to its rigidness a sophisticated vibration insulation is not necessary, in contrary to the practice elsewhere in industrial holography.

The holograms are recorded on photo-thermoplastic film. Interferograms can be generated by double exposure or in real time. They are evaluated with the aid of a commercial holographic image processing system using phase stepping. Owing to its ease in operation and its relatively low cost this camera opens new applications for holographic interferometry in the fields of industrial nondestructive testing.

Introduction

Holographic interferometry is an extremely sensitive and powerful method for non-destructive testing of materials and products by applying an external or internal force. The resulting deformations are detectable down to a fraction of a light wavelength, thus defects can be found. Vibration modes can be examined using time average holography.

However, up to now the industrial applications of holographic interferometry are restricted due to the high price of holographic equipment, the difficulties in aligning and operating such a system which requires trained personnel, and the high sensivity against mechanical vibrations which requires excessive insulation measures.

A way to reduce this sensitivity is to miniaturize the optical setup. With the semiconductor laser a miniaturized souce of coherent radiation suited for holography is available[1-5]. Especially laser diodes emitting in the visible spectral region, which allow the use of usual holographic recording plates and films, could already prove their suitability for holographic interferometry[6].

Description of the setup

Our holographic setup is based on a Toshiba TOLD 9125 laser diode (670 nm, 10 mW). It exhibits an elliptical radiance profile with half divergence angles of 28° and 8°, which allows it to illuminate an object with a size of some cm without any beam expanding optics. Furthermore, the radiation emitted into the larger angle may be used to produce the reference wave. The laser is operated with injection current and temperature stabilization (driver Profile LDC 700 A); it has a specimen-dependent coherence length in the order of some cm, measured with a Michelson interferometer, when operating in a single longitudinal mode. As the longitudinal mode spectrum is higly dependent on temperature and specimen, every laser was tested before use with a grating spectrometer to find out the temperature regimes of single mode operation.

Fig. 1 shows the beam paths. The semiconductor laser is placed below the hologram plane and is oriented with its larger divergence angle horizontally. An electromagnetic shutter in front of it controls the exposure. The only optical component is a plane mirror generating the reference wave. By placing it beneath the object the optical path requirements for coherence are fulfilled sufficiently. The only requirement for the alignment of this mirror is the illumination of the complete hologram area. The holograms are recorded as phase holograms onto photo-thermoplastic film[7] which is available as yard ware in slide film format[8]. This material has a bandpass charecteristic in its spatial frequency transfer function, thus the angle between object and reference wave has to be set dependent on the wavelength, here to about 25°. Hence the reference mirror has to be placed at the edge of the divergent laser radiation bundle. The lower radiance in this direction gives a reference wave amplitude which fits well to the backscattered object wave amplitude of most objects (irradiance ratio of reference wave to object wave in the hologram plane about 5:1).

The holograms are recorded and developped by an automatic photo-thermoplast processor (Rotech HIC 4). The film is transported and sensitized with an electrostatic charge immediately before exposure. Dependent on the object (which usually was spray coated with a white chalk layer) we need an exposure time of several hundred ms. The development is performed using short heat pulses and takes about 1 s; in total the hologram is available within about 15 s. During this process the film is not moved which allows real-time interferometry. Double-exposure holograms with different reference beams[9,10] can be recorded using a second reference mirror placed symmetrically at the other side of the object. An electromagnetic shading element driven by two counteracting spools allows switching off alternatively one of the reference beams during double exposure or the object illumination beam during reconstruction. However, we found it much more convenient to work with real time interferometry as the deformation of the object can directly be observed and controlled by observing the interference pattern. There are no problems with vibrations, stability or drift.

The holograms are evaluated quantitatively using the phase stepping method[11-13]. One of the reference mirrors can be shifted piezoelectrically. Four interference images are fed into a PC 386 and evaluated with a commercial holographic image processing system (Rotech Fringe Expert).

A photograph of the optical setup for real time interferometry is given in fig. 2. Fig. 3 shows the complete system with the control electronics and the PC for image processing.

Examples

As an example, the deformation of a gas pocket lighter under the influence of an internal pressure of some mbar is demonstrated in fig. 4. Dependent on the object and on the deformation itself, we are able to measure deformations in a range between 0,05 μm and 20 μm. As an example for time average holography[14], different vibration modes of a piezoceramic loudspeaker are shown in fig. 5.

178

Work is in progress to adapt the system for electronic speckle interferometry and to complete it with a method to measure the amplitude distribution of vibrating objects.

Conclusion

We present a compact camera for holographic interferometry which could be made very simple in construction, simple in operation and very rigid due to the use of a semiconductor laser. Vibration insulation is not necessary; even under adverse environmental conditions holograms can be recorded reliably. Not at least owing to the low cost of this system holographic interferometry will thus be available for new applications in industrial nondestructive testing.

This research was supported by the Schwerpunktprogramm zur Förderung der Forschung an Fach-hochschulen (Programme for Support of Reseach at Polytechnics) at Baden-Württemberg.

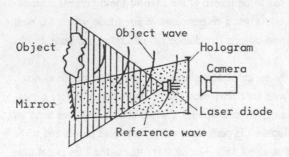

Fig. 1
Optical beam paths

Fig. 2
Optical setup with photo-ther-mo-plast processor
In operation it is shielded against ambient light with a covering cap

Fig. 3
Complete holographic system with PC for image processing.
The optical setup, here with its shielding cover, is located easily accessible on a table above the control electronics.
Two versions are demonstrated, the left, smaller and simpler one for single sheets of photo-ther-moplastic film which are handled manually.

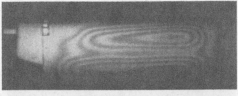

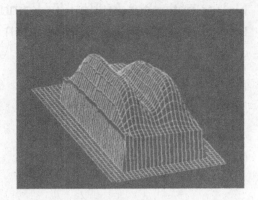

Fig. 4 a) b)   c)
Deformation of a gas pocket lighter under the influence of an internal pressure. The lighter has an internal transverse wall for stiffening. Thus two bulges are formed at each side of it. The maximum deformation is 1.6 μm.
a) Interference fringes
b) Modulo-2π-display of the phase difference between the interfering waves
c) Pseudo-3-dimensional display of the local deformation

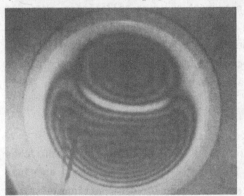
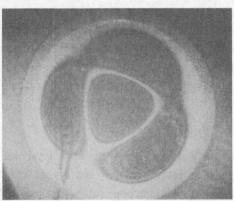

Fig. 5 a)   b)
Time average holograms: Vibration modes of a piezoceramic loudspeaker, sinusoidal driving voltage.
a) Frequency 1.13 kHz, 1,6 V_{pp}
b) 2.62 kHz, 2.6 V_{pp}

References
1. K. TATSUNO, A. ARIMOTO, Appl. Opt. 19, 2096-2097 (1980)
2. M. YONEMURA, Opt. Lett. 10, 1-3 (1985)
3. J. A. DAVIS, M. F. BROWNELL, Opt. Lett. 11, 196-197 (1986)
4. G. C. GILBREATH, A. E. CLEMENT, Opt. Lett. 12, 648-650 (1987)
5. S. HART, G. MENDES, K. BAZARGAN, S. XU, Opt. Lett. 13, 955-957 (1988)
6. B. LAU, E. MATTES, Appl. Opt. 31, 4738-4741 (1992)
7. R. MORAW, In: H. Marwitz (ed.), *Praxis der Holographie* (Expert, Ehningen 1990)
8. Rotech Gesellschaft für Prüftechnik mbH., Bergweg 47, D-83123 Amerang, or Micraudel Électronique et Informatique Appliquées, 93 Rue d'Adelshoffen, F-67300 Schiltigheim
9. G. S. BALLARD, J. Appl. Phys. 39, 4846-4848 (1968)
10. R. DÄNDLIKER, E. MAROM, F. M. MOTTIER, J. Opt. Soc. Am. 66, 23-30 (1976)
11. R. DÄNDLIKER, R. THALMANN, Opt. Eng. 24, 824-831 (1985)
12. P. HARIHARAN, Opt. Eng. 24, 632-638 (1985)
13. B. BREUCKMANN, W. THIEME, Appl. Opt. 24, 2145-2149 (1985)
14. R. J. POWELL, K. A. STETSON, J. Opt. Soc. Am. 55, 1593-1598 (1965)

Neuer optischer Aufbau für kombinierte TV-Holografie- und Shearografie-Anwendungen

A.Ettemeyer, M.Honlet
Ettemeyer Qualitätssicherung
Memminger Str. 72/207, D-89231 Neu-Ulm

1. Einleitung

Speckle-Meßtechniken dienen zur Messung von statischen und dynamischen Verschiebungen bzw. Verschiebungsgradienten und sind als Meßinstrument nur dann vollständig, wenn das optische System von einer Bildverarbeitung und einer quantitativen Bildanalyse unterstützt wird. Hauptvorteil der elektronischen Speckle-Verfahren ist die Möglichkeit, Echtzeitmessungen durchzuführen. Da mit Videotechnik gearbeitet wird, werden nachfolgend die Begriffe TV-Holografie und TV-Shearografie verwendet.

2. TV-Holografie

Die TV-Holografie ist ein unter der Abkürzung ESPI bekanntes, elektronisches Speckle-Interferometer und ermöglicht es, in Echtzeit Verformungsmessungen an einem Bauteil in-plane oder out-of-plane durchzuführen. Bedingung für die Funktionsweise ist das Vorhandensein von Referenzlicht. Die Interferenzmuster, welche das Objekt oder einen Ausschnitt davon überlagern, werden ohne Zeitverzögerung auf einen Fernsehmonitor dargestellt. Der Abstand zwischen den Linien beträgt die halbe Wellenlänge.

3. TV-Shearografie

Die TV-Shearografie ist ein elektronisches Speckle-Scher-Interferometer - auch Shearing-Interferometer genannt - und ermöglicht es, in Echtzeit den Gradienten einer Verformung an der Oberfläche eines Bauteils zu messen. Vorsichtig ausgedrückt kann man auch sagen, daß die Änderungen oder die erste Ableitung der Verschiebung in eine Richtung gemessen wird. Die Empfindlichkeit wird durch Betrag und Richtung der Scherung vorgegeben.

4. Vergleich beider Verfahren

Die TV-Holografie liefert Meßergebnisse, die eine visuell verständliche und zugleich hochgenaue quantitative Auswertung ermöglichen. Ganzkörperbewegungen können jedoch die Meßwerte erheblich beeinflussen und die Empfindlichkeit auf äußere Schwingungseinflüsse ist sehr hoch. In der Praxis ist unter Umständen eine aufwendige Schwingungsisolierung notwendig.

Die TV-Shearografie hingegen liefert Meßergebnisse, die visuell schwieriger nachzuvollziehen sind und für quantitative Aussagen eine deutlich aufwendigere Software zur Auswertung benötigen. Da aber nur der Gradient der Verformung angezeigt wird, ist das Verfahren viel unempfindlicher gegenüber Ganzkörperbewegungen. Dies ist für den mobilen Einsatz ein sehr wichtiges Kriterium, da der Aufwand für die Schwingungsisolierung stark reduziert werden kann.

Aus dem Wunsch nach Flexibilität in der Praxis entstand eine Kombination beider Meßverfahren. Um diese sinnvoll einzusetzen, sollte nur eine Abbildungsoptik verwendet werden. Höchste optische Qualität, die Verwendung des Phasenshift-Verfahrens für die beide Speckle-Verfahren und dies alles untergebracht in einer kompakten, staubdichten Bauweise waren weitere Voraussetzungen.

Bei der TV-Holografie wird das vom Objekt kommende Objektlicht mit dem eingekoppelten Referenzlicht überlagert. Das in der Sensorebene (hier ein CCD-Array) entstehende Interferenzmuster erzeugt kontrastreiche Speckles, welche charakteristisch für den momentanen Zustand der Meßobjekte sind. Durch den Vergleich der momentanen Zustände mit vorher abgespeicherten Intensitätsmustern wird die Information über die Verformung gewonnen.

Bei der TV-Shearografie wird der beobachtete Ausschnitt mit einer speziellen Optik doppelt und zueinander versetzt abgebildet. Tatsächlich werden zwei benachbarte Punkte oder Bereiche vom Objekt in der Sensorebene überlagert. Da diese beiden Bilder miteinander interferieren wird kein Referenzstrahl benötigt.

5. Realisierung einer kombinierten Holografie-/Shearografie-Optik

<u>Abb.1 bis 4</u> zeigt den prinzipiellen Aufbau eines kombinierten TV-Holografie-/Shearo-grafie-Systems. Ausgehend in <u>Abb.1</u> von einer unabhängigen Abbildungsoptik (AO1) wird beispielsweise das Objekt in einer Bildebene (BE1) abgebildet, in der sich normalerweise der photoempfindliche Sensor (CCD1) befindet. Um einen Raum für Manipulationen wie beispielsweise eine Referenzstrahleinkopplung oder eine Schervorrichtung zu schaffen wird ein System (AO2 und AO3) verwendet, welches die zum Bildaufbau beitragenden Strahlen wiederum erst ins Unendliche abbildet und sie dann anschließend wieder zu einem Bild

zusammenführt. Die ursprüngliche Bildebene (BE1) wird somit zur Zwischenbildebene, und das ursprüngliche Bild wird weiter nach hinten verlagert (BE2), wo sich nun der photoempfindliche Sensor (CCD1) befindet.

Zwischen diesen beiden Optiken steht nun ein Raum (R) zur Verfügung, in dem alle Bildpunkte aus der Zwischenbildebene als unendliche Bündel vorliegen. Im Prinzip ist jeder Punkt vom Objekt so gut wie überall in diesem Raum (R) vertreten, und der Durchmesser des jeweiligen Bündels durch die Aperturen begrenzt. Das gesamte optische System hinter der unabhängigen Abbildungsoptik heißt Zwischenbildübertragungssystem (ZBÜS).

Nach Faltung dieses Raumes mit Hilfe von Spiegeln (S1 und S2) wird die Grundvoraussetzung für die realisierte Kombination erzeugt, Abb.2.

Wird nun der Spiegel (S2) in der Achsenmitte geteilt und eine Spiegelhälfte (S2A) in beliebige Richtung verkippt, so werden alle Punkte des Objekts doppelt abgebildet, Abb.3. Damit ist Shearografie ohne zusätzlichen Lichtverlust und mit nur sehr geringem mechanischen Verstellaufwand in beliebiger Richtung möglich. Ist die andere Spiegelhälfte (S2B) auf ein Piezo (P) befestigt, so kann Phasenshift verwendet werden, da ja nur eine der beiden Bildhälften verschoben werden muß. Wird nun dazu noch eine Glasplatte (GP) in diesen Zwischenraum eingesetzt, so kann für ESPI das Referenzlicht eingekoppelt werden. Nach Zuschaltung des Referenzlichts und Rückstellung der verkippten Spiegelhälfte wird mit dieser Anordnung TV-Holografie betrieben.

Für den Simultanbetrieb von TV-Holografie und Shearografie werden die Strahlenbündel mit einem Strahlteiler (ST) aufgeteilt und eine Referenzlichteinkopplung samt zweiten Sensor (CCD2) angeordnet, Abb.4. Nach dem gleichen Prinzip sind mit nur einer Aufnahmeoptik ein-, zwei- oder dreidimensionale Messungen möglich.

Diese Kombination wurde geschützt, in kompakter Form realisiert und wird ständig für vielseitige Meßaufgaben eingesetzt. Z.B zeigt Abb.5 eine Fehleranzeige in einem Flugzeugbauteil mit TV-Holografie und Shearografie. Es handelt sich um eine Decklagenablösung in einem Wabenbauteil.

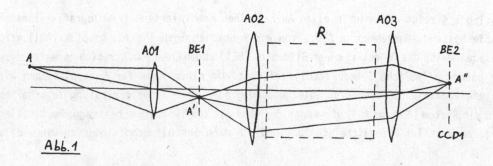

Abb.1

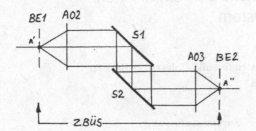

Abb. 2

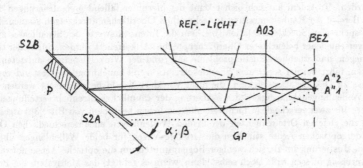

Abb. 3

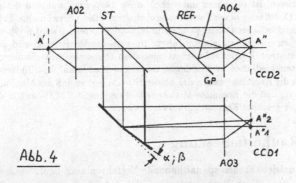

ESPI

SHEARING

Abb. 4

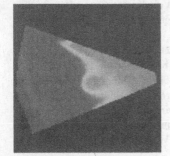

TV-Holografie

Shearografie
Scherrichtung horizontal

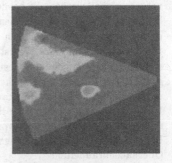

Shearografie
Scherrichtung vertikal

Abb. 5

Specklekorrelation mit einem dichromatischen Fouriertransformationssystem

J. Peters, P. Lehmann, A. Schöne
Institut für Meß-, Regelungs- und Systemtechnik, Universität Bremen
Badgasteinerstr.1-FZB, 28359 Bremen

Zusammenfassung

Vor zwei Jahrzehnten hat Parry [11] vorgeschlagen, die Specklekorrelation zur Bestimmung von Oberflächenrauheiten zu verwenden. Seither ist dieses Verfahren von verschiedenen Autoren untersucht worden. Trotzdem hat sich bisher keine der hierunter fallenden Realisierungen dieses Verfahrens im Bereich der Rauheitsmessung an technischen Oberflächen durchsetzen können. Beim Verfahren der spektralen Specklekorrelation, bei dem die Intensitätswerte des Streulichtes zweier Wellenlängen verschiedener beleuchteter Oberflächenabschnitte korreliert werden, dürfte der tiefere Grund darin liegen, daß erhebliche Justierprobleme aufgrund der Winkeldispersion auftreten. Die Winkeldispersion führt bei der spektralen Specklekorrelation mit zunehmendem Abstand von der optischen Achse zu Dekorrelationseffekten, die nicht von der Oberfläche verursacht werden, sondern von der Wellenlängenabhängigkeit der Skalierung der räumlichen Streulichtverteilungen. In den Streufeldern für zwei verschiedene Lichtwellenlängen liegen bei einem von Bitz [3] vorgeschlagenen Aufbau die gleichen Ortsfrequenzen der beiden Wellenlängen an unterschiedlichen Orten. Lediglich auf der optischen Achse stimmen die Ortsfrequenzen für beide Wellenlängen überein. Dies macht eine Justierung im Bereich weniger Bogenminuten um die optische Achse notwendig, wenn man zuverlässig messen will. Doch selbst dann, wenn es gelingt, das Meßsystem bei Beginn einer Messung so genau zu justieren, ist es nahezu unmöglich, diese Justierung während der gesamten Meßdauer, bei der die Oberfläche bewegt wird, beizubehalten. Die unerwünschte Folge davon ist, daß Intensitäten unterschiedlicher Ortsfrequenzen miteinander korreliert werden [9], was unabhängig von der Oberfläche zu einer Dekorrelation führt. Im vorliegenden Beitrag wird dargelegt, daß mit einem dichromatischen Fouriertransformationssystem eine einheitliche Skalierung der Ortsfrequenzebenen der unterschiedlichen Wellenlängen erreicht werden kann. Ein Bildverarbeitungssystem ermöglicht dann die Kreuzkorrelation der beiden Streubilder, so daß das Maximum unter der üblichen Voraussetzung, daß die Ensemble-Mittelwertbildung nun der Mittelung in der Ortsfrequenzebene entspricht, der gesuchte Korrelationskoeffizient ist.

1 Einführung in die Rauheitsmessung

In der Produktion eines Fertigungsbetriebs stehen spanabhebende Verfahren auch heute noch an erster Stelle. Der Konstrukteur legt im Büro die Sollgeometrie eines Werkstücks fest, mit der die Maße vorgeschrieben werden, die bei der Produktion einzuhalten sind. Damit die Kosten für die Bearbeitung in vorgegebenen Grenzen bleiben, legt der Konstrukteur zusätzlich Toleranzmaße fest. Hiermit erlaubt er der Fertigung Abweichungen von den Sollmaßen.

Die Abweichungen eines Werkstücks vom Sollmaß unterteilen sich nach DIN 4760 [4] in sechs unterschiedliche Ordnungen von Gestaltabweichungen. Die erste Ordnung bezeichnet die Formabweichungen, die zweite Ordnung bezieht sich auf die Welligkeit, die dritte bis fünfte Ordnung beschreiben die unterschiedlichen Arten der Rauheit und die sechste Ordnung beschreibt kristalline Strukturen des Werkstücks. Die quantitative Bestimmung der Gestaltabweichung wird mit zunehmender Ordnung schwieriger. Gegenstand des vorliegenden Vorhabens ist die Bestimmung von Oberflächenrauheiten. Bennett und Mattsson [2] geben einen Überblick über gebräuchliche Verfahren zur Bestimmung von Oberflächenrauheiten.

Die Oberflächenrauheit wird konventionell dadurch bestimmt, daß das Werkstück aus dem Produktionsprozeß genommen und mit einem Tastschnittgerät in einem Labor schwingungsfrei vermessen wird. Hierbei wird die Topographie der Oberfläche des Werkstücks in ein Spannungssignal umgewandelt, das der jeweiligen Oberflächenhöhe proportional ist. Mit Hilfe bekannter statistischer Methoden wird die Oberfläche dann über das Spannungssignal beurteilt. So wird unter anderem der quadratische Mittenrauhwert R_q bestimmt. R_q ist die Standardabweichung der Oberflächenhöhen von ihrer lokalen Mittellinie und entspricht dem englischsprachigen Begriff „rms-roughness". Gl.(1) stellt die numerische Näherung der Definitionsgleichung von R_q nach DIN 4762 [5] für N äquidistante Werte h_i der Oberflächenhöhen dar:

$$R_q = \sqrt{\frac{1}{N-1} \sum_{i=1}^{N} (h_i - \bar{h})^2} \stackrel{N \gg 1}{\approx} \sqrt{\frac{1}{N} \sum_{i=1}^{N} (h_i - \bar{h})^2} \tag{1}$$

mit der mittleren Oberflächenhöhe $\bar{h}$.

Im Gegensatz zu topographischen Verfahren wie dem Tastschnittverfahren erfolgt bei parametrisch optischen Verfahren schon im optischen Bereich eine Transformation der Oberflächenhöhen. Aus dem detektierten Signal läßt sich die Oberflächenhöhe eines diskreten Oberflächenpunktes nicht separieren, vielmehr enthält das detektierte Signal Informationen über die Statistik der Oberflächenhöhen des gesamten beleuchteten Oberflächenbereiches. Der Vorteil von parametrischen Verfahren liegt in der kürzeren Meßzeit und der Möglichkeit, die Messungen berührungslos während des Bearbeitungsprozesses in einem Taktzyklus einer Fertigungsstraße durchführen zu können. Das hier behandelte Verfahren der spektralen Specklekorrelation ist ein solches parametrisches Verfahren.

Parry [11], Bitz [3] und Ruffing [12] beschreiben Verfahren, die mit Hilfe der spektralen Specklekorrelation Oberflächenrauheiten ebener Proben ermitteln. Allen Verfahren gemeinsam ist, daß Justierprobleme aufgrund der Winkeldispersion auftreten. Im Zuge eines Forschungsvorhabens über parametrische Verfahren zur Rauheitsmessung ist es uns gelungen, den Aufbau eines Meßsystems zu entwerfen, mit dem es voraussichtlich möglich sein wird, das Problem der Winkeldispersion bei der spektralen Specklekorrelation mit Hilfe eines dichromatischen Fouriertransformationssystems zu lösen. Im folgenden wird der derzeitige Stand unserer Untersuchungen zu diesem modifizierten Aufbau der spektralen Specklekorrelation dargestellt.

2 Theoretische Zusammenhänge

Bitz [3] und Ruffing [12] benutzen den Kreuz-Korrelationskoeffizienten ρ_{12} zweier statistisch fluktuierender Speckleintensitäten:

$$\rho_{12} = \frac{\langle I_1 I_2 \rangle - \langle I_1 \rangle \langle I_2 \rangle}{\sigma_{I1} \sigma_{I1}} \tag{2}$$

mit:

I_1, I_2 : Speckleintensitäten für unterschiedliche Lichtwellenlängen

σ : Standardabweichungen der Intensitäten

$\langle x \rangle$: Ensemble Mittelwert bzw. Erwartungswert

und leiten hieraus den Kreuz-Korrelationskoeffizienten ρ_{12} voll ausgebildeter Specklemuster einer rauhen Oberfläche der Rauheit R_q in der Fourierebene einer Linse aus den Näherungen von Streufeldannahmen nach Beckmann und Spizzichino [1], der Fourieroptik, beschrieben bei Goodman [8] und zusätzlichen Voraussetzungen hinsichtlich der statistischen Verteilungen der Oberflächenhöhen und des verwendeten Laserstrahles her. Sie erhalten als Ergebnis folgenden Ausdruck für den Korrelationskoeffizienten:

$$\rho_{12} = e^{-R_q^2 (\Omega_1 - \Omega_2)^2} e^{-\frac{L_x^2}{4} ((x_1 - x_2) + 2\pi \Delta m_\xi)^2} e^{-\frac{L_y^2}{4} (2\pi \Delta m_\eta)^2} \tag{3}$$

mit:

$$\rho_{12} \quad : \quad \text{Kreuz-Korrelationskoeffizient} \qquad\qquad \lambda_i \quad : \quad \text{Lichtwellenlänge}$$

$$R_q \quad : \quad \text{Oberflächenrauheit} \qquad\qquad k_i = \frac{2\pi}{\lambda_i} \quad : \quad \text{Wellenzahl}$$

$$\vec{\xi}_i = (\xi_i, \eta_i) \quad : \quad \text{Koordinaten in der Fourierebene (Ortsfrequenzebene)}$$

$$\vec{m}_i = (m_{\xi i}, m_{\eta i}) \quad : \quad \text{Ortsfrequenzen} \qquad\qquad \Theta_{1i} \quad : \quad \text{Einfallswinkel}$$

$$m_{\xi i} = \frac{\cos\Theta_{2i}\xi_i}{\lambda_i f_i}; \Delta m_\xi = m_{\xi 1} - m_{\xi 2} \qquad \Theta_{2i} \quad : \quad \text{Beobachtungswinkel}$$

$$\qquad\qquad\qquad\qquad\qquad\qquad\qquad\qquad L \quad : \quad \text{Laserstrahlradius}$$

$$m_{\eta i} = \frac{\eta_i}{\lambda_i f_i}; \Delta m_\eta = m_{\eta 1} - m_{\eta 2} \qquad L_x = \frac{L}{\cos\Theta_1}$$

$$\Omega_i = k_i(\cos\Theta_{1i} + \cos\Theta_{2i}) \qquad\qquad L_y = L$$

$$\chi_i = k_i(\sin\Theta_{1i} - \sin\Theta_{2i})$$

Die aus der Literatur bekannten Ansätze setzen den Spezialfall voraus, daß in Gl.(3) die Exponenten der zweiten und dritten Exponentialfunktion Null werden, während der Exponent der ersten Exponentialfunktion von Null verschiedene Werte, die von der Rauheit abhängen, annehmen kann. Dieser Spezialfall liegt vor, wenn die Gleichungen:

$$\Theta_{1i} = \Theta_{2i}, \qquad (4) \qquad\qquad \Delta m_\xi = 0, \qquad (5) \qquad\qquad \Delta m_\eta = 0, \qquad (6)$$

erfüllt sind. Für die hier betrachtete spektrale Specklekorrelation wird zusätzlich die Gleichheit der Einfallswinkel gefordert: $\Theta_{11} = \Theta_{12} = \Theta_1$. Damit geht Gl(3) über in [12],[9]:

$$\rho_{12} = e^{-(2(k_1 - k_2)\cos\Theta_1 R_q)^2}. \qquad (7)$$

3 Experimentelle Realisierung

Man versucht nun Meß- und Auswerteverfahren zu finden, mit denen man Schätzwerte r_{12} für die Korrelationskoeffizienten ρ_{12} erhält, um danach mit Gl.(7) die Größe R_q bestimmen zu können. Zur experimentellen Bestimmung eines solchen Schätzwertes r_{12} werden in der Literatur zwei unterschiedliche Aufbauten vorgeschlagen. Bei beiden wird die Oberfläche mit einem parallen Lichtbündel eines Laserstrahls beleuchtet. Als Einfallswinkel wird meistens $\Theta_1 = 45°$ gewählt. In Richtung des geometrischen Reflexionswinkels wird ein Linsen-System installiert, welches das Fernfeld aus dem „Unendlichen" in seine Brennebene „holt". Goodman [8] zeigt, daß diese Eigenschaft einer Linse als Fouriertransformation der „Transmittanzfunktion" eines im Strahlengang vor der Linse befindlichen Objektes beschrieben werden kann. Zur Detektion des gestreuten Lichtes wird in der Brennebene des Linsensystems eine Blende installiert, deren Durchmesser kleiner als der eines Speckles ist und die auf Photomultiplier abgebildet wird, wobei eine Trennung des Lichtes der beiden Wellenlängen mit Hilfe von Interferenzfiltern erfolgt. Zur Erzeugung einer hinreichenden Zahl von Intensitätswertepaaren wird die Oberfläche durch den Meßfleck bewegt [3], [12]. Dieses Verfahren haben wir erprobt [13]. Es stellte sich als höchst problematisch dar, da schon geringe Verdrehungen des Werkstücks im Bereich weniger Bogenminuten eine Dekorrelation der Speckleintensitäten bewirkt, unabhängig von der Oberflächenrauheit. Dies ist darauf zurückzuführen, daß die Bedingungen für die Exponenten nach Gln.(4) bis (6) dann nicht mehr erfüllt sind.

Abhilfe schafft hier, wie im folgenden näher erläutert, ein achromatisches Fouriertransformationssystem. Bei einem achromatischen Fouriertransformationssystem gilt die Bedingung $\lambda_i f_i = \text{const.}$ Damit sind die Gln.(5) und (6) nicht mehr nur für $\xi_i = \eta_i = 0$ erfüllt, sondern für alle $\xi_1 = \xi_2$ und $\eta_1 = \eta_2$. Wird nun als zweite Möglichkeit der Detektion in der Ebene, in der die Blende installiert war, eine CCD-Kamera positioniert, so kann in vielen Fällen die Ensemble-Mittelwertbildung durch eine Mittelwertbildung in der Ortsfrequenzebene ersetzt werden. Zur separaten Detektion der Speckle-muster der unterschiedlichen Wellenlängen kann vor der Ortsfrequenzebene ein Strahlteiler installiert werden, so daß mit Hilfe von Interferenzfiltern jeweils mit einer Kamera das entsprechende Muster aufgenommen werden kann. Verwendet man nur eine Kamera ohne Strahlaufteilung, so ist es nicht möglich, die zueinander gehörenden Specklemuster zeitgleich aufzunehmen. Vielmehr kann man sie nur hintereinander durch Austauschen des Interferenzfilters gewinnen.

Ein achromatisches Fouriertransformationssystem mit Glaslinsen und holographisch optischen Elementen haben schon Finke und Ruffing [7] in Anlehnung an Morris [10] vorgeschlagen. Wir haben einen modifizierten Aufbau entworfen, in dem ausschließlich einfache Glaslinsen benutzt werden, deren Brennweiten der Bedingung $\lambda_2 f_2 = \lambda_1 f_1$ folgen. Dabei wird das an der Oberfläche gestreute Licht zunächst von einem Strahlteiler in zwei gleiche Anteile aufgeteilt, bevor jeder Anteil eine wellenlängen-angepaßte Linse erreicht. Jetzt können jeweils mit einem CCD-Array und dem entsprechenden Interferenzfilter die beiden Specklemuster gleichzeitig detektiert werden, wobei sichergestellt ist, daß die Skalierung in der Ortsfrequenzebene die gleiche ist. Abbildung 1 zeigt den verwendeten Aufbau. Der

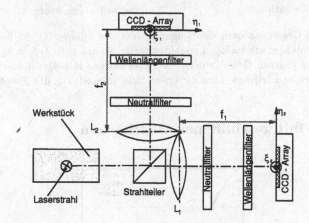

Abbildung 1: Realisierter optischer Aufbau zur dichromatischen Fouriertransformation

Aufbau ist so konzipiert, daß möglichst wenige optische Komponenten in den Strahlengang integriert werden müssen. Die beiden CCD-Arrays, zwei Hitachi Kameras KPM1, sind in der Lage, die von den Linsen mit den Brennweiten f_1 und f_2 erzeugten Specklemuster mit genügend hoher Auflösung zu detektieren. Innerhalb eines Speckles liegen ca. 10 Pixel einer Zeile des CCD-Arrays. Vor den Kameras ist jeweils ein Wellenlängenfilter positioniert, so daß jede Kamera nur eine Wellenlänge des von dem Laser emittierten Lichtes empfängt, und zwar diejenige Wellenlänge, an die die Brennweite der zugehörigen Linse angepaßt ist.

Zum Ausgleich von unvermeidlichen Dejustierungen des Aufbaus kann jetzt, nachdem für eine gleiche Skalierung in der Ortsfrequenzebene gesorgt wurde, eine Kreuzkorrelation der detektierten Bilder vorgenommen werden, um den gesuchten Schätzwert für den Korrelationskoeffizienten als Maximum der Kreuzkorrelationsfunktion aus der Mittelwertbildung in der Ortsfrequenzebene zu bestimmen.

4   Ergebnisse mit einem Spalt

Zunächst sollen Meßergebnisse mit einem Aufbau vorgestellt werden, bei dem der Effekt der Win-keldispersion, daß heißt der unterschiedlichen Abstände der einzelnen Beugungsordnungen für die unterschiedlichen Wellenlängen, nicht beeinflußt wird. Abbildung 2 zeigt das Beugungsmuster eines

Spaltes der Breite 0,2 mm für die beiden unterschiedlichen Wellenlängen $\lambda_1=488$ nm und $\lambda_2=514$ nm, die mit ein und derselben Linse auf einer Zeile des CCD-Arrays erzeugt wurden. Der Effekt der Winkeldispersion ist deutlich erkennbar.

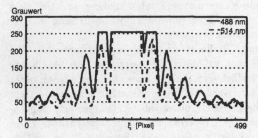

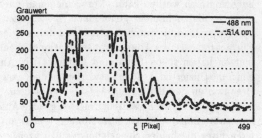

Abbildung 2: Beugungsmuster eines Spaltes für zwei unterschiedliche Wellenlängen bei Verwendung derselben Transformationslinse

Abbildung 3: Beugungsmuster eines Spaltes für zwei unterschiedliche Wellenlängen bei Verwendung wellenlängenangepaßter Linsen

Abbildung 3 zeigt im Gegensatz dazu die Beugungsbilder des Spaltes für die beiden Wellenlängen jeweils mit einer wellenlängenangepaßten Linsenbrennweite, so daß gilt: $\lambda_1 f_1 = \lambda_2 f_2$. Die Beugungsmaxima liegen jetzt an gleichen Orten in der Ortsfrequenzebene. In beiden Bildern überstrahlen die Intensitäten der nullten und teilweise auch der ersten Beugungsordnung den Empfindlichkeitsbereich der CCD- Kamera.

5   Ergebnisse mit geschliffenen Oberflächen

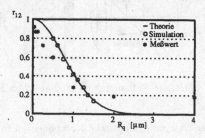

Abbildung 4: r_{12} über R_{qtast} bei $\lambda_1 = 488nm$ und $\lambda_2 = 514nm$

Abbildung 4 zeigt erste Meßergebnisse mit dem in diesem Vortrag vorgestellten Meßaufbau im Vergleich zu Simulationsergebnissen nach der Monte Carlo Methode [9] und dem theoretischen Verlauf gemäß Gl.(7). Als experimentelles Resultat ist der Schätzwert r_{12} des Specklekorrelationskoeffizienten über dem mit einem Tastschnittgerät ermittelten Rauheitskennwert R_{qtast} aufgetragen. Die Abweichungen der experimentellen Werte, die an Rugotest 104 Rauheitsnormalen ermittelt wurden, von der theoretischen Kurve lassen sich dadurch erklären, daß bei der Lasermessung nicht wie beim Tastschnittverfahren nach DIN4774 [6] vorgegangen wird, sondern alle Abweichungen von einer Mittellinie, die durch den Laserstrahlradius festgelegt ist, als Rauheit aufgefaßt werden. Bei den Messungen konnte nachgewiesen werden, daß bei dieser Meßanordnung Dejustierungen im Bereich von ±1 Grad zulässig sind, während bei den ursprünglich vorgeschlagenen Meßanordnungen höchstens wenige Bogenminuten toleriert werden können. Um jedoch den Meßaufbau auch bei stark welligen Oberflächen oder Formabweichungen industriell realisieren zu können, sollte das Meßsystem auch bei noch größeren Dejustierungen funktionssicher arbeiten. In dieser Richtung setzen wir unsere Untersuchungen fort.

Danksagung

Die Ergebnisse, die in der vorliegenden Arbeit dargestellt wurden, konnten in einem von der Deutschen Forschungsgemeinschaft (DFG) unter dem Kurzzeichen Scho344/3- geförderten Vorhaben gewonnen werden.

Literatur

[1] Beckmann, P.; Spizzichino, A.: The Scattering of Electromagnetic Waves from Rough Surfaces. Pergamon Press, Oxford 1963

[2] Bennet, J. M.; Mattsson, L.: Introduction to Surface Roughness and Scattering. Optical Society of America, Washington D.C. 1989

[3] Bitz, G.: Verfahren zur Bestimmung von Rauheitskenngrößen durch Specklekorrelation. Diss. Universität Karlsruhe. Forschrittsberichte der VDI Zeitschriften, Reihe 8, Nr.47 1982.

[4] DIN 4760. Gestaltabweichungen. Beuth Verlag Berlin Juni 1982.

[5] DIN 4762. Oberflächenrauheit. Beuth Verlag Berlin Januar 1989.

[6] DIN 4774. Messung der Wellentiefe mit elektrischen Tastschnittgeräten. Beuth Verlag Berlin Januar 1981.

[7] Finke, T.; Ruffing, B.: Surface Roughness Measurement by Speckle Correlation using an achromatic Fourier Transform System. Journal of Optical Society of America A (eingereicht; Vorabinformation durch Herrn Finke; Universität Karlsruhe, Institut für Mess- und Regelungstechnik mit Maschinenlaboratorium)

[8] Goodman, J. W.: Introduction to Fourier Optics. McGraw-Hill, New York 1968

[9] Lehmann, P.; Schöne, A.; Peters, J.: Simulation der Lichtstreuung an technischen rauhen Oberflächen als Grundlage laseroptischer Rauheitsmeßverfahren. Forschung im Ingenieurwesen-Engineering Research. Bd. 59, Nr. 4, 1993.

[10] Morris, G.M.: Diffraction theory for an achromatic Fourier transformation. Applied Optics, 1981, Vol.20,No.11, pp 2017-2025.

[11] Parry, G.: Some effects of surface roughness on the appearance of speckle in polychromatic light. Optics Communications, Vol. 12, No.1, September 1974, pp. 75-78.

[12] Ruffing, B.: Berührungslose Rauheitsmessung technischer Oberflächen mit Specklekorrelationsverfahren. Diss. Universität Karlsruhe 1987.

[13] Schöne, A.; Peters, J.; Lehmann, P.: Optisches Meßverfahren zur Bestimmung von Oberflächenrauheiten während des Bearbeitungsprozesses. IV. Meßtechnisches Symposium Aachen 10.-11.10.1991.

Oberflächeninspektion mittels topometrischer Verfahren

H. Rein, E. Klaas
Breuckmann GmbH
Torenstr. 13, D-88709 Meersburg

Aufgabenstellung der Topometrie

Die Güte der Oberfläche ist für viele Produkte ein entscheidendes Qualitätsmerkmal und erfordert eine zuverlässige Qualitätskontrolle. Die visuelle bzw. manuelle Oberflächenprüfung - basierend auf subjektiven Bewertungskriterien und somit abhängig von Erfahrung und Einschätzung des jeweiligen Kontrolleurs - wird daher in verstärktem Maße durch automatisierte Prüftechniken ersetzt.

Bildverarbeitende Systeme messen und prüfen heute in vielen Bereichen der industriellen Fertigung auf Maßhaltigkeit, Lage, Orientierung, Vollständigkeit, Farbtreue u.v.m. Die moderne topometrische Meßtechnik eröffnet die 3. Dimension in der Bildverarbeitung. Meßszenen und Prüfobjekte werden bildhaft erfaßt und 3-dimensional vermessen.

- schnell
- berührungslos
- hochauflösend

optoSIS - Erkennung typischer Oberflächenfehler

Das optische Oberflächen-Inspektions-Systems optoSIS (Surface-Inspection-System) arbeitet auf der Basis der topometrischen Meß- und Prüftechnik mittels bildhafter Triangulation. Dabei wird die zu vermessende Szene bzw. das Prüfobjekt mit strukturiertem Licht beleuchtet. Im einfachsten Fall durch Projektion eines hochpräzisen periodischen Gitters. Die auf das Objekt aufprojizierte Linienstruktur wird unter einem Winkel ϑ mittels eines Viewing-Systems aufgenommen und im Bildverarbeitungssystem gespeichert. Die Topometrie-Software berechnet aus diesen "Höhenlinien" die 3D-Information der Meßszene.

Das optoSIS - System gewährleistet eine objektive Analyse 3-dimensionaler Objektoberflächen: schnell, sicher und vor allem reproduzierbar. Die Prüfung typischer Oberflächenfehler umfaßt dabei folgende Punkte:

- Detektion
- Visualisierung
- Vermessung
- Bewertung
- Dokumentation

opto*SIS* - Berührungslose Vermessung von Dellen und Beulen

Entwickelt wurde diese opto*SIS* - Systemausführung zur objektiven und sicheren Erkennung von Dellen und Beulen an lackierten und unlackierten Blechteilen. Solche Fehler sind zum Teil minimale Formabweichungen in der Blechteilgeometrie, welche jedoch bei entsprechendem Lichteinfall sichtbar werden und so den optischen Gesamteindruck der lackierten Struktur stören (z.B. Motorhaube, Fahrzeugtüre usw.). Aufgebaut als flexibles Meßsystem zur Stichprobenkontrolle kann die ganze Bandbreite von Karosserieteilen untersucht werden.

Oberflächeninspektion mit höchster Auflösung

Mit Hilfe der leistungsstarken Blitz-Projektionseinheit werden nacheinander mehrere phasenverschobene Gitter-strukturen aufprojiziert. Als Projektionsmaske finden aperiodisch verlaufende Strichgitter Verwendung, welche mittels Schärfenebenenverkippung nach Scheimpflug, trotz schräger Beleuchtungsrichtung über die gesamte Meßfläche scharf abgebildet werden. Als Referenzgitter wird die CCD-Struktur einer direkt moiré-fähigen Kamera benützt.

Sind die zur Auswertung benötigten phasenverschobenen Bilder eingelesen, erfolgt die Vorverarbeitung zur Gewinnung der 3D-Daten. Nach dem Lösen

Oberflächen-Inspektions-System opto*SIS* zur Detektion von Dellen und Beulen.

der Phasenshiftgleichung für jeden Bildpunkt können die resultierenden modulo-2π Rampen durch den Demodulationsprozeß phasenrichtig zusammengefügt werden. Schon nach wenigen Sekunden steht der gesamte 3D-Datensatz der Meßfläche als Zwischenergebnis zur Verfügung.

An dieser Stelle setzt die Untersuchung auf 3D-Oberflächenfehler an. Das Prüfsystem bietet dafür zwei unterschiedliche Lösungsansätze. Die erste Möglichkeit erfordert einen als Meßreferenz abgespeicherten Datensatz eines fehlerfreien Masterteiles. Bei dem zweiten Ansatz kann auf den sogenannten Gutteilvergleich verzichtet werden.

Gutteilvergleich

Die ermittelten Geometriedaten des Meßobjektes können nun mit den Daten eines zuvor gemessenen Objektes verglichen werden. Dazu ist es notwendig, daß das Meßobjekt in exakt der gleichen Lage wie das bereits Aufgenommene positioniert ist. Bei gegebenen Voraussetzungen entspricht das Differnzhöhenbild beider Datensätze genau der Abweichung zwischen Meß- und Referenzobjekt. Dieses Ergebnisbild beinhaltet alle zur weiteren Auswertung benötigten Daten. Der Vorteil dieser Vorgehensweise liegt in der einfachen Möglichkeit eine Entscheidung ob Gut oder Schlecht zu treffen. Bei bereits fehlerbehafteten

Referenzobjekten ergibt eine Untersuchung mit Hilfe des Formvergleichs entsprechend verfälschte Ergebnisse.

Flächenapproximation

Für viele Anwendungen steht ein fehlerfreies Referenzobjekt nicht zur Verfügung oder es fehlt an einer Aufspannvorrichtung zur exakten Positionierung der Meßobjekte. Eine Untersuchung aufgrund einer Vergleichsmessung macht dann keinen Sinn. Dennoch besteht die Möglichkeit der Weiterverabeitung. Dabei wird aus dem gewonnenen Datensatz, auf Basis eines Verfahrens zur Polynomapproximation, eine Ausgleichsfläche berechnet. Diese Fläche kommt dem idealen Verlauf eines Karosserieteils sehr nahe. Zum Abschluß wird nach lokalen Abweichungen von der berechneten Idealgeometrie gesucht. Der ganze Meß- und Auswertevorgang ist innerhalb von 20 Sekunden abgeschlossen.

Die Grundkonfiguration des Prüfsystems ermöglicht die sichere Vermessung einer Formabweichung in der Größenordnung von 1/20000 der Meßflächenausdehnung. Beispielsweise ergibt sich bei Untersuchung eines Objektbereiches entsprechend dem DIN-A4 Format eine Tiefenauflösung von mindestens 15 Mikrometern. Der Meßbereich des Systems kann durch Wechseln der Gittermodule und Änderung der geometrischen Verhältnisse des Systemaufbaus, beispielsweise durch Vergrößerung des Arbeitsabstandes, in weiten Grenzen den Erfordernissen angepaßt werden. Objekte bis zu einer Größe von 2 Quadratmetern können mit einem Meßvorgang untersucht werden.

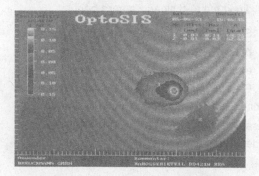

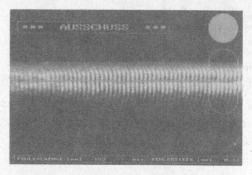

Beispiel einer Untersuchung auf Dellen und Beulen   Beispiel einer Untersuchung auf Einschnürungen

optoSIS - Detektion von Einschnürungen im Preßwerk

Als weitere optoSIS - Ausführung wird ein System zur Erkennung und Klassifikation von Einschnürungen, ebenfalls an Tiefziehteilen, vorgestellt. Einschnürungen sind typische Materialverdünnungen, die während des Preßvorganges vornehmlich bei kleinen Biegeradien oder ungenügend geschmierten Preßwerkzeugen auftreten können. Diese Einschnürungen können sich sofort, oder - gefährlicher noch - zu einem späteren Zeitpunkt, bei Belastung des Bauteiles, bis zu Durchrissen ausprägen. Gerade bei Sicherheitsteilen im Automobil- oder Maschinenbau ist deswegen eine 100%-Kontrolle direkt an der Pressenstraße erforderlich.

Die rauhen Umgebungsbedingungen im Preßwerk, mit starken Erschütterungen sowie hoher Schmutzbelastung, erfordern ein äußerst robust aufgebautes, vollautomatisch arbeitendes

Meßsystem. Die topometrische Meßtechnik basiert hierbei auf einem Projected-Fringes-Verfahren. Daraus resultieren kompakte Abmessungen und geringes Gewicht der Sensorik. Die Handhabung wird erleichtert und eine schnelle Umrüstung auf unterschiedliche Preßteile ermöglicht.

Blitzschnelle Meßwertaufnahme

Kameramodul und Projektionseinheit befinden sich in einem Gehäuse mit fester Basis und festem Beleuchtungswinkel ϑ. Um den Einfluß von Umgebungsschwingungen auszuschalten, wird eine Kombination aus sehr kurzer Beleuchtungsdauer und verkürzter Video-Bildaufnahmezeit angewandt. Die Projektionseinheit besteht aus einer Xenon-Blitzentladungslampe und einem Kondensorsystem zur gleichmäßigen Ausleuchtung des Liniengitters. Zur Abbildung des Liniengitters auf die zu untersuchende Objektstelle bzw. zur Aufnahme der Meßszene werden auf minimale Verzeichnung korrigierte Objektive verwendet. Eine zur Verfügung stehende Auswahl verschiedener Objektivbrennweiten ermöglicht die Anpassung an unterschiedliche Meßfeldgrößen.

Die Beschränkung auf ein einzelnes Auswertebild und eine Blitzbelichtungszeit von 15 Mikrosekunden, erlauben die Meßdatenaufnahme am bewegten Meßobjekt, der Produktionstakt bleibt unbeeinflußt.

Als Bildverarbeitungssystem wird das auf VMEbus-Basis aufgebaute System OPTOVision eingesetzt. Im OPTOVision System mit dem ffp-16 oder dem afp-1024 Prozessor wird dabei nach der Philosophie des "Rechnens mit Bildern" gearbeitet. Das bedeutet, daß der Prozessor für die Bildvorverarbeitung auf komplette Bilder zugreifen kann, die in Bildregistern bis zu einer Größe von 1024×1024 Pixel abgelegt sind.

Die eingesetzte Konfiguration besteht aus der Steuerkarte ffp-16, einer Bildspeicherkarte zur schnellen Verarbeitung von 4 Videobildern mit jeweils 512×512 Pixeln, einer softwarekonfigurierbaren Rechenwerkkarte und einer ebenfalls frei konfigurierbaren 8×8-Filterkarte. Als Host-Prozessor wird eine VME-680X0 CPU eingesetzt. Die Softwarebasis bildet das multitaskingfähige Echtzeit-Betzriebssystem OS-9.

Prozeßbezogene Auswertung im Sekundentakt

Das vom Projektor auf die zu untersuchende Objektstelle aufgeblitzte Streifenbild wird von der Kamera blitzsynchron aufgenommen. Zur Vermeidung von Fehlmessungen werden die eingelesenen Meßdaten auf Vollständigkeit geprüft und nach Bildqualitätskriterien wie z.B. dem Streifenkontrast, beurteilt. Bei nicht ausreichender Qualität des Streifenbildes, beispielsweise durch Defokussierung eines Objektives, oder durch falsch orientierte bzw. fehlende Meßobjekte, wird die Auswertung abgebrochen und die Ursache dafür an die Systemsteuerung weitergeleitet.

Die Anzahl der aufprojizierten Streifen beträgt ca. 1/4 der Pixelanzahl einer CCD-Zeile des Kamerasensors. Damit kann eine sichere Meßwertauswertung nach dem Pseudo-Phasenshift-Verfahren (Trägerfrequenzverfahren) vorgenommen werden.

194

Objektdaten können weiterverarbeitet werden. Durch Differentiation des Oberflächenverlaufs und anschließender Hochpaßfilterung zur Eliminierung der niederfrequenten Formveränderungen werden vorhandene Fehlerstellen herausgearbeitet. Fehlerstellen machen sich also durch einen höheren Gradienten der Steigung gegenüber dem beliebig geformten Meßobjekt bemerkbar. Die automatische Vermessung der detektierten Bereiche erfolgt mit einer Tiefenauflösung von etwa 1/4000 der lateralen Meßfeldgröße. Bei einem Meßfeld in der Größe des DIN-A5 Formats, ergibt dies eine Tiefenauflösung von etwa 50 µm.

Die Bewertung der vermessenen Fehlerstellen basiert auf frei wählbaren Toleranzgrenzen, welche dem System bei Einrichtung auf das jeweilige Prüfobjekt übergeben werden.

Der gesamte Meßvorgang, bestehend aus
⌐ Bildaufnahme
⌐ Vermessung der Fehler nach Lage, Tiefe und Ausrichtung
⌐ Klassifizierung nach Gutteil, Nacharbeit oder Ausschuß
dauert in etwa 1 Sekunde.

opto*SIS* - Zusammenfassung

Durch die Verwendung geblitzter Sensorsysteme kann eine Untersuchung auf Oberflächendeformationen unabhängig vom Umgebungslicht, bei der Einbildauswertung sogar an bewegten Meßobjekten erfolgen. Durch geeignete Hardware, in Verbindung mit problemangepaßter Auswertung, werden Meßzyklen von wenigen Videotakten ermöglicht. Eingesetzt und erprobt zur Prüfung von Tiefziehteilen, verdeutlicht das Meßsystem den hohen Entwicklungsstand topometrischer Online-Techniken. Die Einsatzfähigkeit der Topometrie für eine Vielzahl von Meßaufgaben in der industriellen Produktion ist gegeben.

Einsatz des opto*SIS* - Sensors in der Serienprüfung an Tiefziehteilen

Literatur
B. Breuckmann, E. Klaas: Sensoren und Geräte für die 3D-Meßtechnik. F&E Feinwerktechnik & Meßtechnik 100. 1992.
B. Breuckmann, G. Jansen: Optische 3D-Meßverfahren und Bildverarbeitung. KEM Konstruktion Elektronik Maschinenbau 9. 1992.
P. Gerspacher, H. Rein: Berührungslose Ebenheitsprüfung und Oberflächeninspektion. Praxis der industriellen Bildverarbeitung. Franzis Verlag. 1993.

Topometrische 3D-Koordinatenmeßtechnik

H. Winterberg, F. Halbauer
Breuckmann GmbH
Torenstr. 13, D-88709 Meersburg

1. Einleitung

Die Erfassung von 3D-Koordinaten an komplexen Bauteilen
erfolgt heute überwiegend mit taktilen Sensoren. Den
Vorteilen dieser bewährten Meßtechnik bezüglich Ver-
breitung und Meßgenauigkeit steht der Nachteil der oft
extrem langen Meßdauer gegenüber. Außerdem sind weiche
Materialien (z.B. Plastilinmodelle) mit taktilen Sen-
soren nicht oder nur mit erhöhtem Aufwand zu vermessen.
Mit dem topometrischen 3D-Meßsystem opto*CAM* steht jetzt
eine Computer-gestützte optische Sensorik zur
Verfügung, welche den Vorteil der berührungslosen und
schnellen Meßwerterfassung vereint. Die 3-dimensionale
Vermessung (Digitalisierung) komplexer Objekte aus
beliebigen Ansichten erfolgt innerhalb weniger
Sekunden. Je nach Aufgabenstellung werden dabei die 3D-
Koordinaten von einigen Tausend bis einigen Millionen
Meßpunkten erfaßt. Das zeit- und kostenintensive Messen
mit taktilen Meßsystemen ist so in vielen Einsatz-
bereichen vermeidbar geworden.

2. Topometrische 3D-Koordinatenmeßtechnik

Der opto*CAM*-Sensor arbeitet nach dem Prinzip der bild-
haften Triangulation. Über ein LCD-Display werden
rechnergesteuert Streifensysteme unterschiedlicher
Gitterperiode auf das Objekt projiziert. Die Streifensysteme werden von einer CCD-Kamera erfaßt, die unter
einem definierten Winkel zur Beleuchtungsrichtung angeordnet ist. Mittels eines kombinierten Greycode-

Phasenshift-Verfahrens können so die 3D-Koordinaten der Meßszene bildpunktweise und voneinander unabhängig nach den bekannten Triangulationsgesetzen berechnet werden. Die dazu notwendigen Parameter der Sensorgeometrie werden über neu entwickelte Algorithmen auf einer Kalibriervorrichtung bestimmt. Dabei werden auch Fehler der Abbildungsoptiken oder der Kamera berücksichtigt.

Als Ergebnis der topometrischen 3D-Vermessung steht ein Datensatz von (x,y,z)-Koordinaten (z.B. im VDAFS-Format) zur Verfügung. Bildbereiche, in denen keine sinnvollen Daten vorliegen (z.B. aufgrund von Abschattungen, Löchern o.ä.) werden automatisch erkannt und markiert.

3. Verknüpfung mehrerer Ansichten

Alle Ergebnisdaten beziehen sich auf ein gemeinsames Koordinatensystem, welches fest mit der Position des topometrischen Sensors verknüpft ist. Die Anbindung an ein übergeordnetes Koordinatensystem einer Koordinaten-meßmaschine, eines Roboters oder anderen Handhabungs-systemes kann durch entsprechende Kalibriervorgänge (z.B. Vermessung eines Eichkörpers) erfolgen. Für die Gesamtvermessung eines Objektes können so durch Positionierung von Sensor oder Objekt mehrere Teilansichten mit definierten Bezugskoordinaten aufgenommen und rechnerisch verknüpft werden.

Bild 1 zeigt die Vermessung eines 80 cm langen Urmodelles eines Querträgers in 5 Teilabschnitten. Die Größe der einzelnen Meßfenster betrug ca. 20x25 cm mit einem Überlappungsbereich von ca. 5 cm. Der optoCAM-Sensor war auf einer DEA-Portal-Meßmaschine angebracht.

4. Systemspezifikationen

Bei einem Arbeitsabstand von 60 cm und einem gleichzeitig erfaßbaren Meßvolumen von ca. 240x180x150 mm (BxHxT) erreicht der topometrische 3D-Sensor eine

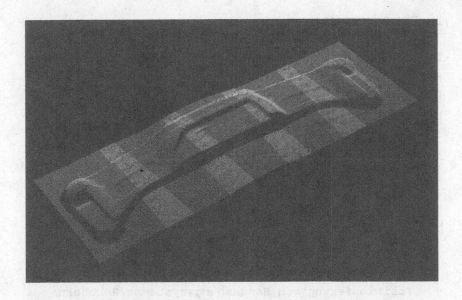

Bild 1: Verknüpfung mehrerer Teilansichten zu einem
        Datensatz

Auflösung von 30μm. Die Merkmalsgenauigkeit liegt bei
±30μm. Sie gibt an, wie gut bestimmte Objektmerkmale
vermessen werden können (z.B. Kanten, Kugeldurchmesser
oder Höhendifferenzen). Die maximale Meßunsicherheit
bezieht sich auf die im Meßvolumen maximal auftretende
Abweichung von tatsächlicher zur gemessener Objektkoor-
dinate und liegt bei ±50μm.
Weitere wichtige Merkmale sind folgend aufgelistet:
-   Die Dauer der Meßaufnahme beträgt
    nur ca. 2 Sekunden.
-   Der Sensor arbeitet mit einer Blitzlampe, so daß
    Umgebungslicht keinen Einfluß auf die Messung hat.
-   Der Sensor zeichnet sich durch einen stabilen
    Systemaufbau, geschützte Bauweise und geringe
    Momente aus, so daß er sich gut für industrielle
    Anwendungen eignet.

5. Einsatzgebiete

Die Einsatzgebiete des topometrischen 3D-Koordinaten-
Meßsystemes opto*CAM* sind aufgrund der hohen
Flexibilität sehr vielfältig. Neben der reinen
Digitalisierung von Freiformflächen kann er auch in
folgenden Gebieten eingesetzt werden:
- Modell-, Formen- und Werkzeugbau
  (z.B Erfassung von Urmodellen)
- Eingangs- und Endkontrolle
  (z.B. Vermessung von Maßhaltigkeiten oder
  Vollständigkeitskontrolle)
- Fertigungsüberwachung und Serienprüfung
  (z.B. bei der Zahnradvermessung)
- Positionierung von Handhabungssystemen/Robotern
  (z.B. für Montageaufgaben)
- Vermessung von Fertigungsstraßen
  (z.B für die Roboter-Offline-Programmierung)
- Verformungsanalyse
  (z.B. bei thermischer Belastung)

6. Literatur

- B. Breuckmann, Topometrische Oberflächenprüfung und
  3D-Koordinatenmeßtechnik, Vision Jahrbuch 1993
- B. Breuckmann, Optische 3D-Meßsysteme für Online-
  Anwendungen, Technisches Messen 57, 1990
- H. Wolf, Codierter Lichtansatz zur schnellen
  dreidimensionalen Bilderfassung, bild & ton 5/6,
  1992, S. 111-114

Determination of Slope and Curvature in Object Deformation with Computer Generated Holograms

Lingli Wang, Theo Tschudi
Institute of Applied Physics, TH Darmstadt,
Hochschulstr. 6, D-64289 Darmstadt, F. R. Germany

Double exposure interferometry is well known to determine mechanical parameters in deformation analysis. Normally it is not possible to determine both, slope and curvature, in the same experiment. We propose a method to determine slope and curvature of object deformation recorded simultaneously on two holographic plates. Between the two exposures, one of the holographic plates is moved along its surface, and the curvature of the deformed object is obtained. The slope and curvature of a clamped square plate on which a concentrated force is applied at the center is mesaured as an example to illustrate the capability of this method.

By using computer generated hologram(CGH), it is easy to get the images of the deformed object. A incident monochromatic plane wave passes through a transparent grating and it is separated into several orders. Only the $\pm$first-order diffraction waves are allowed to pass through the spatial filter, which are reflected by the surface of object. Then the reflected waves go through CGH and there are two images of the object behind the CGH. Putting two holographic plates in the positions of the two images and one grating between one holographic plate and CGH, the slope and curvature of the deformed object can be obtained.

The slope of the deformed object can be obtained as

$$\frac{\partial w}{\partial x} = \frac{md}{4l_2 sec^2\theta} \qquad (1)$$

in which, w(x,y) is the deformation of the object; m is a positive integer; d is the grating pitch; l_2 is the distance between CGH and holographic plate; θ is the angle of incident wave.

The curvature of the deformed object can also be obtained as

$$\frac{\partial^2 w}{\partial x^2} = \frac{m\pi d}{8k\lambda l_2 \cdot \Delta x} \qquad (2)$$

in which, λ is the laser wavelength; Δx is shearing value between the two-exposure, along one direction of holographic plate.

The slope and curvature of a clamped square plate on which a concentrated force is applied at the center is mesaured as an example to illustrate the capability of this method. It is also possible to obtain the curvature and relative slope of one deformed object simultaneously on different holographic plates, taking out the grating between CGH and holographic plate and making two-exposure before and after the object is deformed or between two object's deformed states.

Integration of Projection Moiré with Coordinate Measuring Machines

M.C. Shellabear & A. Rönner
EOS GmbH Electro Optical Systems
W-82152 Planegg/München

1. Introduction - Requirements of Coordinate Metrology and Digitization

In many areas of industry and technology there is a very important requirement to obtain shape information from physical objects. The range of shapes which need to be measured is almost infinite, but can be split into two general classifications:

i) Nominally regular geometric forms, such as planes, cylinders, cones, straight lines, etc. These can be described with simple mathematical functions defined by a small number of measured coordinate points.

ii) Free-form curves and surfaces. These generally need to be defined or approximated using more complex mathematical functions, and typically require a large number of measurement points to describe the shape with acceptable accuracy.

The range of applications is extremely wide, but can be also broadly divided into two areas:

i) Dimensional checking, where the measured data are compared to nominal design data. This is typically used to check the accuracy or deviations of manufactured parts.

ii) Digitization, i.e. the acquisition of shape data where no nominal data exists. This is typically required for transferring the shape of a hand-styled model into a CAD/CAM system, and usually involves free-forms.

Dimensional checking has been used for centuries, and over the last few decades has been developed into a specialized technology known as coordinate metrology, based around the coordinate measuring machine (CMM). Digitization of complex shapes has only become a realistic proposition in recent years, with the availability of high computer power and sophisticated software. The CMM generally consists of a mechanical positioning system having three to six degrees of freedom, carrying a sensor to detect the object surface, and linked to a computer to analyse the results. By far the most widely used type of sensor is the mechanical touch-trigger probe, which sends out an electrical signal whenever it contacts the object surface, causing the computer to read the instantaneous machine position. This is well suited to measuring regular geometric forms by probing a set of individual points. The software can calculate a best-fit geometrical element from the probed coordinates, provided that the contact sensor (typically a sphere) has been calibrated with a known reference to compensate for its finite size. Free-forms can also be measured using touch probes, but with two limitations. First, the requirement to probe a large number of points and reposition the machine between each means that the measurement is usually much slower. Secondly, the irregular surface orientation requires much more complex analysis software to correctly compensate for the sensor offset.

These problems can be reduced or overcome by using other types of sensor. Scanning mechanical probes measure the position continuously whilst moving over the surface and therefore increase the speed, but they still require offset compensation. Optical sensors are of particular interest, as their non-contact operation offers the potential for high-speed data acquisition and also avoids the risk of damaging soft surfaces (for example clay models). Three principal classes of optical sensor have so far been used with CMMs: (i) laser triangulation, which measures the distance to a projected point, set of points or line; (ii) focus detection, which identifies surface points lying at a given distance; and (iii) video imaging, for two-dimensional measurements within the image plane. Of these, laser triangulation is the most widely used for three-dimensional measurement and digitization. Many other optical techniques have also been applied to shape measurement problems, for example photogrammetry (passive triangulation) and laser radar. However in most cases the sensors have been to large and heavy to be mounted on standard CMMs, and have therefore been difficult to integrate within the established coordinate metrology environment.

Projection moiré is an optical technique which offers non-contact measurement of large numbers of surface coordinates simultaneously within a three-dimensional volume. This technique was also previously restricted to laboratory systems due to the size and weight of the sensor, but new developments have produced a compact measurement head which is compatible with standard CMMs.

2. Absolute shape measurement with projection moiré and laser triangulation

Projection moiré has been established for some years as a method for measuring surface shape [REID 1984]. The basic principle is that a grating of parallel bright and dark lines is projected onto a test surface, which is viewed at an angle through a reference grating (Figure 1). The image of the projected grating appears distorted due to the oblique observation angle, and interacts with the reference grating to produce moiré fringes whose position is a function of the surface shape and the optical geometry of the measuring system. When certain geometrical conditions are satisfied, the moiré fringes lie in planes perpendicular to the viewing axis, and therefore correspond to surface contours. By moving the projection grating in a known direction and analysing the change in intensity in the image, it is also possible to calculate the optical phase of the moiré fringes. This is known as the phase-shift method, and enables both the relative height and slope of the surface to be calculated with high resolution at every point in the image. The surface shape can be obtained by combining this height information with the position in the image, and calibrating the sensor to determine its sensitivity in the lateral (x and y) and axial (z) directions.

In order to calculate surface coordinates in a fixed reference system based on the sensor, it is necessary to know the absolute distance to at least one point in the image. This can be achieved by using the well-established technique of laser triangulation [SHELLABEAR 1992]. A laser beam is projected onto the object surface, and observed at an angle in the same way as the projected moiré grating. As the object surface is moved towards or away from the sensor (in the z direction), the laser spot moves

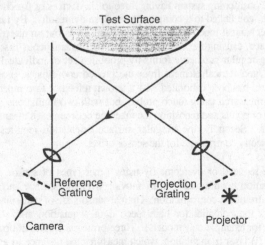

1.   Projection moire principle

from side to side (in the x direction) in the image, with each position corresponding to a unique distance. The absolute distance to the laser spot on the surface can therefore be determined by means of a second calibration. This point can then be used as a reference for calculating absolute x,y,z coordinates at every pixel in the image.

Digitization of surfaces which are larger than the field of view of the moiré sensor requires multiple measurements from different positions. In this case it is necessary to transform the results from each individual measurement into a global coordinate system which is fixed relative to the test object.

3. The EOSCAN 100 measurement system

The EOSCAN 100 is a compact sensor head comprising projection moiré and laser triangulation optics for absolute measurement. Figure 2 shows the functional layout and principal components: CCD video camera, diode laser, grating projector and phase-shift device. Figure 3 shows the sensor head itself. This measures $100 \times 100 \times 35$ mm^3, weighs less than 0.6 kg and (in the version shown) incorporates a Renishaw Autojoint coupling. It can therefore be mounted on almost all types of CMM via the widely-used Renishaw probe-system, which includes 2-axis rotation heads for both manual and CNC systems. Various types of touch-probe and other sensors (including laser triangulaion sensors) can also be mounted on the same system, and can be exchanged and remounted with extremely high reproducibility within a few seconds.

A single cable carrying all the power, control and video signals links the sensor head to a control unit. This in turn is connected to a standard PC containing plug-in boards for the input and output of video and control signals. The standard software operates under Windows, and includes many functions for controlling the measurement procedure and analysing the results. These include:
- visual displays as phase images (contour maps), wire-frame plots or profile sections;
- interrogation of individual coordinates and normal vectors;

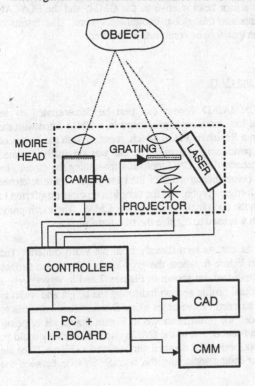

2. EOSCAN principle sketch

204

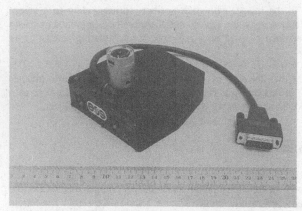

3.   EOSCAN head photo

- image processing, masking and filtering operations;
- automatic detection of edges and form-lines (i.e. lines of maximum local curvature);
- data reduction according to surface curvature, point spacing or specified number of points;
- calculation of best-fit circle or sphere from many measurement points;
- data export in various formats including VDA-FS.

Optional software enables the calculation of B-spline or NURBS surfaces from the point data, with an IGES interface for exporting the results to CAD systems.

A single EOSCAN 100 measurement digitizes a surface area of up to 120x80 mm², delivering up to 390,000 individual coordinate points with an accuracy of 0.1mm. Larger areas and complete objects can be digitized by moving the sensor head around the object using the CMM. The necessary viewing angles are calibrated once using a reference sphere, in much the same way that touch-probes are calibrated. This defines the orientation of the sensor head relative to the CMM, and the EOSCAN software then transforms all measurement points into this global coordinate system. The results of measurements from various different sensors can therefore be combined as required.

4.   Application: Integrating EOSCAN, CMM and CAD

The capabilities of the complete EOSCAN/CMM/CAD system can best be demonstrated by an application example. Figure 4 shows a 300mm long model car, positioned in a CMM fitted with the EOSCAN 100. The sensor head is mounted on a Renishaw PH10 which, together with the 3-axes of the CMM, provides a 5-axis CNC-controlled positioning system. The car was originally designed using a Tebis CAD/CAM system, which also generated the CNC toolpaths for milling the model. Its reference coordinate system is defined by three perpendicular sides of the base-plate. These reference planes were measured using a conventional touch-trigger probe, and the coordinate points transferred to the CAD system to orientate the nominal data in the coordinate system of the CMM. The touch-probe was then replaced with the EOSCAN 100, which was used to digitize the free-form surfaces.

Figure 5 shows a moiré image of the front of the car, as seen directly from the video camera. The result of the phase-shift calculation is shown in Figure 6, where the grey-levels indicate the surface shape. Two examples of the EOSCAN software output are shown in Figures 7 and 8, respectively a wire-frame plot of the whole image and a horizontal profile section indicating the height and width in millimetres. This measurement procedure was repeated twelve times using five sensor orientations to build up the complete surface shape of the model; the result is shown in Figure 9 as a set of point-clouds. These points were exported on-line via a network link to the CAD system, where they could be directly compared to the design data. Figure 10(a) shows the nominal surface, and in Figure 10(b) the measured points are superimposed, with colour-coding indicating the normal distance between the measured coordinates and the nominal surface.

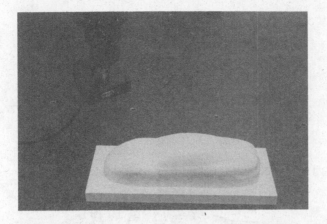

4.   Model car mounted on a CMM with EOSCAN 100

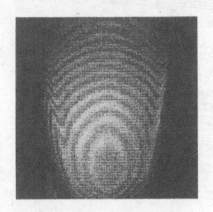

5.   Moiré image of the car bonnet   6.   Phase image of the car bonnet

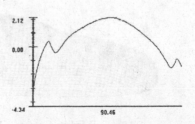

7.   EOSCAN 3D plot of the car bonnet   8.   EOSCAN section through the car bonnet

9. Complete car digitized as point-clouds

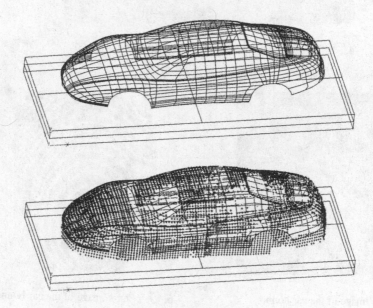

10(a). Nominal CAD design data
10(b). Nominal-actual comparison in CAD

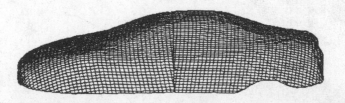

11. B-Spline surfaces calculated from the point data

The generation of surfaces can also be demonstrated using the same measurement data, as shown in Figure 11. In this case, one half of the car has been modelled as two B-spline surfaces, which can easily be imported into third-party CAD systems. The calculated surfaces can be compared with the original measurement points to judge their accuracy, and can also be modified or further processed within the CAD system as required.

5. Summary

The EOSCAN 100 is one of a number of optical techniques which can digitize large quantities of surface points rapidly and without contact. This offers great advantages for some measurement applications, notably where free-form surfaces are concerned. It must however be recognised that there still many applications where such optical technologies cannot compete with the established techniques. The compatibility of the EOSCAN with standard CMMs is therefore of great importance, as it enables various methods to be combined within a single machine and each to be applied to its best advantage. The ability to calculate curves and surfaces, and to efficiently transfer these and the point data to CAD systems is also very important, as this simplifies and accelerates the design and development process.

The examples presented above demonstrate that this technology is already capable of performing sophisticated measurement and digitization tasks. With continuing improvement of the interfaces to both CMMs and CAD systems, the capabilities will certainly increase in the near future.

References:

Reid, G.T., Rixon, R.C. & Messer, H.I., "Absolute and comparative measurements of three-dimensional shape by phase measuring moiré topography", Optics and Laser Technology, **16**, 315-319 (1984).

Shellabear, M.C. & Langer, H.J., "Electro-optic shape measurement using projection moiré", in *Optoelektronische Verfahren in der Koordinatenmeßtechnik*, VDI Seminar, Aachen, 9-10 September 1992.

Vergleich inkrementaler Wegmeßsysteme

G. ULBERS
UPM (Ulbers Präzisions-Meßtechnik)
Lindenbaumstraße 10, D-7730 VS-Weilersbach

Einleitung

Inkrementale Wegmeßsysteme haben sich einen großen Markt erobert, besonders im Bereich von Werkzeugmaschinen, Robotern und Meßmaschinen. Hier sind heute vor allen Dingen noch die inkrementalen Systeme vertreten, bei denen die Weginformation auf einem Teilungsträger (Maßstab oder Teilscheibe) in diskreter Form mit regelmäßigen Markierungen den sog. Inkrementen vorliegt, zwischen denen noch meist eine elektronische Interpolation vorgenommen wird.

Für ein inkrementales Wegmeßsystem brauchen die Inkremente nun nicht in körperlicher Form vorzuliegen, sondern sie können auch als Interferenzstreifen mit einem Interferometer realisiert werden, zwischen denen ebenfalls meist noch eine Interpolation vorgenommen wird.

Inkrementale Meßverfahren mit einem Teilungsträger als Maßverkörperung

Das inkrementale Meßverfahren mit einem Teilungsträger als Maßverkörperung hat sich gegenüber anderen Verfahren, wie z.B. codierte Meßverfahren durchgesetzt, weil große Meßlängen, beliebige Nullpunktwahl, ein wesentlich geringerer Auswertungsaufwand und damit eine größere Zuverlässigkeit erreicht wird. Der Meßwert wird durch elektrisches Auszählen der Perioden oder der nach einer Unterteilung der Perioden gewonnenen Zähl-Inkremente, ausgehend von einem beliebigen Nullpunkt, bestimmt. Zum Wiederfinden dieses Nullpunktes ist auf den inkrementalen Maßstäben häufig eine Referenzspur mit einer oder mehreren Referenzmarken angebracht.

Bei den optischen Meßverfahren zum Abtasten von inkrementalen Maßstäben werden Abbildungs- und interferentielle Meßverfahren unterschieden. Die interferentiellen Verfahren haben allerdings nichts mit den Interferometern zu tun, von denen später die Rede sein wird.

Bei den optisch inkrementalen Abbildungsverfahren fällt Licht durch ein Abtastgitter auf einen Maßstab. Je nachdem, ob es sich um ein Durchlicht- oder Auflichtmeßsystem handelt, fällt es durch diesen hindurch oder wird von diesem wieder zurückgeworfen auf eine Reihe von Fotodetektoren. Auf einen Glasmaßstab ist ein Gitter in Form von lichtdurchlässigen und lichtundurchlässigen Strichen, auf Stahlmaßstäbe in Form von lichtreflektierenden oder lichtabsorbierenden Strichen aufgebracht. Die Teilungsperioden, die letzlich die mögliche Meßauflösung festlegen, liegen bei einigen µm bis einige zehntel Millimeter. Die Teilung wird mit einer Abtastplatte (meist aus Glas), die je nach Abtastverfahren verschieden gestaltet ist, abgetastet. Die Gitterteilung ist herstellungsbedingt nie absolut perfekt, außerdem wird die Teilung beeinflußt durch mechanische Belastung des Maßstabes (Durchbiegung liegend, Dehnung hängend, Deformation durch Befestigung). Ein Beispiel für den Einfluß der Gewichtskraft eines Glasmaßstabes auf die Teilung bei hängender Befestigung zeigt Bild 1. Ein Glasmaßstab von 1,80 m Länge wurde im liegenden und hängenden Zustand mit Hilfe eines Laserinterferometers vermessen. Im liegenden Zustand stimmte die gemessene Abweichung von ±1 µm mit dem Meßprotokoll des Herstellers überein. Im hängenden Zustand ergab sich über einen Meßweg von 1,60 m eine Abweichung von 23 µm. Solch ein Wert liegt weit über dem Fehler eines Laserinterferometers durch Einflüsse der Luft, der bei dem genannten Meßweg brechungsindexkompensiert bei ca. 1-2 µm liegen würde.

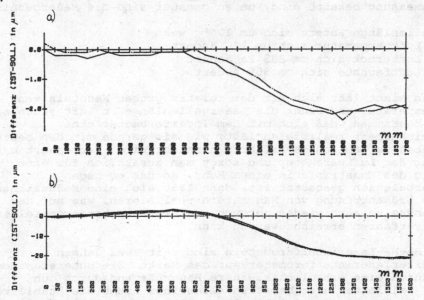

Bild 1: Fehler eines Glasmaßstabs a)liegend b)hängend

Das System der Gitterabtastung ist verschmutzungstolerant bezüglich der Mittelung des Abtastgitters über viele Gitterperioden des Maßstabs. Es ist aber nicht verschmutzungstolerant bezüglich des Abstandes und des Spaltes zwischen Abtastgitter und Maßstab, daher müssen auch inkrementale Systeme gegen Umwelteinflüsse geschützt werden.

Bei Einsatz von Glas bzw. Stahlmaßstäben ist weiterhin deren Ausdehnungskoeffizient mit der Temperatur zu berücksichtigen. Durch die Ausdehnung des Maßstabs ändert sich seine Gitterkonstante, was zu einem Meßfehler führt, wenn das zu messende Werkstück sich nicht mit dem gleichen Ausdehnungs- koeffizienten ausgedehnt hat.

Bei Wegmessungen sollte man das Abbesche Prinzip einhalten, um den
Meßfehler so klein wie möglich zu halten. Dies ist mit einem
Maßstabssystem nur schwer möglich, weil schon die Baugröße dies
nicht zuläßt. Bei einer idealen Meßanordnung sollte sich sowohl
der Maßstab, wie auch das Abtastgitter in der Meßachse befinden,
was nicht möglich ist.

Interferometer

Die meisten für die inkrementale Längenmessung eingesetzten
Interferometer basieren auf dem Michelsonprinzip. Man kann sich
den Wellenzug des Laserlichtes bis zum Meßreflektor in Analogie zu
obigen Ausführungen als Maßstabsgitter vorstellen, daß mit einem
Referenzwellenzug als Abtastgitter abgetastet wird.
Voraussetzung für eine genaue Längenmessung mit einem
Interferometer ist die genaue Kenntnis der Wellenlänge des
verwendeten Lichtes. Die Wellenlänge ist abhängig von der
Lufttemperatur, dem Luftdruck, dem Wasserdampfgehalt und der
Zusammensetzung der Luft. Je genauer diese Parameter bei der
Längenmessung bekannt sind, um so genauer sind die Meßergebnisse.

Die Wellenlänge ändert sich um 10^{-6}, wenn
* die Lufttemperatur sich um 1°C ändert
* der Luftdruck sich um 333 Pa ändert
* die Luftfeuchte sich um 80% ändert

Wie man sieht läßt sich mit der relativ groben Kenntnis von
Lufttemperatur und Druck die Laserwellenlänge in Luft schon so
genau bestimmen, daß sich mit dem Interferometer eine
Meßunsicherheit realisieren läßt, wie sie gerade mit den besten
Maßstäben im Abbildungsverfahren erreicht wird. Verbessert man die
Messung der Luftparameter und sorgt man zusätzlich für eine
Führung des Meßstrahls in einem Rohr, so daß er gegen
Lufturbulenzen geschützt ist, dann läßt sich eine Meßunsicherheit
in der Größenordnung von Nanometern realisieren, was auf der
Maßstabsseite auch wieder nur von den besten interferentiellen
Abtastverfahren erreicht werden kann.

Neben He-Ne-Laserinterferometern sind seit zwei Jahren
Halbleiterlaserinterferometer auf dem Markt. Sie unterscheiden
sich durch eine erheblich kleinere Baugröße und durch den
Preis. Halbleiterlaser besitzen allerdings nicht die Stabilität
von He-Ne-Lasern, deren Wellenlänge bis auf ca. 10^{-7} stabil ist.
Halbleiterlaser erreichen bestenfalls eine Stabilität der
Wellenlänge von 10^{-6}. Die Wellenlänge des Halbleiterlasers ändert
sich mit der Betriebsdauer um $\leq 10^{-5}$ pro Jahr, daher sollte solch
ein System alle halbe Jahr rekalibriert werden. Allerdings hängt
auch die He-Ne-Laserwellenlänge vom Alter ab. Ein Hersteller
schreibt z. B. eine jährliche Rekalibrierung vor.

Die Fa. UPM bietet ein neues Halbleiterlaserinterferometer
erheblich kostengünstiger als He-Ne-Laserinterferometer an. Dieses
hat außerdem den Vorteil, daß es ein trägerfrequentes System und
daher wesentlich weniger anfällig gegen Störungen ist. Der max.
erreichbare Meßweg liegt bei 1 m. Bild 2 zeigt das
Funktionsprinzip.

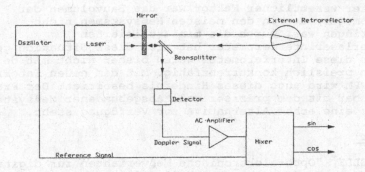

Bild 2: Funktionsprinzip des UPM Halbleiterlaser-Interferometers

Vergleich von Interferometern mit maßstabsgebundenen Meßsystemen

Von Vertretern der maßstabsgebundenen Meßtechnik wird gegen
Interferometer immer wieder das Argument der Empfindlichkeit gegen
athmosphärische Störungen ins Feld geführt. Wenn die Luftparameter
gemessen werden und der Laserstrahl gegen Luftturbulenzen
abgeschirmt wird, trifft dieses Argument nicht mehr zu.
Interferometer haben potentiell einen entscheidenden Vorteil bzgl.
Vermeidung des Abbeschen Fehlers. Der Meßstrahl läßt sich genau in
der Meßachse führen, was mit Maßstäben nicht möglich ist. Ein
weiterer Vorteil des Interferometers ist, daß sich das Inkrement
nicht durch mechanische Einflüsse, wie bei der Befestigung eines
Maßstabes verändern kann. Der Maßstab kann durch die mechanische
Befestigung nicht unerheblich deformiert werden.

Bei Halbleiterlaserinterferometern werden sehr kleine
Meßreflektoren verwendet, verwendet wird z.B. eine
retroreflektierende Kugel mit 2 mm Durchmesser. Diese kleine Masse
kann im Gegensatz zu den relativ schweren Abtastköpfen von
Maßstäben hohen Beschleunigungen ausgesetzt werden, ohne zu
Schwingungen angeregt zu werden.

Hieraus folgt, daß sich mit einem Interferometer mehr Fehler
vermeiden lassen, als mit einem maßstabsgebundenen System und
daraus letzlich potentiell auch eine höhere Meßgenauigkeit des
Interferometers folgt. In der Praxis wird dies auch dadurch
demonstriert, daß oft Laserinterferometer zur Kalibrierung solcher
maßstabsgebundener Systeme eingesetzt werden. Damit dies sinnvoll
ist, sollte solch ein Interferometer eine um eine Größenordnung
kleinere Meßunsicherheit besitzen, was in der Praxis dann wohl
auch gegeben sein muß. Warum man aber nun noch nicht die
Maßstabsysteme durch Interferometer ersetzt hat, wird im folgenden
erläutert.

Wirtschaftlichkeit von Interferometern und Gitter-Meßsystemen

Interferometer lagen bis vor kurzem gegenüber hochpräzisen
maßstabsgebundenen Meßsystemen im Preis wenigstens noch um einen
Faktor zwei höher. Dies verbietet schon oft den Ersatz von
Maßstäben.

Ein zweiter wesentlicher Faktor war das Bauvolumen der
Interferometer, das in den meisten Meßsystemen nicht
unterzubringen war. Durch die miniaturisierten
Halbleiterlaserinterferometer hat sich diese Situation geändert,
aber auch diese Interferometer waren bisher nicht mit den
Maßstäben preislich konkurrenzfähig. Mit dem neuen Interferometer
der Fa. UPM wird auch dieses Hindernis beseitigt. Der Preis ist
vergleichbar mit dem präziser maßstabsgebundener Meßsysteme, so
daß jetzt eine echte Alternative zur Verfügung steht.

Literatur:

/1/ G. NELLE, "Opto-elektronische Meßverfahren zur digitalen
    Längenmessung", Konstruktion 43 (1991) S. 401-410,
    Springer-Verlag

Hologramm-Maßstab zur absoluten Positionserfassung in der Ebene

U. Schilling*, P. Drabarek

Robert Bosch GmbH, Postfach 10 60 50, D - 70049 Stuttgart

*Universität Stuttgart, Institut für Technische Optik, Pfaffenwaldring 9, D - 70569 Stuttgart

1. Einleitung

Bei der Steuerung von modernen elektro-mechanischen Geräten und Maschinen besteht ein Bedarf, Bewegungen von Bauteilen quer zur Beobachtungsrichtung in der Ebene absolut zu vermessen. Interferometrische Meßverfahren sind aufgrund der steigenden Anforderungen an die Präzision dieser Geräte und der dabei geforderten Wegauflösung vorteilhaft.

Vorgestellt wird ein neuartiges interferometrisches Meßverfahren zur absoluten ein- und zweidimensionalen Positionserfassung in der Ebene, das auf der Rekonstruktion einer Welle durch ein Hologramm, den sogenannten Hologramm-Maßstab, beruht. Es besitzt die Positionsauflösung der bekannten inkrementalen interferometrischen Meßverfahren, zeichnet sich jedoch im Vergleich zu diesen durch einen deutlich erweiterten Meßbereich aus, innerhalb dessen die Position eindeutig bestimmbar ist.

Im folgenden wird das Prinzip dieses Verfahrens erläutert. Anschließend wird als Beispiel ein im Labor realisierter eindimensionaler Hologramm-Maßstab vorgestellt.

2. Der Hologramm-Maßstab zur absoluten Positionserfassung

Die auf der Holographie basierende Rekonstruktion von Wellenfronten läßt sich in der in Bild 1 dargestellten Weise für die hochauflösende optische Verschiebungsmeßtechnik einsetzen.

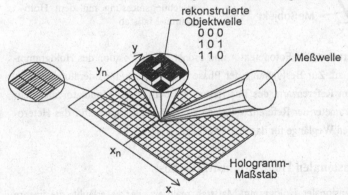

Bild 1: Modellierung einer Meßwelle durch Beleuchtung eines Hologramm-Maßstabs

Eine Meßwelle trifft im Gebiet um den Punkt (x_n, y_n) auf das als Maßstab dienende Hologramm auf. Dabei wird die Phasenfront der Meßwelle durch Beugung verändert und so, abgesehen von Beugungseffekten am Rand der Meßwelle, ein Teil der im Hologramm-Maßstab gespeicherten Objektwelle rekonstruiert. Bei geeigneter Gestaltung des Hologramm-Maßstabs ist eine eindeutige Zuordnung der Phasenverteilung der vom beleuchteten Gebiet rekonstruierten Objektwelle auf die Position eines die Meßwelle emittierenden Meßkopfs relativ zum Hologramm-Maßstab möglich.

Bild 1 zeigt ein Beispiel eines Hologramm-Maßstabs für die zweidimensionale Positionserfassung. Das Hologramm ist aus Elementarfeldern aufgebaut, die jeweils aus Gittern mit identischer Gitterkonstante, aber unterschiedlicher Anfangsphase (0 bzw. π, d.h. um eine halbe Gitterperiode versetzt) bestehen. Die Meßwelle beleuchtet gleichzeitig eine Matrix mehrerer Elementarfelder, so daß, wie in Bild 1 angedeutet, die rekonstruierte Objektwelle eine matrixförmige Phasenfront mit unterschiedlichen Phasenwerten (0, π) besitzt. Die absolute Position ist aus der geometrischen Anordnung dieser Phasenwerte bestimmbar. Die Positionsauflösung ist durch die Gitterkonstante begrenzt.

Eine mögliche Interferometeranordnung zur Verschiebungsmessung mit dem Hologramm-Maßstab ist in Bild 2 skizziert. Das auf dem Meßobjekt angebrachte Hologramm wird so von einer Meß- und einer Referenzwelle beleuchtet, daß die gebeugte Meßwelle (1. Beugungsordnung) und die ungebeugte Referenzwelle (0. Beugungsordnung) interferieren. Eine Fotodetektoranordnung, z.B. ein Fotodiodenarray oder ein CCD-Chip, nimmt das resultierende Interferogramm auf.

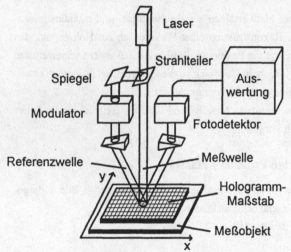

Bild 2: Meßanordnung zur Verschiebungsmessung mit dem Hologramm-Maßstab

Die Auswertung liefert die Phasenverteilung im Fotodetektor und ordnet ihr die Position des Hologramm-Maßstabs relativ zur Meßanordnung zu. Zur Bestimmung der Phase lassen sich in der Literatur beschriebene Verfahren einsetzen /1/. Der im Referenzarm des Interferometers befindliche Modulator dient zur Veränderung der entsprechenden Parameter der Referenzwelle wie z.B. der Lichtfrequenz für das Heterodyn-Verfahren /2,3/ oder der optischen Weglänge für das Phasenschiebeverfahren /4/.

3. Realisierung eines eindimensionalen Hologramm-Maßstabs

Als Labormuster wurde ein eindimensionaler Hologramm-Maßstab realisiert, der es erlaubt, die lineare Verschiebung des Maßstabs relativ zur Meßanordnung absolut und mit hoher Genauigkeit zu bestimmen. Sein grundsätzlicher Aufbau ist in Bild 3 dargestellt.

Der Maßstab besteht aus n nebeneinander angeordneten Spuren, die jeweils Gitterstrukturen enthalten. Die Spuren 1 und 2 besitzen Gitter mit unterschiedlichen Gitterkonstanten. Der mit ihnen erreichbare absolute Meßbereich ist durch die "Schwebungsfrequenz" der Gitter gegeben. Die Spuren 3 - n erweitern den Meßbereich durch eine Digitalkodierung.

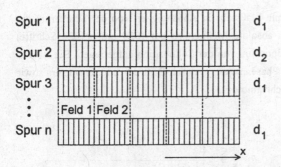

Bild 3: Aufbau des Hologramm-Maßstabs zur eindimensionalen Verschiebungsmessung

Die Funktion des Hologramm-Maßstabs beruht darauf, daß bei Verschiebung eines Gitters mit der Gitterkonstanten d_1 (Spur 1) in x-Richtung für die Lichtphase φ_1 in der 1. Beugungsordnung gilt /5/:

$$\varphi_1 = 2\pi \frac{x}{d_1} \qquad (1)$$

Dabei ist die Phase jedoch nur bis auf Vielfache von 2π eindeutig bestimmbar. Der eindeutige Meßbereich beträgt also eine Periode d_1.

Um den absoluten Meßbereich zu vergrößern, wird ein zweites Gitter mit der Gitterkonstanten $d_2 = d_1 + \Delta$ (Spur 2 in Bild 3) verwendet. Für die Differenzphase φ_{Diff} der an diesen Gittern gebeugten Wellen in Abhängigkeit von der Verschiebung x ergibt sich:

$$\varphi_{Diff} = \varphi_1 - \varphi_2 = 2\pi \frac{x}{D} \qquad \text{mit } D = \frac{d_1 \cdot d_2}{|d_2 - d_1|} \qquad (2)$$

Zur Auswertung wird zunächst gemäß Gleichung (2) die Position x grob bestimmt. Dabei ist der Eindeutigkeitsbereich durch die "Schwebungsfrequenz" D, die Positionsauflösung durch $D \cdot \Delta\varphi / 2\pi$ vorgegeben ($\Delta\varphi$.. Phasenmeßgenauigkeit). Um auf den Eindeutigkeitsbereich der Spur 1 rückschließen zu können, darf die maximale Positionsmeßunsicherheit höchstens d_1 betragen. In einem 2. Schritt wird nach Gleichung (1) die Feinposition ermittelt, so daß die resultierende Positionsauflösung durch das Gitter der Spur 1 mit der Gitterkonstanten d_1 gegeben ist.

Die Spuren 3 - n bestehen aus Gittern, die alle die Gitterkonstante d_1, jedoch in verschiedenen Feldern unterschiedliche Anfangsphase besitzen. Bei Verschiebung des Hologramms in x-Richtung erhält man für diese Spuren ebenfalls den Zusammenhang nach Gleichung (1). Zusätzlich treten an den Sprungstellen der Anfangsphase Sprünge in der Phase des von diesen Spuren gebeugten Lichts auf, so daß die Lichtphase entweder mit der der Spur 1 übereinstimmt oder aber um π verschoben ist. Dies ermöglicht die Einführung einer Digitalkodierung. Im Beispiel in Bild 3 besitzen in Feld 1 die Spuren 3 und n die Anfangsphase 0, beim Übergang zu Feld 2 ändert sich die Phase von Spur 3 auf π.

Zur Demonstration wurde ein Hologramm-Maßstab mit 6 Spuren als Amplituden-Transmissionshologramm realisiert. Die Gitterkonstanten d_1 bzw. d_2 betrugen 8 μm bzw. 8,53 μm, so daß sich D = 128 μm ergab; zusätzlich wurde durch 4 digitalkodierte Spuren der Meßbereich um den Faktor 8 auf 1024 μm erweitert. Um Meßunsicherheiten an den Sprungstellen der Gitter 3 - n auszuschließen, war dabei ein Faktor 2 als Redundanz berücksichtigt.

Um die Funktion des eindimensionalen Hologramm-Maßstabs nachzuweisen, wurde im Labor ein Interferometer ähnlich dem in Bild 2 aufgebaut. Ein Schrittmotor mit der Schrittweite 1 μm verschob den Hologramm-Maßstab in x-Richtung. Die Bestimmung der Phasenverteilung erfolgte mit dem Phasenschiebeverfahren.

216

Bild 5a zeigt die ermittelte Position des Hologramm-Maßstabs in Abhängigkeit von der Schrittzahl des Schrittmotors. Wie bereits oben erwähnt, war der absolute Meßbereich auf 1024 μm (=1024 Schritte) beschränkt. In Bild 5b ist die Abweichung der Meßposition von der Sollposition aufgetragen. Die Abweichungen ergaben sich aus Ungenauigkeiten bei der Phasenmessung, durch Drift des Interferometers sowie durch die Nichtlinearität und das Spindelspiel des Schrittmotors.

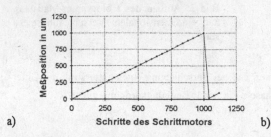
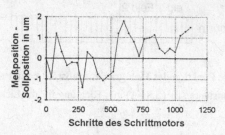

a)   b)

Bild 5: Position des Hologramm-Maßstabs in Abhängigkeit von der Schrittzahl des Schrittmotors. a) Meßposition, b) Abweichung der Meßposition von der Sollposition

4. Zusammenfassung

Mit Hilfe des hier vorgestellten Konzepts eines Hologramm-Maßstabs ist es möglich, die absolute Position eines Objekts in der Ebene zu erfassen. In dem am zu vermessenden Objekt angebrachten Hologramm-Maßstab ist eine im gesamten Meßbereich eindeutige Objektwelle gespeichert. Durch Beleuchtung eines Ausschnitts des Hologramms mit einer Meßwelle wird ein Teil der Objektwelle rekonstruiert. Die interferometrische Auswertung der Phasenfront dieser Welle erlaubt die hochgenaue absolute Bestimmung der Position des Meßkopfs relativ zum Maßstab.

Als Labormuster wurde ein eindimensionaler Hologramm-Maßstab mit 6 Spuren realisiert und in ein Interferometer eingebaut. Das so entstandene Meßsystem erlaubte die absolute Positionsbestimmung in einem Meßbereich von einem Millimeter. Die Auflösung war durch die Gitterkonstante von 8 μm bestimmt.

Durch die Erweiterung des Hologramm-Maßstabs um weitere Spuren läßt sich der Meßbereich beliebig vergrößern, ohne dabei die Positionsauflösung zu beeinträchtigen. Bei Verwendung von insgesamt 12 Spuren ergibt sich beispielsweise ein absoluter Meßbereich von ca. 65 mm.

Literaturverzeichnis

/1/   Tiziani, H.J.: Rechnerunterstützte Laser-Meßtechnik. Technisches Messen 54 (1987) 6: 221-230.

/2/   Dändliker, R.: Heterodyne Holographic Interferometry. In: Progress in Optics, Vol. XVII, ed. Wolf, E., North Holland Publ., Amsterdam 1980: 2-84.

/3/   Drabarek, P.; Pfendler, T.; Winter, K.: Heterodyn-Interferometer zur berührungslosen Weg- und Winkelmessung für industrielle Anwendungen. 9. Int. Kongreß Laser 89, München 1989 (Conf. Proc.: 225-228).

/4/   Creath, K.: Phase-Measurement Interferometry Techniques. In: Progress in Optics, Vol. XXVI, ed. Wolf, E., North Holland Publ., Amsterdam 1988: 351-393.

/5/   Makosch, G.; Schoenes, F.J.: Interferometric method for checking the mask alignment precision in the lithographic process. Applied Optics 23 (1984) 4: 628-632.

Online - Materialflußkontrolle in der automatischen Fertigung durch einen Laserscanner

H. Stettmer, WZL Laboratorium für Werkzeugmaschinen und
Betriebslehre der RWTH Aachen/D

Bei flexiblen Fertigungssystemen steigt die Wahrscheinlichkeit von Materialflußfehlern mit zunehmender Flexibilität. Da Materialflußfehler häufig zu Kollisionsschäden und zum Stillstand von Fertigungsanlagen führen, müssen Kollisionen erkannt werden, bevor sie auftreten.

Die Überprüfung der Werkstückträger alleine reicht nicht aus, da diese falsch bestückt oder Werkstücke während des Transportes verrutschen können. Deshalb wurde am Laboratorium für Werkzeugmaschinen ein Laserscanner entwickelt, der während der Abarbeitung einer Palette durch den Roboter die Bestückung der Palette überprüft (Bild 1). Liegt ein Bestückungsfehler vor, wird der Roboter vor dem Zugriff auf die Palette gestoppt.

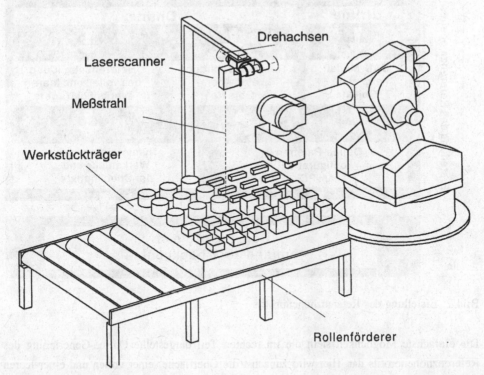

Bild 1: Prinzipieller Aufbau des Laserscanners

Der Lasersensor ist ca. 1,6 m über dem Werkstückträger montiert. Darum muß zur Erfassung des Höhenprofils ein Sensor mit einem Meßbereich bis zu 2 m eingesetzt werden.

Dazu stehen 3 verschiedene Meßprinzipien zur Verfügung:

- Pulslaufzeitmessung
- Triangulation
- Phasenmodulation

Aufgrund umfangreicher Versuche wird hier ein Sensor eingesetzt, der nach dem Prinzip der Phasenmodulation arbeitet. Dieser Sensor läßt im Vergleich mit den anderen Verfahren die Vermessung relativ schlecht reflektierender Objekte zu. Meßfehler lassen sich durch die Streubreite des Meßwertes eindeutig erkennen und auswerten.

Der Laserscanner vermißt jeweils das aktuelle Höhenprofil eines Werkstückträgers. Ein Bestückungsfehler liegt dann vor, wenn die gemessenen Werte von einem vor Arbeitsbeginn erstellten Referenzhöhenprofil abweichen. Zur Erstellung des Referenzhöhenprofils gibt es nach Bild 2 drei verschiedene Möglichkeiten.

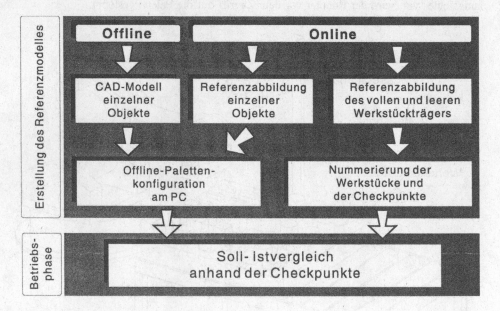

Bild 2: Erstellung des Referenzhöhenprofils

Die einfachste Möglichkeit stellt die im rechten Teil dargestellte Online-Generierung des Referenzhöhenprofils dar. Hier wird zunächst die Oberfläche einer vollen und einer leeren Musterpalette erfaßt und abgespeichert.

Der Nachteil dieses Verfahrens ist, daß zur Erstellung des Referenzhöhenprofils der Produktionsprozeß gestoppt werden muß.

Aus diesem Grunde wurde, wie im linken Teil des Bildes dargestellt, ein System zur Offline-Generierung des Referenzhöhenprofils entwickelt. Ausgehend vom 3D-Geometriemodell, das in der modernen Konstruktion auf CAD-Systemen ohnehin erstellt wird, werden die Höhenprofile für die einzelnen Werkstücke automatisch erstellt. Diese werden mit dem Palettenkonfigurationssystem zu dem Referenzprofil der vollständig bestückten Palette zusammengesetzt. Steht kein CAD-System zur Verfügung, so können die Höhenprofile einzelner Werkstücke durch eine Abbildung mit dem Scanner erzeugt und mit dem System zur Palettenkonfiguration zu einem Höhenprofil zusammengesetzt werden.

Alle spanend zu bearbeitetenden Werkstücke haben maximal 6 Freiheitsgrade. Darum läßt sich die Lage des Werkstückes anhand einiger ausgezeichneter Punkte, sogenannter Checkpunkte, überprüfen. Bild 3 zeigt beispielhaft einige Checkpunkte auf dem Höhenprofil des Werkstückes.

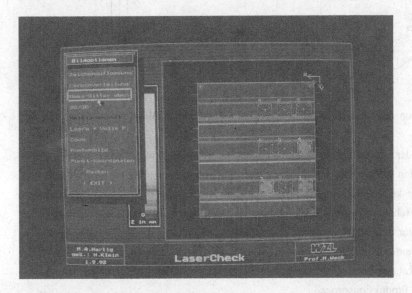

Bild 3: Werkstück mit Checkpunkten

Durch dieÜberprüfung der Werkstücke anhand der Checkpunkte läßt sich die Prüfzeit auf ca. 20 Sekunden reduzieren. Da die Zykluszeit in der spanenden Fertigung großer Teile ist jedoch meist mehr als 1 Minute beträgt. läßt sich die Überwachung der Werkstückposition somit parallel zum Produktionsprozeß durchführen. Der Laserscanner wurde nach dem im Bild 4 dargestellten Konzept in eine flexibele Fertigungsanlage integriert.

Bevor der Roboter auf die Palette zugreift, wird dem zur Steuerung des Laserscanners eingesetzten PC mitgeteilt, auf welches Werkstück der Roboter als nächstes zugreifen wird.

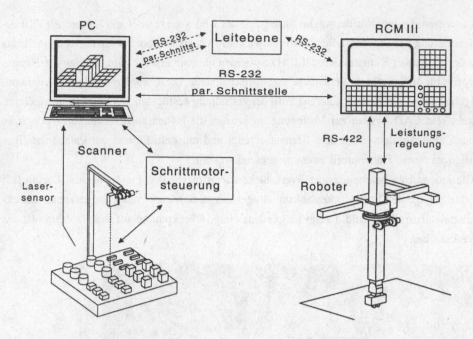

Bild 4: Integration des Laserscanners in eine flexible Fertigungsanlage

Daraufhin wird der Laserstrahl zu den entsprechenden Checkpunkten geführt. Die gemessenen Z-Koordinaten werden mit den Referenzkoordinatensystem abgelegten Koordinaten verglichen. Solange kein Unterschied auftritt, kann der Roboter sein Programm ungehindert fortsetzen. Unterscheiden sich beide Werte, liegt ein Bestückungsfehler vor. Der Roboter muß in diesem Fall vor dem Palettenzugriff gestoppt und auf ein anderes Werkstück umgelenkt werden. Dazu werden vom Laserscanner die Koordinaten eines anderen, gleichartigen Werkstückes an den Roboter übertragen. Der Roboter wird neu gestartet, greift das Werkstück und bringt es in den Produktionsprozeß.

Mit dem hier beschriebenen Laserscanner wurde ein einfach zu bedienendes System entwickelt, das zur Materialflußkontrolle in der automatisierten Fertigung geeignet ist. Da das System die Bestückung der Werkstückträger Hauptzeitparallel kontrolliert, führt die Überwachung der zu keiner Verlängerung der Produktionszeit. Die Kontrolle erfolgt auf der Basis des digitalisierten Höhenprofils, was aus dem CAD-System oder einem eigenen Scannvorgang resultiert. Darum kann der Laserscanner zur Kontrolle beliebig geformter Werkstücke eingesetzt werden.

Anwendung von Lasertriangulationssensoren in der automatisierten Meß- und Prüftechnik

J. Köhn, G. Kröhnert, G. Lensch
NUTECH GmbH, Ilsahl 5, 24536 Neumünster

Es wird der Entwicklungsvorgang eines Meßsystems zur Vermessung von gekrümmten Flugzeugsegmenten dargestellt.

Die NUTECH GmbH baut spezielle Problemlösungen für die Meßtechnik, basierend auf dem Prinzip der Lasertriangulation. In diesem Falle sollen nahezu beliebig geformte und gekrümmte Blechsegmente drei-dimensional erfaßt werden, um die Steuerdaten für Lackierroboter automatisch zu erzeugen. Die Lackierroboter bringen großflächig Maskenlack auf, der, partiell abgezogen, das Einätzen von Konturen ermöglicht. Derartige Bauteile besitzen hohe Festigkeit bei geringem Gewicht; die polierten Oberflächen weisen gute aerodynamischen Eigenschaften auf, ohne diese farblackieren zu müssen.

Das mögliche Bauteilspektrum umfaßt Bauteile mit einer Größe bis zu 3 m x 10 m Kantenlänge. Um ein ausreichend dichtes Meßraster zu erhalten, wurde ein System konzipiert, das aus 14 übereinander angeordneten Meßköpfen besteht mit jeweils 135 mm Abstand, also einer Gesamthöhe von etwa 3 m.

Die Anforderung aufgrund der Bauteile an die Meßköpfe waren:

Meßabstand: 1.250 mm

Meßhub: 1.500 mm

mittlere Auflösung: 3 mm

Das in einem Chargenrahmen befestigte Bauteil fährt in einem Trans-portsystem an der Meßsäule vorbei und erzeugt Taktimpulse, die je-

weils Entfernungsmessungen auslösen. Die Bauteil-Topografie ergibt
sich aus:

- den y-abhängigen Taktimpulsen
- den y-abhängigen Meßkopfpositionen
- den z-Koordinaten aus den Meßwerten

In Abhängigkeit von der Transportgeschwindigkeit ergibt sich eine
Zeit für die Generierung von 14 Meßwerten und deren Übertragung an
einen Leitrechner von max. 133 ms. Dies bedeutet simultanes Messen
der Meßköpfe nach Aufforderung, also ein typisches master-slave
System.

Die Voruntersuchungen der möglichen Bauteiloberflächen auf Glanz und
Rauhigkeit erbrachten Aussagen über die minimal notwendige Laserleistung, die maximal zulässige Krümmung der Oberflächen und die
Laserschutzklasse mit Bewertung der Meßbarkeit von möglicherweise
später hinzukommenden Bauteilen und die Grundlage zur Auswahl eines
geeigneten optischen Zeilensensors.

Die gezeigten extremen Unterschiede der Blechproben (matt bis hoch-
spiegelnd) erfordern eine außergewöhnlich hohe Dynamik des Sensores.

Die Auswahlkriterien der Sensoren wurden dargelegt: Lateraldioden
und CCD-Zeilen waren nicht geeignet, nur mit Diodenzeilen konnten
die Randbedingungen erfüllt werden.

Über die übliche Beschaltung und die Erläuterung der Ausgangssignale
von Diodenzeilen ergab sich die Antwort auf die Frage nach einer
optimalen Auswerteelektronik, die mit nur geringer Peripherie allen
Anforderungen gerecht wird.

High Precision Laser Triangulation by Speckle Decorrelation

Hendrik Rothe[1], Horst Truckenbrodt[2]
[1]: Institute of Technology, Friedrich–Schiller–University Jena, D–07743 Jena, Löbdergraben 32
[2]: Institute of Engineering Optics, Technical University Ilmenau, D–98693 Ilmenau, PSF 327

1. Introduction

Because of their versatility, adaptability and robustness laser triangulation sensors are widely used in industry /1/,/2/. Sources of measurement error in laser triangulation are fluctuations of intensity and coherence of laser light, surface roughness and damage as well as quantum, photonic and thermal noise of the laser source, CCD–sensors and signal processing electronics.

Therefore the ratio of measurement distance to precision in current triangulation sensors is limited to about 500 : 1, which is not enough for many industrial applications. Speckle noise caused by destruction of temporal and spatial coherence of the incoming laser beam is considered to be the main source of measurement error in laser triangulation /3/. This paper deals with a new approach of removing speckle noise from triangulation data based on multivariate statistics and digital signal processing.

2. Objectives

Regarding Figure 1 will ease understanding this paper. The x–axis shows the number of pixels N of

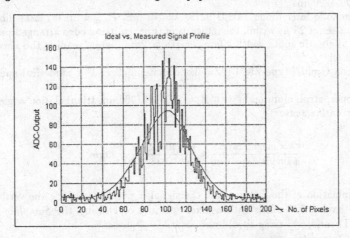

Figure 1: Laser triangulation signals: influence of speckles

a CCD–sensor, while the y–axis displays the sensor response after 8 bit analog–to–digital conversion. In case of an ideal measurement the signal is a smooth *Gaussian*. The real measurement shows speckle noise clearly indicated by gross changes in the signal between two pixels (sampling periods). Because in laser triangulation the maximum of the signal is used as indicator for range measurement speckles cause severe distortions. To circumvent the problems normally averaging over a number of measurements is used for signal reconstruction. Apparently, averaging can not deal correctly with anomalies in the measurement data like the corrupted signal in Figure 1, especially if interpolation (*subpixeling*) is used to determine precisely the maximum of the signal. Of course, with a large number of measurements one can find the population mean of the range. But this requires a large number of measurements. In practice, one needs a method determining an estimate for the range with a maximum of accuracy using a minimum of measurements. In the sense of statistics we have a large *data pool* comprising all possible observations of the "true" signal corrupted with several kinds of noise. Those samples consist of an information carrying part, e.g. a *Gaussian*, and – at least for ranging – meaningless parts, i.e. noise. The part of the signals carrying relevant information is the *same* within all observations of the process. Therefore the parts of the signal representing the ideal *Gaussian* should be significantly *correlated*. If this holds, we should look for an algorithm that is able to put significantly correlated parts of the signal in one data pool and noise in another one. With other words, the approach should handle measurement signals in such a manner, that the correlation matrix of transformed signals consists of ones only. This heuristic approach leads informally to the idea of *speckle decorrelation* by adapted signal processing procedures.

3. Speckle Decorrelation

Speckle noise can be removed by a procedure called *rectification*. It is based on multivariate *Principal Component Analysis* /4/,/5/,/6/. The underlying mathematical procedure and its application to laser triangulation can be found in /7/.

4. Experiments

Setup *Sensor*: Thomson TH 7863ACCA, 2/3", frame transfer type, pixel aperture $(23 \times 23)\mu m$, No. of pixels: $H \times V = 384 \times 288$; with a special clock regime $H \times V = 384 \times 576$ can be obtained

Sensor-Kit: Thomson TH 79KB63, Signal-to-Noise Ratio 50dB, analog output signal, digital signal 8 Bit

Image Processing Board: ITI–VFG 1400–768, pixel synchronous clock, pixel clock $0 \ldots 15$ MHz, 8 Bit analog input, 12 Bit digital input, gain/offset compensation

Integration Time: $20 \ldots 200$ms

Laser Source: monomode laser diode TOLD 9215, $\lambda = 670$ nm, $P_{max} = 3$mW, laser diode clock 20 kHz, i.e. 400 laser pulses of 25 μs within the integration time, emitting edge arranged parallel to the plane of incidence, asigmatic spot: width $40\mu m$, length ≈ 2mm, while reading the sensor the laser source was switched off

Lens: Carl Zeiss Jena GmbH, type ZKM ("Zweikoordinatenmeßgerät"), numerical aperture 0.033, magnification 1.0

Samples: brass, copper, steel, aluminium with $R_a = 200/600/900$nm, triangulation angles 5(5)75°

Average speckle size on the sensor:

$$\frac{\lambda \times \text{magnification}}{\text{numerical aperture}} \approx \frac{0.67\mu m \times 1.0}{0.033} \approx 20\mu m$$

Because the determination of the maximum of the signal is performed along the vertical columns of the CCD–sensor having 576 pixels with a pixel aperture of $11.5\mu m$, the *Sampling Theorem* is obviously fulfilled. Therefore *aliasing* should not occur.

Subpixeling method: fotometric mean

Removing Speckle Noise Now an example with practical dimensions of data shall be considered. We use a range cell of 60 pixels, 8 measurements and a resolution of analog–to–digital conversion of 8 bits.

The measured original data as well as the rectified original data are displayed in Figure 2. Speckle noise can be seen clearly. If averaging would be used, the detected range would have a relatively large variance.

The rectified data are computed using only the first Principal Component. Speckle noise is fully removed.

Figure 3 shows a Box & Whisker Plot of standard deviations of pixelwise averaged original data (SD_{ORIG}) vs. rectified data (SD_{RECT}) as necessary for ranging. The picture indicates the preference of Principal Component Filtering over averaging. The Box & Whisker plot clarifies that 25%–75% of all standard deviation values of rectified data are within a small area below approximately 10 units. Because SD_{ORIG} has a much larger variability there will be severe distortions of in ranging.

Results of Triangulation Ranging Figure 4 shows a plot of typical experimental results. Because

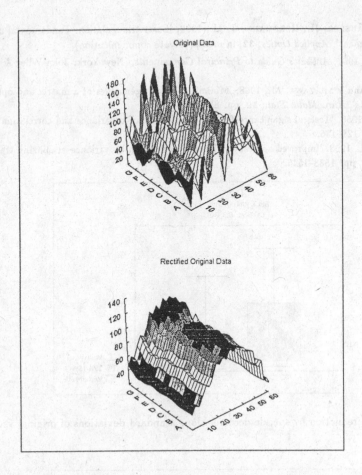

Figure 2: Surface plots: original vs. rectified data

the lens employed has a depth of focus of 5mm, the measurement range is restricted to ±2.5mm. While near zero (2.5mm absolute range) the error is relatively small, it increases at the peripheral regions. But even in case of the largest value of surface roughness (R_a=900nm) the absolute measurement error was always smaller than approximately 2...2.5 microns.

5. Conclusions

In laser triangulation the main source of measurement error are speckles. Usually, the stabilization of triangulation data is done by averaging over several measurements. It was shown that Principal Component Filtering can provide optimal signal reconstruction. In our experimental setup we obtained within a measurement range of 5mm an absolute precision of 2 microns, the ratio of measurement distance to precision was increased to 2500 : 1.

6. References

1. SEITZ, G., TIZIANI, II., LITSCIIEL, R., 1986, '3-D-Koordinatenmessung durch optische Triangulation', *Feinwerktechnik und Meßtechnik*, 94, pp. 423–425.
2. LINSE, V. and SCHMIDBERGER, E., 1991, 'Methoden der berührungslosen Prüfung in der Produktion von Leiterplattenbaugruppen', *Qualität und Zuverlässigkeit*, 36, pp. 270–276.

3. DORSCH, G., HÄUSLER, H., HERRMANN, J. M., 1993, 'Laser triangulation: Fundamental uncertainty of distance measurement', *Applied Optics*, **32**, in press, (private communication).

4. JACKSON, J. E., 1991, *A User's Guide to Principal Components*, (New York: John Wiley & Sons), pp. 80–104.

5. OKAMOTO, M. and KANAZAWA, M., 1968, 'Minimization of eigenvalues of a matrix and optimality of principal components', *Ann. Math. Stat.*, **39**, pp. 859–863.

6. LAWLEY, D.N., 1956, 'Tests of significance for the latent roots of covariance and correlation matrices', *Biometrika*, **43**, pp. 128–136.

7. ROTHE, H. et. al., 1992, 'Improved accuracy in laser triangulation by variance–stabilizing transformations.', *Opt. Eng.*, **31**, pp. 1538–1545.

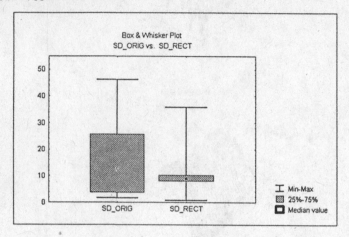

Figure 3: Variance reduction by speckle decorrelation: standard deviations of original vs. rectified data

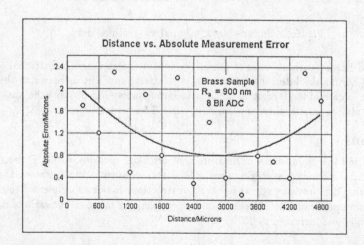

Figure 4: Experimental results: distance vs. absolute measurement error (triangulation angle: 45°)

Geometrische Modellierung und Parameteridentifikation bei Feldmeßverfahren zur optischen Formerfassung

W. Nadeborn, P. Andrä, W. Osten
Bremer Institut für Angewandte Strahltechnik (BIAS)
Klagenfurter Str. 2, D - 2800 Bremen 33

1. EINLEITUNG

Optische Meßverfahren, basierend auf strukturierter Beleuchtung mit Weißlicht oder Laser [HALIOUA / LIU] oder holografischem Contouring [THALMANN / DÄNDLIKER; RASTOGI / PFLUG], erlauben eine berührungslose und feldweise Erfassung der Form und Lage 3-dimensionaler Objekte. Der Meßprozeß untergliedert sich dabei in 2 Etappen:

1. Jedem Meßpunkt wird als primäre Meßgröße ein Phasen- bzw. Phasendifferenzwert zugeordnet.

2. Mit Hilfe eines Geometriemodells werden die primären Meßwerte in 3D-Koordinaten umgerechnet. Zuvor sind die Parameter des Geometriemodells durch Kalibrierung zu ermitteln.

Für den 1. Schritt stehen aus der holografischen Interferometrie stammende bewährte Standardverfahren, wie z. B. Phase-Shifting- oder Fouriertransformationsverfahren [OSTEN], zur Verfügung. Gegenstand dieser Untersuchung ist der 2. Schritt, bei dem in der Praxis oft mit groben Näherungen gearbeitet wird, von dem jedoch - wie gezeigt wird - die erreichbare Koordinatenmeßgenauigkeit in entscheidendem Maße abhängt. Aus Platzgründen konzentrieren sich die anschließenden Darlegungen auf das Streifenprojektionsverfahren, sie sind jedoch auch auf andere Feldmeßverfahren zur optischen 3D-Formerfassung übertragbar.

2. GEOMETRISCHE MODELLIERUNG

Das folgende Geometriemodell geht von dem Konzept der Verschneidung beobachtungsseitiger Hauptstrahlen und projektionssseitig generierter Konturflächen (Flächen konstanter Phase bzw. Phasendifferenz) aus, s. Abb. 1. Es ist vektororientiert und somit nicht abhängig von der Wahl eines speziellen Koordinatensystems.

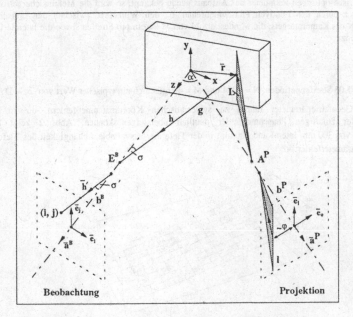

Abb. 1: Geometrische Verhältnisse beim Streifenprojektionsverfahren

Die Eintritts- und Austrittsrichtungen $\overline{h}$ und $\overline{h}'$ des Beobachtungshauptstrahls g sind über die Beziehung

$$\overline{h}' = (\kappa^B - 1)(\overline{a}^B \cdot \overline{h})\overline{a}^B + \overline{h} \qquad\qquad \kappa^B = \tan\sigma / \tan\sigma' = \text{const} \qquad\qquad (1)$$

verknüpft. Daraus läßt sich für g die Geradengleichung

$$\overline{r} = \overline{r}_{E^a} + \lambda\left[\overline{c}_0 + \overline{c}_i(i - i_0) + \overline{c}_j(j - j_0)\right] \qquad\qquad (2)$$

ableiten, wobei die Vektorkoeffizienten $\overline{c}$ Funktionen der Systemkonstanten $\{\overline{a}^B, \overline{e}_i, \overline{e}_j, \kappa^B, b^B, ...\}$ (vgl. Abb. 1) darstellen sowie i und j Pixelindizes und λ einen freien Abstandsparameter bezeichnen.

Jeder Projektionsrasterlinie l ist eindeutig ein Phasenwert $(\varphi - \varphi_0)$ zugeordnet, so daß die Gleichung der korrespondierenden Lichtschnittebene L analog zu (2) als

$$\overline{r} = \overline{r}_{A'} + \nu\left[\overline{d}_0 + \overline{d}_\varphi(\varphi - \varphi_0) + \overline{d}_l\mu\right] \qquad\qquad (3)$$

geschrieben werden kann. Die Vektorkoeffizienten $\overline{d}$ sind Funktionen der Systemkonstanten $\{\overline{a}^P, \overline{e}_\varphi, \overline{e}_l, \kappa^P, b^P, ...\}$. Löst man (2) und (3) als Gleichungssystem bezüglich der Variablen $\{\lambda, \mu, \nu\}$, so erhält man ein Gesamtgeometriemodell der Form

$$\overline{r} = \overline{r}_{E^a} + \left[\overline{c}_0 + \overline{c}_i(i - i_0) + \overline{c}_j(j - j_0)\right] \frac{\tilde{k}_0 + \tilde{k}_\varphi(\varphi - \varphi_0)}{k_0 + k_i(i - i_0) + k_j(j - j_0) + \left[k_\varphi + k_{i\varphi}(i - i_0) + k_{j\varphi}(j - j_0)\right](\varphi - \varphi_0)}, \qquad (4)$$

welches die Objektraumkoordinaten $\overline{r}$ über eine Reihe von konstanten Systemparametern $\overline{c}$ und k, die den Meßaufbau charakterisieren, mit den primären Meßgrößen i, j und φ verbindet.

3. VERFAHRENSEIGENSCHAFTEN

Das Modell (4) erlaubt neben der Berechnung von Koordinaten aus Phasenwerten auch die Ableitung von Verfahrenseigenschaften:

- Sind alle geometrisch-optischen Parameter des Aufbaus genau bekannt, so wird die Meßunsicherheit δz bezüglich der Tiefenkoordinate z durch den relativen Phasenmeßfehler ε, den Winkel α zwischen den optischen Achsen, die Pixelanzahl N x N des Kameratargets, die Mindestanzahl von M Pixeln pro Streifen sowie die laterale Dimension D des Meßobjekts begrenzt:

$$\delta z = \frac{D M \varepsilon}{N \tan\alpha} . \qquad\qquad (5)$$

Für $\alpha \approx 30°$, $\varepsilon \approx 0.05$ Streifenperioden, N = 1024 und M = 4 ergibt sich ein typischer Wert von $\delta z \approx D / 3000$.

- Fehlerbehaftete Geometrieparameter führen zu systematischen Koordinatenmeßfehlern, die in der Praxis die Auswirkungen der zufälligen Phasenmeßfehler deutlich übersteigen können. Abb. 2 zeigt beispielhaft für Meßdimensionen von 300 mm lateral und 100 mm in der Tiefe die ortsvariable Abhängigkeit des Tiefenmeßfehlers δz vom Geometrieparameterfehler $\delta\alpha$

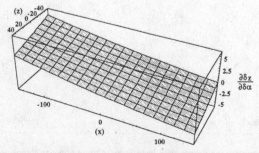

Abb. 2 : Sensitivität des Tiefenkoordinatenmeßfehlers δz gegenüber dem Geometrieparameterfehler $\delta\alpha$ (in mm / °)

- Eine vereinfachte Koordinatenberechnung, die auf der Annahme

$$(z - z_{ref}) \sim (\varphi - \varphi_{ref}) \tag{6}$$

basiert, ist bestenfalls für flache Objekte akzeptabel. Bei größerem Abstand von der Referenzebene weisen die Konturflächen $\varphi - \varphi_{ref} = \text{const}$ nicht zu vernachlässigende ortsvariable Neigungen und Abstände untereinander auf (s. Abb. 3).

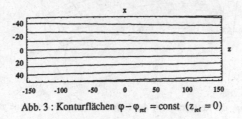

Abb. 3 : Konturflächen $\varphi - \varphi_{ref} = \text{const}$ $(z_{ref} = 0)$

4. KALIBRIERUNG

Anstelle der in der Praxis schwierigen direkten Vermessung aller relevanten Parameter des Meßaufbaus ist eine Identifikation der in das Modell (4) eingehenden vektoriellen und skalaren Koeffizienten $\overline{c}$ und k möglich. Dies kann in einem 2-Schritt-Prozeß unter Benutzung einer präzise gefertigten Ebene als Kalibrierkörper erfolgen:

Zunächst wird mit Hilfe eines linearen Kamerakalibrierverfahrens [TSAI] die äußere Orientierung der Kamera bezüglich zweier unterschiedlicher Stellungen der Kalibrierebene (s. Abb. 4) bestimmt.

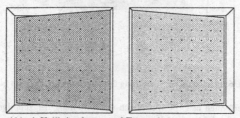

Abb. 4: Kalibrierebenen und Testpunkte

Daraus lassen sich die Geometriekonstanten $\overline{r}_g$, $\overline{c}_0$, $\overline{c}_i$ und $\overline{c}_j$ aus dem Kameramodell (2) sowie, darauf aufbauend, für eine gewisse Anzahl von Testpunkten $\overline{r}(i, j)$ der Abstandsparameter

$$\lambda(i, j) = \frac{\overline{r}(i, j) - \overline{r}_g^B}{\overline{c}_0 + \overline{c}_i (i - i_0) + \overline{c}_j (j - j_0)} \tag{7}$$

ermitteln.

Berücksichtigt man, daß im Gesamtmodell (4) die skalaren Koeffizienten $\tilde{k}_0, \tilde{k}_\varphi, k_0, k_i, k_j, k_\varphi, k_{i\varphi}$ und $k_{j\varphi}$ offenbar nur bis auf einen gemeinsamen Faktor bestimmt werden müssen und sich der Phasenoffset φ_0 - vorausgesetzt, er ist für alle Meßpunkte gleich - in diese Koeffizienten integrieren läßt, so kann man nun in einem zweiten Schritt die folgende, ebenfalls lineare Identifikationsaufgabe

$$\lambda(i_n, j_n)\{k_0' + k_i'(i_n - i_0) + k_j'(j_n - j_0) + [k_\varphi' + k_{i\varphi}'(i_n - i_0) + k_{j\varphi}'(j_n - j_0)] \varphi_n\} - \tilde{k}_\varphi' \varphi_n = 1 , \tag{8}$$

$$(n = 1, ..., N; \quad N \geq 7),$$

bezüglich der entsprechend modifizierten Parameter $\tilde{k}_\varphi', k_0', k_i', k_j', k_\varphi', k_{i\varphi}'$ und $k_{j\varphi}'$ lösen. Hierbei durchläuft der Index n die Menge der ausgewählten Testpunkte $(N \approx 200)$.

Da die Phaseneingangswerte φ_n unvermeidlich fehlerbehaftet sind (ebenso die zuvor kalibrierten Kameraparameter), können auch die Projektionskoeffizienten $\tilde{k}_\varphi', k_0', k_i', k_j', k_\varphi', k_{i\varphi}'$ und $k_{j\varphi}'$ nur mit einer begrenzten Genauigkeit identifiziert werden.

Abb. 5 veranschaulicht für relative Phasenmeßfehler von 0.05 Streifenperioden die Verteilung des dadurch bedingten systematischen Restfehlers bzgl. der Tiefenkoordinate z.

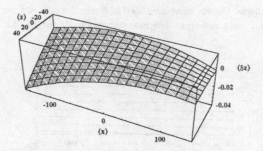

Abb. 5: Verteilung der Restverzerrung der Tiefenkoordinate in der Ebene y = 0 (in mm)

Abb. 6 zeigt beispielhaft die verbleibende systematische Verzerrung der Objektebene $z = 0$.

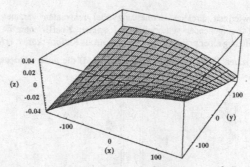

Abb. 6: systematische Verzerrung der Objektebene z = 0 (in mm)

Wenngleich die erhaltenen Fehlerformen hauptsächlich durch die Verteilung der zufälligen Eingangsfehler bedingt sind, bleibt doch ihre Größenordnung in der dargestellten Höhe reproduzierbar.

Somit ermöglicht das entwickelte modellgestützte Kalibrierverfahren die Identifikation der maßgeblichen Geometrieparameter des Meßaufbaus mit einer Genauigkeit, die ausreichend ist, um die verbleibenden *systematischen* Restverzerrungen der ermittelten Koordinaten unter das durch die eingehenden statistischen Phasenmeßfehler gegebene Niveau (vgl. Abschätzung (5)) zu senken.

5. LITERATUR

M. Halioua and H. C. Liu, "Optical three-dimensional sensing by phase measuring profilometry, " Optics & Lasers in Eng., Vol. 11, pp. 185 - 215, 1989

R. Thalmann and R. Dändliker, "Holographic Contouring using electronic phase measurement, " Opt. Eng., Vol. 24, pp. 930 - 935, Nov. / Dec. 1985

P. K. Rastogi and L. Pflug, "Real-time holographic phase organization technique to obtain customized contouring of diffuse surfaces, " Appl. Opt., Vol. 30, pp. 1603 - 1610, May 1991

W. Osten, "Digitale Verarbeitung und Auswertung von Interferenzbildern", pp. 39 - 75, Akademie-Verlag Berlin, 1991

R. Y. Tsai, "A versatile camera calibration technique for high-accuracy 3-D machine vision metrology using off-the-shelf TV cameras and lenses, " IEEE J. Robot. Automat. RA-3(4), pp. 323 - 344, 1987

Novel Technique of Distance Measurement

Tilen Zorec, Božo Vukas

Iskra Elektrooptika d.d., Ljubljana, Rep. of Slovenia

We developed an improved method of distance measurement by laser rangefinder - the method of sampling. Its favourable characteristics are the increased reliability of ranging (reduced false alarm rate), increased number of ranged targets, lower consumption and costs, and exclusive use of standard and easily obtainable components.

INTRODUCTION

A typical counter circuit of the laser rangefinder has until recently been composed of one or more counters, of a start/stop logic that distributes control signals to the counters, and of an output circuit for communication with the surroundings (display control, data transmission etc.). The number of targets ranged by such a circuit depends on the number of counters. The counter logic is regular and repeatable: the basic module is repeated for every decade; after that the entire module chain is repeated for each individual target.

In rangefinders of this type in Iskra Elektrooptika series production the number of simultaneously ranged targets is 1 to 5. To avoid engagement of undesirable near targets (camouflage, bush and the like), the operator sets on the rangefinder the minimum ranging distance which prevents measurements under limits of this distance. A proper selection of the blocked range takes care that the interesting target is always the first or the second one beyond the blocked range.

A counter circuit in the first laser rangefinders consisted of several tens of SSI and MSI integrated circuits. Later on, the thin film hybrid technology enabled miniaturisation of instruments and thereby numerous new applications. Some applications, however, were still not possible due to high consumption and the relatively high cost of hybrids. In the last decade, Iskra Elektrooptika has in conjunction with the Faculty of Electrotechnics of University of Ljubljana developed two monolithic integrated circuits (MV-1 and LUMP), each of which combines the entire counter logic of a typical rangefinder and is capable to range two targets simultaneously.

Both circuits are made in the HCMOS technology where consumption is favourably low (about 30 mA). Each is encapsulated in a standard 40-pin dual-in-line package or built in a hybrid circuit to enable ranging a larger number of targets. Both the LUMP (in a simple manner) and the MV-1 (a little more complicated) allow cascade connections.

Novel Technique: Sampling

To enhance rangefinder reliability and to increase the number of ranged targets, to simplify the process of ranging and to lower the overall costs, we developed a new method of distance measurement: ranging by sampling.

This method employs a ranging circuit that consists of standard components only. It is based on writing the receiver output pulses within a defined ranging interval into a special memory. This memory content is processed by the microprocessor that at the same time carries out also all other control functions in the rangefinder. To build as cheap and as miniature a ranging circuit as possible, composed of the possibly smallest number of (easily obtainable) components, we employ the 8-bit Intel, Philips or Signetics 80C51 family microprocessors. As memory component the 9-bit memory is used, arranged in FIFO (first-in-first-out) structure of either 256, 512, 1K, 2K, 4K, 8K or 16K bit length.

The significant feature of FIFO memories is that their pin layout is the same regardless of memory capacity, which means that identical circuits can be designed for rangefinders of different ranges. They are available in 28-pin dual-in-line packages of 300 or 600 mils width and can be connected directly to the microprocessor data bus.

In miniature rangefinders faster and more expensive memories are employed. Here external components are not used (precision oscillator only). The receiver signal is led directly into one memory bit and other bits neglected.

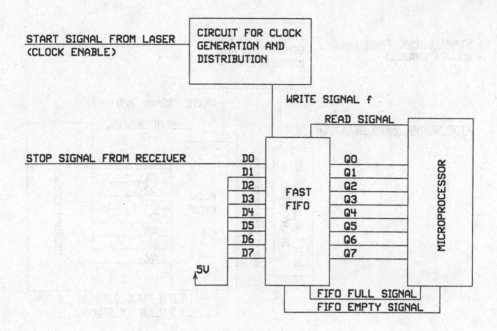

Ranging circuit with the fast FIFO memory

To cut down costs, larger ranging circuits with slower FIFOs (about 100 ns write cycle) should be selected. Unfortunately, for ranging with an accuracy of a few metres they are too slow. But the accuracy can be improved by means of series-parallel conversion obtained through the shift register or delay line. An 8-bit register writes data into memory on every 8 shifts or delays. At 100 ns FIFO write cycle time the resolution is 100/8 = 12.5 ns. The resolution is limited by the register or delay line speed as well.

Cycle time of the faster state-of-the-art FIFO memories is in the 20 ns class. In combination with a shift register or delay line they would provide 2.5 ns rangefinder accuracy, which means under half a metre. For the time being we are satisfied with the 2.5 metres ranging accuracy, so we apply the slower FIFOs in our present sampling method.

Although the FIFOs contain more memory cells (a 512, 1024, ... x 9 structure has 4608, 9216 ... memory cells) than our LUMPs and MV-1s, lower consumption of the ranging circuit can be achieved with FIFO memories than with these two counter circuits. Consumption of a typical FIFO with a nominal active power supply current is 100 mA, and

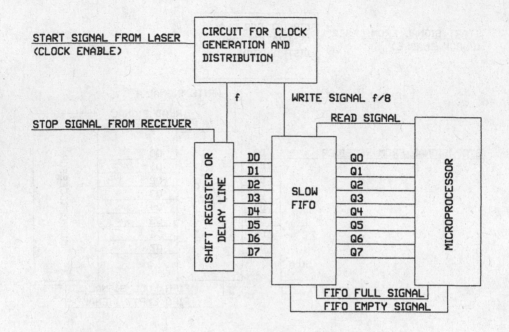

Ranging circuit with the slow FIFO memory

with a nominal stand-by current 12 mA. For a comparison of consumption in LUMP/MV-1 counter circuits and in FIFO memories let us take the example of a conventional laser rangefinder with 2 Hz repetition: ·

FIFO memory:

$Q = U*I*t;$ $\qquad$ $U = 5V;$

Stand-by current: $\quad I1 = 12$ mA; $t1 = 0.4998$ s

Power supply current: $I2 = 100$ mA; $t2 = 200$ μs $+ 5$ ms

$Q = Q1 + Q2 = U*U1*t1 + U*I2*t2 = 29.988 + 2.6 = 32.588$ mJ

LUMP/MV-1:

$Q = U*I*t;$ $\quad U = 5V;$ $\quad I = 30$ mA; $\quad t = 0.5$ s

$Q = 75$ mJ

Consumptions would be equal only in a [(75 - 29.988)/2.6]*2 = 34.6 Hz laser rangefinder. In rangefinders of higher repetition the LUMP/MV-1 counter circuits still consume less. With FIFO memories with a power down mode (power down current from 0.9 to 4 mA) the 34.6 Hz limit can be raised to 55.9 Hz. As the development in the field of FIFO memories is quite rapid, more favourable results in consumption can be expected shortly.

The sampling method of ranging eliminates the problem of detecting only the first few targets within the laser beam since in sampling all targets detected in one measurement are automatically written into memory. E.g., at a range of 10 km and resolution of 40 m (dictated by the receiver) the number of detectable targets is 250. Range blocking is in sampling method defined in the system software. The microprocessor selects per program the detected targets to be indicated on display or to be transmitted to other units.

In the laser rangefinder, information on a target (stop pulse from receiver) lasts normally for 100 ns. We sample with resolution of 2.5 m (16.66 ns), the stop pulse comprises 6 samples. Therefore an interference can hardly be misinterpreted as a target.

Sampling contributes to measurement reliability: the momentary interference duration is well distinguished from a stop pulse.

Ein nichtzählendes Diodenlaser-Interferometer zur Untersuchung von Oberflächentopographien

A. Abou–Zeid, P. Wiese
Physikalisch–Technische Bundesanstalt, 5.101
Bundesallee 100, 38116 Braunschweig

Unter Nutzung der Möglichkeit der Durchstimmbarkeit der Wellenlänge eines Diodenlasers wurde ein neuartiges Interferenzverfahren zur Untersuchung von Oberflächentopographien auf der Basis eines nichtzählenden Interferometers entwickelt. Durch die Modulation des Diodenstroms und die Verwendung von Lock–In–Techniken kann die Wellenlänge so nachgeführt werden, daß sich das Interferometer ständig in einem Interferenzmaximum befindet. Der Verlauf der Steuerspannung zum Nachführen der Wellenlänge entspricht dann dem Verlauf der Oberflächentopographie. Vergleichsmessungen zwischen dem nichtzählenden Diodenlaser–Interferometer und einem hochauflösenden (7,5 nm) zählenden Interferometer bzw. einem Phase–Shift–Interferenzmikroskop zeigen eine gute Übereinstimmung.

1 Meßprinzip

Das Meßprinzip des nichtzählenden Interferometers wird anhand eines Michelson-Interferometers mit ungleichlangen Armen erläutert (siehe Bild 1). Die Phasenlage Φ des Interferenzsignals am Ausgang des Interferometers hängt von der Wellenlänge λ und vom Wegunterschied s zwischen den beiden Interferometerarmen ab.

$$\Phi = \frac{s}{\lambda/2} \qquad (1)$$

Eine Verschiebung des Meßspiegels ändert die Armlängendifferenz s. In der nichtzählenden Betriebsart wird eine Änderung von s durch Nachführung der Wellenlänge kompensiert, so daß die Phasenlage konstant bleibt. Die Meßgröße Δs wird in eine proportionale Wellenlängenänderung $\Delta \lambda$ umgesetzt. Wird eine temperaturstabilisierte Laserdiode als Lichtquelle eingesetzt, so können die Wellenlängenänderungen durch Diodenstromänderungen realisiert und gemessen werden. Es gilt:

$$\Delta s = \left\{ \frac{s_0}{\lambda_0/2} \cdot \frac{1}{2} \cdot d\lambda/di \right\} \cdot \Delta i \qquad (2)$$

Die zu messende Verschiebung Δs ist proportional zur gemessenen Stromänderung Δi. Der Laserdiodenstrom läßt sich nicht beliebig weit durchstimmen, dadurch wird der Meßbereich begrenzt.

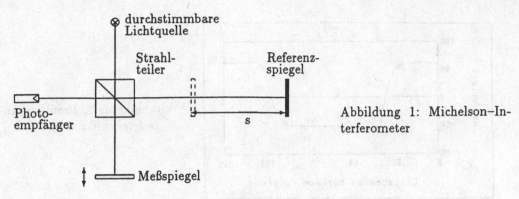

Abbildung 1: Michelson–Interferometer

Die Auflösung des Oberflächensensors wird bestimmt durch die Auflösung bei der Messung von Δi. Eine Verbesserung der Auflösung wird durch die Temperaturabhängigkeit der Wellenlänge der Laserdiode begrenzt [1]. Bei der zur Zeit erreichten Temperaturstabilisierung auf $+/-$ 0,45 mK ist eine Auflösung von 1/4000 des Meßbereiches sinnvoll.

Die Tabelle 1 gibt eine Übersicht über mögliche Kombinationen von Auflösung und Meßbereich, wobei die folgenden Koeffizienten eingesetzt wurden: $d\lambda/di$: 0,007 nm/mA, λ_0: 780 nm, Durchstimmbereich für den Diodenstrom: 10 mA.

Armlängendifferenz s_0	Meßbereich	Auflösung
10 mm	1 μm	0,25 nm
100 mm	10 μm	2,5 nm
200 mm	20 μm	5,0 nm

Tabelle 1: Verschiedene Kombinationen von Auflösung und Meßbereich

Die Festlegung des Interferometers auf ein Interferenzmaximums geschieht mit Hilfe von Lock–In–Techniken. Der Laserdiodenstrom wird sinusförmig moduliert. Befindet sich das Interferometer im Interferenzmaximum, so treten im Photoempfängersignal nur gerade Vielfache der Modulationsfrequenz auf, während die ungeraden Frequenzkomponenten verschwinden. Dies wird mit einem Lock–In–Verstärkers detektiert. In einem geschlossenen Regelkreis wird der Diodenstrom und damit die Laserwellenlänge solange nachgeregelt, bis das Ausgangssignal des Lock–In–Verstärkers zu Null wird. Die Ausgangsgröße des Gesamtsystems ist der Laserdiodenstrom.

2 Meßergebnisse

2.1 Oberflächenmessungen

Zur Untersuchung von Oberflächentopographien wurde ein Michelson–Interferometer (siehe Bild 1) aufgebaut, wobei die zu untersuchende Oberfläche den Meßspiegel des Interferometers darstellt. Mit einem x–y–Verschiebetisches wird die Oberfläche abgerastert.

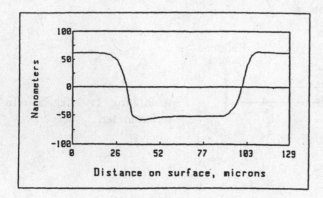

Abbildung 2: Profillinie Interferenzmikroskop

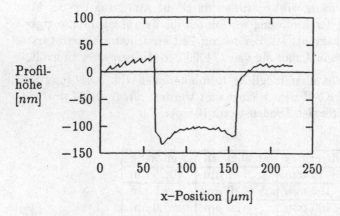

Abbildung 3: Profillinie Diodenlaser–Interferometer

Die Profilhöhe der Oberfläche bestimmt dabei die optische Weglänge im Meßarm des Interferometers.

Als Testobjekt für die Oberflächenmessungen wurde ein Bruchstück eines Silicium–Wafers verwendet. Die Topographie dieses Waferstücks wurde als Referenz mit einem Phase-Shift–Interferenzmikroskop untersucht [2]. In die sehr glatte, ebene Waferoberfläche sind rechteckige Vertiefungen eingeätzt. Die Bilder 2 und 3 zeigen zwei Profilschnitte. Vergleicht man die Profilschnitte miteinander, so zeigt die mit dem Interferenzmikroskop aufgenommene Linie einen glatteren Verlauf. Der Meßwert für die Tiefe des Rechtecks, etwa 120 nm, ist bei beiden Systemen gleich. Damit liefert das Diodenlaser–Interferometer schon recht brauchbare Ergebnisse.

Die Unterschiede in den Meßergebnissen werden durch bislang nicht ausreichend abgeschirmte äußere Einflüsse wie Führungsfehler vom x–y–Scantisch, thermische und mechanische Drift des Aufbaus und Änderungen im Brechungsindex der Luft verursacht.

2.2 Vergleichsmessungen mit einem hochauflösendem zählenden Interferometer

Um die Eigenschaften des nichtzählenden Diodenlaser–Interferometers unabhängig von äußeren Störeinflüßen zu untersuchen, wurde eine Spiegelverschiebung gleichzeitig mit einem hochauflösenden zählenden Interferometer und mit dem Diodenlaser–Interferometer gemessen.

Die Differenz zur Ausgleichsgeraden beider Systeme (Bild 4) zeigt einen parabelförmigen Verlauf. Die Abweichungen von der Nullinie betragen bis zu 0,15 μm und sind im Vergleich zur angestrebten Auflösung von 7,5 nm sehr groß. Bei aufeinanderfolgenden Messungen ergibt sich immer wieder ein parabelförmiger Verlauf, unabhängig von der Bewegungsrichtung des Tripelspiegels. Dieses Verhalten deutet auf eine systematische Abweichung hin, deren Ursache wahrscheinlich in einem nichtlinearen Zusammenhang zwischen der Wellenlänge und dem Diodenstrom zu suchen ist.

Es wurde versucht, mit einem einfachen Ansatz die Nichtlinearität zu eliminieren. Dazu wurde durch die ursprünglichen Meßwerte keine Gerade, sondern eine Parabel gelegt. Die Differenz zu dieser Ausgleichsparabel ist in Bild 5 dargestellt. Die Abweichungen von der Nullinie liegen nun in der Größenordnung der angestrebten Auflösung.

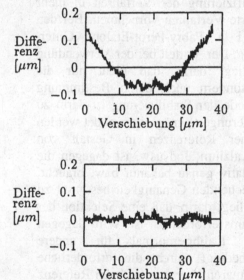

Abbildung 4: Vergleichsmessung mit Heterodyn–Interferometer, Differenz zur Ausgleichsgeraden

Abbildung 5: Vergleichsmessung mit Heterodyn–Interferometer, Differenz zur Ausgleichsparabel

Diese Arbeit wurde von der Deutschen Forschungs Gemeinschaft (DFG) gefördert.

Literatur

[1] *Abou–Zeid, A.:* Diodenlaser in der industriellen Meßtechnik. Technische Mitteilungen, 85 (1992), S. 34–43

[2] *Prettyjohns, Keith N.* und *Wyant, James C.:* Three–dimensional surface metrology using a computer–controlled non–contact instrument. SPIE Vol. 599, Optics in Engineering Measurement (1985), S. 304–308

Absolute Frequenzstabilität eines auf Rb stabilisierten Diodenlasers

A. Abou-Zeid, N. Bader, G. Prellinger
PTB Labor 5.101
Bundesallee 100, D-38116 Braunschweig

Die Verwendung von Diodenlasern in der klassischen interferometrischen Meßtechnik erfordert sowohl die Stabilisierung der Wellenlänge als auch deren Bestimmung. Bezüglich der Stabilisierung läßt sich dabei anhand der verwendeten Referenzen eine grobe Klassifizierung der Verfahren in mehr technisch und in mehr physikalisch orientierte Verfahren vornehmen. Bei den technisch orientierten Verfahren kommen i. a. Fabry-Perot-Etalons, Gitter, Endmaße und Strichmaßstäbe zur Anwendung. Der Vorteil bei der Verwendung eines Strichmaßstabes bzw. Endmaßes liegt darin, daß nicht nur die Stabilisierung der Wellenlänge an Luft, sondern auch deren Bestimmung möglich ist /1/. Ein Vorteil bei dieser Methode der Stabilisierung ist darin zu sehen, daß jeder Diodenlaser, der den Anforderungen genügt, verwendet werden kann. Bei der Verwendung physikalischer Referenzen in Gestalt von Absorptionslinien des Cäsium, Rubidium, Kalzium, Jod usw. ist dagegen die Vakuumwellenlänge bzw. Frequenz i. a. relativ genau bekannt bzw. braucht, falls erforderlich, nur einmal mit der entsprechenden Genauigkeit bestimmt zu werden. Ein Nachteil bei diesem Verfahren liegt darin, daß eine Selektion der Laserdiode bezüglich ihres Emissionsspektrums erforderlich ist. Wird dagegen eine Laserdiode als Referenz bzw. Kalibriernormal für andere Laserdiodensysteme verwendet, so ist, bedingt durch die erforderliche Langzeitstabilität, der Verwendung einer atomaren bzw. molekularen Referenz trotz des zuvor erwähnten Nachteiles der Vorzug zu geben.

Die Doppler-verbreiterten Absorptionslinien des Rubidium

Das Rubidium ist ein Alkalimetall und besteht in seiner natürlichen Isotopenverteilung zu 72,2 % aus dem Isotop ^{85}Rb und zu 27,8 % aus dem Isotop ^{87}Rb . Es weist in den Wellenlängenbereichen von 780.24 und 794.76 nm ausgeprägte Absorptionslinien auf, die sich hervorragend zur Stabilisierung von Diodenlasern eignen. In der Abb.1 ist ein Wellenlängenscan durch Variation des Stromes durch die Laserdiode über die Absorptionslinien im Bereich von 780,24 nm dargestellt. Die beiden inneren Peaks sind dem Isotop ^{85}Rb und die beiden äußeren dem Isotop ^{87}Rb zuzuordnen.

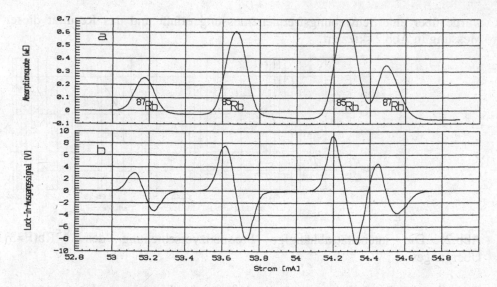

Abb. 1: Die Doppler-verbreiterten Absorptionslinien (Teilbild a) des Rubidiums natürlicher Isotopenverteilung sowie der Verlauf der ersten Ableitung dieser vier Absorptionspeaks (Teilbild b)

Diese vier Absorptionslinien in dem Teilbild 1a bestehen ihrerseits aus mehreren Linien (Tripletts), die, bedingt durch den Effekt der Dopplerverbreiterung, zu jeweils einem Absorptionspeak verschmieren. In dieser Form sind die Absorptionslinien auf Grund der nahezu symmetrischen Verhältnisse (keine Richtungserkennung der Regelabweichung möglich) im Bereich der Absorptionsmaxima als Bezugsgröße für eine Frequenzstabilisierung nicht brauchbar. Ein vorzeichenbehaftetes Signal bezüglich der Regelabweichung liefert die erste Ableitung bzw. jede Ableitung höherer ungeradzahliger Ordnung. In Teilbild 1b ist der Verlauf der bei dieser Stabilisierung verwendeten ersten Ableitung dargestellt.

Die intensitätsabhängige Frequenzverschiebung des Rb-D2 Überganges

Bei hohen äußeren elektrischen Feldern kann eine Verschiebung der atomaren Energieniveaus und somit eine damit gekoppelte Verschiebung der spektralen Lage der Absorptionslinien auftreten. Dieses physikalische Phänomen ist als Stark-Effekt bekannt. Hohe äußere elektrische Felder, in diesem konkreten Fall handelt es sich dabei um Wechselfelder, können durch eine ausreichend intensive Bestrahlung des Mediums mit kohärentem Licht erzeugt werden. Für ein Kalibriernormal ist es unumgänglich, die für eine bestimmte zu erreichende absolute Genauigkeit maximal zulässige Strahlungsflußdichte zu kennen. Dazu wurde die Intensität durch eine Absorptionszelle in einem Bereich von 400 nW/cm^2 bis ca. 6,2 mW/cm^2 variiert. Die aufgetretene Frequenzverschiebung wurde über eine Schwebungsfrequenzmessung erfaßt und das Resultat dieser Messung in Abb.2 skizziert.

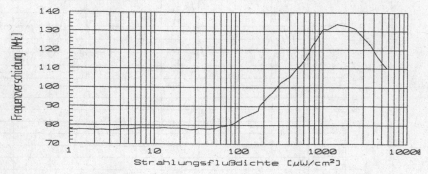

Abb.2: Die intensitätsabhängige Frequenzverschiebung des ^{85}Rb(F=3) Überganges

Das relevante Ergebnis besteht darin, daß, für den betrachteten Übergang, eine Strahlungsflußdichte von ca. 50 µW/cm^2 nicht überschritten werden darf, um eine Frequenzverschiebung zu vermeiden.

Die erreichten Stabilitätswerte eines Rubidium stabilisierten Diodenlasers

Um einen Eindruck von der Verbesserung gegenüber einer Parameter-stabilisierten Laserdiode zu vermitteln, wurden Schwebungsfrequenzmessungen durchgeführt. In der Abb. 3 sind zwei derartige Meßkurven dargestellt. Die Kurve a repräsentiert eine freilaufende, d. h. Parameter-stabilisierte Laserdiode, während die Kurve b das Verhalten eines Rubidium-stabilisierten Diodenlasers widerspiegelt. Beide Messungen wurden, abgesehen von der Art und Weise der Stabilisierung, unter gleichen Bedingungen gewonnen.

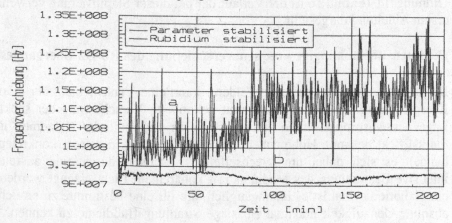

Abb.3: Stabilitätsverhalten einer Parameter-stabilisierten (Kurve a) und einer Rubidium-stabilisierten Laserdiode (Kurve b)

Festzustellen sind im wesentlichen zwei Verbesserungen. Die eine besteht darin, daß die Breite des "Rauschbalkens" , also vergleichsweise kurzfristige Frequenzfluktuationen, deutlich reduziert wurden. Zum anderen wurde das für die gedachte Anwendung entscheidende Kriterium der Langzeitstabilität erheblich verbessert.

Aus der Kurve b wurde auf numerischem Wege die Allen-Varianz /2/ , ein im Bereich der Frequenzstandards allgemein anerkanntes Vergleichskriterium, ermittelt. Abb. 4 enthält das Ergebnis dieser Kalkulation in Gestalt der Wurzel aus der Allen-Varianz. Bei kleinen Zeitkonstanten liegt die Wurzel aus der Allen-Varianz bei ca. $2{,}5 \cdot 10^{-10}$ und steigt zu größeren Zeitkonstanten auf einen Wert von ca. $1{,}3 \cdot 10^{-9}$ an.

Abb.4: Wurzel aus der Allen-Varianz aufgetragen über der Intgrationskonstanten

Eine worst-case Betrachtung in Gestalt des klassischen 3σ Wertes liefert für die relative Stabilität einen Wert von ca. $\pm 6 \cdot 10^{-9}$. Insgesamt kann anhand dieser Betrachtungen die Aussage gemacht werden, daß eine Stabilisierung der Frequenz eines Diodenlasers auf eine Doppler-verbreiterte Absorptionslinie des Rubidiums die gewünschte Verbesserung bezüglich der Langzeitstabilität erbringt.

/1/ Abou-Zeid A., Erdtmann B., Kunzmann H.,Prellinger G.; PTB-Mitt. 101,333
    1991

/2/ Allen David W. Proceedings of the IEEE Vol. 54, No 2 Februar 1966

Diese Arbeit wurde von der Kommission der Europäischen Gemeinschaft (BCR) gefördert.

Laser Radar Parameters for 3D-Measurements

Myllylä, R.[1], Kostamovaara, J.[2], Moring, I[1]., Määttä, K.[2]
1) VTT Optoelectronics Laboratory, P.O.Box 202, SF-90571 Oulu, Finland
2) University of Oulu, Electronics Laboratory, P.O.Box 191, SF-90101 Oulu, Finland

ABSTRACT

A laser radar is capable of creating a range image of a scene by scanning the measurement directions. Significant in the designing of laser devices for various 3D-measurement applications is the choice of parameters such as pulse power, pulse repetition frequency, responsiveness of the receiver, and the number of single shots in a mean value. Differences in these parameters result in significant deviations in range, speed and precision. Other important parameters include the shape and reflectivity of the object, software limitations and calibration methods.

1.   INTRODUCTION

Authors have been faced with the problem of developing range imaging devices based on the laser radar technology for many years /1/. These devices are characterized by a large measurement range, up to several tens of meters. Laser radar technology operates much like conventional radio-frequency radars, measuring the elapsed time between a laser transmission and the reflected return of the pulse. With sophisticated design an accuracy of one mm has been achieved. In addition to the range measurement, two orientation angles, the pan and tilt angles, are measured.

Laser radars have had their first applications in severe environments like in space, and in the military battlefield in smart weapon systems and unmanned autonomous vehicles. They have potential for night vision use as well. With the development of semiconductor lasers, laser radar has found wide applications in the industrial field such as in profiling, positioning, docking, monitoring, tracking, as well as in robotics.

With action distance ranges on the order of tens of meters and accuracies on the order of millimeters, the testing and calibration of the sensor can be problematic. The factors to be tested are mainly precision and accuracy which depend on the range, and to some extent, the field of view, depth of field, and measurement time. Other characteristics of interest when comparing sensors are of course cost, calibration requirements, physical size, environmental requirements, object reflectivity and shape dependence etc.

The next chapter concentrates on the parameters of the laser radar itself, thereafter the influence of the object has been evaluated, and finally the testing and calibration.

2.   LASER RADAR PERFORMANCE CHARACTERISATION

2.1   Optics and signal power

The basic laser radar consists of a transmitter and a receiver as shown in Figure 1. The source of light is usually a laser diode. The source is placed in or near the focal plane of the transmitter objective so that a properly focused beam is projected to the target. The reflected beam is collected by the receiver objective focused at or near the focal plane of the receiver objective, so that part or all of the beam illuminates the detector surface. The received signal power can be estimated using conventional radiometry. Neither coherent and diffraction effects nor aberration phenomena are normally needed to be taken into consideration. This simplifies the calculation without substantial loss of accuracy.

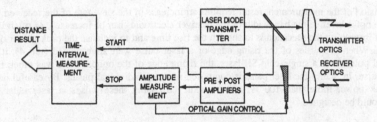

Figure 1. The block diagram of a laser radar.

The signal power (S) estimation is much more sophisticated than the well-known radar equation which for the Lambertian target power distribution is defined by ref. 2.

$$S = \frac{P \tau \rho A}{\pi R^2} ,$$
(1)

where R = radar-target range, P = transmitted pulse power, ρ = the back scattering coefficient of the target, A = aperture area, and τ = optical efficiency. This equation holds when the distance R is large and all power entering the aperture area A, of the receiver, is collected by the sensor. At shorter ranges this is not necessarily the case. In reference 2, a mathematical model and program has been designed using a sophisticated method for approximating the performance of the laser radar system, especially in the determination of the optimum configuration of the optics so that the signal power function covers the specific measurement range.

2.2   Resolution and accuracy

The basic idea in the pulsed time-of-flight laser rangefinding technique is to use intense, impulselike laser pulses in order to minimize the timing uncertainty in the detection process. This enables one to minimize the measurement time needed for the specified resolution. In some cases even a single shot measurement is sufficient.

The timing jitter is proportional to the ratio of the noise of the signal and its slew rate. By some approximations (ref. 3) using leading edge detection, the range jitter is correspondingly

$$\sigma_R = \frac{0{,}35 \cdot c}{2 \cdot B \cdot SNR \cdot \sqrt{n}} ,$$
(2)

where c is the speed of light, B the bandwith of the receiver channel, SNR the peak-to rms signal-to-noise ratio of the timing signal and n the number of averaged measurements.

By using low cost standard commercially available components, a receiver bandwidth of about 100 MHz can easily be realized. With this bandwidth and a signal-to-noise ratio above 10, which is typically needed for reliable detection, single shot resolution is at cm-level. For non-cooperative targets which are common in industrial applications a SNR of 10 can be realized with a laser peak power level of about 10-30 W within a measurement range of 20-30 meters (ref. 3) with the diameter of the receiver optics on the order of centimeters.

As can be seen from Eg. 2, the final resolution of the measurement can be improved by averaging successive measurement results, the improvement being proportional to the square root of the number of averaged results. The measurement rate in some specific applications depends thus on the measurement resolution specified and on the pulsing rate of the laser diode transmitter. The duty factor of the high power semiconductor laser diodes with power level of 10-30 W is typically 0.01-0.1%. A pulse length of 10 ns gives a maximum pulsing rate of 10-100 kHz. This means that the measurement time to reach a measurement resolution of 1 mm, for example, is 10-1 ms, respectively. Laser diodes with lower power levels have higher duty factors and the measurement time is correspondingly decreased. However, at the same time the range is also shortened.

Improvement of the precision of the measurement result requires an increase in the slew-rate of the received light pulse or a decrease in the noise level. If the bandwidth of the receiver electronics can be increased, as often is the case, one possible way of improving the precision is to shorten the rise time and to increase the peak power of the transmitted optical pulse. When properties of the rising edge of a laser pulse were investigated (ref. 4), it was noticed that in the optical pulse from a single-chip SH-laser, the rising edge of the optical pulse was much faster than that of the current pulse. In reality the laser pulse is composed of many individual pulses. By careful design 100 ps pulses, with peak power more than 100 W, can be achieved. Using these pulses, a laser radar with submillimeter accuracy could be designed.

In a single measurement the accuracy is defined by the statistics and the combined effect of the systematic errors. These errors are produced by drifts of the various electronic subblocks of the laser radar such as in the receiver channel and in the time interval measurement. Another important error source is the change in the transversal dimension of overlap between the transmitter and receiver beams as the measurement distance is changed. This results in a corresponding change of the pulse shape of the receiver channel due to the inhomogeniety of the laser beam and the mode dispersion if the receiver diode is pigtailed (ref. 3). In general, the measurement spot size affects the spatial accuracy because the measurement result depends on a weighted average of the reflections from the spot area. The optical stop pulse is formed as a sum of the received pulses over the whole area of the transmitter beam weighted by the beam irradiance and reflectivity distributions (ref. 5). The surface profile of the target alters the arrival time of these pulses at the detector by different amounts, modulating the shape of the received optical pulse collected over the whole transmitter beam area. This results in variations in the stop pulse which leads to errors in the distance measured. The error increases as the beam becomes larger.

The optimum solution from an accuracy point of view is obviously a system with a common receiver channel for both the start and stop pulses to minimize the temperature drift, and with coaxial optics with autofocusing capability so that the measurement spot is well defined and small at all distances.

3. THE EFFECT OF OBJECT SHAPE ON THE ACCURACY

A number of measurements of objects of different kinds were made using the prototype for a commercial laser radar device /5/. The transmitter and receiver spot sizes are 30 mm. The effect of a tilted smooth surface with a homogenous reflectivity was measured. The error remains within ±1 mm when the angle of tilt varies between ± 60°. With larger angles leading/trailing edge timing is better than leading edge timing (constant fraction principle). This seems reasonable because when tilting a homogeneous object, the timing pulses become wider symmetrically on both the leading and trailing sides and, thus, the timing moment remains the same. The error can be notably larger if the reflectivity of the tilted surface varies locally.

Figure 2 shows measurement of the insides and outsides of the 90° corners. The sides of the corners have a constant reflection coefficient. The maximum smoothing error is about one fifth of the spot size. As can be seen in the figure, there is a small increase in the error just before and after the tip of the corner, probably caused by a retroreflection from the tilted surface to the other side and back, because this phenomenon is not present in measurement from the outside corner. The error naturally increases when the angles get smaller.

A large error in the distance measurement can result if a part of the measurement beam goes past the target and a part of the signal received is obtained from another surface behind it, if the distance between the surfaces is smaller than half the pulse length. The error can be tens of centimetres.

How well a hole in a surface can be detected depends on the width of the hole with respect to the spot size and the reflectivity properties of its front and bottom surfaces. Figure 3 shows the measured depth of a 195 mm deep from a 20 mm wide hole with a different bottom and front surface reflectivity. It can be said that the depth of a hole can be measured reliably only if its diameter is larger than the measurement spot.

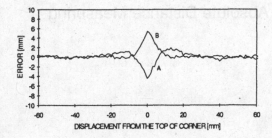

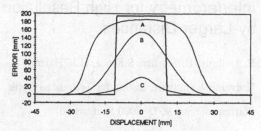

Figure 2. Distance error when measuring an inner corner (A) and outer corner (B).

Figure 3. Measured profiles of a hole. The ratio of the reflection coefficients of the bottom and the front surface is 12:1 (A), 1:1 (B) and 1:12 (C).

4. CALIBRATION AND TESTING

The performance of range imaging devices is determined by many factors, but no common practice exists for testing and evaluating them. Evaluation and comparison of the characteristics and performance of various sensors is important. For the sensor developer, they need to know the state of the art, for manufacturer, they need to be sure of the quality of their products, and for the potential end-user, they need to make correct decisions between alternatives. Some evaluation methods have been described in reference 6.

The performance characterization is based on a reference object with various shapes and reflectivity profiles. A comprehensive 3-D reference point set with good accuracy is also needed. They allow the accuracy and repeatibility of 3D measurement to be tested in the different parts of the 3-D viewing volume of the sensor. The accuracy and repeatibility of distance measurement can be controlled using a straight, accurate reference line. After a reference object has been measured, the results are compared with its known 3-D coordinates or shape, with the deviation being taken to represent the accuracy of the 3-D point measurement and the deviation between successive measurements the precision.

5. DISCUSSION

Recently, industrial applications of laser radar have been realized. Millimeter level accuracy has opened up a lot of new applications. The devices are mainly designed for small series production leading to a high price. The dominating price factors are the opto-mechanics and the complicated testing. The technology has been developed so that industrial mass-production could be possible. Integration of electronics and opto-mechanics will be needed. Promotion of the unification, and if possible the standardization of the testing and evaluation procedures for these types of measurement systems, should be stressed in the future.

6. REFERENCES

1. I. Kaisto, J. Kostamovaara, I. Moring and R. Myllylä, "Laser rangefinding techniques in the sensing of 3-D objects". Proc. SPIE, Vol. 1260, 122-133 (1990).
2. J. Wang, K. Määttä and J. Kostamovaara, "Signal power estimation in short range laser radars". Proc. ICALEO '91, 16-26 (1991).
3. J. Kostamovaara, K. Määttä, M. Koskinen and R. Myllylä, "Pulsed time-of-flight laser rangefinding techniques for industrial applications". Proc SPIE, Vol. 1633, 114-127 (1992).
4. A. Kilpelä, S. Vainshtein, J. Kostamovaara and R. Myllylä, "Subnanosecond high power laser pulses for time-of-flight laser distance meters". Proc. SPIE, Vol. 1821, 365-374 (1992).
5. K. Määttä and J. Kostamovaara, "The effect of the measurement spot size on the accuracy of laser radar devices in industrial metrology". Proc SPIE, Vol. 1821, 332-342 (1992).
6. J. Paakkari and I. Moring, "Method for evaluating the performance of range imaging devices". Proc. SPIE, Vol. 1821, 350-356 (1992).

Interferometry for High Resolution Absolute Distance Measuring by Larger Distances

E. Dalhoff, E. Fischer, S.Kreuz, H.J. Tiziani

Universität Stuttgart, Institut für Technische Optik

Pfaffenwaldring 9, D- 7000 Stuttgart 80

1. Introduction

There are some techniques of absolute remote distance measurement. Some of it deserve the title of absolutness without restriction (triangulation e.g.) but are not useful measuring at distances of some ten metres because of practical reasons. The other techniques have an unambiguity range within the measurement is absolute. Among the incoherent techniques the phase measurement of modulated light seem to have reached some limitation with an resolution of less than 1 mm. This limitation is set by the bandwidth of the photodetector. The coherent techniques operate with two wavelengths to extend the unambiguity range of the classical interferometry. The unambiguity range is then half the synthetic wavelength and can be adjusted quite arbitrarily from some ten micrometers on. If heterodyne detection is involved a relative resolution of the electronic phase measurement of 10^{-4} is achievable. As is valid for the phase measurement of intensity modulated light cascading of 2 or more stages of different synthetic wavelength is necessary in order to improve the relative resolution of the whole system.

The main advantage of the coherent technique is that a resolution below 1 micron is possible. In addition the signal-to-noise ratio (SNR) makes the effort of coherent detection attractive especially working at large distance. In Fig.1 the SNR of the detector signal is depicted for the coherent system discussed in this paper and the incoherent phase measuring technique depending on the power of the light backreflected by the object.

In the following a set-up of a double heterodyne interferometer with an unambiguity range of 100 metres and a resolution of 0.1 mm (corresponding to a relative resolution of 10^{-6}) will be described and experimental results will be presented.

2. Set-Up

The set-up is designed to yield a resolution of 0.1 mm with an unambiguity range of 100 metres. A sketch of it is shown in Fig.2. Light of a monomode laser diode with wavelength λ_1 is frequency-shifted by a 500 MHz acousto-optic modulator (AOM) in order to yield the second wavelength λ_2. This corresponds to a synthetic wavelength of 60 cm. Light of the wavelengths λ_1, λ_2 is used as reference

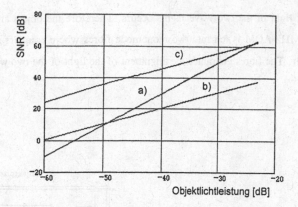

Fig.1 SNR dependent on object light power for a) incoherent phase measuring, b) coherent heterodyne detection on one wavelength (BW 15 MHz), c) double heterodyne detection (BW 15 MHz).

light, while the object light is shifted by means of two additional AOM with f1=80 resp. f2=80.1 MHz in order to provide double heterodyne reception /1/. The object and reference light gets superimposed on a photodetector. Dropping the dc-component the signal after demodulation is given by

$$(1) \qquad i_{det} = |i_{det}| \cdot \cos\left(2\pi \cdot \left(\Delta f \cdot t + \tfrac{2z}{\Lambda}\right)\right)$$

where $\Delta f = |f_1 - f_2|$ and $\Lambda = (\lambda_1 \cdot \lambda_2)/|\lambda_1 - \lambda_2|$. Z is the distance to be measured. Thus $\Delta f = 100$ kHz is the heterodyne frequency of the demodulated signal and the unambiguity range of the measurement is given half the synthetic wavelength $\Lambda/2 = 30$cm. The distance is evaluated from a measurement of the phase of the heterodyne signals of a measuring and a control interferometer with fixed paths. The extended unambiguity range of 100 metres is obtained by computing the results of two measurements with slightly different synthetical wavelength $\Lambda = 58.9$ cm corresponding to a frequency shift of the AOM of 501.5 MHz. In order to maintain the phase resolution of $2\pi/3600$ after more than 300 cycles of the unambiguity range, the relative stability of the synthetic wavelength has to be better than 10^{-6}. Rewriting eq.1 to $\Lambda = c/\Delta v$(where Δv denotes the frequency difference between the two wavelengths) it can be seen that in this set-up the stability of the phase is depending only on the frequency stability of the AOM and not on the frequency stability of the laser diode. This is an advantage of this approach since a relative stability of an electronic oscillator of better than10^{-6} is state of the art.

In addition to the unavoidable noise in the detection process the decorrelation of the two speckle patterns created by the object light when measuring at rough surfaces causes a phase error /2/. The standard deviation of this statistical phase error depends on the wavelength difference, the roughness of the target and its tilt. Measuring at a perpendicular oriented surface with a mean roughness of 10 μm it is calculated to be below 0.1 mm for the set-up in question. This phase error increases when a misalignment between the object light of the two wavelengths occurs. Therefore the object and reference light after the 80 resp. 80.1 MHz AOM is fed into two monomode fibres where couplers split the light for the control interferometer. The fibres guarantee the alignment of the light of the two wavelengths.

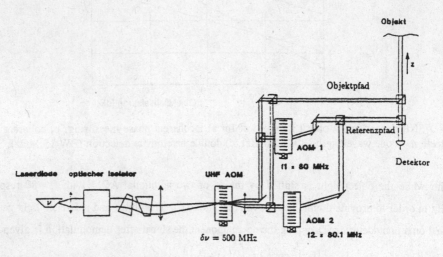

Fig.2: Double heterodyne interferometer set-up.

3. Results

Fig. 3 shows a 30 s stability measurement at a rough surface at 20 m distance. The resolution is 0.08 mm. By now systematic errors reduce the accuracy to about 1 mm. Fig. 4 shows a measurement at 100 metres distance at a retroreflector with an rms-phase-deviation of 0.2 mm. It shows, that a stable phase measurement is obtained at a distance the optical path length of which is about 40 times the coherence length of the laserdiode used.

Literatur:

/1/   Z.Sodnik, E.Fischer, Th.Ittner u. H.J.Tiziani: Two-wavelength double heterodyne interferometry using a matched grating technique.

Applied Optics, Vol.30 No.22, 1991

/2/   Vry,U. et.al.: Higher order statistical properties of speckle fileds and their application to rough surface interferometry

Journal of the Optical Society of America, Vol .3 No.7, 1986

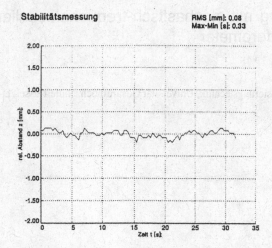

Fig.3: Stability measurement at a rough target at 5 metres distance. $\Delta z_{rms}=0.08$ mm

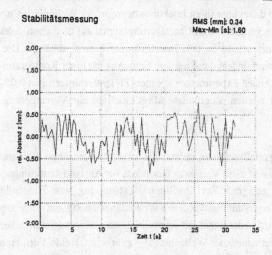

Fig.4: Stability measurement at a cornercube at 100 metres distance. $\Delta z_{rms}=0.34$ mm.

Entfernungsmessung mit stochastisch-frequenzmoduliertem Halbleiterlaser-Interferometer

J. C. Braasch, W. Holzapfel

Institut für Meß- und Automatisierungstechnik, Universität Gesamthochschule Kassel, D-34109 Kassel

1 Zusammenfassung

Ein neuartiges Entfernungsmeßverfahren auf der Basis eines Halbleiterlaser-Interferometers wird vorgestellt. Die Laserdiode wird dabei über ihren Injektionsstrom in der Wellenlänge stochastisch moduliert. Dadurch entsteht ein zeitlich veränderliches Überlagerungssignal auf dem Photodetektor, dessen mittlere Frequenz vom Laufzeitunterschied in den Interferometerarmen abhängt. Die Zielentfernung (Laufzeitunterschied) wird bestimmt, indem diese mittlere Frequenz mit einer Komparator-Frequenzzähler-Kombination gemessen wird. In ersten Experimenten konnten Zielentfernungen von 0,05–1,5 m mit relativen Meßfehlern < 0,5 % erfaßt werden. Begründete Möglichkeiten zur Verringerung des Meßfehlers werden angegeben.

2 Einleitung

Kohärente optische Entfernungsmeßverfahren nutzen die kurze Wellenlänge der Strahlungsquelle als Referenz der Messung. Dadurch sind diese Verfahren besonders für Abstandsbestimmungen im nahen und mittleren Entfernungsbereich geeignet. Zur absoluten Bestimmung von Abständen oberhalb der Wellenlänge sind mehrere Messungen mit verschiedenen Wellenlängen nötig, um eindeutige Meßergebnisse zu erhalten. Zu diesem Zweck wird in der Regel mit deterministisch wellenlängenmodulierten Lasern[1] oder mit mehren Strahlungsquellen verschiedener Wellenlänge[2,3] gearbeitet. Beide Verfahren erfordern einen relativ hohen apparativen Aufwand zur Frequenzmessung bzw. zur Wellenlängenstabilisierung der verwendeten Laser. Das hier vorgestellte Verfahren benötigt wenig Hardware, erlaubt die Kompensation vieler zufälliger Störungen und hat eine weitgehend lineare Kennlinie sowie hohe Auflösung.

3 Prinzip

Die optische Anordnung des rauschmodulierten Laserdioden-Interferometers besteht im einfachsten Fall (Bild 1) aus einem Zweistrahlinterferometer vom Typ Michelson. Die Strahlungsquelle ist eine Monomode-Laserdiode, deren Injektionsstrom durch einen Rauschgenerator mit nachgeschaltetem Tiefpaßfilter stochastisch moduliert wird. Die Amplituden der Strommodulation sind normalverteilt mit einem Effektivwert von nur wenigen Prozent des Schwellenstroms der Laserdiode. In Bereichen ohne Modensprung ist die Frequenzmodulation $\delta\Omega$ (und die Lichtleistungsmodulation) der Laserdiode der Strommodulation δi in guter Näherung proportional.

Die modulierte Strahlung wird am Strahlteiler in eine Referenz- (Index r) und eine Zielwelle (Index z) aufgespalten, die nach der Reflexion am jeweiligen Spiegel auf dem Photodetektor elektrooptisch gemischt werden. Von der Komparator-Frequenzzähler-Kombination wird die mittlere Frequenz des Photodetektorssignals während der Torzeit T bestimmt. Diese mittlere Frequenz ist abhängig von der Zielentfernung z und wird zur Entfernungsmessung genutzt.

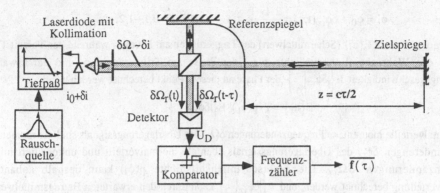

Bild 1: Prinzip des rauschmodulierten Laserentfernungsmessers

4 Theorie

Vernachlässigt man die geringe Leitungsmodulation der Laserdiode, so ist das normierte AC-Detektorsignal U_D des Zweistrahlinterferometers mit stochastisch-frequenzmodulierter Laserquelle:

$$U_D = \cos\left[\Omega_0 \cdot \tau + \int_0^t \{\delta\Omega_r(t') - \delta\Omega_z(t')\}\, dt'\right] = \cos\left[\Omega_0 \cdot \tau + \int_0^t \{\delta\Omega_r(t') - \delta\Omega_r(t'-\tau)\}\, dt'\right] \qquad (1)$$

τ = Laufzeitdifferenz, Ω_0 = mittlere Laserfrequenz, $\delta\Omega$ = momentane Frequenzänderung

Die Varianz der momentanen Frequenz $\delta\Omega_{rz}$ (Überlagerungsfrequenz) des Detektorsignals U_D ist

$$\overline{\delta\Omega_{rz}^2} = \overline{\left(\delta\Omega_r(t) - \delta\Omega_r(t-\tau)\right)^2}. \qquad (2)$$

Mit der Autokorrelationsfunktion $AKF_{\delta\Omega}(\tau)$ des Frequenzmodulationsprozesses

$$AKF_{\delta\Omega}(\tau) = \overline{\delta\Omega_r(t) \cdot \delta\Omega_r(t-\tau)} \qquad (3)$$

folgt bei stationärem Modulationsprozeß für die Varianz der Überlagerungsfrequenz

$$\overline{\delta\Omega_{rz}^2} = 2 \cdot AKF_{\delta\Omega}(0) - 2 \cdot AKF_{\delta\Omega}(\tau). \qquad (4)$$

Für eine normalverteilte Frequenzmodulation $\delta\Omega_r$ ist aufgrund des linearen Zusammenhangs auch die momentane Frequenz $\delta\Omega_{rz}$ des überlagerten Signals normalverteilt. Zwischen dem Betragsmittelwert und dem Effektivwert der Überlagerungsfrequenz besteht deshalb ein fester Zusammenhang.

$$\overline{|\delta\Omega_{rz}|} = \sqrt{\frac{2}{\pi}} \sqrt{\overline{\delta\Omega_{rz}^2}}. \qquad (5)$$

Damit ist [4]:

$$\overline{|\delta\Omega_{rz}|} = \frac{2}{\sqrt{\pi}} \sqrt{AKF_{\delta\Omega}(0) - AKF_{\delta\Omega}(\tau)} \qquad (6)$$

Da die $AKF_{\delta\Omega}$ der Frequenzmodulation durch die Filtercharakteristik vorgegeben ist, kann nach Gl. 6 die Laufzeit τ (Zielentfernung) bestimmt werden, wenn $\overline{|\delta\Omega_{rz}|}$ bekannt ist. Zur Messung von $\overline{|\delta\Omega_{rz}|}$ wird die Komparator-Frequenzzähler-Kombination verwendet.

Das Detektorsignal U_D hat die Form (vgl. Gl. 1):

254

$$U_D \propto \cos(\phi_0 + \delta\phi_{rz}(t,\tau)) \tag{7}$$

ϕ_0 = Offsetphase des Interferometers, $\delta\phi_{rz}$ = momentane Phasenänderung des Überlagerungssignals
Der Komparator schaltet im Nulldurchgang des Detektorsignals an den Phasenniveaus

$$\phi_i = \phi_0 + \delta\phi_{rz}(t,\tau) = \frac{2 \cdot i - 1}{2} \cdot \pi \qquad i = 0,1,-1,2,-2... \tag{8}$$

Die erwartete Anzahl $E\{n_i\}$ (Scharmittelwert) der Triggerungen am Niveau i während der Torzeit T kann anhand der Wahrscheinlichkeitsdichte $p(\phi_i)$ der Phase am Triggerpunkt ϕ_i und der erwarteten Änderungsgeschwindigkeit $E\left\{\left|\delta\dot\phi_{rz}\right|_{\phi_i}\right\}$ der Phase an diesem Punkt berechnet werden[5]:

$$E\{n_i\} = T \cdot p(\phi_i) \cdot E\left\{\left|\delta\dot\phi_{rz}\right|_{\phi_i}\right\} \tag{9}$$

Für normalverteilte momentane Frequenzänderungen $\delta\Omega_{rz}$ des Überlagerungssignals sind die momentanen Phasenänderungen $\delta\phi_{rz}$ des Überlagerungssignals ebenfalls normalverteilt und unkorreliert mit den Frequenzänderungen $\delta\Omega_{rz}$. Die Wahrscheinlichkeitsdichte $p(\phi_i)$ kann deshalb anhand der Normalverteilung berechnet werden, und $E\left\{\left|\delta\dot\phi_{rz}\right|_{\phi_i}\right\}$ kann durch den erwarteten Betragsmittelwert der Frequenzänderung ersetzt werden. Die Scharmittelung kann unter der Vorraussetzung eines ergodischen Prozesses durch Zeitmittelung ersetzt werden. Damit wird aus Gl. 9:

$$\overline{n_i} = T \cdot p(\phi_i) \cdot \overline{|\delta\Omega_{rz}|} \tag{10}$$

Die insgesamt gezählten Triggerungen ergeben sich durch Summierung über alle Stellen i

$$\overline{n} = T \cdot \overline{|\delta\Omega_{rz}|} \cdot \sum_{i=-\infty}^{\infty} p(\phi_i) \tag{11}$$

Die vom Frequenzzähler gemessene mittlere Frequenz ist somit:

$$\overline{f} = \frac{\overline{n}}{2 \cdot T} = \overline{|\delta\Omega_{rz}|} \cdot \frac{1}{2} \sum_{i=-\infty}^{\infty} p(\phi_i) = \overline{|\delta\Omega_{rz}|} \cdot E_f(\phi_0, \sigma_{\phi rz}) \tag{12}$$

Das gesuchte $\overline{|\delta\Omega_{rz}|}$ kann demnach bis auf einen Empfindlichkeitsfaktor E_f (13)

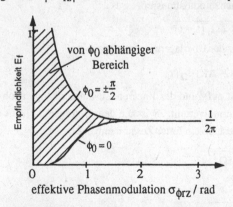

$$E_f = \frac{1}{2} \sum_{i=-\infty}^{\infty} \left[\frac{1}{\sigma_{\phi rz} \cdot \sqrt{2\pi}} \cdot e^{-\frac{1}{2}\left(\frac{\frac{2i-1}{2} \cdot \pi - \phi_0}{\sigma_{\phi rz}}\right)^2} \right]$$

mit der Komparator-Frequenzzähler-Kombination gemessen werden. E_f hängt von der effektiven Phasenmodulation $\sigma_{\phi rz}$, und der Offsetphase ϕ_0 des Interferometers ab (Bild 2). Für effektive Phasenmodulationen $\sigma_{\phi rz}$ größer als $\pi/2$ konvergiert E_f gegen $1/2\pi$. In diesem Bereich liefert die Meßanordnung exakte Ergebnisse, die nicht von ϕ_0 abhängen. $\sigma_{\phi rz}$ kann im Fall der tiefpaßgefilterten Modulation mit $fe \cdot \tau \ll 1$ direkt aus der

Bild 2: Empfindlichkeitsfaktor E_f der Komparator-Frequenzzähler-Anordnung zur Messung von $\overline{|\delta\Omega_{rz}|}$

Frequenzmodulation $\delta\Omega_r$ der Laserdiode berechnet werden[6]:

$$\sigma_\phi = \sigma_{\delta\Omega} \cdot \tau \tag{14}$$

Für übliche Fabry-Perot-Laserdioden kann eine effektive Frequenzmodulation von mindestens $\sigma_{\delta\Omega} = 2\pi \cdot 10$ GHz erreicht werden[1]. Daraus ergibt sich ein kleinster Wert für τ von 25 ps (entsprechend einer Zielentfernung von 3,75 mm). D. h. Abstände > 3,75 mm können ohne Stabilisierung der Offsetphase des Interferometers genau gemessen werden.

5 Experimentelle Untersuchungen

Mit einem einfachen Halbleiterlaser-Interferometer (Bild 1) haben wir Zielentfernungen z von 0,1 m bis 1,5 m gemessen. Zur Strommodulation wurde tiefpaßgefiltertes (10. Ordnung) weißes Rauschen mit einer Eckfrequenz $f_e = 25$ kHz und einem Effektivwert von 0,7 mA verwendet.

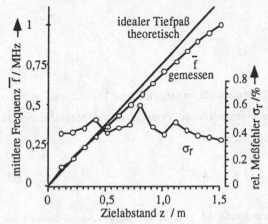

Bild 3: Gemessene Kennlinie und relativer Meßfehler des Rauschmodulierten–Entfernungsmessers.

Innerhalb der Modulationsbandbreite wandelte die verwendete Laserdiode vom Typ Sharp LT015 Stromänderungen in einem konstanten Verhältnis von 2,74 GHz/mA in Frequenzänderungen um und wurde somit in ihrer optischen Frequenz mit 1,92 GHz effektiv moduliert. Unsere Meßergebnisse stimmen gut mit der theoretischen Kennlinie des Entfernungsmessers mit ideal tiefpaßgefilterter Rauschquelle ($f_e = 25$ kHz, $\sigma_{\delta\Omega} = 1,92$ GHz) überein (Bild 3). Der relative Meßfehler (Standardabweichung von 100 Einzelmessungen mit T = 1 s) war nahezu konstant und betrug im Mittel 0,45 %.

6 Ausblick

Im Rahmen eines anderen Beitrags[7] zu dieser Thematik beschreiben wir den Einfluß des Filters im Modulationskanal auf die Kennlinie des E-Messers und auf den Abbruchfehler infolge der begrenzten Mittelungszeit. Dort wird gezeigt, daß das Produkt aus der Eckfrequenz f_e des modulierenden Prozesses und der Torzeit des Zählers T entscheidenden Einfluß auf die Meßgenauigkeit haben. Durch Erhöhung von $f_e \cdot T$ kann der Abbruchfehler der zeitlichen Mittelung und damit der Entfernungsmeßfehler deutlich verringert werden. Die Verwendung eines angepaßten optischen Teleskops ermöglicht auch die Entfernungsmessung mit inkooperativen Zielen[8].

Diese Arbeit wurde mit Mitteln des BMFT unter dem Förderkennzeichen 13 N 5884 gefördert. Die Verantwortung für den Inhalt dieser Veröffentlichung liegt bei den Autoren.

7 Literatur

/1/ G. Beheim, K. Fritsch, Electr. Lett., Vol. 21, 3, pp. 93-94, 1985
/2/ F. Bien, M. Carnac, H. J. Caulfield, and S. Ezekiel, Appl. Opt., Vol. 20, 3, pp. 400-403, 1981
/3/ P. de Groot, Appl. Opt., Vol 30, 25, pp. 3612-3616, 1991
/4/ B. M. Horton, Proceedings of the IRE, 47, pp. 821-828, 1959
/5/ A. Papoulis; "Probability Random Variables and Stochastic Processes", McGraw-Hill, 1991
/6/ K. Peterman; "Laser Diode Modulation and Noise", Kluwer Academic Publishers, 1988
/7/ W. Holzapfel; W. Baetz; J. C. Braasch; Proccedings on SPIE–International Symposium on Optical Instrumentation and Applied Science: Conf. 2000 Current Developments in Optical Design and Optical Engineering III, San Diego 1993
/8/ M. Hölscher, W. Holzapfel; Proccedings on SPIE–International Symposium on Optical Instrumentation and Applied Science: Conf. 2000 Current Developments in Optical Design and Optical Engineering III, San Diego 1993

Absolutinterferometrie mit durchstimmbaren Halbleiterlasern

T. Pfeifer und J. Thiel

Fraunhofer-Institut für Produktionstechnologie IPT

Steinbachstr. 17, D-52074 Aachen

1. Einleitung

Die Absolutinterferometrie mit durchstimmbaren Lasern erlaubt im Gegensatz zur konventionellen Interferometrie mit fester Laserwellenlänge λ Abstände statisch ohne Verschieben des Meßreflektors zu bestimmen [1]. Die ursprüngliche Interferometergleichung:

$$\Delta L = (\Delta m + \Delta \varphi) \frac{\lambda}{2} = \Delta \phi \frac{\lambda}{2} \tag{1}$$

bei der der Verschiebeweg ΔL durch eine Phasenänderung $\Delta \phi$ ausgedrückt werden kann, die sich aus einem ganzzahligen Anteil Δm und der Restphase $\Delta \varphi$ zusammensetzt, besitzt dabei formal dieselbe Gestalt wie die Gleichung für die Absolutinterferometrie:

$$L_{abs} = \Delta \phi \frac{\Lambda_S}{2} \qquad mit: \quad \Lambda_S = \frac{\lambda_1 \cdot \lambda_2}{\lambda_2 - \lambda_1} \tag{2}$$
$$und: \quad \Delta \phi = \phi_1 - \phi_2$$

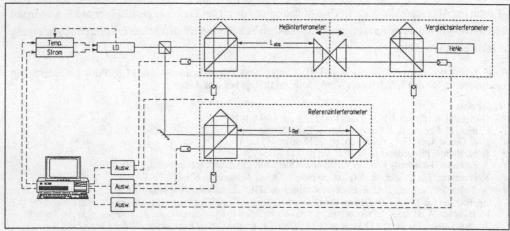

Bild 1. *Meßanordnung zur Absolutinterferometrie*

bei der jedoch anstelle der konstanten Laserwellenlänge λ die synthetische Wellenlänge Λ_S steht. Voraussetzung für dieses Verfahren ist allerdings, daß sich die Wellenlänge kontinuierlich ohne Modensprung von λ_1 nach λ_2 durchstimmen läßt. Halbleiterlaser besitzen neben anderen positiven Eigenschaften heutzutage die größten modensprungfreien Durchstimmbereiche, weswegen sie sich für dieses Meßverfahren besonders anbieten. Ihre Wellenlänge läßt sich dabei um typisch 0,2 nm durch eine Änderung des Betriebsstromes durchstimmen [2].

2. Absolutinterferometrie mit Referenzinterferometer

<u>Bild 1</u> zeigt eine Meßanordnung, bei der der Strahl der verwendeten Laserdiode durch einen Strahlteiler aufgespalten wird und die Teilstrahlen ein Meß- und ein Referenzinterferometer nach der Art von Michelson durchlaufen. Die Verwendung des Referenzinterferometers mit konstanter und bekannter Länge L_{Ref} hat den Vorteil, daß sich der gesuchte absolute Abstand im Meßinterferometer L_{abs} auch ohne exakte Kenntnis der Start- und Endwellenlänge λ_1 bzw. λ_2 auf sehr einfache Art und Weise ermitteln läßt:

$$L_{abs} = L_{Ref} \frac{\Delta\phi_{abs}}{\Delta\phi_{Ref}} \tag{3}$$

Zur Messung der Phasendifferenzen $\Delta\phi_{abs}$ und $\Delta\phi_{Ref}$ werden die Interferenzsignale mittels Photodioden in elektrische Signale umgewandelt und während des Durchstimmens der Wellenlänge jeweils die Nulldurchgänge der Quadratursignale mit Hilfe einer Zählelektronik registriert. Durch eine statistische Auflösungserweiterung, bei der zu Beginn und zum Ende des Wellenlängenshifts die Zählerinhalte des Meß- sowie des Referenzinterferometers überabgetastet werden, lassen sich zahlreiche Phasendifferenzpaare und daraus nach Gl. 3 entsprechend viele Quotienten berechnen, über die während eines einzigen Meßzyklus effizient gemittelt werden kann.

In gegensinniger Anordnung zum Absolutinterferometer (vgl. Bild 1) befindet sich ein kommerzielles HeNe-Laserinterferometer zur Vergleichslängenmessung. <u>Bild 2</u> zeigt das Ergebnis einer Vergleichsmessung, bei der der Abstand des Meßreflektors jeweils nach 10 Messungen in etwa 10 µm-Schritten verändert wurde. Der Grundabstand wurde bei unterschiedlichen Messungen zwischen 10 cm und 70 cm variiert und beträgt in diesem Fall 50 cm. Die durchgezogene Kurve stellt die Meßwerte des Vergleichsinterferometers dar,

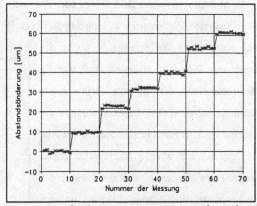

Bild 2. *Vergleichsmessung von 10 µm-Abstandsänderungen (Grundabstand 50 cm)*

258

die Meßpunkte geben die Meßwerte des Absolutinterferometers wieder, die eine Streuung von etwa ± 2 µm aufweisen.

3. Absolutinterferometrie mit Wellenlängenregelung

Wird anstelle des in Bild 1 gezeigten Referenz-interferometers ein Regelinterferometer konstanter Länge verwendet, das eine Regelung der Wellen-länge auf bestimmte Werte erlaubt, läßt sich der Absolutabstand mit Hilfe von Gl. 1 berechnen. Hierbei wird zunächst die Wellenlänge auf den Wert λ_1 zu Beginn eines modensprungfreien Be-reiches geregelt, die Anfangsphase im Meßinter-ferometer φ_1 gemessen und sodann die Wellenlän-ge auf λ_2 am Ende des modensprungfreien Berei-ches durchgestimmt und eingeregelt, wobei wäh-renddessen die Phasenänderung $\Delta\phi$ registriert wird.

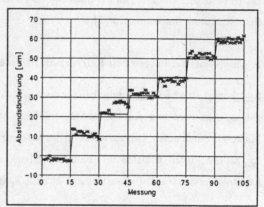

Bild 3. *Vergleichsmessung von 10 µm-Abstands-änderungen (Wellenlängenregelung)*

Bild 3 zeigt eine Bild 2 entsprechende Messung mit der voran beschriebenen Methode. In einem Meßbereich von 10 cm bis 70 cm beträgt die Streuung der Meßwerte etwa ± 10 µm. Vorteilhaft ist bei diesem Verfahren jedoch die Kenntnis der Wellenlänge, die ein relatives Weitermessen im Sinne der herkömmlichen Interferometrie ermöglicht.

4. Absolutinterferometrie mit zwei Halbleiterlasern

In die Meßanordnung aus Bild 1 wird über den ersten Strahlteiler der Strahl einer zweiten Laserdiode mit einer diskreten Wellenlänge λ_3 eingekoppelt (vgl. Bild 4), die so gewählt ist, daß sich mit $\lambda_{1,2}$ und λ_3 eine synthetische Wellenlänge Λ_S ergibt, die mindestens doppelt so groß sein muß, wie die Meßunsicherheit, die mit dem Verfahren der kontinuierlichen Wellenlängenmodulation er-reichbar ist. Hierdurch kann zunächst der Ab-solutabstand auf wenige Mikrometer genau vorgemessen werden, so daß die Periode der synthetischen Wellenlänge bekannt ist und anschließend durch Messung der zu $\lambda_{1,2}$ und

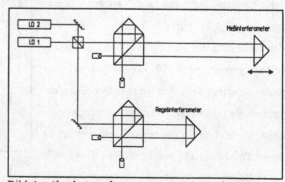

Bild 4. *Absolutinterferometrie mit 2 Laserdioden*

λ_3 gehörigen Restphasen $\varphi_{1,2}$ und φ_3 die Phase der synthetischen Wellenlänge bestimmt werden kann.

Bild 5 zeigt eine Meßreihe, bei der der Absolutabstand von einem beliebigen Grundabstand aus insgesamt über 10 cm verändert wurde. Aufgetragen sind die Differenzen zwischen Vergleichs- und Absolutinterferometer über die logarithmisch dargestellte Verschiebestrecke. Es konnte in diesem Bereich eine Abweichung aller Meßwerte von weniger als 140 nm erreicht werden.

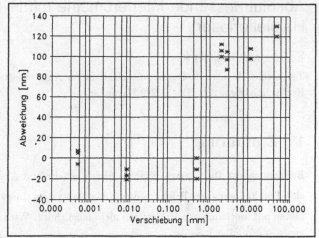

Bild 5. *Vergleichsmessung mit zwei Laserdioden*

5. Zusammenfassung

Die Absolutinterferometrie in der hier dargestellten Form bietet neben dem Vorteil, Abstände statisch ohne Bewegung des Meßreflektors bestimmen zu können, den weiteren Vorteil der intrinsischen Brechzahlkompensation. Aufgrund des geringen Abstandes von Meß- und Referenzinterferometer liegen in beiden nahezu dieselben Umgebungsbedingungen vor, so daß mit diesem Verfahren sämtliche Einflüsse auf den Brechungsindex, wie Temperatur, Druck und Feuchtigkeit, aber auch Einflüsse von Fremdgasen kompensiert werden. Strahlunterbrechungen, die in der herkömmlichen Interferometrie zum Abbruch einer Messung führen, da der Anschluß an die vorangegangenen Meßwerte verloren ist, führen hier lediglich zu einer Unterbrechung der Messung. Der Einsatz von Lasern mit größeren modensprungfreien Durchstimmbereichen wird dabei in Zukunft den Anschluß der Auflösung an die Einzelwellenlänge ermöglichen.

Literatur

[1]   Pfeifer, T.; Thiel, J.: Absolutinterferometrie mit durchstimmbaren Halbleiterlasern. Technisches Messen 60 (1993), S. 185-191.

[2]   Abou-Zeid, A.; Emissionswellenlänge kommerzieller Laserdioden als Funktion ihrer Betriebsparameter. PTB-Mitteilungen 94 (3/1984), S. 163-168.

Absolutmessende Interferometrie mit Hilfe stabilisierter Halbleiterlaser

K. Gerstner, T. Tschudi

Technische Hochschule Darmstadt   Institut für Angewandte Physik
Hochschulstrasse 6   D – 64289 Darmstadt

1   Einführung

Berührungslose Distanzmeßverfahren finden zunehmend Anwendung in vielen Bereichen, wenn eine Beschädigung des Prüflings durch das Meßsystem ausgeschlossen werden soll oder das Antasten mit mechanischen Prüfspitzen nicht möglich ist. Wird eine besonders hohe Auflösung und Genauigkeit gefordert, werden interferometrische Distanzmeßverfahren verwendet. Gegenstand dieser Arbeit ist ein interferometrisches Distanzmeßverfahren, das einen Meßbereich von 125 μm bei einer Genauigkeit von 0.8 μm umfaßt. Der realisierte Meßbereich läßt sich ohne Probleme erweitern. Die Genauigkeit ist durch das Prinzip der Phasenauswertung gegeben und läßt sich durch Einfügen eines Korrekturtermes ebenfalls verbessern. Dieses Meßverfahren ist innerhalb des Meßbereiches absolutmessend, d.h. der Objektabstand wird punktweise gemessen, sobald genügend Intensität in das Interferometersystem zurückreflektiert wird. Auf diese Weise kommt das Interferometer ohne das Anfahren von Referenzpunkten aus. Des weiteren sind Unterbrechungen des Sondenstrahles während des Meßvorganges zulässig. Das Meßverfahren basiert auf dem bekannten Verfahren der Zweiwellenlängeninterferometrie. [1], [2], [3],[4]. Bei diesem Verfahren wird das Interferometersystem von zwei unterschiedlichen, spektral dicht benachbarten Wellenlängen λ_1, λ_2 so durchlaufen, daß die Laserstrahlung dieser beiden Wellenlängen den gleichen optischen Weg nimmt. Die zwei Interferenzfunktionen bei den verschiedenen Wellenlängen werden dann getrennt detektiert und der Phasenauswertung zugänglich gemacht. Den gesuchten Objektabstand erhält man dann aus [5], [6]:

$$x = \lambda_{\text{syn}} \cdot (\Phi_{\lambda 1} - \Phi_{\lambda 2}) = \lambda_{\text{syn}} \cdot \Phi_{\text{syn}} \tag{1}$$

Die sogenannte synthetische Wellenlänge λ_{syn} ergibt sich aus:

$$\lambda_{\text{syn}} = \frac{\lambda_1 \lambda_2}{|\lambda_1 - \lambda_2|} \tag{2}$$

Ein großes Problem bei der Zweiwellenlängeninterferometrie stellt die Lichtquelle dar, die zwei dicht benachbarte Laserlinien mit genügender Stabilität zur Verfügung stellen muß. In dieser Arbeit wird eine derartige Lichtquelle vorgestellt, die mit Hilfe stabilisierter Halbleiterlaser aufgebaut wurde. Darüberhinaus wird ein direktes Phasenauswerteverfahren vorgestellt, das die gewünschte Genauigkeit bei einem kleinen Systemaufwand bietet.

2 Zweiwellenlängenlichtquelle

Die von uns aufgebaute Zweiwellenlängenlichtquelle verwendet zwei Halbleiterlaser des gleichen Typs, deren Parameter Chiptemperatur und Injektionsstrom stabilisiert werden. Die gewünschte Wellenlängendifferenz wird durch verschiedene Betriebstemperaturen und durch Ausnutzen von Exemplarstreuungen erreicht. Beide Laser sind so eingebaut, daß sie in orthogonalen Polarisationsrichtungen emittieren. Mit Hilfe der Parameterstabilisierung wird eine kurzzeitige Stabilität der Wellenlänge von 10^{-6} erreicht, jedoch muß aufgrund spektraler Alterung der Halbleiterlaser eine optische Wellenlängenstabilisierung vorgenommen werden. Im Falle des ersten Lasers wird eine Ankopplung der Wellenlänge an einen hochstabilen konfokalen Resonator (Resonatorlänge 1cm) vorgenommen. Diese Art der Wellenlängenstabilisierung ist bekannt [7] und nicht Gegenstand dieser Arbeit.

Es wäre prinzipiell möglich, die zweite Laserlinie ebenfalls auf diese Weise absolut zu stabilisieren. Dies würde jedoch einen vergleichsweise hohen Aufwand bedeuten und außerdem wäre dann der Wert der synthetischen Wellenlänge festgelegt. In unserem System wird direkt die Wellenlängendifferenz stabilisiert. Beide Laser werden in einen zweiten Fabry Perot Resonator eingekoppelt, dessen Resonatorlänge der halben zu stabilisierenden synthetischen Wellenlänge entspricht. Die Resonatorlänge könnte auch ein ganzzahliges Vielfaches der halben synthetischen Wellenlänge betragen, dann wäre die synthetische Wellenlänge aber nicht mehr eindeutig festgelegt. Wird nun ein Resonatorspiegel gescant, erscheinen die jeweiligen Transmissionsmaxima für beide Wellenlängen gleichzeitig. Wird dieses Regelkriterium verletzt, indem sich die Wellenlänge des zweiten Lasers verändert, erscheint eine Zeitdifferenz zwischen den Transmissionsmaxima beider Wellenlängen. Dieses Fehlersignal wird nun über eine Auswerteelektronik der Temperaturregelung von Laser 2 zugeführt und so die Wellenlänge auf ihren ursprünglichen Wert zurückgeregelt. Abbildung 1 zeigt das Prinzip dieser Stabilisierung.

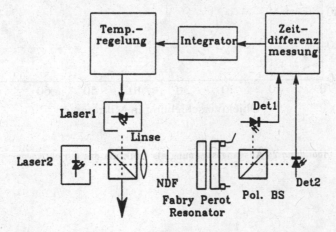

Abbildung 1: Prinzip der Differenzwellenlängenstabilisierung

3 Direktes Phasenauswerteverfahren

Unser Phasenauswerteverfahren bestimmt direkt die Phasendifferenz zwischen den beiden Interferenzsignalen, aus der der gesuchte Objektabstand gewonnen wird. Dazu wird der Referenzweg mit Hilfe eines Phasenmodulators (Piezotranslator, Pockelszelle) über eine Lichtwellenlänge verschoben. Dabei werden beide Interferenzsignale detektiert. Nach Eliminierung des DC-Offsets werden in einer Scanperiode von jedem Interferenzsignal ein gleichsinniger Nulldurchgang detektiert. Dabei entsteht ein Ausgangsimpuls, dessen Länge von der Phasendifferenz der beiden Interferenzsignlae abhängt. Nach einem Integrator erhält man eine Spannung, die von der Phasendifferenz und damit vom Objektabstand abhängt. Da dieses Verfahren keinerlei Berechnungen verlangt, also voll analog arbeitet, kann bei Verwendung eines geeigneten Phasenmodulators eine hohe Meßgeschwindigkeit (etwa 10000 Messungen pro Sekunde) erreicht werden. Möchte man die Genauigkeit von 0.8 μm verbessern muß noch ein Korrekturterm zugefügt werden, der die Lage des ersten zur Detektion verwendeten Nulldurchganges relativ zur Nullage des Phasenmodulators berücksichtigt. Abbildung 2 zeigt die gemessene Objektabstandsänderung als Funktion der realen Objektabstandsänderung. Dazu wurde der Objektabstand mit Hilfe eines abstandsgeregelten Piezosystems verkleinert. Die Sprungstelle indiziert das Ende des Meß- bzw. Eindeutigkeitsbereiches.

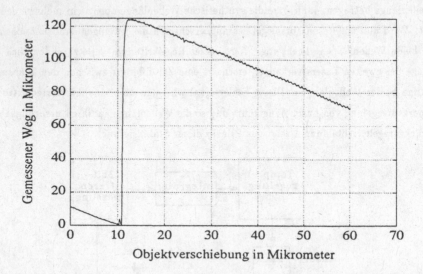

Abbildung 2: Ausgangsspannung als Funktion des Objektweges

4 Zusammenfassung

In dieser Arbeit wurde ein neues Stabilisierungsverfahren vorgestellt, das zur Stabilisierung einer einstellbaren Wellenlängendifferenz dient. Damit ist es möglich die für die Zweiwellenlängeninterferometrie nötigen synthetischen Wellenlängen problemangepaßt zur Verfügung zu stellen. Darüberhinaus wurde ein Phasenauswerteverfahren vorgestellt, mit dem kurze Meßzeiten zu realisieren sind.

Literatur

[1] P. Lam, J.D. Gaskill, J.C. Wyant
Two–wavelength holographic interferometer'„ *Applied Optics, vol. 23,No.18, pp.3079*

[2] A.F. Fercher, H.Z. Hu, U. Vry
'Rough surface interferometry with a two–wavelength heterodyne speckle interferometer'
Applied Optics, Vol. 24, No. 14, pp. 2181

[3] R. Dändliker, R. Thalmann, D. Prongue
'Two–wavelength laser interferometry using superheterodyne detection'
Optics Letters, Vol. 13, pp. 339

[4] C.C. Williams, K.H. Wickramasinghe
'Absolute optical ranging with 200-nm resolution'
Optics Letters, Vol. 14, No. 11, pp. 542

[5] K. Creath
'Step height measurement using two–wavelength phase–shifting interferometry'
Applied Optics, Vol. 26, No. 14, pp. 2810

[6] Y. Cheng, J. C. Wyant
'Two–wavelength phase shifting interferometry'
Applied Optics, Vol. 23, No. 24, pp. 4539

[7] C. Breant, P. Laurent, A. Clairon
Ültra–narrow linewidth of optically self–locked diode lasers'„ *Frequency standards and metrology, Springer Verlag, pp.441*

High-Speed, High-Accuracy, Non-Contact Radius Measurement with Laser Profiler

Li Song, Nathalie Colbert, Peiying Zhu
Servo-Robot Inc.
1380 Graham Bell Street, Boucherville (Quebec) J4B 6H5 CANADA

1. INTRODUCTION

Although optical principles have been sucessfully applied in metrological measuring instruments for many years, the coming of machine vision has transformed this classic domain by combining the traditional and modern optical technologies with new opto-electronic components and computers. The recently developed advanced active 3D vision systems provide the possibility of high-speed, high-accuracy, non-contact shape measurement and they are widely applied in robotic guidance, real-time process control and industry inspection applications. {1} The main features of active 3D vision systems are : very high data sampling rate, high accuracy, application flexibility, high degree of process independence, and reliability to perform shape measurement in 3D coordinates in harsh industrial environments. These features cannot be obtained with traditional metrological instruments nor by 2D machine vision systems.

In the following sections will be discussed an example of a typical application for laser profilers : high-speed, high-accuracy radius measurement. We will briefly describe the active laser profiler and radius measurement set-up. The error sources of original data will be analysed and the algorithm developed for the purpose of eliminating the outliers and to filter the remaining data will be discussed. Finally, we will present a few examples of calculations which will be followed by the conclusions.

2. SERVO-ROBOT'S LASER PROFILER AND RADIUS MEASUREMENT SET-UP

Servo-Robot's active laser profiler consists of two position sensor. One sensor performs range measurement and the other determines the lateral position for each data point. The range sensing is based on active optical triangulation and the profile measurement is achieved by means of an autosynchronized scanning configuration. Figure 1 illustrates the laser profiler (M-SPOT) used to perform the radius measurement and the test conditions.

In order to perform the radius measurement of a partial or complete cylindrical surface, the scanning plan of the laser profiler should be perpendicular to the axis of the cylindrical surface. This orientation condition ensures that the obtained profile is a part of or a complete circle.

Fig. 1

We know, on the one hand, that the optical intensity level on an optical sensor used to make range measurement is a function of the surface condition (orientation and roughness). On the other hand, a stable intensity level on the sensitive area of the optical sensor yields a better accuracy of measurement. Therefore, for the measurement of a metallic curved surface, real-time intensity control was applied. The output power of the laser diode is adjusted to compensate for the variation of light intensity on the optical sensor due to the variation of the surface condition and to keep the intensity at the optimal level.

3. ERROR SOURCE

We deal with three types of measuring errors. The first type of error is the systematic error which is related to the laser profiler used to perform the radius measurement and the measuring set-up which determines the relative position (including the orientation) between the measured part and the laser profiler. This measuring error can generally be eliminated or compensated by adjusting the measuring set-up and by recalibrating the results of the measurements. So, it is not the purpose of this paper to investigate this type of measuring errors.

The second type of error is the measured surface-dependent error. This error is not a zero mean error and it is directly related to the surface condition of the measured part. Two major sources of this type of error are listed below.

- The outliers due to local imperfections of the measured part, such as cracks, scratches and deposits on the surface.
- The range measuring error due to the variation of the reflectivity and the surface angular orientation inside an area which corresponds to the dimension of the projected laser beam on the surface of the measured part.

As it is analysed (2), the local condition of the surface can affect the measured results of the corresponding point. Although this error appears as a random error for different profile measurements, it is repeatable for one particular surface profile and it is not a zero mean error.

The third type of error, which includes the random and zero mean error, can be categorized as follows :

- The random error due to the uncertainty of the measurement relative to the problem of speckles of laser beams.
- The random error caused by the interpolation used for coordinate calculations.
- The random error related to other mechanical, optical and electrical noises.

4. DATA PROCESSING AND RADIUS CALCULATION

A standard least squares method may be used to estimate the radius from the laser profile. However, as mentioned before, the laser profile is subject to various types of random and surface-dependent noise, in this case, some noisy data points will cause large residuals. (A residual is defined as the difference between the data point and the fitted data.) Mathematically, those points are called outliers and all the remaining points are called inliers.

Robust regression methods (3), such as M-estimators, R-estimators, L-estimators, random sample consensus (RANSAC), Least Median Squares method (LMS), Median of the Intercepts (MI), etc., have been developed to reduce the effect of outliers.

The basic ideas behind RANSAC, LMS and MI methods are very similar. Instead of using all points for fitting, they choose some subset of data to estimate the parameters.

The major problem for these robust regression methods is that they are computationally intensive. Therefore, it is hard to apply them in our application since the speed is one of our major concerns. On the other hand, *a priori* information on the measured radius is known in our application. Specifically, we know that the measured radius is in the range $R_g \pm \Delta R$. Based on this knowledge, we have developed the following algorithm.

Let P denote the data set $(x_1,y_1),(x_2,y_2),...,(x_n,y_n)$, that is, the laser profile, where n is the number of sampling points. Let R represent the radius and (x_c,y_c) be the center of a circle. We want to estimate the parameters (R,x_c,y_c) from the data set P.

The algorithm consists of the following four steps which can be divided into two categories of calculation. The first category (Steps 1-3) is to eliminate the outliers, while the purpose of the second (Step 4) is to eliminate zero mean random errors and to make the final calculations.

1) Divide the data set into three subsets P_1,P_2 and P_3 where

$P_1 : (x_1,y_1),(x_2,y_2),...,(x_m,y_m)$ $P_2 : (x_{m+1},y_{m+1}),(x_{m+2},y_{m+2}),...,(x_{2m},y_{2m})$ $P_3 : (x_{2m+1},y_{2m+1}),(x_{2m+2},y_{2m+2}),...,(x_n,y_n)$

where m = [n/3], [ ] denotes the maximum integer of n/3.

2) For each point p_i in the data set P_1,

a) Set counters S and F to zero.

b) Randomly select one point p_j from the data P_2, and another point p_k from the data set P_3.

c) Calculate the parameters (R, x_c, y_c) from these three points (p_i,p_j,p_k).

d) If $R_g - \Delta R < R < R_g + \Delta R$, then increase counter S by 1; otherwise increase counter F by 1.

e) If counter $S \geq T_s$, then the point p_i is considered as an inlier, copy it to data set P' and go to Step 3, otherwise continue.

f) If counter $F > T_f$, then the point p_i is considered as an outlier, go to Step 3, otherwise go back to Step 2 (b).

3) For each point in the data set P_2, proceed as in Step 2, except the second and the third points are randomly selected from the data set P_1 and P_3. A similar procedure is also applied to the data set P_3.

4) The data set P' now contains all inliers, a least squares method is applied to this data set to finally estimate the parameters (R, x_c,y_c).

As one can see, the computation time of outlier elimination depends on two thresholds T_s and T_f. The algorithm at most computes $T_s + T_f$ times for each data point.

5. EXAMPLES OF CALCULATION

Figure 2 shows the profile data of a cylinder obtained by using Servo-Robot's laser profiler. In order to highlight the error in the obtained data, the deviation of the data from a fitted circle is magnified 100 times in Fig. 2 and the following figures. A least squares method is applied to this data set and the calculated radius R_c is 25.05 mm.

In Fig. 3, we set R_g = 25.05 ΔR = 0.05 T_s = 2 and T_f = 28. By applying the algorithm described in previous sections, 10% of data points are eliminated and R_c = 25.041 mm.

In Fig. 4, R_g = 25.05 ΔR = 0.025 T_s = 2 T_f = 28, 22% of data points are eliminated and R_c = 25.040 mm.

In Fig. 5, R_g = 25.05 ΔR = 0.025 T_s = 3 T_f = 32, 32% of data points are eliminated and R_c = 25.036 mm.

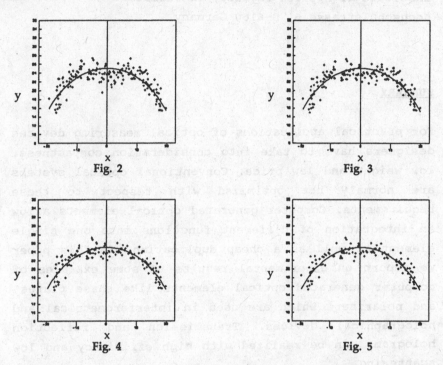

Fig. 2

Fig. 3

Fig. 4

Fig. 5

It is to be noted that the points to be eliminated depend on selected ΔR, T_s and T_f and some inliers could also be eliminated if the thresholds T_s and T_f are not properly determined.

6. CONCLUSION

It is shown that the high-speed, high-accuracy non-contact radius measurement can be performed by using a laser profiler. Two regression methods should be respectively applied to eliminate, first, the outliers due to surface-dependent errors and, second, the zero mean random errors. The good combination of T_s, T_f, ΔR and R_g ensures minimum calculation time and maximum measurement accuracy.

REFERENCES

1. M. Rioux. *Laser Range Finder Based on Synchronized Scanner*, Appl. Opt. 23:3837-3844, 1984.

2. L. Song, J.P. Boillot, G. Lemelin and R.H. Lessard. *3-D Active Vision System with High Lateral Resolution and Large Depth of Field*, Proc. 1822:206-216, 1992.

3. P.J. Huber. *Robust Statistics*. Wiley, New York, 1981.

Computer Generated Optical Elements for Application in Optical Measuring Devices

T.Tschudi, D.Columbus, M.Deininger, K.Gerstner,
J.Hoßfeld, L.Wang
Institute of Applied Physics, TH Darmstadt
Hochschulstrasse 6, D-6100 Germany

SUMMARY

For practical applications of optical measuring devices
designers have to take into consideration compactness,
low weight and low price. Conventional optical systems
are normaly not optimized with respect to these
requirements. Computer generated optical elements allow
an integration of different functions into one single
element as well as a cheap duplication. In this paper
we report on experimental results of some examples of
computer generated optical elements like phase plates,
and polarisers which are used in interferometrical and
holographical devices. Transmission and reflection
holograms can be realized with high efficiency and low
scattering.

1. PHASE PLATES FOR PHASE-SHIFT INTERFEROMETRY

In the well known phase-shift method in interferometry
the reference pass of the interferometer will be
changed usualy in four steps with equidistant phase
steps. On each step, the intensity of the interference
function will be measured. Interpreting the four
measured intensities the phase of the interferometer
pass and with this the object distance can be
determined. With the aid of computer generated
holograms it is possible to get the four nesessary
phase values at the same time. The phase shift will not
longer be applied mechanically but with the help of a

reflection hologram. The reference beam will be split up in four beams with phase difference of 90°. With an other hologram, also the object beam is split into four beams, but without any phase difference. By superposing the four corresponding object- and reference beams, the interference phase can be detected with a quadrant photodiode. Additionally, both of the holograms can be combined with focusing lenses and/or other beam shaping and beam guiding elements.

2. POLARIZERS

A new type of optical element - the polarizing computer generated hologram (PCGH) - may integrate different optical functions like polarizing beam splitting, beam shaping, imaging, polarization forming or conversion, and more in one single element. We have produced a polarization multiplexer/demultiplexer for optical telecommunication using a phase PCGH.

One way to obtain time stable polarizing features of such elements is to change the complex transmission of an optical anisotrop substrate. We present a method to get phase PCGH's which is based on the production of a surface relief structure in a birefringent substrate (refractive indices n_o and n_e). Subsequently the relief structure is filled with a material of refractive index n_e (or n_o). Therefore, only one polarisation component of the incoming light is diffracted by the structure, while the orthogonal one is transmitted without change. Two of such elements in contact or one element with both surfaces prepared may transform the incoming light in amplitude, phase, and polarization and thus act as a general PCGH.

The surface relief may be fabricated using lithographic techniques. Therefore, in contrast to interferometrically obtained polarizing holograms, no photosensitivity of the holographic material is required. In our experiments we used calcite plates cutted parallel to the c-axis. We spin-coated them with photoresist and exposed the calculated hologram structure into the resist. After removal of the exposed parts the resist acted as a mask in the subsequent dry etch process. The complete PCGH was obtained after filling the relief structure with PMMA and gluing together two of such elements.

We will report about two different configurations of polarization multiplexers / demultiplexers which have been build and we show that our approach for the production of phase PCGH can produce elements of high efficiency and low cross-talk, which incorporate different optical functions including polarization dependent ones in a single element. In both cases one polarization conserving fiber carrying two optical signals of orthogonal polarization direction was connected with two fibers each of them carrying one of the signals. In the first system the collimation and focusing of the incoming and outgoing light were performed by GRIN-lenses, while the major task of the PCGH was to act as a polarizing beam splitter. In the second system all three tasks were performed by a single PCGH.

Phase-Conjugating Elements in Optical Information Processing

T. Tschudi, C. Denz, T. Rauch, J.Lembcke
Institute of Applied Physics, TH Darmstadt
Hochschulstrasse 6, D - 6100 Darmstadt

SUMMARY

Parallel optical information processing systems are well suited for information reduction and preprocessing. Three important specifications for such applications have to be fulfiled: High space-bandwidth-product, amplification for signal restoring, and compensation of aberrations. The introduction of phase-conjugating elements into optical information processing systems will help to realize these specifications. The advantages of phase conjugation in optics, like exact counterpropagation and phase reconstruction, are well known, but the number of experimental realized applications is still small. This report gives a small insight into realised applications of phase conjugation. Setups of phase conjugating mirrors in combination with different types of interferometers (Michelson, Sagnac, and Fabry-Perot) will be presented. We report on applications on image subtraction, contrast amplification, phase visualisation, and parallel optical feedback systems. In all our experiments we used $BaTiO_3$ photorefractive crystals and Ar^+-ion laser for pump/signal waves.

1. PCM IN A MICHELSON INTERFEROMETER

A folded setup equivalent to a Michelson interferometer with arms of equal length is used. The light of both interferometer arms is directed by mirrors onto a self-pumped phase comjugating mirror, in which it is focused by a lens. The angular selectivity of self-pumped phase conjugation allows the separation of the two arms in a single volume. Inserting images into the two arms of the interferometer allows the realization of parallel optical logic operations like addition, XOR-operation and subtraction. Most promising application is its use as novelty filter.

2. PCM IN A SAGNAC INTERFEROMETER

The use of optical image processing systems depends on the amount of information channels and the nonlinear coupling between channels. We studied the spatial resolution of a phase-conjugating ring-resonator consisting of a Sagnac interferometer and a phase conjugating mirror with high gain. We also examined the contrast function of a set of incoming signals which depends on the gain of the PC resonator and the feedback ratio of the whole system. We obtained about 10^5 independent channels within our system. The transferfunction of the system is investigated by comparing the power spectrum of the incoming signal and the output signal. Control of coupling strength between channels is possible. We are using ring resonators for the realisation of neural nets and filtering systems.

3. PCM IN A FABRY-PEROT INTERFEROMETER

A Fabry-Perot interferometer of low finesse is used for phase front measurements. While the first mirror of the Fabry-Perot setup is a normal dielectrical one, the second mirror is a self-pumped phase conjugating mirror in which the incoming wave was focused by a internal lens. The interferometer can be regarded as being neutral due to the exact phase front reconstruction via phase conjugation and therefore does not contribute to the interference pattern of the beams. If a phase object is inserted in front of the interferometer into the laser beam, it causes distortions of the wave front, resulting in a change of the interference pattern. This shift can be used to determine the thickness of the phase object. Moreover this setup can prove the quality of phase conjugation.

Photo-thermische Raster-Nahfeld-Mikroskopie an Oberflächen und dünnen Schichten

M. Stopka, R. Linnemann, K. Maßeli, E. Oesterschulze und R. Kassing
Universität Kassel
Institut für Technische Physik
Heinrich–Plett–Str. 40
3500 Kassel, BRD

Kurzfassung

Es werden zwei photo–thermische Meßverfahren vorgestellt, welche die lateral hochaufgelöste Untersuchung thermischer Eigenschaften von Substraten und Dünnschichtsystemen erlauben. Beide Verfahren basieren auf der Anwendung von Meßtechniken der Raster-Nahfeld-Mikroskopie. So wird die Messung der Laser-angeregten Wärmeausdehnung einer Oberfläche mit Hilfe eines Raster–Tunnel–Mikroskops (STM) beschrieben und außerdem ein als Raster–Wärme–Mikroskop (engl.: Scanning Thermal Microscope SThM) bezeichnetes Verfahren vorgestellt, mit dem die Temperatur der Oberfläche gemessen werden kann.

Einleitung

Unter der Anwendung photo–thermischer Meßtechniken zur Bestimmung thermischer Parameter versteht man die Erwärmung eines Festkörpers mit einem periodisch modulierten oder gepulsten Laserstrahl und die Untersuchung thermischer Relaxationsprozesse des Materials. Photo–thermische Meßverfahren sind im allgemeinen kontaktlose und zerstörungsfreie Untersuchungsmethoden. Makroskopische photo–thermische Meßmethoden sind z.B. die Messung der thermischen Wärmeausdehnung der Festkörperoberfläche mit einem Laser–Interferometer [1] oder die Messung der Wärmeabstrahlung der Oberfläche mit einem Infrarot-Detektor.

Die ortsaufgelöste Untersuchung thermischer Eigenschaften mit makroskopischen photo–thermischen Meßverfahren ist unter anderem bei der Deposition dünner Schichten von Interesse. Hier geht es um den Einfluß des Depositionprozesses auf die thermischen Eigenschaften der Schichten. Durch ihre Untersuchung können der Prozeß charakterisiert und Prozeßparameter optimiert werden. Dabei zeigt sich, daß mikroskopische Struktur- und Homogenitätsunterschiede in der Schicht deren makroskopische thermische Eigenschaften beeinflussen können. Eine Feststellung, welche besonders bei der Verwendung dünner Schichten für mikroelektronische, mikromechanische und mikrooptische Bauelemente von Bedeutung ist. Es ist daher notwendig, die thermischen Eigenschaften dünner Schichten auch auf mikroskopischer Skala zu untersuchen, also mit einer größeren lateralen Ortsauflösung als derjenigen makroskopischer Meßverfahren.

Aufgrund der kleinen geometrischen Dimensionen der verwendeten Sensoren, welche die laterale Auflösung bestimmen, ist besonders die Raster–Nahfeld–Mikroskopie [2, 3] zur thermischen Charakterisierung dünner Schichten geeignet. Im folgenden werden zwei Methoden photo–thermischer Raster-Nahfeld-Mikroskopie beschrieben. Eine Methode verwendet ein Raster–Tunnel–Mikroskop (STM) zur hochaufgelösten Messung der periodischen Wärmeausdehnung der Oberfläche eines Festkörpers. Das Raster–Wärme–Mikroskop (SThM) bietet dagegen die Möglichkeit, die Temperaturänderung der Oberfläche der Probe mit Hilfe kleiner, als Spitze geformter Thermoelemente zu messen.

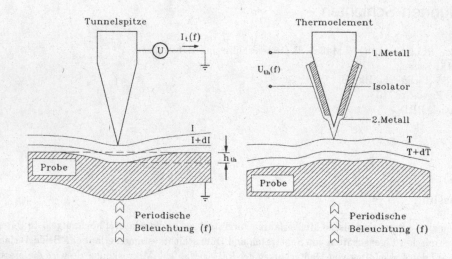

Abbildung 1: Schematischer experimenteller Aufbau des photo–thermischen STM (linke Seite) und des Raster–Wärme–Mikroskops (rechte Seite)

Experimenteller Aufbau

Der Aufbau zur Messung der periodischen Wärmeausdehnung mit einem Raster–Tunnel–Mikroskop ist in Abb. 1 (linke Seite) schematisch dargestellt. Ein He/Ne–Laser (633 nm) mit einer optischen Ausgangsleistung von 7 mW dient als Beleuchtungs– bzw. Heizquelle. Der Laserstrahl wird mit einem mechanischen Chopper periodisch moduliert und mit Hilfe einer konventionellen Fokussierungsoptik auf die Probenrückseite fokussiert. Der Fokusdurchmesser beträgt ungefähr 20 μm. Abb. 1 zeigt die Detektion der Wärmeausdehnung mit einer Wolframspitze von der Probenvorderseite. Neben dieser Transmissionsanordnung kann mit dem verwendeten Aufbau die Wärmeausdehnung aber ebenso von der Rückseite der untersuchten Probe in einer Reflexionsanordnung gemessen werden.

Die periodische Ausdehnung der Oberfläche führt zu einer Modulation des Tunnelstromes. Der Regelkreis des Raster–Tunnel–Mikroskops hält den Tunnelstrom und damit den Abstand der Wolframspitze von der Probenoberfläche konstant. Die Amplitude der periodischen Wärmeausdehnung h_{th} ergibt sich somit aus der Regelspannung des piezoelektrischen Aktuators, welcher den Abstand der Spitze von der Probe einstellt.

Das Raster–Wärme–Mikroskop (SThM) verwendet denselben optischen Versuchsaufbau. In diesem Fall wird jedoch anstelle der Wärmeausdehnung der Oberfläche die laterale Temperaturverteilung auf der Probenoberfläche direkt gemessen. Zu diesem Zweck werden miniaturisierte, als kleine Spitze geformte Thermoelement–Sensoren eingesetzt. Abb. 1 (rechte Seite) zeigt den schematischen Versuchsaufbau des Raster–Wärme–Mikroskops. Die Thermoelement–Sonde besteht aus einer Wolframspitze, welche bis auf den äußersten Spitzenbereich mit einer elektrischen und möglichst auch thermisch isolierenden Schicht bedeckt ist. Durch Deposition einer dünnen Nickelschicht wird ein Thermoelementkontakt an der Spitze hergestellt. Die Temperaturauflösung wird durch die differentielle Thermospannung beider Metalle bestimmt. Die Ortsauflösung hängt von den geometrischen Abmessungen des äußersten Spitzenkontaktes und von dessen Abstand zur Probenoberfläche ab.

Theorie

Der Wärmediffusionsprozess in einem Festkörper wird durch die dreidimensionale Wärmeleitungsgleichung beschrieben, welche für einen homogenen und isotropen Festkörper mit temperaturunabhängigen thermischen Parametern lautet [4]:

$$\triangle T(\vec{r},t) - \frac{1}{\alpha}\frac{\partial T(\vec{r},t)}{\partial t} = -\frac{S(\vec{r},t)}{\kappa}. \tag{1}$$

Hierbei bezeichnen $T(\vec{r},t)$ die Temperatur, κ die Wärmeleitfähigkeit, ρ die Dichte, c die Wärmekapazität, $S(\vec{r},t)$ die Intensitätverteilung der Laserquelle und $\alpha = \kappa/(\rho c)$ die thermische Diffusivität.
Die Annahme eines Gaußschen Intensitätsprofils (TEM$_{00}$-Mode des einfallenden Laserstrahls) für den Quellterm ergibt für die experimentelle Situation eine axiale Symmetrie. Wählt man geeignete Randbedingungen für die Temperatur und den Wärmestrom an jeder Probengrenzfläche, kann die Wärmeleitungsgleichung mit Hilfe einer speziellen Integral–Transformation, der sogenannten Hankel–Transformation nullter Ordnung, gelöst werden [5].
Die thermische Ausdehnung der Probenoberfläche kann bei Vernachlässigung elastischer Deformationen unter der Annahme eines linearen thermischen Ausdehnungskoeffizienten berechnet werden. Für eine exakte Lösung muß zusätzlich die Bewegungsgleichung des betrachteten Materials berücksichtigt werden.

Ergebnisse

Erste Messungen der Wärmeausdehnung mit dem photo–thermischen Raster–Tunnel–Mikroskop erfolgten an hochorientiertem pyrolytischem Graphit (HOPG) in der Transmissionsanordnung. In Abb. 2 (linke Seite) ist die periodische Oberflächenausdehnung als Funktion der Zeit dargestellt. Die Messung erfolgte in diesem Fall über die Analog–Digital–Wandlerkarte eines angeschlossenen Personalcomputers. Man erkennt, daß der Wandler aufgrund seiner diskreten Spannungsauflösung die Auflösung der Rückkopplungspannung des piezoelektrischen Aktuators begrenzt. Aus diesem Grund wurde ein Lock–In–Verstärker eingesetzt und die Amplitude und Phase der thermischen Ausdehnung gemessen. Abb. 2 (rechte Seite) zeigt die Amplitude der thermischen Ausdehnung als Funktion der optischen Leistung des einfallenden He/Ne–Laserstrahls. Wie erwartet, steigt die Amplitude der thermischen Ausdehnung für geringe optische Leistungen linear an.

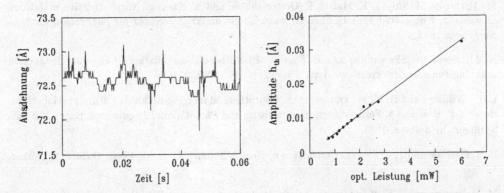

Abbildung 2: Periodischer Zeitverlauf der thermischen Ausdehnung (linke Seite) für eine Modulationsfrequenz von 70 Hz und Amplitude (rechte Seite) der thermischen Ausdehnung der Oberfläche einer Graphitfolie (Dicke 160 μm) als Funktion der optischen Leistung des einfallenden Laserstrahls.

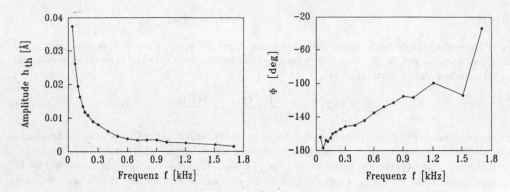

Abbildung 3: Amplitude (linke Seite) und Phase (rechte Seite) der thermischen Ausdehnung der Oberfläche der Graphitfolie als Funktion der Modulationsfrequenz des einfallenden Laserstrahls.

In Abb. 3 ist die Amplitude und die Phase der thermischen Ausdehnung als Funktion der Modulationsfrequenz abgebildet. Rauschmessungen zeigen, daß die Detektion der thermischen Ausdehnung des photo–thermischen RTM auf eine Empfindlichkeit von ungefähr 0.3 pm begrenzt ist, was einer Temperaturauflösung von etwa 10^{-7} K entspricht.

Zusammenfassung

Es wurden zwei Methoden hochauflösender photo–thermischer Raster–Nahfeld–Mikroskopien vorgestellt, welche zur Untersuchung thermischer Eigenschaften von Substraten oder Dünnschichtsystemen auf einer mikroskopischen Skala dienen. Beide Methoden arbeiten kontaktlos und zerstörungsfrei und können als Erweiterung photo–thermischer Meßverfahren in den Sub–Mikrometerbereich betrachtet werden.

Literatur

[1] M. Tochtrop, M. Stopka, K. Maßeli, E. Oesterschulze, and R. Kassing. Nondestructive evaluation of solids and deposited films by thermal wave interferometry. *accepted for publication in Appl. Surf. Sci.*, 1993.

[2] N. M. Amer, A. Skumanich, and D. Ripple. Photothermal modulation of the gap distance in scanning tunneling miccroscopy. *Appl. Phys. Lett.*, 49:137–139, 1986.

[3] C. C. Williams and H. K. Wickramasinghe. Photothermal imaging with sub-100 nm spatial resolution. In P. Hess and J. Pelzl, editors, *Photoacoustic and Photothermal Phenomena*, pages 364–369. Springer, Heidelberg, 1988.

[4] H. S. Carslaw and J. C. Jaeger. *Conduction of Heat in Solids*. Clarendon Press, Oxford, 2. edition, 1959.

[5] M. V. Iravani and H. K. Wickramasinghe. Scattering matrix approach to thermal wave propagation in layered structures. 58:122–131, 1985.

Homodyne Muliphase Sensor

V. Greco, C. Iemmi, S. Ledesma, A. Mannoni, G. Molesini, F. Quercioli

Istituto Nazionale di Ottica

Largo E. Fermi 6, Firenze 50125, Italy

Interferometry provides sensitive techniques for measuring displacements. The classical homodyne approach is based on two channels in close quadrature and bidirectional fringe counting electronics.[1] Working out four counts per fringe, in double pass the resolution is set at $\lambda/8$.[2,3] Better resolution is achieved by digitization of the quadrature signals; the ultimate performance is set by the noise level and by the errors with respect to the nominal conditions.[4,5] The multiphase approach, adding a proper number of measuring channels, allows for the application of robust algorithms which reduce the effect of noise and errors, so improving the actual resolution.[6,7]

A homodyne multiphase displacement sensor is here presented, which operates on a pattern of four signals in quadrature to monitor the displacement of a mirror with a resolution better than 1 nm. The optical setup is based on a Twyman-Green configuration with focusing lenses in both arms and the target and the reference mirrors at the foci (Fig. 1). In one arm a structured delay plate is inserted, with 45° sectors which in double pass induce phase differences of $0, \pi/4, \pi/2, 3\pi/2$ rad (Fig. 2). The resulting structured interference pattern allows for selecting the four signals in quadrature, written as

$$I_i = A_i + B_i \cos \left[\phi + (i-1)\frac{\pi}{2} + \epsilon_i \right] \quad , \tag{1}$$

with $i = 1, 2, 3, 4$; A_i and B_i are the bias intensity and the modulation term, respectively; ϕ is the phase to be determined, and ϵ_i is the quadrature error. With mathematics manipulations it is shown that the phase ϕ can be obtained from

$$\tan \phi = \frac{b}{(I_1 - I_3 - a) \cos \epsilon} \left(\frac{I_4 - I_2 - c}{d} - \frac{I_1 - I_3 - a}{b} \sin \epsilon \right) \quad . \tag{2}$$

The quantities a, b, c, d, ϵ depend on the experimental conditions, and can be determined with a calibration procedure.

Experiments have been carried out with a laboratory prototype of the interferometer. An array of four detectors has been used; the electronics provided digitized signals, read by a desktop computer via IEEE-488 interface. One of the folding mirrors was mounted on a piezoelectric

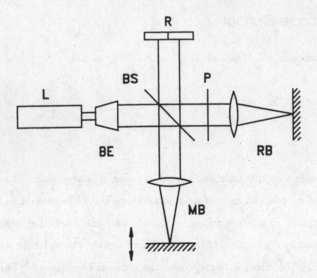

Fig. 1. *Twyman-Green interferometer in focused light, configured to serve as a displacement sensor. L, laser; BE, beam expander; BS, beam splitter; MB, measuring arm; RB, reference arm; P, optical delay plate; R, four detector array.*

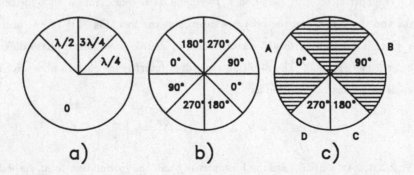

Fig. 2. *Delay plate structure and optical path difference (OPD). a) Single pass OPD. b) Double pass OPD, including the effect of pupil inversion produced by the focused light configuration. c) Selection of four quadrature signals with a shaped mask.*

transducer, also driven via IEEE-488 interface. The minimum step voltage available was 244.2 mV; the corresponding minimum displacement of the mirror, according to the characteristic curve provided by the manufacturer, varied from 2.9 nm at starting voltage up to 6.1 nm at about 500 V applied.

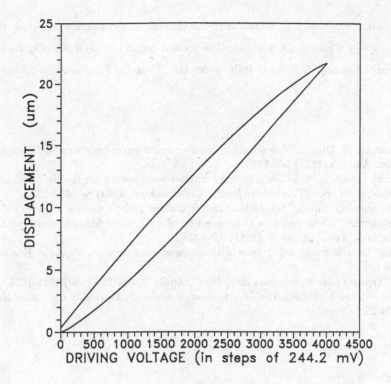

Fig. 3. *Calibration curve of a piezoceramic translator. The displacement in μm is obtained after converting the phase angle difference to length units.*

To evaluate the performance of the interferometer as a displacement sensor, the calibration curve of the piezoelectric transducer has been retraced. A sequence of 4000 unit steps of 244.2 mV has been applied, followed by an equal sequence of reversed steps back to zero voltage. The corresponding cycle is displayed in Fig. 3; the two branches of the curve clearly account for the hysteresis of the piezoceramics. The curve is in fair accord with the one given by the manufacturer.

A further estimate of the performance has been obtained after taking a sequence of data points under steady and unperturbed conditions. The residual noise level typically is of the order of 0.08 nm. Such figure is taken as the actual sensitivity of the sensor.

In conclusion, it has been shown that four signals with quadrature and balancing errors can be handled to obtain the phase in explicit form. A Twyman-Green interferometer in focused light including a structured delay plate has been mounted in laboratory, demonstrating the working principle and achieving sub-nanometric performance.

Acknowledgments.   C.I. and S.L. acknowledge the support of the Consejo Nacional de Investigaciones Científicas y Técnicas (Argentine). The present paper has been partially supported by National Research Council (C.N.R.) of Italy, under the "Progetto Finalizzato" on Electrooptical Technologies.

1. E.R. Peck and S.W. Obetz, "Wavelength or length measurement by reversible fringe counting", J. Opt. Soc. Am. 43 (1953) 505-509.
2. F.T. Arecchi and A. Sona, "Long-distance interferometry with an He-Ne laser", in: Proc. Quasi-Optics, ed. J. Fox (Polytechnic Institute of Brooklyn, 1964) pp. 623-633.
3. S.F. Jacobs and J.G. Small, "Liquid level interferometer", Appl. Optics 20 (1981) 3508-3513.
4. P.L.M. Heydmann, "Determination and correction of quadrature fringe measurement errors in interferometers", Appl. Optics 20 (1981) 3382-3384.
5. K.P. Birch, "Optical fringe subdivision with nanometric accuracy", Precision Engineering 12 (1990) 195-198.
6. L. Mertz, "Optical homodyne phase metrology", Appl. Optics 28 (1989) 1011-1014.
7. M. Hercher, "Ultra-high resolution interferometric sensors", Optics & Photonics News, Nov 1991, pp 24-29.

Vergleichbarkeit von Streulicht- und optisch profilometrischen Messungen an optisch glatten Oberflächen

B. Harnisch
European Space Research and Technology Centre, P.O. Box 299,
2200 AG Noordwijk, The Netherlands

1. Einführung

Optische Hochleistungssyteme werden neben den Abbildungseigenschaften durch das Streulicht charakterisiert. Besonders im Bereich der Hochleistungsoptik ist man deshalb an einer genauen Berechnung des Streulichtes in der Abbildungsebene interessiert. Für diese Berechnungen sind Messungen der Bidirectional- Scattering- Density- Function der optischen Funktionsflächen und Begrenzungsflächen notwendig, die dann mit einer geeigneten Software weiterverarbeitet werden können.

Das Interesse an der Vergleichbarkeit profilometrischer mit Streulichtmessungen resultiert einerseits aus der wesentlich größeren Akzeptanz der Industrie gegenüber profilometrischen Meßgeräten und dem Kostenaufwand, der beispielsweise durch die Messung der BRDF von Funktionsflächen für ein x-ray Teleskop bei Verwendung entsprechender Röntgenstrahlung entstehen würde.

Die Vergleichbarkeit von profilometrischen mit Streulichtmessungen wird dadurch erschwert, daß bei profilometrischen Messungen das Profil der Oberfläche $z(x,y)$ ermittelt wird, z ist hierbei die Profilamplitude an der Oberflächenkoordinate (x,y) und bei Streulichtmessungen die spektrale Leistungsdichte $S(f_x, f_y)$, wobei $f_{x,y}$ die Ortsfrequenz in der jeweiligen Koordinate bezeichnet. Prinzipiell sind Profil und spektrale Leistungsdichte mit Hilfe einer Fouriertransformation ineinander überführbar. Da jedoch beide Messungen jeweils einen begrenzten Bandbereich von Ortsfrequenzen repräsentieren (Übertragungsfunktion), stellt die Ortsfrequenzdarstellung des Oberflächenprofils eine Faltung der Oberfläche mit der Übertragungsfunktion des Meßsystems dar. Will man darüberhinaus von einem gemessenen Ortsfrequenzbereich auf einen benachbarten extrapolieren, muß die Oberfläche durch ein geeignetes Modell charakterisiert werden.

Eine ausführliche Beschreibung aller mathematischen Beziehungen zwischen den profilometrischen und den Streulichtmeßgrößen kann hier nicht wiedergegeben werden, es sei deshalb an dieser Stelle auf zwei wesentliche Arbeiten verwiesen [Stover 1990, Church

1979], die stellvertretend für eine Vielzahl von Veröffentlichungen der jeweiligen Autoren stehen sollen.

Der Übergang vom Oberflächenprofil zur spektralen Leistungsdichte, ist exakt gegeben durch eine Ensemblemittelung statistisch unabhängiger Messungen mit dem Grenzübergang zu einer unendlichen Meßlänge L:

$$S(f_x, f_y) = \langle \lim_{L \to \infty} \frac{1}{L} | \mathcal{F}[z(x, y, L)] |^2 \rangle$$

$$= \langle \lim_{L \to \infty} \lim_{M \to \infty} \frac{1}{LM} | \int_{-L/2}^{L/2} \int_{-M/2}^{M/2} z(x, y, L) e^{i(2\Pi f_x x)} e^{i(2\Pi f_y y)} dx dy |^2 \rangle$$

Der Zusammenhang zwischen BSDF und spektraler Leistungsdichte, der Zugang also vom Streulicht zur spektralen Leistungsdichte der Oberfläche wird nach der Rayleigh-Rice Vektortheorie berechnet zu:

$$BSDF = \frac{(dP/d\Omega_s) \, d\Omega_s}{P_i} = (\frac{16\Pi^2}{\lambda^4}) \cos\Theta_i \cos^2\Theta_s \, Q \, S(f_x, f_y) \, d\Omega_s$$

Hierin ist Θ_i der Winkel des einfallenden Bündels und Θ_s der Winkel des gestreuten Bündels, Q der optische Faktor, der Reflektions- und Polarisationseigenschaften berücksichtigt, dP die in das Raumwinkelelement $d\Omega_s$ gestreute differentielle Leistung relativ zur einfallenden Leistung P_i. Somit ist eine Berechnung der BRDF aus profilometrischen Messungen prinzipiell durchführbar.

Da der Grenzwertübergang in der Messung nicht möglich ist, ist eine exakte Umrechnung des gemessenen Profils in die spektrale Leistungsdichte und die BSDF nicht möglich. Es bieten sich zwei Wege an: Einerseits kann man die Übertragungsfunktionen der Meßgeräte und ihren Einfluss auf die Messung berechnen, eine Entfaltung dieser Übertragungsfunktion durchführen und die Oberflächeneigenschaften anhand der Amplituden der Ortsfrequenzen der spektralen Leistungsdichte miteinander vergleichen. Andererseits besteht die Möglichkeit, die Oberfläche mit geeigneten Modellen zu beschreiben, die Parameter dieser Modelle aus den Meßdaten zu berechnen und miteinander zu vergleichen.

2. Übertragungsfunktionen und vergleichbarer Spektralbereich

Der meßbare Ortsfrequenzbereich der optischen Mikroprofilometrie wird einerseits durch die Meßlänge L und andererseits durch den

Abstand benachbarter Meßpunkte bestimmt. Die Übertragungsfunktionen der Abbildungsoptik und der CCD Matrix führen dazu, daß die Amplituden höherer Ortsfrequenz schlechter als die niederer übertragen werden. In [Church 1985] wird deshalb eine Multiplikation mit der inversen Übertragungsfunktion innerhalb der Bandgrenzen der Messung im Ortsfrequenzraum vorgeschlagen.

Für die Berechnung der rms-Rauheit R_q, die oft zum Vergleich herangezogen wird, hat diese Korrektur allerdings nur einen geringfügigen Einfluß, da realistische Oberflächenmodelle über einen starken Abfall der Amplituden bezüglich der Ortsfrequenzen verfügen und mit dieser Methode nur der hochfrequente Anteil angehoben wird. Der Ortsfrequenzbereich für optisch mikroprofilometrische Messungen wird gegeben durchberechnet aus der Pixelzahl N, der Pixelgröße d'_{pix} und der Vergrößerung des Objektives V_{obj}:

$$f_{min} = \frac{1}{L} \leq f \leq \frac{N}{2}L = f_{max} \quad mit \; L = \frac{Nd'_{pix}}{V_{obj}}$$

Für das Streulichtverfahren kann von einer Beugung an sinusförmigen Oberflächenwellen, die gemäß einer Fourierzerlegung die Oberfläche generieren, ausgegangen werden. Die Übertragungsfunktion ist im gemessenen Ortsfrequenzbereich 1, wenn man sicherstellt, daß bei kleinen Streuwinkeln kein reflektiertes Licht mit gemessen wird. Für senkrechte Inzidenz ergibt sich folgender meßbarer Ortsfrequenzbereich, wobei $\Theta_{min, max}$ der jeweilige Streuwinkel ist:

$$f_{min} = \frac{\sin\Theta_{min}}{\lambda} \leq f \leq \frac{\sin\Theta_{max}}{\lambda} = f_{max}$$

Man beachte hierbei die Abhängigkeit von der Wellenlänge λ. In Tabelle 1 sind die meßbaren Oberflächenwellen (inverse Ortsfrequenzen d = 1/f) für optische Profilometer und die Streulichtmessung bei jeweils $\lambda=0.6328\mu m$ dargestellt.

Meßmethode	V_{obj}	$d_{min} = 1 / f_{max}$	$d_{max} = 1 / f_{min}$
Streulicht		0.63 μm	72.7 μm
Profilometer $d'_{pix} = 13\mu m$ N = 1024	3.2x 12.5x 25x 50x	4.06 μm 0.96 μm 0.52 μm 0.26 μm	4157 μm 980 μm 532 μm 266 μm

3. Beschreibung der Oberfläche durch Modelle

Es existieren mehrere Modelle zur Beschreibung glatter Oberflächen, in denen die Natur der Oberfläche durch einen funktionalen Zusammenhang und entsprechenden Parametern charakterisiert wird, die aus den Meßdaten bestimmt und miteinander verglichen werden können. Die bekanntesten Modelle sind in [Stover 1990] zusammengefaßt.

Ein sehr häufig verwendetes ist das der gaußverteilten Oberflächenamplitude. Eine sogenannte gauß'sche Oberfläche läßt sich durch zwei Parameter, die rms-Rauheit Rq und die Korrelationslänge τ beschreiben. Unglücklicherweise werden glatte Oberflächen durch dieses Modell nur sehr schlecht beschrieben. Beide Parameter, besonders jedoch die Korrelationslänge, hängen stark von der Übertragungsfunktion des Meßgerätes ab und sind somit weitgehend ungeeignet.

Ein zuerst von [Berry 1979] auf der mathematischen Theorie der Fraktale aufgestelltes Modell, das von einer Selbstähnlichkeit der Oberflächen in verschiedenen Ortsfrequenzbereichen ausgeht, charakterisiert glatte Oberflächen in erstaunlich gutem Maße. Die Oberfläche wird durch zwei Parameter, die fraktale Haussdorf-Besicovitch Dimension und einen Lateralparameter, die Topothesie eindeutig beschrieben. Die spektrale Leistungsdichte hat für den eindimensionalen Fall (korrugierte Oberflächen) die Form:

$$S(f_x) = \frac{K_n}{f_x^n}$$

Interessant an diesem Modell ist, daß es sehr gut die mit der Streulichtmethode vermessene spektrale Leistungsdichte von Metallspiegeln, Glasfunktionsflächen und optischen Schichten wiedergibt. Dies ist auch der Grund, das es konsistent mit der BSDF Definition für ein verbreitetes Streulichtanalyseprogramm ist, für das dieses Modell empirisch aus Messungen gewonnen wurde. Dieses Gültigkeit dieses Modells wurde in [Harnisch 1992] anhand profilometrischer Messungen an verschiedenen glatten Oberflächen untersucht.

Diese Arbeit basiert auf Untersuchungen, die an der Friedrich-Schiller-Universität Jena, Sektion Technologie durchgeführt wurden.

[Berry 1979]   J. Phys.   A 12(1979)781-97.
[Church 1979]   Opt. Engn. 18,2(1979)125-36.
[Church 1985]   Opt. Engn. 24,3(1985)388-95.
[Harnisch 1992] Int. Oberflächenkoll. Chemnitz, 03.-05.2.1992.
[Stover 1990]   Optical Scattering, Mc Graw Hill 1990.

Streulichtmessungen beim Laserstrahlschneiden

G.Sepold, B. Heidenreich, Ch. Binroth

BIAS, Bremer Institut für angewandte Strahltechnik

28359 Bremen, Klagenfurter Str. 2

1.   Einleitung

CO_2-Laserstrahlung mit einer Wellenlänge von 10,6 µm kann durch Wärmewirkung Schädigung beim Menschen hervorrufen. Die nach der Unfallvorhütungsvorschrift geltende maximal zulässige Bestrahlungsstärke (MZB) für CO_2-Laserlicht im Dauerstrichbetrieb liegt bei 0,1 W/cm^2 /1/.

Bei der Entwicklung einer großen CO_2-Laserstrahlschneidanlage für den Schiffbau trat in diesem Zusammenhang folgendes Problem auf: Um die Arbeitssicherheit zu gewährleisten, wurde vom Betriebsrat und von der Berufsgenossenschaft gefordert, die gesamte Anlage zu kapseln oder sie an einem von der übrigen Produktion getrennten Sonderstandort aufzustellen /2/. Beides verhindert eine Integration der Laserstrahlschneidanlage in die Produktionslinie des Schiffbaus, die als eine Voraussetzung für den wirtschaftlichen Betrieb der Anlage benötigt wird. Es sollten daher im Anlagenbereich Strahlungsmessungen während des Laserstrahlschneidens durchgeführt und Aussagen erarbeitet werden, ob das Bedienungspersonal oder der Kranführer durch aus dem Bearbeitungsprozeß rückgestreute Strahlung gefährdet sind. Für die Werkstoffe Baustahl, Aluminium und CrNi-Stahl wurden die Bereiche bestimmt, in denen keine Gefährdung von Haut und Augen zu erwarten ist bzw. in denen zusätzliche Schutzmaßnahmen erforderlich sind.

2.   Meßprinzip und -anordnung

Für die Messung der Bestrahlungsstärke E wurde ein sogenanntes elektrisch kalibriertes Radiometer mit einem pyroelektrischen Detektor ausgewählt /3/. Der prinzipielle Aufbau des Meßsystems kann wie folgt beschrieben werden: Die einfallende Strahlung wird mit einem mechanischen Chopper mit 15 Hz moduliert und anschließend auf der Oberfläche eines pyroelektrischen Detektors absorbiert. Der Detektor besteht aus einem Lithium-Tantal-Kristall mit einer Meßfläche von 0,5 cm^2 . Während die optische Strahlung unterbrochen ist, werden synchron über ein elektrisches Signal Heizelemente angesteuert, die sich auf der Detektoroberfläche befinden. Durch die optisch und elektrisch zugeführte Wärme werden zunächst zwei unterschiedliche Detektorsignale erzeugt, die verstärkt und synchron demoduliert werden. Ein nachgeschalteter Verstärker erzeugt aus diesen beiden Signalen ein Differenzsignal, das in einem geschlossenen Regelkreis durch Verändern des Heizstromes auf Null reduziert wird. Die elektrische Leistung zum Betreiben der Heizelemente kann sehr genau gemessen werden und entspricht nach der Nullkompensation der optisch eingestrahlten Lichtleistung. Gemessen und zur Anzeige gebracht wird bei dem Radiometer die benötigte elektrische Heizleistung, um die optische Strahlungswärme zu kompensieren. Das Radiometer wird wie folgt spezifiziert:

- Wellenlängenbereich: UV - IR,
- Leistungsmessung: absolut und pro Fläche,
- Meßbereich: 10^{-6} bis 0,1 W oder $2 \cdot 10^{-6}$ bis 0,2 W/cm^2,
- Genauigkeit: ± 1 % absolut im Bereich 250 nm bis 2 μm,
- Analogausgang.

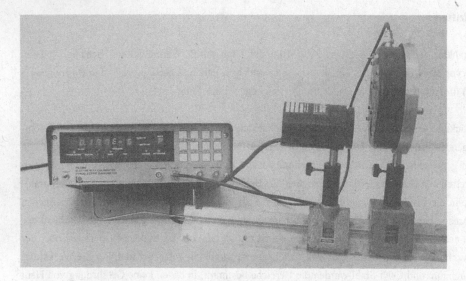

Bild 1: Radiometer zur Messung rückgestreuter Laserstrahlung, v.l.n.r. Anzeigegerät des Radiometers, Dreiecksschiene mit Detektor und Chopper

Der gesamte Meßplatz besteht aus einem Chopper, dem Detektor des Radiometers und dem Anzeigegerät. Der Chopper und der Detektor sind zusammen auf einer Dreiecksschiene montiert und aufeinander einjustiert. Das System kann als komplette Einheit positioniert und an verschiedenen Orten relativ zur Strahlungsquelle angebracht werden, vgl. Bild 1. Für die Ermittlung der Bestrahlungsstärke wurde die DIN VDE 0837 "Strahlungssicherheit von Laser-Einrichtungen" berücksichtigt /4/. Beim Schneiden von Blechen sind mehrere Möglichkeiten der Strahlungsreflexion vorhanden, die jeweils getrennt untersucht wurden /2/. Laserstrahlung kann während des Schneidens aus der Schnittfuge und beim Anschnitt an der Blechkante rückgestreut werden. Ferner wird Strahlung an schrägliegenden Blechen reflektiert, die bei Konturschnitten als Reststücke unter den Schneidtisch fallen und dort als Reflexionsflächen zurückbleiben. Die gemessenen Bestrahlungstärken wurden für praxisrelevante Einstellungen bei der Laserleistung, der Spiegelbrennweite, etc. ermittelt, die für die Bearbeitung der Werkstoffe Schiffbaustahl, CrNi-Stahl sowie einer AlMg-Legierung auf der Schneidanlage in Frage kommen. Die räumliche Zuordnung der Meßpunkte zum Bearbeitungsort wurde mit Hilfe von den Kugelkoordinaten r, ϑ und φ beschrieben. Dabei beschreibt r den Abstand vom Schneidkopf, ϑ den Neigungswinkel zwischen r und der Blechoberfläche sowie φ den Winkel zwischen r und der Schnittfuge.

3. Experimente

3.1 Messung der rückgestreuten Strahlung während des Schneidens

Für die Messung der rückgestreuten Strahlung während des Schneidens wurde das Radiometer parallel zur Schnittfuge ($\varphi = 0°$) angeordnet. Der Beobachtungswinkel betrug $\vartheta = 25°$ zur Blechoberfläche. Vorversuche ergaben maximale Werte unter diesem Beobachtungswinkel für die untersuchten Schiffbaubleche der Dicke 6 mm, 10 mm und 12 mm, vgl. Bild 2.

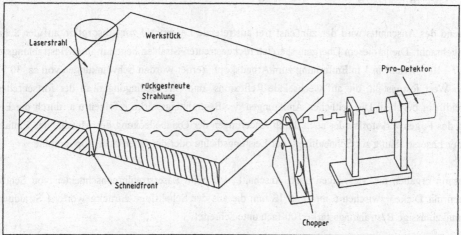

Bild 2:   Meßaufbau für rückgestreute Strahlung beim CO_2-Laserstrahlschneiden

Blechdicke s	6 mm	10 mm	12 mm
Brennweite f	127 mm	127 mm	127 mm
Laserleistung P_L	1900 W	1950 W	1900 W
Schneidgeschwindigkeit v	1,8 m/min	1,1 m/min	0,63 m/min
Bestrahlungsstärke E			
E (r = 1 m)	$0,7 - 0,8 \cdot 10^{-3}$ W/cm^2	$0,4 - 0,6 \cdot 10^{-3}$ W/cm^2 *	$0,1 - 0,2 \cdot 10^{-3}$ W/cm^2 **
E (r = 2 m)	$0,3 - 0,6 \cdot 10^{-3}$ W/cm^2	$0,1 - 0,4 \cdot 10^{-3}$ W/cm^2	$0,08 - 0,15 \cdot 10^{-3}$ W/cm^2
E (r = 3 m)	$0,1 - 0,2 \cdot 10^{-3}$ W/cm^2	$0,05 - 0,15 \cdot 10^{-3}$ W/cm^2	-

* während des Anschnitts steigt dieser Wert auf $0,7 \cdot 10^{-3}$ W/cm^2

** kurzzeitige Spitzenwerte bei $0,25 \cdot 10^{-3}$ W/cm^2

Tafel 1:   Gemessene Bestrahlungsstärke während des Schneidens bei verschiedenen Blechdicken und Beobachtungsabständen ($\vartheta = 25°$, $\varphi = 0°$, Werkstoff: Schiffbaustahl der Güte A, geprimert)

Die Messungen ergaben bei den Schneidversuchen an Schiffbaustahl Bestrahlungsstärken, die in 1 m Entfernung durchgängig unterhalb $1 \cdot 10^{-3}$ W/cm^2 lagen, vgl. Tafel 1. Die maximal zulässige Bestrahlungsstärke von $100 \cdot 10^{-3}$ W/cm^2 wird damit 100-fach unterschritten. Weiterhin wurde beobachtet, daß mit wachsendem Abstand die Bestrahlungsstärke stark abfällt. Sie verringert sich bei einem von 1 m auf 3 m vergrößerten Abstand etwa um den Faktor 10. Mit vergrößerter Blechdicke konnte ebenfalls ein Abnahme der rückgestreuten Strahlung festgestellt werden.

Die höchsten Bestrahlungsstärken wurden normal zur Schnittfuge bei $\varphi = 0°$ ermittelt. Mit steigendem φ-Winkel sinkt die Bestrahlungsstärke; bei $\varphi = 25°$ war nur noch eine sehr geringe Amplitudenüberhöhung gegenüber dem Grundrauschen zu verzeichnen. Die maximal nachweisbaren Werte lagen bei $\varphi = 25°$ um $0,03 \cdot 10^{-3}$ W/cm^2.

3.2 Messung der rückgestreuten Strahlung während des Anschnitts

Während des Anschnitts wird der zunächst frei austretende Laserstrahl zur Absorption auf der Schneidfront gebracht. Die in diesem Übergangsbereich rückgestreuten Strahlen liegen in ihrer Bestrahlungsstärke bei $0,7 \cdot 10^{-3}$ W/cm^2 in 1 m Entfernung zum Arbeitskopf. Ferner wurden Schwankungen von ca. 30 % um diesen Wert festgestellt, die auf wechselnde Reflexions- und Streubedingungen an der fortschreitenden Schneidfront beruhen. Die zeitlichen Änderungen der Bestrahlungsstärke erfolgen u.a. durch die Erwärmung des Fugenwerkstoffs oder auch durch Änderungen der Oxidbedeckung des Metalls, des Einfallwinkels der Laserstrahlung zur Schneidfront, der Leistungsdichte oder der Oberflächentopographie.

Insgesamt ergaben die Messungen beim Anschnitt, daß beim Laserstrahlbrennschneiden von Schiffbaublechen mit Dicken zwischen 6 mm und 12 mm die aus der Schnittfuge zurückgeworfene Strahlung die maximal zulässige Bestrahlungsstärke 100-fach unterschreitet.

3.3 Messung der rückgestreuten Strahlung an schrägliegenden Blechen

Tritt der Laserstrahl durch die schon geschnittene Schnittfuge hindurch, kann der Strahl an einem schrägliegenden Reststück, das unterhalb der Blechschneidauflage liegt, reflektiert werden. Die je nach Oberflächenbeschaffenheit direkt reflektierte bzw. gestreute Strahlung kann zu einer potentiellen Gefährdung des Anlagenbedieners oder des über der Anlage hinwegfahrenden Kranführers werden. Erste Messungen zeigten im Vergleich zum Schneiden und zum Anschnitt vergrößerte Bestrahlungsstärken, so daß für die Messungen an schrägliegenden Blechen auch auf die höherreflektierende Werkstoffe CrNi-Stahl und Aluminium ausgedehnt wurden. Berücksichtigt wurden die Werkstoffe Schiffbaustahl, Güte A mit geprimerter Oberfläche sowie kaltgewalzter CrNi-Stahl (Werkstoff-Nr. 1.4301) und Aluminiumblech (AlMg4,5Mn) beide mit gleichmäßig matter Oberfläche (Ra < 1,9 µm). Der Versuchsaufbau ist in Bild 3 dargestellt.

Während der Messungen an schrägliegenden Blechen wurde eine nahezu 10-fach höhere Bestrahlungsstärken als beim Anstechen und Schneiden nachgewiesen. Jedoch wird bei der Rückstreuung an Schiffbaublech in 1 m Entfernung vom Schneidkopf die zulässige Bestrahlungsstärke 10-fach unterschritten, vgl. Bild 4. Für die entwickelte Laserstrahlschneidanlage bedeutet dies, daß ein reiner Betrieb mit Schiffbaustahl hinsichtlich der rückgestreuten Laserstrahlung ungefährlich ist. Die Anlage kann ohne zusätzliche Sicherheitsvorkehrungen, z.B. in der Schiffbauhalle einer Werft, eingesetzt werden.

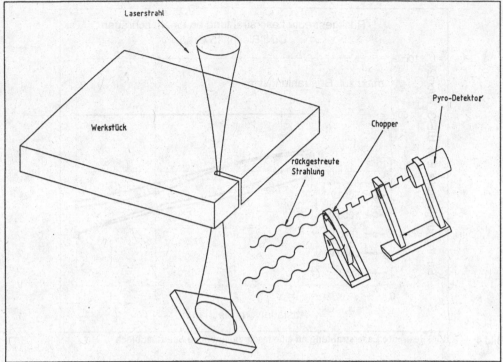

Bild 3: Aufbau zur Messung der Streustrahlung, die an schrägliegenden Blechstücken unterhalb der Blechplatte reflektiert wird

Bei der Bearbeitung von CrNi-Stählen und Aluminiumwerkstoffen muß je nach Oberflächenzustand differenziert werden. Das gleichmäßig matte CrNi-Stahlblech ist ähnlich unkritisch wie Schiffbaustahl, vgl. Bild 5. Dagegen kam es bei einem optisch spiegelnden CrNi-Blech (Ra < 0,4 μm) auch noch in 4 m Entfernung zu einer Überschreitung der zulässigen Bestrahlungsstärke. Bei einem mattgrauen Al-Blech wurde in 4 m Entfernung vom Schneidkopf die zulässige Bestrahlungsstärke 10-fach unterschritten.

Demnach muß beim Schneiden von Aluminiumwerkstoffen und CrNi-Stahl mit einer Überschreitung der maximal zulässigen Bestrahlungswerte gerechnet werden. Es sind daher lokale Schutzvorrichtungen, wie z.B. Stellwände und entsprechende persönliche Schutzausrüstungen, wie Schutzbrillen und -handschuhe bei diesen Werkstoffen vorzusehen.

4. Zusammenfassung

Ziel der Messungen war das Gefahrenpotential für die Anlagenbediener von großformatigen CO_2-Laserstrahlschneidanlagen durch Strahlungsanteile zu bestimmen, die im Bearbeitungsprozeß nicht absorbiert und in den Anlagenbereich rückgestreut werden. Hierfür wurden Messungen mit einem pyroelektrischen

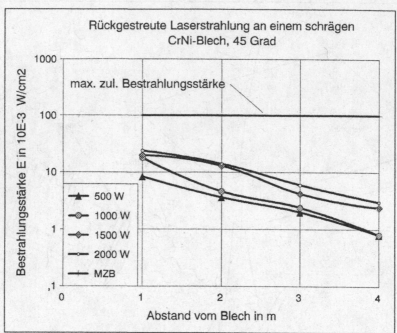

Bild 4: Rückgestreute Laserstrahlung an einem schrägliegenden Schiffbaublech

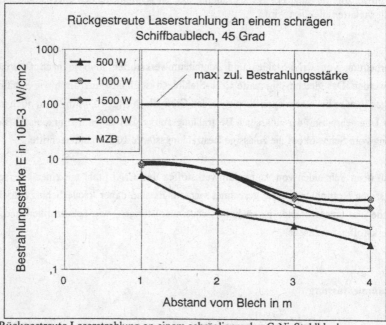

Bild 5: Rückgestreute Laserstrahlung an einem schrägliegenden CrNi-Stahlblech

Radiometer während des Schneidens von Schiffbaustahl, CrNi-Stahl und einer Aluminiumlegierung durchgeführt. Für die Messungen wurden die unterschiedlichen Betriebszustände des Schneidens, wie Anstechen, Schneiden und Rückstreuung an schrägliegenden Reststücken berücksichtigt.

Beim Schneiden wurden die größten Bestrahlungstärken parallel zur Schnittfuge nachgewiesen. Schnitte an Schiffbaustahl ergaben in 1 m Entfernung vom Schneidkopf eine Bestrahlungsstärke, welche die maximal zulässige Bestrahlungsstärke (MZB) um den Faktor 100 unterschreitet. Beim Blechanschnitt steigen die Meßwerte um 30 % gegenüber den Werten beim Normalschnitt an. An schrägliegenden Blechen, die als Reststücke unter der Blechschneidauflage liegen, wurden im Vergleich zum Anschnitt oder zum Normalschnitt größere Werte ermittelt, so daß die Rückstreuung an schrägliegenden Blechstücke im Vergleich das größte Gefahrenpotential aufweist. Die hier gemessenen Werte unterschreiten die maximal zulässige Bestrahlungstärke um den Faktor 10 bei einem Meßabstand von 1 m zum Schneidkopf.

Je nach Oberflächenbeschaffenheit kommt es bei CrNi-Stahlblech zur Streuung oder zur spiegelnden Reflexion. Die Rückstreuung an einem schrägliegenden CrNi-Blech mit einer mattgrauen, kaltgewalzten Oberfläche ergab Bestrahlungsstärken in 4 m Entfernung von der 0,1-fachen maximal zulässigen Bestrahlungsstärke. Für ein hochglänzendes CrNi-Blech wurde der 5-fache Wert der MZB ermittelt. In 1 m Entfernung wurde an einem schrägliegenden mattgrauen Al-Blech der 0,1 bis 1-fache Wert der MZB gemessen.

Aus den Messungen können folgende Schlüsse für den Betrieb der untersuchten Schneidanlage gezogen werden: Für den Werkstoff unlegierter Baustahl bzw. Schiffbaustahl sind keine weiteren Schutzmaßnahmen zum Abschirmen der rückgestreuten Strahlung notwendig. Für die Bearbeitung von CrNi-Stahl und Aluminiumwerkstoffen darf der Anlagenbereich nur mit geeigneter Schutzausrüstung betreten werden.

5.   Literatur

/1/   UVV "Laserstrahlung" VVBG 93), Hrsg. Berufsgenossenschaft für Feinmechanik und Elektrotechnik (Carl Hegemanns Verlag), Köln (1988)

/2/   G. Sepold, B. Heidenreich, Weiterentwicklung der Laserstrahlschneidtechnologie für die schiffbauliche Fertigung, Endbericht zum BMFT-Vorhaben (1.1.89 - 31.3.92) mit FKZ 18S0002B, FDS, Forschungszentrum des Deutschen Schiffbaus, Bericht Nr. 242/1992

/3/   W.M. Doyle, B.C. McIntosh, Implementation of a System of Optical Calibration Based on Pyroelectric Radiometry, Optical Engineering, Nov. 1976

/4/   DIN VDE 0837 "Strahlungssicherheit von Laser-Einrichtungen", Beuth Verlag, Berlin, Feb. 1986

Beurteilung der Wellenfront von CO_2-Lasern mittels phase retrival

Hembd Ch. , Tiziani, H.J.
Institut für Technische Optik
Pfaffenwaldring 9
D-7000 Stuttgart 80

Abstract

Wavefront quality of high power lasers in material processing is among the properties which have not yet been thoroughly investigated. This is particularly due to inherent difficulties in the use of interferometry at very high powers. Nevertheless focus quality is affected by wavefront-distortion as well as mode-distribution. Wavefront errors may be introduced by the laser resonator itself or by thermal effects in the beam guiding system.

Phase retrieval may offer a solution for industrial applications to gain phase information from pure intensity mearurements. First results from CO2 lasers are presented.

Zusammenfassung

Zu den bislang wenig untersuchten Aspekten der Strahlqualität an Bearbeitungslasern gehört die Vermessung der Wellenfront. Dies liegt an grundsätzlichen Schwierigkeiten bei der Anwendung interferometrischer Methoden im Hochleistungsbereich. Die Phasenverteilung ist jedoch neben der Intensitätsverteilung mitentscheidend für eine gute Fokusqualität. Sowohl durch den Laserresonator als auch durch das Strahlführungssystem können Wellenfrontdeformationen verursacht werden.

Wir stellen ein robustes Verfahren vor, um aus Intensitätsmessungen Information über die Wellenfront zu gewinnen. Erste Ergebnisse, die mit dieser Methode an CO_2-Lasern gewonnen wurden, werden vorgestellt und diskutiert.

Einleitung

Die Messung der Phasenverteilung an Hochleistungslasern stellt schon allein auf Grund der hohen Leistungen im Strahl große Anforderungen an die Meßtechnik. Zusätzlich kann das Auftreten höherer Moden und dadurch bedingte partielle Kohärenz interferometrische Messungen erschweren.

Für interferometrische Messungen kann man versuchen, über Strahlteiler ein sehr genaues sample des Strahls zu erhalten. Dies stellt hohe thermische Anforderungen an die Strahlteiler, die sich nicht unter thermischer Belastung verformen dürfen. Eine andere, hier untersuchte Möglichkeit der Wellenfrontmessung stellt die Methode des phase retrieval dar /1-4/, bei der aus Messungen des Intensitätsprofils die Phase ermittelt wird. Dieses Verfahren arbeitet prinzipiell auch noch bei partiell kohärentem Laserlicht. /6/

Grundlagen

In den vergangenen Jahren sind verschiedene Algorithmen entwickelt worden, die aus Intensitätsmessungen Phaseninformation gewinnen. Eine Vielzahl von Algorithmen arbeitet dabei mit Informationen in der Bild und Fourierebene. Für Laserstrahlen erfordert dies eine genaue Auflösung der Fokusstruktur und stellt damit sehr hohe Anforderungen an die Meßtechnik. Zudem sind die Algorithmen in der Regel iterativ und somit enorm zeitaufwendig. Im Gegensatz dazu bietet die Transportgleichung für die Intensität /1,2/

$$dI/dz = \ grad_t \ (\ I \ grad_t \ \phi \ )$$

I: Intensität, ϕ: Phase, $grad_t$ Gradient in tangentialer Richtung

einen Zugang , bei dem der unfokussierte Laserstrahl in zwei Ebenen z1, z2 gemessen wird. Eine geschlossene Lösung der Transportgleichung wurde von TEAGUE /1/ angegeben.

Experimentelle Ergebnisse

Die experimentelle Umsetzung erfolgte mit einer pyroelektrischen Zeile mit 64 Elementen, die durch mechanische Translation ein Feld der Größe 64 x 64 Pixel erfaßt. Bei einer Pixelbreite von 100µm ergibt das ein Bildfeld von 6,4 x 6,4 mm. Die Dynamik beträgt 12 bit.

Zunächst wurde in Simulationsrechnungen mit Gaußstrahlen getestet in welchem Bereich von Strahldurchmesser und Bildabstand dz der Algorithmus korrekte Ergebnisse liefert. Demnach sollte der mit dz anwachsende Kontrast dI/I nicht kleiner als 0,1 werden. Zu große Werte von dz sind dagegen unkritisch . Der Bereich des erfaßten Strahls sollte zumindest den zweifachen Strahldurchmesser (definiert beim $1/e^2$ Wert) umfassen.

Erste Messungen wurden an einem 20 Watt CO_2-Wellenleiterlaser durchgeführt. Der Ausgangsstrahl entspricht in guter Näherung einem TEM_{00}-Mode (EH_{11}-Mode). Die Detektorzeile wurde direkt in der Strahltaille des Lasers positioniert und schrittweise verschoben. Eine berechnete Wellenfront ist in Bild 1 gezeigt.

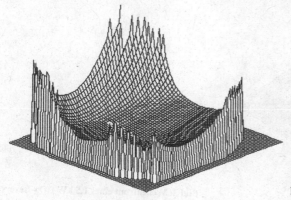

Bild 1: Wellenfront eines 20 Watt CO_2-Lasers

Einen Vergleich der gemessenen Krümmungsradien mit dem theoretischen Verlauf gibt Bild 2. Die durchgezogene Kurve entspricht dem erwarteten Taillenverlauf für einen reinen Gaußstrahl.

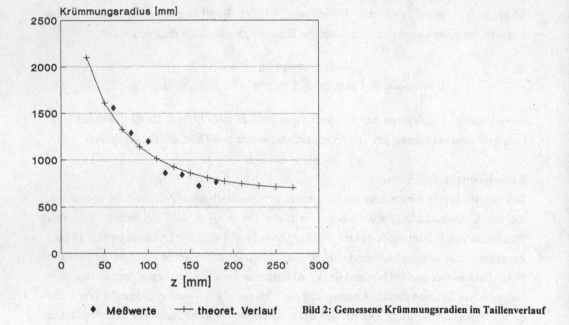

Bild 2: Gemessene Krümmungsradien im Taillenverlauf

Messungen wurden ebenso an einem 1,5 kW CO_2-Laser durchgeführt. Mit Hilfe eines wassergekühlten Reflexionsgitters aus Kupfer wurden 0,2 % der Strahlleistung in die erste Ordnung ausgekoppelt. Die erste Ordnung wurde auf die pyroelektrische Zeile abgebildet. Im rekonstruierten Phasenbild (Bild3) ist eine Struktur zu erkennen, die deutlich vom Grundmode abweicht.

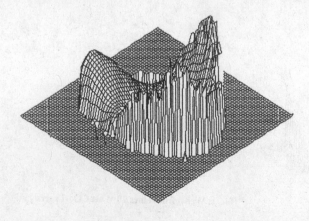

Bild 3: Wellenfront eines 1,5 kW CO_2-Lasers

Die Simulation eines Grundmodes mit ca. 10% Anteil (Amplitude) an TEM_{03}-Laguerre-Mode (Bild 4) gibt näherungsweise Übereinstimmung mit der experimentellen Phasenverteilung.

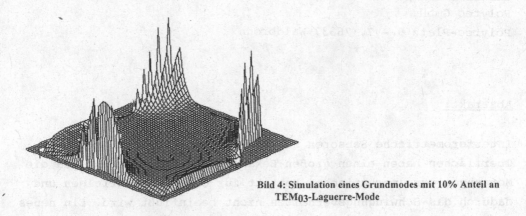

Bild 4: Simulation eines Grundmodes mit 10% Anteil an TEM_{03}-Laguerre-Mode

Danksagung

Die Autoren danken dem "Institut für Strahlwerkzeuge", Stuttgart für die freundliche Unterstützung bei Messungen an einem Hochleistungslaser. Besonderer Dank gilt der "Deutschen Forschungsgemeinschaft" für die finanzielle Unterstützung der Arbeiten im Rahmen des Sonderforschungsbereichs 349.

Literatur:

/1/ Teague, M.R.: Deterministic phase retrieval: a Green's function solution, JOSA 73 (11) 1434-1441 (1983)

/2/ Streibl, N.: Phase Imaging by the Transport Equation of Intensity, Opt. Comm. 49 (1) 6-10 (1984)

/3/ Baltes: Inverse Source Problems in optics, Springer (1980)

/4/ Fienup: Phase Retrieval Algorithms: a comparision, Applied Optics 21 (15) (1982)

/5/ Dooghin, A.; Kundikova, N. D.; Zel'dovich, B. Ya.: Phase retrieval from laser beam intensity profiles, Optics Comm. 91 193-196 (1992)

/6/ Roddier, F. Wavefront Sensing and the Irradiance Transport Equation, Applied Optics 29 (10) 1402-3 (1990)

Scanning Laser Vibrometer

Dipl.-Phys. Martin Feser

Polytec GmbH

Polytec-Platz 5 - 7, 76337 Waldbronn

Abstrakt:

Interferometrische Sensoren zur Analyse der Vibration von
Oberflächen haben einen großen Bekanntheitsgrad erreicht, da die
Messungen ohne mechanischen Kontakt zur Oberfläche erfolgen und
dadurch das Schwingungsverhalten nicht beeinflußt wird. Ein neues
Meßsystem zur flächenhaften Schwingungsanalyse wurde entwickelt,
das die Vorteile interferometrischer Schwingungsmeßtechnik mit
schneller FFT-Signalauswertung und modernster Bildverarbeitung
vereint.

Meßprinzip:

Der Laserstrahl eines Einpunkt Meßkopfes wird über zwei Scan-
spiegel auf die zu untersuchende Fläche gelenkt.

Nach ferngesteuerter Fokussierung, Einstellung der Abtastgrößen
und Festlegung des Meßbereiches wird an jedem Punkt des ausgewählten Meßrasters die Schnelle mittels des Laser Interferometers
und schneller FFT-Signalanalyse ermittelt und gespeichert. Im
optischen Meßkopf befindet sich eine CCD-Kamera, mit der das
Meßobjekt on-line beobachtet wird. Das Objektbild dient der
Festlegung des Meßrasters und der Meßpunkte sowie der
anschließenden Datenpräsentation.

Software

Die Software wurde nach neuesten benutzerfreundlichen Gesichtspunkten entwickelt und stellt einen wesentlichen Bestandteil des
Meßsystems dar. Die Unterteilung in Analysemodus, Flächenscan

und Datenpräsentation entspricht der logischen Abfolge eines Meßzyklus.

Im Analysemodus werden die Signal- und Abtastparameter festgelegt. Zwei Fenster für die Signalanzeige des zweikanaligen Meßsystems sind frei einstellbar. Der Laserstrahl läßt sich frei über das Meßobjekt bewegen und die Signalanzeige erfolgt on-line für einen Meßpunkt. A/D-Frequenzen, Anzahl der Samplepunkte, Triggermodus und verschiedene FFT-Funktionen sind frei wählbar.

Im Programmpunkt Flächenscan wird der Scanbereich und die Anzahl der Scanpunkte mit Hilfe des aktuellen Videobildes festgelegt.

Die während des Scans gleichzeitig ermittelten FFT- und RMS-Daten werden unter dem Menüpunkt "Datenpräsentation" als zwei- oder drei-dimensionale Graphiken dargestellt.

Bei der FFT-Darstellung von bis zu 8 Frequenzbändern kann zwischen Amplitude, Schnelle, Beschleunigung, Phase und Übertragungsfunktion ausgewählt werden. Die Schnelle wird bei den RMS-Daten dargestellt.

Bei der 2-dimensionalen Darstellung wird dem abgespeicherten Objektbild eine Falschfarben- oder Isoliniendarstellung überlagert, die das Schwingungsbild des Meßobjektes repräsentieren. Als Netzgraphik oder farbiges Flächenbild wird bei der drei-dimensionalen Darstellung das Schwingungsverhalten präsentiert.

Anwendungsbeispiele:

In der Automobilindustrie gibt es sehr viele Anwendungen für das Scanning Vibrometer. Das Meßsystem ist vielseitig einsetzbar im Karosseriebau, Motoren-, Getriebe- und Katalysatorenbau. Das flächenhafte Schwingungsbild gibt einen schnellen und präzisen Überblick über das Schwingungsverhalten einzelner Teile im

Betrieb in verschiedenen Lastbereichen. Starke Schwingungsamplitu-
den bei lärmabstrahlenden Oberflächen, z.B. im Karosseriebau,
lassen sich sehr einfach lokalisieren und mit entsprechenden
Maßnahmen beseitigen.

Im Instrumentenbau kann das Meßsystem zur Analyse der zahlreichen
Eigenschwingungen eines Musikinstrumentes eingesetzt werden.
Sowohl dem Verständnis bezüglich des Schwingungsverhaltens als
auch der Klassifikation von Musikinstrumenten können diese
Messungen sehr hilfreich sein.

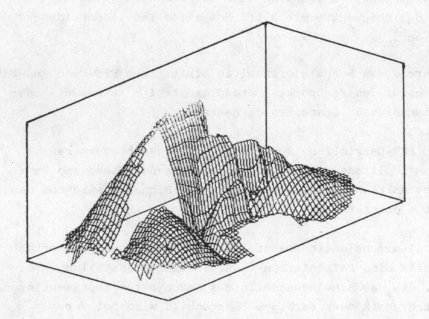

Schwingungsform einer Konzertgeige bei 293 Hz

Weitere Anwendungsmöglichkeiten gibt es im Lautsprecher-
Turbinen-, Maschinen- und Elektromaschinenbau.

Messung von Rayleighwellen zur Untersuchung des Verschleiß-mechanismus an der Oberfläche kristalliner Werkstoffe mit Hilfe der Doppel-Plus-Holographie

C. Henning, D. Mewes

Institut für Verfahrenstechnik, Universität Hannover

Callinstraße 36, D-30167 Hannover

EINLEITUNG

Der wiederholte Aufprall fester oder flüssiger Partikeln auf eine Werkstoffoberfläche führt zum Ver-schleiß durch Erosion [1, 2]. In den Experimenten werden ebene Werkstoffe aus austenitischem Stahl durch den wiederholten Aufprall fester Partikeln geschädigt. Dies führt infolge der einsetzenden Werk-stoffermüdung zur Verfestigung oberflächennaher Werkstoffbereiche, zur Anrißbildung und schließlich zum Materialabtrag. Die durch den Aufprall einzelner oder mehrerer Partikeln hervorgerufene zeitab-hängige Oberflächendeformation führt u.a. zur Ausbreitung von Rayleighwellen an der Werkstoffober-fläche. Diese werden mit Hilfe der holographischen Interferometrie mit einem Doppelpuls-Laser ge-messen. Dabei können die geschädigten Werkstoffbereiche anhand der gestörten Ausbreitung der Ober-flächenwellen erkannt werden. Aufgrund der wie bei Anwendungsfällen kleinen Amplitude der erzeug-ten Rayleighwellen von einigen 10nm wird ein Verfahren der hochauflösenden holographischen Inter-ferometrie eingesetzt.

AUFBAU ZUR AUFNAHME DER INTERFEROGRAMME

Die Aufnahme der holographischen Interferogramme erfolgt mit der Zwei-Referenzstrahl-Technik [3]. Der optische Aufbau für die Aufnahme ist im Bild 1 schematisch dargestellt. Der an dem ersten Strahl-teiler reflektierte Strahl wird mit Hilfe eines leicht dejustierten Michelson-Interferometers in zwei leicht divergente Referenzstrahlen aufgeteilt. Das Umschalten der Laserpulse zwischen den beiden Strahlen-gängen für die Referenzwellen erfolgt mit einer einzigen Pockelszelle. Dazu wird während der Emissi-on des Doppelpulses die Polarisationsrichtung des ersten Laserpulses mit Hilfe der Pockelszelle ge-dreht. Jeweils einer der Referenzstrahlen wird durch eines der beiden sich in den Interferometerarmen befindenden Polarisationsprismen gesperrt. Der Rotator (Polarisationsdreher) dreht die Polarisations-richtung um 45°. Dies bewirkt für beide Referenzstrahlen eine senkrecht polarisierte Komponente der Laserstrahlung, so daß beide mit dem senkrecht polarisierten Objektstrahl interferieren können. Der Treiber für die externe Pockelszelle wird regulär für die Güteschaltung von Pulslasern eingesetzt. Die Elektronik der Hochspannungsversorgung ist somit für einen anderen Anwendungsfall konzipiert: Die an den Kristall angelegte Hochspannung dreht die Polarisationsrichtung um nur 45°, um zur Emission eines Laserpulses für die Dauer von ca. 0,5µs abgeschaltet zu werden. Der Schaltvorgang an der

Pockelszelle muß also mit einer zeitlichen Genauigkeit von wenigen zehntel Mikrosekunden auf den durchtretenden zweiten Laserpuls synchronisiert werden. Die an den Kristall angelegte Spannung wird auf das doppelte des üblichen Wertes erhöht, um die für das Schalten zwischen den Referenzstrahlen erforderliche Drehung von 90° zu erzielen.

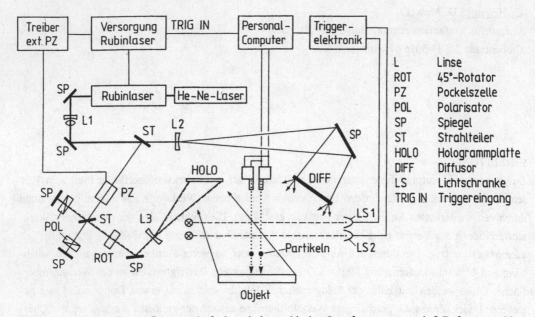

Bild 1: Experimenteller Aufbau zur Aufnahme holographischer Interferogramme mit 2 Referenzstrahlen

STEUERUNG DER AUFNAHME

Die Aufnahme eines Interferogramms von der Oberflächendeformation erfordert noch die Synchronisierung zwischen der Emission des Doppelpulses und dem Auftreffen der Partikel auf die Werkstoffoberfläche. Aufgrund der hohen Ausbreitungsgeschwindigkeit der Rayleighwelle von $\cong 3 \cdot 10^3$ m/s muß der Aufprall wenige 10µs vor dem zweiten Laserpuls erfolgen. Wegen der relativ geringen Geschwindigkeit der Partiteln von 30...40m/s sind die Fluktuationen der Flugzeiten ($\cong 10^{-3}$s) groß im Vergleich zu der geforderten Genauigkeit für die Synchronisation. Für jede Aufnahme eines Interferogramms wird die Geschwindigkeit der Partikel gemessen. Während ihres Fluges wird -in Abhängigkeit von der gemessenen Geschwindigkeit- eine individuelle Verzögerungszeit berechnet, mit der der Laser ausgelöst wird. Die Messung der Flugzeit, die Berechnung der Verzögerungszeit und die Steuerung der Hologrammaufnahme erfolgen über einen Personal-Computer. Die zeitlichen Anforderungen werden durch die Multifunktionskarte im Rechner gewährleistet, die mit einem eigenem Prozessor ausgestattet ist.

REKONSTRUKTION DER INTERFEROGRAMME UND PHASENMESSUNG

Die Interferogramme werden nach der Phasen-Shift-Methode rekonstruiert und ausgewertet [3]. Bildaufnahme, Bildvorverarbeitung und Berechnung der Interferenzphase erfolgen nach den bekannten Phasen-Shift-Gleichungen [3, 4] mit einem selbstentwickelten Bildverarbeitungsprogramm auf PC-Basis. Die somit vorliegende Interferenzphasenverteilung modulo(2π) erfordert eine spezielle Auswertung

(Demodulation). Zur sicheren Demodulation werden die digitalisierten Intensitätsverteilungen der rekonstruierten Interferogramme zunächst gefiltert. Dafür haben sich besonders Binominalfilter bewährt, die als Faltung im Ortsbereich angewendet werden. Wegen der geringen Amplitude der Rayleighwellen führen die durch das Speckle-Rauschen hervorgerufenen Fehler der Intensitätsmessungen zu relativ großen Fluktuationen in der Interferenzphasenverteilung. Dadurch entstehende Inkonsistenzstellen (Änderung der Interferenzphase entlang eines geschlossenen Weges ungleich Null) führen zu Fehlern bei der Demodulation der Interferenzphasenverteilung entlang eines vorgegebenen Weges [5], wie etwa bei zeilenweisem Vorgehen. Die Demodulation erfolgt daher, ähnlich dem von ETTEMEYER et. al. [Ref. [6.15] aus [5]] vorgeschlagenen Verfahren, entlang des Weges minimaler Änderung der Interferenzphase. Da aufgrund der unter dem Betriebssystem MS-DOS herrschenden Beschränkung bei der Speicherplatzverwaltung nicht das gesamte Interferenzphasenfeld gleichzeitig berücksichtigt werden kann, wird das Verfahren nur auf einen Ausschnitt des Feldes angewandt. Sobald ein zusammenhängendes Gebiet demoduliert ist, wird ein noch nicht demodulierter Bereich nachgeladen.

ERGEBNISSE DER INTERFEROMETRISCHEN MESSUNG

Im Bild 2a sind die nach dem Phasen-Shift-Verfahren erhaltenen Intensitätsverteilungen des holographischen Interferogramms einer Oberflächenwelle dargestellt. Die nicht in der Objektebene rekonstruierten Interferogramme sind mit einem Binominalfilter (Filtermaske: 11×11) gefiltert. Die räumliche Auflösung und die Dynamik des Bildes sind durch den Video-Kopierprozessor eingeschränkt. Im Vergleich dazu ist im Bild 2b das auf die Objektoberfläche fokussierte zugehörige Bild des Interferogramms dargestellt.

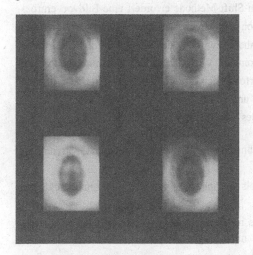 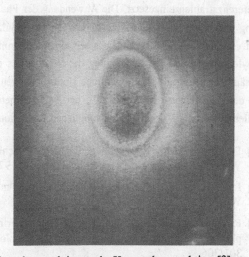

Bild 2a: Rekonstruierte Intensitätsverteilungen, Selbstrekonstruktionen, b: Kreuzrekonstruktion [3]

Bild 3a ist die Grauwertdarstellung der gemessenen Interferenzphase. Die zugehörige Oberflächendeformation wird mit Hilfe des Sensitivitätsvektors berechnet [6] und ist im Bild 3b als Pseudo-3D Darstellung zu sehen. Anhand der Bilder 3a und b ist zu erkennen, daß die gemessene Oberflächendeformation in einem kleinen Bereich deutlich größer ist als im Rest des Bildes. Dieser Bereich fällt mit dem im Bild 2b erkennbaren Bereich der Werkstoffschädigung zusammen. Dort ist die Reflektivität aufgrund der erhöhten Oberflächenrauheit des Materials reduziert.

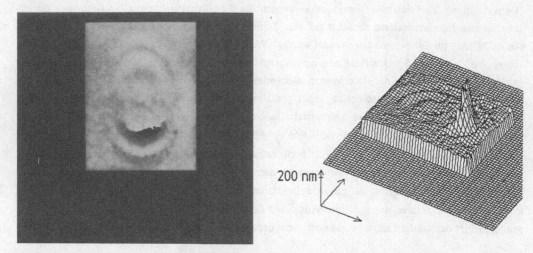

Bild 3a: Gemessene Interferenzphase (Grauwertdarstellung), b: Oberflächendeformation

ZUSAMMENFASSUNG UND AUSBLICK

Mit Hilfe der Zwei-Referenzstrahl-Technik werden holographische Interferogramme von Oberflächenwellen mit kleiner Amplitude ($\approx 10^{-8}$m) aufgenommen, die mit dem Phasen-Shift-Verfahren rekonstruiert werden. Zur Aufnahme der Interferogramme wird eine digitale Steuerung zur Synchronisation der transienten Oberflächenwelle mit der Emission des Doppelpulses und dem Umschalten zwischen den Referenzstrahlen eingesetzt. Die Anwendung der Phasen-Shift-Methode erfordert eine Bildvorverarbeitung und einen speziellen Algorithmus zur Demodulation der berechneten Interferenzphasenverteilung auf einem PC. Die Messungen lassen einen Zusammenhang zwischen der Amplitude der gemessenen Rayleighwelle und der Schädigung des Werkstoffes erkennen. Zur Klärung des Zusammenhangs sind zusätzliche materialwissenschaftliche Untersuchungen erforderlich. Durch Variation der Kristallitgröße an der Oberfläche der Werkstoffe, der Stoßparameter und durch geeignete Werkstoffbeschichtungen sollen langfristig Vorschläge zur möglichen Reduktion des Verschleißes hergeleitet werden.

[1] D. Mewes: Modellvorstellung für den Verschleißmechanismus beim Prallbeschuß kristalliner Werkstoffoberflächen; Dissertation TU Berlin 1970

[2] I. M. Hutchings: A model for the erosion of metals by spherical particles at normal incidence; Wear 70 (1981), 269/281

[3] R. Dändliker, R. Thalmann, J.-F. Willemin: Fringe interpolation by two-reference-beam holographic interferometry: Reducing sensitivity to hologram misalignement; Opt. Comm. 42 (1982) 5, 301/306

[4] W. Jüptner, T. M. Kreis, H. Kreitlow: Automatic Evaluation of Holographic Interferograms by Reference Beam Phase Shifting; Proc. of SPIE 398 (1983), 22/29

[5] W. Osten: Digitale Verarbeitung und Auswertung von Interferenzbildern; Akademie Verlag, Berlin 1991

[6] J. E. Sollid: Holographic Interferometry Applied to Measurements of Small Static Displacements of Diffusely Reflecting Surfaces; Appl. Opt. 8 (1969) 8, 1587/1595

Optical Measuring and Testing System of the Large-Dimensional Rails' Robot

Janusz SZPYTKO, Stefan STUPNICKI
University of Mining and Metallurgy
al. Mickiewicza 30,
30-059 Cracow, Poland

Introduction

Automated overhead crane (robot) is an example of large-dimensional rails' material handling device (LDRMH) working in Computer Integrated Manufacturing (CIM). Robots requirements include necessary accuracy of operation movements, reliability, safety, diagnosis ability.

Possibility of operation motions control (with assumed precision) along a taken trajectory and continuous monitoring of the object technical condition are principal criterions in considerations on robot application in CIM.

The presented noncontact optical measuring and testing system enable control/testing crane/robot planned trajectory (operation movement) and investigation his technical condition. The optical measuring and testing system is based on laser and vision techniques, computer-aided image analysis [1, 2, 3].

Measuring System

Diagnostic measuring system of LDRMH is based on laser and visual technique (CCD camera) and computer image analysis. The system allows for remote both static and dynamic measurements in noncontact and automatic way with accuracy required for this class of devices. The block scheme of the system is presented in Figure 1.

In the measuring system low power (about 1.0 mW) He-Ne lasers are in use. Light beam emitted by a laser emitter has the function of visible

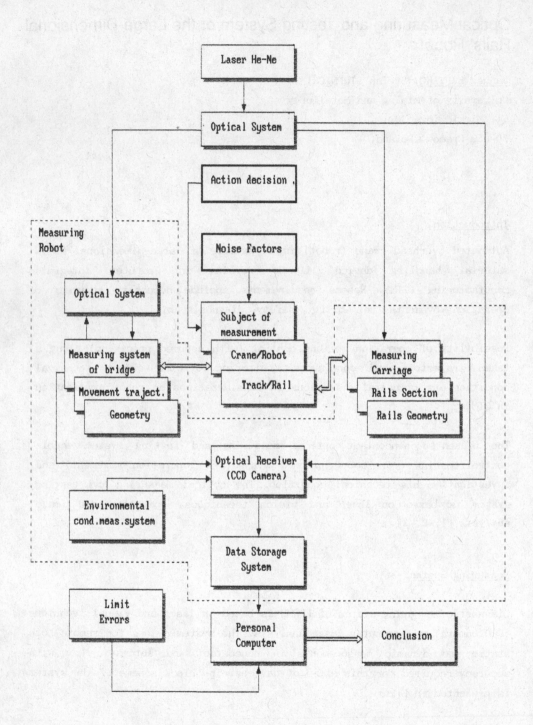

Figure 1. Block Diagram of Crane/Robot Measuring and Testing System.

measurement base conducted in cartesian system. The measurement base is coordinated with the railroad of known theoretical geometric relations and where an LDRMH is moving. Moreover, the laser emitted beam is useful in noncontact measurement of linear and angle displacements, in control of measuring devices operation.

A universal measuring robot of module structure is used in measurement. The main modules of the measuring robot are: measuring system of the bridge, measuring carriage of the track and rails, environmental condition measuring system and optical receiver. Changes of the LDRMH motion trajectory parameters, geometry of railroad and bridge in relation to fixed measurement base are recorded by CCD camera.

The measuring robot can measure and record in each t moment (or in each rail section on its length):
-the LDRMH bridge and rails displacements in LDRMH run,
-geometry of rail: deviations from linearity in the horizontal and vertical plane, width of rail head,
-actual distance of measurement section (point) of rail from the beginning of the rail (the first measurement point),
-geometry of the bridge and bridge travel mechanism of the LDRMH in the horizontal and vertical plane,
-environmental conditions or additional necessary information.

The measurement data are storage in the database system. Using Personal Computer and special authors' image analysis software the technical condition of the crane/robot and crane track/rail can be concluded.

References

1.Szpytko J.: Analysis of Motion Trajectory of Large-Dimensional Rails' Transport Facilities for Diagnostic Needs. International Academic Publishes, pp. 397-402, Beijing, 1991.

2.Szpytko J., Stupnicki S.: The rails' geometry laser measuring system. Polish Patent no. 156986, dated October 22, 1992.

3.Szpytko J. Stupnicki S.: The laser measuring system of the crane bridge movements. Polish Patent no. 156582, dated February 22, 1993.

3D-Lichtschnittsensor zur Biegewinkelerfassung

W. Heckel und M. Geiger
Lehrstuhl für Fertigungstechnologie, Universität Erlangen-Nürnberg
Egerlandstr. 11, D-91058 Erlangen, Tel.: 09131-857140, Fax: 09131-36403

Abstract

Air bending is a valuable manufacturing method in flexible and computer integrated manufacturing systems but adaptive control strategies are necessary to overcome deviations from the desired bending angle caused by material tolerances. This paper describes an industrial application of optical 3-D sensing aiming at an in-process measurement of bending angles in a commercial bending machine. The measurement system is based on the principle of light sectioning. The in-process measurement of the bending angle after release of the workpiece achieves a precision of ± 0,03°. The evaluation time is 0,1 s. The experimental results prove that incremental bending steps can be applied to adjust the bending angle.

1   Ziele

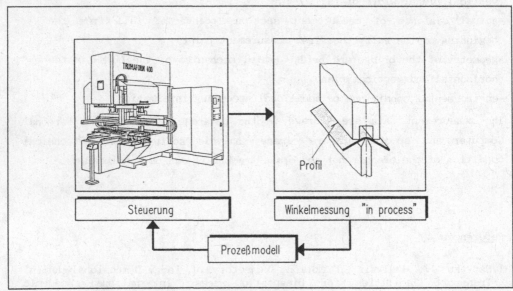

Bild 1:   Adaptive-Control-System zum freien Biegen

Das freie Biegen ist ein für den Einsatz in flexiblen Fertigungssystemen besonders vorteilhaftes Fertigungsverfahren, da mit einem Werkzeug eine große Bandbreite verschiedener Biegewinkel hergestellt werden können. Die erreichbare Winkelgenauigkeit beim freien Biegen wird vor allem durch die Toleranzen der Werkstoffkenndaten (Blechdicke, Fließgrenze, Anisotropie, usw.) auf etwa 1° begrenzt.

Die erforderliche Steigerung der Fertigungsqualität setzt daher den Aufbau eines Prozeßregelkreises zum freien Biegen (schematisch dargestellt in Bild 1) voraus.

2 Das Meßprinzip des Lichtschnittverfahrens

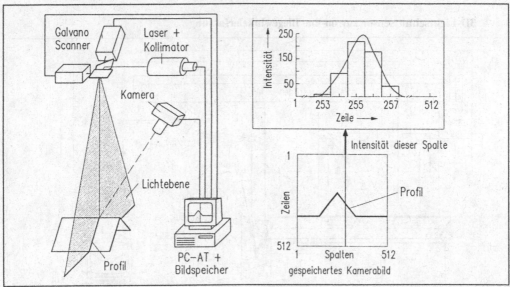

Bild 2:   *Prinzip des Lichtschnittverfahrens*

Der hier vorgestellte 3D-Lichtschnittsensor zur on-line Biegewinkelerfassung bildet die Grundlage für diesen Regelkreis und ähnelt dem Lichtschnittsensor nach [1]. Ein fokussierter Laserstrahl wird mittels eines 2-Achsen-Galvanoscanners während einer Integrationsperiode des Detektors geradlinig über das Werkstück geführt, vgl. Bild 2. Die Schnittlinie der so erzeugten "Lichtebene" mit der Werkstückoberfläche, das "Profil" des Werkstücks, wird mit einer CCD-Kamera unter dem Triangulationswinkel Θ detektiert und zur Auswertung im Bildspeicher des Meßrechners abgelegt. Aus dem Verlauf des Profils im Bildspeicher erhält man mit einer einzigen Messung die 3D-Information der Objektoberfläche entlang des Profils (bestehend aus bis zu 512 Meßpunkten).

In jeder Spalte des Bildspeichers repräsentiert die Position des Profils die 3D-Koordinaten eines Punkts der Werkstückoberfläche. Durch Anwendung eines Interpolationsalgorithmus' kann Profilposition jeder Spalte mit Subpixelgenauigkeit bestimmt werden. Im zweiten Auswerteschritt wird durch die Profildaten der Biegeschenkel je eine Regressionsgerade gelegt. Diese beiden Regressionsgeraden werden im dritten und entscheidenden Auswerteschritt vom Bildspeicher- in das 3D-Weltkoordinatensystem transformiert. Der Schnittwinkel der beiden Geraden in Weltkoordinaten liefert schließlich den gesuchten Biegewinkel des Werkstücks. Die Transformationsformeln wurden in [2] umfassend dargelegt. Die Meßgenauigkeit des biegemaschinenunabhängigen Lichtschnittsensors beträgt ± 0,05° bei einem Arbeitsabstand von 1000 mm [3]. Die Meßzeit incl. Auswertung beträgt rund 0,1 s.

Dieses 3D-Meßsystem nach dem Lichtschnittprinzip ist in der Lage, wesentlich mehr Information als den Biegewinkel des Werkstücks zu liefern. Mit einer Messung erhält man für bis zu 512 Punkte der Werkstückoberfläche entlang des Profils die absoluten 3D-Weltkoordinaten. Das Meßprinzip eignet sich

308

damit hervorragend für eine on-line Konturerfassung als Grundkomponente von Regelkreisen bei einer Reihe weiterer Fertigungsverfahren. Der 3D-Lichtschnittsensor wird daher sicher weitere interessante Anwendungen in der Fertigungstechnik finden. Im weiteren konzentriert sich der Beitrag auf die on-line Biegewinkelerfassung mit dem Lichtschnittsensor.

3   3D-Lichtschnittsensor zur on-line Biegewinkelerfassung

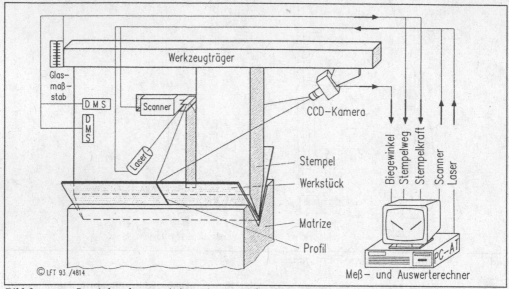

Bild 3:   *Spezialwerkzeug mit integriertem Lichtschnittsensor*

Zur on-line Biegewinkelerfassung wurde ein Sonderwerkzeug mit integriertem Lichtschnittsensor für das freie Biegen entwickelt [4], siehe Bild 3. Dabei wurden Laserdiode, Resonanz-Scanner (aufgrund der festen geometrischen Randbedingungen reicht hier eine bewegte Achse aus) und CCD-Kamera in ein Versuchswerkzeug für die NC-gesteuerte 10-Achsen-Biegemaschine am Lehrstuhl für Fertigungstechnologie integriert. Die definierte Lage des Biegeteils im Bildfeld der Kamera vereinfacht die Kalibrierung der Kamera wesentlich. Über einen Glasmaßstab in der Biegemaschine und Dehnmeßstreifen am Biegestempel ist die gleichzeitige Erfassung von Stempelweg und -kraft möglich. Das gesamte Meßsystem ist auf einem Werkzeugträger der Biegemaschine aufgebaut, so daß jederzeit ein Werkzeugwechsel ohne Veränderung des Lichtschnittsensors möglich ist.

4   Ergebnisse der on-line Biegewinkelerfassung

Bild 4 zeigt den on-line gemessenen Verlauf des Biegewinkels in Abhängigkeit der Eintauchtiefe des Stempels in das Gesenk. Dabei wurde ein Werkstück (St14O3, $s_0 = 1,0$mm) bis zu einer definierten Eintauchtiefe gebogen und anschließend unter laufender Kontrolle des Biegewinkels entlastet. Dieser Be- und Entlastungszyklus wurde bei einem Werkstück mit jeweils um 0,02 mm erhöhter Eintauchtiefe vielmals wiederholt. Die vollständige Entlastung des Werkstücks ist in jedem Zyklus deutlich am Biegewinkelverlauf zu erkennen (siehe Bild 4). Dies belegt, daß der entscheidende Biegewinkel des entlasteten Werkstücks direkt im Gesenk meßbar ist. Bei den nachfolgenden Belastungen des Werkstücks werden bei gleicher Eintauchtiefe identische Winkelwerte wie bei den vorherigen Belastungen erzielt. Mit einer Erhöhung der maximalen Eintauchtiefe um 0,02 mm nimmt der entlastete Biegewinkel um etwa

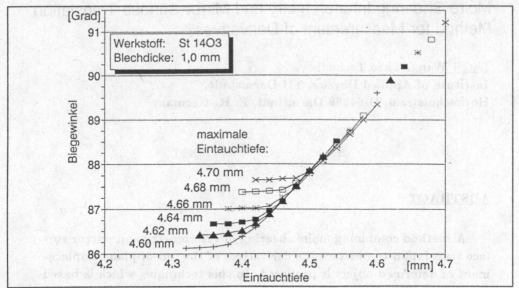

Bild 4:   *On-line Biegewinkelerfassung*

0,3° zu. Diese Veränderung ist eindeutig detektierbar. Die in Bild 4 dargestellten Messungen belegen, daß ein Regelzyklus mit inkrementalen Korrekturbiegungen und jeweils Messung des entlasteten Biegewinkels möglich ist. Die Meßgenauigkeit des on-line Lichtschnittsensors beträgt ± 0,03° bei einer Meß- und Auswertezeit von 0,1 s.

Nachdem der entlastete Biegewinkel direkt gemessen werden kann, ist für eine Prozeßregelung des freien Biegens im Prinzip keine Information über das Werkstück erforderlich. Dennoch werden im industriellen Einsatz wissensbasierte Regelstrategien auf der Basis von a-priori-Wissen von Werkstoff und Werkstückgeometrie evtl. in Verbindung mit einer prozeßbegleitenden Messung der Stempelkraft und des Stempelwegs vorteilhaft sein, um zum einen die Zahl der Winkelmessungen zu minimieren und zum anderen Rückschlüsse auf weitere Biegungen des gleichen Werkstücks ziehen zu können.

5   Danksagung

Die diesem Bericht zugrundeliegenden Forschungsarbeiten wurden von der Deutschen Forschungsgemeinschaft unter dem Aktenzeichen Ge 530/6-1 gefördert. Die Autoren danken für diese Unterstützung.

6   Literaturverzeichnis

[1]   *Häusler, G.; Heckel, W.: Light sectioning with large depth and high resolution. In: Applied Optics, Vol. 27 (1988), No. 24, S. 5165.*

[2]   *Heckel, W.: Use of Laser Metrology for In-process Measurement of Bending Angles. In: Lasers in Engineering, Vol. 2 (1993) No. 1, im Druck.*

[3]   *Geiger, M.; Heckel, W.: 3D-Lichtschnittsensor zur Biegewinkelerfassung. In: Blech Rohre Profile 40 (1993) 3, S. 235-240.*

[4]   *Geiger, M.; Heckel, W.: Biegemaschine zum Biegen flächiger Werkstücke. Patentanmeldung P 43 12 565.4, 1993.*

Moiré Shearing Interferometry and Mirror Surface Translation Method for Measurement of Derivatives

Lingli Wang, Theo Tschudi
Institute of Applied Physics, TH Darmstadt,
Hochschulstr. 6, D-64289 Darmstadt, F. R. Germany

ABSTRACT

A method combining moiré shearing interferometry and mirror surface translation to measure the derivatives of the out-of-plane displacement of deformed object is proposed. In this technique, which is based on grating diffraction, only one grating is used. Instead of the usual inconvenient step of reflection coating of the surface of the object under test, the mirror surface translation method is used. Compared with other techniques, the proposed method is an easy way to obtain the derivatives of the deformed object, because only one ordinary grating (e.g. grating with a pitch of 0.005-0.01 cm) is used and a high sensitivity can be reached. For illustration the derivatives of a clamped steel plate with a concentrated force applied to its center is measured by this method.

1. INTRODUCTION

It is relatively easy to obtain the stress distribution by the correspondence slope or curvature of a deformed object. There are many specialised methods available to identify the slope or curvature of a deformed object. Grating shearing interferometry and mirror surface translation is on such way of obtaining the slope or curvature of a deformed object. However, the object surface, which will be measured normally, must be like a mirror, which is generally not practical. In addition many surfaces are impossible to make mirrorlike even by utilising the technique of metallization, since the surfaces are not smooth enough.

In our work, a mirror surface translated method is used to translate the mirror surface to the object surface which has to be measured. By using grating shearing interferometry, the slopes of a clamped steel plate with a concentrated force at its center are measured. The theoretical values agree with the experimental results. The mirror surface translated method makes grating shearing interferometry easy to measure the slope of deformed objects.

2. PRINCIPLE AND EXPERIMENT

A incident monochromatic plane wave passes through a transparent grating and is separated into several orders. Only the $\pm$first-order diffraction waves are allowed to pass through a spatial filter, which are reflected by the surface of object and finally reach on observing plane.

Supposing that the incident wave is expressed by $exp2\pi i(\theta x)/\lambda$, the transmittance amplitude of the grating is $T(x) = \sum_{n=-\infty}^{\infty} A_n exp2\pi inx/d$. d is the grating pitch. When the object is in initial state, the first exposure is made. The intensity distribution on plane P is

$$I_1 = 2a_1^2 + 2a_1^2 cos2\pi(2/d)(x - z_p\theta) \tag{1}$$

in which, a_1 is a constant; z_p is the distance from grating to observing plane. θ is the angle of incidence. When the object is deformed by the deformation $w(x,y)$, the second exposure is made. Letting $\partial w/\partial x = [w(x + b) - w(x - b)]/(2b)$, the intensity distribution of the second exposure on plane P will be

$$I_1 = 2a_1^2 + 2a_1^2 cos\frac{4\pi}{d}(x - z_p\theta - 4l_2\frac{\partial w}{\partial x}sec^2\theta), \tag{2}$$

where, $b = \frac{\Delta}{d} = \frac{2l_2\lambda}{d}sec^3\theta$, which is a half shearing value between the ± 1 diffraction orders; λ is the wavelength; l_2 is the distance between the object and the observation plane.

After the second-exposure, the total intensity distribution on plane P will be

$$F(x,y) = I_1 + I_2 = 4a_1^2 + 4a_1^2 cos\frac{4\pi}{d}(x - z_p\theta - 2l_2\frac{\partial w}{\partial x}sec^2\theta)cos\frac{4\pi}{d}(2l_2\frac{\partial w}{\partial x}sec^2\theta). \tag{3}$$

We know that the intensity distribution of the moiré pattern may be obtained by assuming that the observing system is a low-pass filter which cuts off the components whose spatial frequency is higher than that of the grating. Assuming that the function $\frac{\partial w}{\partial x}$ varies slowly, the fringes of the moiré pattern are given by the set of maxima when the sign of the term of $cos[4\pi l_2(\frac{\partial w}{\partial x})sec^2\theta]/d$ is positive, as

$$\frac{\partial w}{\partial x} = \frac{md}{4l_2 sec^2\theta} \qquad (4)$$

Where, m is a positive integer. The object which has been measured is a circle steel plate. The radius of the plate is 4 cm and the modulus of elasticity $E = 205,947$ MPa. Poisson's ratio $\mu = 0.28$. The steel plate is clamped with a concentrated force applied at the center. A 5mW He-Ne laser and a grating with a pitch of 0.00847cm are used. The mirror surface is translated onto the steel plate.

3. CONCLUSIONS

By using the grating shearing interferometry and mirror surface translated method, it becomes easy to measure the slope distribution of a deformed object with only ordinary grating, e.g., the grating with a pitch of 0.005-0.01cm, and with a high sensitivity. Although it is possible to achieve high sensitivity(the sensitivity is equivalent to the method in this paper) by using one beam shearing moiré interferometry, a grating with a pitch of 0.000167 has to be used. The method used in this paper is a simlpe way of obtaining the slope distribution, especially for surfaces that are not smooth enough. However, utilisation of this method is only in its infancy. Many problems have to be solved. For example, the quality of the fringe patterns is not yet good enough.

4. ACKNOWLEDGMENTS

The work was supported by the Alexander von Humboldt Foundation. The authors express their thanks to Prof. D. Z. Yun and J. B. Dang for valuable discussion and help.

Experimental Investigation of the Optical Characteristics of Diode Lasers in CW- and High-Frequency Pulsed Operation Modes for LDA Applications

H. Wang, D. Dopheide, V. Strunck

Physikalisch-Technische Bundesanstalt (PTB), Department for Fluid Mechanics

Bundesallee 100, D-38116 Braunschweig

1. Introduction

As more wavelengths and higher powers become available, diode lasers continue to find new applications. Their costs, compactness, efficiency and modulation capability make them an ideal tool for many applications. However, laser Doppler anemometry (LDA), laser spectroscopy and other fields require lasers with spectral purity and high optical output power [DAMP, 1988; DOPHEIDE et al, 1988 & 1989; STRUNCK et al, 1992].

The emission characteristics of diode lasers are mainly characterized by the optical output power and the dependence of the emission wavelength on the current and temperature of diode lasers (i.e. mode card). The optical output power of a diode laser can be enhanced by using the high-frequency pulsing technique [STRUNCK et al, 1992; DOPHEIDE et al, 1990]. The index-guided GaAlAs single stripe diode lasers usually show a single longitudinal mode emission. But this holds true only in certain operating temperature and current ranges of diode lasers (operating points).

2. CW- and high-frequency pulsed (HF-pulsed) operation mode of diode lasers

As shown in Fig. 1, the characteristic relationship between output power and injected current of a diode laser reflects the parameters which influence the emission characteristics of a diode laser. In the CW operation mode, the optical characteristics can be described by the relationship between the wavelength λ, temperature T, current I, and output power P. While in the HF-pulsed operation mode, a duty cycle of laser pulses is additionally needed to describe the pulse characteristics of a diode laser:

$$\text{Duty cycle} = (1/f_{\text{pulse}} - \tau_{\text{pulse}})/\tau_{\text{pulse}} \tag{1}$$

3. Typical emission characteristics of various diode lasers

Experimental set-up   Fig. 2 shows the set-up used for measuring mode cards and light output power of diode lasers. The frequency of the pulse generator is fixed at 80 MHz during experiments (i.e. f_{pulse} = 80 MHz), because this is the frequency used as a local oscillator in PTB's HF-pulsed LDA [STRUNCK et al, 1992].

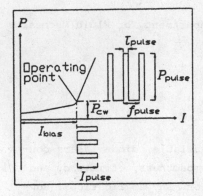

Fig. 1 Characteristic relationship between output power P and injection current I (i.e. I_{bias} + I_{pulse}) of a diode laser in CW- and HF-pulsed operation modes. The current I is fixed in this case. An increase of the bias current I_{bias} decreases the pulse current I_{pulse}. The meaning of the symbols in the figure is as follows:
I_{bias} is the bias current when operating
  in the HF-pulsed mode,
I_{pulse} is the pulse current,
P_{CW} is the pure optical power of
  continuous laser light significant for I_{bias},
f_{pulse} is the frequency of laser pulses,
P_{pulse} is the peak power of HF-pulsed laser light,
τ_{pulse} is the width of laser pulses.

In order to study the influences of duty cycles on the output power of diode lasers, various duty cycles are obtained by varying the pulse width of the pulse generator. Based on the reason explained later, a duty cycle of about 2 is chosen for measuring the mode cards in the HF-pulsed operation mode, unless otherwise specified. The variable attenuator controls the pulse current. The whole test system is programmed for automatic measurements of diode lasers in the CW operation mode. Diode lasers in the HF-pulsed operation mode are measured by manually varying the operating parameter in relatively large steps.

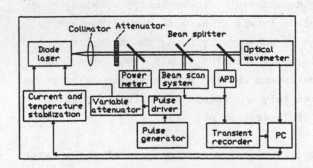

Fig. 2 Block diagram of the experimental set-up used for measuring mode cards and light output power of diode lasers.

Mode card analysis   Typical mode cards of a diode laser consist of four individual cards: 1). λ-I-T card — dependence of the emission wavelength of a diode laser on the current in a certain range of temperatures; 2). λ-T-I card — dependence of the emission wavelength of a diode laser on the temperature in a certain range of currents; 3). λ-I card — dependence of the emission wavelength of a diode laser on the current at constant temperature; and 4). λ-T card — dependence of the emission wavelength of a diode laser on the temperature at constant current. The last two cards are discussed in more detail in the following sections.

Various diode lasers from Hitachi have been selected and measured. For comparison, the dependence of the emission wavelengths of these diode lasers on current and temperature is shown in Fig. 3.

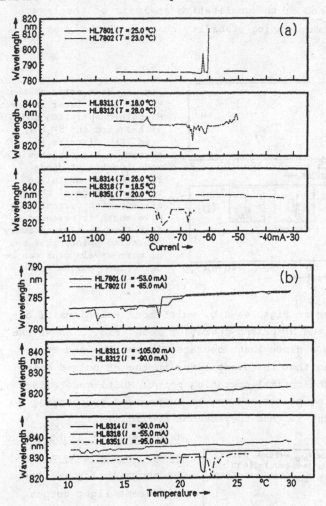

Similar to the Figs. 3a & b, Figs. 4a & b show the dependence of the emission wavelength on the diode laser's temperature and current in the HF-pulsed operation mode. Here, the discontinuities are caused by the multi-mode emission of diode lasers because side modes become comparable to the main mode and several wavelengths are appearing in the pulsed wavemeter's output (refer to Figs. 4a, 4b & 5). The condition preferred in LDA applications is a current and temperature range where most of the optical power contributes to only one mode (i.e. mono-mode emission at $T = 25.0$ °C and $I_{bias} = 76.0$ mA in Figs. 4a & b).

In the CW operation mode, it was observed that the mode jumps vary rapidly from one mode to the other, often without multi-mode emissions during mode hopping. In contrast to this, the mode changes

Fig. 3 Mode cards of various diode lasers at the optimal operating points in the CW operation mode: a) $\lambda-I$ card and b) $\lambda-T$ card. The legends indicate the type designation of diode lasers and its related optimal operating current I and temperature T. The diode lasers from Hitachi are investigated because they are used successfully in the PTB's diode laser-based CW and HF-pulsed LDAs [DOPHEIDE et al, 1988, 1989 & 1990; STRUNCK et al, 1992].

slowly from one mode to the other in the HF-pulsed operation mode, and multi-mode emissions are observed during mode hopping. It takes a few seconds for the diode laser to achieve a final equilibrium and to show a smaller spectral change if its operating point is changed in the HF-pulsed operation mode.

It should be pointed out that 1) different diode lasers have different mode
cards, even if they are of the same type and from the same manufacturer;
2) no significant change of the mode cards was observed during the long-
term usage of diode lasers; and 3) the qualitative analysis of the laser
beam shows that the laser beam pointing stability changes slightly as
operating parameters change.

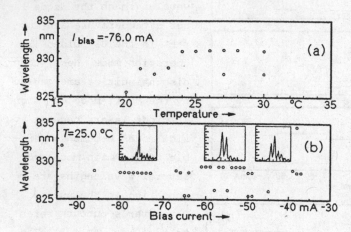

Fig. 4 Mode cards of the
HL8314 diode laser in the
HF-pulsed operation mode:
a) $\lambda-I$ card and b) $\lambda-T$
card. The diagrams in-
serted in Fig. 4b show the
mode spectra of diode
laser emissions in the
relevant range of the op-
erating bias current. The
wavelength is measured
from time to time at every
operating point, thus two
or more wavelengths can be
determined during mode
hopping.

Spectral analysis   As shown in Figs. 4a & b, multi-mode emissions of diode lasers are observed in the HF-pulsed operation mode. Fig. 5 shows the
spectral comparison between a diode laser operated in the CW- and HF-
pulsed modes. It can be seen that the diode laser in the HF-pulsed operation mode is monomodal at the optimal operating point. Multi-mode emissions occur at some operating points and can be suppressed by changing
bias current and temperature, so that they become negligible.

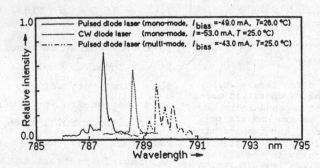

Dependence of light output
power on the duty cycle
As shown in Fig. 6, the
different light output
powers of a diode laser at
different duty cycles are
measured using the
apparatus shown in Fig. 2.
It can easily be seen that
the peak power of a diode
laser can be increased
many times through the
pulsing technique compared

Fig. 5 Typical emission spectra of the HL7801
diode laser at different operating points in
CW-and HF-pulsed operation modes.

to the CW operation mode. For example, the peak power of a diode laser
pulse at a duty cycle of about 2 (f_{pulse} = 80 MHz) is nearly twice that of

the pulse at a duty cycle of about 1, providing the average power in both cases is about 4.0 mW (HL7801 is used). The shorter the trigger pulse of the pulse driver, the stronger the output peak power. However, a duty cycle of about 2 to 3 is recommended for f_{pulse} = 80 MHz, since the time response of diode lasers is limited. The shorter laser pulses need very strong current pulses and their pulse shapes become noisy.

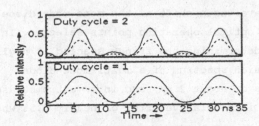

Fig. 6 Comparison of the light output of an HL7801 diode laser operating at various duty cycles (f_{pulse} = 80 MHz). All full lines are adjusted by varying the attenuation to correspond the same peak output power of the HF-pulsed diode laser; the same applies to the dashed lines. The difference between the peak output powers represented by the full and dashed lines is about a factor 2. For simplicity, I_{bias} is set to zero. In the CW operation mode, the diode laser has a CW power of 4.3 mW at a temperature of 25.0 °C and a current of −53.0 mA.

	Duty cycle	Line type	Average power (mW)	Estimated I_{pulse} (mA)
Upper part	2	full	5.1	− 75.0
		dashed	2.6	
Lower part	1	full	8.0	− 70.0
		dashed	4.5	

4. LDA applications

HF-pulsed LDA   In order to investigate the influence of diode lasers' side modes on the fringe pattern, the experimental set-up in Fig. 2 is modified according to Fig. 9, which actually shows one component HF-pulsed LDA — parts of PTB's two component HF-pulsed LDA [STRUNCK et al , 1992].

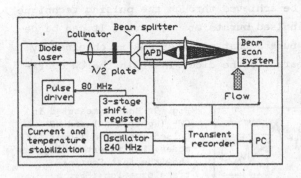

Fig. 7 Block diagram of the experimental set-up used for measuring the fringe pattern and burst signals. The local oscillator operates at 240 MHz and toggles a three-stages shift register. The shift register is made in such a way that only one of its three outputs is set at a time, giving pulses with a frequency of 80 MHz. One output of the shift register is fed into the pulse driver. The transient recorder is externally triggered at 240 MHz for coherent sampling [DOPHEIDE et al, 1990].

Influence of multi-mode emissions of diode lasers on the fringe pattern
The measured fringe patterns of the LDA probe volume are shown in Fig. 8. It can be easily deduced that the side modes of diode lasers can reduce the visibility of fringe patterns, because the side modes impair the coherence length of the diode laser beam and make the alignment very difficult.

318

Referring to Figs. 3a & b and 4a & b, mode jumps may result in a wavelength shift by up to 10 nm. This in turn causes a relative error of up to 1.2% when the fringe spacing is determined according to the following equation:

$$\Delta x = \lambda / [2 \bullet \sin(\varphi)] \qquad (2)$$

Here: Δx is the fringe spacing, λ is the laser wavelength and φ is the half angle between laser beams.

With respect to Figs. 3a & b, the wavelength shift of the mono-mode diode laser is smaller than 0.1 nm at the optimal operating points selected. In such a case, the relative error on determining the fringe spacing is negligible. As shown in Fig. 5, the emission spectrum of a multi-mode diode laser has a typical wavelength range of about 2 nm, and in an extreme case, could be about 3 to 4 nm, resulting in a relative error of nearly 0.5% when the fringe spacing is determined.

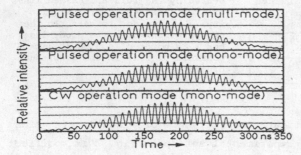

Fig. 8 Fringe pattern measured at different operation modes and points. Differences among the modulation depths of various fringe patterns can be seen. The mode spectra of the diode laser and their related operating points are shown in Fig. 7. In order to obtain the best quality of the fringe patterns, the optical arrangement is adjusted for each operating point and mode.

SNR analysis   Burst signals are measured with the apparatus shown in Fig. 7. The result is presented in Fig. 9. An increase of the signal intensity with a factor of about 2 can be achieved through the pulsing technique when signal intensities of CW and pulsed bursts are compared. It can be deduced from the Fourier spectra of bursts that a better SNR is obtained by the HF-pulsed diode lasers if compared to the diode laser in the CW operation mode.

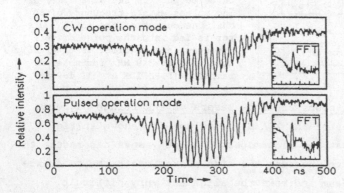

Fig. 9 Bursts measured in CW- (upper left) and HF-pulsed operation (lower left) modes. The inserted diagrams show the fast Fourier transformation (FFT, a logarithmic scale of the intensity is used) of the related burst. Both vertical scales of burst signals are comparable. $f_{pulse} = 80$ MHz. Duty cycle = 1.

By pulsing the HL8312 diode laser at a duty cycle of 3 (f_{pulse} = 80 MHz), it was observed that the intensity of HF-pulsed burst signals can be five times higher than that of CW burst signals. In this case, both the CW light power in the CW operation mode and the average pulse power in the HF-pulsed operation mode are about 4.0 mW. The average pulse power of 4.0 mW is chosen because of limitations of the pulse driver. Such high intensity enhancements ensure the good SNR of the HF-pulsed LDA.

The spectral emission behaviour and the SNR can be further improved by varying the pulse current superimposed on the bias current. This is, however, observed only qualitatively due to the complexity of operating the pulse driver (i.e. emission spectra versus bias current, pulse current and temperature have to be measured).

5. Discussion and conclusion

The mode cards of GaAlAs semiconductor lasers are experimentally investigated in detail in both the CW- and the HF-pulsed operation mode. This is very important for applying diode lasers in diode laser-based LDAs, because the mode cards of diode lasers differ and the relative errors on determining the fringe spacing could increase up to 1.2% if the operating points of a diode laser are not appropriate. The optimal operating points of diode lasers insure the visibility of the fringe pattern, and allow the accurate determination of the Doppler frequency.

The optical output power in the HF-pulsed operation mode is strongly enhanced compared with the CW operation mode and leads to a better SNR. The peak power of PTB's HF-pulsed diode laser is usually twice its average power. Clearly separated laser pulses at duty cycles of about 2 to 3 enable the optimal use of coherent sampling to be made. An optimization of the duty cycle is recommended because of the limited time response of diode lasers and the optimal use of the pulse current.

The light emission of a multi-mode diode laser has a short optical coherence length. As a result, adjustment is difficult and the fringe pattern obscures. It is recommended that the optical alignment should be done for the specified operating parameters because of beam pointing stability.

References
S. Damp, Proc. 4th Inter. Symposium on Application of Laser Anemometry to Fluid Mechanics, Instituto Superior Tecnico, Lisbon, paper 5.4 (1988).
D. Dopheide, G. Taux, M. Faber, Experiments in Fluids, 6, 297-298 (1988).
D. Dopheide, H. J. Pfeifer, M. Faber, G. Taux, J. of Laser Applications, 1, 40-44 (1989).
D. Dopheide, V. Strunck, H. J. Pfeifer, Experiments in Fluids, 9, 309-316 (1990).
V. Strunck, D. Dopheide, M. Rinker, 6th Intern. Symposium on Applications of Laser Techniques to Fluid Mechanics, July 20th to 23rd 1992, Lisbon, Portugal, Conference proceedings, paper 11.2 (1992).

Photoelastischer Kraftsensor mit Proportionalverhalten und Resonanzverstärkung

U. Neuschaefer-Rube, W. Holzapfel

Institut für Meß- und Automatisierungstechnik, Universität Gesamthochschule Kassel, D-34109 Kassel

1 Zusammenfassung

Ein neuartiges, laseroptisches Kraftmeßverfahren basierend auf dem photoelastischen Effekt wird beschrieben. Durch eine erweiterte Polariskopanordnung in Verbindung mit einem speziellen Signalauswerteverfahren liefert der Sensor als kraftproportionales Ausgangssignal ein digitalisiertes Zeitintervall τ. Die nach der Theorie zu erwartende Meßauflösung von ca. $4 \cdot 10^{-5}$ bei einer Meßzeit von ca. 4 ms bestätigt sich im Experiment. Der Nichtlinearitätsfehler ist dabei kleiner als 10^{-3}. Um auch kleine Kräfte messen zu können, wird die Empfindlichkeit des photoelastischen Sensorelementes durch beidseitige Verspiegelung und Resonanzankopplung bzw. durch Miniaturisierung (Einsatz einer Sensor-Faser) gesteigert. Mögliche Anwendungen des Sensors werden diskutiert.

2 Einleitung

Bei bekannten photoelastischen Kraftsensoren ist das doppelbrechende Sensorelement entweder innerhalb /1/ oder außerhalb /2/ eines Laserresonators oder in einem Interferometer /3, 4/ angeordnet. Die resonatorexternen Anordnungen sind dabei besonders einfach aufgebaut. Hierbei befindet sich das Sensorelement zwischen zwei Polfiltern (Polarisator, Analysator). Die Abhängigkeit der transmittierten Lichtleistung von der zu messenen Kraft ist allerdings nichtlinear und periodisch. Nur mit Hilfe einer Verzögerungsplatte läßt sich eine Kennlinie erreichen, die über einen gewissen Bereich näherungsweise linear ist. Weiterhin führen Leistungsdriften der verwendeten Lichtquelle zu entsprechenden Driften des Sensorausgangssignals. Durch eine Erweiterung des Aufbaus und Quotientenbildung /5/ läßt sich dies zwar kompensieren, es ist aber ein erhöhter apparativer Aufwand erforderlich.

Der im folgenden beschriebene Kraftsensor vermeidet die Nachteile der Polariskopanordnung. Sein Ausgangssignal ist ein Zeitintervall τ, daß sich proportional zur Kraft F ändert und unabhängig von Lichtleistungsdriften ist.

3 Meßprinzip

Wirkt auf ein photoelastisches Sensorelement eine Kraft F, wird es doppelbrechend. Eingestrahltes linear polarisiertes Licht der Wellenlänge λ spaltet in zwei Anteile parallel zu den mechanischen Hauptspannungen (Hauptachsen) auf. Nach dem Durchstrahlen des Sensorelementes weisen die Teilwellen die Phasendifferenz Δ_S auf, die sich proportional zur Kraft F ändert.

$$\Delta_S = E_S \cdot F + \Delta_{S0} \qquad \text{mit} \quad E_S = \frac{360°}{\lambda} \cdot \frac{G}{b} \cdot C_0 \ . \qquad (1a, b)$$

Neben der kraftinduzierten Phasendifferenz ist die Offsetphasendifferenz Δ_{S0} vorhanden, die durch eine Offsetbelastung des Sensorelementes und durch eingefrorene Eigenspannungen verursacht wird. Der Geometriefaktor G berücksichtigt die Form des Sensorelementes. Bei runden Sensorelementen gilt /6/: $G = 8/\pi$. Die photoelastische Konstante C_0 ist vom verwendeten Sensormaterial abhängig. Da die Empfindlichkeit E_S umgekehrt proportional zur Breite b des Sensorelementes ist, kann durch eine Miniaturisierung die Meßempfindlichkeit erhöht werden. Für einen hochempfindlichen Kraftsensor bieten sich wegen ihres extrem kleinen Durchmessers (b ≈ 125 µm) daher Glasfasern als Sensorelemente an. Gering doppelbrechende (LoBi) Fasern besitzen Offseteigenschaften, die denen von Glaselementen vergleichbar sind ($\Delta_{S0} < 1°$).

Um die Empfindlichkeit des Sensorelementes zu steigern, besteht neben der Miniaturisierung die Möglichkeit der Verspiegelung der Stirnflächen und Resonanzbetrieb /7/. Wird die Lichtwellenlänge λ auf die Dicke t des Sensorelementes abgestimmt ($2 \cdot t = n \cdot \lambda$, n: ganze Zahl), führen Mehrfachdurchläufe des Lichts dazu, daß die resultierende Phasendifferenz $\Delta_{S,Res}$ größer als die Phasendifferenz Δ_S des Sensormaterials ist. Es gilt /8/:

$$\Delta_S = 2 \cdot \arctan\left(\frac{1+R}{1-R} \cdot \tan\left(\frac{\Delta_M}{2}\right)\right) = V \cdot \Delta_M \quad \text{mit} \quad V \approx \frac{1+R}{1-R} \quad \text{für kleine } \Delta_S. \quad (2)$$

Der resultierende Fehler dieser linearen Approximation ist für $\Delta_S < 10°$ kleiner als 0,25%. Für die Resonanzabstimmung ist ein Regelkreis /9/ erforderlich (vgl. Bild 1). Hierbei wird ein Teil des Lichts hinter dem Sensorelement mit einem Wollaston-Prisma in die Anteile P_x und P_y parallel und senkrecht zur Kraftrichtung aufgeteilt und deren Lichtleistungsdifferenz zu Null geregelt.

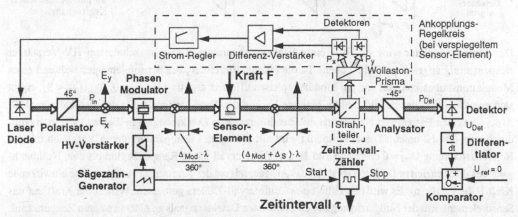

Bild 1: Blockschaltbild des Kraftsensors.

Die verwendete optische Anordnung (Bild 1) stellt eine Erweiterung der bekannten Polariskopanordnung um einen Phasenmodulator dar. Das Sensorelement und der Phasenmodulator sind zwischen zwei Polfiltern (Polarisator, Analysator) angeordnet. In die Anordnung wird das Licht einer Laserdiode eingestrahlt. Der (elektrooptische) Phasenmodulator erzeugt eine der Ansteuerspannung U_{Mod} proportionale Phasendifferenz Δ_{Mod}. Der Detektor erfaßt die transmittierte Lichtleistung P_{Det} und liefert eine hierzu proportionale Spannung U_{Det}.

Werden die Hauptachsen des Phasenmodulators und des Sensorelementes gleich ausgerichtet (S = M = 0°) und für die Polfilter die Azimute P = 45° und A = -45° gewählt, gilt für die vom Detektor gemessene Lichtleistung P_{Det}:

$$P_{Det} = P_{in} \cdot \sin^2\left(\frac{\Delta_S + \Delta_{Mod}}{2}\right). \quad (3)$$

322

Durch die gleiche Ausrichtung von Phasenmodulator und Sensorelement addieren sich in Gl. (3) deren Phasendifferenzen. Die gemessene Lichtleistung P_{Det} hängt somit von der eingestrahlten Lichtleistung P_{in} und der Phasendifferenzsumme $\Delta_S + \Delta_{Mod}$ ab.

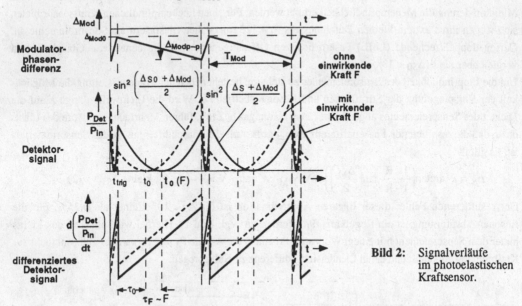

Bild 2: Signalverläufe im photoelastischen Kraftsensor.

Der Phasenmodulator wird mit Hilfe eines Sägezahngenerators mit nachgeschaltetem HV-Verstärker rampenförmig angesteuert (Bild 2 oben). Ist die Phasendifferenz Δ_S des Sensorelementes während eines Modulationshubes konstant und die Modulatorphasendifferenz negativ ($\Delta_{Mod0} < 0$, $\Delta_{Modp-p} < 0$), ergibt sich gemäß Gl. (3) der in Bild 2 Mitte gezeigte Verlauf des Detektorsignals. Da das Detektorsignal für $\Delta_{Mod} = -\Delta_S$ den Extremwert Null besitzt, hat das differenzierte Detektorsignal in diesem Fall einen Nulldurchgang (Bild 2 unten). Dieser Zeitpunkt wird mit Hilfe eines Komparators durch Vergleich mit der Referenzspannung $U_{ref} = 0$ ermittelt (Bild 1). Das Zeitintervall τ vom Rampenbeginn bis zum Nulldurchgang des differenzierten Detektorsignals ist das Ausgangssignal des Kraftsensors und hat ohne einwirkende Kraft F die Länge τ_0. Es wird mit Hilfe eines Zeitintervall-Zählers gemessen. Wirkt eine Kraft auf das Sensorelement, tritt der Nulldurchgang des differenzierten Detektorsignals zu einem späteren Zeitpunkt auf. Das Zeitintervall verlängert sich hierdurch um τ_F. Aufgrund der rampenförmigen Ansteuerung des Phasenmodulators ist die Dauer τ_F der Phasendifferenz Δ_S und damit der zu messenden Kraft F proportional. Driften der eingestrahlten Lichtleistung haben keinen Einfluß auf den Kompensationszeitpunkt und damit auf das Sensorausgangssignal. Für das Zeitintervall τ gilt

$$\tau = \tau_F + \tau_0 \qquad \text{mit} \qquad \tau_F = \frac{E_S \cdot T_{Mod}}{\Delta_{Modp-p}} \cdot F \ . \qquad (4a, b)$$

4 Experimentelle Untersuchungen

In einem Versuchsaufbau gemäß Bild 2 wurden mit dem beschriebenen Verfahren Gewichtskraftmessungen durchgeführt. Als Sensorelemente dienten Sensor-Scheiben (ø25 mm · 1 mm aus Quarzglas) mit unterschiedlichen, beidseitigen Beschichtungen (R = 79%, R = 49%, AR) bzw. eine Sensor-Faser (York LB 800). Zur Ermittlung der jeweiligen statischen Sensorkennlinien (Bild 3) wurden die resultierenden Änderungen des Sensorausgangssignals τ bei Belastung des Sensorelementes durch kalibrierte Massen ermittelt. Die Ansteuerung des Phasenmodulators erfolgte dabei sägezahnförmig mit dem Modulationshub $\Delta_{Modp-p} = -23{,}6°$ und der Rampendauer $T_{Mod} = 3{,}65$ ms.

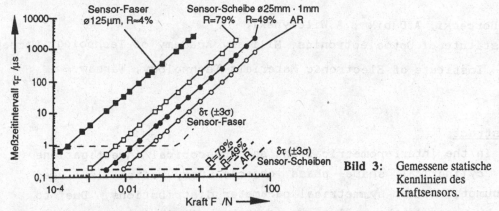

Bild 3: Gemessene statische Kennlinien des Kraftsensors.

Durch eine Verspiegelung des Sensorelementes und durch die Miniaturisierung wird die Kennlinie des Sensors durch die gesteigerte Meßempfindlichkeit nach links verschoben. Gegenüber einer entspiegelten Sensor-Scheibe konnte die Empfindlichkeit des Sensors durch Verspiegelung um den Faktor neun und mit einer Sensorfaser um den Faktor 187 gesteigert werden. Dies entspricht den theoretischen Erwartungen nach Gl. (1b) und (3). Die Linearität der Kennlinien (max. Fehler, bezogen auf den Meßbereich) war mit einer Sensor-Scheibe besser als 10^{-3}. Mit einer Sensor-Faser wurde, bedingt durch eine nicht ideale Lasteinleitung nur ein Nichtlinearitätsfehler von 1,6% erreicht.

Ebenfalls in Bild 3 enthalten ist das gemessene Rauschen $\delta\tau$ des Zeitintervalles. Es beträgt bei kleinen Kräften und einer Sensor-Scheibe ca. 25 ns. Wird der $\pm3\sigma$-Bereich von $\delta\tau$ als Meßunsicherheit zugrunde gelegt, entspricht dies einer relativen Auflösung von ca. $4 \cdot 10^{-5}$. Mit einer Sensor-Faser beträgt das Zeitintervallrauschen bei kleinen Kräften ca. 150 ns. Dies entspricht einer relativen Auflösung von $2{,}5 \cdot 10^{-4}$. Verursacht wird der erhöhte Rauschpegel bei einer Sensor-Faser und bei großen Kräften durch seismische Störungen, die zu einem verrauschten Sensoreingangssignal führen.

5 Ausblick

Aufgrund der hohen Meßempfindlichkeit sind Anwendungen des Meßprinzips besonders zur Messung kleiner Kräfte denkbar. Die Meßbandbreite kann dabei über T_{Mod} der jeweiligen Meßaufgabe angepaßt werden. Mit großer Meßbandbreite sind Anwendungen bei der Messung kleiner Beschleunigungen, z. B. in der Inertialnavigation und der Schwingungsmeßtechnik möglich. Dabei würde die Unempfindlichkeit des Sensors gegenüber magnetischen und elektrischen Störfeldern eine Beschleunigungsmessung auch in Umgebungen ermöglichen, in der z. B. piezoelektrische Meßaufnehmer nicht einsetzbar sind. Eine weitere Anwendung des vorgestellten Meßprinzips könnten empfindliche Kraftmeßdosen sein, die heute üblicherweise mit Hilfe von DMS und einem nachgeschalteten Trägerfrequenzmeßverstärker realisiert werden.

6 Literatur

/1/   Holzapfel, W.; Settgast, W., Applied Optics, Vol. B 28, No. 21, Nov. 1989, pp. 169-172

/2/   Martens, G., Technisches Messen, Vol. 53, No. 9, Sept. 1986, pp. 331-338

/3/   Tventen, A. B.; Dandridge, A.; et. al., Electronics Letters, Vol. 16, No. 22, Oct. 1980, pp. 854-855

/4/   Buchholtz, F.; Kersey, A.; Dandridge, A., Electronics Letters, Vol. 22, No. 2, Oct. 1986, pp. 75-76

/5/   Bertholds, A.; Dändliker, R., Applied Optics, Vol. 25, No. 3, Feb. 1986, pp. 340-343

/6/   Wolf, H., Spannungsoptik, Springer Verlag, Berlin, 1976

/7/   Dändliker, R.; Mayestre, F., Applied Optics, Vol. 28, No. 11, June 1989, pp. 1995-2000

/8/   M. Born, M.; Wolf, E., Principles of Optics, Pergamon Press, Oxford, 1980

/9/   Braasch, J. C.; Holzapfel, W., Electronics Letters, Vol. 28, No. 9, April 1992, pp. 849-850

New Method of the Asymmetrical Phase Objects Investigation by the Laser Quasi-Tomography

M.Borcecki, A.Dubik[*], A.Wilczynski

Institute of Optoelectronics, Military Academy of Technology, Warsaw,

[*] - Institute of Electronic Materials Technology, Warsaw.

Abstract

In the interferometrical and spectroscopical investigations of cylindrical shape phase objects very often is used assumption about symmetrical parameter distributions. Due to it by means of the well known (symmetrical) Abel inversion one can obtain local parameters from experimental data easy enough. However, such ideal situation in reality exists very seldom so that calculated values of sought local parameters are not correct.

Presented method allows to find local parameters of asymmetrical (or symmetrical) phase objects more correctly. In the measurements the computer diagnostics arrangement based on the CCD camera was used [1]. The camera consists of the linear image sensor: CCD-123 type of Fairchild Firm [2]. Real values of sought parameters were calculated by specially prepared numerical code CCDABEL [1].

The method was tested in the laboratory practice and additionally by means the test functions simulating experimental data - for which one can obtain symulated local parameters in analitical way.

Introduction

In the studies of phase objects (plasmas,...) it is usually desirable to analyze the spatial distribution of some parameters (density, intensity,...). In these investigations interferometry and spectroscopy are often used; however, the directly obtained data is not a sought local value but an integrated one along the optical path (see Fig.1). To obtain the spatial distribution, it is necessary to solve the integral equation:

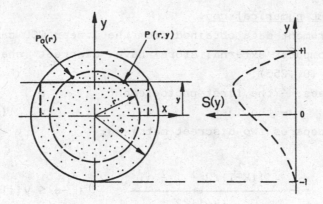

Fig.1. (a) Illustration of the geometrical relationships between the variables: x, y, r, S(y), P(r,y). (b) A typical example obtained by optical experiment, e.g., interferometry or spectroscopy.

$$S(y) = \int_{-\sqrt{a^2 - y^2}}^{+\sqrt{a^2 - y^2}} P(r,y)dx \qquad (1)$$

where   a,r – maximal and current plasma radius,

$$r = (x^2 + y^2)^{1/2}, \quad r \in [|y|, a].$$

S – integrated value along the optical path,

P – real local value (sought).

If the investigated object is assumed to be axial symmetric, the numerical solution method is based on solving the well known Abel inversion:

$$P(r) = -\frac{1}{\pi} \cdot \int_{r}^{a} \frac{S'(y)dy}{(y^2 - r^2)^{1/2}} \qquad (2)$$

where S' – the first derivative of the S function

(in this case for $|y| \leq a$; $P(r,y) = P(r)$)

The CCDABEL numerical code

I) Measurement data obtained from the linear CCD camera (kept in microcomputer external store) have discreet character of byte type (0..255).

User chooses: - the first photoelement p_1,
- the last -,,- p_K (p_K < 1728)

CCDABEL prepares two discreet matrixes:

$$y[i] = - \frac{[(k-1)/2 + 1 - i]}{(k-1)/2} \quad \{ -a \leq y[i] \geq +a \}$$

(3)

$$S_c(y[i]) = S_c[i] - \text{data registered by photoelement "i" of the CCD camera.}$$

where i - a current number of chosen photoelement,
$k = p_k - p_1 + 1$ (number -,,-).

II) In analysis we assumed that the phase object is symmetrical in the direction of the x axis, but in common case may be asymmetrical in direction of the y axis (Fig.1). The asymmetrical local value P(r,y) is represented by the next equation [3]:

$$\boxed{P(r,y) = g(y) \cdot P_o(r)}$$

(4)

(in common case $P(r,y) \neq P_o(r)$)
where g(y) - the function of asymmetry ;
$P_o(r)$ - the even component of P(r,y);

$$S(y) = 2 \cdot \int_o^{\sqrt{a^2-y^2}} P(r,y)dx = 2 \cdot g(y) \cdot \int_o^{\sqrt{a^2-y^2}} P_o(r)dx = g(y) \cdot S_o(y)$$

(5)

where $S_o(y) = 2 \cdot \int_o^{\sqrt{a^2-y^2}} P_o(r)dx$

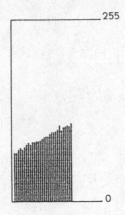

Fig.2. Typical state of registered optical signals in 30 successive photoelements of the CCD camera.

Calculating function g(y) and $P_0(r)$:

$$S(y) = S_0(y) + S_1(y)$$

$$S_0(y) = (S(y) + S(-y))/2 \qquad (6)$$

$$S_1(y) = (S(y) - S(-y))/2$$

where $S_1(y)$ — odd component.

From (5) and (6), the function of asymmetry can be calculated as follows:

$$g(y) = 1 + S_1(y)/S_0(y) \qquad (7)$$

The data registered by the CCD camera have discreet character besides they have considerable experimental scatter (see Fig.2). Therefore, the experimental data must be smoothed. The method involves dividing the data into a number of segments going out from the center [4]. A least – squares polynomials are then fitted to each segment in order to obtain smoothed functions g(y) and $S_0(y)$.

III)
After calculating values of symmetrical (mean) $P_0(r)$ by solving the well known Abel inversion [4] the real local values of P(r,y) we obtain from the equation (4).

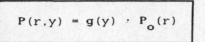

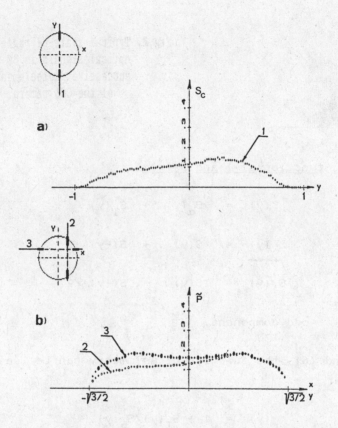

Fig.3. Registered distribution of integrated values S(y) (a)
and calculated local values P(r,y) (b).
1 - experimental data,
2,3 - reconstructed local value distributions adequately
for $x_{const} = 0.5$ and $y_{const} = 0.5$

Conclusions

The presented method for observing asymmetrical integrated
data in which asymmmetry exists normal to the direction of
observation is reliable, and its validity was verified by
using artificial test data (test functions). Furthermore, the
numerical code CCDABEL was applied succesfully to experimental
data from which valid local value were obtained.

The method gives more correct results than the methods assuming asymmetry of phase object in every case. Of course, the full tomography gives the best results but that last method is very complicated. So in many cases the proposed method (The quasi – tomography) is much simpler and obtained reasults are correct enough.

In the near future we plan to use the quasi – tomography by means two laser beams – along axis x and axis y (see Fig.1).

References
1] A.Wilczynski, Doctor's Thesis, Cracow, 1992.
2] CCD Imaging and Signal Processing Catalog and Applications Handbook, Fairchild Weston Schlumberger CCD Imaging Division, 1987.
3] Y.Yasutomo, K.Miyata, S.Himeno, T.Enoto, Y.Ozawa, IEEE Transactions on Plasma Science, vol.PS–9,No.1, 1981.
4] C.J.Cremers, R.C.Birkebak, Applied Optics, vol.5, No.6, 1966.

On Line Statistical Analysis for Laser Gauging Systems

S Chatterjee and **G G Sarkar**

R&D Division, MECON, Ranchi, INDIA
Pin : 834002. Fax (91)(651)(300708)

Introduction

The Laser Non-Contact Gauging System (LNCGS) is an instrument designed for fast and accurate non-contact measurement using Laser beam as a probe. In this system, the Laser beam is made to 'scan' a particular region at a high speed [1,2]. The scanning beam is focussed onto a photodiode by a collecting lens. When the scanning beam is interrupted by an object, the photodiode senses the shadow, and generates a pulse whose width is proportional to the linear dimension of the object which is intercepted by the plane of scanning beam . This may be measured and processed by suitable electronic circuits to estimate the dimension.

The LNCGS consists of two major subsystems : optical and electronic. The optical subsystem consists of a laser source, a scanning and a receiving optics. The electronic subsystem consists of a pulse shaping and measuring circuits and a computer, called 'Instrument Controller' [3] which is responsible for the relevant computations required to deliver the result.

The LNCGS is suitable for use as an on-line process control instrument on the shop floor, where there are a large number of fluctuating sources of error which may not be fully compensated at any time. Moreover, the optical components, exposed to the environment, deteriorate with time. All these indicate that the performance specifications for any LNCGS may change with time.

Hence, it is desired to have a system which can provide a self check on measurement and also increase user confidence.

The present paper proposes an innovative method of improving reliability of Laser Non Contact Measuring System in the shop floor by using on-line statistical techniques. This has been implemented by the authors in the system designed by them.

Statistical Analysis

The process of averaging eliminates most of the random errors of measurement. Hence, almost all LNCGS systems are designed to generate the final result after averaging over several scans. Typically, over 100 scans are used for tool room instruments, where the update rates are slow. However, if the update rate has to be fast then the number of samples averaged must be less.

Accuracy is improved as more and more samples are used in the averaging process. However, as already indicated, using more samples is at the cost of a lower update rate which may not be acceptable in an on-line situation. Hence, it is desirable to have a pre-processor which can enhance the efficiency of the averaging process. Also it is desirable to have feed back on the averaging process itself so that the measurement may be monitored continously. The later objective may be achieved by using other statistical parameters, like, coefficient of variation and skewness while the earlier objective may be achieved using a scheme of outlier rejection.

Outliers

Any real time on-line measurement system would be subject to purely random one-time events which give rise to a spurious and fictitious reading [4].These events are beyond the control of the instrument controller. Data generated by such events are called 'Outliers' [5]. If outliers are present in the averaging process, they introduce a major inaccuracy as they donot follow any distribution. Hence outliers must be eliminated before the averaging process.

The problem of rejection of outliers involves examining a set of data to identify the possible outliers, and then eliminating them. For the purposes of identification, an apropriate criterion must be selected. Since an outlier looks no different from a valid data, the only apropriate criterion which may be used in this case to separate the two must be a statistical criterion. The Chauvenet's criterion [5] has been used effectively to reject the outliers in the instrument designed by us.

Statistical Monitoring

Any measuring instrument is subject to three types of errors [6] *viz*, Gross error, Systemic error and Random error. It is expected that an instrument, that has been otherwise well designed and properly set up, to be free of gross and systemic errors. Random errors follow a Normal distribution. Conversely, it may be concluded that if data generated from a measurement fits into a normal curve, it must only be subject to random errors, which may be effectively eliminated by averaging. Hence, by examining how well a set of measured data confirm to a normal distribution, we are in a position to monitor the measurement process. This may be termed 'Statistical Monitoring'.

Statistical monitoring may be easily carried out by calculation of other statistical parameters, such as coefficient of variation [9] and skewness [10]. In the present context, an increase in the coefficient of variation would indicate a deterioration of the optical components or the measurement environment. Skewness indicates a loss of alignment or other major faults in the system.

Statistical Monitoring may be exploited in two ways. The user may set limits for the various parameters beyond which the instrument flags an error. Or the statistical monitoring may be used to dynamically optimise the update rate of the system for the given environment.

It may be emphasised that statistical monitoring does not replace the conventional hardware selfchecks incorporated in any system. Statistical monitoring seeks to detect error which would not normally be detected by the hardware self check routines.

Implementation

The online statistical analysis proposed for the LNCGS involves considerable computational overheads. Such a system could not have been possible without the current advances in microprocessor electronics. The main issue in the design of the instrument controller [3] has been a trade off between the dynamic performance desired and the statistical monitoring.

Conclusion

Online statistical analysis incorporated in the LNCGS designed by us promises to provide an innovative way to improve the robustness of such systems particularly in shopfloor environments. It may also be used to improve and optimise the dynamic performance of such systems under various conditions. Incorporation of such statistical techniques will enhance the effective use of Laser Non Contact Gauging Systems on the shop floor.

Acknowledgement

The authors would like to thank the authorities of MECON for providing all the facilities to carryout the work and present this paper. The authors would also like to thank Mr. Vikas Thakur for preparing the manuscript.

References :

1. Charshan S.S. *Laser in Industry*, Van Nostrand Reinhold Co., 1972
2. Harry J E, *Industrial Lasers and their Applications,* Mc Graw Hill Book Co, UK,1974
3. Sutap Chatterjee, G G Sarkar, *A Computer Controller for Laser Non Contact Gauging System,* Innovative Applications in Computing, Tata Mc Graw Hill Publishing Co. Ltd., 1992, P.375-382.
4. Cooper W.D., *Electronic Instrumentation and Measurement Techniques,* Prentice Hall of India Pvt. Ltd., New Delhi, 1976, P. 7 - 14.
5. Kennedy John B and Neville Adam M., *Basic Statistical Methods for Engineers and Scientists,*Harper and Row Publishers, New York, 1985, p 233- 236.
6. Kennedy John B and Neville Adam M., *Basic Statistical Methods for Engineers and Scientists,* Harper and Row Publishers, New York, 1985, p 34,52.

Analyse der out-of-plane und in-plane Dehnungen mit Hilfe der Shearografie

W. Steinchen, L. X. Yang, M. Schuth, G. Kupfer
Universität Gh-Kassel, Labor für Spannungsoptik und Holografie, Mönchebergstr. 7, 3500-Kassel

Einleitung

Die in den letzten Jahrzehnten entwickelten kohärent-optischen Meß- und Prüfverfahren wie die holografische Interferometrie, die Speckle Interferometrie usw. spielen wegen der Ganzfeldbetrachtung, der Berührungslosigkeit und der Hochempfindlichkeit in der Verformungsmessung und der zerstörungsfreien Werkstoffprüfung eine immer größere Rolle. Mit der Holografie ist es bereits gelungen, die out-of-plane Verformungen direkt zu messen. Aber es ist schwierig, mit der holografischen Interferometrie oder mit anderen kohärent-optischen Meßtechniken direkt auf der zu prüfenden Objektoberfläche die Dehnungen berührungslos zu messen. Die im Bereich der kohärent-optischen Meßtechnik von Y.Y. Hung entdeckte Shearografie bietet diese Möglichkeit. Sie kann aber nicht nur die out-of-plane Dehnungen, sondern unter bestimmten Bedingungen auch die In-Plane Dehnungen messen. In dieser Dokumentation wird die Ermittlung der out-of-plane und in-plane Dehnungen mit der Shearografie erläutert und diskutiert.

Prinzip der Shearografie

Die Shearografie ist ein optisches Verfahren, bei dem die Kohärenzeigenschaft des Laserlichts ausgenutzt wird (siehe Bild 1). Ein Bauteil wird mit einem Laserstrahl beleuchtet, der durch spezielle optische Elemente auf die Größe des zu untersuchenden Bereiches aufgeweitet wird. Das diffus von der Bauteiloberfläche reflektierte Laserlicht wird durch ein Shearelement beobachtet, das im Objektiv einer Kleinbildkamera oder einer CCD - Videokamera integriert ist. Dadurch bewirkt man, daß sich das reflektierte Objektwellenfeld dupliziert und geringfügig verschoben überlagert. Die beiden Wellenfelder stellen wechselseitig Objekt - bzw. Referenzwelle dar und man erhält ein Mikrointerferenzmuster, das sich z. B. auf einem speziellen Film speichern läßt. Analog zur Holografie werden in der Shearografie im Doppelbelichtungsverfahren von zwei Objektzuständen aufgenommene Mikrointerferenzmuster überlagert. Es entstehen ebenfalls Makrointerferenz-streifen, die jedoch nicht im Vergleich zur Holografie als Höhenlinien der Verformung interpretiert werden können, sondern die die Richtungsableitung der Verformung in

Shearrichtung kennzeichnen. D.h. die Makrointerferenzstreifen bei der Shearografie sind Linien konstanter Steigung der Verformung und stellen somit den Gradient der Verformung in Shearrichtung dar.

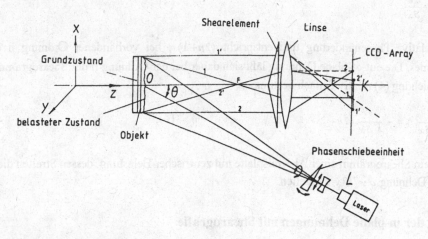

Bild 1. Prinzip der Shearografie

Die prinzipielle theoretische Erläuterung der Shearografie kann in der Literatur [1-3] nachgeschlagen werden. Von Seiten der optischen Theorie läßt sich beweisen[1], daß die Interferenzstreifen den Ort der relativen Phasenänderung $\Delta = (2n+1)\pi, \pm n = 0,1,2\cdots$ darstellen. Aus der geometrischen Theorie ist die relative Phasenänderung Δ wie folgt definiert:

$$\Delta = \frac{2\pi}{\lambda}\left(A\,\frac{\partial u}{\partial x} + B\,\frac{\partial v}{\partial x} + C\,\frac{\partial w}{\partial x}\right)\delta x \qquad (1)$$

Darin sind: λ = Wellenlänge des Lasers (u,v,w) = Verformungsvektor
    δx = Shearabstand auf dem   A, B, C = Geometrische Größen
        Objekt in x - Richtung   des Versuchsaufbaus

Die Interferenzstreifen der Shearografie spiegeln somit den Ort des Gradienten an den Stellen $\Delta = (2n+1)\pi$ der Verformung wider. Wird das Shearelement um 90° gedreht, so entsteht ein Shearabstand in y - Richtung. Aus ∂x und δx wird in der Gleichung (1) ∂y und δy.

Ermittlung der out-of plane Dehnung mit der Shearografie

Wenn die Größe des Objektes im Vergleich zu den Abständen vom Laser bis zum Objekt und von der Kamera bis zum Objekt (vgl. Abb.1) sehr klein ist, kann die Gleichung (1) wie folgt umgeschrieben werden[3]:

$$\Delta = -\frac{2\pi}{\lambda}\left[\sin\theta\,\frac{\partial u}{\partial x} + (1+\cos\theta)\,\frac{\partial w}{\partial x}\right]\delta x \qquad (2)$$

Darin ist: θ = der Winkel zwischen der Beleuchtungs- und der Beobachtungsrichtung (siehe Bild 1).

Richtet man den Winkel θ möglichst gegen Null aus, so erhält man $\sin\theta \approx 0$ und $\cos\theta \approx 1$. Aus Gleichung (2) wird:

$$\Delta = -\frac{4\pi\delta x}{\lambda}\frac{\partial w}{\partial x} \tag{3}$$

Δ ist die relative Phasenänderung und entspricht $(2n+1)\pi$ bei vorhandener Ordnung n des Shearogrammes. Die out-of-plane Dehnung läßt sich daher bei der Ordnung n des Shearogrammes nach der Gleichung (4) relativ einfach bestimmen.

$$\frac{\partial w}{\partial x} = -\frac{\lambda}{4\pi\delta x}\Delta = -\frac{\lambda}{4\delta x}(2n+1) \tag{4}$$

Bild 2 zeigt ein Shearogramm einer Al-Kreisplatte mit zentrischer Belastung, dessen Streifen die out-of-plane Dehnung $\partial w/\partial x$ darstellen.

Ermittlung der in-plane Dehnungen mit Shearografie

Wenn der Laser L und die Kamera K entsprechend wie im Bild 1 dargestellt positioniert sind, d.h. $\angle LOK$ liegt in der xz - Ebene (siehe Bild 1), so erhält man die Gleichungen für Δ_1 und , Δ_2 der Gleichungsgruppe (5). Wenn der Laser L und die Kamera K in der yz - Ebene liegen, d.h. $\angle LOK$ ist in der yz - Ebene, so erhält man die Gleichungen für Δ_3 ,und Δ_4[3].

$$\Delta_1 = -\frac{2\pi}{\lambda}\left[\sin\theta\frac{\partial u}{\partial x} + (1+\cos\theta)\frac{\partial w}{\partial x}\right]\delta x \quad \text{(x-Shearrichtung, } \angle LOK \text{ in xz - Ebene)}$$

$$\Delta_2 = -\frac{2\pi}{\lambda}\left[\sin\theta\frac{\partial u}{\partial y} + (1+\cos\theta)\frac{\partial w}{\partial y}\right]\delta y \quad \text{(y-Shearrichtung, } \angle LOK \text{ in xz - Ebene)} \tag{5}$$

$$\Delta_3 = -\frac{2\pi}{\lambda}\left[\sin\theta\frac{\partial v}{\partial x} + (1+\cos\theta)\frac{\partial w}{\partial x}\right]\delta x \quad \text{(x-Shearrichtung, } \angle LOK \text{ in yz - Ebene)}$$

$$\Delta_4 = -\frac{2\pi}{\lambda}\left[\sin\theta\frac{\partial v}{\partial y} + (1+\cos\theta)\frac{\partial w}{\partial y}\right]\delta y \quad \text{(y-Shearrichtung, } \angle LOK \text{ in yz - Ebene)}$$

Ist der Winkel zwischen der Beleuchtungs- und der Beobachtungsrichtung nahezu 90°, d.h. $\sin\theta \approx 1$ und $\cos\theta \approx 0$, kann die Gleichungsgruppe (5) wie folgt vereinfacht geschrieben werden:

$$\Delta_1 = -\frac{2\pi}{\lambda}\left[\frac{\partial u}{\partial x} + \frac{\partial w}{\partial x}\right]\delta x \quad \text{(x-Shearrichtung, } \angle LOK \text{ in xz - Ebene)}$$

$$\Delta_2 = -\frac{2\pi}{\lambda}\left[\frac{\partial u}{\partial y} + \frac{\partial w}{\partial y}\right]\delta y \quad \text{(y-Shearrichtung, } \angle LOK \text{ in xz - Ebene)} \tag{6}$$

$$\Delta_3 = -\frac{2\pi}{\lambda}\left[\frac{\partial v}{\partial x} + \frac{\partial w}{\partial x}\right]\delta x \quad \text{(x-Shearrichtung, } \angle LOK \text{ in yz - Ebene)}$$

$$\Delta_4 = -\frac{2\pi}{\lambda}\left[\frac{\partial v}{\partial y} + \frac{\partial w}{\partial y}\right]\partial y \quad \text{(y-Shearrichtung, } \angle LOK \text{ in yz - Ebene)}$$

Wenn das Objekt eine in-plane Belastung erfährt, sind normalerweise $\partial w/\partial x$ und $\partial w/\partial y$ im Vergleich zu $\partial u/\partial x$, $\partial u/\partial y$, $\partial v/\partial x$ und $\partial v/\partial y$ klein, d.h., die analog Gleichung (6) erzeugten Shearogramme entsprechen $\partial u/\partial x$, $\partial u/\partial y$, $\partial v/\partial x$ und $\partial v/\partial y$. Es ist daher möglich, jede Komponente des in-plane Dehnungstensors zu beobachten und abzuschätzen. Bei Objekten, die eine einfache Geometrie besitzen und eine einfache Belastung erfahren, kann man Anteile des in-plane Dehnungstensors mit der Shearografie auswerten. Bild 3 zeigt ein Shearogramm mit Querkraftbiegung, dessen Streifen die in - plane Dehnung $\partial u/\partial x$ darstellen.

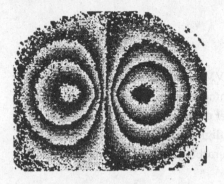

Bild 2. Out-of-plane Dehnung $\partial w/\partial x$ einer zentrisch belasteten Kreisplatte

Bild 3. In-plane Dehnung $\partial u/\partial x$ eines Kragbalkens

Zusammenfassung

Die Shearografie wie von Y.Y. Hung entwickelt, wurde bisher ausschließlich für die out-of-plane Dehnungsmessung verwendet. Jedoch läßt sich anhand der theoretischen Grundgleichung der Shearografie beweisen, daß das Verfahren ebenso für die in-plane Dehnungsmessung geeignet ist. Dies eröffnet der experimentellen Meßtechnik eine Vielzahl von Einsatzmöglichkeiten. So kann die Verwendung von DMS bei vielen Dehnungsmessungen entfallen. Neben der eigentlichen Größe der Dehnung liefert die Shearografie einen weiteren dominanten Vorteil. Es läßt sich anhand des Verlaufs der Interferenzstreifen eindeutig die Aussage treffen, wo und in welcher Richtung die maximalen Dehnungen auftreten. Mit Hilfe den Materialkenngrößen und den geometrischen Abmessungen können somit Bauteile optimiert werden. In Bezug auf die ständig steigenden Anforderungen in der Bauteilentwicklung bietet die Shearografie somit völlig neue Wege im Bereich der Meßtechnik.

Literatur

1. HUNG, Y. Y., "Shearography: A new optical Method for Strain Measurement and Non-destructive Testing", Opt. Eng., 21 (3), 391 (1982)
2. ETTEMEYER, A., "Shearography: ein optisches Verfahren zur zerstörungsfreien Werkstoffprüfung", Technisches Messen, 58 (1991), 6, S. 247-252.
3. YANG, L. X., SCHUTH, M., Arbeitsbericht Nr. 2, "Verhältnis zwischen der relativen Phasenänderung Δ und dem Gradienten der Verformung bei der Shearografie", Labor für Spannungsoptik und Holografie, Universität Gh - Kassel, (1992) (unveröffentlicht)

Anwendung optischer Strukturen in der elektronischen Datenverarbeitung

K.H.Schmidt, W.Waidelich,
Institut für med. Optik der LMU
Barbarastr. 16, 8 München 40

Es wird eine neue Datenverarbeitungsmaschine (DVM) vorgestellt, bei der die Daten nicht mehr wie bei einer von Neumann Maschine sequentiell sondern parallel verarbeitet werden. Die Struktur der DVM ähnelt der einer Abbildung mit einer Linse. Bei zwei Operanden ist die Verarbeitungsgeschwindigkeit um den Faktor 4 höher als bei von Neumann Rechnern. Bei der parallelen Verarbeitung von mehr Operanden erhöht sie sich entsprechend.

Fast alle bisherigen Datenverarbeitungsmaschinen basieren auf der von Neumann Architektur [1,2], bei der die Anweisungen an den Prozessor und die zu bearbeitenden Daten in einem logisch zusammenhängenden Speicher [3] untergebracht sind. Die Abarbeitung der Programme erfolgt durch einen Prozessor, der an den Speicherbaustein angeschlossen ist. Die Inhalte der über den Befehlszeiger vorgegebenen Adressen werden als Anweisungen interpretiert, anhand derer der Inhalt der Speicherzellen im Speicher sowie der Befehlszeiger verändert werden (Abb.1). Das Lesen, Verarbeiten und Abspeichern der Daten erfolgt sequentiell [4,5].

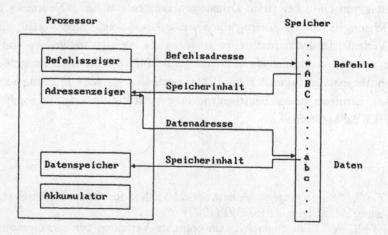

Abb.1 Aufbau eines von Neumann Rechners

Tab. 1 zeigt den Programmablauf anhand der Operation

$$c_n = a_n * b_n \tag{1}$$

wobei bei modernen Prozessoren 3 Schritte eingespart werden können [4,5].

1) Befehl * lesen,
2) Adresse A_n lesen,
3) $a_n := [A_n]$ lesen und a_n in den Akkumulator laden
4) Adresse B_n lesen,
5) $b_n := [B_n]$ lesen,
6) Adresse C_n lesen, und $a_n * b_n$ berechnen
7) $[C_n]$ mit $a_n * b_n$ beschreiben.

Tab. 1 Programmschritte zur Bearbeitung von Gl.1.

Grundlage für die Informationsverarbeitung in der Optik ist die Phasenverschiebung in Medien unterschiedlicher Ausbreitungsgeschwindigkeit und die Superposition der Teilwellen in der Aufnehmerebene [7].

Betrachtet man die Struktur moderner Bildverarbeitungssysteme [6], dann ergeben sich Parallelen zur einer optischen Abbildung.

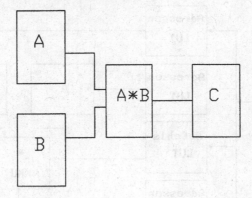

Abb. 2 Vereinfachtes System zur Bildverarbeitung

Die Daten aus den Speichern A und B stehen wie bei einer Linse parallel zur Verfügung, die Verknüpfung * entspricht der Superposition, der Speicher C der Aufnehmerebene.

Die Verarbeitung erfolgt parallel. Bei Verwendung einer Pipelinestruktur laufen die folgenden Prozesse gleichzeitig ab:

1) $[A_n]$ lesen und $[B_n]$ lesen,
2) $a_n * b_n$ berechnen
3) $[C_n]$ mit $a_n * b_n$ beschreiben.

Beim Aufbau nach Abb. 2 ist die Operation * und die Reihenfolge der Operanden in den Speichern A bis C fest vorgegeben.Der Index n steht für die Adresse der Operanden und entspricht der Zeit t bei einer optischen Abbildung.

Die Phasenverschiebung in der Optik entspricht einer zeitliche Verschiebung des Signals. In der Datenverarbeitung läßt sich das durch einen Adressenoffset bzw. noch allgemeiner durch eine indirekte Adressierung realisieren. Zudem wird für die Operation * eine Tabelle eingeführt, mit der verschiedene Funktionen in Abhängigkeit von der Adresse n gewählt werden können.

Abb. 3 zeigt den Aufbau eines neuen Datenverarbeitungsmaschine (DVM), bei dem die Adressen der Speicher A, B und C sowie die Funktion * über Look-up-Tabellen (LUT) ausgewählt werden.

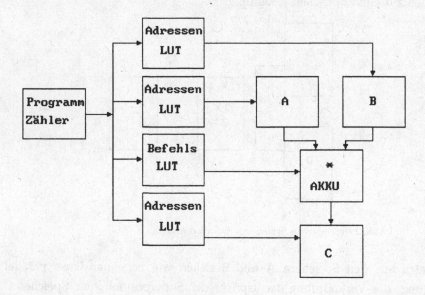

Abb. 3 Vereinfachter Aufbau der DVM

Die Daten stehen in den Speichern A und B, das Ergebnis wird in C geschrieben. Die Programmierung erfolgt über die vier Look-up-Tabellen. Mit der Pipelinetechnik können alle Prozesse parallel ausgeführt werde. Eine Operation benötigt demnach nur die Zeit für einen Speicherzugriff.

Am Beispiel einer Matrixmultiplikation soll die Einsatzmöglichkeit des Verfahrens in der Numerik demonstriert werden:

$$c_{ik} = \sum_{j=1}^{n} a_{ij}\, b_{jk} \qquad\qquad (2)$$

a_n, b_n und c_n werden über die LUT's wie folgt adressiert. Über den Akku kann das Ergebnis der Operation * zwischengespeichert und mit dem Befehl '+' aufsummiert werden. Der Befehl 'R' setzt den Akku nach Auslesen des Wertes auf 0 zurück.

Schritt	A	*	B	Akku	C
1	a_{11}	x	b_{11}	+	-
2	a_{12}	x	b_{21}	+	-
3	a_{13}	x	b_{31}	+	-
4	a_{14}	x	b_{41}	+	-
.	.	.	.	.	.
n	a_{1n}	x	b_{n1}	R	c_{11}
n+1	a_{21}	x	b_{21}	+	-
n+2	a_{22}	x	b_{22}	+	-
.	.	.	.	.	.
n^3	a_{nn}	x	b_{nn}	R	c_{nn}

Tab. 3 Anordnung der Operanden und Operationen in den Look-up-Tabellen zur Bearbeitung von Gl.2

Literatur:

[1] Turing: Intelligence Service, Hrsg. B. Dotzler, K.Kittler, Brinkmann&Bose,Berlin (1987)

[2] M.L.Minsky,S.A.Papert: Percetrons,MIT Press, Second Printing, (1988)

[3] U.Tieze Ch.Schenk: Halbleiterschaltungstechnik, Springer Verlag Berlin (1974)

[4] Intel: Microprozessors, Intel corporation,(1990).

[5] Motorola: 16-Bit Mikro Prozessor Benützer-Handbuch, Motorola INC (1980).

[6] Data Translation: User Manual for DT2861 Arithmetic Frame Grabber, Data Translation (1987)

[7] Bergmann,Schaefer: Optik, Band III, 7.Auflage, De Gruyter, Berlin-New York (1978)

Congress C

Congress-Chairmen: G. Herziger, H. Weber and E.W. Kreutz

Laser in der Fertigung
Laser in Production

Qualitätssicherung in der Laserstrahlschweißtechnik

G. Deinzer, P. Hoffmann, M. Geiger
Lehrstuhl für Fertigungstechnologie, Universität Erlangen-Nürnberg
Egerlandstr. 11, D-91058 Erlangen,
Tel: 09131-85-7140, Fax: 09131-36403

1   Einleitung

Der CO_2-Hochleistungslaser ist aufgrund seiner Verfahrens-, Mengen- und Produktflexibilität aus der modernen Fertigung nicht mehr wegzudenken. Selbst für Anwendungen in der Großserienfertigung bietet das zuverlässige aber auch teure "Werkzeug" Laserstrahl Vorteile. Allen Schweißanwendungen gemeinsam ist, daß mit dem Laser nur dann hohe Bearbeitungsqualitäten erzielt werden, wenn die Vorbereitung der Fügekanten mit hoher Präzision erfolgt und die Prozeßparameter, insbesondere Laserleistungsflußdichte und Schweißgeschwindigkeit, exakt eingehalten werden. Die Qualifizierung der Lasertechnik für weitere Anwendungen geht deshalb mit der Entwicklung von qualitätssichernder Systemtechnik einher. In diesem Beitrag wird über neue Entwicklungen bei prozeßoptimierenden Optiken sowie zur temperaturgeführten Prozeßregelung berichtet. Die nachfolgend beschriebenen Ergebnisse wurden an einer handelsüblichen Laserstrahlwerkzeugmaschine in Portalbauweise gewonnen: Die Strahlquelle ist ein hochfrequent angeregter CO_2-Laser mit bis zu 2,2 kW Ausgangsleistung. Die Führungsmaschine ist in Hybridbauweise ausgeführt. Die Linearachse X bewegt den Werkzeugtisch, während die Linearachsen Y und Z durch sogenannte fliegende Optiken realisiert sind, d.h. die Strahlweglänge zwischen Auskoppelfenster der Strahlquelle und Fokussieroptik des Bearbeitungskopfes variiert mit der Position des Bearbeitungskopfes in der von Y- und Z-Achse aufgespannten Arbeitsebene um bis zu 2,75 m.

2   Adaptive/prozeßoptimierende Optiken

In Kooperation mit einem Hersteller von prozeßoptimierenden Systemen wurden deformierbare Spiegel in die Strahlführung der o.g. Werkzeugmaschine integriert und auf prozeßoptimierende Anwendungen untersucht [1]. Die für die nachfolgenden Anwendungen als Einkanalsystem ausgeführte Optik hat eine elastisch deformierbare Spiegelmembran aus einer Kupferlegierung. Ein rückseitig zentrisch angreifender Piezoaktuator erzeugt nahezu sphärische Oberflächen. In Abhängigkeit des Aktuatorhubs stellen sich Brennweiten zwischen $-\infty$ und -2 m ein. Entscheidend für die mit der Optik erzielbaren Effekte ist der Einbauort in die Strahlführung. Wird in einer bewegten Strahlführung die Optik zwischen Resonatorauskoppelfenster und Strahltaille des Hauptstrahls eingebaut, ermöglicht die Beeinflussung der Strahl-ausbreitung durch die adaptive Optik ein Konstanthalten der Ausleuchtung der Fokussieroptik. Dadurch kann ein konstantes Bearbeitungsergebniss im gesamten Bearbeitungsraum gewährleistet werden. Genauere Ausführungen dazu finden sich in der Literatur [1,2]. Wird die prozeßoptimierende Optik so in den Strahlengang integriert, daß die Strahlweglänge zwischen prozeßoptimierender Optik und Fokussieroptik gering ist, so sind große Divergenzänderungen des Laserstrahls realisierbar ohne die Ausleuchtung der Fokussieroptik wesentlich zu beeinflussen [Schutzrecht angemeldet: 3]. Es stellt sich eine Fokuslagenshift Δz_F ein, bei näherungsweise gleichbleibender Strahlkaustik. **Bild 1** skizziert den Einbaufall in der Versuchsanlage. Die Optik ist anstelle eines Planspiegels in den Bearbeitungskopf integriert. In den Versuchen konnte bei einer Brennweite f_1 von 150 mm bis zu 10 mm Fokusshift Δz_f erzielt werden, ohne daß eine meßbare Verschlechterung der Strahlqualität im Fokus auftrat.

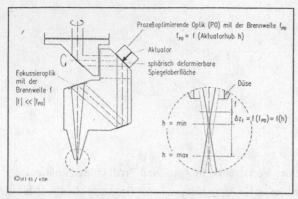

Bild 1: :   Einbaufall prozeßoptimierende Optik für eine variable Fokusverschiebung Δf_z

Diese quasi-masselose Verstellung der Fokuslage eröffnet neue Möglichkeiten in der Bearbeitung stark konturierter Bauteile. Mußte bislang der Brennfleck mechanisch nachgeführt werden, entweder über die Maschinenachsen oder eine Zusatzachse im Bearbeitungskopf, steht mit der prozeßoptimierenden Optik eine Strahlnachführung zur Verfügung, die neben kürzesten Stellzeiten den Vorteil eines verbesserten Zugangs zum Werkstück bietet. Ziehflansche mit kleinen Radien z.B. stellten bislang größte Anforderungen an die Dynamik der Führungsmaschine und an das Geschick des Programmierers, Kollisionen zwischen Bearbeitungskopf und Werkstück zu vermeiden [2,4]. Mit der prozeßoptimierenden Optik ist es möglich eine für den Bearbeitungskopf leichter ausführbare Verfahrbewegung zu realisieren und trotzdem die für das Bearbeitungsergebnis optimale Fokuslage einzuhalten. **Bild 2** zeigt Einschweißungen an einem Testwerkstück. Die Aufgabe bestand darin eine 8 mm tiefe "Versteifungssicke" zu überschweißen. Die Bearbeitungsbahn wurde geradlinig programmiert. Bei konstant gehaltener Fokuslage zur Düsenunterkante reichten die Intensitäten des Laserstrahls im Bereich der Vertiefung nicht aus, den Schweißprozeß fortzusetzen (siehe Fotografie der Blechunterseite). Unter Einsatz der prozeßoptimierenden Optik - sie wurde bei dieser Anwendung als zusätzliche NC-Achse programmiert - wurde der Brennfleck der Werkstückkontur nachgeführt und die Qualität der Schweißnaht entlang der Bearbeitungsbahn ist konstant.

3   Temperaturgeregeltes Laserstrahlschweißen

Die Technologie des Laserstrahlschweißens wird in der Automobilfertigung u.a. zum Fügen von Karosserieblechen im Überlappstoß eingesetzt. Eine weitere Anwendung ist das Verschweißen von sog. Karosseriefalzen an Tür- und Motorhauben. Da es sich in beiden Fällen um das Fügen von Tiefziehblechen handelt, die herstellungsbedingt mit Maßabweichungen behaftet sind, sind die Bearbeitungs-

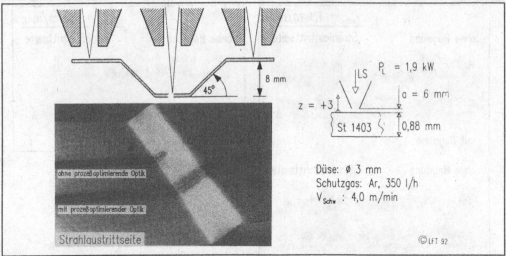

Bild 2: Fokusnachführung bei profilierten Werkstückoberflächen mittels prozeßoptimierender Optik

parameter an die lokalen Abweichungen des Bauteils hinsichtlich Fügespalt, die Absorption beeinflussende Oberflächenschichten bzw. sonstiger Werkstoffinhomogenitäten zu adaptieren. Im folgenden soll für den Anwendungsfall Schweißen am Karosseriefalz ein Verfahren zum temperaturgeregelten Laserstrahlschweißen vorgestellt werden. Ziel der Regelung ist es, eine Einschweißung in das mittlere der drei Bleche zu garantieren, eine Durchschweißung des dritten Bleches, das die Außenhaut einer Automobilkarosserie darstellt, zu verhindern.

Die Vorrichtung zur Regelung des Laserstrahlschweißprozesses besteht aus einem Infrarot-Pyrometer, einem PID-Regler und einer eigenentwickelten elektronischen Schaltung zur Aufbereitung des Meßsignals. Als Stellgröße wurde die Laserleistung, die direkt über die Anregungsamplitude der Hochfrequenzanregung angesteuert wurde, herangezogen. Das an der Blechunterseite aufgenommene Tempertursignal wird in den PID-Regler, mit dem die Regelcharakteristik eingestellt werden kann, eingelesen. Das Signal des Reglers wird in einer eigenentwickelten Schaltung aufbereitet und direkt in die Hochfrequenzanregung eingespeist. Da keine softwaretechnische Signalverarbeitung erforderlich ist, arbeitet die Regelung mit einer, für den Laserstrahlschweißprozeß ausreichen hohen Grenzfrequenz. **Bild 3** zeigt Aufnahmen von Schweißproben, die bei jeweils unterschiedlicher Geschwindigkeit ohne und mit Temperaturregelung laserstrahlgeschweißt wurden. Im Bildausschnitt links unten sind deutlich die beim ungeregelten Schweißprozeß auftretenden Anlauffarben zu erkennen, die durch den Einsatz der Temperaturregelung vermieden werden konnten. Der Bildausschnitt rechts oben zeigt ein Bearbeitungsergebnis bei einer hohen Schweißgeschwindigkeit. Für den ungeregelten Prozeß kam es zum kurzzeitigen Abreißen des Schweißplasmas und damit zu einer schlagartigen Verringerung der Einschweißtiefe. Beim geregelten Laserstrahlschweißen wird durch eine Leistungsanpassung der Schweißprozeß kontinuierlich aufrecht erhalten, wodurch z.B. beim Überlappschweißen von zwei Blechen eine ausreichende Einschweißung in das untere Blech gewährleistet werden kann.

348

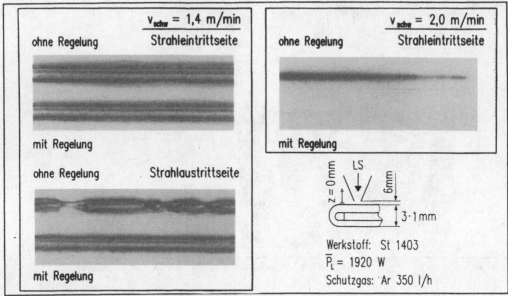

Bild 3: Temperaturgeregeltes Laserstrahlschweißen am Karosseriefalz

4 Zusammenfassung und Ausblick

Die vorgestellten Ergebnisse sind ein wichtiger Beitrag für die weitere Qualifizierung der Laserstrahl-schweißtechnik im industriellen Einsatz. Sowohl die prozeßoptimierende Optik als auch der temperatur-geregelte Schweißprozeß dienen der Qualitätssicherung selbst an Bearbeitungsaufgaben, die bislang für den Laser als kritisch oder nicht machbar galten. Dank der prozeßoptimierenden Optik lassen sich Bauteile größter Abmessungen aber auch komplizierte Strukturen prozeßsicher verschweißen. Mit dem temperaturgeregelten Laserstrahlschweißprozeß steht ein Verfahren zur Verfügung, das mit Blick auf Maßabweichungen in der Fügekantenvorbereitung fehlertolerant ist. Bei variablen Fügespaltweiten im Bereich von 0 bis 0,3 mm werden sowohl für Falzverbindungen definierte Einschweißtiefen garantiert als auch die sichere Durchschweißung für Überlappverbindungen.

5 Literaturverzeichnis

[1]   SCHOTTELIUS, H.U.; GEIGER, M.; HOFFMANN, P.: Adaptive Optik für die prozeß-optimierte Fokussierung in Laserbearbeitungsmaschinen. In: Waidelich, W. (Hrsg.): Laser in der Technik. Berlin: Springer, 1992, S. 652-656

[2]   HOFFMANN, P.: Verfahrensfolge Laserstrahlschneiden und -schweißen -Prozeßführung und Systemtechnik in der 3D-Laserstrahlbearbeitung von Blechformteilen. München: Hanser, 1990 (Reihe Fertigungstechnik Erlangen Bd. 29). Erlangen, Universität, Fertigungstechnik, Diss., 1991

[3]   Schutzrecht P 42 17 705.7 (1992-06-01). Hoffmann, P.; Schuberth, S.; Kozlik, C.; Geiger, M.: Verfahren und Einrichtung zur Materialbearbeitung mittels eines Hochenergie-Laserstrahles sowie Verwendung dieser Einrichtung

[4]   HAFERKAMP, H.; BENECKE, R.: Laserschweißen > >um die Ecke< <. In: Blech Rohre Profile 38 (1991) 4, S. 304-307

Thin Film Growth by Pulsed Laser Deposition

Ian W Boyd

Electronic & Electrical Engineering
UNIVERSITY COLLEGE LONDON
Torrington Place, London WC1E 7JE, UK

Abstract

The use of Pulsed Laser Deposition (PLD) to grow a variety of thin multicomponent films, including dielectric, protective, and superconducting layers, is described. The underlying principles governing the laser-target interaction and removal of material from the irradiated surface are briefly reviewed. Additionally, particular advantages and disadvantages of the technique are discussed, and future directions proposed.

Introduction

Considerable attention has been extended to the application of laser radiation to initiate or enhance thin film growth over the past decade [1]. Dielectric, semiconductor, metal and superconductor films have been grown by a variety of processing modes involving photonic or thermal (or both) reactions induced by the quantised energy. One such technique currently attracting enormous interest is Pulsed Laser Deposition (PLD), particularly for multi-component thin film growth. Although dating back more than 28 years [2], the field burgeoned during the late 1980's principally because of the discovery the new families of layered cuprate superconductors. In fact, some of the best quality superconducting films currently available have been prepared by PLD.

The subject of PLD is paper briefly introduced and reviewed here, indicating the numerous advantages and less desirable effects associated with the technique. The application towards the growth of interface dielectrics and protective layers suitable for a range of multilayer structures with semiconductors will be described, as well as the use of sandwiching methods to prepare BiPbSrCaCuO and PbSrYCaCuO super-conducting films. Finally, the potential for reactive PLD will be summarised.

Background and Theory

Pulsed Laser Deposition is a conceptually simple technique involving the collection on a substrate of material removed from a pulsed laser-irradiated target. It offers many attractions. Material transfer is congruent (i.e. stoichiometry is preserved), atomic flux is easily regulated, novel layers and multilayers may be readily grown, and it is simple and relatively inexpensive. Over the years a wide variety of descriptive terms have been

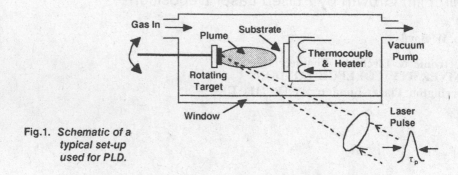

Fig.1. *Schematic of a typical set-up used for PLD.*

associated with PLD, which has been referred to as laser evaporation [2-6], laser assisted deposition and annealing (LADA) [7] laser flash evaporation [8], laser assisted sputtering [9], laser MBE [10], hydrodynamic sputtering [11], laser ablation [12-14] laser ablation deposition [15], and laser evaporation deposition (LEDE) [16]. Figure 1 shows schematically a simple PLD arrangement for film growth.

To make the process as efficient as possible, such that energy is not lost due to carrier or thermal diffusion during absorption, short laser pulses should be used at a wavelength strongly absorbed by the material. Lasers operable in the Q-switched mode, therefore, the Nd:Yag (1.064 μm, 532nm, 355nm, 266nm), ruby (694nm), and excimers (XeCl at 308nm, KrF at 248nm, and ArF at 193nm), can deposit large amounts of energy into a thin surface region in the target. With the recent performance improvements , e.g. reliability and stability, the excimer and Nd: YAG lasers have in particular been most widely used in PLD. The typical energies involved in PLD are close to those required to heat the target materials beyond their melting temperatures and initiate significant evaporation. Material ejection is accompanied by the formation of a plasma or plume of emitted particles just above the target surface and an easily recognisable snapping sound as the velocity of some of the species exceeds the speed of sound in the immediate environment. However, a clear understanding of the underlying non-equilibrium mechanisms involved is extremely difficult to formulate, and only a few papers have addressed this complicated subject [17, 18].

It is instructive to model the energy input to the system as a conventional laser heating problem. By taking temperature averaged values of the usual optical and thermal constants, one can obtain order of magnitude estimates of the fluence required to melt YBCO. For 248nm and 1064 nm light, these are calculated to be at least 70 J/cm^2 and 260 mJ/cm^2 respectively, and values of 140 and 560 mJ/cm^2 are estimated to be the *minimum* fluences necessary to energise the thermal removal process [19]. In reality, good ablation will not necessarily occur near the threshold, but at higher incident energy levels. These estimates are nevertheless in good agreement with the reported observations [20 - 23]. A more complete model clearly must take into account the proper non-linear behaviour of the optical and thermal properties with temperature, as well as the precise extent of the laser beam in space and time. Furthermore, it is known that material already begins to leave the irradiated surface during the pulse, and so interactions between the photons and the ablated plasma must also be considered.

It is well known that at high temperatures thermionic emission of ionic species (and electrons) occurs at surfaces. Traditionally, the flux of these species has been estimated using the Richardson-Dushman (Langmuir-Saha) equations, respectively. In the presence of large photon intensities, however, such standard relationships have to be modified to accommodate additional phenomena such as photo-excitation and photo-ionisation and their consequential effects. When electrons are present near the surface they strongly couple the electromagnetic radiation and are accelerated and collide with any plasma ions or any nearby solid or gas phase atoms (inverse bremsstrahlung). The electron population can then be increased and excited further, causing extremely rapid cascade ionisation. The collective properties of the plasma can thus strongly determine what happens to the remainder of the incident light. The energy absorbed in the plasma is rapidly shared amongst the individual particles, raising their kinetic energy, i.e. temperature, and shielding the surface from further exposure to the laser beam. If the optical absorption of the plasma falls, its increased transparency will enable further surface exposure to re-populate the plasma and again increase absorption and surface shielding. Thus, it could be deduced [24] that an instantaneous equilibrium plasma density, once formed, would attempt to self-regulate against any changes in density, temperature or irradiance. The plasma temperature can be as much as 20,000K whilst the surface temperature of the underlying solid is only a modest amount above the evaporation point (ca. 2500K). Although one can achieve a fair indication of the levels of energy required for ablation to occur by a simple thermal approach,.the mechanisms by which material leaves the surface and their subsequent interactions and behaviour, are much more complex than can be predicted using thermal evaporation.

For laser ablation using visible or UV wavelengths it has been shown [25] that within a cone of acceptance the angle of incidence of the incident laser beam does not affect the trajectory of the ejected material which is virtually always perpendicular to the irradiated surface. Analysis of the deposits obtained by short wavelength laser ablation reveals two components to the thickness contour [25, 26]. The expected broad $\cos \theta$ profile is present, but a highly peaked deposit, whose profile can usually be fitted by a $\cos^n \theta$ curve, where $9 < n < 11$, is superimposed. Increasing the gas pressure used during deposition de-emphasises this peak and produces n-values closer to 4-6 [19, 26]. Such profiles have also been reported for laser ablation by 1064 nm radiation for both ms [27] and ns pulses [28]. Interestingly, for shorter wavelengths but at fluences near the threshold for material removal only the pure $\cos \theta$ form is observed while a $\cos^{10} \theta$ deposition is found only at higher fluences [25]. One way to homogenise the profile is to tilt the rotating target during ablation in order to "spray" the deposit across the substrate. At an angle of wobble of about 30° can produce a uniform film over several square cms [29]. However, Rutherford Back-Scattering (RBS) studies have shown that while the central portions of the $\cos^{10} \theta$ films grown by ablation are essentially stoichiometric the outer edges (containing the $\cos \theta$ component only) are not [25], an observation which may limit this tilting approach to film homogenisation.

In attempting to explain the overall shape of the deposits obtained in PLD, Kelly and Rothenberg [11] have proposed hydrodynamic sputtering which also assists in crater formation on the irradiated surface and confines the ablated material to a $\cos^4 \theta$ profile. The power-value could, however, be increased by enhancing the number of collisions amongst the particles. Others have discussed the effects of shock waves, surface superheating, and sub-surface explosions [30-33]. Singh et al. [17] have modelled a plasma expanding isotropically into the vacuum and have been able to obtain similar atomic profiles to those mentioned above. In their case, however, they prefer to describe the growth contour as a Gaussian profile. The consequences of these observations appear to suggest that there is an unavoidable thermal component to the removal process which contains non-stoichiometric matter.

The most spectacular accompaniment to the ablation growth of films is the colourful plume that forms between target and substrate. It occupies a volume containing high densities of electrons, excited and ionised atoms, and molecular clusters, all moving at very high velocities away from the surface. Its formation and content are subject to intense investigation using e.g. Laser Ion Mass Spectroscopy [34] and optical emission spectroscopy [35]. The phenomena are reminiscent of dielectric breakdown, caused by the highly localised electric field strengths at the surface. Early theories suggested that stoichiometric clusters were ejected. In fact, post-deposition examination of some films reveals large particulates (tens of microns) formed during the growth. Either through their inherent thermodynamic instability, or by collision with themselves or any surrounding molecules, or interaction with the remnants of the laser beam, these then decay into smaller components prior to reaching the substrate.

Many neutral species are also known to form, apparently more so when longer wavelength lasers are used (i.e. 1064 nm rather than 193 nm) [22]. This is most likely due to the restricted variety of interactions available to the less energetic photons, thus eliminating, for example, the possibility of photo-ionisation of the species in the plasma. Electron impact ionisation or cascade ionisation will also be less and so generally the plasmas associated with the shorter wavelength ablation will be the most energetic and reactive. With CO_2 laser ablation, very efficient inverse bremsstrahlung photon-coupling can be achieved in the plasma. Particle velocities can thus be much higher than for 248 nm ablation [36], though it is not clear why stoich-iometric ablation can only be achieved at 10.6µm for low and not high fluences [37].

Because of the usual irradiation geometries used, the plasma, consisting of electrons, ions, atoms, neutrals, molecules, and clusters and particulates of varying sizes, will initially be thin and flat extending over the area of the surface exposed to the laser beam. Soon after its formation and until the end of the laser pulse, it can be considered to be isothermal, with temperatures exceeding 10^3K (7.5eV). Immediately after irradiation, the amount of material augmenting the plasma will drop considerably. It will then expand preferentially away from the target, this direction being where the greatest density gradients are, into the surrounding vacuum, accelerating the species to top speeds of several km/s (up to 1000 times the usual velocities encountered during thermal evaporation). This expansion can push any existing gases away from the target setting up a pressure wave. As the particles slow down and lose energy, they give rise to characteristic UV and visible emission patterns that can be studied in-situ. Such spectroscopic studies have revealed the predominance of excited elemental and monoxide species, in agreement with the mass spectrometry results, although atomic collisions within the plasma may lead to the formation of more intricate structures. The mechanisms behind the formation of certain small cluster ions are very complex, and clearly require considerable study [36].

The explosive velocity distributions associated with the particles leaving the ablated surface have been measured directly using Time-of-Flight (TOF) measurements [38] to be about 2 km/s in 100 mtorr pressures [38-40] while velocities up to 2 x 10^6 cm/s for particles some 7 cm away from the target have been measured using an ion probe [39].

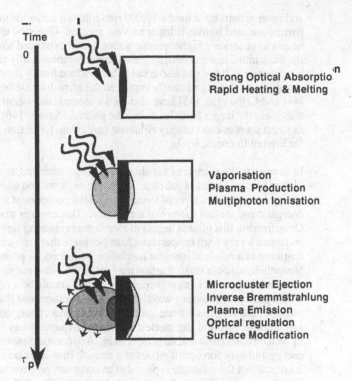

Time
0

Strong Optical Absorption
Rapid Heating & Melting

Vaporisation
Plasma Production
Multiphoton Ionisation

Fig.2. *Simple representation
of the principle steps
operating in PLD.*

Microcluster Ejection
Inverse Bremmstrahlung
Plasma Emission
Optical regulation
Surface Modification

τ_p

The velocity distributions compare well with the formation of supersonic molecular
beams in chemical studies, where the velocities can be described by [14].

$$f(v) = Av^3 \exp (m(v-v_o)^2 / 2kT_s) \qquad (1)$$

Singh et al. [17] have modelled many plume characteristics by treating the partially
ionised plasma as an adiabatically expanding high temperature, high pressure gas, and
have shown that each ablated species accelerates and achieves a velocity dependent upon
its mass and the incident fluence applied (i.e. the plasma temperature induced). While
their model predicted the asymptotic velocity of the ionic species would be proportional to
the square root of their molecular weight, this did not agree with observations, as they
suggested, because of possible interactions of different species within the plume.

A rotating substrate has been used by Dupendent et al. to gather particles of different
speeds formed PLD [12]. As well as the rapid (2 km/s) species mentioned above much
slower components with velocities peaked around 100-200 m/s were found. These
slower species resulted in lumps (from 100's of nm to 10's of μm) forming on the film
surface, with generally the slowest (≈50 m/s) resulting in the largest lumps. In general,
more particulates tend to be formed at higher fluences and for picosecond rather than
nanosecond ablation. Several techniques have been tried to eliminate these undesirable
droplets. By far the most successful method applied to date involves target rotation.
This ensures that the remnants of any previous ablation event, often resolidified off-
stoichiometric mixtures, are not re-irradiated and ablated toward the substrate. Barr [41]

and other groups have used a 10,000 rpm rotating deflector that selectively deflects slow particulates and transmits faster moving species. Gaponov et al. [42] used two colliding beams to generate a higher pressure zone which expanded towards the substrate, which the transmitted beams containing the heavier lumps actually missed. A negative substrate bias of several 100V can also considerably reduce heavy clusters that are similarly charged, and this significantly improved the growth of Ge by PLD [43]. Koren et al. [44] and Chiba et al. [45] have also used a second laser beam during PLD to excite and dissociate the larger particles within the plasma. Sankur [46] has found that using melts as target sources can virtually eliminate particulate formation for the case of both Ge and for limited fluences, B_2O_3.

In summary, the picture of the ablation process, schemed as a series of key events in Fig. 2, may be considered as follows. Under the extreme and sudden absorption of optical energy in a very thin layer of the sample, all the component atoms are ejected in an energetic and violent rupture of the surface. This ensures stoichiometry is conserved. Once formed, this plasma begins to absorb the remaining light in the incident laser pulse and attain a very high temperature. Many clusters that are ejected most likely rapidly fragment as a result of inherent instability, collisions, or photonic interactions. Nevertheless, some small fraction may reach the substrate surface. These can be all but eliminated by careful target preparation (which should be as dense as possible) together with a judicious choice of wavelength, pulse duration and fluence, surrounding gas pressure, and target-substrate geometry. After undergoing collisions with themselves and with background gas species, the plasma expands away from the surface and emits its own characteristic and detectable light. After some distance the particles loose kinetic energy and may form particulates or a smooth film layer on a nearby substrate. Thus, it is critical that the substrate is placed at an optimum position with respect to the plume. Finally, it may also be essential to introduce a rapidly rotating chopper operating synchronously with the laser into the path of the plasma, to eliminate the most slowly moving particles. This is because in some cases, the parameters required to minimise the particulate formation may not necessarily be *identical* to those necessary for optimum quality of film grown.

Experimental Approach

In this section, only some brief examples of PLD will be presented. A vast number of layers can in fact be grown by this method [19]. A typical experimental set-up, as already shown in Fig. 1, basically consists of an evaporation chamber with a laser beam and directing optics positioned outside the cell. The target pellet is commonly irradiated at an angle of 45-60°. The substrate upon which the film grows is usually positioned parallel to the target,although fundamentally, this need not be the case, and often can be heated to sufficiently high temperatures to help promote epitaxy. To help achieve optimum stoichiometric transfer during irradiation, the target is commonly rotated, or the impinging beam is scanned across the target. Although pressures around 10^{-6} torr are common, precise amounts of O_2, N_2O, N_2, or other gases may be pumped through the chamber during deposition. The beam fluences applied for superconductor ablation are around 1-4 J/cm^2, although occasionally higher values are used, and to date most of the lasers used operate at 10Hz, although in principle higher rates would be desirable. Typical deposition rates are around 0.5-5 Å/pulse, and in 10-20 minutes films around 1 μm in thickness can be grown.

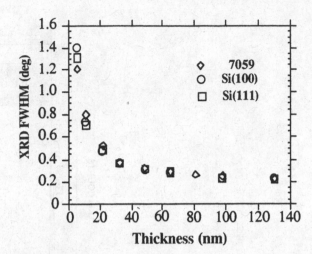

Fig.3. *Thickness dependence of FWHM of XRD 001 peak of CeO2 films grown on various substrates.*

Depending upon the nature of the target irradiated and more importantly the properties of the host substrate (thermal and chemical characteristics) and its temperature, the deposited films may be amorphous, or crystalline. The degree of lattice and thermal expansion mismatch is just as important as in any heterogeneous film growth. One also must ensure that chemical reactivity at the interface is not detrimental to the growing film. Substrates of strontium titanate ($SrTiO_3$), magnesium oxide (MgO), and yttrium stabilised zirconia (YSZ) have been host to amongst the best laser grown thin film superconductors reported to date. Although slightly better properties have occasionally been achieved with Al_2O_3 and $LiNbO_3$ at 500°C thermally-induced stress cracks can hamper reproducibility. To further improve the quality of superconductor films, cheaper and better substrates than $SrTiO_3$ are needed, particularly if microelectronic integration is desirable. There is presently a great deal of investigation into the possible use of buffer layers that might in future enable successful combination of the superconducting films with Si devices. Crucial amongst the criteria to be observed are:

(a) Intermixing and chemical reaction at the superconductor-semiconductor interface
(b) Thermal mismatch between the two components
(c) Strain introduced by mismatch in the lattice constants of the two materials

Initial attempts to grow films on Si required high temperature (850°C) to crystallise the films resulting in cracking, interdiffusion and chemical reaction at the film-substrate interface, leading to T_c's of only 45K for films as thick as 5-8μm [47]. One solution, the growth of buffer layers on the Si has been investigated in recent years. Because of its good lattice match and dielectric properties, we have studied CeO_2 growth by laser ablation. Both substrate temperature and oxygen pressure are important in determining the crystalline quality of the films [48]. Figure 3 shows how the crystalline quality of the substrate does not seem to be crucial, as a gradual narrowing of the FWHM of the 100 peak of the XRD pattern with film thickness is seen to occur independently of whether the substrate was Si (111), Si (100), or 7059 glass. As a novel approach to assisting

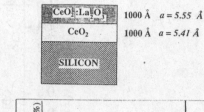

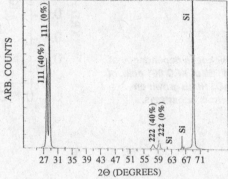

Fig.4. *XRD of undoped and doped ceria showing lattice engineering by PLD.*

optimised lattice matching between different crystal systems, differently doped buffer layers can be grown. Figure 4, for example, shows how CeO_2 doped with 40% La_2O_3 can change the lattice constant of the buffer by around 2.5%. Such sequential layering is exceedingly straightforward in PLD.

The YBCO system is by far the most intensely studied of the superconducting cuprates and the growth of such films on crystal substrates is well established. Conditions used by the vast majority of groups are based on the work of Inam et al. [49]. Although a variety of modifications have been reported, typical conditions include the use of fluences of $1.5 \ J/cm^2$ at 248nm, around 100 mTorr of oxygen, and a substrate heated to between 700-750°C. Such layers can even be grown to a high degree of quality, on a range of buffer layers such as YSZ or CeO_2, also grown by PLD on substrates such as c-Si.

The BiSrCaCuO system is much more difficult to prepare and process. The homologous series $Bi_2Sr_2 Ca_nCu_{n+1}O_x$ contains three members at n = 0, 1 & 2, with superconducting transitions of 20K, 80K and 110K. These are structurally interrelated and differ only by the insertion of the $CaCuO_2$ accommodated by extra layers and are often described by the number of Cu-O planes, i.e. 1 (n=0), 2 (n=1), or 3 (n=2). The chemistry of these phases is unfortunately very complex and isolation of pure single phase ceramic samples has proven to be a major challenge. The 2223 (n=2, T_C=110K) phase, for example, is only stable over a 10K range near its melting point. In general, the thermodynamically stable 2212 (n=1 T_C =80K) phase is preferentially synthesised and the highest T_C phase is then generated by the decomposition of this phase.

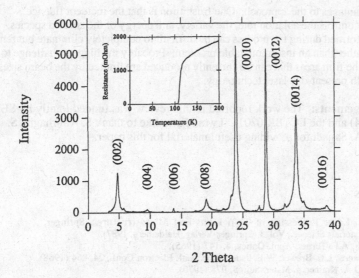

Fig.5. *XRD and resistivity of multilayer-grown Bi(Pb)SrCaCuO films.*

Many attempts to grow layers of BiSrCaCuO have nonetheless been reported, although the level of activity is substantially lower than that for $Ba_2YCu_3O_7$. The incorporation of lead (Pb) on to the bismuth site promotes the formation of the 2223 phase over 2212 and nearly single phase material can be produced by very long annealing times (~ 100 hours) in the presence of a Pb vapour. Alternate PLD of BiSrCaCuO with PbO layers in a sandwich structure has been shown to promote the 2223 phase [50]. Figure 5 shows the X-ray diffraction pattern and electrical resistivity of such a multilayered film, annealed for 15 hours at 854°C. This method has the advantage that it enables independent control over the Pb content in the film and the temperature cycling. A similar approach has been adopted to grow the first films of PbSrYCaCuO, by making sandwiches of PbO and semiconducting SrYCaCuO [51]. The ability to fabricate novel materials by PLD thus opens up a vast new area of exciting research.

The methods of sandwiching and reactive growth in O_2 shown here are only a few of the myriad of possibilities available with PLD. A further example of reactive PLD is evident in the growth of TiN, obtained by ablating pure Ti metal in nitrogen. To date, other approaches have used less dense and more porous TiN targets and often report undesirable incorporation of oxygen impurities. Our method produces crystalline films at 450°C. These are found to be essentially oxygen-free and have smooth mirror-like surfaces, which are certainly amongst the best yet reported [52].

In this brief review, the process of PLD has been described. Although not yet well understood, a simple outline of the perceived mechanisms has been presented in terms of absorption, particle ejection, and plume dynamics. PLD has tremendously exciting prospects and offers many advantages over several traditional deposition technologies. Despite these attractions, there is general agreement that it will not displace any of these, but rather, will introduce increased scope for thin film growth. There are nevertheless

358

some disadvantages to the approach. One frustration is that the incident fluence collectively controls the ablation rate, the variety, and energy of the ablated species. Particulates formed during the process can be difficult to completely eliminate but remain a mischief rather than an insurmountable problem. Probably the biggest challenge to PLD is that the film areas that can be currently produced are limited by the beam sizes available with present-day laser technology.

Acknowledgements: The work reported in this review was funded jointly by SERC (GR/H13154) and the EC (BR-0201). I would also like to thank S. Amirhaghi, S. Naqvi, and A. Sajjadi for providing useful material for this paper.

References

1. I.W. Boyd, **Laser Processing of Thin Films and Microstructures**, Springer Series in Materials Science., Vol. 3 (Springer-Verlag, Heidelberg, 1987).
2. H.M. Smith, A.F. Turner, Appl. Optics, **4**, 147 (1965).
3. P.D. Zavitsanos, L.E. Brewer, W.E. Sauer, Proc. Natl. Electron Conf., 24, 864 (1968).
4. V.S. Ban, D.A. Kramer, J. Mater. Sci., 5, 978 (1970).
5. H. Sankur, J.T. Cheung, J. Vac. Sci. Technol., A1, 1806 (1983).
6. See reference 1, pp280-281.
7. J. T. Cheung, T. Magee, J. Vac. Sci. Technol., A1, 1604 (1983).
8. C. Cali, V.Daneu, A. Orioli, S. Riva-Sanseverino, Appl. Opt., 15, 1327 (1976).
9. M.I.Baleva, M.H.Maksimov, S.M.Metev, M.S.Sendova, J.Mater.Sci. Lett., 5, 533 (1986).
10.J.T. Cheung, J. Madden, J. Vac. Sci. Technol., B5, 705 (1987).
11.R. Kelly, J.E.Rothenberg, Nucl.Inst.& Methods in Phys Res., B7/8, 755 (1985) & refs therein.
12.R. Srinivason, V. Mayne-Banton, Appl. Phys. Lett., 41, 576 (1982).
13.D. Dijkkamp, T. Venkatesan, X.D. Wu, S.H. Shaheen, N. Jisrawi, Y.H. Min-Lee, W.L. McLean, M.Croft, Appl. Phys. Lett., 51, 619 (1987).
14.J.E. Andrew, P.E. Dyer, D. Forster, Appl. Phys. Lett., 43, 717 (1983).
15 H. Dupendant, J.P. Gavigan, D. Givord, A. Lienard, J.P. Rebouillat, Y. Souche, Appl. Surf. Sci., 43, 36 (1989).
16.H.S. Kwok, D.T. Shaw, Q.Y. Ying, J.P. Zheng, S. Witanachchi, E. Petrou, H.S. Kim, Proc. SPIE, 1187, 161 (1989).
17. R.K. Singh, J. Narayan, Phys. Rev. B, 41, 8843 (1990).
18. J.C.S. Kools, T.S. Baller, S.T. de Zwart, J. Dieleman, J. Appl. Phys., 71, 4547 (1992).
19.F. Beech, I.W. Boyd, "Laser Ablation of Electronic Materials", in **Photochemical Processing of Electronic Materials**, ed., I.W. Boyd, R.B. Jackman, Academic Press, London, 1992, pp.388-432.
20.D. Dijkkamp, T. Venkatesan, X.D. Wu, S.A. Shaheen, N. Jisrawa, Y.H. Min-Lee, W.L. McLean, M. Croft, Appl. Phys. Lett., 51, 619 (1987).
21.J. Narayan, N.Biunno, R.Singh, O.W.Holland,O.Auchiello, Appl.Phys.Lett., 51, 1845 (1987).
22.H.S. Kwok, P. Mattocks, L.Shi, X.W. Wang, S. Witanachchi, Q.Y. Ying, J.P. Zheng, D.T. Shaw, Appl. Phys. Lett., 52, 1825 (1988).
23.N.Savva, K.F.Williams, G.M.Davis, M.C.Gower, IEEE J. Quant. Elect., 25, 2399 (1989).
24.A. Caruso, R. Gratton, Plasma Phys., 10, 867 (1968).
25.T. Venkatesan, X.D. Wu, A. Inam, J.B. Wachtman, Appl. Phys. Lett., 52, 1193 (1988).
26.M. Brown, M. Shiloh, R.B. Jackman, I.W. Boyd, Appl. Surf. Sci., 43, 382 (1989).
27.L. Lynds, B.R. Weinberger, G.G. Peterson, H.A. Krasinski, Appl. Phys.Lett., 52, (1988).
28.W. Marine, M. Peray, Y. Mathey, D. Pailharey, Appl. Surf. Sci., 43, 377, (1989).
29.A. Sajjadi, K. Kuen-Lau, F. Saba, F. Beech, I.W. Boyd, Appl. Surf. Sci., 46, 84 (1990).
30.A.M. Bonch-Bruevich, Y.A. Imas, Sov. Phys. Tech., 12, 1407 (1968).
31.F.P. Gagliano, U.C. Paek, Appl. Opt., 13, 274 (1974).
32.P.E. Dyer, A. Issa, P.H. Key, Appl. Phys. Lett., (1990).
33.R.K.Singh, P. Tiwari, J. Narayan, SPIE Proc 1187, 182 (1989).
34.A. Mele, D. Consalvo, D. Stranges, A. Giardini, R. Teghil, Appl. Surf. Sci., 43, 398 (1989).
35.O. Auciello, S. Athavale, O.E. Hankins, M. Sito, A.F. Schreiner, N. Biunno, Appl. Phys. Lett., 53, 72 (1988).

36.P.E. Dyer, RD. Greenough, A. Issa, P.H. Key, Appl. Surf. Sci., 43, 387 (1989).

37.S. Miura, T. Yoshitake, T.Satoh, Y.Miyasaka, N.Shohata, Appl. Phys. Lett., 52, 1008 (1988).

38.J.P. Zheng, Z.Q. Huang, D.T. Shaw, H.S. Kwok, Appl. Phys. Lett.,54, 280 (1989).

39.D.N. Mashburn, D.B. Geohegan, Proc.SPIE 1187, 172 (1989).

40.Yu. A. Bykovskii, S.M. Sil'nov E.A. Sotnichenko, B.A. Shestakov, Sov, Phys. JETP, 66, 285 (1987).

41.W.P. Barr, J. Phys. E, 2, 1024 (1969).

42.S.V. Gaponov, A.A. Gudkov, A.A. Freeman, Sov. Phys. Tech. Phys.,27, 1130 (1982).

43.D. Lubben, S.Barnett, K.Suzuki, S.Gorbatin, J.E. Greene, J.Vac.Sci.Technol., B3, 968 (1985).

44. G. Koren, R.J. Baseman, A. Gupta, M.I. Lutwyche, R.B. Laibowitz, Appl. Phys. Lett., 56, 2144 (1990).

45. H. Chiba, K. Murakami, O. Eryu, K. Shihoyama, T. Mochizuki, K. Masuda, Japan. J. Appl. Phys. (1991).

46.H. Sankur (Private communication).

47.X. D. Wu, A. Inman, T. Venkatessan, C. C. Chang, E.W. Chase, P. Barboux, J. M. Tarascon, and B. Wilkens, Appl. Phys. Lett., 52, 754 (1988).

48.S. Amirhaghi, F. Beech, M. Vickers, P. Barnes, I.W. Boyd, Elec. Lett., 27, 2304 (1991).

49.A. Inman, M.S. Hedge, X.D.Wu, T. Venkatesan, P. England, P.F. Micelli, E.W. Chase, C.C. Chang, J.M. Tarascon, and J.B. Wachtman, Appl. Phys. Lett., 53, 908 (1988).

50.A. Sajjadi, S. Kilgallon, F. Beech, F. Saba, I.W. Boyd, Elec. Lett., 27, 345 (1991).

51. S.H.H. Naqvi, F. Beech, I.W. Boyd, Elec. Lett., 27, 430 (1991).

52. V. Craciun, I.W. Boyd, Mat. Sci & Eng (B), 18, 178 (1992).

Laser Processing of Ceramics

M.F. Modest
Department of Mechanical Engineering
The Pennsylvania State University
University Park, PA 16802

Abstract

Various silicon compound ceramics, in the form of monolithic materials (such as Si_3N_4 and SiC), whisker composites [like Al_2O_3 (matrix)–SiC (whiskers)] and continuous-fiber composites [such as SiC (matrix)–SiC (fibers)] are considered excellent candidates for high-temperature, high-strength applications. Because of the extreme hardness of these materials, lasers are becoming a preferred machining tool for them. Experimental investigations have demonstrated that nearly all ceramics can be efficiently drilled, scribed or cut with a laser, although massive problems remain that are poorly, or not at all understood. These problems include thermal stress (resulting in micro-cracking, macro-cracking or shattering), redeposition of evaporated or liquified material, poor surface finish, undesirable hole and groove tapers, etc.

This paper gives an overview of (*i*) our theoretical work on predicting drilling, scribing and cutting of ceramics, (*ii*) our experimental work on CO_2-laser drilling, scribing and cutting of compound ceramics such as Si_3N_4 and SiC, and continuous fiber composites such as C-SiC and SiC-SiC, and (*iii*) a comparison between experiment and theory.

1 Introduction

A number of different lasers are available for laser materials processing, each with different characteristics which may make a particular laser most suitable for a particular application. The most commonly used lasers for materials processing are the carbon dioxide (CO_2) laser (operating at a wavelength of around $10.6\,\mu m$), the neodymium-YAG (Nd:YAG) ($1.064\,\mu m$) and various excimer lasers (0.2 to $0.6\,\mu m$). Their characteristics vary primarily with respect to power levels, wavelength of operation, focal spot size, and pulsing capabilities.

Laser reactions can be classified into those which are governed mainly by pyrolytic (thermal) or mainly by photolytic (chemical) processes. In photolytic processing the laser light breaks chemical bonds directly; such reactions are generally achieved with short-wavelength lasers, such as excimer lasers. In pyrolytic laser-induced processing the laser serves as a heat source and gives rise to thermochemical reactions. In spite of their thermal character, such laser-driven reactions (*e.g.*, melting, structural transformations, amorphization, formation of metastable phase, etc.) may be very different from those traditionally initiated by a conventional heat treatment. Any of the above lasers may be used for pyrolytic processing.

A number of experimenters investigated laser drilling of materials, primarily of metals, e.g.,[1-8]. Today, one of the primary industrial applications of laser drilling is the drilling of cooling holes in high-temperature alloys in combustors and turbine blades. These alloys may soon be replaced by even tougher composite ceramics with their own need for cooling holes. Experimental investigations on laser cutting and grooving have been reported for many different materials. In the electronics industry laser machining has been used in the cutting of alumina substrates and the separation of silicon wafers into complex shapes[9-13]. Scribing of alumina was considered by Paek and Zaleckas[11]. They compared the performance of CO_2 and Nd-YAG lasers, noting that the YAG laser gave a smoother edge (since, because of its shorter wavelength, it can be focussed better) but left unwanted debris on the surface (since the absorption coefficient is lower at short wavelengths, resulting in subsurface evaporation and explosive removal of particles rather than evaporation).

Detailed grooving experiments, using CW CO_2 lasers, have been carried out by Wallace[14, 15] and Yamamoto and Yamamoto[16] for silicon nitride, and by DeBastiani et al.[17] for silicon carbide. These studies shed some light on the respective laser ablation mechanism, showing that Si_3N_4 decomposes into liquid silicon and gaseous nitrogen, while SiC decomposes into various gases without liquid formation. The studies also discussed how the strength of the materials was affected (indicating decreases in bending strength of up to 70%). Meiners et al.[18] investigated the machinability of alumina, silicon carbide and silicon nitride with a pulsed 600 W $Nd{:}YAG$ laser, measuring removal rates, surface roughness and onset of cracking. They determined that only Si_3N_4 could be machined satisfactorily with a pulsed YAG laser. Cutting of alumina pieces with a CO laser was compared to CO_2 laser cutting as reported by Maisenhälder[19]. Under otherwise identical conditions the cutting rate with a CO laser was approximately 35% greater, apparently due to the fact that alumina has a relatively low absorption coefficient at the CO wavelength, resulting in internal absorption. Smith et al.[20] experimentally determined the maximum feedrate with which alumina plates can be cut with a CW CO_2 laser (without fracturing), as function of plate thickness and laser parameters. They showed that fracture conditions can be related to the magnitude of maximum temperature gradients as predicted from calculations. Tönshoff and Gonschior[21] investigated the influence of several laser processing conditions on thermal stress damage during CO_2 laser cutting of alumina wafers. They found that water cooling produced minor improvements, while crack-free cutting was obtained if the ceramic was preheated to 1100°C. Maruo et al.[22] investigated plasma formation above Si_3N_4 when irradiated by KrF excimer and by CO_2 lasers. They observed strong plasma formation for excimer irradiation, but none with CO_2 irradiation, probably because of the higher power density obtained with the short pulses of the KrF laser. Wallace et al.[23, 24] demonstrated the feasibility of several shaping operations based on overlapping of grooves, including turning, threading, and milling with various silicon compound ceramics. Chryssolouris et al.[25, 26] showed how a dual-beam set-up may be used to mill relatively large chunks of material out of metals and ceramics. An up-to-date review of various laser machining operations has been given by Chryssolouris[27].

2 Theoretical Modeling of Laser Drilling of Ceramics

Most of the theoretical work on laser-processing heat transfer to date has centered on the solution of the classical heat conduction equation for a stationary or moving semi-infinite solid. Cases with and without phase change and for a variety of irradiation conditions have been studied. The simplest case without phase change, where a body is heated over its boundary surface, has been addressed by Carslaw and Jaeger[28], White[29, 30], Rykalin et al.[31], Paek and Gagliano[32], Ready[33], Brugger[34], Maydan[35, 36] and Modest and Abakians[37]. The problem is considerably more complicated when phase change takes place. Soodak[38], Landau[39], and Dabby and Paek[40] studied the problem of melting with complete removal of melt for a stationary solid. To model the laser drilling process von Allmen[3] found a quasi-one-dimensional solution which showed considerable agreement with experiments.

CW laser drilling of ceramics and anisotropic continuous fiber ceramic composites has been investigated by Ramanathan and Modest[41]. Assuming constant thermal properties, but different thermal conductivities in the plane of the fibers (k_r) than normal to them across the thickness of the workpiece (k_z), the axisymmetric transient heat conduction equation governing the process is

$$\rho c \frac{\partial T}{\partial t} = \frac{k_r}{r} \frac{\partial}{\partial r}\left(r \frac{\partial T}{\partial r}\right) + k_z \frac{\partial^2 T}{\partial z^2}, \tag{1}$$

where it is assumed that the ceramic is opaque, i.e., the laser energy is absorbed by a very thin surface layer. Thus, assuming single step material removal at temperature T_{re}, and convection and radiation losses to be negligible[42], the boundary conditions for Eq.(1) are

$$r \to \infty : \quad T \to T_\infty, \tag{2a}$$

$$z = L : \quad \frac{\partial T}{\partial z} = 0, \tag{2b}$$

$$z = s(r, t): \qquad \underbrace{\alpha \overline{H}}_{\text{Absorbed laser energy}} = - \underbrace{\left(k_r \frac{\partial T}{\partial r} \hat{\jmath} + k_z \frac{\partial T}{\partial z} \hat{k} \right) \cdot \hat{n}}_{\text{Conduction into medium}} + \underbrace{\rho \Delta h_{re} \frac{\partial s}{\partial t} \hat{k} \cdot \hat{n}}_{\text{Energy for ablation}}, \qquad (2c)$$

where $s(r, t)$ is the local hole depth, $\hat{\jmath}$, $\hat{k}$ and $\hat{n}$ are the unit vectors in the r- and z-directions and normal to the surface, respectively. The energy required to remove a unit mass of material (by ablation, decomposition, micro-explosions, etc., depending on the ceramic) is Δh_{re}, α is the surface absorptance at the laser wavelength, and $\overline{H}$ is the total local laser irradiation, consisting of a direct component and one due to multiple reflections from the hole's surface. If only diffuse reflections are considered, and if ablated/ejected material does not interfere with the laser beam, $\overline{H}$ follows from[43]

$$\overline{H}(r, z) = \overline{H}_0(r, z) + (1 - \alpha) \int_{\text{hole}} \overline{H}(r', z') \, dF_{dA-dA'}, \qquad (3)$$

where $dF_{dA-dA'}$ is the view factor between ring elements on the hole's surface (cf. Fig. 1). For a Gaussian beam the direct irradiation is

$$\overline{H}_0(r, z) = \frac{2P}{\pi w^2(z)} \exp\left(-\frac{2r^2}{w^2(z)} \right) \hat{k} \cdot \hat{n}, \qquad (4)$$

where P is the total power of the laser beam, and $w(z)$ is the local beam radius, related to that at the beam waist, w_0, which is located at $z = -W$:

$$w^2(z) = w_0^2(z) \left[1 + \left(\frac{(W + z) M^2}{\pi w_0^2 / \lambda} \right)^2 \right] \qquad (5)$$

and $M^2 > 1$ is a beam quality factor describing how rapidly the beam diverges as compared to a perfectly-Gaussian beam.

One final condition is required to determine hole shape $s(r, t)$. For a fixed removal temperature this condition is

$$z = s(r, t): \quad \frac{\partial s}{\partial z} = 0 \quad \text{if} \quad T < T_{re},$$
$$\frac{\partial s}{\partial z} > 0, \quad T = T_{re} \quad \text{otherwise}, \qquad (6)$$

i.e., no ablation occurs until the surface temperature reaches T_{re}, after which T remains at T_{re} and excess energy is used for removing material.

Ramanathan and Modest[41] solved the above set of equations approximately, evaluating conduction losses through an integral method. Some typical results are given in Fig. 2 for a hole drilled with a 500 W CO_2 laser using a 12.7 cm focal-length lens ($w_0 = 150\,\mu m$, $M^2 = 2.7$) focused $W = 8$ mm above a 2.1 mm thick plate of C/SiC composite (long graphitic carbon fibers embedded in β-SiC matrix material). The diameter of the hole can be varied by using different focal positions W. Figure 2 shows that there is substantial improvement in taper between 0.39 s and 0.73 s; however, after approximately twice the drill-through time, no significant additional improvement in hole taper can be expected.

3 Modeling of Laser Scribing and Cutting of Ceramics

Laser scribing or shaping with a moving laser was modeled by Modest and Abakians for irradiation by a CW laser with a Gaussian distribution onto an opaque[42] or semi-transparent[44] semi-infinite solid. They assumed one-step evaporation of material (without beam interference), parallel laser beams, negligible reflection effects and small heat losses (in order to employ a simple integral method for conduction losses). They showed that a semi-transparent solid should behave like an opaque material, unless the absorption coefficient, κ, becomes small, $\kappa w_0 < 40$, which corresponds to absorption coefficients of $\kappa \simeq 2000\,\text{cm}^{-1}$ (a fairly large value). The assumption of a parallel laser beam was relaxed by

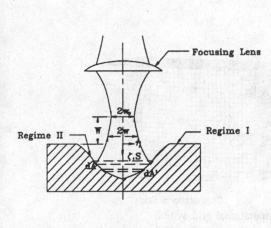

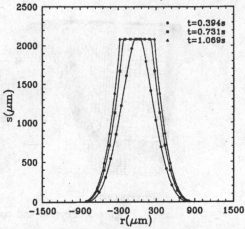

Figure 1: Schematic of laser drilling

Figure 2: Calculated hole profiles for C/SiC, $W = 8\,$mm and $P = 500\,$W

Biyikli and Modest[45], who investigated the effects of beam focusing and focal plane position. The effects of variable properties and multiple-pass scribing, using the same basic model, were treated by Ramanathan and Modest[46,47]. Roy and Modest[48] addressed one of the weakest assumptions made in these studies by including three-dimensional conduction effects and showed that a more accurate treatment of the conduction losses has a considerable effect on the size of the groove formed by evaporation, especially if scanning velocity is low (resulting in higher conduction losses). However, the surface and bulk thermal properties were assumed constant. The boundary element formulation used in[48] to solve the three-dimensional conduction equation for a moving body is unsuitable for variable property conduction within moving bodies. Therefore, in another paper by Roy and Modest[49] and two by Bang and Modest[50,51], the boundary element method is replaced by a finite-difference formulation with a boundary-fitted coordinate system. The transient form of the governing equation is used so that time-varying laser irradiation conditions can be readily accommodated in the future. Augmenting Eq.(1) to include movement of the specimen (with constant velocity u) and variable (but isotropic) properties leads to

$$\frac{\partial h}{\partial t} + u\frac{\partial h}{\partial x} = \nabla \cdot (\alpha_H \nabla h), \quad h = \int_{T_0}^{T} c\,dT, \tag{7}$$

where h is enthalpy and T_0 is a reference temperature. Similarly, the boundary conditions for this three-dimensional problem change to

$$x - \pm\infty, \quad y \to \pm\infty, \quad z \to +\infty: \quad h = h_\infty, \tag{8a}$$

$$z = s(x,y,z): \quad \alpha\overline{H} = -\hat{\mathbf{n}} \cdot (\rho\alpha_H\nabla h) + (v_n - u\hat{\mathbf{i}} \cdot \hat{\mathbf{n}})\rho\Delta h_{re}, \tag{8b}$$

$$v_n - u\hat{\mathbf{i}} \cdot \hat{\mathbf{n}} = 0 \text{ if } h < h(T_{re}),$$

$$> 0 \text{ if } h = h(T_{re}), \tag{8c}$$

where it was assumed that the specimen is of semi-infinite thickness (scribing). Equations (4) and (5) remain valid for the scribing process, of course, as does Eq.(3) (with "hole" replaced by "groove surface" in the integration), provided reflections are perfectly diffuse. In the case of specular reflections beam tracing must be employed as described in detail by Bang and Modest[50,51].

Roy and Modest[49] solved Eqs.(7) and (8) numerically by transforming equation and boundary conditions to a nondimensional boundary-fitted coordinate system symbolically given by

$$\xi = \xi(x,y,z,t), \quad \eta = \eta(x,y,z,t), \quad \zeta = \zeta(x,y,z,t), \quad \tau = \frac{2\alpha_{H,re}\,t}{w_o^2}, \tag{9}$$

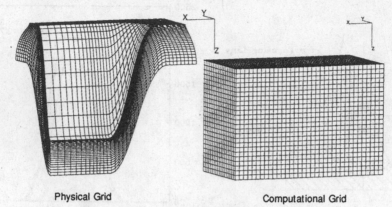

Physical Grid   Computational Grid

Figure 3: Physical and computational grid system

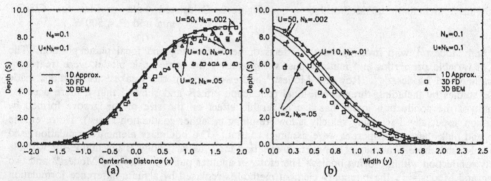

(a)   (b)

Figure 4: Comparison of 1-D and 3-D results (parallel beam, constant properties) (a) groove development along centerline, (b) fully developed groove cross-section

which maps the time-varying, heat-affected zone near the groove's surface to a fixed cube with uniformly spaced nodes as indicated in Fig. 3. The transformed equations were finite-differenced and solved with an approximate factorization algorithm. Because of the strong nonlinearities caused by the groove formation, the quasi-steady solution was found with a time-iterative scheme using the transient form of the governing equations. Some typical resulting groove cross-sections, for the case of constant properties and non-reflective surfaces, are shown in Fig. 4, comparing the present boundary-fitted finite-difference solutions (FD) to the boundary element method of Roy and Modest[49] (BEM), and the quasi-one-dimensional model of Modest and Abakians[42] (1D approximation). The nondimensional parameters in this figure are defined as

$$U = \frac{w_o u}{\sqrt{2}\alpha_{H,re}}, \quad N_k = \frac{\pi w_o \rho \alpha_{H,re}(h_{re} - h_\infty)}{\sqrt{2}\alpha_{re}P}, \quad N_e = \frac{\pi w_o^2 \rho u \Delta h_{re}}{2\alpha_{re}P}, \tag{10}$$

where P is total laser power and y and s are normalized by $w_o/\sqrt{2}$, the $1/e$-radius of the Gaussian beam. It is seen that the quasi-one-dimensional model gives good results for large values of U (scan velocity $u \gg$ diffusion velocity α_H/w_o; i.e., conduction losses are small), but grossly overpredicts groove depth for slow scan velocity/high thermal diffusivity materials.

Effects of beam trapping through multiple reflections within the groove have been modeled by Bang and Modest[50,51]. They divided the material surface into a number of rectangular patches using a

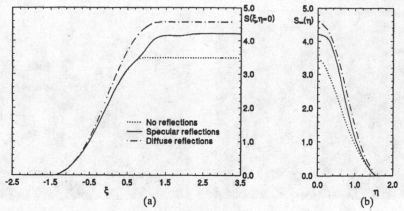

Figure 5: Groove cross-section development for different reflection models ($U = 10$, $N_k = 0.01$, $N_e = 0.01$, parallel beam, $\alpha = 0.3$) (a) groove development along laser scan direction (at centerline), (b) fully developed groove cross-section

bicubic surface representation method. The net radiative flux for these patch elements is obtained by view factor theory (diffusely reflecting surface) or standard ray tracing methods (specularly reflecting surfaces). Three different types of beam trapping have been addressed: a randomly polarized laser beam with (*i*) diffuse and (*ii*) specular reflections at the surface, and (*iii*) a linearly or circularly polarized laser with specular surface reflections obeying Fresnel's relationships. For a polarized laser beam the changing state of polarization of the electric field after reflection was included in the ray tracing method. It was observed that diffuse reflections always carve out deeper grooves than specular reflections, which tend to cause one or more humps along the centerline due to focusing effects, as shown in Fig. 5. It was observed that if the electric vector of the linearly–polarized beam points along the laser scanning direction a narrow and deep groove is obtained; conversely, a wider shallower groove forms. It was also found that a circularly polarized beam results in a similar cross-section as the unpolarized case for the tested parameters. Finally, it was found that a circularly-polarized beam can be used to cut a near-optimum groove. Beyer and Petring[52] also reported that no difference is found between linear and circular polarization in metal-cutting experiments. This has important industrial consequences, since maintaining a constant-orientation linear polarization is difficult, if not impossible, to maintain when cutting/scribing along non-straight paths.

Performing these three-dimensional calculations tends to be CPU time intensive, and the model is limited to scribing of semi-infinite bodies (no through-cutting). Ramanathan and Modest[53] solved Eqs. (7) and (8) approximately for specimens of finite thickness (without considering multiple reflections), using an integral method for the temperature variation across the thickness of the specimen. This two-dimensional approach, while more CPU intensive than the quasi-one-dimensional model, is an order-of-magnitude faster than 3-D calculations and is always rather accurate, even for slow laser scan velocities. An example of applying this 2-D model to through-cutting is given in Fig. 3, which shows the differences of the fully-developed kerf cross-section for a number of plate thicknesses. The results clearly show that there is only a minor difference between truncating the infinite thickness results (at the appropriate thickness) and the results from the cut-through

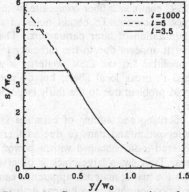

Figure 6: Groove cross-sections for slabs of different thickness, ℓ ($U = 2.83$, $N_k = 0.035$ and $N_e = 0.1$)

Figure 7: Drilling through 3.5 mm thick *C/SiC* composite; power $P = 1500$ W, $f = 6.35$ cm lens, drill time $t = 1$ s, focused above surface at $W = 4$ mm

Figure 8: Drilling through 3.5 mm thick *C/SiC* composite; power $P = 1500$ W, drill time $t = 2.7$ s, $f = 6.35$ cm lens, oblique incidence at 70° off-normal, focused at 6 mm away from hole top

model. A similarly small effect was noticed for the development of the groove depth along the center-line up to the point where the material is completely cut through. The slight increase in the bottom kerf width is due to the accumulation of heat at the corners of the bottom kerf which increases the cutting action at these nodes. Since the differences between the scribing and cutting results were found to be so small, it was concluded that it is always sufficient to predict through-cutting by truncating scribing predictions (which are much easier and faster to obtain).

4 Experimental Holes and Grooves

A well-focused CO_2 laser beam can drill holes into virtually any material, provided the laser has enough power and the absorptance of the material is sufficiently high; this includes all ceramics. However, to obtain holes of the desired diameter and depth, and with acceptable taper is a long, tedious optimization process that, hopefully, can be aided by a good theoretical model. In addition, there may be redeposition of liquid or evaporated material at the hole walls, and thermal stresses within the heat-affected zone with its extreme temperature gradients may cause microcracks, macrocracks or even shattering of the material. Figures 7 and 8 show some examples of holes drilled into the *C/SiC* continuous-fiber composite, with the aim to generate 500 μm diameter oblique holes in 3.5 mm thick specimen. To obtain holes of such high quality and uniform diameter required considerable optimization of laser parameters. The oblique hole in Fig. 8 is actually considerably more uniform that it appears due to the difficulty to cross-section the hole precisely through its mid-point. Unlike monolithic *SiC* the *C/SiC* material is not very susceptible to thermal shock: apparently, microcracks tend to break local fibers, but do not propagate through the material. Also, redeposition is not a great problem due to the fairly large hole diameter (i.e, the debris is easily blown out by the auxiliary jet).

Scribing and cutting of ceramics faces the same problems as drilling, i.e., kerf width and taper, redeposition and damage due to thermal stresses. Figure 9 shows a typical, unoptimized groove in sintered α-*SiC* obtained with a power of $P = 1200$ W and $u = 7.5$ cm/s, using a 6.35 cm focal length lens. Typically, a large crack is observed almost parallel to the groove surface, a substantial distance into the matrix material, apparently caused by thermal stresses. Part of the material above the crack is redeposition showing columnar grains on the surface[17], but it is not clear how much since redeposition (mostly β-*SiC*) is nearly indistinguishable from the original material. The columnar grains detected by deBastiani et al.[17] are more distinct (due to their use of a strong jet of argon shielding gas),

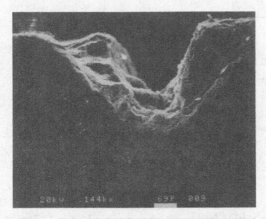

Figure 9: Typical cross-section of grooves formed in sintered α-SiC ($P = 1200\,W$, $u = 7.5\,cm/s$, $f = 6.35\,cm$ lens)

Figure 10: Groove cross-section in hot-pressed silicon nitride ($P = 600\,W$, $u = 5\,cm/s$, $f = 6.35\,cm$ lens)

clearly showing SiC formation by vapor deposition. This, in turn, indicates that the SiC is removed by decomposition into gases (no liquids). The present micrograph shows a layer of silica covering the columnar grains, indicating that some of the silicon gases reacted with air, since no strong jet was used. Figure 10 depict a similar groove for hot-pressed Si_3N_4 ($P = 600\,W$, $f = 6.35\,cm$, $u = 5\,cm/s$), with redeposition in the form of blobs of resolidified liquid silicon indicating that, indeed, Si_3N_4 decomposes into liquid silicon and nitrogen.

For both materials, SiC and Si_3N_4, relatively large chunks of matrix material broke loose during sectioning. This is apparently due to thermal stress damage in the form of micro-cracks throughout the heat-affected zone. This damage is more severe in SiC (probably due to its higher removal temperatrue and, consequently, stronger gradients and stresses), but is also present in Si_3N_4. In contrast, the continuous fiber ceramics suffer only little thermal stress damage during laser processing: four-point bending strength is decreased by about 20%, as compared to up to 70% for Si_3N_4[16] and more for monolithic SiC. Apparently, thermal stress breaks local fibers, but these micro-cracks do not propagate into the material.

5 Comparison of Theory with Experiment

Comparison of hole cross sections with experiment is extremely difficult since sectioning of small, experimentally drilled holes at precisely their midpoint is almost impossible. Figure 4 compares the theoretically predicted hole cross-sections with the holes drilled into the C/SiC composite as shown in Fig. 8, with material properties estimated as reported by Ramanathan and Modest[41]. Agreement is very good considering the poor knowledge of physical properties. Experimental holes generally show less taper than predicted by theory, apparently a beneficial side effect of the (otherwise undesirable) non-Gaussian laser beam profile. A simpler, less time consuming effort is to compare predicted and measured drill-through times. This was done for C/SiC and SiC/SiC composites.[41] Again, agreement between theory and experiment was rather good. Figure 12 shows a comparison of experimentally

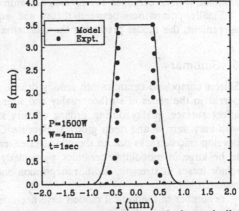

Figure 11: Experimentally and theoretically determined cross-sections of holes drilled through the C/SiC composite; $P = 1500\,W$, drill time $t = 1\,s$, $f = 6.35\,cm$ lens, focused above surface at a distance of $W = 4\,mm$

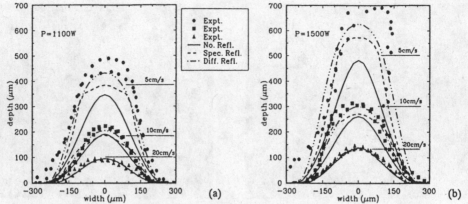

Figure 12: Groove cross-sections in hot-pressed silicon nitride for several scanning speeds; $f = 12.7\,$cm lens, (a) $P = 1100\,$W, (b) $P = 1500\,$W

obtained groove cross-sections with those predicted by the three-dimensional model with multiple reflections[49,51] for hot-pressed silicon nitride. In this figure the "heat of removal" was estimated from the JANAF tables[54] to be $\Delta h_{r_e} \simeq 6206\,$kJ/kg and the absorptance was fixed at 0.15, chosen to give best agreement between theory and experiment. This value appeared reasonable, since the absorptance of Si_3N_4 ranges between 0.19 at room temperature and 0.26 at $1000°C$[55], while the absorptance of liquid silicon—which partially covers the surface during decomposition—is reportedly $\simeq 0.10$[56]. Agreement between experiment and theory is very good. Figure 12 seems to indicate that reflections are primarily diffuse (perhaps due to violent motion within the liquid silicon layer). However, uncertainty of physical properties and of the beam profile could also be factors.

Making similar comparisons for sintered α-SiC was not so successful, overpredicting groove depths and removal rates by about a factor of four, although qualitative trends are predicted well, in particular multiple reflection effects for unpolarized[51] and polarized irradiation[50]. Also, very recent measurements of Ramanathan and Modest[57] have determined the absorptance of Si_3N_4 to be $\simeq 0.45$ during decomposition (for a slightly differently prepared specimen). Using this value in Fig. 12 would also overpredict groove depths by a factor of three. On the other hand, cut cross-sections for the continuous-fiber ceramics are predicted reasonably well.

Finally, comparison between theory and experiment for a polarized laser beam also gave good agreement, the model predicting the qualitative trends perfectly.

6 Summary

Silicon compound ceramics are readily drilled, scribed and cut with a CO_2 laser, although problems persist in the areas of surface quality and strength. Redeposition from liquid and/or vapor negatively affect surface quality during drilling of very small diameter holes and during scribing and cutting with very narrow and deep grooves. Because CO_2 laser machining is a pyrolitic process, all ceramics develop micro-cracks due to thermal stresses, causing a loss in strength. These losses of strength tend to be large in monolithic ceramics, particularly SiC, while continuous-fiber ceramics suffer only very minor losses in strength. Both, redeposition and thermal stresses still require considerable additional research.

A number of theoretical models have been developed, which have the potential to accurately predict the outcome of laser drilling, scribing and cutting operations. To date the accuracy of these models remains hampered somewhat due to the uncertainty in and/or lack of knowledge of important material and plume properties, such as surface absorptance (as function of temperature and surface reactions), ablation temperature and mechanism, the energy required to remove a unit mass of material, and the nature and amount of beam blockage by the plume. Again, additional research is required in this area. Finally, the theoretical models need to be further developed to incorporate such effects as ablation governed by rate equations, laser pulsing and laser-plume-surface interactions, and to predict thermal stresses and redeposition during laser machining.

References

[1] Bar-Isaac, C. and Korn, U., "Moving Heat Source Dynamics in Laser Drilling Processes", *Journal of Applied Physics*, Vol. 3, 1974, pp. 45–54.

[2] Bar-Isaac, C., Korn, U., Shtrikman, S., and Treves, T., "Thermal Structure of the Evaporation Front in Laser Drilling Processes", *Journal of Applied Physics*, Vol. 5, 1974, pp. 121–125.

[3] von Allmen, M., "Laser Drilling Velocity in Metals", *Journal of Applied Physics*, Vol. 47, 1976, pp. 5460–5463.

[4] Hamilton, D. C. and James, D. J., "Hole Drilling with a Repetitively-Pulsed TEA CO_2 Laser", *Journal of Physics D: Applied Physics*, Vol. 9, 1976, pp. L41–L43.

[5] Hamilton, D. C. and Pashby, I. R., "Hole Drilling Studies with a Variable Pulse Length CO_2 Laser", *Optics and Laser Technology*, Vol. 11, 1979, pp. 183–188.

[6] Anthony, T. R., "The Random Walk of a Drilling Laser Beam", *Journal of Applied Physics*, Vol. 51, 1980, pp. 1170–1175.

[7] Polk, D. H., Banas, C. M., Frye, R. W., and Gragosz, R. A., "Laser Processing of Materials", *Industrial Heat Exchangers*, 1986, pp. 357–364.

[8] Chan, C. and Mazumder, J., "Materials Removal by Vaporization and Liquid Expulsion Using a Concentrated Heat Source", In *Proceedings of the Second Joint ASME/JSME Thermal Conference*, Vol. 3, 1987, pp. 283–292.

[9] Lumley, R. M., *American Ceramics Bulletin*, Vol. 48, 1969, p. 850.

[10] Longfellow, J., *American Ceramics Bulletin*, Vol. 52, 1973, p. 513.

[11] Paek, U. C. and Zaleckas, V. J., "Scribing of Alumina Material by YAG and CO_2 Lasers", *American Ceramics Bulletin*, Vol. 54, 1975, pp. 585–588.

[12] Saifi, M. and Borutta, R., "Optimization of Pulsed CO_2 Laser Parameters for Al_2O_3 Scribing", *American Ceramics Bulletin*, Vol. 54, 1975, pp. 986–989.

[13] Belland, R., *Electron. Packaging and Production*, Vol. 15, 1975, p. 56.

[14] Wallace, R. J., *A Study of the Shaping of Hot Pressed Silicon Nitride With a High Power CO_2 Laser*, PhD thesis, University of Southern California, Los Angeles, CA, 1983.

[15] Wallace, R. J., Bass, M., and Copley, S. M., "Curvature of Laser-Machined Grooves in Si_3N_4", *Journal of Applied Physics*, Vol. 59, 1986, pp. 3555–3560.

[16] Yamamoto, J. and Yamamoto, Y., "Laser Machining of Silicon Nitride", In *International Conference on Laser Advanced Materials Processing -- Science and Applications*, High Temperature Society of Japan, Japan Laser Processing Society, Osaka, Japan, 1987, pp. 297–302.

[17] DeBastiani, D., Modest, M. F., and Stubican, V. S., "Mechanisms of Reactions During CO_2-Laser Processing of Silicon Carbide", *Journal of The American Ceramic Society*, Vol. 73, No. 7, 1990, pp. 1947–1952.

[18] Meiners, E., Wiedmaier, M., Dausinger, F., Krastel, K., Masek, I., and Kessier, A., "Micro Machining of Ceramics by Pulsed Nd:YAG Laser", In *Proceedings of ICALEO '91, Laser Materials Processing*, Vol. 74, San Jose, CA, 1991, pp. 327–336.

[19] Maisenhälder, F., "Materials Processing by CO Lasers", In *Laser Advanced Materials Processing -- LAMP '92*, Vol. 1, Nagaoka, Japan, 1992, pp. 43–50.

[20] Smith, R. N., Surprenant, R. P., and Kaminski, D. A., "Fracture Characteristics of an Aluminum Oxide Ceramic During Continuous Wave Carbon Dioxide Laser Cutting", In *Proceedings of ICALEO '91, Laser Materials Processing*, Vol. 71, San Jose, CA, 1992, pp. 337–347.

[21] Tönshoff, H. K. and Gonschior, M., "Reduction of Crack Damage in Laser Cutting of Ceramics", In *Laser Advanced Materials Processing -- LAMP '92*, Vol. 1, Nagaoka, Japan, 1992, pp. 645–650.

[22]Maruo, H., Miyamoto, I., and Ooie, T., "Processing Mechanism of Ceramics with High Power Density Lasers", In *Laser Advanced Materials Processing -- LAMP '92*, Vol. 1, Nagaoka, Japan, 1992, pp. 293–298.

[23]Copley, S. M., Bass, M., and Wallace, R. J., "Shaping Silicon Compound Ceramics With a Continuous Wave Carbon Dioxide Laser", In *Proceedings of the Second International Symposium on Ceramic Machining and Finishing*, NBSA Publ. No. 562, 1979, pp. 283–292.

[24]Copley, S. W., Wallace, R. J., and Bass, M., "Laser Shaping of Materials", In Metzbower, E. A., ed., *Lasers in Materials Processing*, ASME, Metals Park, Ohio, 1983.

[25]Chryssolouris, G. and Bredt, J., "Machining of Ceramics Using a Laser Lathe", *International Ceramics Review*, April 1988, pp. 70–72.

[26]Chryssolouris, G., Bredt, J., Kordas, S., and Wilson, E., "Theoretical Aspects of a Laser Machine Tool", *ASME Journal of Engineering for Industry*, Vol. 110, 1988, pp. 65–70.

[27]Chryssolouris, G., *Laser Machining, Theory and Practice*, Springer Verlag, New York, 1991.

[28]Carslaw, H. S. and Jaeger, J. C., *Conduction of Heat in Solids*, Oxford University Press, 2nd ed., 1959.

[29]White, R. M., "Elastic Wave Generation by Electron Bombardment or Electromagnetic Wave Absorption", *Journal of Applied Physics*, Vol. 34, 1963, pp. 2123–2124.

[30]White, R. M., "Generation of Elastic Waves by Transient Surface Heating", *Journal of Applied Physics*, Vol. 34, 1963, pp. 3559–3567.

[31]Rykalin, N. N., Uglov, A. A., and Makarov, N. I., "Effects of Peak Frequency in a Laser Pulse on the Heating of Metal Sheets", *Soviet Journal of Quantum Electronics*, Vol. 12, 1967, pp. 644–646.

[32]Paek, U.-C. and Gagliano, F. P., "Thermal Analysis of Laser Drilling Processes", *IEEE Journal of Quantum Electronics*, Vol. QE-8, 1972, pp. 112–119.

[33]Ready, J. F., *Effects of High Power Laser Radiation*, Academic Press, New York, 1971.

[34]Brugger, K., "Exact Solutions for the Temperature Rise in a Laser Heated Slab", *Journal of Applied Physics*, Vol. 43, 1972, pp. 577–583.

[35]Maydan, D., "Fast Modulator for Extraction of Internal Laser Power", *Journal of Applied Physics*, Vol. 41, 1970, pp. 1552–1559.

[36]Maydan, D., "Micromachining and Image Recording on Thin Films by Laser Beams", *Bell System Technical Journal*, Vol. 50, 1971, pp. 1761–1789.

[37]Modest, M. F. and Abakians, H., "Heat Conduction in a Moving Semi-Infinite Solid Subjected to Pulsed Laser Irradiation", *ASME Journal of Heat Transfer*, Vol. 108, 1986, pp. 597–601.

[38]Soodak, H., *Effects of Heat Transfer Between Gases and Solids*, PhD thesis, Duke University, Durham, NC, 1943.

[39]Landau, H. G., "Heat Conduction in a Melting Solid", *Quarterly of Applied Mathematics*, Vol. 8, 1950, pp. 81–94.

[40]Dabby, F. W. and Paek, U.-C., "High-Intensity Laser-Induced Vaporization and Explosion of Solid Material", *IEEE Journal of Quantum Electronics*, Vol. QE-8, 1972, pp. 106–111.

[41]Ramanathan, S. and Modest, M. F., "CW Laser Drilling of Composite Ceramics", In *Proceedings of ICALEO '91, Laser Materials Processing*, Vol. 74, San Jose, CA, 1992, pp. 305–326.

[42]Modest, M. F. and Abakians, H., "Evaporative Cutting of a Semi-Infinite Body With a Moving CW Laser", *ASME Journal of Heat Transfer*, Vol. 108, 1986, pp. 602–607.

[43]Modest, M. F., *Radiative Heat Transfer*, McGraw-Hill, New York, 1993.

[44]Abakians, H. and Modest, M. F., "Evaporative Cutting of a Semi-Transparent Body with a Moving CW Laser", *ASME Journal of Heat Transfer*, Vol. 110, 1988, pp. 924–930.

[45]Biyikli, S. and Modest, M. F., "Beam Expansion and Focusing Effects on Evaporative Laser Cutting", *ASME Journal of Heat Transfer*, Vol. 110, May 1988, pp. 529–532.

[46]Ramanathan, S. and Modest, M. F., "Effect of Variable Properties on Evaporative Cutting with a Moving CW Laser", In *Heat Transfer in Space Systems*, Vol. HTD--135, ASME, 1990, pp. 101–108.

[47]Ramanathan, S. and Modest, M. F., "Single and Multiple Pass Cutting of Ceramics With a Moving CW Laser", In *Proceedings of the XXII ICHMT Intl. Symposium on Manufacturing and Materials Processing*, Dubrovnik, Yugoslavia, 1990.

[48]Roy, S. and Modest, M. F., "Three-Dimensional Conduction Effects During Evaporative Scribing with a CW Laser", *Journal of Thermophysics and Heat Transfer*, Vol. 4, No. 2, 1990, pp. 199–203.

[49]Roy, S. and Modest, M. F., "Evaporative Cutting with a Moving CW Laser — Part I: Effects of Three-Dimensional Conduction and Variable Properties", *International Journal of Heat and Mass Transfer*, 1993, accepted for publication.

[50]Bang, S. Y. and Modest, M. F., "Evaporative Scribing with a Moving CW Laser - Effects of Multiple Reflections and Beam Polarization", In *Proceedings of ICALEO '91, Laser Materials Processing*, Vol. 74, San Jose, CA, 1992, pp. 288–304.

[51]Bang, S. Y., Roy, S., and Modest, M. F., "CW Laser Machining of Hard Ceramics — Part II: Effects of Multiple Reflections", *International Journal of Heat and Mass Transfer*, 1993, accepted for publication.

[52]Beyer, E. and Petring, D., "State of the Art in Laser Cutting with CO_2 Lasers", In *9th International Congress on Applications of Lasers and Electro-optics*, 1990, pp. 199–212.

[53]Ramanathan, S. and Modest, M. F., "CW Laser Cutting and Scribing of Composite Ceramics", In *Laser Advanced Materials Processing -- LAMP '92*, Vol. 2, Nagaoka, Japan, 1992, pp. 625–632.

[54]Chase, J. M. W., Davies, C. A., Downey, J. J. R., Frurip, D. J., McDonald, R. A., and Syverud, A. N., eds., *JANAF Thermochemical Tables*, National Bureau of Standards, Washington, DC, 1985.

[55]Roy, S., Bang, S. Y., Modest, M. F., and Stubican, V. S., "Measurement of Spectral, Directional Reflectivities of Solids at High Temperatures Between 9 and 11 μm", In *Proceedings of the ASME/JSME Engineering Joint Conference*, Vol. 4, 1991, pp. 19–26.

[56]Preston, J. R., Sipe, J. E., and Van Driel, H. E., "Phase Diagram of Laser Induced Melt Morphologies on Silicon", In *Materials Research Society Symposia Proceedings*, Vol. 51, Materials Research Society, Pittsburg, Pennsylvania, 1985, pp. 137–142.

[57]Ramanathan, S. and Modest, M. F., "Measurement of Temperatures and Absorptances for Laser Processing Applications", In *ICALEO '93*, Orlando, FL, 1993, submitted for publication.

In-Process-Diagnose beim Laserhärten rotationssymmetrischer Bauteile

K. Dickmann und J. Gröninger
Laserlabor des Fachbereichs Physikalische Technik, FH Münster
Stegerwaldstr. 39, D-48565 Steinfurt

Das Laserhärten rotationssymmetrischer Bauteile bzw. von Teilbereichen komplexer Werkstücke (z.B. Haupt- und Pleuellagersitz von Kurbelwellen) gewinnt in der Praxis zunehmend an Bedeutung, wenn die Randschichthärtung zur Erhöhung der Verschleißfestigkeit nur partiell durchgeführt werden soll. Schwankende Laserstrahl- sowie weitere Prozeßparameter, Materialinhomogenitäten, unterschiedliche Oberflächenbeschaffenheiten und variierende Bauteilgeometrien führen jedoch zu einem unkontrollierten Einfluß auf das Bearbeitungsergebnis.

1. Stand der Technik

Verfahren zur In-Process-Diagnose beim Randschichthärten mit Hochleistungslasern wurde unter Verwendung verschiedener Sensoren umfangreich untersucht /1-6/. Da die Oberflächentemperatur einen entscheidenden Einfluß auf die Härtebahnbreite und -tiefe sowie den Härtewert hat, wurden diese in der überwiegenden Anzahl o.g. Arbeiten als Wert für die Diagnose herangezogen. Aufgrund der einfachen Handhabung sowie der hohen zeitlichen Auflösung (< 1 ms) wird die Temperaturmessung im allgemeinen berührungslos mit Pyrometern durchgeführt. Die bekannten Verfahren zur In-Process-Diagnose und Konzepte zum Aufbau von Regelkreisen beziehen sich überwiegend auf das Randschichthärten ebener Bauteile unter Verwendung von CO_2-Hochleistungslasern. In der letzten Zeit ist jedoch auch der Einsatz von Festkörperlasern für das Randschichthärten von Interesse geworden /7/. Die folgenden Ergebnisse beziehen sich auf entsprechende Arbeiten an rotationssymmetrischen Bauteilen unter dem Einsatz von Nd:YAG-Lasern. Typische Problemstellungen beim Laserhärten von Kurbelwellen sind bspw. das Randschichthärten im Überlappungsbereich beim Anschlußstoß bzw. nebeneinanderliegender Bahnen oder das Härten im Bereich von Bohrungen oder Radien.

2. Versuchstechnik

Die Randschichthärtungen wurden an einem repräsentativen Versuchswerkstoff, Vergütungs-Wellenstahl 42CrMo4, mit einem 400 Watt-Nd:YAG-Laser (Typ VEGA, KLS 322) vorgenommen. Für die Temperaturmessungen wurde ein Pyrometer mit einem Meßbereich von 300 bis 1400° C und einer hohen Anstiegszeit von 0,3 ms eingesetzt.

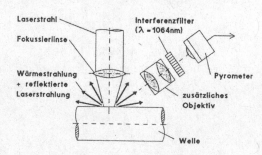

Bild 1: Pyrometeranordnung

Unter Verwendung eines zusätzlichen Objektivs konnte der Durchmesser des Meßflecks auf < 1 mm reduziert werden. Reflektierende störende Laserstrahlung wurde mit einem 1064 nm-Interferenzfilter unterdrückt. Bild 1 zeigt die prinzipielle Anordnung des Pyrometers.

3. Ergebnisse

Voruntersuchungen haben gezeigt, daß bereits bei einer Laserleistung des o.g. Nd:YAG-Lasers von nur 200 Watt Einhärtetiefen bis zu 0.3 mm bei einer Bahnbreite von 3,0 mm möglich sind. (Es soll ausdrücklich darauf hingewiesen werden, daß unter Verwendung dieses Lasersystems nicht der Härteprozeß an sich im Vordergrund stand, sondern ausschließlich dessen Diagnose. Die erzielten Resultate sind jedoch prinzipieller Natur und sollten daher auf andere Laser mit höheren Leistungen übertragbar sein.)

3.1 Anschlußstoß bei der Überlappung umlaufender Härtebahnen

Der Anschlußstoß beim Schließen einer umlaufender Härtebahn an rotationssymmetrischen Bauteilen führt zu Anlaßeffekten der bereits gehärteten Bahn mit einem reduzierten Härtewert oder sogar zu Aufschmelzungen im Überlappungsbereich. Bild 2 zeigt diesen Effekt anhand des Temperaturverlaufs in der Überlappungszone deutlich. Je nach Grad der Überlappung stellt sich ein Temperatursprung bis zu 300° C (bei $\alpha = -7.5°$, siehe Bild 2) ein, der auf die Vorwärmung der Ersthärtung und veränderte Absorptionseigenschaften des bereits gehärteten Materials zurückgeführt werden kann.

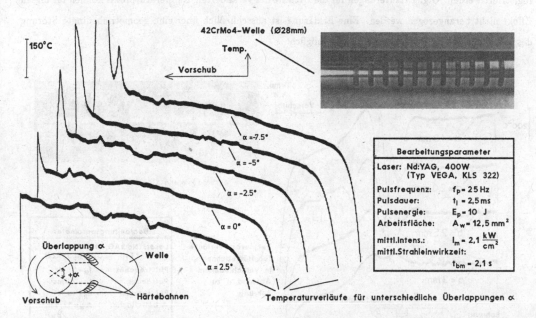

Bild 2: Pyrometrische Erkennung des Anschlußstoßes im Bereich der Überlappung

3.2 Überlappung parallel angeordneter Härtebahnen

Mit dem zur Verfügung stehenden Laseraggregat konnten Spurbreiten von ca. 3mm erzielt werden; mit anderen kW-Hochleistungslasern sind 15 mm heute problemlos möglich. Je nach verfügbarer Laserleistung ist für das Härten von Lagersitzen bspw. an Kurbelwellen die Anordnung paralleler Härtebahnen mit der Forderung nach einem gleichbleibenden Härteresultat über den gesamten Bahnquerschnitt erforderlich. Dieses Problem stellt sich auch beim Anschluß von Härtebahnen in den Radien an den gehärteten Lagersitz dar. Mit Hilfe des Pyrometers konnte ein mit zunehmender Überlappung höherer Temperaturverlauf der zuletzt gehärteten Bahn deutlich detektiert werden. Hiermit verbundene Auswirkungen auf die Querschnittsfläche der Randschichthärtung wurden in /6/ untersucht. Das beobachtete Temperaturverhalten wird auch hier durch die oben erläuterten Zusammenhänge erklärt.

3.3 Einfluß von Geometrieänderungen auf den Härteprozeß

Der Einfluß durch die begrenzte Bauteilabmessung beim Härten von ebenen Flächen oder an schneidenden Bauteilgeometrien und dem damit verbundenen Wärmestau beim Strahlein- und -austritt in das Werkstück wurde in /6,2,3/ untersucht. Aufgrund der Geometrie rotationssymmetrischer Härteflächen tritt dieser Effekt hier jedoch nicht auf. Ein Geometrieeinfluß stellt sich bspw. beim Randschichthärten von Lagersitzen in Kurbelwellen in der Umgebung von Ölbohrungen oder im Bereich der Radien ein. Bild 3 zeigt eine Auswahl pyrometrisch ermittelter Temperaturverläufe im Bereich von Bohrungen (ø4mm) an einer präparierten Welle für unterschiedliche Abstände a der Härtebahn zur Bohrung. Für Abstände a < 3 mm kann unter den hier vorliegenden Versuchsbedingungen im Bohrungsbereich ein deutlich beeinflußtes Temperaturprofil registriert werden. O.g. Erläuterungen für die Ursache des veränderten Temperaturprofils können für diesen Effekt nicht herangezogen werden. Eine Erklärung ist ausschließlich über eine geometriebedingte Störung des Wärmeflusses im Bereich der Bohrung möglich.

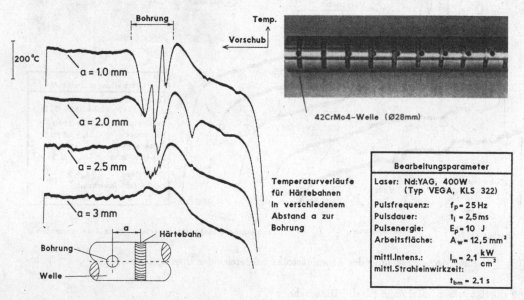

<u>Bild 3:</u> Pyrometrische Detektion von Geometriestörungen (Bohrungen)

4. Konsequenzen für den Einsatz in Regelkreisen

Die Ergebnisse haben gezeigt, daß durch den Prozeß oder die bauteilgeometriebedingten Einflüsse auf den Härteprozeß beim Randschichthärten rotationssymmetrischer Bauteile mit pyrometrischen Temperaturmeßverfahren sicher erkannt werden können. Weitere hier nicht vorgestellte Arbeiten haben darüber hinaus gezeigt, daß auch andere Einflüsse, wie bspw. variierende Oberflächenbeschaffenheit mit diesem Verfahren detektiert werden können. Wenn die hier aufgeführten Untersuchungen auch an einem für Kurbelwellen untypischen Versuchswerkstoff durchgeführt wurden, so bestehen an der Übertragbarkeit auf andere Werkstoffe aufgrund der prinzipiellen Natur der Ergebnisse keine Zweifel. Beim Aufbau eines geschlossenen Regelkreises lassen sich die Temperatursignale des Pyrometers als Ist-Größe in den Regelprozeß zurückführen. Als Stellgröße wird die Laserleistung oder der Vorschub herangezogen.

Hinweis: Bei den hier vorgestellten Arbeiten handelt es sich um Teilergebnisse eines noch laufenden Forschungsverbundvorhabens "Lasertechnik" des Landes NRW.

5. Literatur

/1/   W. König, H. Willerscheid: Qualitätsprüfung beim Laserstrahlhärten anhand von Oberflächenkenngrößen, Laser & Optoelektronik 23 (2) (1991), S. 38-43

/2/   W. König, G. Herziger, H. Willerscheid, K. Wissenbach: Temperaturmessung beim Laserstrahlhärten, Laser-Magazin 1 (1988), S. 16-22

/3/   A. Drenker, L. Böggering, K. Wissenbach, E. Beyer: Adaptive Temperaturregelung beim Umwandlungshärten mit CO_2-Laserstrahlung, Laser & Optoelektronik 23 (5), 1991, S. 48-53

/4/   R. Rosenthal: Online-Diagnose bei der Laseroberflächenbearbeitung durch thermographische Systeme, Laser & Optoelektronik 22 (5), 1990, S. 44-49

/5/   H.W. Bergmann: Kontrolliertes Laserstrahlhärten durch online-Prozeßkontrolle, BMFT-Föderkennzeichen FKZ 13N5445

/6/   W. König, F. U. Meis, C. Schmitz-Justen: Härten mit dem Laserstrahl, VDI-Z. Bd. 128, 1986, Nr. 1/2, S. 27-31

/7/   C. Meyer-Kobbe: Randschichthärten mit Nd:YAG- und CO_2-Lasern, Dr.-Ing. Dissertation/Hannover, 1990

Online Process Control in Laser Beam Welding by Means of Power Regulation in RF-Excited CO$_2$-Lasers

B. Seidel, W. Sokolowski, J. Beersiek, W. Meiners and E. Beyer,
Fraunhofer-Institut für Lasertechnik,
Steinbachstr. 15, 52074 Aachen, FRG

Abstract

In order to make laser beam welding more economical, investment and operating costs for laser welding systems have to be reduced and process reliability and weld quality have to be optimized by choice of appropriate process parameters and process control. The most important criterion for quality of continous weld seams is the attainability of non-interrupted fusion over the whole seam length. The appearance of plasma shielding in laser beam welding is a cause of process interruptions and therefore non-fulfilment of this criterion.

This paper presents a method of control called PSC (plasma shielding control / ler) for avoidance of plasma shielding. The system is based on a two-level controller evaluating the intensity of a spectral line emitted by laser-induced plasma above the workpiece. The controlled variable of the system is the laser power. Three applications for PSC concerning steel and aluminium welding in various laser power ranges are discussed. PSC can be used in order to increase quality assurance, to achieve quality improvement or to reduce process gas costs, thus achieving a reduction in operating costs.

Zusammenfassung

Um die Wirtschaftlichkeit des Laserstrahlschweißens zu steigern, sind die Investitions- und Betriebskosten für Laserschweißanlagen zu reduzieren sowie Prozeßsicherheit und Schweißqualität durch geeignete Prozeßparameterwahl und Prozeßkontrolle zu optimieren. Wichtigstes Kriterium für die Qualität kontinuierlicher Schweißnähte ist die Erreichbarkeit unterbrechungsfreier Verbindung der zu fügenden Teile über die gesamte Nahtlänge. Das Auftreten abschirmender Plasmen beim Laserstrahlschweißen ist die Ursache von Prozeßunterbrechungen und damit Nichterfüllung des wichtigsten Kriteriums.

In dieser Veröffentlichung wird ein Regelungsverfahren, PSC genannt, zur Unterdrückung von abschirmenden Plasmen vorgestellt. Das Reglersystem basiert auf einem Zweipunktregler, dessen Regelgröße die Intensität einer aus dem laserinduzierten Plasma oberhalb des Werkstückes emittierten Spektrallinie ist. Stellgröße des Systems ist die Laserleistung. Es werden drei Anwendungen für den PSC dargestellt, welche das Schweißen von Stahl und Aluminium in unterschiedlichen Laserleistungsbereichen betreffen. Der PSC kann zur Steigerung der Qualitätssicherheit, zur Verbesserung der Qualität und zur Einsparung von Prozeßgaskosten und damit zur Reduzierung der Betriebskosten angewendet werden.

1.Introduction

Plasma shielding, which means a strong absorption of the incident laser beam above the work-piece, is a well known phenomenon in laser beam welding /FOWLER/PIRRI/BEYER/. Plasma shielding is responsible for process interruptions and thus for welding faults. The conditions for plasma shielding, which is determined by the electron density, are mainly influenced by laser intensity and process gas. Due to the time constants for regulation of laser intensity, which are smaller than those used for controlling process gas parameters, the PSC (plasma shielding control / ler) is based on regulation of laser intensity via laser power control. The effect of the PSC-method is considered under various process gas parameters in steel and aluminium welding.

2.Method of control

The basic idea of PSC is to use the fact that there is a strong correlation between the beginning of plasma shielding and an increase in the intensity of a nitrogen spectral line

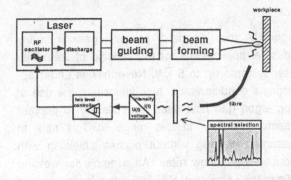

Figure 1: blockdiagram of plasma shielding controller

measured above the workpiece. A block diagram of the PSC is given in fig.1. Light-guiding from interaction zone to sensor can be achieved using fibre or by optical components in the laser beam path. The monitoring signal, intensity of a nitrogen spectral line, causes an immediate interruption of laser power on exceeding an adjustable threshold. Once the monitoring signal has decreased to beneath a second threshold the original laser power is reestablished by the controller. The

duration of power interruption is typically below one ms so that the melting and welding process is not influenced. In order to control the laser power by amplitude of rf-excitation in µs-range a direct line to the rf-oscillator of a rf-excited CO_2-laser is necessary.

3.PSC (plasma shielding controller) applications

3.1 Replacement of helium by argon in process gas for welding with 10 kW class lasers

Helium is preferred as process gas for laser beam welding with high power CO_2 lasers in order to control the laser induced plasma above the workpiece. Investigations on supplying gas-mixtures with portions of helium replaced by argon have shown the possibility of reducing gas costs /KALLA/. The limit for an argon portion in helium/argon gas-mixtures for tolerable seam quality was found to be about 35%. Experiments using PSC under similar welding conditions have demonstrated the possibility of replacing a further 20% helium by argon and giving higher process reliability with respect to plasma shielding. Fig.2 shows an example. For given welding parameters a process gas-mixture of 50% helium and 50% argon leads to a rough weld surface and extreme fluctuations in welding depth (see seam and longitudinal section fig.2). These effects are caused by frequently appearing plasma shielding, demonstrated by the peaks in spectral line intensity (fig.2 left). At the same welding and gas parameters the controlled process under PSC (fig.2 right) produces sound welds with nearly constant welding depth.

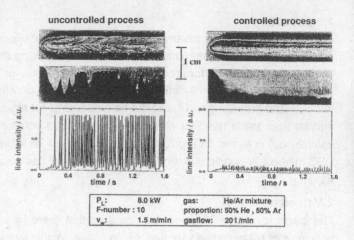

Figure 2: Surface, longitudinal section of seam and monitoring signal of welding process without PSC (left) and with PSC (right).

3.2 Reduction of process gas consumption

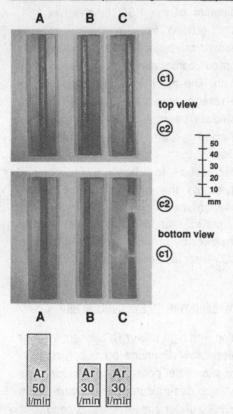

Figure 3: Welding with high and reduced argon process gas flow rate. A) without PSC, B) with PSC, C) without PSC. Areas c1 and c2: interruption of full penetration welding by plasma shielding.

Argon is the standard process gas in many industrial applications for the welding of steel with laser systems up to 5 kW. Nevertheless under appropriate conditions, i.e. high intensities, the use of high argon gas flow rates is necessary to prevent plasma shielding. In this range PSC is able to guarantee welding without plasma shielding with reduced argon flow rates. An example for welding 5.0mm thick sheets of STE355 with a laser power of 5.0kW, welding speed of 2.3m/min, F-number 10 and focal radius 0.5mm is demonstrated in fig. 3. Seam (A) is produced without PSC under argon gas supply by off-axis nozzle (inner diameter 4mm) with flow rate of 50 l/min. Specimen (B) is welded with reduced argon gas flow (30 l/min) and PSC. The weld with identical parameters without PSC, seam (C), shows two areas where full penetration welding is interrupted by plasma shielding. In this example a reduction of process gas consumption of 40% by PSC is possible without diminished seam quality.

To achieve quantitative evaluation the controller activity is determined for various machining parameters. Controller activity is defined as percentage of total duration of laser power interruptions by PSC in relation to total process duration. Fig. 4 shows a diagram of controller activity versus argon flow rate under given nozzle arrangement and focussing conditions. As ascertained in the authors investigations, controller activities higher than approximately 0.5% lead to a processing mode characterized by melt ejection and therefore rough weld seams. This means that possible process gas reduction can be determined by finding the gas flow rate with controller activity 0% and the flow rate for controller activity 0.5%. From fig. 4 the largest reduction potential by PSC is found at high F-numbers. In the case of F 8 the argon gas flow can be reduced from about 60l/min to approximately 40 l/min. About 5 to 7 l/min of argon can be saved in the case of F 4.

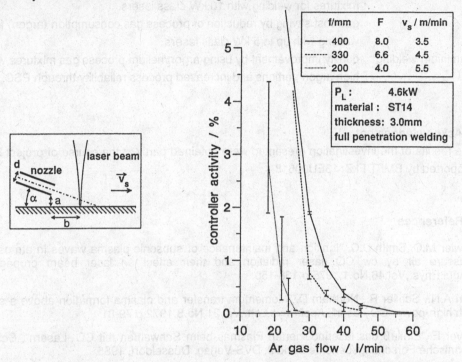

Figure 4: nozzle arrangement (left) (a=5mm, b=10mm, d=4mm, α=30°) and diagram of controller activity (=percentage of total duration of laser power interruption by PSC in relation to total process duration) versus argon gas flow rate under various focussing conditions.

3.3 Quality improvement in aluminium welding

Depending on the alloy, focal point position of laser beam and nozzle type different gas flows and types of process gas are applied to achieve sound aluminium welding /MATSUMURA/BERKMANNS/. Recent investigations /BERKMANNS/ have shown that high argon portions in argon/helium gas-mixtures lead to improved seam quality. Melt ejection and crater formation are suppressed and energy coupling is enhanced. The optimum of improvement is achieved near the threshold to plasma shielding so that uncontrolled processing contains the risk of plasma shielding and thus of process interruption. PSC is necessary for process-reliability with an optimum of seam quality improvement and process efficiency /BEERSIEK/.

4. Conclusion

A method of control called PSC (plasma shielding control) has been developed for CO_2 laser beam welding. The application range mainly covers the field of efficient process gas supply to the interaction zone of laser beam and workpiece as well as quality assurance.

Depending on the material the motivations for using PSC are:

steel welding: gas cost saving by replacement of helium by argon in process gas-mixtures for welding with 10 kW class lasers.

gas cost saving by reduction of process gas consumption (argon) for welding with up to 5 kW class lasers.

aluminium welding: quality improvement by using argon/helium process gas mixtures with high argon portions and increased process reliability through PSC.

5. Acknowledgement

The results of the investigation presented were obtained partly in the course of project EU194 supported by BMFT FkZ: 13EU00618

6. References

Fowler M.C.,Smith D.C.,"Ignition and maintenance of subsonic plasma waves in atmospheric pressure air by cw CO_2 laser radiation and their effect on laser beam propagation", J.Appl.Phys.,Vol.46,No.1,1975,p 138-150

Pirri A.N., Schlier R., Northam D.,"Momentum transfer and plasma formation above a surface with high-power CO_2 laser", Appl.Phys.Lett.,Vol.21,No.3,1972,p 79-81

Beyer E.,"Einfluß des laserinduzierten Plasmas beim Schweißen mit CO_2-Lasern", Schweiß-technische Forschungsberichte Band 2, DVS-Verlag, Düsseldorf, 1985

Kalla G., Beyer E.,"Influence of the working gas parameters on the weld quality using high power lasers", Proc. of LASER '91, p 500-504

Matsumura H., et al.,"CO_2 laser welding characteristics of various aluminium alloys", Proc. LAMP'92, p 529 -533

Berkmanns J., Behler K., Herziger G.,"Schweißen von aushärtbaren und nichtaushärtbaren Aluminiumlegierungen", BMFT-Abschlußbericht F+E-Vorhaben 13N5655, 1993

Beersiek J., Seidel B., Berkmanns J., Behler K.,"Process control in high power laser beam welding with argon/helium process gas mixtures" Proc. LASER'93

Tiefenregelung beim Laserabtragen

Dr. G. Eberl, M. Kuhl, M. Reisacher,   P. Abels, A. Drenker, S. Geißler,
P.Lagner   W. Nöldechen
MAHO Aktiengesellschaft, 8962 Pfronten Inst. f. Lasertechnik, 5100 Aachen

Einleitung

Das Abtragen mittels Laserstrahl bietet dem Werkzeug- und Formenbauer ein
neues Bearbeitungsverfahren.
Das LASERCAV® Programmier-System LCPS zum automatischen Generieren des
CNC-Programmes verarbeitet die Geometriedaten der Gesenke, die mit gängigen CAD-Systemen konstruiert wurden. Die Gesenke werden in einzelne
Schichten zerlegt, die Schichten schraffiert und damit die Spur des Laserstrahls über dem Material berechnet.
Um die durch das LCPS vorgegebene Schichtstärke zu gewährleisten und damit
hohe Formgenauigkeiten zu erzielen, ist eine Tiefenregelung notwendig.

Depth Control for Material Removal by Laser. Stock removal by means of a
laser beam places a new machining process at the disposal of tool and
mould makers. The LASERCAV® Programming System LCPS for automatic generation of the CNC programme processes the geometric data of dies that have
been designed by currently used CAD systems. The LCPS splits up the die
into individual layers. The path of the laser beam over the layer and the
depth control data are then computed. A depth control is necessary to ensure that the layer thickness specified by the LCPS is maintained in order
to produce highly accurate shapes.

1. Stand der Technik

Das Laserabtragen ist eine neue Bearbeitungstechnologie, die konventionelle Verfahren ergänzt - wie z.B. das Erodieren, das Hochgeschwindigkeitsfräsen oder chemische Verfahren zur Strukturierung von Werkzeugen und Formen.

Das Laserabtragen eignet sich besonders zur Herstellung von Gesenken mit
filigranen Geometrien und zur Einarbeitung von Strukturen (z.B. Leder-
strukturen) in Formen.

Schicht für Schicht trägt der gebündelte Laserstrahl Material ab. Typische
Schichtstärken und Zeilenabstände liegen bei 0,1 mm.

Die automatische Generierung der 5-Achsen-NC-Programme für die Laser-Ab-
tragmaschine LASERCAV® erfolgt über ein Programmiersystem, so daß die 3-D-
und Freiformflächenbearbeitung möglich ist.

Die Formgenauigkeit der Gesenke hängt stark davon ab, wie genau die über
das Programmiersystem eingestellte Schichtstärke eingehalten wird.
Im gesteuerten Betrieb waren daher bisher nur Genauigkeiten im Zehntel-
Millimeter-Bereich möglich /1, 2, 3/.

Inzwischen steht ein Tiefensensor, der während der Bearbeitung die Schichttiefe mißt, zur Verfügung, so daß über einen Tiefenregler der CO_2-Laser so angesteuert werden kann, daß die tatsächliche Schichtstärke der vorgegebenen Schichtstärke entspricht. Damit werden deutliche Verbesserungen der Formgenauigkeit erreicht (Hundertstel-Millimeter-Bereich).

2. Die Programmierung der Maschine

2.1 Das LASERCAV® Programmier-System LCPS

Der Schlüssel zur Anwendung der Maschine ist die Programmierung. Dabei gibt es zwei prinzipiell unterschiedliche Wege. Die Programmierung an der Steuerung ist für einfache Programme zweifellos der direkteste Weg. Die Steuerung der Abtragmaschine bietet leistungsfähige G- und M-Funktionen, mit denen Programme z.B. zum Schneiden direkt an der Maschine erstellt werden können. Naturgemäß ist es sehr schwierig, 5-achsige Programme ohne Hilfsmittel an der Maschine zu erstellen.
Ebenso resultiert aus dem geringen Werkzeugdurchmesser beim Abtragen die Schwierigkeit, daß gewöhnlich eine große Anzahl von Bearbeitungssätzen, die praktisch nicht mehr manuell festgelegt werden können, nötig ist. Deshalb gehört zum LASERCAV®-System ein speziell auf die Technologie abgestimmtes Programmierwerkzeug.

Das Programmiersystem LCPS läuft auf einem von Maschine und Steuerung unabhängigen Rechner und speist die Programme über die serielle Schnittstelle in die Maschine ein. Standardmäßig wird momentan mit der Maschine ein PC486 als Hardwareplattform mit UNIX als Betriebssystem für das Programmiersystem geliefert. Je nach Anwendungsfall können 2D- oder 3D-CAD-Daten von Konstruktions- oder Zeichenprogrammen in das Programmiersystem übernommen werden. Für den Transfer von ebenen Zeichnungen wird eine DXF-Schnittstelle angeboten.
3-dimensionale Daten können über die weitverbreitete STL-Schnittstelle (STL=Stereolithographie) von nahezu allen heute üblichen CAD-Systemen übernommen werden. Bild 1 zeigt die LCPS-Bildschirmdarstellung eines zu bearbeitenden Werkstücks.

2.2 Bearbeitung von Gesenken

Ähnlich wie beim Fräsen wird das Material Spur für Spur und Schicht für Schicht abgetragen (Bild 2). Bei der Laserbearbeitung gelten allerdings völlig andere technologische Bedingungen, denen im Programmiersystem Rechnung getragen wird. Der Laserstrahl ist im Gegensatz zu einem Fräser kein geometrisch konstantes Werkzeug. Seine Wirkung auf das Material hängt von der aus der Bearbeitung der letzten Schicht resulierenden Geometrie und einer Reihe anderer Einflußgrößen ab. Aus Bild 2 ist der schichtweise Abtrag ersichtlich. Es zeigt die Bearbeitungsspuren, die für den Abtrag einer Schicht nötig sind.

2.3 Bearbeitung von Oberflächen

Für Gravieraufgaben können 2-dimensionale Zeichnungen mittels spezieller Funktionen auf 3-dimensionale Oberflächen übertragen werden. Abwicklungs- und Projektionsverfahren ermöglichen es beispielsweise, einen Schriftzug auf einen Rotationskörper oder eine Freiformfläche zu bringen. Eine Besonderheit des Programmiersystems bilden die Texturierfunktionen, mit deren Hilfe verschiedene Zufallsmuster zur Strukturierung von Werkstückoberflächen erzeugt werden können.
Außerdem steht eine Option zur Übertragung von Pixelbildern auf Werkstückoberflächen zur Verfügung.

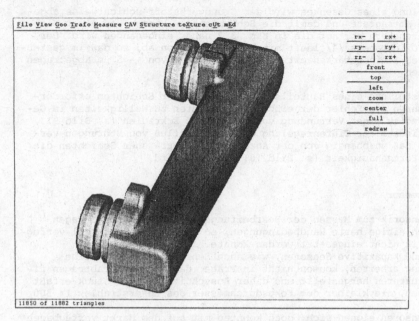

File View Geo Trafo Measure CAV Structure teXture cUt mEd

rx- rx+
ry- ry+
rz- rz+
front
top
left
zoom
center
full
redraw

11850 of 11882 triangles

Bild 1. LCPS-Bildschirmdarstellung eines Werkstückes.

Bearbeitungsspuren
für eine Schicht

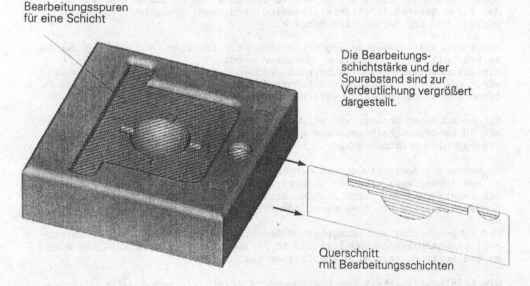

Die Bearbeitungs-
schichtstärke und der
Spurabstand sind zur
Verdeutlichung vergrößert
dargestellt.

Querschnitt
mit Bearbeitungsschichten

Bild 2. Laserabtragen: Spur für Spur und Schicht für Schicht.

3. Tiefenregelung zum formgenauen Laserabtragen

Zur Herstellung eines Gesenkes wird der Grundwerkstoff schichtweise abgetragen. Die Abtragrate und damit die Schichtstärke hängen von einer Vielzahl von Parametern ab, auf die in Kap. 3.2 näher eingegangen wird. Der Prozeß des Oxidspanens /4/ läuft mit hoher Präzision ab, so daß im gesteuerten Betrieb Schichtstärken mit einer Genauigkeit von +- 5 µm abgetragen werden können.

Ist zur Herstellung eines Bauteils der Abtrag vieler Schichten erforderlich, so führen die Fehler der einzelnen Schichten zu Welligkeiten im Gesenkgrund und zu einer Verrundung von Kanten und Eckradien (s. Bild 9). Durch den Einsatz der Tiefenregelung wird der Einfluß von Störungen vermindert, so daß unabhängig von der Anzahl der abgetragenen Schichten die geforderte Formgenauigkeit (s. Bild 10) erreicht wird.

3.1 Tiefensensor

Für eine Sensorik zum Messen der Bearbeitungstiefe beim Laserabtragen stellen sich einige harte Randbedingungen, so daß eine kommerziell verfügbare Sensorik nicht eingesetzt werden konnte.
Induktive und kapazitive Sensoren, wie auch Sensoren, die über eine Schallmessung arbeiten, kommen nicht in Frage, da durch Laserabtragen filigrane Strukturen hergestellt und daher Ausschnitte im Werkstück erfaßt werden müssen, die kleiner dem Fokusdurchmesser des Laserstrahles (< 300 µm) sind.
Optische Sensoren eignen sich; doch konnten mit auf dem Markt verfügbaren optischen Sensoren die geforderten Genauigkeiten nicht erfüllt werden.

Im Folgenden sind die Anforderungen an einen für das Laserabtragen geeigneten Tiefensensor beschrieben.
Unter Berücksichtigung dieser Anforderungen wurde in Zusammenarbeit mit der Firma Steinheil Optronik, Ismaning, ein neuer optischer Sensor entwickelt und zur Serienreife gebracht.

Damit die Abtragmaschine möglichst produktiv arbeitet, muß _während_ der Laserbearbeitung, d.h. on-line, gemessen werden.
Ein zwischengeschalteter Meßzyklus, währenddessen nicht abgetragen wird, würde die Zeit für die Bearbeitungsaufgabe erhöhen und es müßten große Datenmengen gespeichert und verarbeitet werden.

Ein neues Sensorkonzept war notwendig, da die Bearbeitungstiefe on-line mit 10 µm-Genauigkeit gemessen werden muß, ohne den Arbeitsraum der Abtragmaschine einzuschränken.

So wurde die Sensoreinheit in den Bearbeitungskopf integriert. Mit einem 90°-Parabolspiegel wird der Laserstrahl gebündelt. Gemessen wird parallel zum Laserstrahl und durch die Bearbeitungsdüse hindurch mit einer Beobachtungswellenlänge von etwa einem Zehntel der Bearbeitungswellenlänge (900 nm).
Der Fokussierkopf mit integriertem Sensor baut so kompakt, daß er universell einsetzbar und sowohl an die in /1/ beschriebene Vertikal- als auch an die Horizontalmaschine adaptierbar ist.

Die Leistungsfähigkeit des Tiefensensors zeigt sich schon darin, daß die Meßwerterfassung innerhalb weniger Mikrosekunden möglich ist.

In Bild 3 ist das Tiefensignal aufgenommen, das sich ergibt, wenn mäanderförmig abgetragen und der Bearbeitungskopf in Schritten von 20 µm vom Werkstück wegbewegt wird, nachdem eine Laserspur abgetragen ist. Die 20

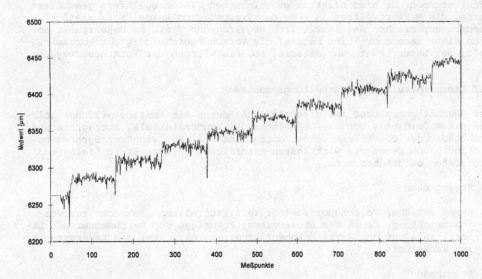

Bild 3. Tiefensignal beim Verfahren des Bearbeitungskopfes in 20 μm-Stufen.

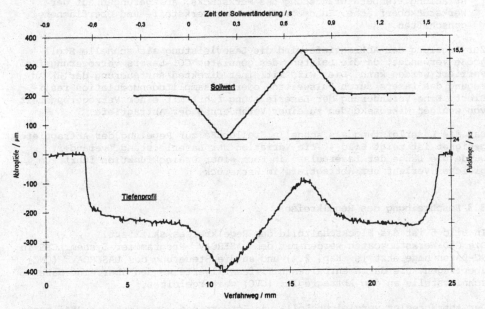

Bild 4. Steuerbarkeit des Bearbeitungsprozesses - Änderung der Pulslänge des HF-Generators und resultierende Bearbeitungstiefe.

µm-Stufen sind deutlich zu sehen. Eine absolut geglättete Kurve kann sich nicht ergeben, da hier nicht im geschlossenen Tiefenregelkreis gearbeitet wurde. Die Meßwertspitzen am Ende der Treppen sind keine "Meßausreißer", sondern entsprechen der tatsächlich abgetragenen Tiefe am Umkehrpunkt am Ende einer Laserspur. Hier beträgt die Vorschubgeschwindigkeit kurzzeitig Null; der Laser trägt mehr Material ab, als während der Vorschubbewegung.

3.2 Steuerbarkeit des Bearbeitungsprozesses

Der Bearbeitungsprozeß ist gekennzeichnet durch ein kontinuierliches Ablösen eines Oxidspans von der Oberfläche des Grundmaterials. Abtragrate und Rauhigkeit der Oberfläche werden durch eine Vielzahl von Prozeßgrößen beeinflußt, von denen die wichtigsten nachfolgend unterteilt in Stell- und Störgrößen aufgeführt sind.

a) Stellgrößen

Laserleistung, Vorschubgeschwindigkeit, Zustellung senkrecht zur Vorschubrichtung, Druck des Prozeßgases, Fokuslage und Durchmesser des Laserstrahls auf der Werkstückoberfläche, Anstellwinkel des Laserstrahls zum Werkstück.

b) Störgrößen

Schwankungen der Leistungen und der Intensitätsverteilung der Laserstrahlung, Temperaturänderung des Werkstücks, Ablagerungen auf der Werkstückoberfläche, Inhomogenitäten der Werkstoff- und Oberflächeneigenschaften.

Zur Regelung der Abtragtiefe wird die Laserleistung als schnelle Stellgröße verwendet, da die Leistung des gepulsten CO_2-Lasers verzögerungsfrei variiert werden kann. Dies wird mit einer direkten Ansteuerung der HF-Anregung des Lasers durch Pulsweiten- oder Pulsamplitudenmodulation realisiert. Eine Veränderung der Laserleistung führt mit einer Verzögerungszeit von wenigen Mikrosekunden zu einer Veränderung der Abtragtiefe.

Daß die Laserleistung als schnelle Stellgröße zur Regelung der Abtragtiefe geeignet ist zeigt Bild 4. Die Variation der Laserleistung (verändert wurde die Länge der Laserpulse) in Form einer Dreieckfunktion führt zum gleichen Verlauf der Abtragtiefe im Werkstück.

3.3 Beschreibung des Regelkreises

In Bild 5 ist das Blockschaltbild des Regelkreises skizziert. Die CAD-Werkstückdaten werden mit dem LASERCAV® Programmier-System LCPS in NC-Daten umgesetzt (s. Kap. 2.1) und an die Steuerung des LASERCAV® (CNC) übertragen. Die darin enthaltenen Tiefensollwerte werden über eine Datenschnittstelle an den Abtragregler (CVC) weitergeleitet.

Der Abtragregler vergleicht Soll- und Istwert und generiert die Stellgröße für die Laserleistung. Der programmierbare Pulsgenerator (PPG) setzt diese Stellgröße in eine Pulsfolge zur Ansteuerung der Strahlquelle um, so daß sich die erforderliche Abtragrate ergibt.

Da der Meßfleck des Tiefensensors auf der Oberfläche des Oxidspans liegt, ist die Messung der erreichten Abtragtiefe unter dem Oxidspan erst beim Abtragen der nächsten Schicht möglich.

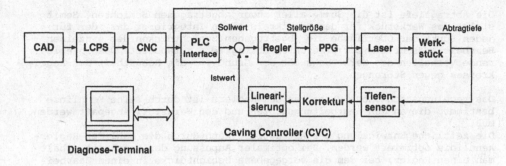

Bild 5. Meß- und Regelsystem für das Abtragen mit Laserstrahlung.

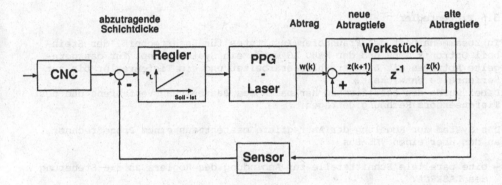

Bild 6. Ersatzschaltbild für den Regelkreis zum Abtragen mit Laserstrahlung.

Die Meßwerte des Sensors werden an den Abtragregler übertragen; daraus werden durch eine Korrektur- und Linearisierungsfunktion Tiefenmeßwerte ermittelt.

3.4 Modell für die Tiefenregelung

Dem Regelkreis liegt ein vereinfachtes regelungstechnisches Modell zugrunde.
Das Werkstückvolumen wird schichtweise mit konstanter Schichtdicke abgetragen, wobei im Folgenden der vollständige Abtrag einer Schicht als ein Bearbeitungsschritt bezeichnet wird. Da die Messung der erreichten Tiefe an jedem Punkt der zu bearbeitenden Oberfläche in einem Bearbeitungsschritt genau einmal erfolgt, ist ein zeitdiskretes Modell für den Regelkreis zweckmäßig.
Das in Bild 6 dargestellte Ersatzschaltbild des Regelkreises beschreibt die Entwicklung der Abtragtiefe für einen Punkt der Oberfläche zeitdiskret von einem Bearbeitungsschritt zum nächsten.

Die Abtragtiefe an einem Punkt der Oberfläche ergibt sich aus der im vorangegangenen Bearbeitungsschritt erreichten Abtragtiefe und der im aktuellen Bearbeitungsschritt abgetragenen Schichtdicke.

Die Abtragtiefe ist die Summe aller zuvor abgetragenen Schichten. Somit
stellt das Werkstück für den Regelkreis einen Integrierer dar, der für
jeden Punkt der Oberfläche die abgetragenen Schichtdicken der einzelnen
Bearbeitungsschritte zur erreichten Abtragtiefe aufaddiert. Das integie-
rende Verhalten des Werkstücks ist der Grund für die Robustheit des Regelkreises gegen Störungen.

Das regelungstechnische Verhalten des Reglers ist durch seine Kennlinie
bestimmt, die an den Bearbeitungsprozeß und den Werkstoff angepaßt werden
muß.
Die zeitliche Entwicklung der Abtragtiefe kann durch die Wahl der Reglerkennlinie optimiert werden. Bei optimaler Anpassung der Kennlinie erhält
man einen Regler, bei dem die vorgegebene Schichtdicke in einem Bearbeitungsschritt abgetragen wird.
Im praktischen Einsatz zeigt sich, daß eine Abweichung der Reglerkennlinie
vom Optimum für den geregelten Abtragprozeß unkritisch ist.

3.5 Tiefenregler

In Zusammenarbeit des Fraunhofer Institutes für Lasertechnik, der Stein-
heil Optronik GmbH und der MAHO AG wurde ein Systemkonzept für das gere-
gelte Abtragen mit Laserstrahlen erarbeitet und ein Tiefenregler bis zur
Serienreife entwickelt.
Dabei wurde den spezifischen Merkmalen des Bearbeitungsverfahrens und des
Tiefensensors Rechnung getragen.

Das System zur Regelung der Abtragtiefe besteht aus einem Prozeßrechner,
an den über einen VME-Bus

- eine parallele Schnittstelle zur Anbindung des Reglers an die Steuerung
  des LASERCAV$^{®}$,

- ein programmierbarer Pulsgenerator zur Steuerung der Laserleistung und

- eine Auswerteeinheit für den Tiefensensor

angeschlossen sind.

Die Funktionen des Systems zum geregelten Abtragen sind durch folgende
Software-Module auf dem Prozeßrechner realisiert:

a) Aufbereitung der Sensormeßwerte
   Die Meßwerte des Tiefensensors werden eingelesen und linearisiert.

b) Kommunikation mit der Steuerung des LASERCAV$^{®}$
   Die Steuerung des LASERCAV$^{®}$ sendet mittels schneller M-Befehle die Sen-
   sor- und Reglerparameter an den Prozeßrechner und empfängt in umgekehr-
   ter Richtung Meßwerte sowie Status- und Fehlermeldungen.

c) Steuerung der Laserleistung
   Der programmierbare Pulsgenerator erzeugt Rechteckpulse zur Steuerung
   der Laserleistung mit einer maximalen Frequenz von 5 MHz und einer
   zeitlichen Auflösung von 100 ns. Dabei wird der Laser so angesteuert,
   daß ein optimales Zündverhalten des Lasers erreicht wird.

d) Diagnose und Testfunktionen
   Die Inbetriebnahme, Wartung und Instandsetzung des Meß- und Regelsy-
   stems wird durch ein Diagnoseterminal unterstützt.

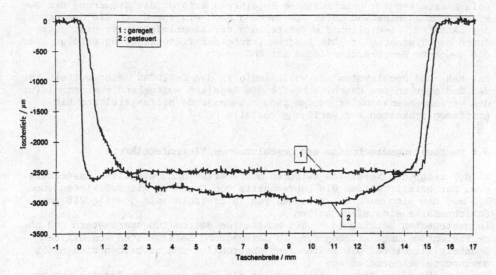

Bild 7. Querschnitt zweier Gesenke mit Solltiefe von 2,5 mm zum Vergleich von gesteuertem und geregeltem Laserabtragen.

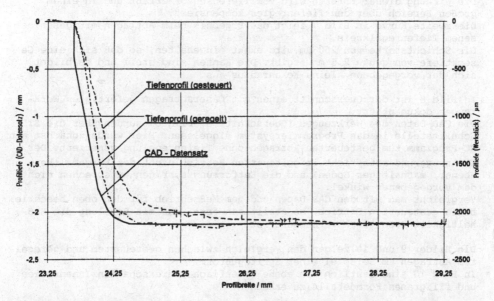

Bild 8. Querschnitt eines Werkzeuges zum Vergleich von CAD-Daten, gesteuertem und geregeltem Laserabtragen.

Beim Bearbeiten mit geschlossenem Regelkreis erfolgt die Steuerung der Bearbeitungsmaschine und deren Überwachung nach wie vor durch die Steuerung des LASERCAV®. Lediglich die Verstellung der Laserleistung erfolgt jetzt durch den Tiefenregler. Die Programmierung der Tiefenregelung erfolgt über die gewohnte Benutzeroberfläche der CNC.

Das Meß- und Regelsystem ist vollständig in den LASERCAV® eingebettet. Für den Bediener an der Maschine bleibt die Regelung weitgehend verborgen. Für den Verfahrensentwickler werden jedoch umfassende Hilfsmittel und Eingriffsmöglichkeiten zur Verfügung gestellt.

3.6 Bearbeitungsergebnisse mit geschlossenem Tiefenregelkreis

Bild 7 zeigt den Vergleich zwischen gesteuertem und geregeltem Laserabtragen. Dargestellt werden die Querschnitte von Gesenken mit Solltiefen von 2,5 mm, die sich nach dem Abtragen von 10 Schichten mit jeweils 250 µm Schichtstärke einstellen sollen.
Im gesteuerten Betrieb (d.h. mit konstanten Bearbeitungsparametern und ohne Tiefenregelung) ergeben sich im Gesenkgrund Welligkeiten von einigen Zehntelmillimetern. Die Kanten sind abgeschrägt und es bilden sich nicht gewünschte Eckenradien aus.
Ursache für diese Formabweichungen ist die Summierung von Formfehlern von Schicht zu Schicht, die mit zunehmender Schichtstärke größer werden, thermische Effekte und veränderten Abtragmechanismen an Gesenkwänden bei zunehmender Gesenktiefe.

Die Wirkung dieser Effekte wird vom Tiefensensor erfaßt und in einem großen Bereich über den Tiefenregler kompensiert.
Die zweite Kurve in Bild 7 zeigt das Ergebnis beim Abtragen mit geschlossenem Tiefenregelkreis.
Die Schichtstärke von 250 µm wird exakt eingehalten, so daß sich eine Gesenktiefe von genau 2,5 mm ergibt. Die Kanten sind steil und es bildet sich der vorgegebene kleine Eckenradius aus.

In Bild 8 ist der Querschnitt eines mit Laserabtragen gefertigten Werkzeuges dokumentiert.
Die CAD-Daten des Werkzeuges (Querschnitt s. Bild 8) wurden über die STL-Schnittstelle in das Programmiersystem eingelesen. Hier wurde zunächst ein NC-Programm zum gesteuerten Abtragen ohne Tiefenregelung generiert. Der Werkzeuggrund entspricht im gesteuerten Betrieb nicht der Vorgabe (kein ebener, maßhaltiger Boden) und die Entformungsschrägen entsprechen nicht dem vorgegebenen Winkel.
Vergleicht man mit den CAD-Daten und dem Meßschrieb für den oben beschriebenen gesteuerten Betrieb, so stellt man fest, daß mit Regelung die Maßhaltigkeit wesentlich verbessert ist.

Die Bilder 9 und 10 zeigen den Vergleich zwischen gesteuertem und geregeltem Abtragen am Beispiel einer Spritzgußform:
In Bild 10 sind deutlich die ebene Bodenfläche, die scharfen Innenradien und filigranen Formdetails zu erkennen.

In Analogie zur Funkenspaltregelung beim elektro-thermischen Verfahren Erodieren (EDM) hat die Tiefenregelung für die Weiterentwicklung des Laserabtragens eine zentrale Bedeutung.
Durch Weiterentwicklung der Tiefenregelung wird sich die Maßhaltigkeit der Gesenke weiter verbessern lassen.

Bild 9. Gesteuertes Laserabtragen - Ausschnitt einer Spritzgußform.

Bild 10. Geregeltes Laserabtragen - Ausschnitt einer Spritzgußform.

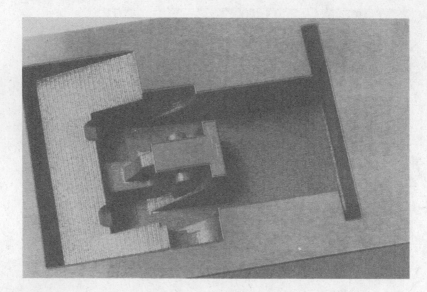

Bild 11. Spritzgußform hergestellt durch Laserabtragen.

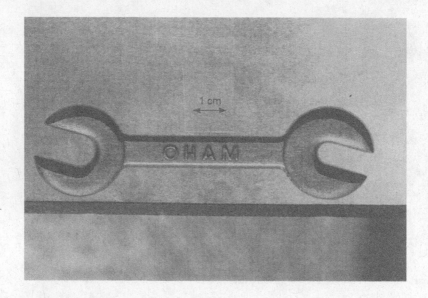

Bild 12. Schmiedegesenk hergestellt durch Laserabtragen.

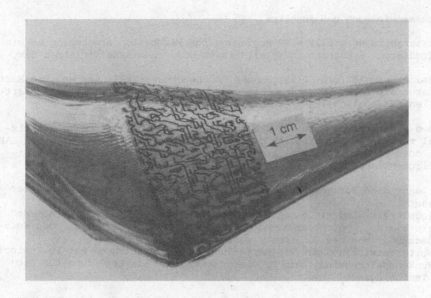

Bild 13. Einarbeiten von Einwachsstrukturen in eine Prothese.

4. Einsatzbeispiele

4.1 Spritzgußform und Schmiedegesenk

Ein Ziel des Laserabtragens ist die Vereinfachung der Herstellung filigraner Formen. Statt einer mehr oder weniger großen Zahl von Elektroden beim Funkenerodieren, ist nur ein einziges Werkzeug - der Laserstrahl - nötig. Gerade bei dünnen Stift- und Stegelektroden ist mit hohem Verschleiß zu rechnen, der beim Laserwerkzeug prinzipiell nicht auftritt. Der Aufwand für die Elektrodenherstellung entfällt vollständig. Vom konstruierten Datensatz zum fertigen Gesenk (Bilder 11 und 12) ist nur mehr ein Bearbeitungsschritt erforderlich. Ein interessanter Nebeneffekt ist bei geeigneter Wahl der Werkstoff- und Bearbeitungsparameter die Laserhärtung der Oberfläche.

4.2 Prothesentexturierung

Eine besondere Art der Oberflächentexturierung stellt das Aufrauhen von Prothesen dar (Bild 13). Das Ziel der Bearbeitung ist, eine bessere Verankerung der Prothese im Knochen zu erreichen. Wird die Oberfläche physiologisch geeignet aufgerauht, so kann das Knochengewebe in die feinen Oberflächenstrukturen hineinwachsen und stellt damit eine gute Verbindung zwischen Implantat und Knochen sicher. Im Programmiersystem finden sich Möglichkeiten, verschiedene Zufallsstrukturen zu generieren und diese auf die Oberfläche der als CAD-Datensatz vorgegebenen Prothese zu übertragen.

5. Zusammenfassung und Ausblick

Das Laserabtragen stellt eine Ergänzung für Bearbeitungstechnologien wie das Erodieren, Hochgeschwindigkeitsfräsen oder chemische Verfahren dar.

Filigrane Gesenke (z.B. Prägegesenke) können hergestellt werden. Die Härte des Materials macht keine Probleme: das Laserabtragen von Glas bzw. Hart-metallen ist möglich.

Mit dem Programmiersystem für die Abtragmaschine (LCPS) steht ein Hilfs-mittel zur Verfügung, das eine direkte und automatische Umsetzung von CAD-Daten oder Bildverarbeitungsdaten in ein 5-Achsen-NC-Programm zur Frei-formflächenbearbeitung realisiert.

Damit sehr hohe Formgenauigkeiten erzielt werden, wird ein Tiefenregler eingesetzt.
Der Regelkreis wird mit einem Tiefensensor, der auf einem optischen Meß-prinzip basiert, geschlossen und die Regelabweichung direkt auf den Hoch-frequenzgenerator des CO_2-Lasers zurückgekoppelt.
Die Weiterentwicklung der Tiefenregelung zum Laserabtragen hat in Analogie zur Funkenspaltregelung beim Erodieren eine zentrale Bedeutung für die neue Technologie.

Mit einigen Arbeitsbeispielen wurde verdeutlicht, daß das Laserabtragen inzwischen einen technologisch hohen Stand erreicht hat und das neue Ver-fahren in den unterschiedlichsten Marktbereichen die Lösung von schwieri-gen Bearbeitungsaufgaben bringt.

Literatur

/1/ Ahlers, R.-J.; Eberl, G.: Neue Laserpraxis im Werkzeugbau - Freiform-flächen abtragen in Metall, Graphit und Hartstoffen, Laser-Praxis, Sonderteil in Hanser-Fachzeitschriften Mai 1992.

/2/ Eberl, G.; Hildebrand, P.; Kuhl, M.; Sutor, U.; Wrba, P.: Neue Ent-wicklungen beim Laserabtragen - Material Removal Using the Laser - New Developments, Laser und Optoelektronik 24(4)/1992.

/3/ Ahlers, R.-J.; Eberl, G.: New Laser Machining Process for Tool Makers - Stock Removal from Free Form Surfaces on Metal, Graphite and Hard Materials, European Production Engineering EPE, 16(1992)4.

/4/ Eberl, G.; Hügel, H.; et al.: Laserspanen - eine neue Technologie zum Abtragen, Laser und Optoelektronik 25(3)/1993.

/5/ Lässiger, B.; Treusch, H.G.; Abels, P.; Nöldechen, W.; Petring, D.; Beyer, E.: Verfahrensentwicklungen zum definierten Formabtrag mit Laserstrahlung, Laser in der Technik - Proceedings LASER 91, Springer-Verlag.

/6/ Abels, P.; Eberl, G.; et al.: Tiefenregelung beim Laserabtragen, Ta-gungsband LASER 93, München, 6/93, (in Vorbereitung).

/7/ Dausinger, F.; Meiners, E.; Masek, F.; Wiedmaier, M.: Materials Pro-cessing with YAG-Lasers integrated in a Turning Center, Proceedings ECLAT 92, Göttingen, 10/92.

/8/ LASERCAV®: registriertes Warenzeichen der MAHO Aktiengesellschaft, Pfronten.

Real Time Process Monitoring with an Integrated Fibre Monitor

CHRIS PETERS[a] JULIAN D C JONES[b] & DAONING SU[b]

[a]Lumonics Ltd., Cosford Lane, Swift Valley, Rugby, Warks, England
[b]Physics Department, Heriot-Watt University, Edinburgh, Scotland.

ABSTRACT

Real-time monitoring and process control for consistent quality and productivity in laser materials processing is likely to become a requirement in the 1990's to fully exploit the flexible manufacturing capability of laser systems in industry. We report on a method of in-situ process monitoring via the cladding power in an Nd:YAG laser system equipped with fibre optic beam delivery. The technique detects light reflected, scattered or emitted from the workpiece during processing. The light signal carries information about the quality of the process. Experimental results are reported for laser cutting and drilling using the optical cladding power monitor which show that real-time feedback control of the process in manufacturing by this technique is practical. Further studies are in hand to determine whether the approach can also be applied effectively to laser welding.

FIBRE CLADDING MONITOR (FCM)

A schematic diagram of the Fibre Cladding Monitor (FCM) is shown in Figure 1. The FCM consists of a $1000\mu m$ optic fibre with a smaller $50\mu m$ monitor fibre coupled to its cladding.
Photodiodes at each end of the monitor fibre are used to detect light passing into the Monitor fibre from the cladding of the main fibre.
Thus light from the laser, and the workpiece can be monitored during laser processing.

RESULTS OF PROCESSING TRIALS

Trials have been carried out to assess the performance of the FCM for inprocess monitoring, the trials included detection of focal point position, percussion drilling and cutting.
All trials were carried out using an industrial 250 Watt pulsed Nd:YAG laser (Lumonics JK702).

The results of processing trials are summarised below:

Figure 2 shows the relative amplitude of the return signal against focal position relative to the surface of an aluminium target. It can be seen that when the beam focus position is coincident with the plane of the target surface, the return signal is a minimum.

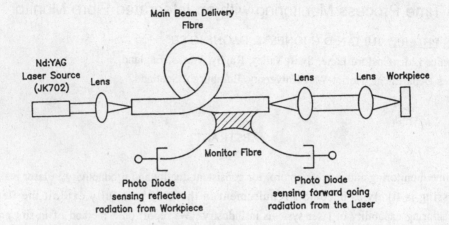

Figure 1   Schematic diagram of Fibre Cladding Monitor used for inprocess monitoring

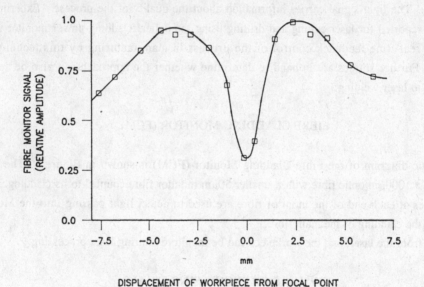

Figure 2.   Trace showing relative amplitude of return signal against focal position relative to work piece surface

Figure 3 shows the trace of the FCM return signal during a multiple shot percussion drilling operation. The trace shows that after 5 shots the signal shows a decrease in amplitude, which corresponds to the full development of the hole through the plate.

Figure 4 shows two traces of the return signal taken during cutting trials. Trace (a) shows the pulse shape of the return signal for a successful, fully penetrating cut made at a speed of

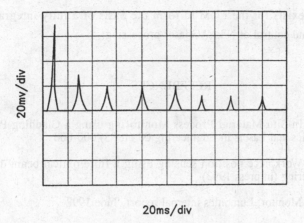

Figure 3.　　FCM return signal during a multishot percussion drilling operation

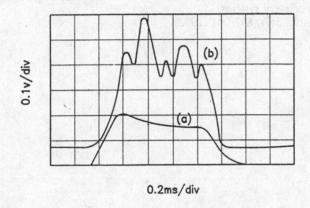

Figure 4.　　Return signal during cutting trials. Trace (a) for fully penetrating cut.
　　　　　　　Trace (b) for cut which is failing due to excessive cutting speed

300mm/min. Trace (b) shows the pulse shape of the return signal after the cutting speed had been increased to 400mm/min to induce cut failure. The increase in signal amplitude and the occurrence of spikes on the peak of the pulse is clear indication of the cut failure.

CONCLUSIONS

Trails with the Fibre Cladding Monitor have demonstrated it's suitability for inprocess monitoring of drilling and cutting operations; work is in-hand to assess it's suitability for laser welding.

Significant potential exists for the FCM to form the basis of a fully integrated system for inprocess monitoring and control of Nd:YAG laser processing.

REFERENCES

1.   Daoning et al, In-Situ Material Process Monitoring using a Cladding Power Detection Technique, Optics and Lasers in Engineering 00 (1993) 000-000

2.   Daoning et al, Workpiece position sensing using a fibre optical beam delivery system, Optical Engineering (in press 1993)

3.   Peters C, Fibre Monitor, Lumonics internal report, June 1992

4.   Daoning Su, A.A.P. Boechat, J.D.C. Jones and C.L.M. Ireland, Euro Patent Publication No. 0507 483 A1 (1992)

5.   A.A.P. Boechat, D. Su and J.D.C. Jones Bi-directional Cladding Power Monitor for fibre-optic beam delivery systems.
Meas. Sci. Technol., 3(9), 897-091 (1992)

Qualitätserzeugung in der Laserstrahlschneidbearbeitung durch Einsatz eines anwendungsgerechten CAD/CAM-Systems

Prof. Dr.-Ing. Dr.-Ing. E.h. H. K. Tönshoff, Dipl.-Ing. R. Kader
LASER ZENTRUM HANNOVER e.V.
Hollerithallee 8
30419 Hannover

Einleitung

Die Produkt- und Produktionsqualität in der Laserstrahlschneidbearbeitung hängt neben den Leistungsmerkmalen der Laserwerkzeugmaschinen wesentlich von den arbeitsvorbereitenden Maßnahmen ab. Die Leistungsfähigkeit einer Laserstrahlschneidanlage kann unter wirtschaftlichen und qualitativen Gesichtspunkten nur dann voll genutzt werden, wenn sie mit einer effektiven NC-Programmierumgebung ausgestattet ist.

Das Werkzeug Laserstrahl zeichnet sich gegenüber den konkurrierenden Trennverfahren wie Stanzen oder Nibbeln durch seine Geometrieflexibilität aus. Daher wird der Laserstrahl besonders dann sinnvoll eingesetzt, wenn komplexe und komplizierte Geometrien **[Bild_1]** in kleinen Stückzahlen geschnitten werden müssen.

Auf der einen Seite steigt mit sinkenden Stückzahlen das Verhältnis von Programmierzeit zu Fertigungszeit. Auf der anderen Seite sind die Investitionskosten von Laserstrahlschneidmaschinen immer noch sehr hoch. Daher muß die Zeit für die NC-Programmerstellung minimiert, und die Auslastung der Laseranlagen optimiert werden. Dem Programmierer sollte daher mit einer geeigneten Programmierperipherie ein komfortables und leistungsfähiges Werkzeug zur Verfügung gestellt werden, mit dessen Hilfe er in der Lage ist, innerhalb kürzester Zeit NC-Programme für komplexeste Werkstückgeometrien mit hinsichtlich Bearbeitungsqualität und Wirtschaftlichkeit optimal eingestellten Prozeßparametern zu generieren. Dadurch werden nicht nur die Programmierzeiten und die durch Testläufe verursachten Maschinenstillstandszeiten verringert. Auch vor- und angrenzende Bereiche wie Vor- und Nachkalkulation sowie Angebotserstellung und

400

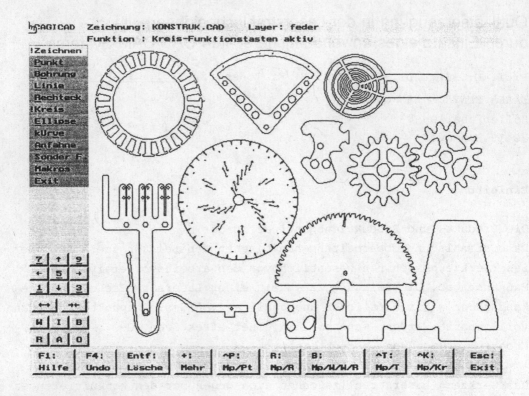

[Bild_1] Komplizierte Geometrien hinsichtlich des Programmieraufwandes
(C) LZH, EGLI, LACURA, MEKO

Auftragsverwaltung können beschleunigt und exakter durchgeführt
werden.

**Erzeugen der Bearbeitungsqualität durch optimale Einstellung der
Prozeßparameter im NC-Programm**

Neben der Geometriebeschreibung ist in der Laserstrahlschneidbearbeitung die Festlegung und Optimierung der Prozeßparameter **[Bild_2]** zur
wirtschaftlichen Realisierung eines geforderten Bearbeitungsergebnis-
ses sehr anspruchsvoll und zeitaufwendig.

Im Gegensatz zu konventionellen Trennverfahren ist das Laserstrahlschneiden ein thermisches Verfahren. In kritischen Bereichen, die sich
durch geringe lokale Materialvolumina auszeichnen, kann ein Wärmestau
entstehen. Daher sind bei der Festlegung der Prozeßparameter kritsche
Konturen besonders zu berücksichtigen.

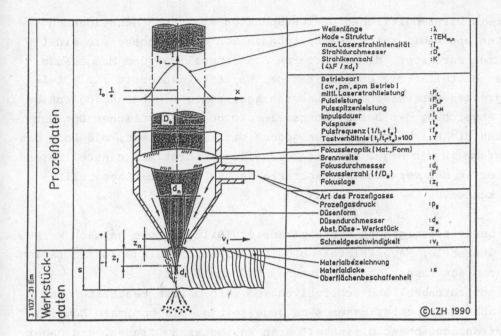

Wellenlänge	:λ
Mode – Struktur	:TEM$_{m,n}$
max. Laserstrahlintensität	:I$_o$
Strahldurchmesser	:D$_o$
Strahlkennzahl (4λF / πd$_f$)	:K
Betriebsart [cw , pm , spm Betrieb]	
mittl. Laserstrahlleistung	:P$_L$
Pulsleistung	:P$_{LP}$
Pulsspitzenleistung	:P$_{LM}$
Impulsdauer	:t$_i$
Pulspause	:t$_o$
Pulsfrequenz (1/t$_i$ + t$_o$)	:f$_p$
Tastverhältnis (t$_i$/t$_i$+t$_o$) ×100	:τ
Fokussieroptik (Mat.,Form)	
Brennweite	:f
Fokusdurchmesser	:d$_f$
Fokussierzahl (f/D$_o$)	:F
Fokuslage	:z$_f$
Art des Prozeßgases	
Prozeßgasdruck	:p$_g$
Düsenform	
Düsendurchmesser	:d$_n$
Abst. Düse – Werkstück	:z$_n$
Schneidgeschwindigkeit	:v$_f$
Materialbezeichnung	
Materialdicke	:s
Oberflächenbeschaffenheit	

Prozeßdaten

Werkstück-daten

3 1137 - 31 Em

©LZH 1990

[Bild_2] Einflußgrößen für den Laserstrahlschneidprozeß (C) LZH 90

Grundsätzlich sollte die Abarbeitungsreihenfolge der Konturen innerhalb eines Werkstückes oder eines Schachtelplanes zeitlich optimiert
sein ("Traveling Salesman Problem"). Jedoch muß bei der Bestimmung der
Reihenfolge berücksichtigt werden, daß zwei nacheinander abzuarbei-
tende Konturen einen dünnen Steg einschließen können. In diesem Fall
besteht wiederum die Gefahr einer durch Wärmestau verursachten thermischen Schädigung des Steges. Falls dies durch Parametereinstellung
nicht vermieden werden kann, muß an dieser Stelle die Abarbeitungsreihenfolge änderbar sein.

Kritische Geometrien im Hinblick auf durch mangelnden Wärmeabtransport
hervorgerufene thermische Schädigung des Werkstückes sind neben schma-
len Stegen auch spitz zulaufende Winkel und kleine Radien. Problematisch dabei ist, kritische Werkstückgeometrien in Abhängigkeit des
Werkstoffes, der Materialdicke und der verwendeten Laseranlage zu erkennen und an diesen Stellen geeignete Prozeßparameter einzustellen.
Diese Einstellung sollte sowohl unter Zugriff auf eine Technologiedatenbank, als auch grafisch interaktiv möglich sein.

Unter rein wirtschaftlichen Gesichtspunkten sind hohe Schneidgeschwindigkeiten und damit kurze Bearbeitungszeiten anzustreben. Die einstellbaren Parameter, die geforderte Schnittqualität und Maßgenauigkeit des Bauteils, die Komplexität der Werkstückgeometrie , die Leistungsgrenzen der verwendeten Laseranlage, sowie die technologisch bedingte Abweichung der Bearbeitungsreihenfolge vom zeitlichen Optimum schränken die Bearbeitungszeiten jedoch ein. Hinzu kommt, daß über das NC-Programm in der Regel nur ein Teil aller theoretisch einstellbaren Parameter an der verwendeten Laserstrahlschneidanlage eingestellt werden können.

Die zu berücksichtigenden Prozeßparameter **[Bild_2]**, die je nach Werkstoff, Geometrie, Materialdicke und installierter Anlagenkonfiguration (bestehend aus Laser, Lasersteuerung, CNC sowie Peripherieeinrichtungen) und schließlich vom geforderten Bearbeitungsergebnis abhängen, beeinflussen sich gegenseitig. Die optimale Kombination der anlagenabhängig einstellbaren Parameter zu finden, und dabei ein vernünftiges Verhältnis von Bearbeitungsqualität und Wirtschaftlichkeit zu erreichen, ist daher als ein wesentliches Problem bei der Laserstrahlschneidbearbeitung anzusehen. Diese Einstellungen sollten daher von einem NC-Programmiersystem unbedingt unterstützt werden. Hierzu eignet sich der Einsatz von Technologiedatenbanken. Die Erfahrung zeigt jedoch, daß die anzusetzenden Prozeßparameter für ein definiertes Schnittergebnis z.T. starken Schwankungen unterliegen. Dies liegt u.a. am Verschleiß der optischen Komponenten, an schwankenden Umgebungs- und Werkstoffeinflüssen und nicht zuletzt an Justagefehlern, die sich durch häufiges Auswechseln und Umrüsten des Bearbeitungskopfes ergeben. Es sollte daher an jeder Stelle der NC-Programmgenerierung ein sinnvolles Maß an manuellen Eingriffsmöglichkeiten gewährleistet sein, um die Voreinstellungen aus einer Technologiedatenbank ggf. zu korrigieren. Dies gilt insbesondere für die Festlegung der Bearbeitungsreihenfolge, dem Zuweisen von Prozeßparametern an beliebige Konturen und Konturelemente (speziell bei kritischen Konturen) sowie die Erstellung von Schachtelplänen.

Neben der optimalen Einstellung der Prozeßparameter sorgen generalisierte Prozessoren für eine reproduzierbare Bearbeitungsqualität auf unterschiedlichen Laserstrahlwerkzeugmaschinen. Spezielle Programmiermethoden wie Unterprogrammtechniken gewährleisten die

Erstellung übersichtlicher NC-Programme. Die Maßhaltigkeit der Werkstücke kann u.a. durch Schnittfugenkompensation und Fahren mit Außenschleifen definiert werden. Wesentliche Bestandteile eines modernen CAD/CAM-Systemes sind NC-Simulatoren, Generatoren für Nestingpläne, Methoden zur Vor- und Nachkalkulation, Auftragsverwaltung und-verfolgung, sowie die Geometrieerzeugung (CAD). Schließlich haben die Komfortabilität, Benutzerfreundlichkeit und das Layout sowie die notwendige Einarbeitungszeit in ein System einen großen Einfluß auf die Benutzerakzeptanz, was sich letztlich auf die Qualität der Bedienereingaben und die Arbeitseffektivität auswirkt. Gerade bei kleinen und mittelständischen Lohnfertigern liegt aufgrund einer typischerweise geringen Arbeitsteilung eine hohe Produktionsverantwortung und Aufgabenvielfalt bei den Mitarbeitern. Diese sollen durch die Anwendung von CAD/CAM-Systemen sowohl bei der Geometriebeschreibung als auch bei der Festlegung der technologischen Parameter in optimaler Weise unterstützt und motiviert werden. Dies setzt neben den funktionalen Möglichkeiten eine einfache und effektive Bedienung der Systeme voraus.

Das *LASER ZENTRUM HANNOVER e.V.* hat ein CAD/CAM-System für die Laserstrahlschneidbearbeitung (CAGILA) in enger Zusammenarbeit mit industriellen Anwendern des Systemes entwickelt.

Durch Einführung dieses anforderungsgerechten CAD/CAM-Systemes können die genannten Firmen die NC-Programmierung ihrer Laserstrahlschneidanlagen jetzt effektiver und komfortabler durchführen. Zudem wurde die Wettbewerbsfähigkeit gesteigert, da nun auch komplizierte und umfangreiche Werkstückgeometrien wirtschaftlich erstellt werden. Die NC-Programme werden im Vorfeld der Fertigung prozeßparameteroptimiert. Dies vermeidet umfangreiche Testläufe an der Laserstrahlschneidanlage und erhöht dadurch deren Auslastung und Verfügbarkeit für die Fertigung.

Durch eine hohe Benutzerakzeptanz aufgrund der einfachen Bedienung des Systemes bedurfte es nur einer kurzen Einarbeitungsphase des Bedienpersonals, dessen Motivation zusätzlich gesteigert wurde. Somit war das neue Verfahren in der NC-Programmierung bereits nach wenigen Tagen etabliert und als effektives Routinewerkzeug in den Produktionsablauf integriert.

404

Literatur

/ 1/ Gonschior, M.; Kader, R.
"Problemlos Anpassen – NC-Programmierung für die Metallbearbeitung mit Laser"
Industrieanzeiger 41/1990, S. 90-91

/ 2/ Gonschior, M.; Kader R.
"Simulation von NC-Programmen für die Lasermaterialbearbeitung"
Laser Magazin 3/90, S. 14-19

/ 3/ Gonschior, M.
"Exploiting CAD/CAM for Better Laser Processing"
European Machining, Jan/Feb '91, S. 65-69

/4/ Gonschior, M.
"CAD/CAM für die Lasermaterialbearbeitung"
Technica, Nr.9, 1990, S. 46-49

/5/ Tönshoff, H.K.; Emmelmann, C.; Gonschior, M.
"Konturangepaßte NC-Programmierung für die Lasermaterialbearbeitung"
VDI-Z 132, 1990, Nr.7, S. 43-46

/6/ Spur, G.; Krause, F.L.; u.a.
"CAD-Methoden zur technologischen Planung und Programmierung von Laserschneidanlagen"
ZwF 83, 1988, S. 409-414

/7/ Warnecke, H.J.; Hardock, G.
"Systemvergleich, Prüfwerkstück zur Beurteilung von Laser-schneidanlagen"
Laser, Juni 1989, S. 22-27

/8/ Gonschior, M.
"CAD/CAM-System für die Lasermaterialbearbeitung"
Industrie Anzeiger 35/36/1990, S. 39-40

/ 9/ Geiger, M.; Kolléra, H.
"Werkstattnahe Programmierung von Laserstrahlschneidmaschinen"
Vortrag im Rahmen des "International Computer Science Meeting miroCAD'93" in Miskolc, Ungarn

/10/ Hoffmann, M.
Entwicklung einer CAD/CAM-Prozeßkette für die Herstellung von Blechbiegeteilen, Reihe Fertigungstechnik Erlangen, Nr.28 Bericht aus dem Lehrstuhl für Fertigungstechnologie (Prof. Geiger), Hanser: München Wien 1991

/11/ Tönshoff, H.K.; Kader, R.
NC-Programmierung für die Laserstrahlschneidbearbeitung
Blech Rohre Profile, April 1993

Einsatz eines voll 3D-fähigen schnellen Sensorsystems zur Nahtverfolgung beim Laserstrahlschweißen

W. Trunzer, H. Lindl, H. Schwarz

Institut für Werkzeugmaschinen und Betriebswissenschaften (iwb)
Technische Universität München, Karl-Hammerschmidt-Str. 39, 85609 Aschheim

Einleitung

Der Laser gewinnt in der Produktionstechnik aufgrund seiner Vorteile und seiner vielfältigen Einsatzmöglichkeiten zunehmend an Bedeutung. Die Haupteinsatzgebiete in der Materialbearbeitung sind das Schneiden und Schweißen. Schneidanwendungen sind derzeit Stand der Technik und sowohl in 2D- als auch in 3D-Anwendungen vertreten. Beim Schweißen treten jedoch häufig Probleme hinsichtlich der vom Laserprozeß geforderten Genauigkeiten auf. Um das Laserschweißen in der automatisierten Produktion einsetzen zu können, ist oftmals massive Spanntechnik erforderlich, um zumindest lagetolerante Schweißnähte wie die Überlappnaht verschweißen zu können /1/.

Ein anderer Weg, die geforderten Toleranzen einzuhalten, führt zum Einsatz von Nahtfolgesensorsystemen, die ein selbständiges Auffinden und Verfolgen von Schweißnähten ermöglichen.

Schweißnahtfolgesensoren

Bei Nahtfolgesensoren handelt es sich um intelligente Sensorsysteme, die in der Lage sind, Konturen zu vermessen und daraus die Schweißnaht extrahieren. Im wesentlichen werden hierfür optische Systeme eingesetzt.

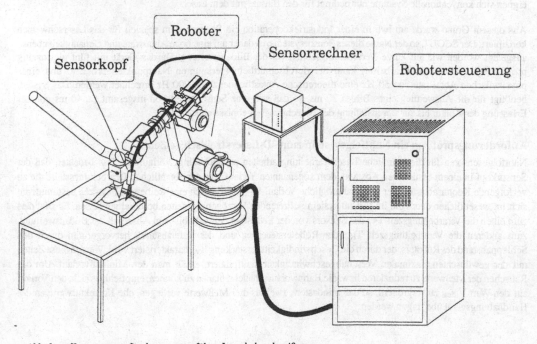

Abb. 1:   Komponenten für das sensorgeführte Laserbahnschweißen

406

Für die Vermessung der Kontur wird im allgemeinen das Triangulationsverfahren verwendet.

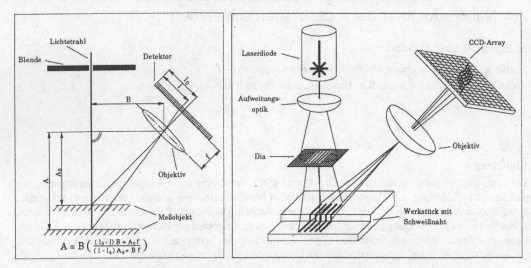

Abb. 2:   Prinzip des Triangulationsverfahrens (links) und Schema eines Mehrstreifenlichtschnittsensors (SCOUT)

Nahtfolgesysteme stehen seit mehreren Jahren zur Verfügung. Sie wurden jedoch für den großen Markt der Schutzgasschweißapplikationen WIG/MIG/MAG konzipiert. Dies hat zur Folge, daß die Systeme bei einer Meßauflösung von teilweise nur 0.5 mm für maximale Schweißgeschwindigkeiten bis ca. 2.5 m/min ausgelegt sind. Beim Laserschweißen werden aus technologischen und wirtschaftlichen Gründen oftmals Schweißgeschwindigkeiten von mindestens 5 m/min gefordert, bei maximalen Abweichungen je nach Nahttyp von teilweise nur 0.1 mm. Somit eignen sich konventionelle Systeme nur bedingt für den Einsatz mit dem Laser.

Aus diesem Grund wurde am iwb in einer Industriekooperation ein Sensorsystem speziell für das Laserschweißen konzipiert. Der SCOUT, so der Name dieses Sensorsystems, ist daher auf eine hohe Abtastrate und Genauigkeit ebenso ausgelegt worden wie auf kurze Verarbeitungszeiten bei der Bildauswertung. Da das Gerät mit fünf gleichzeitig projizierten Lichtschnitten arbeitet, kann bei durchschnittlich 4 erkannten Nahtpunkten pro Bild und einer physikalischen Abtastrate von 50 Hz eine theoretische Abtastfrequenz von 200 Hz errechnet werden. Das System benötigt für die Auswertung eines Bildes 20 ms, so daß sich eine Sensortotzeit von insgesamt ca. 40 ms von der Erfassung der Kontur bis zur Bereitstellung der Berechnungsergebnisse einstellt /2/.

Anforderungsprofil an ein Nahtfolgesystem zum 3D-Laserstrahlschweißen

Nahtfolgesensoren für die räumliche Laserbearbeitung arbeiten prinzipbedingt vorlaufend. Das bedeutet, daß der Sensorkopf in einem bestimmten Abstand, dem sogenannten Vorlauf, vor der eigentlichen Bearbeitungsstelle die zu verfolgende Kontur abtastet. Der minimal mögliche Vorlauf V_{min} wird durch eine Reihe von Faktoren bestimmt, die sich im wesentlichen durch die im Gesamtsystem auftretenden Verzögerungszeiten beschreiben lassen. Es sind dies zum einen die Verarbeitungszeit T_{ST} des Sensors von der Erfassung der Naht bis zum Abschluß der Bildauswertung, zum anderen die Verarbeitungszeit T_{RC} der Robotersteuerung und der Bahnabstand, hervorgerufen durch den Schleppabstand des Roboters, der durch die Geschwindigkeitsverstärkung K_v charakterisiert wird. Werden diese Zeiten mit der gewünschten maximalen Vorschubgeschwindigkeit multipliziert, erhält man den Minimalvorlauf. Um das Rauschen der Meßwerte zu reduzieren bzw. die Bahn vorausschauend planen zu können, empfiehlt es sich, den Vorlauf auf den Wert V_{ideal} zu vergrößern, so daß mindestens zwei bis drei Meßwerte vorliegen, ehe Korrekturwerte an das Handhabungsgerät übertragen werden.

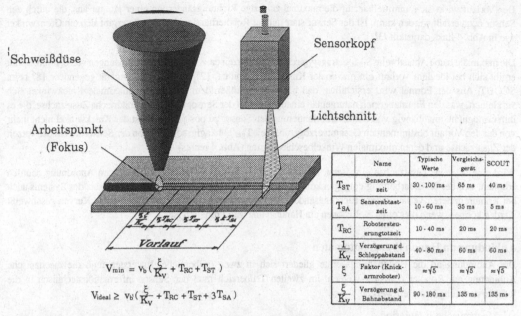

Schweißdüse

Arbeitspunkt
(Fokus)

Sensorkopf

Lichtschnitt

$$V_{min} = v_b \left(\frac{\xi}{K_V} + T_{RC} + T_{ST} \right)$$

$$V_{ideal} \geq v_b \left(\frac{\xi}{K_V} + T_{RC} + T_{ST} + 3T_{SA} \right)$$

	Name	Typische Werte	Vergleichs-gerät	SCOUT
T_{ST}	Sensortot-zeit	30 - 100 ms	65 ms	40 ms
T_{SA}	Sensorabtast-zeit	10 - 60 ms	35 ms	5 ms
T_{RC}	Robotersteu-erungtotzeit	10 - 40 ms	20 ms	20 ms
$\frac{1}{K_V}$	Verzögerung d. Schleppabstand	40 - 80 ms	60 ms	60 ms
ξ	Faktor (Knick-armroboter)	$\approx \sqrt{5}$	$\approx \sqrt{5}$	$\approx \sqrt{5}$
$\frac{\xi}{K_V}$	Verzögerung d. Bahnabstand	90 - 180 ms	135 ms	135 ms

Abb. 3: Einflußgrößen, die den Vorlauf eines Nahtfolgesensorsystems bestimmen

So berechnet sich beim SCOUT der minimale Vorlauf zu 26 mm, der ideale Vorlauf zu 28 mm bei einer Bearbeitungsgeschwindigkeit von 8 m/min (133 mm/s), ein Vergleichsgerät mit derzeit typischen Daten erreicht einen Vorlauf V_{min} von ca. 30 mm und V_{ideal} 44 mm bei gleicher Verfahrgeschwindigkeit.

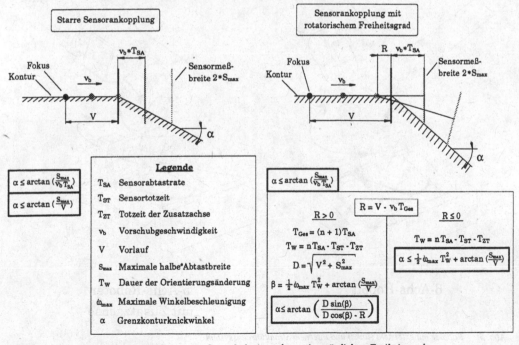

Abb. 4: Der Grenzkonturwinkel bei starrer Sensorbefestigung bzw. mit zusätzlichem Freiheitsgrad

Der Vorlauf wirkt sich unmittelbar auf die maximal zulässige Richtungsänderung einer Kontur aus, die durch den Sensor noch erfaßt werden kann. Ist der Sensor starr mit der Roboterhand verbunden, so ergibt sich ein Grenzwinkel wie in Abb. 4 links dargstellt /3/.

Die maximale halbe Abtastbreite S_{max} beträgt typischerweise derzeit ca. 8 - 10 mm bei Auflösungen von 0.1 mm. Somit ergibt sich bei idealem Vorlauf ein maximaler Knickwinkel von ca. 12° beim Vergleichsgerät gegenüber 18° beim SCOUT. Aus der Formel wird ersichtlich, daß bei sich vergrößerndem Vorlauf der maximale Knickwinkel sich verkleinert, was den Einsatzbereich naturgemäß einengt. Verfügt der Sensor über eine rotatorische Zusatzachse, die es ihm ermöglicht, unabhängig von der Strahlorientierung den Sensor zu positionieren, ist der Knickwinkel nicht mehr von der den Vorlauf bestimmenden Gesamtverzögerungszeit T_{ges} abhängig, sondern von der Sensortotzeit, der Totzeit der Zusatzachse und deren maximalen Winkelbeschleunigung (Abb. 4 rechts)

Wird eine schnelle Zusatzachse eingesetzt, kann sich der Grenzwinkel gegenüber der starren Anordnung deutlich erhöhen, was der 3D-Bearbeitung entgegen kommt. Weiterhin erlauben Zusatzachsen den Einsatz des Systems auch bei Konturen, deren Verlauf starke Richtungsänderungen aufweisen, so daß sogar geschlossene Kurven geschweißt werden können, wenn dies ohne Kollisionen am Handhabungsgerät möglich ist.

Anbindung an Handhabungsautomaten

Die Anbindung an die Handhabungsgeräte gliedert sich in zwei Teilbereiche. Im ersten muß die mechanische Anbindung am Roboter realisiert werden, im zweiten Teilbereich muß der Sensor informationstechnisch in die Robotersteuerung integriert werden.

- Mechanische Integration

Für die 3D-Bearbeitung realer Bauteile ist es zwingend notwendig, den Sensor in seiner Position verändern zu können, ohne dabei die Strahlorientierung bzw. die Fokuslage zu beeinflussen.

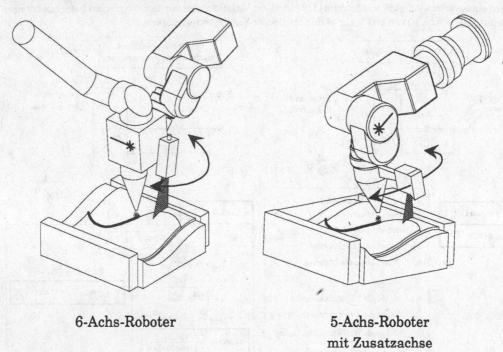

6-Achs-Roboter   5-Achs-Roboter
mit Zusatzachse

Abb. 5: *Roboter mit externem (links) und internem Strahlführungssystem*

In der Praxis werden für die 3D-Laserbearbeitung häufig Roboter mit interner Strahlführung und 5 Achsen eingesetzt, da unter der Annahme eines rotationssymmetrischen Strahls 5 Achsen für die Positionierung und Orientierung des Laserstrahles ausreichen. Um den Vorteil der freien Positionierung des Sensors an diesen Anlagen nutzen zu können, ist hier zumindest eine Zusatzachse erforderlich. Eine maximale Beweglichkeit bietet hier eine um die Laserstrahlachse angebrachte rotatorische Zusatzachse. Sie gewährleistet, daß der Sensor die Schweißnaht stets vorlaufend abtasten kann. Zusatzachsen haben in der Regel den Vorteil, daß sie sehr schnell auf Bewegungsvorgaben reagieren können, was die Dynamik des Gesamtsystems erhöht. Setzt man zur Handhabung einen 6-Achs-Roboter ein, bietet dieser an sich bereits einen Freiheitsgrad, der für die Sensorpositionierung genutzt werden kann. Somit ergibt sich auch nicht die Problematik der Integration der Zusatzachse in die Robotersteuerung. Als 6-Achs-Roboter eignen sich Standardindustrieroboter, die hinsichtlich ihrer Bahnführungsgenauigkeit optimiert sind.

- Informationstechnische Integration

Auf dem Gebiet der informationstechnischen Integration von Nahtfolgesystemen werden derzeit sehr unterschiedliche Konzepte verfolgt. Allen gemeinsam ist, daß die Bildauswertung und die damit verbundene Datenreduktion im Sensorrechner geschieht.

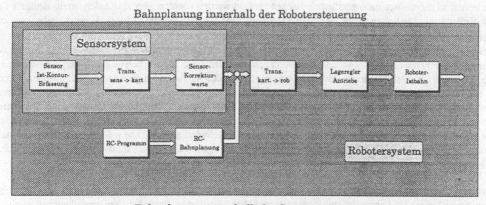

Abb. 6:   Zwei unterschiedliche Strategieen zur Integration eines Nahtfolgesensors in die Robotersteuerung

Bei der Planung eines Gesamtsystems sollte grundsätzlich darauf geachtet werden, daß die Abtastrate des Sensors etwa gleich groß oder höher ist als die Taktrate, mit der die Robotersteuerung die Daten verarbeiten kann. Da die Sensoren immer leistungfähiger werden, insbesondere was die Abtastrate betrifft, tritt derzeit die Schnittstellenproblematik in den Vordergrund. Mit der vorzugsweise eingesetzten seriellen Schnittstelle stoßen moderne Sensorsysteme sehr schnell an die Grenzen, was die Datenübertragungsrate betrifft. So behilft man sich oftmals mit speziellen Schnittstellen, die eine schnelle, echtzeitfähige Punkt-zu-Punkt-Verbindung zwischen Sensorrechner und Robotersteuerung erlauben. Von einer Standardisierung schneller Schnittstellen für zeitkritische Anwendungen, die eine einfache und universelle

Integration ermöglichen würden, ist man jedoch noch weit entfernt. Einen erfolgversprechenden Ansatz beschreiten die unterschiedlichen Feldbussysteme, die einerseits eine deutlich höhere Datenübertragungsrate als die konventielle serielle Schnittstelle erlauben, deren Echtzeitfähigkeit aber für besonders zeitkritische Anwendungen wie die Nahtfolgesensorik oftmals noch nicht genügt.

Praktische Erfahrungen

Das iwb beschäftigt sich seit mehreren Jahren unter anderem mit den Aufgaben der Laserintegration in die rechnergestützte Fertigung. Hierfür wurden im Versuchsfeld des iwb drei Laserbearbeitungszellen aufgebaut. Als Handhabungsgeräte für die Lasermaterialbearbeitung werden 6-Achs-Standardindustrieroboter eingesetzt, die entweder über ein externes Spiegelstrahlführungssystem oder eine Glasfaser mit den Bearbeitungslasern gekoppelt sind /4/.

An diesen Anlagen wurden umfangreiche Untersuchungen zum sensorgeführten Laserstrahlschweißen durchgeführt. Für die ersten Versuche wurde ein System eingesetzt, das die Schweißnaht bei einer Auflösung von ±0.2 mm mit max. 10 Hz bei einem nutzbaren Sichtfeld von ca. 50 mm abtastete. Dabei konnte der Sensor keine Orientierungsänderungen verarbeiten, so daß nur schwach gekrümmte Bahnen verfolgt werden konnten. Hier zeigte sich bereits deutlich, daß ohne Orientierungsänderung keine 3D-Nahtverfolgung möglich ist. In der nächsten Versuchsreihe kam ein Sensorsystem mit einer Abtastrate von 16 Hz und einer Auflösung von ±0.1 mm zum Einsatz, das jedoch nur noch ein nutzbares Sichtfeld von ca. 20 mm besaß. Da mit diesem System eine Orientierungsänderung möglich war, konnte der Nachteil des kleineren Sichtfeldes kompensiert werden, so daß man dieses System bereits als 3D-tauglich bezeichnen kann. Aufgrund der relativ niedrigen Abtastrate konnten Schweißungen bis zu 5 m/min durchgeführt werden, jedoch nur bei relativ glatten Kurvenverläufen. Durch den Einsatz des Sensorsystems SCOUT konnte die maximale Schweißgeschwindigkeit bei guter Nahtqualität auf 8 m/min erhöht werden. Noch höhere Geschwindigkeiten waren aufgrund der begrenzten Laserleistung nicht möglich, doch zeigten Ritzuntersuchungen, daß auch bei 10 m/min und mehr die Führungsgenauigkeit ausreicht. Bei diesen hohen Geschwindigkeiten kann jedoch der Roboter bei stark konturierten Schweißnähten mit Ecken und kleinen Radien den Vorgaben des Sensors aus dynamischen Gründen nicht mehr folgen.

Zusammenfassend kann festgehalten werden, daß mit dem SCOUT inzwischen ein Sensorsystem auf dem Markt ist, das ein zuverlässiges 3D-Laserbahnschweißen auch bei hohen Vorschubgeschwindigkeiten ermöglicht und somit die Einsatzgebiete des Lasers um wesentliche Bereiche erweitern wird.

Ausblick

Eine Nahtform, die prädestiniert ist, sensorgeführt mit dem Laser verschweißt zu werden, ist die Kehlnaht am Überlappstoß. Sie tritt beispielsweise an allen Türen, Motorhauben und Heckdeckeln im Automobilbau auf. Wenn es gelingt, mit dem Laser die Falze dichtzuschweißen, können hierdurch eine Reihe bisher zur verrichtender Arbeitsgänge wie kleben, fixieren und dichten durch das Laserschweißen wirtschaftlich ersetzt werden. Die ersten vielversprechenden Untersuchungen in dieser Richtung wurden bereits am iwb durchgeführt. Selbst bei beschichteten Blechen konnte eine optimale Nahtqualität bei einer definierten Einschweißtiefe erreicht werden, ohne dabei die Blechaußenseite zu verletzen, was für das anschließende Lackieren von ausschlaggebender Bedeutung ist.

Literaturverzeichnis:

/1/   J. Hornig, Lasermaterialbearbeitung bei BMW, European Laser Marketplace '92, 1992

/2/   W. Trunzer, H. Schwarz, H. Lindl, Ein Scout weist den richtigen Weg, S. 8, Produktion Nr. 18, 1993

/3/   G. Pritschow, A. Horn, K. Grefen, Dynamisches Verhalten und Grenzen sensorgeführter Industrieroboter mit vorausblickendem Sensor, S. 155, Robotersysteme 8,1992

/4/   F. Garnich, Laserbearbeitung mit Robotern, Dissertation TU München, 1992

Process Diagnosis and Control for Removal Processes with Excimer Laser Radiation

W. Barkhausen, M. Wehner, K. Wissenbach, Fraunhofer Institut für Lasertechnik, Steinbachstraße 15, D-52074 Aachen, Federal Republic of Germany

E. W. Kreutz, Lehrstuhl für Lasertechnik, RWTH Aachen, Steinbachstraße 15, D-52074 Aachen, Federal Republic of Germany

1. Introduction

Many basic aspects of surface treatment with Excimer laser radiation have been investigated and it has been shown that the processing results are affected by variations of the processing variables as well as by irregularities of the workpiece /1,2/. Therefore the development of on-line process control systems is of great importance for an increase in reproducibility and further acceptance of Excimer laser processing in production line. To ensure high machining quality and short machining time the processing result and all significant processing parameters has to be analyzed and controlled in realtime.

One promising technique of surface treatment is scale removal to improve the corrosion resistance of austenitic steel /3/. For this application the varying thickness and composition of the scale layer require an adaption of the processing parameters (fluence and number of laser pulses per site) to the local surface conditions. A process diagnosis based on Plasma Emission Spectroscopy (PES) is build up for this application. Another application of interest is selective removal of corrosion layers on historical glass paintings to increase light transmission. Because the corrosion layers have varying composition and thickness a process diagnosis with a probe laser and an optical sensor is used to measure the transmittivity of the layer.

A process control system is build up for real time diagnosis of removal processes with Excimer laser radiation. The system is applied to descaling of steel. For selective removal of corrosion layers on glass a closed loop control has been developed to guarantee a high processing quality without damaging the glass.

2. Fundamentals of the process control system

A block diagram of a system for controlled laser surface treatment is shown in fig. 1.

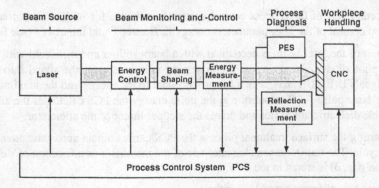

Fig. 1: Schematic diagram of the laser control system.

The control system is based on a process computing system (PCS) and has components for on-line energy measurement, energy control, beam shaping, on-line process diagnosis, laser control and workpiece handling /4/.

The control system is working in four operation modes:

1. fixed processing values - preset by the user
2. constant laser parameter
3. on-line process diagnosis
4. closed loop control.

Ref. 1: In this mode the number of irradiated laser pulses per surface element is preset by controlling pulse repetition rate and processing speed. The beam shape is formed by variable apertures. The optimal process parameters have to be defined by the user.

Ref 2: The automatic control of laser parameters like pulse energy, fluence or intensity is realized with a closed loop control consisting of a measurement device and a stepper motor driven variable attenuator. The laser parameters such as pulse energy are measured on-line and undesired changes caused by varying laser output energy during long time processing or changes in transmission of the optical system e.g. as a function of the repetition rate can be compensated by adaption of the attenuators transmittance.

Ref 3: Diagnosis of the process can be done by

- measurement of reflectivity of the surface to detect laser induced changes of surface morphology and / or selective removal of surface layers

- analysis of laser induced plasma with PES for qualitative measurement of the vapor composition

- measurement of transmittivity for analysis of layer thickness on transparent workpieces.

Measurement and data processing are realized in real-time which allows subsequent analysis of the processing result.

Ref. 4: To optimize the laser parameters according to the local surface conditions the PCS has to compare the measurement results with internally stored values for the actual set of laser parameters. If the measured values are out of range the fluence on the workpiece and/or the number of irradiated laser pulses are adapted.

3. Energy control

One important part of the process control system is a device for on-line measurement and closed loop control of the laser parameters energy E, fluence F and intensity I (see fig. 2).

A small part of the laser beam is decoupled with a beam splitter and measured with two UV-sensitive photodiodes. An electronic preprocessing unit transforms the signal into two DC-voltage signals U(E), U(I max), which are proportional to the energy and the maximum intensity of the laser pulse /1/. To control e.g. the pulse energy the PCS calculates the adjustment of a variable dielectric attenuator and drives the stepper motor of the attenuator.

Before starting the surface treatment process the PCS carries out an automatic attenuator calibration cycle. The transmittance characteristic of the attenuator as a function of stepper motor position (fig. 3) is stored in the PCS.

Fig. 4 shows two different control algorithms:

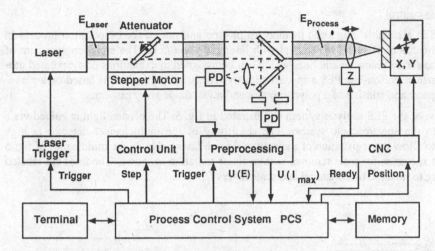

Fig. 2: Operation diagram of a control system for processing parameters. Pulse energy, repetition rate and workpiece handling are controlled by PCS.

a) "P"- Controller, damped control characteristic
b) "Dead Beat" - Controller, instantaneous control characteristic,

both realized as software modules on the PCS. The graphs show the output energy of the control system at a laser repetition rate of 10 Hz. The preset value for process energy (E Process) is changed after a certain number of laser pulses to show the different response characteristics. Depending on the process requirements different control algorithm can be adapted to the PCS.

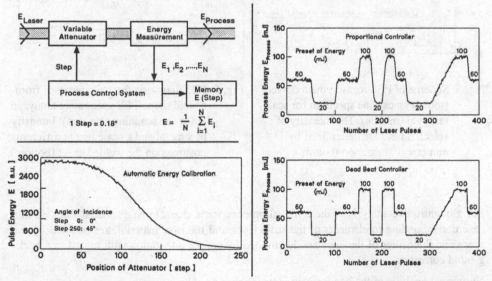

Fig. 3: Scheme of attenuator calibration (top). Transmitted energy of attenuator as a function of attenuator position (bottom).

Fig. 4: Output energy of the PCS with "protional" (top) and "dead beat" (bottom) controller for changing preset values.

414

4. Process control

The PES is a suitable diagnostic method for on-line analysis of the evaporation process in surface treatment or selective removal with laser radiation /5,6/. The emission spectrum of the plume shows characteristic lines which can be assigned to excitations of evaporated atomic elements. For on-line PES a spectroscopy system is built up, which is based on a personal computer and consists of a polychromator and a fast diode array camera.

A scheme of the PES analysis system is illustrated in fig. 5. The plasma light is guided via a fiber into the spectroscopic system. The sensitivity of the multichannel detector is high enough to allow the registration of a spectrum in one laser pulse. In the middle part of fig. 5 a typical spectrum for scale removal with excimer radiation is shown. The lines are labeled according to metallic base material and scale layer /7/.

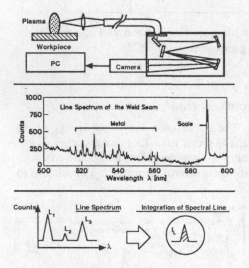

Fig. 5: Scheme of PES measurement device (top). Typical line spectrum for scale removal (middle). The intensity of selected lines are determined by numerical integration (bottom).

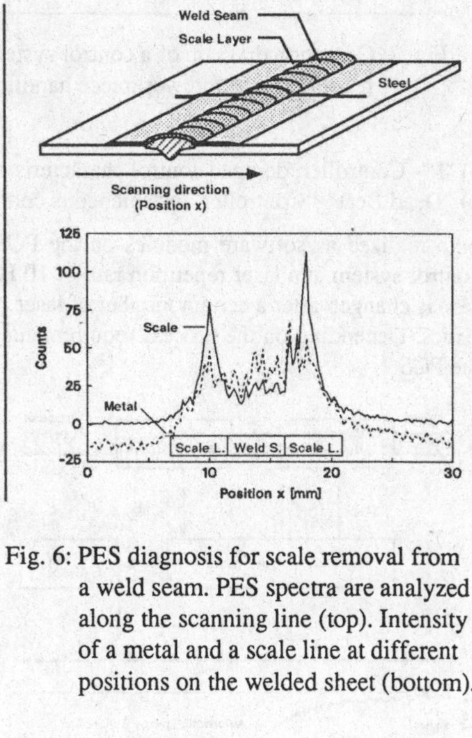

Fig. 6: PES diagnosis for scale removal from a weld seam. PES spectra are analyzed along the scanning line (top). Intensity of a metal and a scale line at different positions on the welded sheet (bottom).

For a quantitative analysis of the removal process some characteristic spectral lines, in this case corresponding to elements of the scale layer and the base material, are selected by the user. The intensities of the lines are determined by simple integration with regard to a background correction.

Numerical analysis of the line intensity data is carried out by
- time development of single lines, showing different composition of the plume (fig. 6),
- comparison of measured intensities with reference values or
- calculation of intensity ratios between different lines.

Local changes in surface finish as well as varying absorption of the laser beam are detected. By comparing the actual measurements with experimental determined reference values closed loop control can be realized.

A typical on-line measurement result by PES is shown in fig. 6. The laser beam is scanned perpendicular to the weld seam starting on the sheet material. The full graph corresponds to the line intensity of Na in the scale layer and the dotted graph shows the intensity of Cr-line from the base material. It can be clearly seen, that there is a higher amount of scale in the vicinity of the weld seam (two maxima). On the other hand the line intensity of Cr increases because the scale and the roughness of the weld seam leads to higher absorption of laser radiation in this region. As a result the processing quality decreases. Therefore an adaption of the laser fluence and number of laser pulses per site by the control system is necessary when reaching the scale. The optimization of the closed loop control based on the PES analysis is the aim of further activities.

Corrosion layers on historical glass paintings are a result of longtime chemical reactions with ambient atmosphere and have varying composition and thickness. For selective removal of these layers a closed loop control is build up (fig. 7, upper part).

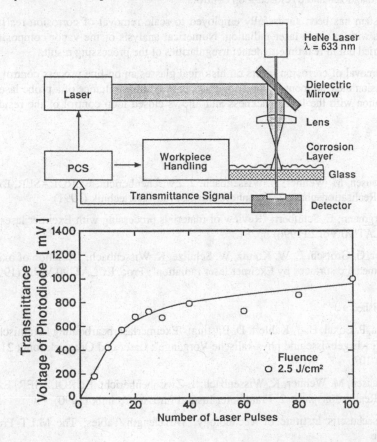

Fig. 7: Scheme of a closed loop control for removal of corrosion layers on glass (top). Transmittance signal of the HeNe laser for successive laser pulses (bottom). The increasing detector signal indicates the decreasing layer thickness.

In order to analyze the removal process in real time the transmission of a HeNe - Laser beam through the glass is measured with a photodiode. The transmittance signal correlates with the thickness of the corrosion layer. Depending on the desired thickness of the removed layer (transmittance signal) PCS controls the number of laser pulses at the actual position and moves the glass automatic to the next position. Fig. 7 (bottom part) shows a typical measurement: the transmittance signal increases because of the decreasing layer thickness as a function of laser pulses.

Summary

In order to improve processing quality and processing speed in surface treatment or removal processing with Excimer laser radiation a control system has been developed. The control system is based on a process computing system (PCS) and consist of a closed loop control for laser energy, fluence and intensity and an on-line process diagnostic using Plasma Emission Spectroscopy (PES) or transmission measurement. The PCS support laser processing with different operation modes a) fixed processing values, b) constant laser parameters, c) on-line process diagnosis and d) closed loop control.

The control system has been sucessfully employed to scale removal of corrosion resistant austenitic steels with excimer laser radiation. Numerical analysis of the vapor composition with PES is carried out in real time to detect irregularities of the processing result.

For selective removal of corrosion layers on historical glasses an on-line process control based on transmission measurement is build up. The measured transmittance of a probe laser is in good correlation with the layer thickness and allows closed loop control of the residual layer thickness.

References

/ 1 / W. Barkhausen, M. Wehner, K. Wissenbach: 2. Zwischenbericht, EUROLASER, Excimerlaser, Realisationsphase 2; Fraunhofer Institut für Lasertechnik (1991)

/ 2 / H. W. Bergmann, E. Schubert: "Review of materials processing with Excimer lasers"; Proc. ECLAT'90 Vol.2 (1990) pp. 813

/ 3 / M. Wehner, G. Grötsch, E. W. Kreutz, W. Schulze, K. Wissenbach: "Ablation of oxide layers on metallic surfaces by Excimer laser radiation"; Proc. ECLAT'90 Vol.2 (1990) pp. 917

/ 4 / to be published

/ 5 / U. Sowada, P. Lokai, H.-J. Kahlert, D. Basting: "Excimerlaserbearbeitung keramischer Werkstoffe - Ergebnisse und physikalische Vorgänge"; Laser und Optoelektronik 21(3) (1989) pp. 107

/ 6 / W. Barkhausen, M. Wehner, K. Wissenbach: 1. Zwischenbericht, EUROLASER, Excimerlaser, Realisationsphase 2; Fraunhofer Institut für Lasertechnik (1990)

/ 7 / e.g.: Massachusetts Institute of Technology, Wavelength Tables; The M.I.T Press, Cambridge

State of the Art and Prospects of Electron Beam and Laser Beam Welding

Georges SAYEGH

Dr. Sc.; Dr. Eng.; Prof. at E A P V (Paris)
Club Laser de Puissance (Arcueil-France)

INTRODUCTION :

No other welding process have created so much scientific and technical interest as Electron beam and laser beam. The wide variety of industrial techniques involved in the processes interests many branches of physics : Electro technics, Electronics, Quantum physics, Optics, Computer science and Metallurgy. It is not surprising therefore to see that Electron beam (EB) and laser beam (LB) occupy a prime place in the R&D of welding processes. Of course the commercial turnover with these processes remains quite low when compared to conventional welding processes ; but their potential prospects appear very promising, especially when considering flexible production and rational utilization of energy.

Our interest here concentrates only on welding applications and more specifically on multi-kilowatt continuous beams.

In the first part, I shall analyse the peculiar characteristics of the heat source generated by EB and LB and the resulting consequences.

In the second part, I shall consider the situation of these processes in the range of welding techniques and analyse the different technologies and versions operating in production. Next I shall present applications in various industries.

In the third part, I shall analyse the technical economic aspects of the use of EB and LB.

Lastly a forecast will be made of the prospects for the processes and the trend to integrate them into flexible manufacturing centers.

I - PECULIAR CHARACTERISTICS OF LB AND EB - CONSEQUENCES

The heat source created by LB and EB welding is characterised by :

a) The high energy density at the impact point of the beam,
b) The high accuracy with which the heat source can be controlled in movement and modulated in power,
c) The necessity of using automated controls through CNC systems.

The high energy density at the impact point-consequences

Thanks to the specific characteristics of LB and EB sources, it is possible to concentrate them onto a very small surface, producing energy densities about 1000 times higher than the conventional welding heat sources. The beam instantaneously vaporises the metal at the impact point and creates a capillary which is filled with plasma type metal vapors. The calorific energy of the beam is transmitted to the workpiece throughout its thickness, unlike conventional processes which transmit heat by conduction from the top surface of the workpiece. When the beam is displaced the liquid metal lining the wall of the capillary flows inwardly and solidifies forming the joint.

The beam -metal interaction has very important consequences on the geometry of the welded joint and on its metallurgical characteristics :

. The heat transfer from the beam across the capillary to its walls produces deep and parallel sided narrow welds.
. The energy input of EB and LB welding is much lower than conventional processes,
. The thermal distorsion and residual stresses after welding are very small ; permitting to produce finished parts which are used as welded,
. There is high temperature gradient during welding and a high cooling rate between the molten zone and the base metal,
. EB and LB are fusion welding. Because of the high cooling rates, the hardness and the toughness of mild and low alloys steels joints may pose problems. Addition of filler material or modification of the microalloying chemical composition of steels could be a solution to the problem,
. Fatigue behavior of LB and EB joints are very acceptable provided they are free from defects such as : cracks, undercut, large porosities.

The high precision of the impact point and power modulation

Photons and electrons can be displaced at very high speeds with mirrors or electromagnetic fields. Because of the very small diameter of the impact point, it is possible to control with high accuracy the focus of the beam in such a way that the applied energy on the workpiece surface follows a given distribution corresponding to best welding conditions. Additionally the power transported in the beam can be modulated very easily with very short time response. Thus it is possible to control the input power into the workpiece in space and in time.

Flexibility and automation

A high degre of automation is needed to take advantage of these processes. Computer controls are employed to ensure all the operating functions of the equipment, or the flexible line such as :
. sequential and logical functions of the welding cycle,
. servo position functions of the workpiece,
. process operating functions (monitoring and controlling of welding parameters, maintenance, production sequence...)

2 - SITUATION OF LB AND EB IN INDUSTRIAL PRODUCTION

EB welding was invented simultaneously in FRANCE and GERMANY around 1956. Multikilowatt laser beam welding was introduced in production with litlle success around 1973 and it developed really in the 78-80 to attain today a high level of expansion.

For both processes, the advantages were rapidly recognised by the potential users and by the system manufactures. Consecutive improvements in the technology make it possible to produce machines satisfying the needs of diversified and very demanding customers.

Applications of EB and LB cover today a whole range of industries such as : Aeronautics, automobile, nuclear, energy, electrical appliances. These welding processes are met :

. in joining very expensive components (jet engines) as well as very cheap ones (sensers),
. in mass production (gears) as well as in unit ones (pressure vessels)
. in welding small sized parts (transducers) as well as large components (off-shore tubes)
. in welding thin components(sawblades) as well as very heavy sections (vapor generater)
. in welding ordinary steels (structural steels) as well as exotic metals (Titanum).

In EB welding, the electron generator (gun) could be placed inside of the vacuum chamber or outside. If the vacuum level should be less than 10^{-2} Pascal in the gun, it can be 1 Pascal around the seam and even at the atmospheric pressure when welding is achieved in air.

In laser welding, parts to be joined are supplied to one or more working stations and the beam is transported to the stations by appropriate mirrors for CO_2 lasers and through fiber optics for YAG lasers. The flexibility of the beam by time-sharing or space-sharing represents a big advantage in flexible production lines.

Today the applications of laser welding are mostly concentrated in the automobile industry and the future of multikilowatt YAG lasers transported with fiber optics is very promising especially in car body welding composed of thin sheets.

The applications of EB welding are mainly in aeronautics and mechanical components where high power levels are required (> 10 kW), but some applications are also found in car industry.

3 - TECHNICAL AND ECONOMIC ASPECTS OF EB AND LB WELDING

The main advantages of EB and LB compared with conventional processes result from the specific characteristics of the heat sources presented in §1.

LB presents the following advantages over EB :
. LB does not need vacuum,
. LB is not disturbed by stray magnetic fields,
. LB can be employed in areas of difficult access,
. LB welding does not produce X rays,
. LB can be used very easily in time-sharing basis.

However LB presents the following disadvantages when compared to EB welding :

. energy efficiency is low for lasers (3% for YAG and 15% for CO_2)
. high operating cost,
. limited possibilities for beam oscillation and focus variation,
. high investment cost for high power beams.

For 5 to 6 mm thick steels, EB and LB produce similar penetrations when the operating parameters are similar. Beyond 5 to 6 mm, the increase in penetration with LB is less than for EB. This is attributed to the presence of plasma at the impact point which perturbates the beam material interaction. When laser welding is achieved in vacuum, the penetration of LB is quite similar to that of EB. Thus the plasma created in vacuum during laser welding is completely transparent to the CO_2 beam.

The choice of an industrial organisation of a new production technique is dictated by one or several of the following considerations :

. the application can be achieved only by the new technique,
. the new technique is cost effective (lower cost/unit),
. for equal cost/unit, better technical characteristics of the product (quality, lifetime.....),
. for equal cost/unit, better operating conditions and job interest.

These rules are valid when considering use of EB or LB welding. It should be pointed out that the cost/unit in an industrial application is difficult to analyse because most users consider this informationas confidential.

Comparing EB and LB investments for a given production which can be achieved by the two processes, one can draw the following results :

a) when the application can be realised with a beam power lower than 5kW, the laser process is more profitable,
b) when the application can be realised with a beam power higher than 10 kW, EB process looks more economical,
c) when the application needs a beam power of 5 to 10 kW, the two processes are competitive and a detailed analysis is neccessary to make the choice between the two.

Additionnally, factors such as : flexibility of the production, reliability of the system, selling promotion ... might play an important role in the decision.

4 - PROSPECTS OF THE EB AND LB WELDING PROCESSES

The annual market for EB equipment is about 150 units. It is about the double for multikilowatt laser equipment. Some applications originally reserved to EB (car industry) have been transfered during the last years to LB.

What are the improvements which can be brought to EB and LB in order to render them still more acceptable in industrial production ? The answer can be resumed in : Realibility, Flexibility and Economy. Here are some directions to follow in order to reach the target :

. improve power stability of the beam in space and in time,
. exploit extensively the use of CNC in maintenance, operating conditions and quality control of the equipment,
. improve real time seam tracking during welding,
. understand more deeply and use the results of beam interaction during welding,
. reduce the investment and the operating costs by simplifying the technology and improving the efficiency of the processes.

The following prospects can be identified :

a) In mass production (automobile), LB will continue to take progressively the place of EB in mechanical parts (gears) and will be used extensively in car body assembly,
b) In aeronautics and space (jet engines) EB continue to occupy a good place as joining technique. With the availability of high power lasers (> 10 kW), LB can modify the position of EB,
c) In energy and heavy industries where high beam powers are involved (15 - 50 kW), EB have a predominant position. Nevertheless because of the multiple advantages of LB welding, programmes using high power lasers associated with robotics (20 kW and more) are under R&D or exploited in production in shipbuilding, off shore or heavy industry.

CONCLUSION

In a 30 years period, EB and LB acquired a prime place in the range of industrial joining processes.

Associated with high degree of automation and CNC, these processes become more and more reliable, flexible, reproduceable and high quality production tools. Thus mass production industries, aeronautics, energy and heavy industries will continue to give a high priority to these processes.

The Electron Beams and the Laser Beams are competitive in certain domains, they remain quite complementary in others where they can co-exist harmoniously for the benefit of industry.

Leistungssteigerung bei der Materialbearbeitung durch Verfahrensintegration

F. Treppe, C. Hermanns, A. Werner, A. Zaboklicki

Fraunhofer-Institut für Produktionstechnologie, IPT
Steinbachstr. 17, D-52074 Aachen

Die Vorteile des Lasereinsatzes in der Fertigungstechnik können nur dann genutzt werden, wenn dessen Anwendungsmöglichkeiten und die Leistungsfähigkeit der Verfahren bekannt sind. Es werden neue Verfahrensansätze aufgezeigt, die die Integration von Laserbearbeitungsverfahren in konventionelle fertigungstechnische Abläufe (z.B. Zerspantechnik, Umformtechnik) beinhalten. Darüber hinaus wird anhand von Beispielen die Nutzung der daraus resultierenden technologischen Potentiale diskutiert.

1. Einleitung

Industrieunternehmen stehen in der heutigen Zeit zuhnemend im Spanungsfeld von markt-, umwelt- und gesellschaftspolitischen Entwicklungen. Um weiterhin auf nationalen und internationalen Märkten bestehen zu können und damit den Unternehmenserfolg auch zukünftig zu sichern, müssen sich die Unternehmen dem Wetbewerb mit neuen Strategiekonzepten stellen. Wesentliche Ansatzpunkte für neue Strategien bieten die optimierte Organisations- und Arbeitsgestaltung in der Produktion sowie die Weiterentwicklung der eingesetzten Produktionstechnologie /1/. Einen interessanten Ansatz zur Innovation und damit zur Leistungssteigerung in der Produktionstechnologie bieten die Laserstrahlverfahren in Verbindung mit konventionellen Bearbeitungstechniken.

2. Verfahrensintegration

Wegen seiner hohen Flexibilität, der guten Führ- und Regelbarkeit eignen sich der CO_2- und der Nd:YAG-Laser zur Einbindung in andere Bearbeitungsmaschinen. Eine solche Integration kann sowohl in einer konventionellen Dreh-, Fräsmaschine oder in einem Robotersystem erfolgen /2/; damit erhöht sich die mit dem Anlagenkonzept realisierbare Anzahl an Bearbeitungstechniken, *Bild 1*. Neben den konventionellen Werkzeugen (z.B. Drehmeißel, Fräswerkzeug, Preßstempel) steht zusätzlich der Laserstrahl als thermisches und verschleißfreies Werkzeug zur Verfügung. So können neben den klassischen Laserstrahlverfahren Schneiden und Schweißen weitere Techniken, wie z.B. das Laserstrahloberflächenveredeln (Härten, Beschichten, Legieren) bzw. das Laserstrahlumformen von metallischen Halbzeugen genutzt werden.

Die zeitliche Reihenfolge des Einsatzs der integrierten Verfahren in einer Werkzeugmaschine bei der Durchführung einer Bearbeitungsaufgabe führt zu einer Aufteilung in *simultane* und *sequentielle* Nutzung der vorhandenen Werkzeuge (Verfahren). Bei der simultanen Verfahrensfolge werden verschiedene Werkzeuge (Dreh- bzw. Fräswerkzeug und Laserstrahl) wie z.B. beim laserunterstützten Warmzerspanen gleichzeitig genutzt. Dagegen werden bei der sequentiellen Fertigung die einzelnen Bearbeitungstechniken nacheinander eingesetzt.

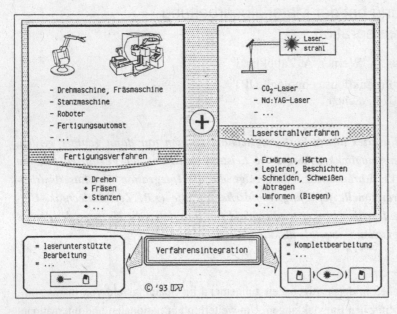

Bild 1: *Konzepte zur Verfahrensintegration*

Hierdurch kann beispielweise auf einer Bearbeitungsmaschine und in einer Aufspannung eine Komplettbearbeitung von Drehbauteilen mit folgender Fertigungsfolge: Drehen -> Laserstrahlhärten -> Hartdrehen durchgeführt werden. Der Vorteil einer derartigen durchgehenden Fertigung liegt vor allem in der Einsparung von Rüstzeiten.

Die realisierte Einbindung des CO_2-Laserstrahls in einer Fräsmaschine ist in **Bild 2** schematisch dargestellt. Der Laserstrahl wurde unter Beachtung der geometrischen und kinematischen Randbedingungen der Werkzeugmaschine integriert. Das Strahlführungssystem des CO_2-Lasers

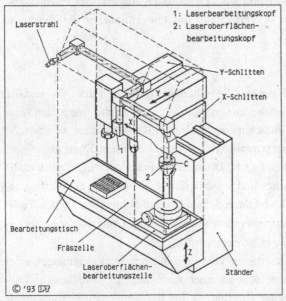

besteht hierbei aus Umlenkspiegeln. Dagegen kann ein Nd:YAG-Laserstrahl mit einem flexiblen Lichtleitfaserkabel an eine Bearbeitungsstelle geführt werden. In beiden Fällen ist eine variable Positionierung des fokussierten Laserstrahls dierekt vor dem Zerspanwerkzeug möglich. Der Laserbrennfleck auf der Werkstückoberfläche kann über das Strahlformungssystem als Kreis, Ellipse oder Rechteck ausgebildet werden. Die Integration des Laserstrahls in einer Werkzeugmaschine dient folgenden Zielen: zum einen, zusätzliche Fertigungsverfahren, wie z.B. Härten, Legieren, Beschichten oder Schweißen in den Funktionsumfang der

Bild 2: *Einbindung des Laserstrahls in einer Fräsmaschine*

Maschine einzugliedern und zum anderen, die konventionelle Zerspanung durch eine partielle Erwärmung des Werkstoffs vor dem Schneideneingriff zu unterstützen.

2.1 Simultane Nutzung unterschiedlicher Werkzeuge

Die Problematik des simultanen Einsatzes verschiedener Werkzeuge wird am Beispiel des laserunterstützten Fräsens diskutiert. Bei der laserunterstützten Warmzerspanung wird mit einem Laserstrahl der Werkstückstoff während des Zerspanvorgangs im Bereich der Zerspanstelle erwärmt, wobei die Wärmeeinbringung auf den Spanungsquerschnitt beschränkt bleibt. Durch die Relativbewegung zwischen Werkstück und Werkzeug wird ein bestimmter Bereich des Werkstoffes kontinuerlich erwärmt und spanend abgetragen.

Infolge der Temperaturerhöhung im Bereich der Zerspanstelle kommt es zur Veränderung der physikalischen Eigenschaften, wie z.B. Festigkeit und Härte, die zur Plastifizierung des Werkstoffes und somit zur deutlichen Verbesserung der Zerspanbarkeit führen. Dadurch ergeben sich folgende Prozeßvorteile: Reduzierung der Zerspankraft, Senkung des Werkzeugverschleißes, höhere Zeitspanvolumina /3/.

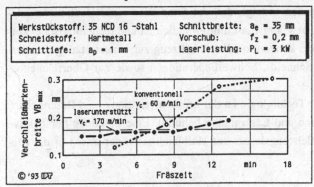

Bild 3 zeigt im Vergleich den Verschleißmarkenvortschritt für das konventionelle und das laserunterstützte Fräsen. Trotz der um den Faktor 3 höheren Schnittgeschwindigkeit beim Fräsen mit Laserunterstützung ist eine deutliche Reduzierung des Werkzeugverschleißes zu verzeichnen.

Bild 3: Werkzeugverschleiß beim Fräsen

Das Bearbeitungspotential der Warmzerspanung beschränkt sich dabei nicht nur auf metallische Legierungen, sondern bietet auch für hochfeste Ingenieurkeramiken interessante Perspektiven. So kann Siliziumnitrid-Keramik, die bislang nur mit geometrisch unbestimmter Schneide zerspanbar ist, mit Laserunterstützung durch Drehen bearbeitet werden. Die dabei erzeugte Oberfläche weist Schleifqualität auf /4/.

2.2 Sequentielle Nutzung der Laserstrahlverfahren

Aufgrund hoher Verfahrensflexibilität bietet die Lasertechnik vielseitige Möglichkeiten bei der Komplettfertigung von Blechbauteilen. Neben den schon etablierten Verfahren Laserstrahlschneiden und -schweißen können neue Techniken, wie das Laserstrahlumformen eingesetzt werden. Das Prinzip des Laserstrahlumformens basiert auf der Einbringung von laserinduzierten thermischen Spannungen in eine definierte Umformzone /5/. Die Herstellung der Kanten des in **Bild 4** gezeigten Würfels erfolgte durch mehrere Überläufe über die gleiche Biegelinie. Der Biegewinkel beträgt im Durchschnitt 5° je Überlauf.

424

Die Möglichkeit zur Formgebung von Blechen mit Hilfe der Laserstrahlung erweitert das Spektrum konventioneller Umformtechniken. Berücksichtigt man die derzeit realisierbaren Umformgrade und Formelemente, so kann das Laserstrahlumformen nach Abschluß der Prozeßentwicklung für die Prototypenherstellung und die Kleinserienfertigung eine interessante Alternative zu den bestehenden Techniken sein.

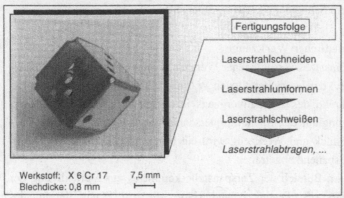

Bild 4: *Bearbeitungsbeispiel für den Einsatz von Laserstrahlverfahren*

3. Zusammenfassung und Ausblick

Die bisherigen Forschungsergebnisse haben gezeigt, daß die Integration eines Hochleistungslasers in eine Werkzeugmaschine (Dreh- bzw. Fräsmaschine) möglich ist und technologische Vorteile bringt. Der Laserstrahl kann wie ein konventionelles Werkzeug zur Unterstützung des Zerspanprozesses oder bei den typischen Schneid-, Schweißoperationen sowie zur Oberflächenbehandlung genutzt werden.

Ein Hemmnis bei der Einführung dieser Technologie ist der große Investitions- und Raumbedarf, den die Kombination Werkzeugmaschine und Laseranlage benötigt. Eine mögliche Perspektive zeigen Entwicklungsarbeiten des Hochleistungslasers auf Halbleiterbasis, der direkt in die Maschine integriert werden könnte /6/.

Das diesem Bericht zugrundeliegende Teilvorhaben wird mit Mitteln der Europäischen Gemeinschaft im Rahmen des Brite/EuRam-Projektes "Laser Assisted Machining" (Projekt-Nr.: BE-3366, Kontrakt-Nr.: BREU-0155) gefördert.

4. Literatur

/1/ Autorenkollektiv: Wettbewerbsfaktor Produktionstechnik, Hrsg.: AWK-Aachener Werkzeugmaschinen Kolloquium, VDI-Verlag, Düsseldorf 1990.

/2/ M. Weck; C. Hermanns: Laserintegration in Werkzeugmaschinen. VDI-Z 135 (1993), Nr.5, S.38-41.

/3/ W. König; F. Treppe; A. Zaboklicki: Laserunterstütztes Fräsen. VDI-Z 134 (1992), Nr.2, S.43-48.

/4/ W. König; A. Zaboklicki: Laserunterstützte Drehbearbeitung von Siliziumnitrid-Keramik. VDI-Z 135 (1993), Nr.6, S.34-39.

/5/ W. König; M. Weck; H.-J. Herfurth; H. Ostendarp; A. Zaboklicki: Formgebung mit Laserstrahlung. VDI-Z 135 (1993), Nr.4, S.14-17.

/6/ Autorenkollektiv: Wettbewerbsfaktor Produktionstechnik, Hrsg.: AWK-Aachener Werkzeugmaschinen Kolloquium, VDI-Verlag, Düsseldorf 1993.

Theoretical Aspects of Laser Material Processing

P. Kapadia
Department of Physics, University of Essex
Wivenhoe Park, Colchester, Essex CO4 3SQ U.K.

1.   Introduction

In the development of mathematical models to study laser welding it has hitherto been usual to consider the keyhole and weldpool regions separately. Actually these are two powerfully interacting facets of the problem. Accordingly the group at the University of Essex consisting of Dr. Dowden, Dr. Ducharme, Mr. M. Glowacki and myself have produced an integrated keyhole and weldpool model for the laser welding of thin metal sheets in which Dr. Ducharme played a leading part.

The work was first presented at ICALEO, USA[1]. For further details of this model it would be appropriate to refer to that paper. It is the object of this paper to extend this model and apply it to a range of materials such as iron, titanium, aluminium and copper.

2.   Inverse bremsstrahlung and Fresnel absorption in the keyhole

For steel at a translation speed of 10 mm/sec the calculated keyhole profile is almost perfectly cylindrical with laser power absorbed almost entirely by inverse bremsstrahlung processes. This is therefore a test of the validity of the inverse bremsstrahlung absorption process. Table 1 compares experimentally and theoretically calculated power absorption in the keyhole and excellent agreement is found between these two values at 10 mm/sec. The radius of the top of the keyhole is determined by the radius of the laser beam in each case (see Figure 1).

At translation speeds of 20 mm/sec the keyhole shape is perfectly conical with almost vertical sides. As the translation speed increases, the shape at the top of the keyhole becomes gentler emphasising the increasing effect of Fresnel absorption. The absorption due to inverse bremsstrahlung is insensitive to the keyhole radius and the increase in absorption with increasing translation speed arises almost entirely from Fresnel absorption. Table 1 demonstrates close agreement between experiment and theory.

The case of titanium is closely analogous to that of steel except that titanium ionises more readily and for this reason is associated with somewhat higher inverse bremsstrahlung absorption. This is shown in Table 2.

A comparison of the theoretical and experimental values for the power absorbed at a translation speed of 80 mm/s for titanium in Table 2 indicates a discrepancy. This deviation is very likely to reflect the very serious defects in the current knowledge of the standard parameters that appear in the Fresnel absorption coefficient particularly their variation with temperature which is very poorly known at present. It strongly emphasises the need for such measurements to be made urgently.

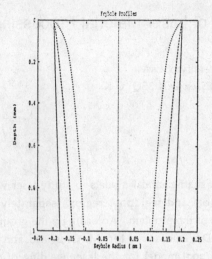

Figure 1. The keyhole radius as a function of depth for three different values of
the weld translation speed (a) 20 mm/s, (b) 40 mm/s and (c) 60 mm/s.

Table 1 – Steel-power absorbed

Translation speed	Power in weldpool	Power absorbed in the keyhole
10 mm/s	359 watts	369 watts
20 mm/s	501 watts	501 watts
40 mm/s	707 watts	732 watts
60 mm/s	914 watts	901 watts

Comparison of (a) power absorption calculated from
experimental weldpool shapes and (b) total bremsstrahlung
and Fresnel absorption in the keyhole produced by a 2.0 kW
continuous CO_2 laser beam of 0.29 mm radius in a 1 mm
thick sheet of steel.

Table 2 – Titanium - power absorbed

Translation speed	Power in weldpool	Power absorbed in the keyhole
20 mm/s	450 watts	449
40 mm/s	561 watts	629
80 mm/s	735 watts	863

Comparison of (a) power absorption calculated from
experimental weldpool shapes and (b) total bremsstrahlung
and Fresnel absorption in the keyhole by a 2.0 kW laser
beam of 0.29 mm radius in a 1 mm thick sheet of titanium.

The variation of the absorption coefficient in the plasma with temperature has a peak that is highest for aluminium and lowest for copper with that for iron lying in an intermediate position but the curves for all the metals considered again have the same general shape. This is also reflected in the variation of the refractive index with temperature in the keyhole for these metals which produces significant distortions of photon paths in the keyhole for CO_2 laser light.

Turning next to the Fresnel absorption process on the wall of the keyhole, the electrical resistivity can be plotted against temperature for the various metals under consideration up to their boiling point. It must, however, be emphasised that these parameters are not known with significant accuracy. The resistivity of iron rises rapidly with temperature subject of course to the limitations of our knowledge of this parameter whereas that for copper rises very gently with temperature. The resistivity of aluminium is somewhat greater than that of copper and the curve for aluminium lies above that of copper. The other aspect of interest in Fresnel absorption is the dependence of the Fresnel absorption coefficient on the angle of incidence of the laser light. This quantity is nearly constant and rises rapidly at about an angle of incidence of 80° and then falls away rapidly. This ensures that photon paths at nearly grazing incidence in the keyhole are particularly important for Fresnel absorption processes. Further, these considerations suggest that the Fresnel absorption is poorest for the most highly electrically conducting metals such as copper.

3.   The problem of welding copper with a continuous CO_2 laser

These considerations explain why aluminium is much harder to weld with a continuous CO_2 laser than iron and titanium and would require a suitable beam profile for the laser. They also provide an explanation as to why it seems to be not possible to weld copper with a continuous CO_2 laser. It seems that the very weak Fresnel absorption for copper on a flat surface would make it very hard to form a keyhole in the first place. Also the size of the keyhole would appear to be so small that the power density of laser light entering the keyhole would be very low. This in itself would make it hard to sustain a keyhole and the force due to surface tension which varies inversely as the keyhole radius would tend to close the keyhole completely. The calculated keyhole profiles for steel, titanium, aluminium and copper support these considerations with steel behaving like titanium. Copper produces unrealistically narrow keyholes. These effects may also be estimated very approximately using a moving line source formula. For an absorbed power of 600 watts the behaviour of the keyhole radius for different translation speed is estimated in the table for iron, aluminium and copper.

Table 3 – Variation of keyhole radius with translation speed

	Translation speed of 20 mm/s	Translation speed of 60 mm/s	Translation speed of 100 mm/s
Iron	0.597	0.199	0.119
Aluminium	0.420	0.140	0.084
Copper	0.132	0.045	0.027

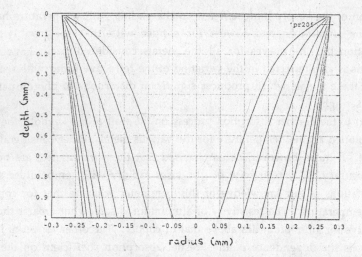

Figure 2.   Variation of the keyhole profile with laser power for steel starting from
an inner profile with a laser power of 400 watts and rising in steps of
200 watts to a laser power of 2 kW. Some of the keyhole profiles for
low laser powers are unlikely to be realised in practice.

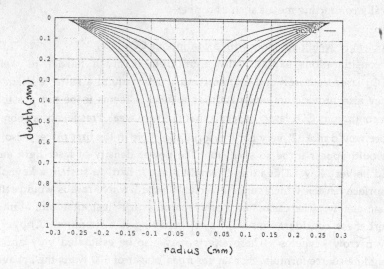

Figure 3.   Variation of the keyhole profile with laser power for copper starting
from an inner profile with a laser power of 400 watts and rising in
steps of 200 watts to a laser power of 2 kW. The keyhole profiles are
all likely to be unrealised in practice for this case of copper.

The keyhole profiles for iron and copper are displayed in Figures 2 and 3. The inner
profiles start at laser powers of 400 watts and rise in steps of 200 watts to 2 kW in each case.
The inverse bremsstrahlung contribution is included in each case. This model can predict the
threshold for the formation of the keyhole, but these considerations are not included in the

present paper. For steel some keyhole profiles will not be realised whereas for copper they are all unlikely to form.

The most favourable situation for the laser welding of copper with a continuous CO_2 laser, even if a keyhole could be formed, would be expected to occur for the lowest practical translation speeds of about 20 mm/s as shown in Figure 3.

In this Figure the keyhole profile shows a severe narrowing at the bottom of the keyhole compared to the corresponding keyhole profiles for iron and titanium. The integrated model predicts that in laser welding more power is absorbed nearer the top of the keyhole than further down. Ablative pressure effects are, therefore, largest near the top of the keyhole and smaller further down. On the other hand, the keyhole radius decreases with depth. Since the excess pressure in the keyhole is inversely proportional to the keyhole radius, it is clear that the excess pressure is largest near the bottom of the keyhole. These opposing effects generate a highly unstable situation. The radiative measurements made on the keyhole clearly point to the presence of such instabilities active even in the case of iron and titanium with a continuous CO_2 laser. In the case of copper this instability is much more severe and could well prove to be prohibitive.

4, Efficiency of laser welding

The efficiency of laser welding can be quantified by considering the power absorbed divided by the incident laser power. This is plotted against the translation speed.

Figure 4 demonstrates that both the cases of 2 kW and 3.25 kW laser powers the relationship between the absorbed laser power and the translation speed is nearly linear over the greater part of the range. The graph for a 2 kW laser power terminates at around 60 mm/s because of a failure to achieve penetration in the material. It is therefore of particular interest to

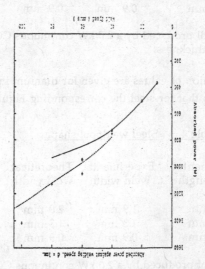

Figure 4. A graph of the absorbed laser power against weld speed to illustrate the concept of efficiency in laser welding. The graphs are for laser powers of 2 kW and 3.25 kW and the points are experimentally measured values.

consider possible ways of averting this limitation. This also suggests that the use of lasers of higher power is restricted by the absorption that can be achieved in a thin metal sheet. The use of higher laser powers, however, is likely to permit higher translation speeds, thereby assisting the efficiency of the laser welding process. This also invites the use of higher power lasers as a desirable exercise to further test and, if necessary, extend the model discussed here.

5. Shape of the weldpool

The next quantity of interest is the shape of the weldpool. Figure 5 for steel displays the theoretically calculated and experimentally measured weldpool shapes which are seen to agree extremely well. The somewhat more tapered experimental shape of the weldpool in the rear of the keyhole is likely to arise from surface convective heat losses which are not taken into account in the theory. Table 3 for the case of steel displays the variation of the shape of the weldpool with translation speed and includes a comparison of experimentally measured and mathematically calculated weldpool widths as well as corresponding keyhole radii for a 2.0 kW continuous CO_2 laser incident on a thin sheet of 1 mm thickness.

Table 4 – Steel weldpool shape

Translation speed	Experimental weld length	Experimental weld width	Theoretical weld width	Theoretical keyhole radius
10 mm/s	3.1 mm	1.6 mm	1.7 mm	0.28 mm
20 mm/s	3.0 mm	1.4 mm	1.4 mm	0.27 mm
40 mm/s	2.8 mm	0.9 mm	1.0 mm	0.24 mm
80 mm/s	3.3 mm	0.9 mm	0.9 mm	0.24 mm

Geometry of weldpool produced by a 2.0 kW continuous CO_2 laser in a thin sheet of 1 mm thickness.

A corresponding selection of values are given for titanium in Table 5. As the weldpool shape is largely analogous to that for steel the corresponding Figure is not reproduced in this paper.

Table 4 – Steel weldpool shape

Translation speed	Experimental weld length	Experimental weld width	Theoretical weld width	Theoretical keyhole radius
20 mm/s	4.3 mm	1.9 mm	2.0 mm	0.28 mm
40 mm/s	3.3 mm	1.3 mm	1.3 mm	0.21 mm
80 mm/s	2.8 mm	0.9 mm	0.9 mm	0.17 mm

Geometry of weldpool produced by a 2.0 kW continuous CO_2 laser in a thin titanium sheet of 1 mm thickness.

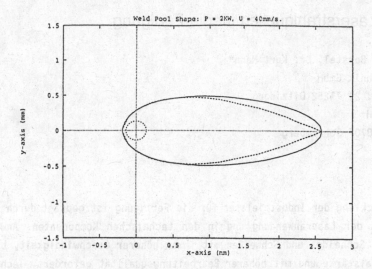

Figure 5.   The weldpool shape for a laser power of 2 kW and a translation speed of 10 mm/s. The solid curve is the theoretical weldpool shape and the broken curve has been determined by experiment. The dashed circle centered at the origin of the (x,y) co-ordinate system represents the keyhole.

These considerations demonstrate the versatility and predictive power of the integrated keyhole and weldpool model for laser welding encompassing a wide range of processing conditions and materials. The extension of these ideas offers rich promise for a variety of future developments in generating a deeper understanding of these processes – a necessary preliminary step to defining more precisely optimal conditions for laser welding and economical based practical industrial applications free from costly trial and error approaches.

References

[1]   R. Ducharme, P. Kapadia, J. Dowden, K. Williams and W.M. Steen (1992) "An integrated mathematical model of the keyhole and weldpool in the laser welding of thin metal sheets" Proceedings of the Laser Institute of America, Oct. 25-29, 1992, Vol. 75, Laser Material Processing, ICALEO (1992) pp.176-186.

Acknowledgements

We are happy to acknowledge the experimental work done by Miss K. Williams and Professor W.M. Steen (Liverpool) and the SERC of the U.K. as well as Air Products plc for funding support for the integrated model. We also acknowledge funding support from EUREKA 194 to extend the work to a range of materials.. We would also like to thank Professor D. Schuöcker (Vienna), Professor Krutz (LLT Aachen) and the organisers of the Munich conference.

Trends bei Laserstrahlquellen für die Fertigung

Dr. Michael von Borstel, Dr. Kurt Mann*
TRUMPF Lasertechnik GmbH
Johann-Maus-Str. 2, 71252 Ditzingen
*HAAS-Laser GmbH
Postfach 572, 78707 Schramberg

Die Weiterentwicklung der Industrielaser für die Fertigung ist geprägt durch ständige Innovationen in der Laseranwendung und in den technischen Komponenten. Anwendungsseitig wird das Schneiden und Schweißen mit immer höherer Geschwindigkeit, bei immer größerer Materialstärke und mit höherer Bearbeitungsqualität gefordert. Technologieseitig beeinflußt eine Reihe verbesserter und neuer Komponenten die Bauweise und Ausführung der Laseraggregate.

1. Trends bei CO_2-Lasern

Folgende Schwerpunkte kennzeichnen heute die Entwicklung von CO_2-Laserquellen:
o   Die Laserleistungen steigen nachhaltig, bei gleichzeitiger Verbesserung der Strahlqualität.
o   Die Laseraggregate werden immer kompakter, der Integrationsgrad steigt, auch der von Laser und Maschine.
o   Die Laser werden "intelligent".
Gleichzeitig konnte in den letzten Jahren das Preis-/Leistungsverhältnis der CO_2-Laser deutlich verbessert werden, und dies wird auch zukünftig ein wichtiges Entwicklungsziel sein.

1.1 Höhere Laserleistung und höhere Strahlqualität

Bei den weitaus wichtigsten Anwendungen des CO_2-Lasers, dem Schneiden und Schweißen, sind Laserleistung und Strahlqualität gleichermaßen wichtig für die erzielbaren Resultate. Die Anforderungen an beide Parameter werden auch zukünftig zunehmen.

Beim Schneiden wird die Möglichkeit zum Trennen immer dickerer Bleche, in noch besserer Qualität gefordert. Heute ist das Schneiden von 20 mm dickem Stahl auf Serienmaschinen verfügbar, und künftig können sicher 25 bis 30 mm getrennt werden.

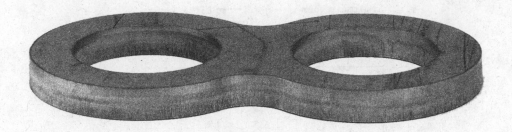

Bild 1: Hochdruck-Laserschnitt in 10 mm Edelstahl

Dabei ist neben einer hohen Laserleistung eine gute Strahlqualität entscheidend. Sie erlaubt eine optimale Anpassung der Strahlkaustik an die Schneidfront.

Auch das stark zunehmende sauerstofffreie Schneiden von Edelstahl und Aluminium, mit Einsatz von Inertgas unter erhöhtem Druck, stellt hohe Leistungsanforderungen, und auch hier ist für das erzielbare Resultat neben der Laserleistung eine hohe Strahlqualität entscheidend. So kann heute Edelstahl von bis zu 10 mm Dicke oxidfrei und ohne Gratbildung getrennt werden.

Ein besonders eindrucksvolles Beispiel für die Bedeutung der Strahlqualität stellt das Hochgeschwindigkeits-Laserschneiden dar. Bleche von 0,2 bis 0,5 mm Dicke, zum Beispiel Elektrobleche, können mit einer Geschwindigkeit von bis zu 250 m/min getrennt werden.

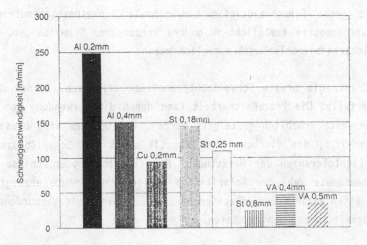

Bild 2: Hochgeschwindigkeitsschneiden mit dem CO_2-Laser
Quelle: Maschinenfabrik Georg Kreuztal/FHG-ILT

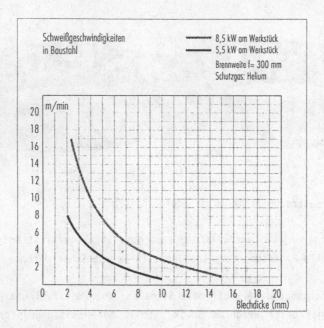

Bild 3: Schweißgeschwindigkeiten in Baustahl (TLF 12000turbo)

Auch beim <u>Laserschweißen</u> ist der Trend zu höherer Leistung unverkennbar. Seit kurzem sind industrietaugliche Laseraggregate bis ca. 12 kW verfügbar. Mit diesen Lasern werden heute Schweißgeschwindigkeiten von 6 m/min in 6 mm Baustahl erreicht.

Zukünftig werden mit Sicherheit noch höhere Laserleistungen verlangt. Eine hohe Strahlqualität ist auch hier wichtig. Sie erlaubt einen kleineren Brennfleckdurch-messer und damit große Einschweißtiefen und schmale Schweißnahtgeometrien. Die schmale Schweißnahtgeometrie ermöglicht besonders verzugsarme Schweißungen, wie sie bisher nur mit Elektronenstrahlschweißen möglich waren.

Auch für Anwendungen, die breite Schweißnahtgeometrien erfordern, bringt die hohe Strahlqualität Vorteile. Die Prozeßsicherheit kann durch die Verwendung von Optiken mit längerer Brennweite erheblich gesteigert werden, die Optiken sind besser gegen Verschmutzung geschützt, die flachere Strahlkaustik macht den Schweißprozeß unemp-findlich gegen Lagetoleranzen der Werkstücke. Neue Entwicklungsergebnisse zeigen, daß auch durch geeignete Wahl der Polarisation breite Schweißnähte erzeugt werden können, ohne auf einen kleinen Brennfleckdurchmesser und die damit verbundene höhere Schweißgeschwindigkeit verzichten zu müssen.

Beim Schweißen von Aluminium wurden in letzter Zeit durch die Kombination von hoher Strahlqualität und hoher Laserleistung bedeutende Fortschritte erzielt. Im Applika-

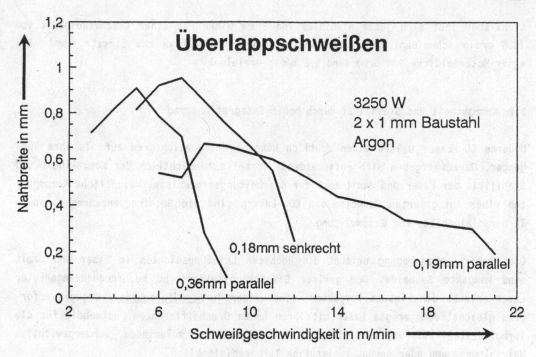

Überlappschweißen

3250 W
2 x 1 mm Baustahl
Argon

0,18mm senkrecht

0,36mm parallel

0,19mm parallel

Bild 4: Schweißnahtbreite in Abhängigkeit von der Schweißgeschwindigkeit für
verschiedene Laserstrahlparameter
Quelle: Dr. Dausinger, Universität Stuttgart IFSW

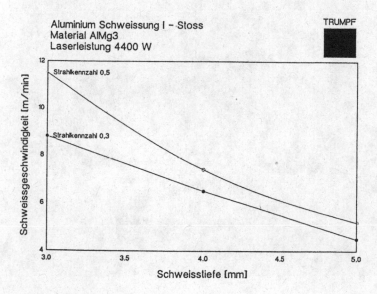

Aluminium Schweissung I – Stoss
Material AlMg3
Laserleistung 4400 W

TRUMPF

Strahlkennzahl 0,5

Strahlkennzahl 0,3

Bild 5: Schweißgeschwindigkeiten in Aluminium

tionslabor läßt sich jetzt Aluminium von 3 mm Dicke mit einer Geschwindigkeit von 11,5 m/min schweißen, wobei eine Laserleistung von 4,4 kW zum Einsatz kommt. Bei einer Materialdicke von 5 mm sind 5,2 m/min erzielbar.

1.2 Kompaktheit und Stabilität durch hohen Integrationsgrad

Moderne CO_2-Laser weisen einen deutlich höheren Integrationsgrad auf als ihre Vorgänger. Daraus ergeben sich entscheidende Vorteile hinsichtlich der Kompaktheit und Stabilität der Laser und damit auch der Bearbeitungsergebnisse. Wesentliche Komponenten einer integrierten Bauweise von CO_2-Lasern sind die Hochfrequenzanregung sowie Turboradialgebläse zur Gasumwälzung.

Die Hochfrequenzanregung erlaubt die höchsten Leistungsdichten im Laser und damit eine kompakte Bauweise. Der geringe Gasströmungswiderstand hochfrequenzangeregter Laser erlaubt den Einsatz kompakter, einstufiger Turboradialgebläse. Dagegen erfordern gleichstromangeregte Laser mit ihren hohen Druckdifferenzen, notwendig für die Turbulenzstabilisierung der Entladung, zur Gasumwälzung voluminöse, schwergewichtige Wälzkolbenpumpen oder große, zweistufige Turboradialgebläse.

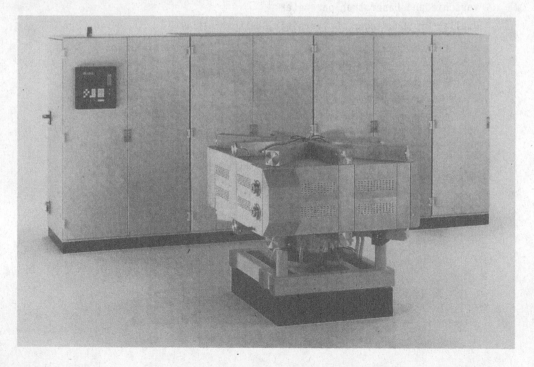

Bild 6: Schneid- und Schweißlaser mit 12 kW Strahlleistung (TRUMPF TLF 12000turbo)

Durch die Kombination dieser Elemente mit der quadratischen Faltung des Resonators erreichte TRUMPF einen besonders kompakten und stabilen Aufbau der Laser. Laser mit Leistungen bis zu 12 kW sind jetzt in dieser Technologie als Serienprodukte verfügbar.

Es zeigte sich, daß die Hochfrequenztechnologie auch und gerade bei höheren Laserleistungen entscheidende Vorteile gegenüber der Gleichstromtechnologie aufweist. Während sich die Hochfrequenzanregung problemlos im Multikilowatt-Bereich einsetzen läßt, wird es bei zunehmender Laserleistung mit der Gleichstromanregung immer schwieriger, eine stabile, kontrollierbare Entladung zu erreichen. Deshalb wird, bei Strahlleistungen von 2 kW und darüber, von nahezu allen Laserherstellern weltweit die Hochfrequenztechnologie eingesetzt. Auch bei Laserleistungen unter 2 kW setzen die Hersteller zunehmend auf die Hochfrequenztechnologie, da sie unbestritten die gleichförmigere, stabilere und besser ansteuerbare Laseranregung ist.

Ein anderer Ansatz für kompakte, integrierte Laseraggregate sind die diffusionsgekühlten oder Wand-stabilisierten CO_2-Laser. Bei diesem Lasertyp wird die im Lasergas entstehende Verlustwärme per Diffusion über die Wand des Entladungsraumes abgeführt, so daß auf eine leistungsstarke Gasumwälzung verzichtet werden kann. Preislich werden die diffusionsgekühlten Laser besonders interessant, wenn es gelingt, die Laser "sealed off" zu betreiben, d. h. mit einem abgeschlossenen Entladungsraum ohne Gasaustausch. Diese "sealed off" Bauart findet zur Zeit selbst bei diffusionsgekühlten Lasern kleiner Leistung keine Anwendung, da die Langzeit-Leistungsstabilität dieser Laser nicht ausreichend ist.

Um genügend Wärmeaustausch zu ermöglichen sind geringe Wandabstände erforderlich. Die Entladungsräume der diffusionsgekühlten Laser werden daher in Plattenform oder in koaxialer Bauweise ausgeführt. Die Entwicklungsaufgaben bestehen darin, für dieses Anregungsvolumen einen angepaßten Resonator mit gutem Wirkungsgrad, hoher Strahlkennzahl, symmetrischem Strahlprofil und akzeptabler Optikbelastung zu finden.

Da einerseits die Lösung dieser Aufgaben mit zunehmender Laserleistung immer schwieriger wird, andererseits die verlangten Laserleistungen in den letzten Jahren stetig gestiegen sind und auch weiter steigen, konnte sich das Bauprinzip des diffusionsgekühlten Lasers, obwohl seit über einem Jahrzehnt an verschiedenen Stellen verfolgt, bisher in der Praxis nicht durchsetzen.

1.3 Intelligenter Laser

Der "intelligente" Laser - die Kombination von Laser, Sensorik und Steuerungstechnik - ist von den genannten Entwicklungs-Schwerpunkten sicher derjenige, der heute noch am weitesten in den Kinderschuhen steckt, zugleich aber auch mit Sicherheit der zukunftsträchtigste. Als Stichpunkte seien genannt: geschlossene Regelkreise für Laserleistung, Strahlqualität, Strahlverlauf und Fokuslage; Laserleistungs- und Polarisationsanpassung an die Bearbeitungsgeometrie; On-Line-Pulsformung, abhängig von Bearbeitungsart, -stand und -qualität; Selbstdiagnose- und Expertensysteme.

Offene Lasersteuerungen lassen die Kommunikation mit beliebigen Sensorsystemen zu. Die Reaktionszeiten moderner Lasersteuerungen liegen im ms-Bereich. Einzelne Regelgrößen, z. B. die Hochfrequenzleistung von hochfrequenzangeregten Lasern lassen sich mit Zeitkonstanten von 1 sec regeln.

Die Schwachpunkte liegen heute eindeutig in den Bereichen der Sensorik und der adaptiven Komponenten. Ansätze gibt es viele, jedoch wenige industriereif und kostengünstig. Als Beispiele seien genannt: Strahldiagnose- und Strahllagesensoren, Nahtverfolgungssysteme, Durchstech- und Tiefensensoren, Sensoren zur On-Line-Kontrolle von Bearbeitungsergebnissen (Schnittqualität, Schweißnaht, Härte), adaptive Optiken.

2.   Trends bei Nd:YAG-Lasern

Nd:YAG-Laser unterscheiden sich von CO_2-Lasern vor allem durch ihre kürzere, im nahen Infraroten liegende Wellenlänge. Sie können dadurch Bunt- und Edelmetalle bearbeiten. Vor allem aber läßt sich ihr Licht problemlos über Glasfaserkabel leiten. Ein Handicap gegenüber den CO_2-Lasern war jedoch lange Zeit die thermische Linse, die eine schnelle Entwicklung zu höheren Leistungen verhinderte. Für Laser bis ca. 100 W konnte der Einfluß der thermischen Linse mit speziell angepaßten Resonatoren so gering gehalten werden, daß eine Auswirkung nur minimal zu bemerken war. Bei höheren Leistungen war die thermische Linse nur schwer in den Griff zu bekommen. Lösungsansätze wie Slab-Laser oder "Stand-by"-Betrieb konnten sich aus preislichen und handhabungstechnischen Gründen nicht durchsetzen. Erst mit der konsequenten Anpassung des Laserlichtkabels an die Laser waren die Auswirkungen der thermischen Linse zu eliminieren. Heute sind Nd:YAG-Laser bis 2 kW verfügbar, und dies wird sicherlich nicht die Obergrenze bleiben.

Neben der Tendenz zu höherer Leistung werden bei Nd:YAG-Laser ähnliche Trends wie bei CO_2-Lasern beobachtet:

o   Verbesserung der Strahlqualität
o   Verbesserung des Strahlverhaltens
o   cw-Betrieb
o   Diodenanregung

2.1 Auswirkungen der Strahlqualität

Die Strahlqualität ist bei der Materialbearbeitung mit Nd:YAG-Lasern ein entscheidender Parameter, der je nach Applikation höher als die Laserleistung bewertet werden muß. Versuche mit Lasern gleicher Leistung haben gezeigt, daß bei gleichen experimentellen Bedingungen eine Verdoppelung der Strahlqualität auch eine Verdoppelung der Bearbeitungsgeschwindigkeit bringen kann. Darüber hinaus liefert ein Nd:YAG-Laser mit guter Strahlqualität einen großen Arbeitsabstand und eine große Schärfentiefe. Ein großer Arbeitsabstand verhindert eine schnelle Verschmutzung der Optik, ein ganz wesentlicher Punkt z. B. beim Schweißen von verzinktem Blech. Eine große Schärfentiefe erhöht die Toleranz im Abstand zwischen Bearbeitungsoptik und Werkstück. Dies macht sich besonders bei der 3-D Bearbeitung mit Roboter positiv bemerkbar. Die zur Zeit mit einem 2 kW Nd:YAG Laser erreichte Strahlqualität liegt bei 25 μm · rad (1/4 θ · do). Die derzeit erreichbare Leistungsdichte ermöglicht selbst cw-Lasern das Schweißen von Aluminium.

2.2 Verbesserung des Strahlverhaltens

Die Strahleigenschaften, wie Intensitätsverteilung, Divergenz oder Laserleistung, sollten bei einem Laser, der in der Fertigung eingesetzt wird, unabhängig von äußeren Gegebenheiten sein. Dies verlangt eine leistungsunabhängige Strahlqualität und eine Eliminierung von Alterungseffekten, wie z. B. Lampenverschleiß.

Ein Schritt auf dem Wege zur leistungsunabhängigen Strahlqualität ist mit der Lichtleitung über Glasfaser zu machen. Aber erst wenn auch die Divergenz konstant gehalten wird, ist das Ziel einer konstanten Strahlqualität erreicht.

Die Forderung, Alterungseffekte auszugleichen oder Schwankungen in der Leistung zu kompensieren, läßt sich mit einer Laserleistungsregelung erfüllen. Die Laserleistung wird dabei permanent gemessen, mit dem Sollwert verglichen und ggf. nachgeregelt. Die Zykluszeit liegt dabei unter 0,1 ms. Die Resultate der Laserleistungsregelung sind lineare Kennlinie, exakte Reproduzierbarkeit von beliebig geformten Laserpulsen und eine im Bereich der Regelgrenze mögliche Kompensation der Lampenalterung.

2.3 cw-betriebene Nd:YAG Laser

Mit der Verfügbarkeit höherer Laserleistungen wird der cw-betriebene Nd:YAG-Laser auch für die Materialbearbeitung sehr interessant. Gegenüber dem gepulsten Nd:YAG-Laser weist er sowohl beim Schweißen als auch beim Schneiden deutlich höhere Bearbeitungsgeschwindigkeiten auf. Dabei steht weniger die Bearbeitung von dicken Materialien im Vordergrund als vielmehr die Bearbeitung von dreidimensionalen Teilen mit Materialdicken von einigen Millimetern. Denkt man z. B. an das 3D-Schneiden von Karosserieblech, so ist schon bei einem 400 W cw-Laser nicht mehr der Laser, sondern der Roboter die geschwindig-keitsbegrenzende Komponente. Mit Laserleistungen bis zu 2 kW kann der cw-Nd:YAG-Laser auch ganz beachtliche Schweißgeschwindigkeiten erreichen. Bei Einsätzen in der Automobilbranche, wo typischerweise 2 Millimeterbleche verschweißt werden, sind Schweißgeschwindigkeiten bis zu 4 m/min. möglich.

Wie beim CO_2-Laser, so geht auch beim Nd:YAG-Laser der Trend vor allem aus Preisgründen zu kompakt aufgebauten Lasergeräten mit Laser, Versorgung, Steuerung und Kühlung in einem Gehäuse. Das Laserlicht wird mit einem oder mehreren Laserlichtkabeln zum Bearbeitungsort geleitet.

Bild 7:

1 kW cw-Nd:YAG-Lasergerät mit konstanter Strahlqualität, Laserleistungsregelung und zwei 600 µm Laserlichtkabeln

2.4 Diodengepumpte Nd:YAG-Laser

Der Wirkungsgrad der lampengepumpten Festkörperlaser und die für die thermische Linse verantwortliche hohe Wärmeeinbringung lassen nach Alternativen beim optischen Pumpen suchen. Eine zur Zeit hoch gehandelte, aber noch im Entwicklungsstadium stehende Alternative ist das Pumpen des Nd:YAG-Kristalls mit Laserdiodenarrays. Diese Art der Anregung verspricht im Vergleich zur Lampenanregung eine deutliche Effizienzsteigerung, bessere Strahlqualität und eine höhere Standzeit. Außerdem kann die Baugröße des Lasergerätes reduziert werden. Der Einsatz dieses Anregungsprinzips scheitert jedoch zur Zeit noch am Preis und an der Verfügbarkeit der Diodenarrays.

Obwohl CO_2- und Nd:YAG-Laser in den letzten Jahren sowohl in Leistung, Strahlqualität als auch im Aufbau und der Handhabung deutliche Verbesserungen erfahren haben, ist auch in den kommenden Jahren mit einem hohen Grand an Innovation zu rechnen.

Measurement and Reproduction of Microgeometries with Lasers

H.K. Tönshoff, J.Mommsen, L.Overmeyer
Laser Zentrum Hannover e.V.
Hollerithallee 8, 3000 Hannover 21

1. Introduction

New developments in microelectronics and micromechanics require adequate production technologies to produce small structures and parts. UV laserbeams are well-suited to meet the requirements in absorption behaviour, spot size and thermal influence. Therefore, this paper presents a system and its working technique to rebuild microstructures based on laseroptically scanned surface data.

2. Micromachining system

At present, material processing with excimer lasers is almost restricted to applications such as drilling, cutting, material removal and surface modification of different materials /TÖNS92/. Essentially, the applied system technology consists of simple optical arrangements, fixed apertures for beam shaping, a workpiece handling system and the excimer laser.

The integration of the modules excimer laser, workpiece handling, flexible mask, energy control and beam handling system in a multiaxis CNC-controlled micromachining unit makes a new way of flexible micromachining possible which includes besides the dynamic material removal also an automatic NC-programming. Figure 1 shows the schematic set-up of a complex system for structuring technical surfaces with excimer lasers.

3. Generation of geometry data for the micro material removal

The data base for the micro material removal processes to be executed can be CAD-data or data from an optical surface measurement system, used for the material removal of geometrically undefined structures /OVER92/. The system works non-tactilly, so that damage caused by the measurement is prevented. The laser beam emitted from a laser diode is focussed at the surface of the object to be analysed. If the distance between the focusing optics and the object changes, the optical system brings the surface into focus again. A position measuring element combined with the moveable focusing lens generates an output which is proportional to the surface distance. During the

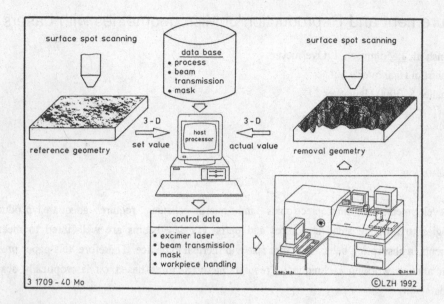

Figure 1: Control system for 3-dimensional structuring of technical surfaces with excimer-laser

measurement the object to be analyzed is moved line by line beneath the sensor. As a result, a convertable data is stored and can be computed for different analysis methods.

Due to changing qualities of the scanned surface, some data could be faulty, for example caused by an overshoot amplitude, which then have to be manually or automatically adjusted. For this purpose the measuring data are visualized as gray scale-, contour- or false color photo and filter algorithms or an interactive correction of the measured data are used. The corrected data are the three-dimensional index data for the following formation of the NC-program and the material removal process.

4. NC-programming for micromachining with excimer lasers

For a rational material removal it is necessary to combine the parameter surface position, the removal depth and the square dimension of each removal very flexibly. This means to move the workpiece in the x-y-plane parallel to the surface of the material and to vary the number of pulses, the frequence and the beam cross-section with a flexible mask.

In order to rebuild three-dimensional microstructures from data measured by a surface measurement system which includes the information position and depth, two different methods have been developed:

- Output Separated from Movement (OSM)

The three-dimenssional surface to reproduce is seperated into slices where each slice has the height of one pulse´s removal depth. Outlining, that different dimensions of the mask cross-section are used, the cavities of each slice have to be removed with a minimum of pulses. Index data is computed to remove the material layer by layer, starting with the largest mask scale and ending with the smallest one for each layer. Significant for the OSM- method is, that the workpiece is moved into the calculated position and then the pulse is fired while the workpiece stays.

- Position Synchronized Output (PSO)

With the PSO method outputs - and so material removal - is effected simultaneously to the movement of the workpiece. Because the movement is not limited to a direction which is parallel to the edges of the mask cross-section, but also diagonal, a better removal quality is achieved. This means an overlapping arrangement of material removal cross-sections. It does not result in overlapping edges of cross-sections directly removed one after another. The combination of the position of the workpiece, the output signal and the mask cross-section for the PSO method is totally different to the OSM method:
- The volume to be removed is not devided into separate layers
- Coordinates of same height values form one diagonal. Therefore, a special coordinate system was defined.
- The overlap of the removal cross-sections within one diagonal can be determined by the software. This is different to the OSM method where the overlap is only random.

The two methods have in common that cavities are filled up to the upper layer, in succession from the biggest to the smallest mask removal surface.

5. Machining of complex microstructures

Micromachining with excimer lasers enables the reproduction of complex structures in many materials with a high precision. But often it is necessary to structure large areas like for example embossed rollers. At the Laser Zentrum Hannover e.V. a multiaxis CNC-controlled micromachining unit (ELPEC-μ) was constructed which makes complex surface treatments possible.

One example for three-dimensional micromachining with excimer lasers is shown in fig. 2. The shown surface was scanned optically and rebuilt in the polymere material Araldit using the PSO method. The smooth grid structure that can be seen has to be improved. To avoid these structures the mathematic algorithms for index data generation and the overlapping factor have to be optimized.

444

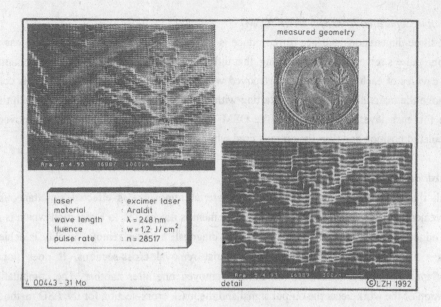

Figure 2: Material removal of a 3d-microstructure

6. References

[HERZ91]   Herziger, G.; Wester, R.: Abtragen mit Laserstrahlung; Laser und
            Optoelektronik; 23(4) (1991), pp. 64-69

[OVER92]   Overmeyer, L.; Dickmann, K.: Dynamischer Autofokussensor zur
            dreidimensionalen Mikrostrukturerfassung; Technisches Messen TM, 1/1992

[TÖNS92]   Tönshoff, H.K.; Mommsen, J.: Process of   generating three-dimen-
            sional microstructures with excimer lasers. Proceedings of Eclat
            '92,Göttigen, 4th. Europ. Conf. on Laser Treatment of Materials

Mikrobearbeitung von Silizium

M.Alavi, A. Schumacher, H.-J. Wagner

Institut für Mikro- und Informationstechnik der
Hahn-Schickard-Gesellschaft für angewandte Forschung e.V.
D-78052 Villingen-Schwenningen, F.R.G.

Abstract

Mikromechanische Systeme bestehen häufig aus dreidimensionalen mechanischen Strukturen in Silizium, deren Formen und Abmessungen zumindest in einer Dimension so klein sind, daß feinmechanische Formgebungsverfahren nicht sinnvoll eingesetzt werden können. Zur Herstellung derartiger Strukturen werden in der Regel photolithographische Verfahren und anisotropes Ätzen eingesetzt. Form und Abmessungen der Mikrostrukturen werden hierbei durch Maskierschichten auf der Waferoberfläche festgelegt. Die maximal realisierbare Strukturtiefe und damit das Aspektverhältnis werden jedoch im wesentlichen durch die kristallographische Orientierung des Siliziumsubstrats bestimmt. Die in $<110>$-Silizium herstellbaren V-Gräben sind beispielsweise durch ein Aspektverhältnis von 0.35 festgelegt. Dieses Verhältnis ist durch die Kristallstruktur von Silizium vorgegeben und kann bei konventionellen Verfahren durch die Prozeßparameter nicht beeinflußt werden. In dieser Arbeit wird ein neuartiges Verfahren vorgestellt, das durch Kombination der Laser-Mikrobearbeitung (Nd:YAG-Laser) eines photolithographisch strukturierten Siliziumwafers mit nachfolgendem anisotropen Ätzen, die Herstellung einer Vielfalt neuer Strukturen mit definierten Berandungsflächen und variierbaren Aspektverhältnissen ermöglicht. Durch lokale, definierte Zerstörung der ätzresistenten {111}-Ebenen innerhalb der mit Laserstrahlung bearbeiteten Zone und nachfolgendes anisotropes Ätzen können vertikal zur Oberfläche angeordnete Mikrokanäle in $<110>$-Silizium und teilweise geschlossene lateral zur Oberfläche angeordnete Mikrokanäle in $<110>$- und $<111>$-orientiertem Silizium hergestellt werden. Diese Basisstrukturen finden Anwendung bei der Realisierung von z.B. Mikrowellen-Bandpaßfiltern, Transportsystemen für Gase und Flüssigkeiten zur Chipkühlung bzw. biochemischen Analytik, Haltevorrichtungen für mikrooptische Komponenten und Mikroschlitzen zur elektrischen und optischen Kontaktierung von Sensorarrays. Mit diesem neuartigen Fertigungsverfahren Laser-Mikrobearbeitung/Anisotropes Ätzen können ebenfalls Balkenstrukturen ohne Verwendung von Ätzstoppschichten realisiert werden, die als Resonatorelemente für frequenzanaloge Kraftsensoren oder in Verbindung mit einer monolithisch integrierten Membranstruktur als Basiselement eines mikromechanischen Drucksensors dienen.

Absorption Behaviour of Solder Pastes in Laser Soldering

Christoph HAMANN[☆], Hermann KEHRER[*], Hans-Georg ROSEN[☆], Clemens SCHERER[☆]

☆   Siemens AG, Otto-Hahn-Ring 6, D - 8000 Munich 83

*   BLV Licht- und Vakuumtechnik GmbH, Münchner Str. 10, D - 8019 Steinhöring near Munich

Abstract

The aim of these investigations is to examine various solder pastes for their absorption behaviour with respect to laser beam of wavelength of 1.06 µm. It is intended to varify the importance of absorption of flux over metallic composition and reactivity of flux.

Because absorbed energy cannot be measured directly the signals corresponding to the fractures of laser beam transmitted through and reflected from a specimen are measured simultaneously in a calibrated setup to gain information on the absorption behaviour of solder pastes. Four different fluxes and three different metallic compositions were investigated. The pastes and the fluxes were spread on glass plates to allow for transmission measurements. In addition, temperature of paste on board was measured with a thermocouple attached to the board to show that results of the rather theoretic setup of the absorption measurements can be transfered to real printed wire boards. Best coupling of energy was achieved with solder pastes containing a flux with an absorption coefficient of about $\tau = 7 \cdot 10^{-3}$ μm^{-1}.

Introduction

Laser beam soldering is a single step procedure in which one solder joint is made after the other /Hamann/. In order to be economically competitive laser beam soldering has to be fast. This involves the problem of heating up a mixture of organic compounds - the flux - having a evapouration temperature below 200°C and metallic particles generally melting above 200 °C with a high intensity energy beam. Additionally, the organic component of the solder paste has a higher absorption coefficient than the metal /Greenstein/.

The absorption of the solder pastes is calculated by measuring the laser beam power and subtracting the reflected and the transmitted power signals. The experimental set-up is shown in fig. 1. The solder paste specimens are irradiated by a Nd:YAG laser. The resulting test signals are recorded by fast-response photodiodes. The photodiodes (1 to 4) are located at the focal point of collimating lenses and at the focal point of a semi-ellipsoid (8), respectively. The test signals are recorded by a four-channel digital transient recorder (10) and transferred directly to a PC (11) for subsequent processing. The paste specimens (9) are aligned by means of a He-Ne laser (6) and a video camera (5). As a result of the geometry of the ellipsoid, the reflected signal is measured at two points:

- the diffusely reflected component is measured at the outer focal point of the ellipsoid (2),
- the directly reflected component which leaves the ellipsoid through the inlet aperture is separated from the beam path with the aid of a beam splitter and measured with a second diode (1).

Four series of experiments were conducted using the fluxes and pastes of tab. 1. The measurement results were assessed with regard to the following points:

- power densities for examining the absorption of fluxes were selected such that no visible damage could be detected in the flux material,
- soldering was considered to be completed when a single solder ball was formed from the melt.
- in comparing the absorption coefficients, their values were taken at the same pulse energy at different power densities,

The first measurements gave the absorption of the flux alone, see fig. 2. The absorption coefficient τ is derived from the equation

$$I = I_0 * e^{-\tau d}$$

The individual values are displayed in tab. 2. In the second series, different alloys using the same flux were examined. The other two series of measurements allow a comparison to be made of the behaviour of particular pastes using different types of flux material. The solder pastes were tested as soon as the paste specimens with a size of 2×2 mm² were printed onto a glass plate by means of a stencil.

1st series of measurements: Fluxes

F-SW 26/602
F-SW 26/603
F-SW 32/308
F-SW 33/018

2nd series of measurements:

Sn59 Pb35.5 Bi3.9 In1.6	F-SW 26/602
Sn62 Pb36 Ag2	F-SW 26/602
Sn63 Pb37	F-SW 26/602

3rd series of measurements:

Sn59 Pb35.5 Bi3.9 In1.6	F-SW 26/602
Sn59 Pb35.5 Bi3.9 In1.6	F-SW 32/308

4th series of measurements:

Sn62 Pb36 Ag2	F-SW 26/602
Sn62 Pb36 Ag2	F-SW 26/603
Sn62 Pb36 Ag2	F-SW 32/308
Sn62 Pb36 Ag2	F-SW 33/018

Tab. 1: Solder pastes and fluxes investigated (provided by Metallwerke Goslar)

Fluxes	F-SW26/602	F-SW26/603	F-SW32/308	F-SW33/018
$\tau[10^{-3}*\mu m^{-1}]$	$9.5 \pm 5\%$	$7.7 \pm 4\%$	$7.1 \pm 4\%$	$3.2 \pm 5\%$

Tab. 2: Absorption coefficient of different fluxes

The time of solder ball formation can be observed from the transmission signal. At the beginning of processing, the transmission through the 100 µm thick paste specimens is still very low. The mean thickness of the balls in the untreated paste is 40 to 50 µm, i.e. two to three layers of solder balls lie on top of each other. The transmitted signal is therefore strongly divergent due to multiple reflections. A strong rise in the transmission signal indicates that the solder balls coalesce to form a bigger single solder ball. The steeper the rise in the signal, the faster is the formation of the solder ball.

The diffuse reflected signal which was measured at the second focus of the ellipsoid provides most information on the melting onset of solder pastes. The reflected laser power is weak at the beginning but increases as soon as the paste starts to melt. The solder balls in the paste coalesce and form a molten surface. At this time, the diffuse reflected laser power is at its maximum. A continued irradiation with the laser leads to the slow formation of a single ball due to the surface tension of the melt. This process can be observed very effectively with the diffusely reflected signal so that this signal was primarily examined for comparing the melting of different solder pastes.

For the comparative examination of the pastes, two significant points were determined in the time curve of the diffuse reflected signal. The first characteristic melting point is that at which the diffuse reflected signal begins to rise strongly. This point, which has the largest curvature on the rising edge of the curve, is designated as t_{c1} and indicates the onset of melting. The second characteristic point t_{c2} is that with the largest curvature on the trailing part of the curve after the maximum of the diffuse reflected signal. These two points represent the time characteristic of the paste from the onset of melting through to the formation of a single ball.

In addition, temperature of paste on board was measured with a thermocouple attached to the board under the point of interaction to show that results of the rather theoretic setup of the absorption measurements can be transfered to real printed wire boards.

Results

The comparison of pastes of different alloys using the same flux showed small differences only. The time at which solder balls were formed was approximately the same, merely the absorbed energy differed slightly. It was shown that this can be attributed to the different reflection coefficients of the solder alloys, as could be seen from the amplitude of the diffuse reflected laser power.

Two types of alloys were used in the comparison of pastes with different fluxes. It was established that in this case the chemical reactivity and absorption of the flux are the only key factors determining the formation of solder balls whereas the reflection coefficient of the alloys plays a subordinate role. As an example, fig. 3 shows the absorption of SnPbAg2 with different fluxes.

The importance of chemical reactivity can be particularly clearly seen from the melting behaviour of the solder Sn59Pb35.5Bi3.9In1.6. The low melting point BiIn-alloy was added to increase heat conduction within the paste. Whilst its combination with flux F-SW26/602 produced the same result as most of the other pastes, it showed clear differences when combined with flux F-SW32/308. When combining this Bi-In alloy with flux F-SW32/308 (halogen-free activators), the formation of a single solder ball takes even longer than for solders without that alloy. In this combination, individual solder balls are formed first which then coalesce to a single ball after a relatively long time.

A similar effect was observed for the alloy Sn62Pb36Ag2. The combination of this alloy with flux F-SW33/018 (halogen-free activators) required a very long time for forming a solder ball, see fig. 4, even longer than the previous combination. An additional factor in this case is that flux F-SW33/018 has the lowest absorption coefficient of all the fluxes investigated. It can thus be seen that in addition to chemical reactivity the absorption of the flux is another important factor in solder ball formation.

Furthermore it was found that the main factor determining the onset of melting is the laser intensity. The time of solder ball formation, on the other hand, depends strongly on the chemical reactions and on the absorption coefficient of the flux. Differing soldering results due to the laser parameters used could only be found with compositions containing halogen-free flux. Paste SnPbAg2 in combination with flux F-SW33/018 showed shorter times for solder ball formation with increasing laser intensity, see fig. 5. A behaviour opposite to this was observed for paste SnPbAg2 with flux F-SW32/308. Pastes containing organic activators and without halogen additions generally require longer times for solder ball formation. It thus follows that the chemical components contained in the flux, especially the activators, play an important role in the formation of solder balls.

It was generally observed that the soldering of pastes with flux containing halogen is not critical under the present experimental conditions. The results of soldering for these pastes as well as the values determined for the characteristic times and the absorption coefficients showed practically no dependence on the laser parameters.

Despite the low absorption of paste Sn62PbAg2 with F-SW33/018, the temperature was as high as for paste with the F-SW26/603 which absorbs twice as much. This may be due to absorption of the underlying solder of the pad, which absorbs transmitted laser power. On the other hand, paste F-SW26/602, which has the highest absorption, showed the lowest on-pad temperature. The reason is that the laser power is absorbed in the top layer of the paste, and the temperature in the pad rises due to thermal conduction of the paste, which is lower than that of the metal.

The highest temperature, indicating highest energy absorption, is exhibited by pastes that have an absorption of about $7 \times 10^{-3} \ \mu m^{-1}$.

This work was conducted under BRITE / EURAM 3341.

Literature

Chr. Hamann, H.-G. Rosen, C. Scherer, Laserlöten mit dem CO_2- und Nd-Laser, Proc. Laser '87, p. 553
M.Greenstein, Optical absorption aspects of laser soldering for high density interconnects, Applied Optics, 1 Nov. '89, Vol. 28, No. 21, p.4595

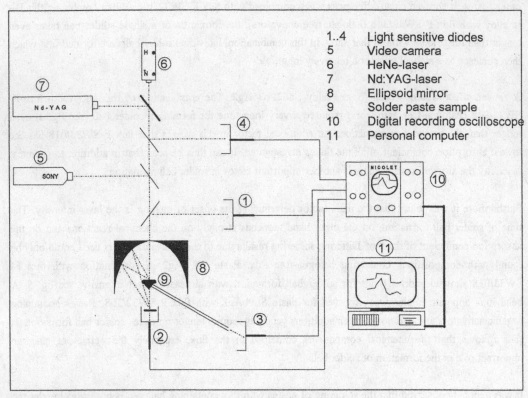

1..4	Light sensitive diodes
5	Video camera
6	HeNe-laser
7	Nd:YAG-laser
8	Ellipsoid mirror
9	Solder paste sample
10	Digital recording oscilloscope
11	Personal computer

Fig.1: Experimental setup for determination of absorption of solder pastes

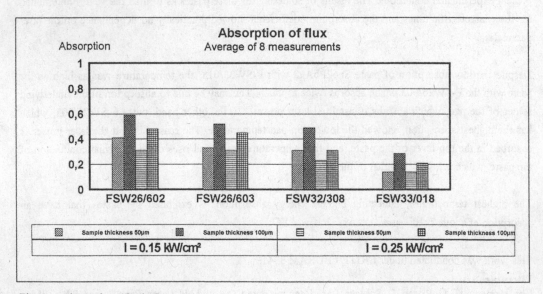

Fig. 2: Absorption of sole flux

451

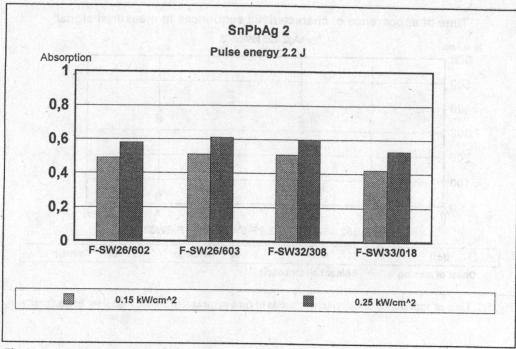

Fig. 3: Absorption of solder pastes (same alloy, different flux) and equal pulse energy

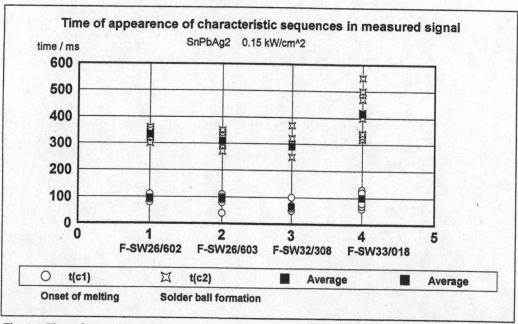

Fig. 4: Time of appearance of characteristic points of time in measured signal; low laser beam density

452

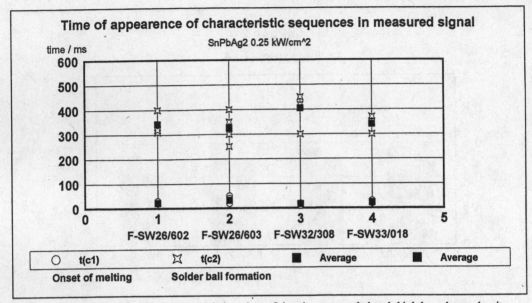

Fig. 5: Time of appearance of characteristic points of time in measured signal; high laser beam density

Fertigung von Medikamentenpumpen mittels Lasertechnik

J. Eggersglüß, G. Lensch
NU-TECH GmbH
Ilsahl 5, D-24536 Neumünster

Die Neuentwicklung einer implantierbaren Infusionspumpe aus dem Werkstoff Titan forderte präzise Füge- und Bohrverfahren. Durch Musterschweißungen und Absprachen in der Entwicklungsphase ist es zur lasergerechten Konstruktion vieler Baugruppen der Pumpe gekommen. Zur Zeit werden vier Laserbearbeitungsverfahren zur Fertigung der Pumpe eingesetzt.

CO_2-Laser Schneidtechnik wird an ebenen Blechteilen und an Tiefziehteilen eingesetzt, wobei ein Sensorschneidkopf zwingend ist, um Toleranzen der Tiefziehteile auszugleichen. Nachträgliches Einschweißen von Bauteilen entlang der geschnittenen Bohrungen fordert eine hohe Schnittkantenqualität.

Drei verschiedene Filterstützsiebe werden aus Titanblech mittels Laserbohr- und Laserschneidtechnik hergestellt. Max. 1.800 Bohrungen mit dem Durchmesser 0,12 mm werden pro Sieb auf einer Ringfläche eingebracht. Für die Bohrbearbeitung wird ein 35W Nd-YAG-Laser der Fa. Lumonics mit sehr guter Strahlqualität eingesetzt. Bei vorgegebenem Bohrbild wird eine Bohrgeschwindigkeit von 4,5 Bohrungen pro Sekunde erreicht.

Um Qualitätsrichtlinien zu erfüllen und die Handhabung der Pumpe zu erleichtern, werden einige Bauteile der Pumpe laserbeschriftet.

Den größten Anteil der Laserbearbeitung nimmt das Laserschweißen ein. 16 verschiedene Bauteile mit insgesamt 30 Nähten werden pro Pumpe geschweißt. Die Anforderungen an die einzelnen Schweißnähte sind sehr unterschiedlich. Es geht von der reinen Festigkeit über Flüssigkeitsdicht, Schweißen unter Reinraumbedingungen bis zur Dichtigkeitsforderung von $5 \cdot 10^{-10}$ mbar·L/s, nachgewiesen im Heliumlecktest.

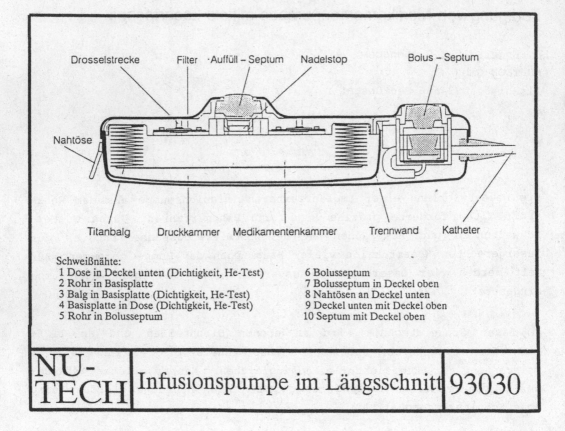

Schweißnähte:
1 Dose in Deckel unten (Dichtigkeit, He-Test)   6 Bolusseptum
2 Rohr in Basisplatte   7 Bolusseptum in Deckel oben
3 Balg in Basisplatte (Dichtigkeit, He-Test)   8 Nahtösen an Deckel unten
4 Basisplatte in Dose (Dichtigkeit, He-Test)   9 Deckel unten mit Deckel oben
5 Rohr in Bolusseptum   10 Septum mit Deckel oben

NU-TECH	Infusionspumpe im Längsschnitt	93030

Die Schweißnähte werden an 3 Stationen mit einem 100W Nd-YAG Laser der Fa. Haas ausgeführt. Die erste Station ist eine 4-Achs-Anlage mit der die meisten Schweißnähte bearbeitet werden. Die Medikamentenkammer wird mit einer Schweißnaht an der zweiten Station verschlossen. Es wird in einer Handschuh-Box unter definierter Heliumatmosphäre geschweißt. Hier kommt Lichtleitertechnik mit TV-Beobachtung der Schweißstelle zum Einsatz.

Die dritte Station, an der Schweißaufgaben erledigt werden, ist eine Strahlablenkeinheit, die in einer Laminar Flow-Box angebracht ist, um das Bauteil vor Schmutz und Staub zu schützen. An dieser Station werden auch die Beschriftungsaufgaben ausgeführt.

Probleme bei den Schweißaufgaben bilden die Tiefziehteile mit ihren Toleranzen, die teilweise zu große Schweißspalten ergeben. Eine weitere Schwierigkeit bilden kleine Bauteile, die wärmeempflindlich

sind, weil sie Kunststoffteile enthalten, die in der Nähe der Schweißnähte liegen. Bisher gibt es noch eine Freihandschweißung der Nahtösen an der Außenseite der Pumpe. Dieses Teil soll jetzt so geändert werden, daß eine halbautomatische Schweißung in einer Vorrichtung erfolgen kann.
Die Zuführung der Bauteile in die Schweißvorrichtungen erfolgt in allen Fällen von Hand.

Bei Stückzahlen von ca. 1.000 Pumpen pro Jahr rentiert sich eine vollautomatische Schweißung mit automatischer Zuführung der Bauteile nicht. Konventionelle Schweißverfahren wie WIG oder auch Mikroplasma hätten bei der Mehrzahl der Schweißnähte Schwierigkeiten, diese herzustellen.

Die Lasertechnik bietet durch ihren flexiblen Einsatz ein wertvolles Fertigungsverfahren zur Herstellung der Infusionspumpen.

Adhesion Pretreatment of Polyolefin with Nd:YAG Laser

G. Habenicht
Lehrstuhl für Fügetechnik
Technische Universität München

S. Tsuno
SUNSTAR
Engineering INC.
Osaka/Japan

We have developed a new technique of surface treatment
using an infrared laser. The treatment is very precise and
is adaptable for automation. Our important conclusion is
that this treatment offers not only high potential adhesion
but it also points to possible new kinds of treatment.
With these treatments we can develop new applications for
bonding polyolefins.
Current treatment of polyolefins can be classified mainly
in two ways. One way is the introduction of functional
groups in the surface. The other way is the improving of
the adhesion properties by application of a primer as a
special coating. But the adhesion enhancement is limited.
We present a third way. The primer is coated on the carbon
black-pigmented polyolefin, and the primer is radiated by
the YAG-laser. Then the beam is absorbed on the boundary,
and the primer and polyolefin are "welded". This generates
the strong bonding. The waves of the YAG-laser are not
absorbed by the primer. Therefore the YAG-laser is well
suited for a primer/laser treatment.
In order to make the treatment effective, the suitable
primer "PP2" has been selected. The primer "PP2" has a
polyolefin chain and reactive functional groups. With the
primer/laser treatment, shear strength values up to 5 N/mm2
were achieved on polyethylene using 1-comp. Polyurethane
adhesive.

Laser Micro-Welding and Micro-Melting for Connection of Optoelectronic Micro-Components

M.Becker,R.Güther,R.Staske, R.Olschewsky,

Ferdinand-Braun-Institut für Höchstfrequenztechnik im Forschungsverbund Berlin

Rudower Chaussee 5, D-12489 Berlin-Adlershof, FRG

H.Gruhl and H.Richter

Forschungsinstitut der Deutschen Bundespost TELEKOM

Am Kavalleriesand 3, D-64295 Darmstadt, FRG

Abstract: Laser-melting and laser-welding as micromachining methods promise for microoptics high precision connection with long time stability without post correction.

Introduction: In this contribution we will describe a possibility of high precision fastening of submounts, fibres and chips of optoelectronics by laser-microwelding without alignment post-correction. With help of the impulse-solidstate laser (λ= 1.06 μm) serveral materials can be welded or melted. In respect with the solution of handling and packaging problems the mechanical dimension of submounts is designed this high as necessary, thus low as possible. The small geometrical dimensions of submounts support the long term stability, in respect with offset, and low cost assembly. Laser microwelding with inclusion of vacuum-welding is widely known (DORN et al., BÜTTGENBACH , GRUHL and BECKER, GÜTHER et al.). In this paper we discuss how to combine fastness with offset-precision smaller than 0.5 μm . This is required for fastening of single mode components of optoelectronics.Essential parameters we used for handling and assembling are listed in the following table:

parameter of	feature
laser	- pulsed solid state laser (λ_{Nd} =1.06μm) - simultaneous performance all of the welding points
micromachining	- beam transmission via optical fibre (core diameter < 200 μm) - diameter of focus (< 100 μm) - angle of incidence of light (~ 45°)
work piece	- diameter of fibre submounting (< 1 mm) - geometry of slit (fastening groove < 20μm) - surface quality (flatness < 5 μm) - homogeneous structure of material without imperfections
material	copper,kovar,silicon,GaAs,glass fibre

In the next table we show a matrix with materials which were sucessfully connected by laser welding (x) or laser melting ([x]) in this work:

	copper	kovar	silicon	GaAs	glass fibre
copper	x	x			[x]
kovar		x	x		[x]
silicon			x	x	x
GaAs				x	
glass fibre					

In Fig. 1 we show different steps of fastening by laser welding based on Si. Figs. 2-4 deal with a new vaccum laser welding device. The comparison between air welding and vacuum welding is given. Melting quality for polished and etched Si- and GaAs-surfaces is tested in Figs 5 and 6 and Fig. 4 shows the generation of double focus.

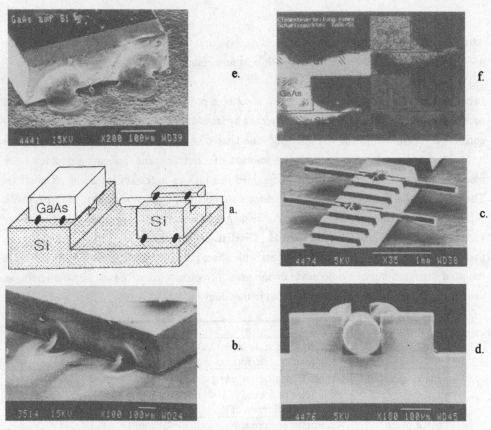

Fig. 1   Steps for Si-based fastening of micro-optical components by laser welding (Nd):

    a.   Scheme of a chip-fibre coupling

    b.   Si-Si-welding by double focus

    c.   Fibres on a grooved Si-submount (submount: VOGES,KLEIN,Uni Dortmund)

    d.   Front view of c.

    e.   GaAs-chip on Si-submount

    f.   Cut of a welding point of e.: Element analysis:on the top:left:Si;right:Ga,

                at the bottom: left:Si and GaAs;right:As.

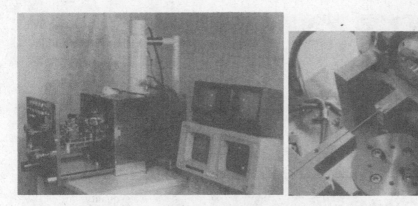

Fig. 2   Device for vacuum laser welding with REM-and light observation. At right: detail.

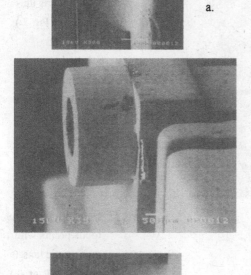

a.

b.

c.

Fig. 4   Laser welding of
two kovar sub-
mounts in air
(same position
as in Fig. 3 b)

Fig. 3   Vacuum laser welding of two kovar submounts
(submounts: OTT,WAGNER,FhG/IPM,Freiburg)
a. Upper welding point of b.
b. Two welding points can be seen in the angle between cyclindrical submount
and a rectangular submount surface.
c. Lower welding point of b.

460

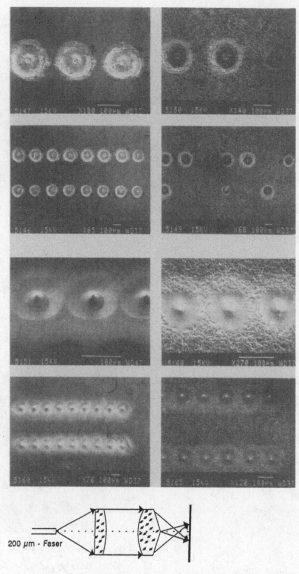

Fig. 5 GaAs-melting with
double focus illu-
mination (Fig.7):
- on the left:
polished surface:
constant quality
- on the right:
etched surface:
variations of
quality

Fig. 6 Si-melting with
double focus illu-
mination (Fig.7):
- on the left:
polished surface:
constant quality
- on the right:
etched surface:
constant quality

200 µm - Faser

Fig. 7 Generation of two
focus points with
a prismatic lens
(GÜTHER et al.)

For fruitful cooperation we thank to Prof.Wagner,Mr.Ott, FhG/IPM Freiburg, Prof.Voges, Dr.Klein, Uni Dortmund,Dipl.-Phys.Olivier,TELEKOM,Berlin and Dipl.-Metall. Klein,FBH.

References:
[1] DORN,L.,GRUTZECK,H.,JAFARI,S.,'Schweißen und Löten mit
        Festkörperlasern',Springer,Berlin,1992
[2] BÜTTGENBACH,S.,'Mikrofügen mit Laserstrahl',in: 1.Fachgespräch AVT-FIOS,
        Nov.1989,VDI/VDE Technologiezentrum,Berlin
[3] GRUHL,H.,BECKER,M.,'Verfahren zum direkten Fixieren einer Lichtleitfaser auf einem
        Faserhalter mittels Laserschweißen',Patent appl. P 41 40 283.9 (1991)
[4] GÜTHER,R.,RICHTER,H.,GRUHL,H.,'Strahlaufspaltungs- und Fokussierungs-
        optik',Patent appl. P 42 30 224.2(1992)

Herstellung hartmetallähnlicher Verschleißschutzschichten durch Laserauftragsschweißen von Hartstoff-Hartlegierungs-Pulvergemischen

St. Nowotny, A. Techel, A. Luft, W. Reitzenstein

FhE-Institut für Werkstoffphysik und Schichttechnologie

Helmholtzstraße 20, 8027 Dresden

Einführung

Für den Verschleißschutz abrasiv und erosiv belasteter Bauteile des Motoren- und Triebwerksbaus sowie der Umformtechnik und Werkzeugindustrie werden häufig Hartmetalle in Form von aufgelöteten gesinterten Formkörpern oder thermisch gespritzten Schichten eingesetzt. Bedingt durch den Herstellungsprozeß sind die auf diese Art verarbeiteten Hartstoffe spröd und porös und weisen nur eine begrenzte Haftfestigkeit zum Substrat auf. Es besteht somit Bedarf an hartmetallähnlichen Schutzschichten, die neben einer ausreichenden Verschleißfestigkeit über eine höhere Duktilität und größere Schichtfestigkeit sowie eine metallurgische Bindung zum Substrat verfügen.

Für das lokale Beschichten von metallischen Bauteilen mit hartmetallähnlichen Werkstoffen ist das Laserauftragschweißen sehr gut geeignet, da die Verwendung von Pulver als Zusatzwerkstoff nahezu beliebige Mischungen aus Hartstoffen und metallischer Matrix ermöglicht und die Hartstoffauflösung in der Schmelze über die Variation der Schmelzbadtemperatur und -lebensdauer beeinflußt werden kann. Zur Erzielung von riß- und porenfreien Beschichtungen mit einem hohen Hartstoffanteil ist jedoch eine Optimierung des Verfahrens und der Werkstoffe vorzunehmen.

Lösungsweg

Das Beschichten erfolgte mit einem CO_2-Laser vom Typ Rofin Sinar RS 6000 bei einer Leistung zwischen 3,0-4,2kW und defokussiertem Laserstrahl auf Vergütungsstahl. Vorversuche mit verschiedenen Hartstoff-Hartlegierungsverbunden haben ergeben, daß sich mit Kobalt umhülltem Wolframkarbid (Amperit 522.2; H.C. Starck Berlin) in Verbindung mit einer duktilen, borarmen Nickelhartlegierung

(NiCrobor 20; VAUTID-Verschleißtechnik Stuttgart) gute Ergebnisse hinsichtlich Verarbeitbarkeit, Rißfreiheit und Verschleißfestigkeit erzielen lassen /1/. Es wurde agglomeriertes Hartstoffpulver der Fraktion 45-90μm mit einem mittleren Durchmesser von 75μm verwendet. Die Teilchen haben eine unregelmäßige, blockige Form und bestehen aus mehreren WC-Kristallen. Eine vollständige Umhüllung der WC-Agglomerate mit Kobalt ist nicht immer gewährleistet. Die Nickelhartlegierung lag als gasverdüstes Pulver mit kugeliger Teilchenform und einem mittleren Teilchendurchmesser von 85μm vor (Abb.1). Das Mischen der Pulver erfolgte unmittelbar vor der Bearbeitungsstelle mit Hilfe eines speziell für die Lasermaterialbearbeitung mit kontinuierlicher Pulverzufuhr entwickelten Pulvermischkopfes. Mit diesem Pulvermischkopf kann die Gefahr der Entmischung von Pulvern unterschiedlicher Dichte ausgeschlossen und die überschüssige Fördergasmenge abgeschieden werden. Hartstoffverteilung, Gefüge und Erstarrungsstruktur der erzeugten Schichten wurden metallographisch und elektronenmikroskopisch in Verbindung mit EDX-Mikroanalyse

Abb. 1: Elektronenmikroskopische Aufnahme des verwendeten Pulvergemisches WC/Co-NiCrB

untersucht. Zur Charakterisierung des abrasiven Verschleißverhaltens der Schichten wurden Verschleißversuche sowohl mit gebundenem Abrasivkorn (Schleifpapier, SiC, mittlere Korngröße 70μm) als auch mit losem Abrasivkorn (SiC, mittlere Korngröße 115-320μm) gegen Stahl 100Cr6 durchgeführt.

Ergebnisse

Die in einem bestimmten Parameterfeld erzielten Beschichtungen mit Hartstoffgehalten bis zu etwa 50Vol% waren durchweg riß- und nahezu porenfrei (Abb. 2). Die Erfahrungen zeigen, daß rißfreie und porenarme Schweißraupen mit Hartstoffgehalten über 50Vol% nur schwer realisierbar sind /2/. Abb. 3 zeigt als Beispiel das Gefüge einer Beschichtung, die durch einen hohen Hartstoffgehalt, eine homogene Hartstoffverteilung und eine geringe Hartstoffauflösung gekennzeichnet ist.

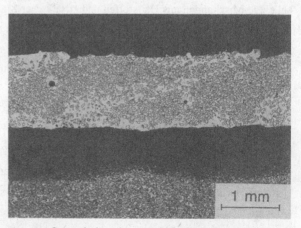

Abb. 2: Querschnitt einer WC/Co-NiCrB-Beschichtung (Spurüberlappung 43%)

Die agglomerierten Hartstoffteilchen sind fest in die Nickelmatrix eingebettet und gleichmäßig verteilt. Die Metallmatrix ist dendritisch erstarrt. Das Kobalt wird vollständig vom Nickelmischkristall aufgenommen, daneben entsteht ein Eutektikum aus Nickel und einer borreichen Phase, vermutlich Ni_3B, wodurch die Härte der Matrix gesteigert wird.

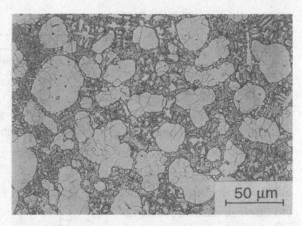

Abb. 3: Mikrostruktur der WC/Co-NiCrB-Beschichtung

Die Mikro- sowie die quantitative Gefüge-analyse ergaben, daß man bei den mit niedriger Intensität ($<1,2x10^4W/cm^2$) erzeugten Beschichtungen von einem nahezu vollständigen Erhalt der WC-Teilchen ausgehen kann. Mit zunehmender Intensität ist eine geringe Auflösung der Hartstoffteilchen zu verzeichnen. Demzufolge erscheint es nicht zweckmäßig, den Hartstoffgehalt im Pulver wesentlich über 50Vol% zu erhöhen, da die zum Beschichten der hartstoffreicheren Pulver notwendige höhere Intensität die Auflösung der Hartstoffpartikel in der Schmelze forciert und somit einer Steigerung des Hartstoffgehaltes in der Schicht entgegenwirkt. Im Intensitätsbereich von 1,0-$1,5x10^4W/cm^2$ ist der Anteil an wiederausgeschiedenen WC-, W_2C- und spröden Mischkarbidphasen jedoch vernachlässigbar, wodurch sich die geringe Rißneigung erklärt.

Die Ergebnisse der Verschleißuntersuchungen zeigen, daß für eine beanspruchungsgerechte Bauteil-beschichtung die wirksame Abrasivkorngröße berücksichtigt werden muß. Im Schleifpapiertest (Abb.4) wurde der höchste Verschleißwiderstand an der Schicht gemessen, die mit einer Intensität von $1,2x10^4W/cm^2$ aufgetragen wurde und einen Hartstoffgehalt von 50Vol% aufwies. Aus dem Vergleich der unterschiedlichen Beschichtungen geht hervor, daß bereits ein Hartstoffanteil von etwa 50Vol% ausreicht, um ähnliche Verschleißeigenschaften wie ge-

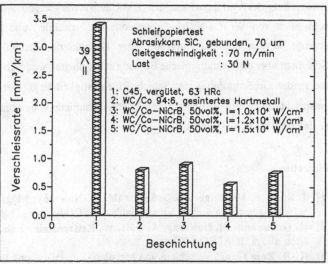

Abb. 4: Verschleißraten der WC/Co-NiCrB-Beschichtungen im Vergleich zu gesintertem WC/Co-Hartmetall und vergütetem Stahl

464

sintertes WC/Co-Hartmetall zu erzielen.

Unter den Bedingungen grobkörnigen Abrasivverschleißes mit losem Abrasivkorn und einem metallischen Gegenkörper ist der Volumenabtrag der Verbundschicht deutlich größer als der der reinen Hartlegierungsschicht (Abb.5). Diese Tatsache ist offensichtlich durch den Effekt zu erklären, daß der Angriff der wesentlich größeren Abrasivkörner zu einem vollständigen Ausbrechen der Hartstoffteilchen führt und diese ihrerseits als Abrasiv wirken und die Belastungsbedingungen weiter verschärfen /3/.

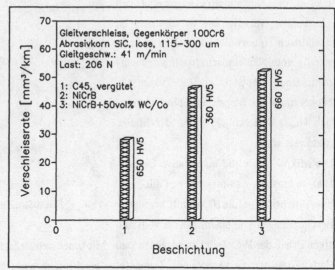

Abb. 5: Verschleißraten einer hartstoffhaltigen und einer hartstofffreien NiCrB-Beschichtung im Vergleich zu vergütetem Stahl

Zusammenfassung

Auf der Grundlage der erzielten Optimierung der Werkstoffe und Verfahrensparameter ist die industrielle Anwendung des Verfahrens möglich. Die Untersuchungsergebnisse zeigen, daß durch das Laserauftragschweißen von WC/Co-NiCrB-Pulvergemischen rißfreie und porenarme Verschleißschutzschichten erzeugt werden können. Aufgrund einer kontrollierten Grundwerkstoffanschmelzung verfügen die Schichten über eine metallurgische Bindung zum Substrat, wodurch eine Haftfestigkeit gewährleistet wird, die in der Größenordnung der Schicht-Bruchfestigkeit liegt. Der Verschleißwiderstand der Schichten entspricht unter der Bedingung der Übereinstimmung von Abrasiv- und Karbidkorngröße dem des gesinterten WC/Co-Hartmetalles.

Literatur

[1] **J. Shen, F. Dausinger, B. Grünenwald, St. Nowotny**: Möglichkeiten zur Optimierung der Randschichteigenschaften eines Einsatzstahles mit CO2-Lasern. Laser und Optoelektronik 23(6)/1991

[2] **R. Gassmann, St. Nowotny, A. Luft, W. Reitzenstein, J.Shen**: Laser Cladding of Hard Particles Rich alloys. ICALEO 1992

[3] **K.-H. Zum Gahr**: Verschleiß von Metallen, Keramiken und Polymeren im Vergleich. Reibung und Verschleiß bei metallischen und nichtmetallischen Werkstoffen, DGM-Verlag 1989

Laser Surface Hardening of Heat Treatable Steels with CO_2- and Nd:YAG Lasers

A. Lang, H. Stiele, D. Müller, R. Jaschek, J. Domes, R. Bierwirth, H.W. Bergmann

Universität Erlangen-Nürnberg
Forschungsverbund Lasertechnologie Erlangen
Lehrstuhl Werkstoffwissenschaften 2, Metalle
Martensstr. 5, D-91058 Erlangen

Abstract

The present paper describes a comparison between Nd:YAG- and CO_2-laser beam hardening of several steels such as C 45 and 42 CrMo 4. Hardness measurements, microstructures and residual stresses are reported. The investigations show that the Nd:YAG-laser is favoured for smaller geometries whereas the CO_2-laser is suited for large surface applications due to the high output power.

Introduction

Recently Nd:YAG-lasers with output powers of 1 - 2 kW have been made commercially available. This enables surface treatments like hardening or melting by Nd:YAG-lasers. The shorter wavelength compared to the CO_2-laser allows fibre transmission and a better absorption behaviour without an additional coating. An exact measurement of the surface temperature can be made when no coating was used in the hardening process. A further disadvantage of the coating is the bad integration in an automated production line /1/. At least fibre transmission enables hardening of complex geometries and inaccessible parts /2/.

Experimental Set-up

Heat treatable steels like 42 CrMo 4 and C 45 (dimensions 200 x 40 x 20 mm) quenched and tempered for a homogenous carbon distribution and two lasers, Nd:YAG- and CO_2-laser, with a 3-axis handling system were used for the experiments. The CO_2-laser beam hardened samples are graphite-coated, whereas the Nd:YAG-laser beam hardened samples were only ground. CO_2-laser experiments were carried out with a controlled temperature, while the Nd:YAG-laser equipment provides only a temperature recording. The following table contains the two beam sources of the applied lasers.

Beam source	CO₂-laser	Nd:YAG-laser
Mean power	4 kW, cw	1 kW, pw
Beam transmission	Mirror	Fibre (1 mm)
Beam shaping	Facet optic (10 x 10 mm)	Collimator lens (f = 80 mm) Focusing lens (f = 80 mm)
Spot dimension	Square (10 x 10 mm)	Circle, diameter 6 mm

Results and Discussion

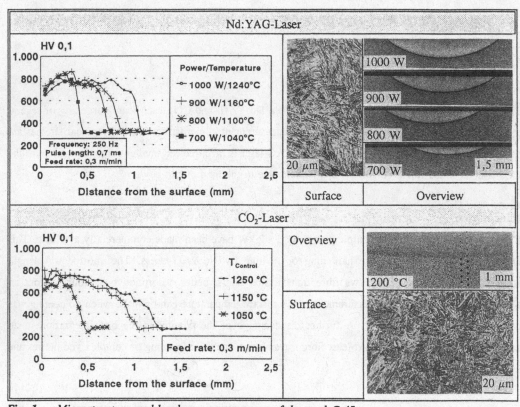

Fig. 1: *Microstructure and hardness measurement of the steel C 45*

The hardness measurement and the microstructure of the steel type C 45 can be seen in Fig. 1. A hardening depth of 1 mm can be achieved with both lasers. The reduction of hardness at the surface could be an effect of self tempering which arises very easily due to the high martensite start temperature of this steel type. A decarburization of the edge or retained austenite can not be detected. The semicircle form of the Nd:YAG-laser beam hardened areas results from the Gauss-distribution and the round spot which generates a decreasing beam affecting time from the middle of the track to the edge. A significant difference between the two laser processes is demonstrated by the hardness profile which decreases in the case of the CO_2-laser slowly to the value of the bulk material whereas the Nd:YAG-laser treated sample first shows a constant hardness and then a rapid reduction of the hardness.

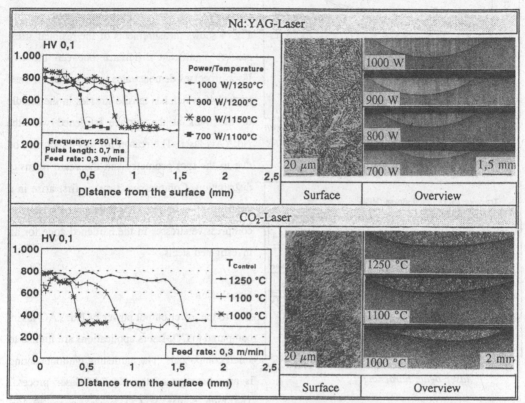

Fig. 2: *Microstructure and hardness measurement of the steel 42 CrMo 4*

Fig. 2 outlines the hardening results for the steel type 42 CrMo 4. A distinct difference between the two laser systems is the hardening depth which results in the case of the CO_2-laser nearly 1,6 mm whereas the Nd:YAG-laser beam hardened sample reaches, with a similar surface temperature, only a value of 1 mm. The scattering hardness values at the surface of the Nd:YAG-laser treated samples are induced by segregations in the material which can be seen in the microstructure. The CO_2-laser treated samples do not show a segregation. This is due to the higher beam affecting time which leads to a homogenous microstructure and a reduced hardness at the surface.

Fig. 3 demonstrates a comparison of the achieved hardening depth for both laser systems. Due to an improved beam coupling at the steel surface of the Nd:YAG-laser radiation a significant lower intensity is necessary to achieve the same hardening depth than by CO_2-laser beam hardening. A calculation of the electric efficiency I_{el} (Nd:YAG-Laser 5,5 %, CO_2-Laser 10 %, manufacturer data) shows that the

468

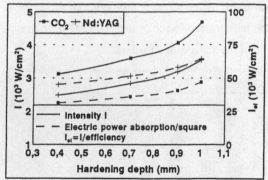

Fig. 3: *Comparison of the efficiency between Nd:YAG- and CO₂-laser beam hardening*

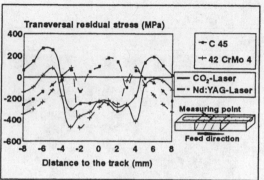

Fig. 4: *Residual stresses after Nd:YAG and CO₂-laser beam hardening*

energy consumption by CO_2-laser beam hardening is lower compared to the Nd:YAG-laser for a specified hardening depth.

Fig. 4 shows measurements of the residual stresses. A significant difference between the two laser processes demonstrates the result of the steel type C 45 which has tensile stresses in the middle of the track at the surface hardened with the Nd:YAG-laser. The reason for this effect may be due to the spot geometry and the beam intensity distribution. Compressive stresses first arise in a depth of 40 μm. The CO_2-laser process leads to compressive stresses in the hardened zone for all investigated steels.

Conclusion

Due to the lower output power of Nd:YAG-lasers compared to CO_2-lasers applications are limited to small components. The hardening without coating is an advantage of the Nd:YAG-laser process. The high power of CO_2-lasers up to 20 kW allows the hardening of large areas or round geometries like crankshafts without an overlapping.

Acknowledgement

These investigations were financially supported by the German Ministry of Research and Technology BMFT (Förderkennzahl: 13N6021-8) and the companies Rofin Sinar Laser GmbH (Hamburg), Messer Griesheim GmbH (Puchheim) und Thyssen HOT (Nürnberg).

References

/1/ Th. Rudlaff, F. Dausinger: Effects of transformation hardening without absorptive coatings using CO₂- and Nd:YAG-lasers, ed. B.L. Mordike, Laser treatment of materials, Göttingen 1992 (ECLAT 1992), P. 313 - 318.

/2/ H.K. Tönshoff, C. Meyer: Randschichthärten mit dem kW-Festkörperlaser, HTM 44, 1989, P. 32 - 36.

Surface Treatments of Components for the Automotive Industry Using Excimer Lasers

K. Schutte[1], E. Schubert[2], H.W. Bergmann[1]
[1]Forschungsverbund Lasertechnologie Erlangen (FLE), Universität Erlangen-Nürnberg, Institut für Werkstoffwissenschaften 2, Martensstraße 5, D-8520 Erlangen
[2]ATZ-EVUS Sulzbach-Rosenberg, Bereich III, Außenstelle Vilseck, Rinostraße 1, D-8453 Vilseck

Introduction

Fundamental investigations of the surface treatment of metals using excimer lasers have been carried out in recent years. In this paper 3-dimensional materials processing of parts for the automotive industry, like cylinder blocks, is reported and the technical advantages are outlined.

Today laser applications in automotive industry are reliable technological processes, which are implemented in manufacturing lines. Commonly most of the applied lasers are cw CO_2-lasers with optical mirror systems or alternatively Nd:YAG-lasers with a beam guiding system using fibre optics /1/. Excimer lasers are industrially used for marking, drilling of multi-layer circuit boards and similar electronic applications. In this paper it is shown that excimer lasers are a suitable tool for motor car applications. Machining processes (e.g. cutting, grinding and polishing) generate a technical surface of the metal different to an ideal crystal. The metallic substrate is covered by several layers of a thickness from a few nanometres to about one millimeter. One of the advantages of an excimer laser treatment of these surfaces is the small penetration depth of the laser light due to the short wavelength (from 193nm up to 351nm) and the short pulse duration (typically 10 to 100ns) compared to the lasers commonly used in car industry. The main process reported in literature (/2/,/3/,/4/) is the removal of thin surface layers (like deformation layers, organic adsorbates, metallic oxides or grinding particles, etc.). The possibility of modifying surfaces without a heat affected zone in the substrate reveals the excimer laser machine to an innovative and excellent working tool. Another reason for the use of an excimer laser treatment is the avoidance of chemical etching techniques. In near future pollution problems will become much more important for the automotive industry. This will lead in particular to higher production costs for components and has to be considered.

Experimental Set-up

A work station for excimer laser surface treatments of materials has been designed. The applied beam source was the Siemens Excimer laser XP2020 with an optical power of 2J/pulse, an average power of

40W using XeCl (wavelength 308nm) as laser active medium /5/. The rectangular cross-section of the excimer laser beam is different compared to the typical rotationally symmetric beam profiles of commercial CO_2-Lasers. An optical mask projection system is used to generate a homogeneous intensity profile and shape of the laser pulse on the treated samples surfaces, see figure 1. The installed handling system allows the motion controlling via personal computer in three linear axes (x,y,z) and two rotational axes (u,v), see fig. 1. Using special programmes and arrangements of axes a machining of typical components, like cylinder blocks or camshafts, from the automotive industry with complex three dimensional geometries could be achieved. Possible dimensions for working pieces are diameters of about 200mm for outside treatments and minimal diameters of 50mm for

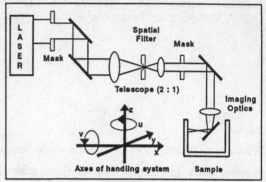

Fig. 1 Schematic drawing of laser optics using a mask projection system and available motion axes.

inside treatment. Enhanced process and laser safety is provided by an on-line process and quality control by means of spectral differential reflectometry /6/. The sample is scanned during laser treatment by a fibre optic transmitted measurement beam while the spectral response of the treated surface is detected. The host computer is able to decide if the reflection signal is the same as a master signal preprogrammed by the user. The process control was successfully tested by optical etching of cast iron /7/.

Experimental Results

The experiments described below are the opening of graphite spherulites or laminae in cast iron surfaces and the smoothing of components. The aim is to increase the lifetime of motor parts and to reduce the noise for motion components. Cylinder liners used in car production are often made out of pearlitic-ferritic cast iron. It is well known that opened graphite spherulites can act as an oil reservoir. Classical processes remove deformation layers by chemical means. It was found that optical etching occurs using a contact free optical process by illuminating the surface with an excimer laser. Depending on the machining procedure (e.g. honing the cylinder bore) number of tools generate deformation layer thicknesses up to 10 μm. For such large values a removal process by UV-light is not efficient in terms of costs and machining time. Therefore the authors examined first the influence of the initial surface roughness on the rate of opened graphite by treating several cast iron samples. Figure 3 traces the laying-bare of graphite and change in roughness by excimer laser treatment of nodular cast iron as a function of the samples roughness when applying one specific power flux density and a specific number of pulses. Collecting all data for the investigated different flux densities makes it possible to determine a processing window for the optical etching process mentioned above, see fig. 3.

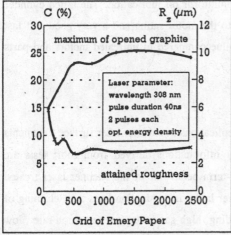

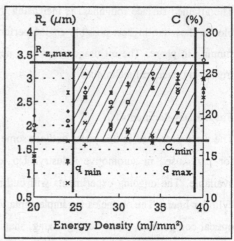

Fig. 2 Laying-bare of graphite and change of roughness by excimer laser treatment of nodular cast iron; 2 pulses each

Fig. 3 Processing window for excimer laser surface treatment of nodular cast iron; 2 pulses each , 500 - 2400 grit

Using the results from test samples the optimum process parameters could be applied to cylinder bores commercially produced by our partner of the german automotive industry. Detailed studies enable also a process window to be found for technological components and industrial relevant surface treatments including the advantages of a process and quality control. Several cylinder blocks with different honing qualities were illuminated to investigate the behaviour of the modified surfaces running under service conditions. Figure 4 b-d show SEM pictures of the untreated and treated surface areas before and after running the cylinders on an engine-test rig of about six and sixty hours.

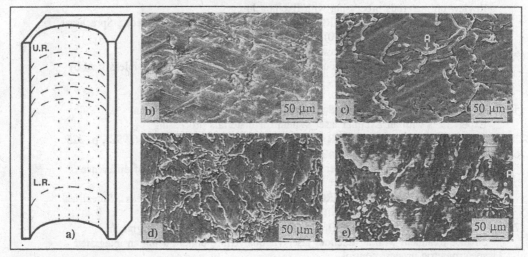

Fig. 4 Excimer laser treatment of cylinder liner surfaces:
a) cylinder bore b) honed surface c) laser treated surface d) after 6,5h-test e) after 60h-test

472

The attained surface roughness is increasing the active volume for oil lubrication. The tested cylinder blocks show the highest possible engine performance up to the motor rating and a very good and fast running-in phase. The oil usage drops to a neglectable value compared to untreated motors and parts treated by chemical techniques.

Future Developments

The work presented in this paper outlines some possible applications of excimer laser surface treatments for parts used in automotive industry. Up to now only informations derived from short tests are available. The ongoing experiments will outline the long-term behaviour of the excimer laser treated cylinder liners. The obstacles for implementation of excimer lasers in production lines are : cleaning of optical components, beam homogenizing, shaping and guiding, high gas costs, low repetition rate , low average power, maintenance and reliability of electrical components. The acceptance for such lasers will rise only if the mentioned problems are solved in the future.

References

/1/ D.M. Roessler
New laser processing developments in the automotive industry
The Industrial Laser Annual Handbook, 1990 edition, D. Belforte and M. Levitt eds.,
PennWell Publ. Co. Tulsa (1990), p. 109

/2/ E. Schubert, H.W. Bergmann
The influence of exposure time, wavelength and repetition rate on the quality of
surface treatments of metals with Excimer laser
OPTO ELEKTRONIK MAGAZIN, Vol. 5, No. 3 (1989), p. 334

/3/ H.W. Bergmann, E. Schubert
Review on materials processing with Excimer lasers
Proc. ECLAT '90 Erlangen (FRG), H.W. Bergmann and R. Kupfer eds., Sprechsaal
Publishing Group Coburg (1990), p. 813

/4/ A.S. Bransden, J.H.P.C. Megaw, P.H. Balkwill, C. Westcott
Metal surface treatment and reduction in pitting corrosion of 304L stainless steel by
excimer laser
J. Modern Optics, Vol. 37, No. 4 (1990), p. 813

/5/ K. Schutte, S.M. Rosiwal, E. Schubert, R. Queitsch, H.W. Bergmann
Experiences with an Excimer laser system for materials processing
Proc. of 10th Int. Cong. LASER '91 Munich, Springer Verlag Berlin (1992), p. 325

/6/ S.M. Rosiwal, H.W. Bergmann
Surface treatments with Excimer laser and quality control by means of difference
reflectometry
Proc. ECLAT '90 Erlangen (FRG), H.W. Bergmann and R. Kupfer eds., Sprechsaal
Publishing Group Coburg (1990), p. 895

/7/ H. Hitzler, Ch. Pfleiderer, K.O. Greulich, S.M. Rosiwal, H.W. Bergmann
Surface treatment with an Excimer laser via quartzfibre bundle using an integrated
spectrometric process control
Proc. of 10th Int. Cong. LASER '91 Munich, Springer Verlag Berlin (1992), p. 364

Laser Beam Material Processing Using a Combination of a High Power CW-CO$_2$-Laser and a TEA-CO$_2$-Laser

R. Jaschek, H.W. Bergmann

Friedrich-Alexander-Universität Erlangen-Nürnberg
Forschungsverbund Lasertechnologie Erlangen (FLE)
Lehrstuhl Werkstoffwissenschaften 2, Metalle (WTM)
Martensstr. 5, D-91058 Erlangen

1. Abstract

The radiation of a CO$_2$-laser is highly reflected at metallic surfaces. A plasma, generated by a TEA-CO$_2$-laser, can be used to increase the absorption of a cw-CO$_2$-laser during laser beam transformation hardening without any harmful melting of the surface. An increased case depth can be achieved during laser beam transformation hardening of a heat treatable steel without preoxidizing of the metal surface or using additional coatings. The radiation of a 18 kW cw-CO$_2$-laser was formed by a facetted mirror and superimposed by a TEA-CO$_2$-laser pulse with an energy of 6 J. The induced plasma interacts and modifies the surface of the heat treatable steel resulting in a higher absorption of the 18 kW cw-CO$_2$-laser radiation. A linear polarized beam irradiated under the Brewster angle shows an enhanced absorption. Superimposing a TEA-CO$_2$-laser pulse with an energy of 180 mJ to a linear polarized 4 kW CO$_2$-laser beam irradiated under an angle of 70° an even increased case depth can be observed.

2. Introduction

During laser beam transformation hardening an increased case depth and also an increased hardened volume can be achieved using dielectric coatings (e.g. an oxide layer) or coatings of isolated particles (e.g. graphite or ceramic suspension) [1,2], but also by specific pulse shaping [3] or the use of a linear polarized laser beam irradiated under the Brewster angle [4]. An increased case depth and hardened volume could be achieved by a combination of a cw-CO$_2$-laser beam and a TEA-CO$_2$-laser beam. The two beams are superimposed using a copper mirror with a diffraction grating which allows beam focusing with the same optics [5]. This "two laser beam technique" could be first verified for perpendicular incidence of both beams. No additional coatings are required and no harmful oxidation can be observed. A very thin melted layer (thickness smaller than 10 μm) is responsible for the increased absorption [5]. The plasma dynamics can be observed with short time photography and spectroscopy [5,6].

3. Experimental set-up

The investigations were carried out using two commercial TEA-CO_2-lasers. Table 1 summarizes the characteristic parameters of these TEA-CO_2-lasers.

Table 1: Specifications of the TEA-CO_2-lasers

model	pulse energy	P	P_{max}	repetition rate	pulse duration (FWHM)	cross section of raw beam
M1	180 mJ	18 W	3,6 MW	100 Hz	50 ns	8 x 8 mm^2
M2	6 J	75 W	71 MW	10 Hz	70 ns	25 x 25 mm^2

The experimental set-up is shown in Fig. 1. The model M1 TEA-CO_2-laser with a pulse energy of 180 mJ was combined with a 4 kW cw-CO_2-laser (Eurolas 4000, Messer Griesheim) which beam was linear polarized and focused with a concave mirror (focal point: 150 mm). The angle of incidence was 70° respectively. The model M2 TEA-CO_2-laser with a pulse energy of 6 J was combined with a 18 kW cw-CO_2-laser beam (UTIL SM41-18) shaped with a facetted mirror. The focal area was 12 x 12 mm^2 and partly covered by the plasma generated by the high power TEA-CO_2-laser pulse. The incidence of the cw-CO_2-laser beam was perpendicular to the surface of the workpiece. The heat treatable steel DIN C 45 (No. 1.1730) used for these experiments was pre-heat-treated to obtain a fine distribution of carbon inside the workpiece, which was grounded and cleaned with acetone prior to each experiment. The shielding gas nitrogen avoids an oxidation of the steel surface.

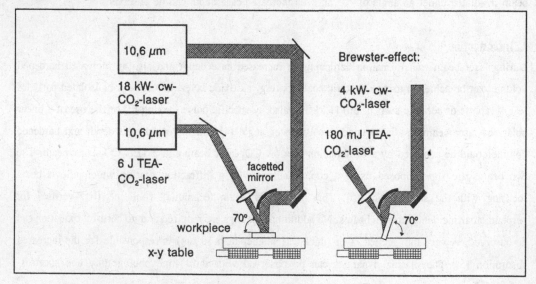

Fig. 1 *Experimental set-up for the "two laser beam technique"*

4. Experimental results and discussion

Fig. 2 and Fig. 3 show the case depths for different power of the $cw-CO_2$-laser radiation and the two different experimental set-up of the "two laser beam technique".

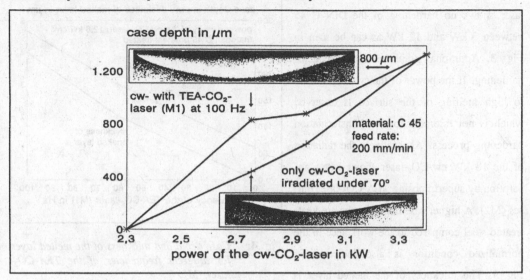

Fig. 2 *Achieved case depths using a combination of a linear polarized $cw-CO_2$-laser beam irradiated under an angle of 70° with a 180 mJ $TEA-CO_2$-laser beam (M1)*

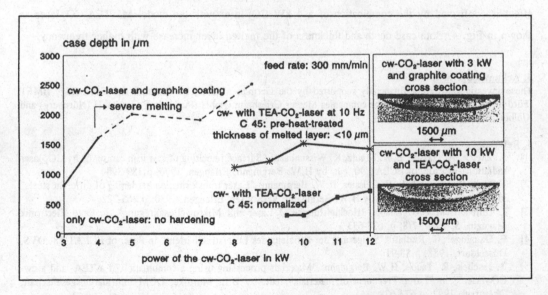

Fig. 3 *Achieved case depths using a combination of a 18 kW $cw-CO_2$-laser beam with a 6 J $TEA-CO_2$-laser beam (M2)*

With a 4 kW $cw-CO_2$-laser and a linear polarized laser beam under an angle of 70° a hardening of the steel is possible even at low power due to the increased absorption based on the Brewster effect. With

476

the additional TEA-CO_2-laser (M1) at a frequency of 100 Hz even an increased case depth is observed with a larger hardened volume detected from the cross sections of the hardened tracks seen in Fig. 2.

The irradiation only with the 18 kW cw-CO_2-laser shows no hardening of the DIN C 45 between 3 kW and 12 kW as can be seen in Fig. 3. Additional graphite coatings enables hardening. If the power of the cw-CO_2-laser is to high melting of the surface is observed which is not tolerable for the transformation hardening process. Alternatively the radiation of the 18 kW cw-CO_2-laser shows better absorption by superimposing TEA-CO_2-laser pulses (M2). A higher case depth for the pre-heat-treated steel compared to the workpiece in the normalized condition is also indicated in Fig. 3. The thickness of the melted layer is still below 10 μm. The limited repetition rate

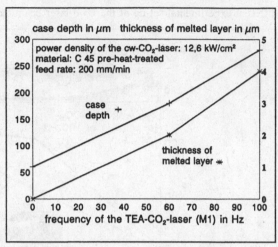

Fig. 4:   Case depth and thickness of the melted layer for different frequencies of the TEA-CO_2-laser (M1)

of 10 Hz for this type of TEA-CO_2-laser leads to a reduced coupling of the cw-CO_2-laser radiation. This effect is confirmed by the combination of a 4 kW CO_2-laser with the model M1 TEA-CO_2-laser as shown in Fig. 4. Both case depth and thickness of the melted layer increase with higher frequency.

5. Acknowledgement
These investigations were financially supported by the German Ministry of Research and Technology BMFT (Förderkennzahl: 13N5936) and the companies Messer Griesheim GmbH (Puchheim), Geat mbH (Nürnberg) and Kleiber (Erlangen).

6. References
[1]   A. Zwick, A. Gasser, E.W. Kreutz, K. Wissenbach: "Surface remelting of cast iron camshafts by CO_2-laser radiation" in Proc. of ECLAT'90, edt. by H.W. Bergmann, Erlangen, 1990, p.389-398

[2]   J. Bach, R. Damaschek, E. Geissler, H.W. Bergmann: "Laser transformation hardening of different steels" in Proc. of ECLAT'90, edt. by H.W. Bergmann, R. Kupfer, Erlangen, 1990, p.265-282

[3]   J. Uhlenbusch, Z.B. Zhang: "Hochleistungs-CO_2-Laser mit Mikrowellenanregung", in Opto Electronik Magazin, 5, 1989, 7/8, p. 628-633

[4]   F. Dausinger, T. Rudlaff: "Steigerung der Effizienz des Laserstrahlhärtens" in Proc. of ECLAT'88, DVS, Düsseldorf, 1988, p.88-91

[5]   R. Jaschek, R. Taube, H.W. Bergmann: "Materials processing using a combination of a TEA- and a cw-CO_2-laser", in "Laser Treatment of materials", edt. by B.L. Mordike, DGM Informationsgesellschaft, Oberursel, 1992, p.673-678

[6]   H.W. Bergmann, R. Jaschek: "Untersuchungen zum Absorptionsverhalten von Kurzpulslaserstrahlung bei flächiger Bestrahlung von dünnen Oberflächenschichten", in "Strahl-Stoff-Wechselwirkungen bei der Laserstrahlbearbeitung", edt. by M. Geiger and F. Hollmann, Meisenbach Verlag, Bamberg, 1993, p. 21-26

The Laser Assisted Manufacture of Tailored Protective Coatings on Turbine Blades of Steel and Titanium

W. Amende, A. Coulon, W. Kachler
MAN Technologie AG, Munich (D), GEC Alsthom, Belfort (F), MAN Energie GmbH, Nürnberg (D)

1. Introduction

The predominant wear of turbine blades in the low pressure section of steam turbines is based on the impingement of small water droplets, which come into being in the expanding water steam.

A conventional measure against the droplet erosion consists in the application of brazed edges, especially of cobalt base alloys. The new procedure, described in this paper, concerns the fusion welding of the coating material by means of a high power laser beam as the thermal instrument (1). In this case, coating and process should fullfill the following preconditions:

- A high clad adhesive strength without corrosive infiltration
- A small thermal load of the component for its low distortion
- In the interest of a high erosion resistance: The microstructure should show finely distributed carbides.

The preconditions mentioned above are satiesfied when hard facing is performed by means of the laser beam.

2. Involved Materials and Processing

In the considered work, steam turbine blades of steel 1.4939 were cladded with a cobalt base alloy. Titanium blades were provided with a β- titanium coating. The chemical composition of the involved claddings are given in table 1.

Co base alloy

element	Cr	C	Si	Mn	Mo	Fe	Ni	W	Co
w%	28	1,1	1,1	0,01	0,02	0,3	0,34	4,65	bal

β-titanium

element	Al	Sn	Zr	Mo	Cr	Fe	Ti
w%	5	2	4	4	2	1	bal

Table 1: The chemical composition of the applied coating materials

The amount of TiC was graded in a range up to 20 %.
The powdery coating material was delivered continuously into the spot of the laser beam. The beam source was a CO_2 laser with nominal power up to 5 kW. With careful coordination of the processing data, uniform beads were manufactured. Adjacent overlapping beads resulted in cladded areas.

3. The Cladding of Turbine Blade Edges

Fig. 1 shows the cross section of a cladding on steel 1.4939. In order to achieve a finely crystalline structure, the clad is built up of two layers with a thickness of 1 mm each. The hardness is mainly determined by the local amount of carbides.

A further influence in the processing of the mentioned material combination is given by the amount of iron, which is absorbed from the substrate material. With an increasing iron amalgamation, the hardness of the cobalt rich clad is reduced. However, with careful development of the processing data, the iron absorption is limited to an amount of 5 % approx. With this precondition a layer hardness of 500 HV 0,5 is achieved with sufficient certainty.

With laser assisted hard facing, also in case of a blade thickness of 6 mm only, the depth of the heat affected zone could be limited to approximately 1 mm. The hardness profile showed a hardness of approximately 400 HV in the martensitic area. In the hardened condition the component is subjected to an increased susceptibility to stress corrosion cracking. This is why, subsequent heat treatment, 640 °C / 2 h for instance, was carried out.

The alternative application of titanium alloys for blade material allows longer blades and therefore an increased steam cross section. The protective cobalt base coatings, effectively applied on steel blades, are not suited for the treatment of titanium alloys, because both materials form a rather brittle fusion zone. For this reason, in this case, the coating material is similar to the substrate in order to ensure a ductile transition zone. Some β-titanium alloys have proved a good resistance against drop erosion (2). Their erosion resistance is improved by means of finely distributed inclusions of titanium carbide.

As in the case of cladding with cobalt base alloys, the titanium base coating material was delivered in powder form. The TiC containing powders consisted of a mechanical mixture of β-titanium and carbide. The grain size of the carbide powder amounted to 5- 20 microns.

The well known high reactivity of titanium with the atmosphere (e.g. O_2, H_2 and N_2) requires careful protection of the melting process with a shrouding gas. Consequently, the powder delivery as well as the melting zone and the cooling beads completely were completely kept under an argon shield.

The microstructure of a laser cladding on Ti Al6 V4, enriched with titanium carbide, is shown in _fig. 2_ . The uniform distribution of the carbide particles is of major importance. Furthermore, the cooling rate must be high in order to achieve the titanium in its β-modification. Coarse α-grains cause a decreased cladding ductility. A high toughness, however, is considered as an absolute necessary precondition for good resistance against droplet erosion.

The clad hardness amounted to 380 - 400 HV 0,5. .The coating hardness can be changed by means of a heat treatment. The increase in hardness and strength depends on the precipitation of finely distributed α-particles from the β solid solution. Their formation begins in coherence to the matrix lattice. During the further ageing process they pass over to the semicoherent and the non-coherent condition. With the ageing of laser coatings at 600 °C the hardness development , shown in **fig. 3** , was evaluated. The reduction in hardness, observed after passing a hardness maximum, is explained with an overageing of the material.

The turbine blades with a length of 900 mm , were cladded in the head area (fig. 4)over a length of 200 mm with a cobalt base alloy, given in table 1. The clad thickness on the edge as well as on the vaulted flank amounted to 2 mm. A particular attention was paid to the transition between the hard faced area and the uncladded substrate. This transition zone acts as a geometric and a metallurgical notch, which may affect the fatigue strength badly. For this reason, continuous transitions were produced by sloping the powder delivery rate, in order to reduce the geometric notch effect. Corresponding test samples , provided with a similar cladding, had proved, that in this manner the fatigue strength was not deteriorated to an inadmissible amount. The beveling of the clad area at its bottom edge served the same purpose.

Porosity and bonding faults were not admitted, because these failures in general initiate fatigue cracking and stress corrosion cracking. The distortion of the blades, in all instances, amounted to less than 1,2 mm, and therefore it remained within the allowance.

In a test facility of Dornier GmbH, the drop erosion resistance of the coatings was investigated under conditions similar to the operating load. The essential test data are given in table 2 .

droplet size	1,2 mm
relative velocity between	
sample and droplets	600 m/s
angle of impact	90°
tested sample area	104 mm ^{2}

Table 2: Data of the drop erosion tests on laser claddings of cobalt base alloys on steel 1.4939

The gravimetric material loss in comparison to test results of various protective coatings is plotted against the exposure time in fig. 5 . The results can be summarized as follows: With regard to the incubation time and the erosion rate , the laser claddings proved superior to the compared coatings. The reason is thought to be the most fine and uniform distribution of the carbide phase.

4. Conclusions

Protective coatings have been applied on the leading edges of steam turbine blades by means of a high power laser beam. The blade materials were steel 1.4939 and Ti Al6 V4.

The claddings had a favourable microstructure and a low amalgamation with the substrate. The thermal load of the components was comparably low, so that a good accuracy to gauge was achieved. The reliable metallurgical fusion ensures a high adhesive strength. Further characterization of the coatings highlighted properties capable of extending service life.

From the technical and the economical point of view, the laser assisted cladding proved a most promising alternative for the manufacture of local hard facings on steam turbine components.

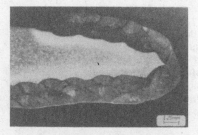

Fig. 1: The cross section of a laser cladded leading edge of a steel turbine blade

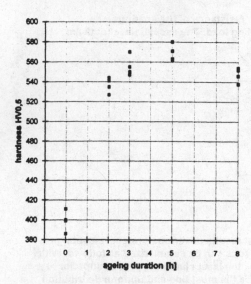

Fig. 2: The microstructure of a laser cladding consisting of β-titanium with 20 % TiC on a substrate of Ti Al6 V4

Fig.3: The hardness in laser claddings of β-titanium after an ageing treatment at 600 °C

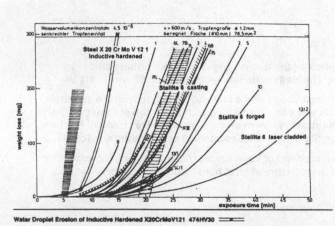

Water Droplet Erosion of Inductive Hardened X20CrMoV121 474HV30

Comparison with Scatterbands of Cobalt-Base Alloys (Stellite 6) and Tempered 12%Cr - Steels

Fig. 4: The laser cladded leading edges of large steam turbine blades

Fig. 5: The gravimetric material loss of various protective coatings under drop erosion

Acoustic Emission, Magnetic and Wear-Resisting Performances Research of Laser Quenching

Zhang KuiWu.Tao XueBin.Zhai ShaoYan
Beijing Machine Tool Research Institute.Beijing.China

In this papaer, Some properties of laser quenched.such as fragility. wear-resisting and magnetic performance.and their influencing factors wereinvestigated.so as to be reference to design of parts with magnetic and mechanical specifications.

Test   Condition

1.Acoustic emission test

The materials of specimens are 20 steel and 45 steel.the size of specimensare $140 \times 15 \times 8$mm. Specimens were tested by using DUNEGAN/ENDECU acoustic emittermade in USA. with 3-point crankle method and the laser quenching face down.

2.Wear-resisting test.

Materials of Specimens are 20 steel.45 steel and 42CrMo steel.thesize of specimens are $25 \times 10 \times 6$mm. The opposing wear block is whole quenched and low-temperature tempered GCr15 steel with hardness HRC 60, size 35×10mm.Specimens were weared on Amsler abrasion testing machine.the wear-resisting is judged with the block volume weared.Testing conditions are listed as follows: wearing pressure:500 N. rotation rate: 200 rpm.lubricating oil:20 machine oil mixed with 0.8% Cr_2O_3.time interval of measuring: 1.5 hours.

3. Magnetic test

The circular specimens are made of 20.20Cr and 42CrMo steel with outer diameter30mm. inner diameter 26mm.specimens are tested on CL8 soft magnetic acoustic frequency magnetic characteristic instrument connected with plotter.Testing parameters are listed as follows: Magnetic field intensity: Ho=5 Oc/cm.Magnetic induction intensity: Bo=2100 GS/cm.Primary magnetizing current :Ip=0.5A. Primary winding: N1=650(winding).Secondary winding:N2=150(winding).Secondary voltage: Vb=4 ∽ 4.5V.

4.Laser quenching

All specimens.coated by self-made 1005GW laser absorber (unless special declaration).were laser quenched on 1.5 KW CO_2 CNC laser heat treatment equipment made in china with output power P=1000W for quenching.

Results and Discussion of Test

1.Acoustic emission test

Surfaces of specimens were laser quenched with different scanning speed.Effects of quenching quality. laser absorber. number of quenching track and tempering temperature on character of acoustic emission are shown in Tab 1.

(1).If the surface of specimen is melted during laser quenching. its fracture strength and displacement will be lower than that of normal quenched specimen.The fracture strength of 20 and 45 steel will decrease 9.5% and 9% respectively comparing with normal quenched.

(2).The breaking load of specimens coated by 1005GW laser absorber is higher than that of specimens coated by phosphide laser absorber (reference to group II No.S020 and S001). The main reason is that phosphide in the coating can infiltrate into specimen during laser quenching. and diffuse along the higher active energy crystal boundary.so as to increase the tendency of fragility.[1

(3).Being low-temperature tempered after laser quenching. the fracture strength and displacement increases as tempering temperature increases. The specimens of 20 and 45 steel. tempered at 250° C.yield with no crackle occured on the surface.

(4).When specimens were crankled. all crackles occured on the surface.are perpendicular to those scanning line.Crackles occured on 20 steel were limited within laser quenching tracks. and no specimen breaked during testing. However, when crackles occured on the quenching tracks of 45 steel specimen.it spreaded immediately across Matrix resulting breaking. See Fig.1

(5).The fracture strength and displacement of specimens with one laser quenching track (width of track 4mm) is less than that of specimens with two.

482

Table 1   Acoustic emission test

Group	Material	Number of specimen	B1	B2	B3	B4	B5	B6		
I	20 Steel	Number of specimen	B1	B2	B3	B4	B5	B6		
		Scanning speed (mm/min)	1200	1200	1200	1200	1200	1200		
		Quality of hardening	fine	fine	fine	fine	fine	Surface melted		
		Number of scanning line	2	2	2	2	1	2		
		Tempering temp.(°C)	——	150	200	250	——	——		
		Fracture strength (MPa)	89.4	92.7	101.6	yield	63.4	80.9		
		Elastic displacement (mm)	0.89	1.01	1.07		0.86	0.83		
	45 Steel	Number of specimen	5016	5017	5018	5024		5026		
		Scanning speed (mm/min)	1900	1900	1900	1500		1020		
		Quality of hardening	fine	fine	fine	fine		Surface melted		
		Number of scanning line	2	2	2	2		2		
		Tempering temp(°c)	——	150	200	250		——		
		Fracture strength (MPa)	109.7	115.3	124.2	yield		96.8		
		Fracture displacement (mm)	1.16	1.26	1.49			0.93		
II	45 steel	Number of specimen	5001	5002	5014	5013	5007	5005	5020	5019
		Laser absorber	phosphide absorber						1005GW	1005GW
		Quality of hardening	fine	fine	fine	fine	fine	surface melted	fine	fine
		Number of scanning line	2	2	2	2	1	1	2	2
		Tempering temp (°C)		150	200	250				200
		Breaking load(KN)	16.6	19.4	22 yield	21.3 yield	16.8	14.2	17.9	17.7 yield

Fig.1 Acoustic emission specimen
upper:laser quenched 45 steel
lower:laser quenched 20 steel

2. Wear-resisting performance

Results of wearing test are listed   in Tab 2.contrasting with whole quenched 45 steel specimens.

Table 2   Wear--resisting

Group	Material	Record of Heat Treatment	Surface Hardness HV0.1	Relative Wear-resisting
I	45	Whole quenching	685 693	1
II	45	Normal laser quenching U=2000mm/min	765 743 767 768 660	2.31
III	45	Normal laser hardening U=2000 mm/min Tempering at 150° C 200° C 250° C	559 575 655	1.38 1.27 1.12
IV	45	Laser hardening U=600 mm/min Surface melted slightly Tempering at 200° C	563 591	1.38 1.38
V	20	Normal laser hardening U=1450 mm/min Not enough heating of laser quenching U=2000 mm/min	604 383	3.09 1.32
VI	42CrMo	Normal laser hardening U=1800 mm/min Tempering at 200° C	816 706 593	2.92 2.77 2.63

(1).The wear properties of normal laser quenched specimens of 45 steel (group II) are 1.3 times higher than that of whole quenched specimens. fewer scrathing tracks on the surface of laser quenched specimens after being weared.

(2).When tempered at 150°C .200°C and 250°C (group III).the wear-resistancing of laser quenched 45 steel decreased. but it is still higher than that of specimen in group I.We can see from the mention above. for the purpose of improving toughness, low-temperature tempering is necessary after laser quenched.but for wear-resisting.the tempering is not necessary.

(3).For the laser quenched surface slightly melted of 45 steel.its wear-resisting is decreased.but higher than that of whole quenched.its stability against tempering is better than that of normal laser quenched.

(4).The wear-resisting of normal laser quenched 20 steel specimens (group V) is higher than that of laser quenched 45 steel specimens. perhaps this is because of its higher toughness (reference to the acoustic emission test).It should point that. the laser quenching technology of 20 steel should be controlled strictly.otherwise the result will not be good.

(5).The wear-resisting of laser quenched 42CrMo steel specimens is 1.9 times higher than that of whole quenched 45 steel specimens. 60% higher than that of laser quenched 45 steel specimens.the stability against tempering of 42CrMo steel is better than that of 45 steel.its wear- resisting decreased a little when tempered at 200° C.

3.Magnetic test

The hysteresis loop of three kinds of steels. each of which with 3 kinds of microstructure.are shown in Fig.2 and some magnetic parameters measured are listed in Tab 3.

raw material   laser quenching   whole quenching

Fig.2 hysteresis loop

Table 3   The result of magnetic test

	Material	40 Cr	42 CrMo	20 Steel
Hc (Oe)	Raw material	16	18.5	13
	Laser quenching	20	21	15
	Whole quenching	22	21.5	20
Br (Gs)	Raw material	8190	7140	9870
	Laser quenching	7980	7560	10710
	Whole quenching	9870	10920	11760
Bs (Gs)	Raw material	10500	10500	14910
	Laser quenching	11550	11340	15750
	Whole quenching	12810	13860	15540
μ	Raw material	256.1	244.2	331.9
	Laser quenching	176.3	175.8	297.2
	Whole quenching	312.4	346.5	296

(1).The order of materials is 20.40Cr and 42CrMo steel. the order of which microstructure is raw material.laser quenched and whole quenched. their coercive force (Hc) is in an acending order. For 20 steel,the Hc of laser quenched specimen is more lower than that of whole quenched specimen.

(2).The magnetic saturation intensity Bs of 20 steel is highest of all. the magnetic saturation intensity of 40Cr steel is the same as that of 42CrMo.The descending order of Bs is whole quenched.laser quenched (Bs of 20 steel is similar for two quenching ways) and raw material.Bs of laser quenched 20 steel is highest of all.

(3).The order of materials is 20.40Cr and 42CrMo.the order of which microstructure is whole quenched. laser quenched (except 40Cr) and raw material.their residual magnetic induction Br is in an descending order.

(4).The order of material is 20.40Cr and 42CrMo steel.and the microstructure order of 42CrMo is whole quenched.raw material and laser quenched.their magnetic conductivity is a descending order.the μ of laser quenched 40Cr steel has been descended 43.5% and 31.1% comparing with the whole quenched and raw material.For 42CrMo.it has been descended 49.2% and 28% in same cases. the μ of 20 steel is similar for the two quenching ways.has been descended only about 10% comparing with raw material.

Conclusion

1.Suitable laser absorber.correct processing of laser quenching and proper Low-temp.tempering can help to increase fracture strength and displacement deform-resisting performance of structure steel.
2.The wear-resisting of laser quenched 45.42 CrMo and 20 steel is respectively 1.3. 1.9 and 2 times higher than that of the 45 whole quenched (the wear-resisting of 20 steel is very sensitive to the will descend the wear-resisting.Among them.the stability against tempering of 42CrMo is better.no obvious change when being tempered at 200° C
3.Among all specimens hardened.the laser quenched 20 steel specimen has smallest Hc.highest Bs and little loss of μ.

Appreciation

Tianjing Machine Tool Electeric General Works supported our magnetic test.
621 Research Institute assisted us in acoustic emission test.

Reference

〈1〉. 张魁武、赵杰、激光淬火裂纹、金属热处理、1992、(6):47
〈2〉. Zhang KuiWu. Chang XiaoHui.Huang ZhuXiu.Shao BaoRu. The components of Electrimagnetic clutch with Laser Handening. will be published in 1993.
〈3〉. [日] 近角聪信等编、韩俊德、杨膺善译. 磁性体手册、冶金工业出版社出版. 1985年11月。

$Y_3Al_5O_{12}$:Nd and YAlO_3:Nd Active Elements with Improved Paramters

A.Ya.Neiman
Ural State University
620083, Ekaterinburg, Russia

On the basis of comprehensive investigations of defect structure, mass- and charge transport phenomenas and solid state reactions mechanisms /1-3/ a new technique of thermochemical treatment of high temperature crystals used for solid state lasers $Y_3Al_5O_{12}$, YAlO$_3$, $Gd_3Sc_xGa_{5-x}O_{12}$, $Y_3Sc_xGa_{5-x}O_{12}$ and others have been developed in order to improve their functional properties and increase the rate life.

Highly effective coating on the side surface and original technique of it deposition for the protection of laser elements from the damage by UV-radiation of pumping source have been worked out. The coating can absorb UV-radiation in desired spectral range 0.2 - 0.5 mcm depending on composition. The laser elements with protective coating allows to operate in continuously operating and different pulse conditions. It is very important that the coating provides the increase of laser efficiency in 1.25 times for the devices operated in single pulse conditions.

A solid state method for preparation the transparent-profile side surface with desired relief parameters (solid state etch process) have been produced for the lasers operated in continuous conditions. The method provides: removal of thin surface layer with crazes after mechanical abrasion treatment, increase of optical transparency of side surface in pumping range and decrease of generation threshold up to 10...15 %.

The coatings and etch process could be used in manufacturing of laser elements with different geometry and sizes.

The coatings have an excellent adhesion to the element surface, they are highly stable to the mechanical and temperature actions and stresses. They also have good resistivity to the cooling liquids, even aggressive (water, alcohol, Si-organic liquids, inorganic media).

Indexes of basic properties of $Y_3Al_5O_{12}$:Nd AND $YAlO_3$:Nd
Active Elements with special treatment of side surface.

Characteristic of Active Element	UV-stability	Laser efficiency (in relation to element without treatment)	FUNCTIONAL PARAMETERS:	
			PULSE regime, maximal Radiation Energy	Continuous regime, Change of Laser threshold
STANDARD $Y_3Al_5O_{12}$:Nd	$<10^2$	1.0	Active Element:size,mm 6.3x80 0.28 J 5.0x50 0.17 J	
With UV- & IR- active coating	$>10^6$	1.25	6.3x80 0.40 J 5.0x50 0.24 J	
With transparent-profile side surface	$<10^2$	1.15		decrease on 10...15 %
STANDARD $YAlO_3$:Nd	$<10^2$	1.0	3.0x50 0.14 J	
With UV- & IR- active coating	$>10^6$	1.25	3.0x50 0.25 J	
With transparent-profile side surface	$<10^2$	1.15		decrease on 10...15 %

The key characteristics are shown in the Table. In table the UV-stability means the number of pumping pulses (20 J) without decrease Radiation Energy when laser operates without any pumping source filtration. The cutting wavelength of side surface coating is 0,42 mcm.

One can see that new techniques provide substantial improvement of basic functional properties of laser rods.

Beside this new thermochemical approach reveals some additional possibilities. For example we developed a kind of solid state welding and used it for production of experimental laser system which cools by means of spontaneous heat transfer. Another novel direction of thermochemical treatment application is production of multy-friquency active elements. These new complex elements can effective operate on two-, three- and even four fixed wavelengthes.

Literature

1. A.Ya.Neiman et al. Doclady akademii nauk (SSSR), 1978,v.240, N4, 876-879.

2. A.Ya.Neiman et al. J.Inorganic Chemistry (USSR), 1980, v.25, N9, 2340-2345.

3. A.Ya.Neiman et al. J.Physical Chemistry (USSR), 1985, V.59, N9, 2360-2361.

Phase Transformations by Nanosecond Iodine Laser Pulses

I. Sárady[+], C. Magnusson[+], L-Y. Wei[+],
and
M. Chvojka[*], B. Králiková[*] and J. Skála[*]

[+]Luleå University of Technology, S-951 87 Luleå, Sweden

[*]Czech Academy of Sciences, Institute of Physics,
Na Slovance 2, CSR-180 40 Prague 8

1. Summary

The interaction of intense laser radiation with a metal surface in vacuum is a complicated process which involves surface melting, vaporisation, plasma formation, plasma heating and finally a breakdown or explosion of the superheated plasma. The explosion of the plasma causes a high-pressure shock wave, which acts as an impact on the material. The generated stress wave propagates and is absorbed in the material, causing elastic and plastic deformation, formation of dislocations, slip bands or other microstructural changes. Of the materials tested, an austenitic Hadfield manganese-alloyed steel was found to have the strongest tendency for a martensitic phase transformation, induced only by plastic deformation. An Iodine photo-dissociation laser [1] emitting at the wavelength of 1.315 µm was used for the experiments. The pulse duration was typically 1 ns and the pulse energy could be varied between 1-15 Joules.

2. Experimental treatment of target material

A plano-convex single lens with a focal length of 175 mm and free aperture $\varnothing 100$ mm was used to focus the beam to a spot size of approximately 0.2 mm diameter. The lens served as the window to the vacuum chamber but it could be moved up and down inside a sealed cylinder, allowing optimum focusing. The diameter of the focused spot was in the order of 0.2 mm. The energy and power density in the focused spot were greater than 1×10^4 Jcm^{-2} and 1×10^{13} Wcm^{-2} respectively. The martensitic transformation has been confirmed using optical, scanning- and transmission electron microscopy. Calculations of impact/momentum, peak pressure and peak temperature were performed using a one-dimensional finite difference method with moving boundaries [5,6].

Comparison of the form of the craters created in these experiments showed differences from the craters created by Q-switched Nd:YAG- or Ruby laser pulses [6]. This indicates that the interaction mechanism between the material and the Iodine laser pulse was different. In the case of 15-25 ns, 0.1-0.5 J Q-switched Nd:YAG- or Ruby laser pulses, the craters were almost conical. With the 1 ns, 5 J Iodine laser pulses the craters were like a section of a sphere (figure 1 and 2).

3. Microscopic examination of the samples

A number of observation techniques were used to evaluate the craters produced, including:

1. Optical Microscopy (OM), with Differential Interference Contrast (DIC) for measurement of the crater depth, molten layer thickness and for detection of microstructural changes.
2. Scanning Electron Microscopy (SEM), with Secondary and Backscattered Electron Imaging (SEI and BEI) for enhanced topographic contrast in microstructural changes.
3. Transmission Electron Microscopy (TEM) for detailed microstructural examination.

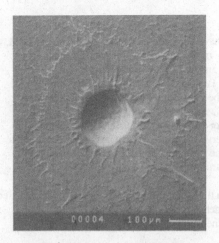

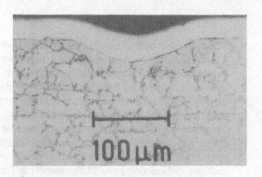

Figure 2. The crater cross-section;
(Irradiated specimen with Ni-coating).
OM, 250X. It seems likely that
a part of the molten material was ejected
from the crater by the high pressure.
The molten layer is just a few μm thick.

Figure 1. The crater created by the
Iodine laser irradiation. SEM, 100X.

The electropolished samples for TEM were investigated first by optical microscopy. Here, a rough estimation of the thickness of the sample and a pre-selection of them was possible. About half of the prepared samples were suitably thin for TEM-investigations.

The microstructural changes, i.e. twinning and martensite formation in the irradiated samples could be observed both in OM and in SEM and TEM. All of the TEM-specimens prepared from irradiated samples had clearly detectable amounts of microstructural changes/transformation and an increase in dislocation density, but in none of the unirradiated dummies were similar features found.

Figure 3.   Figure 4.   Figure 5.

Figure 3. TEM image of a specimen, prepared from an irradiated area. Martensite systems and an
increase in dislocation density compared with figure 4 can be observed.

Figure 4. TEM-image of a specimen, prepared from a non-irradiated area. The structure is completely
different, compared with the irradiated samples.

Figure 5. TEM dark field image of another irradiated sample, containing twin systems.

490

4. Discussion and conclusions

The microstructural changes in the irradiated samples have been confirmed by OM, SEM and TEM. These changes induced were similar to those caused by rapid deformation - e.g. drop-forging or by explosion loading - and are mainly composed of twins and an increase in dislocation density. The density of the changes decreased with increased depth, i.e. distance from the irradiated surface. The driving force for the transformation has been found to be the strongest in the case of Iodine laser pulses, compared with the samples irradiated by other laser sources [5, 6].

The crater depth and the crater formation mechanism for the samples irradiated by the Iodine laser pulses was completely different from those earlier observed: The crater form is a cap or section of a sphere, which can be explained by the assumption of the formation of a plasma ball, containing multiply ionised atoms and a large number of free electrons.

The scenario can be imagined as follows: Very early in the pulse, the surface of the target is vaporised and further heated until it becomes ionised, forming a dense metal plasma ball [2].

The plasma formation can be imagined stepwise as in the following figures below:

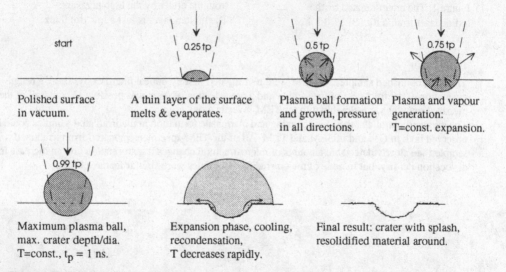

start	0.25 tp	0.5 tp	0.75 tp
Polished surface in vacuum.	A thin layer of the surface melts & evaporates.	Plasma ball formation and growth, pressure in all directions.	Plasma and vapour generation: T=const. expansion.

0.99 tp

Maximum plasma ball, max. crater depth/dia. T=const., t_p = 1 ns.	Expansion phase, cooling, recondensation, T decreases rapidly.	Final result: crater with splash, resolidified material around.

Figure 6. The suggested model for the formation of a plasma ball and its expansion. The ejection of molten material occurs continuously, until the surface is re-solidified.

As the described plasma ball is formed and expands, a shock wave will be generated in the material. The pressure amplitude must be higher than the tensile strength of the material in a volume around the impact where microstructural changes can be expected. The energy transfer from the laser pulse to the target surface occurs not directly but via the plasma ball. The electrons in the plasma ball absorb laser energy in a way similar to that of inverse bremsstrahlung. The interaction between the plasma ball and the underlying material is through intense, broadband radiation. By this, the crater form as a section of a sphere can be explained [3].

The initial shock wave propagates with the velocity of sound, 6×10^3 m/s. The propagation of the melt front is much slower, about 2×10^3 m/s so the transformation reaches far in the material before the surface melts, evaporates or re-condenses. This explains the presence of transformed material in a large volume around the crater.

At the very high levels of flux described here - $1.6 \cdot 10^{13}$ Wcm^{-2} for the Iodine laser - the amount of melting and boiling involved is minimised because any vapour generated is immediately and extensively ionised.

The opaque plasma produced prevents the majority of the pulse energy reaching the substrate and also sets up a high energy shock wave within the material (see fig. 6). The result is an interaction where the material removal process is minimised and the microstructural changes or shock hardening effects are maximised. In this case the pressure exerted by the plasma on the substrate can be calculated by a modified FDM-model for the non-equilibrium case to exceed 100 GPa [4]. (See Table 1).

Table 1. Calculated and measured values for Iodine laser pulses.
Pulse parameters: Pulse energy, $E_p > 5$ J; Pulse duration, $t_p = 1$ ns;
Wavelength $\lambda = 1{,}315$ nm; Power density, $P_d = 1.6 \cdot 10^{13}$ Wcm^{-2}

Description	Symbol	Calculated	Measured	Unit
Plasma temperature	T_n	120×10^3	-----	K
Maximum pressure	P	110×10^9	-----	Pa
Plasma initiation time	t_i	0.17	-----	ns
Plasma end time	t_e	17.7	-----	ns
Max. evap. front velocity	v_v	1.7×10^3	-----	m/s
Mass loss vapour	m_v	5.32	~ 5.0	µg
Mass loss liquid	m_l	2.66	~ 2.5	µg
Depth of molten layer	d_m	0.7	≈ 1.0	µg

5. Acknowledgements

The authors express their thanks to the Swedish Institute and STUF for financial support.
Personal thanks to Mr. J. Grahn for the assistance with SEM-investigations and Dr. J. Powell for discussions and advice.

(Keywords: Laser pulses, Iodine, Shock hardening, Phase transformation, Hadfield steel).

6. References

1. Chvojka, M. et al.: 100 GW Pulsed Iodine Photodissociation Laser System PERUN I. Czech J. of Phys. B 38, (1988).
2. Gilath, I.; Eliezer, S. and Bar-Noy, T.: Hemispherical shock wave decay in laser-matter interaction. Laser and Particle Beams (1993) pp. 221-225.
3. Goldman, S.R.; Dingus, R.S.; Kirkpatrick, R.C.; Kopp, R.A.; Stover, E.K. and Watt, R.G.: Laser-matter Interaction at intensities of 10^{12} W/cm^2 and below, SPIE, Vol. 1279, Laser-Assisted Processing II, Proceedings 13-14 March 1990, ISBN 0-8194-0326-1
4. Dingus, R.S.: Laser-ablation processes. SPIE Vol. 1627 Solid State Lasers III, 1992, pp 388-395.
5. Meijer, J.; Sárady, I. and van Sprang, I.: Shock hardening experiments on austenitic Hadfield steels by high-intensity TEA CO$_2$ laser pulses. Proc. ICALEO'87, pp253-259.
6. Sárady, I.; Magnusson, C.F.; Wei, L-Y. and Meijer, J.: Phase Transformations by High-Intensity Sub-Microsecond Laser Pulses; Proc. ICALEO'92, pp 228-236.

Legierungsentwicklung von Laserbeschichtungen auf Eigenbasis

R. Haude, R. Wilkenhöner, A. Weisheit, D. Burchards, B. L. Mordike

Institut für Werkstoffkunde und Werkstofftechnik der TU - Clausthal

Agricolastr. 6, D - 38 678 Clausthal - Zellerfeld

Die beim Laserbeschichten auftretende Problematik umfaßt vorallem Riß- und Porenprobleme sowie Haftungsprobleme, desweiteren soll eine gewünschte Schichtdicke bei möglichst geringer Nachbearbeitung - also eine möglichst glatte Oberfläche - hergestellt werden. Die Schichtdicke und die Oberflächenbeschaffenheit werden weitgehend durch die Prozeßparameter bestimmt. Die Riß-, Poren- und Haftungsprobleme werden durch die spezifischen Eigenschaften der hier eingesetzten Werkstoffe wie, Wärmeleitfähigkeit, Wärmekapazität und Temperaturausdehnungskoeffizient, gesteuert. Da Stähle aufgrund ihres günstigen Preises und ihres Eigenschaftsprofiles die meistverwandten Grundwerkstoffe für Laserbeschichtungen darstellen, wurde eine Beschichtung auf Eisenbasis entwickelt, um die Riß-, Poren- und Haftungsproblematik zu minimieren.

Ziel dieser Untersuchung war die Legierungsentwicklung einer Laserbeschichtung auf Eisenbasis, in die unterschiedliche Gehalte an Chrom-, Vanadium- und Titankarbid eingemischt wurden.

Als Legierungsmatrix diente ein, aus eigener Herstellung stammendes, Stahlpulver, mit einer Korngrößenverteilung von 25 bis 200 µm. Diesem wurden unterschiedliche Gehalte an Chrom-, Vanadium- und Titankarbid beigemischt und anschließend getrocknet. Die weitere Bearbeitung erfolgte ein- oder zweistufig. Bei der zweistufigen Bearbeitung wurde die Beschichtung mit einem geeigneten Binder versetzt und dann in definierter Schichtdicke pastenförmig auf den Grundwerkstoff aufgetragen. Nach einer anschließenden Trocknung an Luft oder im Ofen erfolgte die Laserbearbeitung dieser Beschichtungen an einem 400 W

Nd:YAG - Laser der Firma LUMONICS. Die einstufigen Versuche erfolgten an einem 5 kW CO_2 - Laser der Firma HERAEUS , wobei das Pulver simultan mit einem Tandem - Pulverförderer der Firma METCO bereitgestellt wurde.

Die Gefüge der Proben zeigten je nach Prozeßparametern und Karbidanteil unterschiedliche Phasenanteile. Dabei können die primären Karbide durch hohe Leistungsdichten teilweise oder vollständig aufgeschmolzen werden und sich danach als fein verteilte Karbide wiederausscheiden, wie dies die folgende Abbildung 1 zeigt.

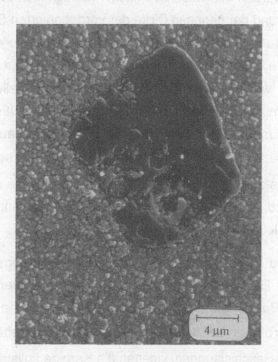

Abb. 1: Rasterelektronenmikroskopische Aufnahme einer Beschichtung aus 50 Gew.-% TiC und Stahlpulver.Das große Titankarbidteilchen wurde bereits geringfügig angeschmolzen, die kleinen Partikelchen sind wiederausgeschiedene Karbide.

Über die eingebrachte Streckenenergie und den Karbidanteil lassen sich die Härte und die Duktilität in einem weiten Bereich einstellen. Zunächst läßt sich durch die Streckenenergie regeln, ob die Karbide primär erhalten bleiben oder aufgeschmolzen werden und sich dann wieder als feinverteilte, kleine Karbide ausscheiden, dadurch lassen sich direkt Härte und Zähigkeit regeln. Desweiteren kann in der Ausgangsmischung der Karbidanteil vergrößert werden, so daß die Härte ebenfalls direkt erhöht wird, wenn die Karbide erhalten bleiben. Dadurch lassen sich bei 10 Gew.-% Cr_3C_2 Härten um 500 HV 0,3 erzielen, mit 20 Gew.-% Cr_3C_2 um 600 HV 0,3 und bei 50 Gew.-% TiC und VC Härten von über 1000 HV 0,3.

Bei der Laserbearbeitung der pastenförmig aufgebrachten Schichten am Nd:YAG - Laser wurden maximale Oberflächenrauhigkeiten zwischen 12 und 50 µm erreicht. Aufgrund der geringen Dicke der vordeponierten Schicht mußten relativ niedrige Leistungen gewählt werden, um die Karbide zu erhalten. Aufgrund der geringen Dicke der Beschichtungen war es allerdings auch möglich Feinbeschichtungen von weniger als 50 µm Schichtdicke herzustellen. Der Nd:YAG - Laser kombiniert mit dem pastenförmigen Auftragen empfiehlt sich besonders für die Herstellung dünner Beschichtungen, die sehr partiell aufgebracht werden können. Am CO_2 -Laser lassen sich hingegen auch großflächigere Beschichtungen mit Dicken bis zu 3 mm herstellen, indenen die Karbide aufgrund größerer Pulvermengen und einer demzufolge geringeren Energieeinbringung fast vollständig erhalten bleiben.

Zusammenfassend läßt sich sagen, daß durch die Legierungsentwicklung einer Beschichtung auf Eisenbasis durch die Artgleichheit bei der Verwendung auf Stählen die Riß-, Poren- und Haftungsprobleme deutlich verringert werden konnten. CO_2-Laser größerer Leistungen eigneten sich dabei eher zur Herstellung dickerer, flächiger Beschichtungen, indenen die Karbide vollständig erhalten bleiben (siehe Abb. 2). Nd:YAG - Laser geringerer Leistungen empfehlen sich in Verwendung mit dem pastenförmigen Auftragen zur Produktion von Feinbeschichtungen insbesondere für partielle Anwendungen.

Hinsichtlich des Preises ist diese Legierungszusammensetzung sicherlich sehr attraktiv, da der Grundpreis für Eisenbasislegierungen deutlich unter dem von

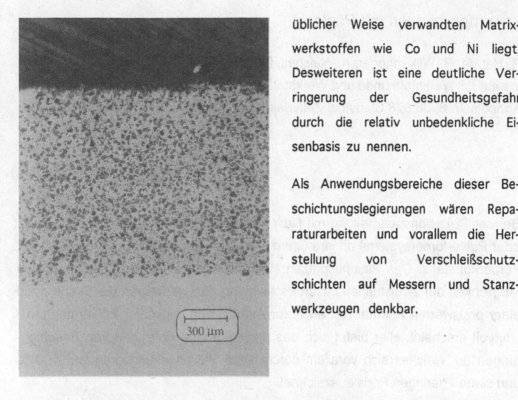

üblicher Weise verwandten Matrix-werkstoffen wie Co und Ni liegt. Desweiteren ist eine deutliche Ver-ringerung der Gesundheitsgefahr durch die relativ unbedenkliche Ei-senbasis zu nennen.

Als Anwendungsbereiche dieser Be-schichtungslegierungen wären Repa-raturarbeiten und vorallem die Her-stellung von Verschleißschutz-schichten auf Messern und Stanz-werkzeugen denkbar.

__Abb. 2:__ Lichtmikroskopische Aufnahme einer Beschichtung aus 50 Gew.%-TiC mit 50 Gew.%- Stahlpulver. Die Karbide sind vollständig erhalten geblieben und ho-mogen verteilt.

Laserbeschichten mit Pasten

R. Haude, R. Wilkenhöner, A. Weisheit, D. Burchards, B. L. Mordike
Institut für Werkstoffkunde und Werkstofftechnik der TU Clausthal
Agricolastr. 6, 38 678 Clausthal-Zellerfeld

Bei der Präzisionsbearbeitung mit Nd:YAG-Lasern stoßen herkömmliche Draht- oder Pulverfördersysteme an ihre Grenzen bezüglich der Fördermengen und der Förderkonstanz. Das Plasmaspritzen stellt hingegen eine sehr kostenintensive Möglichkeit der zweistufigen Laserbeschichtung dar, weswegen die Suche nach einer preiswerten Alternative - auch zur Aufbringung minimaler Pulvermengen - sinnvoll erscheint. Hier bietet sich das pastenförmige Auftragen von Beschichtungen an, welches sich vorallem durch einen Pulverwirkungsgrad von 100 % und seinen geringen Preis auszeichnet.

Ziel dieser Untersuchung war es zunächst einen geeigneten Binder bezüglich Viskosität, Sedimentation, Kantenstabilität und Trocknung zu finden, der weitergehend durch geeignete Verfahren in definierter Schichtdicke aufgetragen werden sollte. Schließlich sollten an dem am Besten geeigneten Binder die Laserparameter optimiert werden.

Als Ausgangsstoffe dienten vier Binder: zwei metallorganische Binder, ein wässriges System und Wasserglas. Diese Binder wurden zunächst mit Stellit 6 Pulver gemischt, durch eine einfache Abziehvorrichtung in unterschiedlichen, definierten Schichtdicken auf das Grundmaterial, einen unlegierten Baustahl, aufgetragen und getrocknet. Dem schloß sich die Laserbearbeitung mit einem 400 W Nd:YAG-Laser der Firma LUMONICS an. Die so hergestellten Beschichtungen wurden bezüglich ihrer Schichtqualität, ihres Gefüges, ihrer Rauhigkeit, Härte und Phasenzusammensetzung untersucht. Schließlich wurden an dem am Besten geeigneten Binder mit der günstigsten Schichtdicke die Laserparameter optimiert.

<u>Abb. 1:</u> Eine Stellit 6 Einzelspur , bei der rechts und links die Paste mit dem Pulver zu erkennen ist.

Bei der Auswertung der Ergebnisse zeigte sich das wässrige System den drei anderen Bindern hinsichtlich Kantenstabilität und Schichtqualität als überlegen. Eine Aufbringung der Pasten in Schichtdicken um 50 bis 100 µm erwies sich als vorteilhaft. Aufgrund der raschen Erstarrung bei der Laserbeschichtung konnten mit dem Stellit 6 Pulver Härten von 800 HV 0,3 erzielt werden. Den Übergang von zellularer zu dendritischer Erstarrung im Bereich der Oberfläche, an der höhere Abkühlgeschwindigkeiten herrschten, zeigt die folgende Abb. 2.

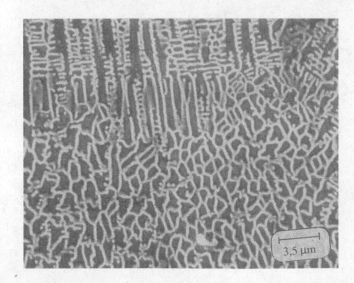

<u>Abb. 2:</u> Rasterelektronenmikroskopische Aufnahme des Gefüges einer Stellit 6 Probe nach der Laserbehandlung.

Die Aufmischung mit dem Grundwerkstoff konnte auf etwa 6 % reduziert werden.

Als Zusammenfassung der Resultate läßt sich sagen, daß das pastenförmige Auftragen von Beschichtungen eine preiswerte Alternative zum Plasmaspritzen darstellt. Es eignet sich besonders für Anwendungen in Kombination mit dem Festkörperlaser und kann dabei insbesondere zur Herstellung von Feinbeschichtungen um 50 μm eingesetzt werden. Desweiteren kann es sehr gut partiell angewandt werden.

Mögliche Einsatzgebiete sind das Auftragen von Beschichtungen beim zweistufigen Laserbeschichten von Verschleiß- und Korrosionsschutzschichten. Ein weiteres Anwendungsgebiet stellt das Laserlegieren dar.

On-Line Monitor for the CO_2-Laser Welding Process

M. Jurca
M. JURCA Optoelektronik, Rodgau/D

1. INTRODUCTION

Generally the quality requirements of the laser welding process are strongly dependent on the function and properties of the workpiece. This makes the definition of an on-line Laser Welding Monitor difficult. The laser welding process represents the equilibrium of several different physical mechanisms acting and partially disturbing each other permanently. Besides the ambient process parameters are also influencing and disturbing this equilibrium. For detecting and classifying these kinds of disturbances it is necessary to measure more than one independent process parameter or signal.

The comparison of the results of different **welding monitoring methods** under industrial welding conditions has shown that the following methods are usable for an on-line monitoring system:

♦ Plasma signal monitoring (including UV-light detectors at different wave-lengths and other sensors with similar output signals),
♦ Spatter size evaluation,
♦ Temperature measurements
  (especially for overlap welds),
♦ Acoustic emission processing
  (only a few special applications).

Some of these signals are ambiguous with respect to the monitoring of the welding quality. Only the combination of two or more signals reduces substantially the number of "false alarms".

2. THE SOLUTION

The here presented monitoring method (see figs. 1 and 2) is based on the detection and processing of the plasma plume UV-light emission **(P-detector)** and the evaluation of the size of the occurring welding spatters **(S-detector)** by detecting their near-infrared light emission.

3. LASER WELDING MONITOR LWM 900

The heart of the LWM 900 is a PC-like computer. The interactive operation with the LWM is made via dialogue boxes on a VGA-monitor and a membrane protected keypad on the LWM-front side (see fig. 3).

The installation of such a system, except the wiring and mounting, takes place under production conditions without disturbing it. After an initial **set-up** all signals of five welds will be memorised and processed to a REFERENCE signal during the **teach mode**. If the quality of the welded parts during this phase is acceptable, the **automatic welding** process can be started. By comparing the actual signals with the memorised REFERENCE data it is possible to detect small welding failures and the drift of some welding process parameters. Simultaneously serial data from laser beam diagnostic devices can be monitored and memorised for documentation purposes.

LWM 900 - SYSTEM FEATURES

- A filter bank processes the electrical spectrum of the P-signal in real-time up to 10 kHz.
- An Automatic Gain Control circuitry in the P-signal processing path avoids the degradation of the monitoring resolution for different signal amplitudes.
- In the TEACH-mode 5 welds are taken to build a REFERENCE for the actual process.
- The TEACH-mode operation optimises automatically the LWM-parameters and considers the workpiece geometry,
- Up to 8 different REFERENCES can be loaded automatically, if different work-pieces have to be welded.
- In the AUTOMATIC-mode the actual signals are compared in real-time with the REFERENCE.
- The NIR-spatter detection is combined with the P-signal processing, because no other signal contains the information on spatters.
- Failure processing occurs within 0.5 s after the welding process using FUZZY-LOGIC routines.
- All signals are related to the actual welding nozzle position.
- Simple dialogue based operation is used.
- Welding process drift is displayed over 24 h.
- 100% welding process documentation is available .
- Additional sensor inputs are available.
- Adjusting and calibration tools are available as options.
- Remote system and process diagnostic as well as general customer support via MODEM are possible.

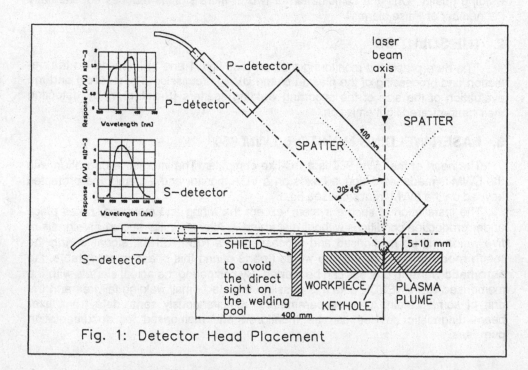

Fig. 1: Detector Head Placement

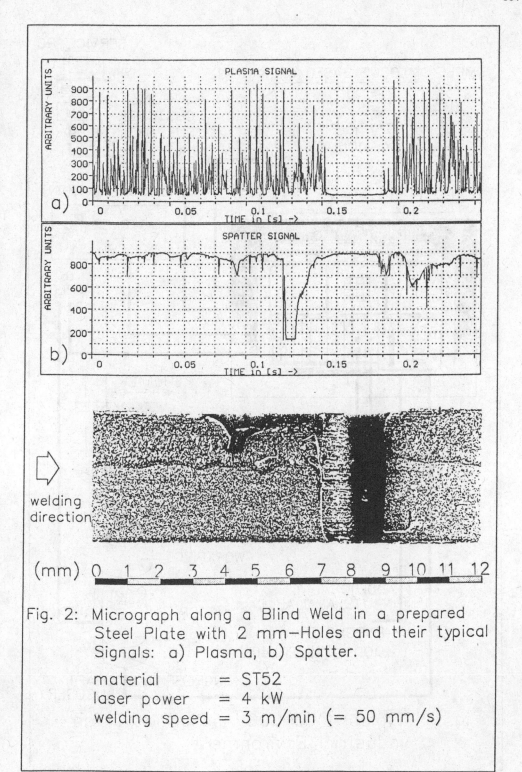

Fig. 2: Micrograph along a Blind Weld in a prepared
Steel Plate with 2 mm—Holes and their typical
Signals: a) Plasma, b) Spatter.

material   = ST52
laser power   = 4 kW
welding speed = 3 m/min (= 50 mm/s)

502

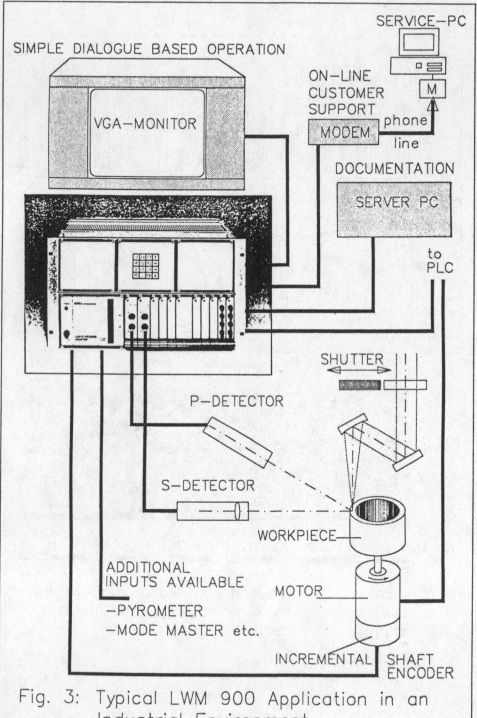

SIMPLE DIALOGUE BASED OPERATION

SERVICE-PC

ON-LINE
CUSTOMER
SUPPORT

VGA-MONITOR

M

MODEM

phone
line

DOCUMENTATION

SERVER PC

to
PLC

SHUTTER

P-DETECTOR

S-DETECTOR

WORKPIECE

ADDITIONAL
INPUTS AVAILABLE

MOTOR

-PYROMETER
-MODE MASTER etc.

INCREMENTAL SHAFT
ENCODER

Fig. 3: Typical LWM 900 Application in an
Industrial Environment

Aspects for Quality Assurance with a Plasma-Monitoring System During Laser Beam Welding

K. Zimmermann, R. Klein, R. Poprawe
Thyssen Laser-Technk GmbH, 52074 Aachen

1. Introduction

Significant advancement has been achieved in welding by laser radiation in the last years considering the technological characteristics of the process and its applicability for industrial manufacturing. Due to the high quality of laser energy and its high degree of reproducibility many applications have been developed and are in operation today. However, general manufacturing technology tends to demand not only high manufacturing quality but emphasis is brought to the control of the quality increasingly. This tendency has been taken up by many institutes and promising results on fundamental aspects of on-line quality control during laser welding have been demonstrated [1,2]. Especially the analysis of the laser induced plasma seems to be an important key to monitor the process quality on-line. Yet, the research results so far involve complex equipment and are realized mainly on a laboratory scale.

In this paper a plasma-monitoring system shall be described in its characteristics of operation. Emphasis has been laid on the correlation between welding quality and observed characteristic data of the plasma behaviour. Of major importance are the parameters

- intensity of the laser induced plasma
- temporal behaviour of the plasma intensity
- detection of plasma-interrupts and their correlation to the process quality

In summary the described procedure and device are capable of monitoring the laser welding quality in terms of pores and holes with significant accuracy especially for the application of welding automotive sheet metal of a typical thickness of 0.8 mm. Zinc-coated as well as uncoated material has been investigated.

2. Experimental set up

When welding with high power lasers the metal vapour plasma above the keyhole has vital influence on the coupling of the incident laser energy into the work piece. The fluctuations in appearance of the plasma are closely related to the laser welding process itself. Through analysing the fluctuations in duration and intensity in suitable frequency bands, momentary information on the process can be gained. Thus, a differentiating monitoring technique evaluating the produced weld seam quality seems to be feasible. For the outlined purpose an experimental set up, shown in Figure 1, is briefly described in the following. A deep-drawing sheet metal of the grade St14 has been used for all experiments. The sheet thickness is 0.8 mm. The employed laser source was a 2.5 kW CO_2-laser with a TEM_{10} mode structure. To obtain a signal of the plasma emitted light two methods have been examined:

- In the first case, a multi glass fibre bundle is attached to the outside of the welding nozzle. The fibre having a direct view on the plasma is equipped with a gas jet, protecting the optical surface against contamination. In addition the fibre is both transmitting the plasma light to the diode via an interference filter (450 nm, 10 nm bandwidth) and itself absorbs reflected laser radiation. With a 6 mm diameter fibre bundle enough plasma light could be collected to place the inlet up to 200 mm away from the processing zone.

- Another method is collecting the emitted plasma light through the welding head via a focusing scraper mirror. Since the wavelengths of interest are not efficiently reflected by plane copper mirrors molybdenum or other similar coatings should be utilised. This method is specially suitable for facilities offering very few space around the welding head.

From the diode, having its peak intensity at 500 nm, the signal is amplified and transported to the controller unit, where ahead of further processing the signal is analog/digital transformed at a sampling frequency of 10 kHz.

3. Monitoring technique

Initial motivation for development of a new monitoring technique was the detection of long interrupts in the laser induced plasma leading for instance to holes. This standard can be considered state of the art and commercial systems are available. These systems basically are able to find major errors in the weld quality with a digital evaluation method but fail to offer a detailed view on the welding process itself yielding quantitative data on the process quality. In this paper the development of a more differentiating monitoring technique is investigated adding to the phases "green" for excellent quality and "red" for rejection a third "yellow" phase, were the process tends to get unstable but still is producing good quality parts. In this phase the facility operator should be supplied with enough information by the plasma monitoring system to re-stabilise the process.

Provided the monitoring criteria can be derived with a reasonable cost-benefit ratio in terms of later realisation in software running on a PC-board, the following charact. data have been chosen (Fig. 2):

1. Plasma interrupts grouped in three categories
2. Plasma flashes grouped in two categories
3. Average plasma intensity

Interrupts or flashes in the plasma are emphasised events to the frequent fluctuations bearing the potential of a high information content. This has been worked out in prior programmes [3], but in distinction to the observation of single events an additional information content is attainable through investigating their duration length and quantity of appearance. Fluctuations remaining under a definable plasma intensity threshold are identified as an interruption and a timer is subsequently started. The elapsed time until the momentary intensity level has exceeded the set threshold again is measured and afterwards sorted into three different categories. The three categories are defined by two independent adjustable times T_1 and T_2:

Category 1: $t < T_1$
Category 2: $T_1 < t < T_2$
Category 3: $T_2 < t$

Every detected interrupt is registered in one of the categories where the corresponding counter is incremented. After an adjustable time period the counter positions are read out for the subsequent evaluation process. Plasma flashes, originating from laser induced absorption waves with partial shielding of the surface i.e. when using argon as assist gas with insufficient flow rates or other incorrect parameters are basically registered in the same manner as mentioned above. In this work, the different flash lengths are grouped in two categories identified by a longer or shorter duration then the adjustable time T_3 respectively. Finally, the average integral intensity of the plasma emitted light is utilized and evaluated in terms of exceeding the adjustable upper and lower limits I_{max} and I_{min}. The so far mentioned monitoring features are programmed in a real-time assembler routine running on a 386 PC-board framed by an interactive user surface written in Pascal-language.

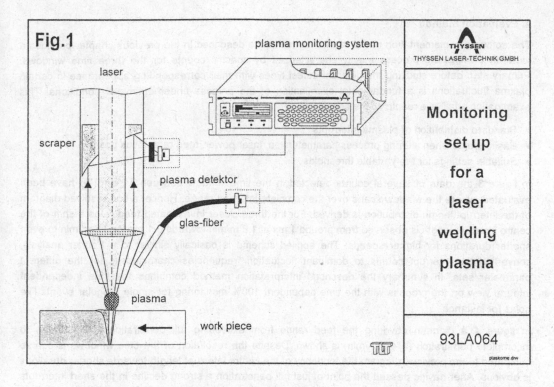

Fig.1

plasma monitoring system

THYSSEN
THYSSEN LASER-TECHNIK GMBH

laser

scraper

plasma detektor

glas-fiber

plasma

work piece

Monitoring
set up
for a
laser
welding
plasma

93LA064

plaskome.drw

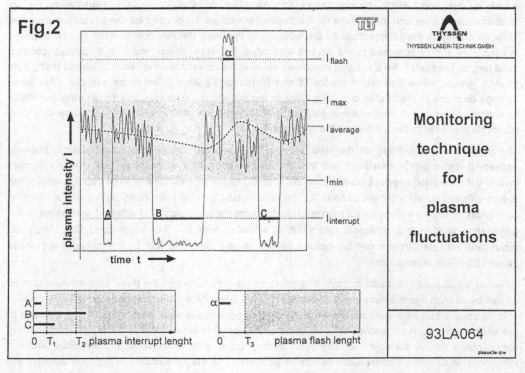

Fig.2

THYSSEN
THYSSEN LASER-TECHNIK GMBH

I_{flash}

I_{max}

$I_{average}$

I_{min}

$I_{interrupt}$

plasma intensity I

time t

α

A   B   C

A
B
C

0   T_1 T_2 plasma interrupt lenght

α

0   T_3 plasma flash lenght

Monitoring
technique
for
plasma
fluctuations

93LA064

plaaus3e.drw

4. Evaluation method

The software implementation of the monitoring technique described in the previous chapter provides a view on the welding process mainly characterized by specific counts for the three time windows. Primary step before studying the different defect types with their corresponding appearance in certain plasma fluctuations is a fundamental examination of the process under optimized conditions. This examination shall give results to the aspects of:

- Standard distribution of plasma interrupts
- Basic effects when altering process parameters i.e. laser power, feed rate, focus position
- Suitable settings for the variable thresholds

In figure 3 the data of several counts detected in the time window between T_1 and T_2 have been evaluated, where the window scans over the complete interrupt range. Hence, a fine resolved diagram of the interrupt length distribution is derived. For the three closer investigated feed rates a shift of the centre interrupt length is observed from around 1 ms at 1.8 m/min towards 0.1 ms for 3.6 m/min to even shorter durations for higher speeds. The applied scheme is basically similar to a Fourier analysis, converting the interrupt counts to dominant fluctuation frequencies characteristic for the different parameter sets. In summary the described interpretation method combines the time independent integral view on the process with the time dependent 100% monitoring for single irregular events like holes for instance.

In figure 4 a diagram covering the feed range from exceeding full penetration (< 2.4 m/min) to incomplete penetration (> 4.2 m/min) is shown. Despite the resolution of the time windows is coarse compared to the previous diagram the tendency of the centre interrupt length towards shorter durations is obvious. After having passed the point of just full penetration a strong decline in the short interrupts count is detected, which clearly marks the transition to the incomplete penetration region. Subsequently, upper and lower limits for the three channels are introduced for the evaluation purpose. Through changing the time settings for the windows or the limits the monitoring result can be adjusted in sensitivity and be matched to the desired application (i.e. laser power, material, thickness, coating, focal length, feed rate). As a suitable parameter set to achieve a monitoring result similar to traffic light phases "green, yellow and red" 1 ms for T1 and 10 ms for T2 were found to be suitable. With these settings changes in the counts only take place in the two shorter time windows for altering but stable conditions whereas the third channel only responds to irregular events with their corresponding long interrupts that directly have effect on the weld quality.

The plasma rate is defined as the ratio between the observed time with ambient plasma intensity exceeding the interrupt threshold and the absolute elapsed time in percent. It is derived through inverting the counted interrupt times. Since the coupling of the offered laser energy into the work piece is less efficient during interrupt periods [4], the plasma ratio can be interpreted as a general coupling coefficient. In Figure 4 the plasma ratio is evaluated from data of welds with different feed rates. A full penetration weld with a sufficient root width is achieved with 3 - 3.6 m/min feed rate. Thus, an incomplete root penetration can be avoided for the utilized parameter set by controlling the process around 50 - 70% plasma rate.

Different focus positions seem to have little effect on the ratio between the three time window counts, as can be seen in figure 5, whereas the average integral intensity strongly responds on a focus position shift. In square butt joint weld applications an excellent edge preparation and a gap free positioning of the sheets are important prerequisites for good quality joins. However, this is an aspect emphasizing the demand for a suitable monitoring technique, since both mentioned prerequisites are not influenceable or difficult controllable factors respectively. The diagram shown in figure 6 displays the

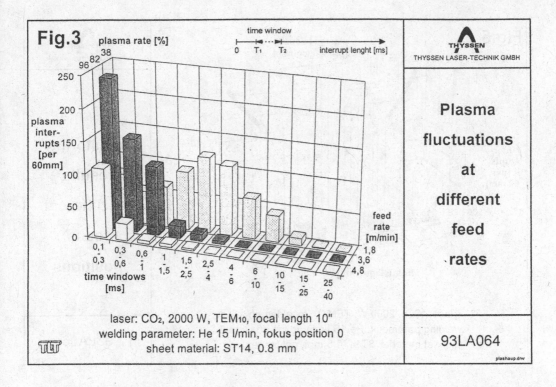

Fig.3

plasma rate [%]

time window

0   T₁   T₂   interrupt lenght [ms]

THYSSEN
THYSSEN LASER-TECHNIK GMBH

Plasma

fluctuations

at

different

feed

rates

laser: CO₂, 2000 W, TEM₁₀, focal length 10"
welding parameter: He 15 l/min, fokus position 0
sheet material: ST14, 0.8 mm

93LA064

plashaup.drw

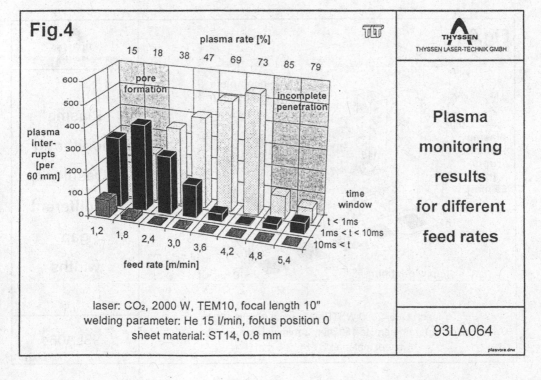

Fig.4

plasma rate [%]

THYSSEN
THYSSEN LASER-TECHNIK GMBH

Plasma

monitoring

results

for different

feed rates

laser: CO₂, 2000 W, TEM10, focal length 10"
welding parameter: He 15 l/min, fokus position 0
sheet material: ST14, 0.8 mm

93LA064

plasvore.drw

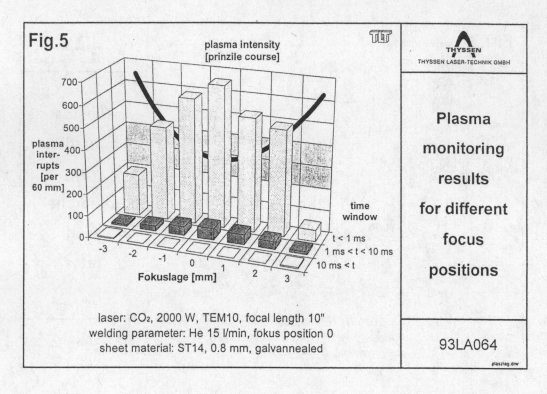

Fig.5

plasma intensity
[prinzile course]

plasma
inter-
rupts
[per
60 mm]

700
600
500
400
300
200
100
0

Fokuslage [mm]

-3 -2 -1 0 1 2 3

time
window

t < 1 ms
1 ms < t < 10 ms
10 ms < t

laser: CO₂, 2000 W, TEM10, focal length 10"
welding parameter: He 15 l/min, fokus position 0
sheet material: ST14, 0.8 mm, galvannealed

THYSSEN
THYSSEN LASER-TECHNIK GMBH

Plasma

monitoring

results

for different

focus

positions

93LA064

plaszlag.drw

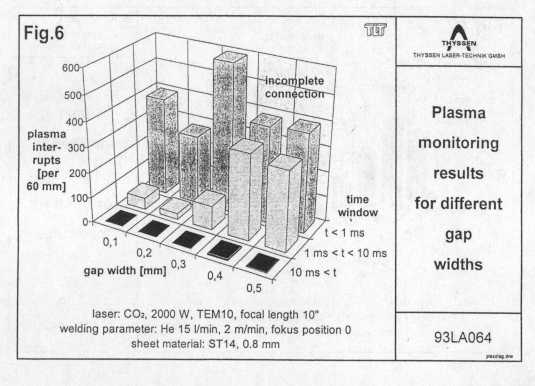

Fig.6

plasma
inter-
rupts
[per
60 mm]

600
500
400
300
200
100
0

incomplete
connection

gap width [mm]

0,1 0,2 0,3 0,4 0,5

time
window

t < 1 ms
1 ms < t < 10 ms
10 ms < t

laser: CO₂, 2000 W, TEM10, focal length 10"
welding parameter: He 15 l/min, 2 m/min, fokus position 0
sheet material: ST14, 0.8 mm

THYSSEN
THYSSEN LASER-TECHNIK GMBH

Plasma

monitoring

results

for different

gap

widths

93LA064

plaszlag.drw

effect of an increasing gap width on the interrupt counts. Starting from zero gap a strong influence of the width on the characteristic count ratio and amount is observed. Thus, the monitoring is accomplished by the previously introduced upper and lower count limits. The same correlation is valid for the monitoring of the joint of top and bottom sheet of an overlap weld. From a certain gap width on where no junction is achieved the typical counts change immediately.

In summary the described evaluation methods for the individual effects all have a deviation from the characteristic count ratio or number of counts in common. This represents the key effect utilized for the operation of the plasma monitoring system.

5. Performance specifications for the plasma monitoring system

Apart from realizing the outlined evaluation process for the monitoring results the succeeding aspects for a plasma monitoring system have to be taken into account:

- 100% on-line monitoring
- User surface for the operator
- Compatibility to industrial standards
- Options for further steps towards quality control

To establish a true 100% on-line monitoring the basic structure of the software is organized in 10 kHz cycles (100 µs) around the sampling routine occupying near 60 µs CPU-time. The evaluation routines are split in several cycles running in loops in the spare time after the sampling routine. The update of the display or the I/O-ports and the read out of the keyboard buffer is performed in the same manner. The user surface is divided into the monitoring and the editing mode. At first the system is completely controlled via interface. Hence, the monitoring process is started and subsequent results are transmitted to the installation controller. The editing mode having a password protection offers a menu-driven adjustment possibility for all monitoring variables.

The operator can choose between the monitoring of short welds of defined length and virtually continuous welds. For the short welds the monitoring process is started once and the reached count levels are evaluated after finishing the complete seam length. This offers an integral view on the produced quality having no local resolution. When the monitoring is frequently intermitted, the determined counts are directly evaluated and finally reseted to commence the next cycle. This technique is utilized for longer or continuous welds where a local resolution of defects is required. In both cases, the determined monitoring and evaluation data can be stored on the built in Floppy Disk or are sent to the printer port for documentation purposes. The data is also available at the serial RS232 Interface in prospect of further network integration and evaluation steps towards quality control.

6. Summary

A plasma monitoring system is described and demonstrated in its operation with examples offering a highly differentiating view on the laser welding process. In principle, the stability limits for hole- and pore-formation as well as the degree of penetration are covered. The data evaluation adds to the phases "green" for excellent quality and "red" for rejection a third "yellow" phase, were the process tends to get unstable but still is producing quality parts in a given tolerance band. Therefore, the facility operator has the opportunity to re-stabilise the process prior to malefunction of the production system. Through changing the evaluation variables the monitoring result can be adjusted in sensitivity and be matched to various application.

510

References:

[1] W. Gatzweiler, D. Maischner, E. Beyer; "On-line plasma diagnostics for process-control in welding with CO_2 lasers"; Fraunhofer Institut für Lasertechnik

[2] G. Herziger; "The Industrial Laser Annual Handbook 1986"; eds. D. Belforte and Lewitt, Penn Well Books (Laser Fokus)

[3] Y. Arata; N. Abe, T. Oda; "Beam Hole Behaviour during Laser Beam Welding"; L. I. A. Vol. 38; Proc. ICALEO '83

[4] E. Beyer, A. Gasser, W. Gatzweiler, W. Sokolowski; "Plasma Fluctuations in Laser Welding with cw-CO_2-lasers"; Proc. ICALEO '87

Laserstrahlschweißen von Aluminiumlegierungen mit KW Nd:YAG-Lasern

E.U. Beske und J. Schumacher
Laser Zentrum Hannover e.V.
Hollerithallee 8   D-30419 Hannover

1. Einleitung

Aluminium hat sich mittlerweile als Konstruktionswerkstoff hoher Festigkeit und Korrosionsbeständigkeit etablieren können. Das Haupteinsatzgebiet liegt im Bereich der dünnblechverarbeitenden Industrie und der Luftfahrtindustrie. Aufgrund seiner geringen Dichte und der damit verbundenen Energieeinsparung durch Gewichtsverminderung gewinnt dieser Werkstoff auch im Bereich der Automobilindustrie zunehmend an Bedeutung. In diesem Umfeld stellt das Laserstrahlschweißen eine interessante Möglichkeit zum Fügen von Aluminiumlegierungen dar /1/, da traditionelle Fügetechniken wie z.B. das Widerstandspunktschweißen nur sehr begrenzt einsetzbar sind. Mit den in neuerer Zeit verfügbaren mittleren Leistungen von 2 kW und erheblich höheren Pulsspitzenleistungen können Nd:YAG-Laser mit dem Vorteil einer äußerst flexiblen Strahlführung über Lichtwellenleiter auch für dreidimensionale Fügeaufgaben genutzt werden /2, 3/. Vorteilhaft gegenüber den CO_2-Lasern ist weiterhin eine erhöhte Absorption durch Aluminiumwerkstoffe /4/. Eine Abschirmung des Fügebereiches von der Laserstrahlung durch Plasmabildung wurde für Nd:YAG-Laser bisher nicht beobachtet /5/.

In diesem Beitrag werden die Randbedingungen für einen derartigen Fügeprozeß beschrieben. Die z.T. problematische Porenbildung und Rißentstehung beim Schweißprozeß werden anhand von Untersuchungen zum Überlappstoßschweißen an den Werkstoffen AlMg0.4Si1.2, AlMgSiCuMn und AlMg3 charakterisiert. Abschließend werden Methoden zur Vermeidung dieser Schweißnahtimperfektionen aufgezeigt.

2. Probleme beim Laserstrahlschweißen von Aluminiumlegierungen

Die beim Laserstrahlschweißen von Aluminiumlegierungen häufig auftretende Porenbildung stellt keine laserspezifische Besonderheit dar, sondern kann auch bei konventionellen Schweißverfahren beobachtet werden /6/. Aus mehreren Untersuchungen ist bekannt, daß diese Porenbildung zum großen Teil wasserstoffinduziert ist /7, 8/. Allerdings bestehen derzeit unterschiedliche Theorien woher der Wasserstoff aufgenommen wird, der dann beim Erkalten der Schmelze aufgrund abnehmender Löslichkeit zur Entstehung von Poren im Schweißnahtgefüge führt. Grundsätzlich kann Wasserstoff aus der umgebenden Atmosphäre oder aus der den Aluminiumwerkstoff umgebenden Oxidschicht aufgenommen werden. Voraussetzung für die erste Annahme ist eine nicht ausreichende Abschirmung des Schweißbades.

Je nach Porengröße und deren Verteilung im Nahtquerschnitt wirken sich die Poren negativ auf die erzielbare Schweißnahtfestigkeit aus. Daher wurden unterschiedliche Me-

thoden zur Vermeidung der Porenbildung untersucht, die im einzelnen in Kap. 4 dargestellt werden.

Noch erheblich kritischer bezüglich der erzielbaren Dauerschwingfestigkeiten ist für
viele Aluminiumlegierungen die Rißbildung. Gerade die für das Laserstrahlschweißen
mit gepulsten Nd:YAG-Lasern charakteristischen enorm hohen Abkühlgeschwindigkeiten
der Schmelze induzieren z.T. eine sehr ausgeprägt Rißbildung bei Legierungen mit vergleichsweise geringer Menge an Korngrenzeneutektikum /9/. Ursache ist die bei fallenden Temperaturen in der Schmelze abnehmende Menge an Eutektikum bei ansteigendem Volumen der bereits erstarrten Körner aus Legierungsbestandteilen. Durch den hohen Wärmeausdehnungskoeffizienten kommt es während der Abkühlung gleichzeitig zu einer Veränderung der Lage der Körner. Die dabei auftretenden Verschiebungslücken müssen durch
die noch flüssigen Bestandteile aufgefüllt werden, da diese den geringsten Verfor-
mungswiderstand aufweisen. Die zwangsläufig auftretenden Werkstofftrennungen können
nur zu einem Teil durch das nachfließende Korngrenzeneutektikum ausgefüllt werden. Es
entstehen so Fehlstellen, die sich bei weiterer Erstarrung und Abkühlung des Werkstoffes zu Rissen aneinanderreihen. In Kap. 5 werden unterschiedliche Methoden zur
Vermeidung der Rißbildung und Ergebnisse zu Untersuchungen mit einer Vorwärmung des
Werkstoffes vorgestellt.

3. Laserstrahlquelle

Im Rahmen der beschriebenen Untersuchungen wurde ein gepulster Nd:YAG-Laser der Firma
Lumonics eingesetzt, der eine mittlere Ausgangsleistung von bis zu 1,5 kW durch eine
Parallelschaltung von drei Kavitäten erzielt. Die Strahlführung erfolgt über drei
einzelne Stufenindex-Lichtwellenleiter mit einem Kerndurchmesser von 800 μm zu einem
Bearbeitungskopf. Hier werden die drei Teilstrahlen kollimiert und auf einen gemeinsamen Brennpunkt auf der Werkstückoberfläche fokussiert. Vorteilhaft für das Laserstrahlschweißen von Aluminiumwerkstoffen ist bei dieser Strahlquelle, daß vergleichsweise hohe Pulsenergien bis zu 60 J bei Pulsspitzenleistungen im Bereich zwischen 5
und 10 kW bei relativ hohen Pulsfolgefrequenzen zur Verfügung stehen. In diesem Para-
meterbereich lassen sich qualitativ hochwertige Durchschweißungen im Überlappstoß bei
einer Blechdicke von 2 X 1.5 mm erzielen. Die Pulsfolgefrequenzen liegen dann immer
noch bei ca. 50 Hz, so daß sich bei ausreichender Überlappung der einzelnen Laserpulse Vorschubgeschwindigkeiten im Bereich größer 1 m/min. ergeben.

4. Methoden zur Vermeidung der Porenbildung

Die Untersuchungen zur Porenbildung wurden an drei unterschiedlichen Aluminiumlegierungen (AlMg0.4Si1.2, AlMgSiCuMn, AlMg 3) durchgeführt. Eine Modifikation des Bearbeitungskopfes ermöglichte den vollständigen Abschluß der Schweißstelle von der umgebenden Atmosphäre. Die Porenbildung kann durch diese Maßnahme jedoch nur unwesentlich
verringert werden. Offensichtlich wird ein großer Anteil des Wasserstoffes in den Poren durch die den Werkstoff umgebende Oxidschicht induziert, die im Überlappstoß ohne
Durchschweißung dreifach in der Schweißnaht aufgeschmolzen wird.
Erheblich größeren Einfluß auf die Porenbildung haben dagegen die Parameter der Laserpulse. Bei vergleichbar hoher Einschweißtiefe konnte eine Porenbildung durch For-

mung der einzelnen Laserpulse weitgehend unterdrückt werden (Pulsformung). Der leistungsführenden Puls wird dazu mit einem sogenannten "Nachsektor" verlängert. Dies ermöglicht am Pulsende während der bereits einsetzenden Verringerung der Temperatur der Schmelze eine erhöhte Ausgasungszeit für den Wasserstoff. Allerdings führt diese Vorgehensweise auch zu einer Verringerung der Energie des Laserpulses, die für den Aufschmelzvorgang genutzt werden kann. Daher muß i.a. die Pulsfolgefrequenz verringert werden, um eine gleichbleibende Einschweißtiefe gegenüber dem Prozeß ohne Pulsformung zu erreichen. Die Vorschubgeschwindigkeit verringert sich, da eine Überdeckung von mindestens 50% des minimalen Fokusradius pro Puls einzuhalten ist, um eine homogene Verbindung zwischen den beiden Blechen sicherzustellen.

5. Methoden zur Vermeidung der Rißbildung

Wesentlichem Einfluß auf die Entstehung von Heißrissen haben einerseits die Legierungsbestandteile und andererseits das Schweißnahtvolumen. Aufgrund des Anforderungsprofiles wurde mit allen drei Legierungen ein Überlappstoß bei einer Blechdicke zwischen
2 x 1,25 mm und 2 x 1,5 mm angestrebt, bei dem die Verbindungsnahtbreite mindestens der einfachen Blechdicke entspricht.

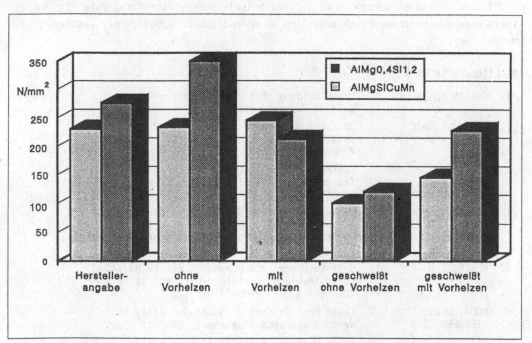

Bild 1: Einfluß des Vorheizens auf die Zugfestigkeit nach DIN 50145

Zur Vermeidung der Rißbildung stehen unterschiedliche Methoden zur Verfügung, denen gemeinsam ist, daß der zeitliche Temperaturgradient in der Abkühlphase zwischen zwei aufeinanderfolgenden Laserpulsen verringert wird:

- Vorwärmen des Werkstoffes im Bereich der Schweißnaht auf Temperaturen über 250°C. Die Wirksamkeit dieser Methode wurde für zwei Werkstoffe nachgewiesen (Bild 1)
- Einsatz eines Lasers mit erhöhter mittlerer Leistung (> 2kW) aufgrund eines höheren Tastverhältnisses bei ausreichender Pulsspitzenleistung (> 3 kW). Diese Strahlquellen sind derzeit noch nicht verfügbar, mit einer Markteinführung ist aber in den nächsten Jahren zu rechnen.

Andere Methoden basieren auf einer Änderung der Zusammensetzung der Schmelze. Dies erfordert die direkte Zuführung von Zusatzmaterial in Form von Pulver oder Draht in die Schmelzzone:

- Vergrößertes Volumen des Korngrenzeneutektikums durch Zusatzwerkstoff mit niedrigschmelzenden Legierungslementen.
- Bei Legierungen, die Magnesium enthalten, Zusatzwerkstoff mit erhöhtem Magnesiumanteil, so daß dieses bei niedrigen Temperaturen verdampfende Element in ausreichender Menge im Schweißnahtgefüge vorhanden ist.

Die Entscheidung, welche der Methoden einzusetzen ist, wird wesentlich durch die Randbedingungen wie beispielsweise Bauteilgeomtrie, Zugänglichkeit etc. beeinflußt und kann nicht allgemeingültig entschieden werden. Im Vorfeld ist grundsätzlich abzuklären, ob nicht durch ein verändertes Nahtvolumen oder einen anderen Werkstoff eine Rißbildung vermieden werden kann. So sind beispielsweise Stumpfstoßnähte für den untersuchten Blechdickenbereich erheblich unempfindlicher bezüglich der Entstehung von Heißrissen.

6. Literaturverzeichnis

/1/	HUNTINGTON, C.A.	Laser Welding of Aluminium Alloys Welding Journal, 4/1983
/2/	TÖNSHOFF, H.K. BESKE, E.U.	3D-Laser Material Processing by kW Nd:YAG-Lasers, Optical Fibre and Robotics, Proc. of ISATA, Vienna, 1990
/3/	Bütje, R. Meyer, C. Bödecker, V.	Festkörperlaser im kW-Bereich, Teil 4 Industrie Anzeiger 110/1988 Nr. 59/60
/4/	Tönshoff, H.K. Beske, E.U. Schumacher, J.	Untersuchungen zur Energieeinkopplung beim Schweißen metallischer Werkstoffe mit kW- Festkörperlasern, Vortrag DPG-Tagung Greifswald, März 1993
/5/	BESKE, E.U.	Untersuchungen zum Schweißen mit kW Nd:YAG- Laserstrahlung, Fortschr.-Ber. VDI Reihe 2 Nr. 257 VDI-Verlag, Düsseldorf 1992
/6/	MOON, D.W. ETZBOWER, E.A.	Laser Beam Welding of Aluminium Alloy 5456 Welding Research Supplement, Februar 1983
/7/	MASUMOTO, I. KUTSUNA, M.	Laser Welding of AA5083 Aluminium Alloy DOC. IV-566-91
/8/	N.N.	Fügen mit CO_2-Hochleistungslasern 3. Zwischenbericht, Fördernummer 13 N 5651 0, 1990, Dornier Luftfahrt GmbH
/9/	WEIßMANTEL, H. JUNG; R.	Rißbildung von Aluminium beim Schweißen mit dem Laserstrahl, Feinwerktechnik und Meßtechnik 88, 1980

Prozeßkontrolle beim Laserstrahlschweißen von Aluminium mit Ar/He-Arbeitsgasgemischen

J. Beersiek, B. Seidel, J. Berkmanns, K. Behler, E. Beyer
Fraunhofer-Institut für Lasertechnik, Steinbachstr. 15, 52074 Aachen

Zusammenfassung:

Die Prozeßentwicklung zum Laserstrahlschweißen von Aluminium Werkstoffen hat gezeigt, daß eine wesentlich genauere Anpassung des Arbeitsgasgemisches aus Ar und He an die Laserleistung, Strahlintensität und die Bearbeitungsgeschwindigkeit erforderlich ist, um eine optimale Nahtqualität zu erreichen, als bei Stahl. Die besten Ergebnisse hinsichtlich Nahtqualität und Energieeinkopplung werden erzielt, wenn der Prozeß nahe am Bereich der Plasmaabschirmung geführt wird. Eine hohe Prozeßsicherheit und damit eine gesicherte Bearbeitungsqualität kann mittels zeitaufgelöster Analyse und Kontrolle des laserinduzierten Plasmas erreicht werden.
Mit spektroskopischen Methoden werden Kontrollsignale generiert, die zur Regelung des Schweißprozesses genutzt werden können. Ein Beispiel zeigt,daß dieser Regelkreis das Auftreten abschirmender Plasmen sicher verhindern kann.

Einleitung:

Die Variationsbreite der Bearbeitungsparameter für das Laserstrahlschweißen von Aluminium (Fokuslage, Prozeßgasführung, Vorschubgeschwindigkeit, Laserleistung, etc.), innerhalb der eine gesicherte Prozeßführung und eine ausreichende Nahtqualität erzielt wird, sind wesentlich kleiner als bei Stahl. In der industriellen Anwendung können durch auftretende Abweichungen innerhalb der Fertigungstoleranzen oder fertigungstechnische Vorgaben (3D- Anwendung) Fehler induziert werden. Die Sicherung eines hohen Qualitätsstandards, wird durch die On- line Kontrolle des Bearbeitungsvorganges unterstützt.

Stand der Prozeßentwicklung:

Es hat sich herausgestellt, daß mit dem Einsatz eines He/Ar - Arbeitsgasgemisches, bei sonst konstanten Prozeßparametern, sowohl die Energieeinkopplung, wie auch die Nahtqualität optimiert werden kann /BERKMANNS, MATSUMARA/. Im Vergleich zu Schweißungen mit Stickstoff als Arbeitsgas ergab sich auch eine verbesserte dynamische Schwingfestigkeit der Schweißnähte. In Bild 1 ist beispielhaft die Querschnittsfläche der Schweißnaht in Abhängigkeit von dem Mischungsverhältnis des He/Ar Arbeitsgases beim Schweißen von Al99,5 aufgetragen. Die unter dem Diagramm dargestellten Querschliffe entsprechen der Kurve bei einer Laserleistung von 5kW.
Aus diesen Ergebnissen ist ersichtlich, daß die maximale Querschnittsfläche, d.h. die optimale Energieeinkopplung, mit Schweißparametern nahe an der Grenze zum Bereich der Plasmaabschirmung erzielt wird. Darum ist bei einer optimalen Prozeßführung durch Veränderung von Bearbeitungsparametern während des Bearbeitungsprozesses (Bauteilgeometrie, Fokuslage, Kantenversatz, lokale Variation des Arbeitsgasflusses etc.) das Auftreten von Plasmaabschirmungen nicht auszuschließen. Dieses Problem bildet den Ansatz für das in Bild 2 beschriebene Kontroll- und Regelungskonzept.

Funktionsprinzip:

Das Licht des Bearbeitungsplasmas wird über einen Lichtleiter einem Bandpaßfilter zugeführt. Dieser selektiert eine Spektrallinie aus dem Spektrum des beobachteten Lichtes. Die Intensität dieser Spektrallinie wird mittels eines Photomultipliers oder einer Photodiode in äquivalente elektronische Signale umgewandelt, die dann zur Weiterverarbeitung genutzt werden können. Die zeitlichen Veränderungen von Spektrallinien des Laser induzierten Plasmas zeigen ein sehr komplexes Verhalten, daß von Prozeßparametern und deren Schwankungen während des Prozesses abhängt. Für eine Prozeßkontrolle müssen die, für eine konkrete Überwachung notwendigen Informationen aus den Intensitätsschwankungen selektiert werden. Im Falle eines abschirmenden Plasmas ist die Generierung eines binären Kontrollsignals mit den Zuständen "high" (Plasmaabschirmung liegt vor) und "Low" (erwünschter Bearbeitungs-

Bild 1:

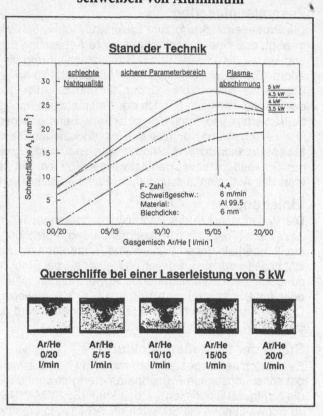

Variation des Arbeitsgases beim Laserstrahl-schweißen von Aluminium

prozeß) sinnvoll. Es kann durch den Vergleich der Intensität einer Spektrallinie mit zwei Schwellwerten entsprechend den Zuständen "high" und "low" erzeugt werden. Die Wahl der Schwellwerte muß an den Bearbeitungsparametern und den Charakteristika des Sensorikproblems (Abstand des Lichtleiters von der Wechselwirkungszone, Bandpaßfilter, Detektor, usw.) orientiert werden. Dabei ist zu beachten, daß die auftretenden Frequenzen des Plasmasignals im Bereich von 1kHz bis 10 kHz liegen. Bei dem heutigen Stand der Technik kann als relevante Prozeßgröße für ein regelndes Eingreifen nur die Laserleistung mit entsprechenden Frequenzen geschaltet werden.

Ergebnisse:

Aus Bild 1 wird deutlich, daß ein Zusammenhang zwischen der Zündung eines abschirmenden Plasmazustandes und dem Anteil von Argon im Arbeitsgas besteht. Neben Argon- Spektrallinien sind im Spektrum eines abschirmenden Plasmas, Spektrallinien von atomaren nicht ionisierten Stickstoffatomen zu finden, Bild 3 / POUEYO/.

Das Auftreten von atomaren Stickstoff- Spektrallinien kann nur dadurch erklärt werden, daß N_2- Moleküle aus der Umgebungsatmosphäre dissoziieren. Dies kann aufgrund der Dissoziationsenergie von Stickstoff (7,6 eV) /HERZ-BERG/ einen Großteil der abschirmenden Wirkung dieses Phänomens ausmachen. Eine quantitative Analyse der Vorgänge ist Inhalt laufender Untersuchungen. Deshalb wurden Spektrallinien des Argon (810 nm) und des atomaren Stickstoff (746,8 nm) auf ihre Tauglichkeit zur Detektion abschirmender Zustände untersucht. Bild 4 zeigt exemplarisch 2 Schweißprozesse, die mit gleichen Parametern ohne und mit Nutzung des Kontrollsignals zur Prozeßregelung durchgeführt worden sind.

Aufgetragen sind jeweils 3 Meßkurven die den zeitlichen Verlauf des Bearbeitungsprozesses anhand einer Argon- bzw. einer Stickstoff- Spektrallinie und des daraus generierten Kontrollsignals wiedergeben. Die Prozeßparameter sind so eingestellt worden, daß sie an der Grenze zu dem Bereich der Plasmaabschirmung entsprechend Bild 1 anzusiedeln sind. Der ungeregelte Prozeß verdeutlicht die Wirkung der Abschirmung auf das Schweißergebnis. In diesem Fall läuft das Kontrollsignal nur als Vergleichsgröße mit und hat keinen Einfluß auf die Laserleistung.

Bild 2:

Meß- und Regelkonzept

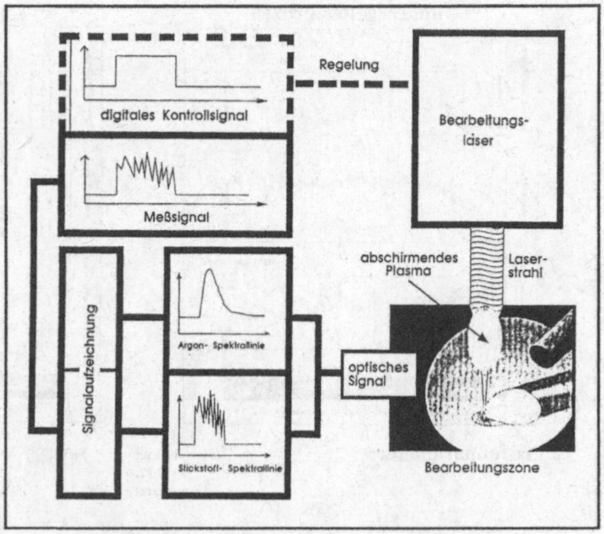

Bild 3:

Charakteristische Spektrallinien eines abschirmenden- und eines Bearbeitungsplasmas

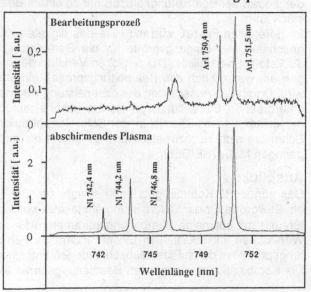

Bild 4: **Vergleich von ungeregeltem und geregeltem Laserstrahlschweißen von Aluminium**

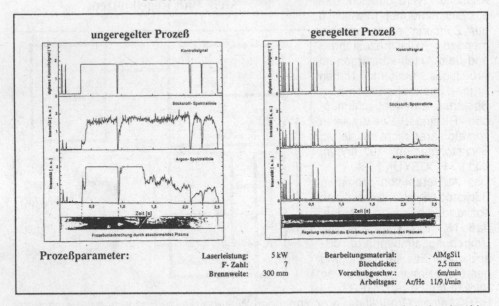

Der Schweißprozeß startet auf dem Werkstück. Nach etwa 0,4 s wird ein abschirmendes Plasma induziert, welches den Schweißprozeß vollkommen unterbricht. Dies ist mit einem deutlichen Anstieg des Signalpegels der Stickstoff- bzw. Argon-Spektrallinie verbunden. Dieser Anstieg schaltet das digitale Kontrollsignal, welches einen Bearbeitungsfehler anzeigt. Eine halbe Sekunde später bricht das abschirmende Plasma zusammen und der Bearbeitungsprozeß setzt wieder ein. Die Signalpegel der Spektrallinien fallen ab und schalten das Kontrollsignal wieder zurück in den "low"- Zustand. Das abschirmende Plasma wird jedoch sofort wieder gezündet, sodaß der Bearbeitungsprozeß nur zu einem Bearbeitungspunkt auf dem Werkstück führt.

Im geregelten Prozeß wird mit Hilfe des digitalen Kontrollsignals die Laserleistung geschaltet. Als Eingangsgröße für die Generierung des Kontrollsignals dient die Stickstoff- Spektrallinie. Die zeitlichen Verläufe der beobachteten Spektrallinien zeigen an, wann durch den Bearbeitungsprozeß ein abschirmendes Plasma induziert wird. Durch das Ausschalten der Laserleistung wird der Abschirmeffekt jedoch sofort unterbrochen. Die Zeitdauer, in der die Laserleistung zurückgenommen wird, ist dabei kleiner als 1ms. Dies ist in der Schweißnaht aufgrund der Relaxationszeiten der Schmelze nicht nachzuweisen und beeinflußt die mittlere Laserleistung nur in sehr geringem Maße /SEIDEL/.

Ausblick:

Das generierte Kontrollsignal kann sowohl zur On- line Prozeßkontrolle, wie auch als Stellgröße für die Laserleistung eingesetzt werden. Dies ermöglicht die Prozeßführung mit optimalen Bearbeitungsparametern für das Schweißen von Aluminium Werkstoffen im industriellen Einsatz indem es Fehler, die während des Bearbeitungsprozeßes durch Plasmaabschirmungen entstehen, anzeigt und verhindert.

Die Kombination aus optimalen Bearbeitungsparametern und die Verhinderung von

Fehlern durch Plasmaabschirmungen ist eine Verfahrensverbesserung des Laserstrahlschweißens von Aluminium, sowohl in Hinsicht auf die Nahtqualität, als auch auf die Qualitätssicherung. Als natürliche Folge ergibt sich daraus eine verbesserte Wirtschaftlichkeit.

Literatur:

Berkmanns J., Behler K., Herziger G., "Schweißen von aushärtbaren und nicht aushärtbaren Aluminiumlegierungen", BMFT- Abschlußbericht F+E- Vorhaben 13N5655, 1993

Matsumara H., et al., "CO_2- laser welding characteristics of various aluminium alloys", Proc. LAMP'92, 529- 533

A.Poueyo, G.Deshors, R. Fabro, A.M. De Frutos, J. Orza, " Study of laser induced plasma in welding conditions with continuous high power Co_2-lasers", Proc. LAMP'92

Herzberg G., "Spectra of Diatomic Molecules", 1950, Van Nostrand Reinhold Company

Seidel B, Sokolowski W., Beersiek J., Meiners W., Beyer E., "Online process control in laser beam welding by means of power regulation in RF- excited CO_2- Lasers" Proc. LASER'93

The Visualization of the Gas Flow During the Welding Process

D. Maischner, D. Becker, G. Funke, E. Beyer

Fraunhofer-Institut für Lasertechnik
Steinbachstr. 15, D-52074 Aachen

1 Introduction

The effectiveness of the welding process and the quality of the weld seam are strongly influenced by the following parameters:

- special type of process gas
- gas flow
    - configuration of the nozzle
    - flow rate
    - laser induced plasma

The gas flow and the influence of the plasma and the melt on the gas flow can only be partially defined by mathematical models or by measurements of quantified parameters. We have proved the effectiveness of the laser light sheet method to describe the gas flow and the welding process qualitatively. Correlations between the gas flow, the welding process and the quality of the weld seam are demonstrated. The aim is to get a deeper understanding of the welding process as well as to optimize the effectiveness of the process and the quality of the weld seam by optimizing the gas flow with respect to the processing parameters.

2 Laser Light Sheet Method

2.1 Principle

The principle of the laser light sheet method can be described as follows. The gas flow is charged with tracers. The tracers are illuminated by the beam of a sounding laser, which is widened in one direction in space using cylindrical lenses. Part of the laser radiation scattered by the tracers is monitored by a camera. The visible movement of the tracers describes the streamlines projected onto the laser light sheet plane. Preconditions are sufficient scattering of the laser beam by the tracers and sufficient tracing behaviour which means sufficient ability of the tracer particles to follow the gas flow.

2.2 Applicability During the Welding Process

The applicability of the laser light sheet method during the welding process requires the adaptation of the following components to the process [1]:

- tracer
- sounding laser
- mode of observation

2.2.1 Tracer

The chosen tracer is boron nitride on the basis of the subsequent criteria:

- tracing behaviour
- temperature increase of the tracer
- scattering of the light of the sounding laser

Tracing behaviour

The tracing behaviour is estimated by calculating the limit frequency of the gas flow which the tracer particles can still follow sufficiently, say with an accuracy of 5%. The calculation of the limit frequency is based on the assumption that the movement of the tracer in the gas flow is only influenced by the stokes-friction and the inertial force. We computed a Fourier transform of the equation of the gas and tracer velocity to find a solution to the equation of motion. The tracing behaviour can be described by the amplitude ratio and the phase angle of the Fourier-transformed gas and tracer velocities.

Helium is used as process gas. In this case boron nitride particles (maximal diameter: 1 μm) can follow the gas flow sufficiently up to a limit frequency of 7 kHz.

Temperature increase of the tracer

The temperature increase of the tracer is mainly due to absorption of part of the power of the processing laser. The scattering and absorption of the laser beam by the tracer is described by the Mie-theory. For a given tracer particle the Mie-theory yields the intensity distribution of the scattered laser light and the amount of beam power absorbed by the tracer.

The temperature increase depends on the velocity of the tracer and the location of passage in the processing laser beam. On the worst assumption that the velocity of the tracer is typical 40 m/s and that it crosses the middle of the CO_2 laser beam (process gas: helium; maximum diameter of the tracer: 1 µm), we calculate an increase in temperature of 1000 °C. The temperature increase is considerably lower than the sublimation temperature of 3000 °C. The temperature increase of the tracer by the plasma and the power of the sounding laser can be neglected.

Scattering of the light of the sounding laser

The light scattered by different tracers at 90° to the laser light sheet plane is monitored with a camera. Of the tracers examined only boron nitride particles showed sufficient scattering of the laser light for the given test set-up.

2.2.2 Sounding Laser

An argon-ion laser with a maximum power of 9 W emitted at a wavelength of 514.5 nm is used. For our realization of the laser light sheet method a minimum power of 7 W is used. In this case sufficient intensity of the sounding laser beam is scattered by the tracers towards the camera. The minimum necessary power is defined by the size of the observation volume and the sensitivity of the observation mode.

2.2.3 Observation Mode

A black-and-white video camera with temporal resolution of 1000 pictures per second is used. To prevent an irradiation of the scattered light by the plasma an interference filter is added to the camera. The filter is transmissive for the wavelength of the sounding laser, with a bandwidth of 1 nm.

3 Results

In welding of aluminium faultless weld seams can be achieved for a wide area of nozzle positions. In overlap welding of zinc-coated steel, the adjustment of the nozzle has an important influence on the quality of the weld seam, especially if the zinc layer is in the joint face and if there is no or only a small gap between the workpieces.

We have proved that there is a correlation between the necessary adjustment of the nozzle and the influence of the plasma on the gas flow during the welding process. The plasma flow and the gas flow can be described by the 'jet-theory'. We calculated the influence of the plasma flow on the gas flow on the basis of the dynamic pressures. For zinc coated

Figure 1: optimal nozzle position   Figure 2: misalignment of the nozzle

Parameters:

material	: aluminium	special type of process gas	: helium-argon gas mixture
thickness of workpiece	: 2,5 mm	flow rate	: 20 l/min
CO_2 laser power	: 4,5 kW	diameter of nozzle	: 6 mm
welding speed	: 4 m/min		

Figure 3: optimal nozzle position   Figure 4: misalignment of the nozzle

Parameters:

material	: zinc-coated steel	special type of process gas	: helium
thickness of workpiece	: 3 mm	flow rate	: 30 l/min
thickness of zinc layer	: 7.5 µm	diameter of nozzle	: 4 mm
CO_2 laser power	: 4,5 kW		
welding speed	: 3 m/min		

steal we found an significant influence of the plasma flow on the gas flow while for aluminium there is almost no influence of the plasma. The experiments with the laser light sheet method have validated the theoretical calculation.

Figure 1 and figure 2 show snap shots of the gas flow and the plasma during welding of aluminium. In case of an optimal nozzle position (fig. 1) as well as in case of an adjustment of the nozzle in the welding direction (fig. 2) the gas flow passed the plasma nearly unaffected. Faultless weld seams are produced.

In welding of zinc coated steal (fig. 3 and fig. 4) a deflection of the gas flow by the plasma is observed for the given processing parameters. As a consequence of the misalignement of the nozzle (fig. 4) there is a formation of pores and holes in the weld seam.

The laser light sheet method can be used to avoid these errors and to allow an accurate positioning of the nozzle.

As a consequence of variations of the keyhole diameter and variations of the plasma flow velocity the influence of the plasma gas flow strongly fluctuates during the welding process. In addition to the influences shown above, the laser light sheet method allows to describe the dynamic interactions between the plasma flow and the gas flow.

Reference

[1] Gerhard Funke;
     Laser-Lichtschnitt-Verfahren zur Untersuchung des Prozeßgaseinflusses beim Tiefschweißen mit CO$_2$-Laserstrahlung;
     Diplomarbeit am Lehrstuhl für Lasertechnik, Mathematisch-Naturwissenschaftliche Fakultät der Rheinisch-Westfälischen Technischen Hochschule Aachen

Eigenschaften des keyholes und des laserinduzierten Plasmas beim Schweißen mit CO$_2$-Laser

Dipl.-Phys. H. Lindl[+], Dipl.-Phys. A. Widl[*]

[+] Institut für Werkzeugmaschinen und Betriebswissenschaften (iwb) der TU München,
Karl-Hammerschmidt-Str. 39, 85609 Aschheim, Deutschland
[*] Mannesmann Pilotentwicklungsgesellschaft mbH, Thomas-Dehler-Str. 18, 81737 München, Deutschland

1. Einleitung

Während des Schweißens mit CO$_2$-Laser formiert sich bei ausreichender Laserintensität ein "keyhole". Geometrische und dynamische Eigenschaften des Schmelzbades in Abhängigkeit der Laserleistung und des Schutzgasflusses lassen sich mit Hilfe eines zweiten Laserstrahls bestimmen, der fokussiert und ortsauflösend durch das keyhole hindurchtritt. Dies wurde für den Fall von unbeschichtetem Dünnblech mit Argon als Schutzgas durchgeführt. Es lassen sich charakteristische Schwingungen der Schmelzbadoberfläche beobachten, deren Frequenzen im wesentlichen unabhängig von unterschiedlichen Einflußgrößen wie Vorschubgeschwindigkeit, Laserleistung und Tastfrequenz des Bearbeitungslasers sind. Gleichzeitig durchgeführte ortsaufgelöste spektroskopische Untersuchungen des laserinduzierten Plasmas an FeI-Linien zeigen ein vom Radius im keyhole und von der Laserleistung nur schwach abhängiges Temperaturverhalten. Die mittlere Temperatur des Plasmas beträgt dabei 0.6 eV.

2. Versuchsaufbau

Zur Untersuchung des keyholeradius wurde zusammen mit dem CO$_2$-Laserstrahl, der einen Fokusradius von etwa 310 μm besitzt, ein Meßlaserstrahl (Nd:YAG gepulst, HeNe kontinuierlich) durch das keyhole geführt, der auf einen Radius kleiner als 30 μm fokussiert wurde (Bild 1). Mit Hilfe einer beweglichen Optik, die auf

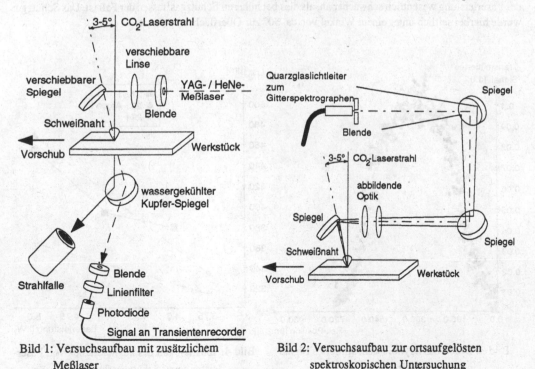

Bild 1: Versuchsaufbau mit zusätzlichem Meßlaser

Bild 2: Versuchsaufbau zur ortsaufgelösten spektroskopischen Untersuchung

einer Linearachse verfahren werden kann, kann das keyhole senkrecht zur Vorschubrichtung abgetastet werden. Das Licht des Bearbeitungslasers und des Meßlasers wird über einen weiteren vor Spritzern geschützten und wassergekühlten Umlenkspiegel unterhalb des Werkstückes umgelenkt. Während das CO_2-Laserlicht in eine Strahlfalle läuft, kann der zeitliche Verlauf der Intensität des Meßlaserlichtes über eine Photodiode mit entsprechendem Linienfilter beobachtet werden. Die Photodiodensignale wurden mit Hilfe eines 10 MHz-Transientenrecorders aufgezeichnet. Als Werkstück wurde Stahl mit Blechstärken von 2 und 3 mm verwendet, das mit einer Blindschweißnaht durchschweißt wurde.

Auch das sekundär emittierte Prozeßlicht wurde untersucht, und zwar einerseits ebenfalls mit Hilfe zweier Photodioden, mit denen über die beiden Umlenkspiegel das Licht von oben und unten gleichzeitig beobachtet werden kann, andererseits mit Hilfe eines Gitterspektrographen, mit dem über eine stark vergrößernde optische Abbildung eine ortsaufgelöste Messung der Elektronentemperatur im keyhole durchgeführt wurde (Bild 2).

3. Keyholeradius

Ein Maß für den keyholeradius erhält man beispielsweise über das ortsabhängige transmittierte Signal des gepulsten Nd:YAG-Lasers. In Bild 3 ist das Integral des Signales der einzelnen Pulse über der Position des Meßlaserstrahles aufgezeichnet. Während einer Schweißung wird dabei der Meßlaserstrahl kontinuierlich über das keyhole hinwegbewegt, so daß man eine über die Schweißnaht gemittelte Information erhält. Als tatsächliches Maß für den keyholeradius kann dann beispielsweise die Breite der Kurve bei der Hälfte des Maximums genommen werden.

Betrachtet man dieses Maß für den keyholeradius in Abhängigkeit einzelner Prozeßparameter, so ist ein deutliches Anwachsen des Radius mit der Laserleistung festzustellen (Bild 4). Der CO_2-Laserstrahl war bei diesen Untersuchungen auf die Blechoberfläche fokussiert. Liegt eine teilweise Abschirmung der Laserstrahlung vor, wie dies im Bild 4 bei einem Schutzgasfluß von 22 l/min Argon der Fall ist, was anhand des Schweißergebnisses durch die Ausbildung eines Nagelkopfes erkannt werden kann, so steigt der Radius mit der Laserleistung wesentlich schwächer an, als dies bei höheren Schutzgasflüssen der Fall ist. Das Schutzgas wurde hierbei seitlich unter einem Winkel von ca. 30° zur Oberfläche zugeführt.

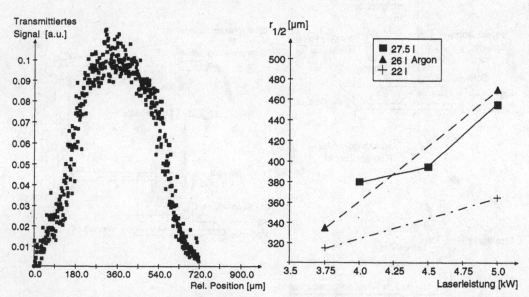

Bild 3: Durch das keyhole transmittiertes Signal in Abhängigkeit der Position

Bild 4: Abhängigkeit des keyholeradius von Laserleistung und Schutzgasfluß

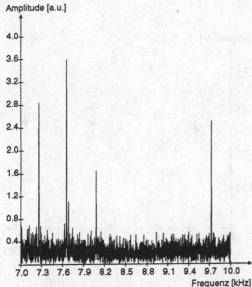

Bild 5:Charakteristische Peaks im Frequenzspektrum zwischen 7 und 10 kHz

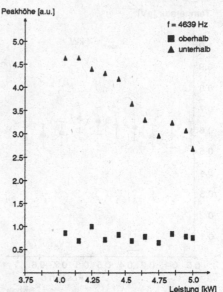

Bild 6:Abhängigkeit der relativen Höhe eines Peaks bei 4640 Hz bei Beobachtung der Ober- bzw. der Unterseite des keyholes

4. Dynamik des keyholes

Beobachtet man den zeitlichen Verlauf des sekundär emittierten Prozeßlichtes, so stellt man nach einer Fouriertransformation charakteristische Peaks im Frequenzspektrum fest, die auch in Körperschallmessungen während des Schweißens eindeutig identifiziert werden konnten /1/. Die Lage dieser Peaks im Frequenzspektrum hängt nur schwach von den Prozeßparametern wie beispielsweise der Laserleistung ab, ihre relative Höhe ist jedoch sowohl von Prozeßparametern abhängig als auch von der Seite von der aus beobachtet wird (von oben bzw. unten) und von der radialen Position im keyhole. Bild 5 zeigt vier dieser Peaks im Bereich von 7 bis 10 kHz, die aus dem zeitlichen Verlauf des Prozeßlichtes stammen. Bei einigen Peaks ist deren relative Höhe stark davon abhängig, ob von die Oberseite oder die Unterseite des keyholes beobachtet wird. Die Amplitude einer solchen Linie ist in Bild 6 in Abhägigkeit der Laserleistung dargestellt, die von der Oberseite aus nicht beobachtbar ist, auf der Unterseite hingegen sehr stark ausgeprägt ist und mit

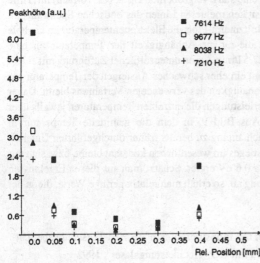

Bild 7:Radiale Abhängigkeit und Lokalisierung an der Schmelzbadoberfläche

zunehmender Laserleistung deutlich abfällt. Betrachtet man zusätzlich die radiale Abhängigkeit dieser Schwingungen mit Hilfe eines fokussiert durch das keyhole geführten und kontinuierlich betriebenen HeNe-Lasers, so ist zu erkennen, daß in diesem Fall die Schwingungen am Rand des keyholes, d.h. an der Schmelzbadwand lokalisiert sind. Sie treten also sowohl im Prozeßlicht als auch an der Schmelzbadoberfläche auf.

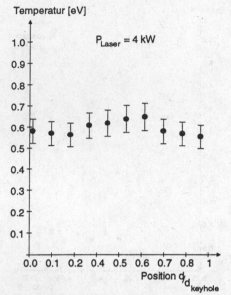

Bild 8: Radialer Verlauf der Elektronentempera-
tur im keyhole

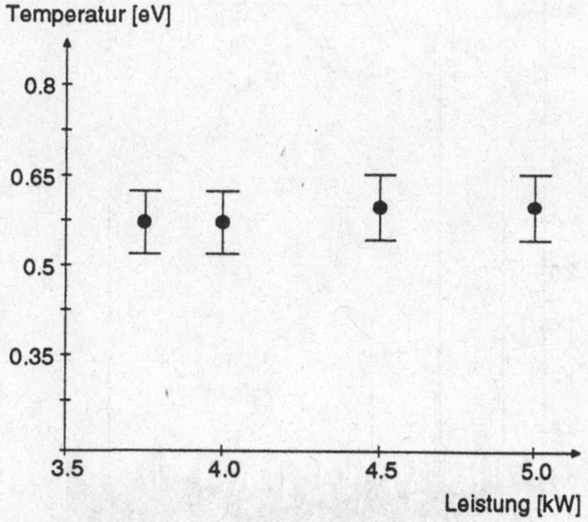

Bild 9:Mittlere Temperatur im keyhole in Abhän-
gigkeit der Laserleistung

5. Temperatur im keyhole

Das nach oben hin austretende Prozeßlicht wurde über eine stark vergrößernde Optik (ca. 70-fach) mit Hilfe eines Gitterspektrographen beobachtet. Aus den Intensitäten mehrerer Linien des optischen Spektrums im Bereich zwischen 400 und 450 nm kann mit Hilfe von Boltzmann-Plots die Elektronentemperatur im keyhole ortsaufgelöst bestimmt werden /2,3,4/. In Bild 8 ist die radiale Abhängigkeit der Temperatur für eine Laserleistung von 4 kW bei einem Schutzgasfluß von 27.5 l/min Argon unter seitlicher Zuführung mit einem Winkel von 30° zur Oberfläche dargestellt. Man erkennt ein eher schwaches Ansteigen der Temperatur zur Mitte hin, das jedoch im wesentlichen innerhalb der Genauigkeit des verwendeten Verfahrens bleibt. Daher wurde für die Untersuchungen bei verschiedenen Laserleistungen die einzelnen Temperaturen jeweils über den gesamten Querschnitt des keyholes gemittelt. Aus Bild 9, in dem die gemittelte Temperatur in Abhängigkeit der Laserleistung dargestellt ist, zeigt sich analog zu bereits früher durchgeführten Untersuchungen /2/, daß die Temperatur trotz des Leistungsanstieges im wesentlichen konstant bleibt, bzw. sich nur schwach ändert und eine mittlere Temperatur von etwa 0.6 eV ergibt. Schätzt man aus diesen Ergebnissen den Absorptionskoeffizienten für inverse Bremsstrahlung ab, so erhält man relativ geringe Werte, die in der Größenordnung bis zu 0.5 cm^{-1} liegen.

6. Literatur

/1/ J. Milberg, "Untersuchungen zum 3D-Bearbeiten mit Industrierobotern", BMFT-Forschungsbericht Nr.13N5547/0 im Verbundprojekt "3D-Bearbeiten mit CO_2-Hochleistungslaser", 1992

/2/ F. Garnich, H. Lindl, "Plasma Spectroscopy in Laser Beam Welding", In: W. Waidelich (ed.), Laser in der Technik, S.424, Berlin, Springer-Verlag 1992

/3/ W. Sokolowski et. al., "Spectroscopic Study of Laser Induced Plasma in the Welding Process of Steel and Aluminum", Proc. SPIE '88, Vol. 1020, S. 96 (1988)

/4/ H.R. Griem, "Plasma Spectroscopy", New York, McGraw-Hill 1964

Evaporation in Deep Penetration Welding with Laser Radiation

D. Becker, W. Schulz
Lehrstuhl für Lasertechnik , RWTH Aachen
Steinbachstr.15, D - 52074 Aachen

Abstract

In deep penetration welding most of the molten material does not cross the keyhole as vapour but has to be accelerated around it. The driving force for this acceleration is assumed to be given by the ablation pressure of evaporating material. The required evaporation rate and the power needed for the evaporation process itself are calculated. It is shown that although the pressure gradients in the melt increase drastically with the processing velocity, the power needed for evaporation is negligible in a wide range of parameters.

Introduction

In deep penetration welding with laser radiation the molten material is evaporated partially and a keyhole is formed, in which the laser radiation is absorbed efficiently. The permanent evaporation of material is required to maintain the keyhole against the capillary force. Most of the material is not evaporated but flows around the keyhole /1/. The driving force for the flow of molten material around the keyhole may be given by the ablation pressure due to evaporation or by tension at the keyhole walls induced by the Marangoni effect due to a temperature variation along the keyhole-melt interface. Since at the interface evaporation takes place, the temperature is very close to the equilibrium evaporation temperature and the Marangoni effect can be neglected /2/. Therefore the evaporation rate can be calculated once the required driving forces are known. Given the local evaporation rate the heat flow that is needed for the evaporation process itself can be calculated. Since a part of the evaporated material recondensates at the rear front of the keyhole, part of the required power for evaporation is coupled back into the material. The total required power P_{tot} then contains the power P_λ that has to be coupled into the material in order to reach the evaporation temperature at the keyhole wall and the power P_{vap} that is needed for the evaporation process itself. P_λ is known quite well /3/ whereas P_{vap} and the power ratio $\beta = P_{vap} / P_{tot}$ is calculated in this paper.

Method

The shape of the molten region and the flow field in it is calculated two-dimensionally in a plane perpendicular to the beam direction. The molten region is bounded by two interfaces. The first one, the liquid-solid interface, is iterated by calculating the temperature fields inside and outside the melt pool. The boundary condition for heat flow at this interface, including the latent heat of fusion, is used to determine the position of the interface. The second interface, the keyhole wall, is assumed to be a fixed circle of radius r_o. Given the molten region, the flow field can be calculated and the pressure distribution at the gaseous-liquid interface can be determined. Since the pressure is assumed to be due to evaporation, the evaporation rate can now be calculated. For this purpose an evaporation model /4/ is used, in which the evaporation rate and the ablation pressure is determined from given temperature of the surface and pressure above the surface, namely inside the keyhole. The surface temperature has to be somewhat higher than the equilibrium evaporation temperature in order to have a non-zero evaporation rate. The pressure inside the keyhole is assumed to be the ambient pressure plus the pressure due to capillary forces. With the evaporation model the local surface temperatures and evaporation rates can be calculated. These values are used again as boundary conditions for the calculation of heat and mass flow in the melt and the whole procedure is then carried out again until it converges to the steady state. The calculations of temperature- and velocity-fields are done numerically with a finite volume method.

The input parameters are the thermophysical properties of the material, the keyhole radius r_0 and the processing velocity v_o. In all calculations the material is assumed to be a mild steel, the other parameters are indicated.

Results

In Fig. 1 a calculated melt pool is shown. The velocities v in the melt are shown in a grey scale plot. The dark regions at the side parts of the keyhole indicate that velocities much greater than the processing velocity only occur in the narrow channel between keyhole and liquid-solid interface. The narrow tip at the end of the melt pool forms in such a way that the latent heat of fusion that is set free at the recristallisation process can be dissipated into the ambient material.

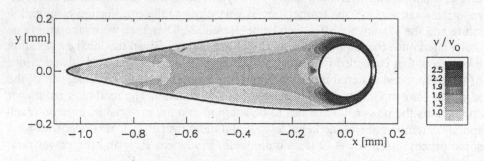

Fig. 1 : Melt pool shape and velocities in the pool (r_0 = 0.1 mm , v_0 = 6 m/min)

In Fig. 2 the difference Δp between pressure in the melt at the gaseous-liquid interface and the ambient pressure is plotted versus the angle. The pressure decreases from a maximum value at the front to a minimum at the side part of the keyhole, where the velocities are highest and rises again to a local maximum at the rear part of the keyhole. This maximum is lower than that at the front part because the melt is partially slowed down by friction at the solid walls. The surface temperatures belonging to such ablation pressures are in the range of $\pm 2\%$ of the equilibrium evaporation temperature.

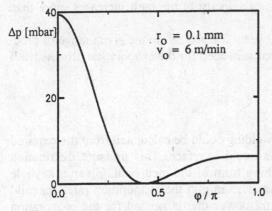

Fig. 2 : Pressure in the melt at the keyhole wall

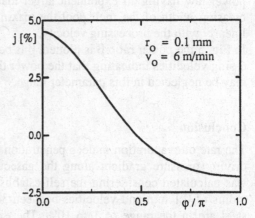

Fig. 3 : Evaporation rate j at the keyhole wall

In Fig. 3 the normalized evaporation rate j is shown. At the front 4% of the incoming material is evaporated, 96% of it flows around the keyhole. At the rear part of the keyhole the evaporation rate is negative, indicating a recondensation of material into the melt. The integration of the evaporation rate along the interface results in a total material loss of 1.7%. That means that if the keyhole remains a cylinder as assumed, then not all the evaporated material will recondensate but some of it has to be transported through the keyhole into the environment. The power ratio β for this example is 2.9%

These calculations are carried out for a variation of processing velocities. The results are shown in Fig. 4 and 5.

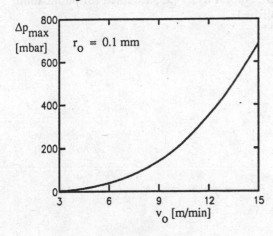

Fig. 4 : maximum pressure drop versus processing velocity

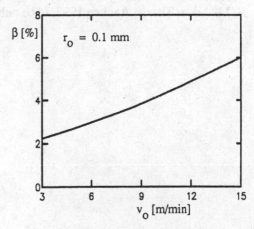

Fig. 5 : Power ratio β versus processing velocity

In Fig. 4 the pressure drop Δp_{max} in the melt along the keyhole wall from maximum pressure at the front to the minimum pressure at the side is plotted versus the processing velocity. The pressure drop increases strongly with the processing velocity, namely with a power law having an exponnent larger than 2. This is due to the fact, that with the decreasing width of the melt pool the maximum velocity in the melt increases more than linearly with the processing velocity.

In Fig. 5 the power ratio β is plotted. β is between 2% and 6% in the given range of processing velocities indicating that the power that is needed for the evaporation process itself may be neglected in this parameter range.

Conclusion

The rate of evaporation in deep penetration welding could be calculated from the required driving pressure gradient along the gaseous-liquid interface. This pressure distribution was calculated considering the self establishing form of the melt pool. Given a keyhole radius of 0.1 mm and velocities between 3 and 15 m/min the evaporation rates for mild steel are in the range of 1 to 10%. The extra power that is needed for the evaporation process in this range does not exceed 6% of the power that has to be coupled into the material in order to reach the evaporation temperature and can therefore be neglected compared with the total required power. The pressure drop in the melt rises up to 0.7 bar in this parameter range. This indicates that for high processing velocities the hydrodynamical pressures exceed the capillary pressures drastically which may lead to instabilities in the formation of the gaseous-liquid interface

Literature

/1/ M. Funk, Thesis, Aachen 1993, to be published

/2/ D. Becker, Thesis, Aachen 1993, to be published

/3/ W. Schulz, D. Becker, J. Franke, R. Kemmerling, G. Herziger : Heat conduction
losses in laser cutting of metals, J. Phys. D : Appl. Phys 1993, accepted for publication

/4/ M. Aden, Thesis, Aachen 1993, to be published

Anwendung der Grauwert-Theorie auf die Datenverarbeitung beim Laserstrahlschweißen

Y.H. Xiao, National Centre of Laser Technology, P.O.Box 8511, Beijing, China,100015
T.C. Zuo, Beijing Polytechnic University, Pingleyuan 100, Beijing, China,100022

1. Einleitung

Der Laserstrahl zeichnet sich durch günstigen Tiefschweißeffekt aus. Jedoch zeichnet sich auch durch die starke Reflektion der Laserstrahl von metallischen Oberflächen und das von metalldampf geformte Plasma aus. Verschiedene Laseranlagen haben verschiedene Eigenschaften. Eine wichtige Voraussetzung für die Schweißqualität von Hochleistungslaserstrahlen ist dashalbe, ob man die verschiedenen physikalischen Faktoren richtig beherrschen kann.

Selbstverständlich hoffen wir auf eine beste Nahtstelle beim Schweißen. Hierfür hat man viele Forschungsarbeiten daran gemacht: z. B. die Rechnung nach der physikalischen Grundformel (es geht um die Eigenschaften von Laserstrahlen, die Thermocharaktere des geschwweißten Materials, die Laserleistung und die Wirkungsdauer), damit die geometrische Form der Nahtstelle voraus berechnet werden kann. Noch dazu hat man versucht, daß durch die Messungen der geometrischen Form von einer Menge Nachstellen die besten Schweißparameter mit Sensoren ausgewählt werden. In diesem Artikel geht es auch um die Zusammenhang zwischen den geometrischen Formen der Nahtstelle und den Schweißparametern.

Hierbei schlagen wir eine mathematisch analytische Methode - **Grauwert-Theorie** vor, dadurch kann man mit relativ wenigen Testen die Zusammenhängen zwischen den verschiedenen Experiments- parametern und der Schmelztiefe und -breite finden und ihr mathematisches Modell gründen. Damit kann man auch eine Einflußordnung von allen Paramentern auf die geometrische Form der Nahtstelle finden. Hierfür macht man eine Reihe Versuche, aus vielen Faktoren, die die unterschiedlichen Einflüsse auf eine Sache haben, die Einflußstufe zu ordnen.

2. Untersuchungsmethode

Beim Hochleistungsschweißen gibt es viele Faktoren, die den Einfluß auf die Schmelztiefe und -Breite haben. Folgende Parameter wurden hier als diskutierte Objekte ausgewählt:
- Laserstrahlleistung (Kurzzeichen P)
- Schweißgeschwindigkeit (V)
- Brennenweite der Fokusssieroptik (R)
- Fokuslage (F)
- Durchflußmenge des Schutzgases (Q)
- Spritzwinkel des Schutzgases (W)

Bei den Untersuchungen verändert sich jedesmal nur ein Parameter regelmäßig, die anderen Parameter sind jedoch in den Normalbedingungen befestigt. Nach dem Schweißen messen wir die Schnitttiefe und - breite jedesmal.

3. Analysenmethode

Eine neue mathematische Methode -Grauwert -Theorie wird hier in Anwendung gebracht, um die unbekannten Daten abzuleiten und das mathematische Modell zu geründen.

Sei $\qquad [x^{(0)}(i)] = [x^{(0)}(1), x^{(0)}(2), x^{(0)}(3), \cdots]$
eine stochastische Datenfolge.

Laß die Daten addieren $\quad x^{(1)}(i) = \sum\limits_{s=1}^{i} x^{(0)}(s)$

erhalten wir dann eine neue Datenfolge:

$$[x^{(1)}(i)] = [x^{(1)}(1), x^{(1)}(2), x^{(1)}(3), \cdots]$$

$$= [x^{(0)}(1), \sum\limits_{s=1}^{2} x^{(0)}(s), \sum\limits_{s=1}^{3} x^{(0)}(s), \cdots]$$

Wenn die Additionsmale genug sind, verwandelt sich die Folge in die unstochastische und kann sich eine Exponentialkurve annähren.

Für die folgende Datenfolge:

$$[x_k^{(0)}(i)] \qquad \begin{matrix} k = 1, 2, 3, \cdots h \\ i = 1, 2, 3, \cdots N \end{matrix}$$

haben wir die entsprechende Folgen von Erstordnung:

$$[x_k^{(1)}(i)] = \sum\limits_{s=1}^{i} x^{(0)}(s) \qquad \begin{matrix} k = 1, 2, 3, \cdots h \\ i = 1, 2, 3, \cdots N \end{matrix}$$

und die entsprechende hochsubstrahierte Datenfolge:

$$[a^{(j)}(x_k, i)] \qquad \begin{matrix} i = 1, 2, \cdots N \\ j = 3, 4, \cdots I \\ k = 1, 2, \cdots h \end{matrix}$$

$$a^{(0)}(x, i) = x^{(0)}(i)$$

$$a^{(1)}(x, i) = x^{(0)}(i) - x^{(0)}(i-1)$$

$$\vdots \qquad\qquad \vdots$$

$$a^{(j)}(x, i) = a^{(j-1)}(x_k, i) - a^{(j-1)}(x_k, i-1)$$

$$A = \begin{vmatrix} -a^{(n-1)}(x_1^{(1)}, 2) & -a^{(n-2)}(x_1^{(1)}, 2) & \cdots & -a^{(1)}(x_1^{(1)}, 2) \\ -a^{(n-1)}(x_1^{(1)}, 3) & -a^{(n-2)}(x_1^{(1)}, 3) & \cdots & -a^{(1)}(x_1^{(1)}, 3) \\ \vdots & \vdots & & \vdots \\ -a^{(n-1)}(x_1^{(1)}, N) & -a^{(n-2)}(x_1^{(1)}, N) & \cdots & -a^{(1)}(x_1^{(1)}, N) \end{vmatrix}$$

$$C = \begin{vmatrix} -1/2 [x_1^{(1)}(2) + x_1^{(1)}(1)] & x_2^{(1)}(2) & \cdots & x_h^{(1)}(2) \\ -1/2 [x_1^{(1)}(3) + x_1^{(1)}(2)] & x_2^{(1)}(3) & \cdots & x_h^{(1)}(3) \\ \vdots & \vdots & & \vdots \\ -1/2 [x_1^{(1)}(N) + x_1^{(1)}(N-1)] & x_2^{(1)}(N) & \cdots & x_h^{(1)}(N) \end{vmatrix}$$

$$a = [a_1, a_2, \cdots, a_n, : b_1, b_2, \cdots, b_{h-1}]^T$$

$$Y_N = [a^{(n)}(x_1^{(1)}, 2), a^{(n)}(x_1^{(1)}, 3), \cdots, a^{(n)}(x_1^{(1)}, N)]^T$$

(Hier T bedeutet die Umkehrung der Matrix)

Nach der Methode des kleinsten Quadrades haben wir:

$$\grave{a} = [(A : C)^T (A : C)]^{-1} (A : C)^T Y_N.$$

Die Elemente in $\grave{a}$ sind die Koeffiziente vom folgenden Differenzengleichung:

$$\sum\limits_{i=0}^{h} [a \, d^{(n-i)}(x^{(1)}) / d j^{(n-i)}] = \sum\limits_{i=0}^{h} b_i \, x_{i+1}^{(1)}$$

4. Untersuchungsergebnisse und Rechnung

Unter der obigen Untersuchungsbedingungen haben wir die Experimente in Serien durchgeführt, und die Daten werden mit der obigen Analysenmethode verarbeitet. Daraus ergeben sich die folgenden Ergebnisse:

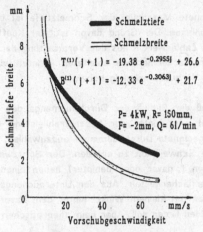

$$T^{(1)}(j+1) = -19.38\, e^{-0.2955j} + 26.6$$
$$B^{(1)}(j+1) = -12.33\, e^{-0.3063j} + 21.7$$

P= 4kW, R=150mm, F=-2mm, Q= 6 l/min

Der Einfluß der Schweißgeschwindigkeit (V)

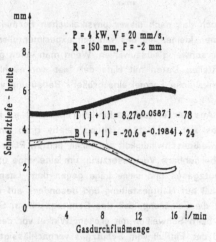

P = 4 kW, V = 20 mm/s, R = 150 mm, F = -2 mm

$$T(j+1) = 8.27\, e^{0.0587j} - 78$$
$$B(j+1) = -20.6\, e^{-0.1984j} + 24$$

Einfluß der Durchflußmenge des Schutzgases (Q)

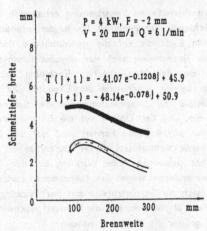

P = 4 kW, F = -2 mm
V = 20 mm/s Q = 6 l/min

$$T(j+1) = -41.07\, e^{-0.1208j} + 45.9$$
$$B(j+1) = -48.14\, e^{-0.078j} + 50.9$$

Einfluß der Brennweite der Fokusieroptik

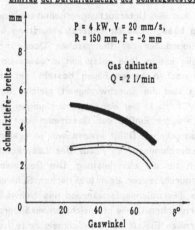

P = 4 kW, V = 20 mm/s, R = 150 mm, F = -2 mm

Gas dahinten
Q = 2 l/min

Einfluß des Strahlwinkels des Schutzgases

Um die statische Zusammenhang zwischen allen Faktoren zu finden, wird das folgende statishe Modell benutzt:

$$x_1^{(O)}(j) = \sum_{i=1}^{h} b_i\, x_{i+1}^{(O)}(j) + b_o \qquad (\text{hier } N = 0 \text{ und } h = h)$$

Nach der Rechnung haben wir dann:

$$T \sim +1.015\,P - 0.095\,V - 0.45\,R - 0.428\,F + 0.48\,Q$$
$$B \sim +0.70\,P - 0.735\,V - 0.15\,R - 0.176\,F - 0.53\,Q$$

Daraus erhalten wir schließlich die Einflußordnung von aller Faktoren auf die geometrische Form der Schweißnahtstelle beim Laserstrahlschweißen.

Für Schmelztiefe T: $\quad P \longrightarrow V \longrightarrow Q \longrightarrow R \longrightarrow F$

Für Schmelzbreite B: $\quad V \longrightarrow P \longrightarrow Q \longrightarrow F \longrightarrow R$

5. Diskusion

(1) Die Einflüsse der Laserstrahlleistung, der Brennenweite, der Fokussieroptik und der Fokuslage auf die Schweißnahtstaltung wurden viel diskutiert, und ein physikalisches Modell wird schon gegründet:

$$T_{max} = K P^{2/3} (\alpha_0 d_0)^{-1/3} Q^{-2/3} V^{-1/3}$$

Jedoch das nach dieser physikalischen Formel gerechnete Maxsimum der Schmelztiefe ist im allgemeinen kleiner als das von den experimentellen Ergebnissen. Der Grund davon ist, der Koeffizient K sehr schwerig auszuwählen. Wenn man einen genauen Zahlenwert K aus der Verarbeitung der experimentellen Daten mit Hilfe der hier vorgeschlagenden Rechenmethode ableiteten könnte, würde die physikalische Formel eine größere Bedeutung von der Vorausberechnung haben.

(2) Aus der erwähnten Einflußordnung ist es klar, daß der Einfluß der Durchflußmenge des Schutzgases auf die Nahtgestaltung relativ groß und steht erst hinter der Laserstrahlleistung und der Schweißgeschwindigkeit auf einen dritten Platz. Eine geeignete Durchflußmenge auszuwählen, ist eine unübersehbare Voraussetzung, um eine enge und tiefe Schweißnaht zu erzielen. Der Spritzwinkel des Schutzgases und seine Lage gegen dem Laserstrahlen (davor oder dahinter) haben einen großen Einfluß auf Nahtgestaltung und besonders auf die Oberflächengestalt. Aus den Untersuchungen ergibt sich die Folgerung, daß das Schutzgas beim Schweißen mit kleinem Winkel hinter dem Laserstrahlen und beim Schweißen mit großem Winkel vor dem Strahlen liegen soll. Aber beim theoretischen Analyse wird der Einfluß vom Schutzgas vernachlässigt.

(3) Die aus den Untersuchungsergebnissen und durch mathematische Verarbeitung erhaltende Einflußordnung basagt, daß die Laserstrahlleistung und die Schweißgeschwindigkeit (d. h. die Intensität und die Aktionsdauer der thermischen Quelle) die größten Einflüsse auf die geometrische Form der Nahtstelle haben. Dies entspricht in gewissen Sinne der Berechnung nach der physikalischen Formel. Aber zwischen beiden Fällen besteht ein Untersied. Bei der physikalischen Rechnung haben die Leistung und die Geschwindigkeit gleiche Einflußstufe: beiden erscheinen als Exponent 2/3 ($T \sim P^{2/3} V^{-2/3}$). Bei den Untersuchungen ist der Einfluß der Laserleistung auf die Schmelztiefe größer als der Einfluß der Geschwindigkeit auf die Tiefe, und der Einfluß auf die Schmelzbreite ist gerad umgekehrt. Dafür können wir so erklären: bei der physikalichen Formel wird nur den Dampfbildungsdurchmesser im Fokus des Laserlichtbündels berücksichtigt und dieser Durchmesser bezieht sich nur auf die Laserleistung. Die Geschwindigkeit hat deswegen keine Wirkung auf den Dampfbildungsdurchmesser. Beim praktischen Schweißen ist es anderes: wegen der thermischen Leitfähigkeit und des Thermoumlaufs während des Schweißens hat sich die Schmelzbreite mit der Wirkungsdauer der thermischen Quelle offensichtlich verändert und daraus ergibt sich, daß die Schweißgeschwindigkeit einen großen Einfluß auf die Schmelzbreite hat, sogar größer als die Laserleistung.

(4) Die Einflüsse der Fokuslage und der Brennenweite der Fokussieroptik landeten auf die letzten Plätze, und ihre Einflüsse auf die Schmelztiefe und auf die Schmelzbreite sind gerad umgekehrt. Der Einfluß der der Fokuslage auf die Oberflächenschmeltzweite ist größer als der Brennenweite.

Literatur

1. *R. Rothe, K. Teske, G. Sepold*
   Das Laserpreßschweißen - ein neues Verfahren. Laser Magazin N.4, 29,11. 1985
   S.24/26 Magazin- Verlage Kronberg
2. *Deng Ju-long*
   Greywert Systems. Feb. 1985. China.

Laserstrahlschweißen von Sonderwerkstoffen
- Erweiterung des Anwendungsspektrums -

M. Dahmen[1], J. Berkmanns[2], K. Behler[2], E. Beyer[2]

[1]Lehrstuhl für Lasertechnik der Rheinisch-Westfälischen Technischen Hochschule Aachen
[2]Fraunhofer-Institut für Lasertechnik   Steinbachstraße 15, D-52074 Aachen

Zur Erweiterung des Werkstoffspektrums und der Erschließung neuer Einsatzfelder für das Laserstrahlschweißen werden Ergebnisse von Untersuchungen der Schweißeignung von Nichteisenmetallen, wie Reintitan und Nickelbasiswerkstoffen, dargestellt. Sie gewähren Einblick in die Metallurgie der Verbindung und erlauben Aussagen über die Gebrauchseigenschaften geschweißter Bauteile. Anwendungsgrenzen sowie neue Entwicklungs- und Einsatzmöglichkeiten werden aufgezeigt.

1. Einleitung

Das Fügen mit CO_2-Laserstrahlung hat sich seit den ersten Versuchen in den Siebzigerjahren inzwischen etabliert. Die spezifischen Eigenschaften, wie der konzentrierte Energieeintrag, ermöglichen einerseits ein verzugsarmes Schweißen und einen Schutz des Werkstückes vor thermischer Schädigung, andererseits eröffnen sich Möglichkeiten, Werkstoffe zu bearbeiten, die bisher mit konventionellen Verfahren nicht oder nur unter erschwerten Bedingungen zu fügen sind.

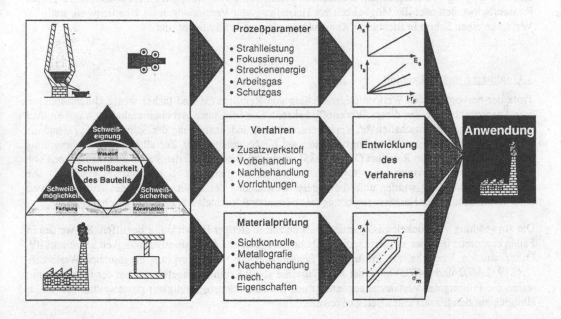

Bild 1: Schweißbarkeit von Metallen mit besonderer Gewichtung der Schweißeignung

Gemessen an der Häufigkeit der Anwendung von Stahlwerkstoffen wird auch das Laserstrahlschweißen

·heute bevorzugt von der stahlverarbeitenden Industrie eingesetzt. Zahlenmäßig die höchsten Anwendungszahlen weisen die Automobilindustrie, der allgemeine Maschinenbau und die Elektroindustrie auf /1/. Der Einsatz der Lasertechnik in der schweißtechnischen Fertigung weiterer Industriezweige erfordert daher die Prüfung der Schweißeignung branchenüblich eingesetzter Werkstoffe. Ausgehend von einer Beschreibung des derzeitigen Standes der Technik, werden Ergebnisse zielgerichtet durchgeführter Versuche (Bild 1) zum Schweißen von Reintitan (Luft- und Raumfahrttechnik, Apparatebau) und Nickelbasis-Feinblechen (Apparatebau) dargestellt.

2. Stand der Technik

Den Stand der Technik stellen derzeit das Laserstrahlschweißen von Stahlwerkstoffen sowie von Aluminium und seinen Legierungen dar.

Im Rahmen der Stahlverarbeitung liegt das Hauptanwendungsgebiet beim Schweißen von Feinblechen in der Automobilfertigung. Durch die schmale Schweißnaht werden trotz werkstoffbedingter Aufhärtung die Tiefzieheigenschaften des Materials nicht beeinträchtigt. Die Dauerfestigkeiten von Bauteilen sind aufgrund der Ausbildung von Längsverbänden gegenüber konventionell gefertigten Komponenten deutlich erhöht /2/.

Laserstrahlschweißen von Aluminiumwerkstoffen war bisher Gegenstand der Forschung und steht kurz vor seiner Einführung in die breite industrielle Anwendung. Bei diesen Werkstoffen können die besonderen Vorteile des Laserstrahlschweißens in hohem Maße ausgenutzt werden, da hier hohe Schweißgeschwindigkeiten erreicht werden, die auch mit Einsatz von Zusatzwerkstoff über denen von konventionellen Verfahren liegen /3/. Die kontinuierliche Zufuhr von Zusatzmaterial während des Schweißprozesses ermöglicht die Steuerung der Nahteigenschaften in weiten Grenzen /4/. Im Hinblick auf die Fertigung von Halbzeugen, wie das Schweißen von Strangpreßprofilen im kombinierten Prozeß Pressen/Schweißen oder die Möglichkeit zur Herstellung von Verbindungen im Überlappstoß, stellt das Verfahren einen Schritt in Richtung "Konstruktionselement Schweißnaht" dar.

3. Grundlagen und Zielsetzung

Trotz der hervorragenden Werkstoffeigenschaften von Reintitan /5/ sind bisher wenig Untersuchungen zum Laserstrahlschweißen dieses Werkstoffs bekannt. Die bekannten Arbeiten enthalten Angaben über die mechanischen Eigenschaften /6/, Verfahrensparameter und Ausbildung der Schweißnähte /7/ und die Auswirkung verschiedener Reinigungsverfahren auf die Nahtqualität /8/. Zur allgemeinen Untersuchung der Schweißeignung von Reintitan (Ti2, 3.7035) wurden an Blechen mit der Stärke 3mm, ausgehend von der Verfahrensentwicklung, die mechanischen Eigenschaften der Schweißnähte ermittelt. Die Verfahrensentwicklung wurden unter besonderer Gewichtung der Schutzwirkung des Arbeitgases und einer Untersuchung der Kantenvorbereitung durch Laserstrahlschmelzschneiden /9/ durchgeführt.

Die Anwendung von Nickelbasislegierungen ist derzeit in stetiger Entwicklung begriffen. Sie werden in Fällen eingesetzt in denen es gleichzeitig auf hohe Festigkeit und Korrosionsbeständigkeit ankommt /10/. Daher·wurden Versuche zur Prüfung der Schweißeignung dreier typisch eingesetzter Werkstoffe (2.4819, 2.4602 und 2.4605) mit einer Materialstärke von 0,5 mm durchgeführt. Ziele der Untersuchung waren die Prüfung der Verfahrenstauglichkeit und der Korrosionsbeständigkeit geschweißter Proben im Hinblick auf den Einsatz unter Heißgaskorrosion.

4. Ergebnisse

4.1 Laserstrahlschweißen von Reintitan

Hinsichtlich der optischen und der thermophysikalischen Eigenschaften (a = $8,14 \cdot 10^{-6}$ m²/s, T_m = 1667°C) wurde eine sehr gute Schweißeignung festgestellt /11/, in Übereinstimmung mit den Angaben der DVS-Richtlinie 2713. Aufgrund der hohen Affinität gegenüber Sauerstoff und Stickstoff ist ein Oxidationsschutz an Ober- und Unterraupe erforderlich. Der Prozeßverlauf ist dem beim Schweißen von hochlegierten Stählen ähnlich und erfordert keine zusätzlichen Maßnahmen zur Steuerung des laserinduzierten Metalldampfplasmas. Typische Prozeßparameter und Querschliffe der Schweißnähte sind in Bild 2 dargestellt /12/.

Die mit hoher Streckenenergie (E_s = 69J/mm) geschweißte Naht weist eine ausgeprägte Trapezform auf (Bild 2 links). Die Wärmeeinflußzone ist durch Kornvergröberung gekennzeichnet, der aufgeschmolzene Bereich weist einen hohen Anteil an Umwandlungsgefüge auf. Wird mit einer angepaßten Fokussierung geschweißt, bildet sich eine Schweißnaht mit fast parallelen Flanken aus. Die Streckenenergie kann reduziert werden (34,5 J/mm). In der WEZ ist nur noch eine leichte Kornvergröberung festzustellen. Die Erstarrung erfolgt stengelig (Bild 2 rechts). Der dargestellte Querschliff weist einen von der Wurzel in die Schweißnaht hineinragenden Bereich martensitischer Umwandlung auf, der auf auf die Aufnahme von Sauerstoff infolge fehlenden Wurzelschutzes zurückzuführen ist /11/.

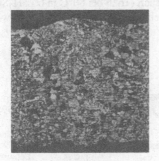 

P_L = 2300 W	P_L = 2300 W
v_s = 2 m/min	v_s = 4 m/min
r_F = 255 µm	r_F = 120 µm
F = 11	F = 5,5
Arbeitsgas Ar 30 L/min	Arbeitsgas He 30 L/min
Wurzelschutz He	kein Wurzelschutz

Bild2: Nahtform von Schweißnähten in Titan (3.7035 s = 3mm)

Die mechanischen Eigenschaften der geschweißten Proben entsprechen denen des Grundwerkstoffs. Die in DIN 17869 E angegebenen Werte der Dehngrenze $R_{p0,2}$ werden sowohl bei gefrästen als auch bei laserstrahlgeschnittenen Stoßkanten mit 361MPa bzw. 420MPa überschritten. Die Zugfestigkeiten betragen 468MPa bzw. 480MPa. Die höheren Werte bei der Fügevorbereitung durch Laserstrahlschmelzschneiden resultieren aus einer Verfestigung der Schweiße durch Sauerstoffaufnahme infolge einer prozeßbedingten Verrundung der Schnittflanken. Die Bruchdehnung fällt von 33% auf 22%. Die Schweißnähte besitzen ausreichende Festigkeiten. Durch Bestimmung des Sauerstoffäquivalentes kann die Tauglichkeit des Schmelzschneidens als Vorfertigungsschritt nachgewiesen werden. Das Sauerstoffäquivalent wurde zu maximal 1150ppm (entsprechend ca.

230HV10) ermittelt. /11/

Die schleppend angeordnete Zufuhr von Argon kombinierte die Wirkung des Arbeitsgases mit der eines Schutzgases. Aus dem vorher gesagten ist die Notwendigkeit eines Wurzelschutzes abzuleiten, der im Rahmen der Versuche durch eine fliegend angeordnete Düse gewährleistet werden konnte. Hauptanforderung an die Arbeitsgase, und in der beschriebenen Versuchsanordnung an die umgebende Atmosphäre, ist Wasserfreiheit, da durch Wasserstoffaufnahme die Randbereiche der Schweißnaht und die WEZ stark verspröden /12/.

4.2 Laserstrahlschweißen von Nickelbasis-Feinblechen

Die Verfahrensprüfung wurden unter Verwendung eines längsgeströmten, hochfrequenzangeregten CO_2-Hochleistungslasers mit einer Nennleistung von 6 kW durchgeführt. Die flexible Leistungssteuerung dieser Strahlquellenart ermöglichte das Schweißen von Überlappverbindungen in einem praktisch handhabbaren Bereich der Vorschubgeschwindigkeiten um 10 m/min.

Die Schweißeignung aller drei Werkstoffe erwies sich als gut. Poren oder Risse wurden in keiner Probe festgestellt. Das Gefüge ist aufgrund der hohen Abkühlraten im Randbereich der aufgeschmolzenenen Zone zellulär und wird zur Schweißnahtmitte hin ausgeprägt dendritisch /13/.

Die Korrosionseigenschaften wurden in einem Kochversuch (168h bei 80°C in 60%H_2SO_4 + 2,5%HCl + 0,2%HF+ 0,5%Asche) durchgeführt. Bild 3 gibt die Ergebnisse für 2.4602 und 2.4605 wieder.

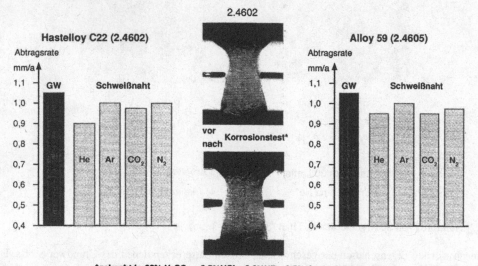

Bild 3: Korrosionsverhalten und Ausbildung von Schweißnähten in Feinblechen aus 2.4602 und 2.4605

Bei allen drei Werkstoffen zeichnete sich ein vergleichbares Bild der Ergebnisse ab. Grundsätzlich ist festzustellen, daß die Schweißnaht eine höhere Korrosionsbeständigkeit aufweist als der Grundwerkstoff.

Dies trifft sowohl auf die Oberflächen als auch auf die Zwischenlage, die zusätzlich der Spaltkorrosion ausgesetzt ist, zu. Mögliche Ursachen hierfür sind in der gerichteten Erstarrung der Schweiße, verbunden mit einer Kornfeinung, und in einem Abbrand unerwünschter Begleitelemente /14/, nicht aber der Legierungselemente, durch den Aufschmelzprozeß zu sehen /13/.

4. Zusammenfassung

Anhand von Schweißversuchen an Reintitan und ausgewählten Nickelbasislegierungen konnten die Eignung des Verfahrens sowie die Eignung dieser Werkstoffe zum Schweißen mit Laserstrahlung nachgewiesen werden. Die Ergebnisse lassen erkennen, daß die Gebrauchseigenschaften geschweißter Bauteile durch die Eigenschaften der Schweißnaht nicht beeinträchtigt werden. Unter Berücksichtigung der Branchenspezifität der Werkstoffe ergibt sich die Möglichkeit dem Laserstrahlschweißen Anwendungsbereiche im Apparatebau und in der Luft- und Raumfahrtindustrie zu erschließen.

5. Literatur

/1/ Meyer, A.: *Lasertechnik 2000 - Laseroszillatoren und Lasersysteme in der industriellen Materialbearbeitung*; Prognos Ag, Basel 1989

/2/ Behler, K. et al.: Laserstrahlschweißen von Karosserieblechen - Ermittlung der mechanisch-technologischen Eigenschaften; *DVS-Berichte* 146, 165 (1992)

/3/ Berkmanns, J. u. K. Behler u. G. Herziger: *Fügen mit CO_2-Hochleistungslasern "Schweißen von aushärtbaren und nicht aushärtbaren Aluminiumlegierungen"*; BMFT-Forschungbericht, FKZ 13 N5655, Aachen 1993

/4/ Berkmanns, J. et al.: Mechanisch-technologische Eigenschaften laserstrahlgeschweißter Aluminiumverbindungen; *DVS-Berichte* 146, 222 (1992)

/5/ Buijs, N.W.: Titaan, een goed alternatief (1); *Metaal en Kunststof* 17, 40 (1991)

/6/ Metzbower, E.A. et al.: Mechanical poperties of laser beam welds; *Welding-Journal* 63, 39-s (1984)

/7/ Denney, P.E. u. E.A. Metzbower: Laser beam welding of titanium; *Welding-Journal* 68, 342-s (1989)

/8/ Bergmann, H.W. u. Juckenrath u. M. Cantello: Laser welding of cp-Ti and Ti Al6 V4; in S.K. Gosh: *Laser 4, High Power Lasers in Metal Processing*; IITT-International, Gournay-sur-Marne 1989

/9/ Hugo, T.: *Laserstrahlschmelzschneiden von Reintitan*; Diplomarbeit, Fachhochschule Aachen, 1993

/10/ Betteridge, H.: *The Nimonic Alloys*; Edward Anold, London 1974

/11/ Rinkens, T.: *Laserstrahlschweißen von Reintitan*; Diplomarbeit, Fachhochschule Aachen, 1993

/12/ Song, Y.-A.: *Thermische Verfahrenstechnik technisch reinen Titans unter besonderer Berücksichtigung des Laserstrahlschneidens und -schweißens*; Studienarbeit, LLT RWTH Aachen, 1992

/13/ Bohlen, R: *Untersuchungen zur Anwendbarkeit des CO_2-Laserstrahlschweißens von Dünnblechen aus Nickelbasislegierungen für Plattenwärmeaustauscher*; Diplomarbeit, LLT RWTH Aachen, 1992

/14/ Uhlig, G. et al.: Laserstrahl-Schmelzschweißen nichtrostender Stähle; *Bänder Bleche Rohre* 4/1992, S. 60

Laserstrahlschweißen verzinkter Bleche
Einfluß von Fügespalt und Zinkschichtdicke

U. Bethke [1], D. Päthe [1], F. Trösken [2], P. Zopf [1]

1 INPRO Innovationsgesellschaft für fortgeschrittene Produktionssysteme in der Fahrzeugindustrie mbH
   Nürnberger Straße 68/69, D-1000 Berlin 30
2 Krupp Forschungsinstitut GmbH
   Münchner Straße 100, D-4300 Essen 1

Kurzfassung

Der Einsatz verzinkter Bleche ist in der Automobilindustrie als Korrosionsschutzmaßnahme weit verbreitet. Aus produktionstechnischen Gründen ist dabei die Überlappnaht die häufigste Fügeform. Das kleine Schmelzbadvolumen beim Laserstrahlschweißen ohne Zusatzdraht erlaubt nur sehr kleine zulässige Fügespalte. Aus spanntechnischer Sicht ist bei der Überlappnaht ein technischer Nullspalt am einfachsten zu realisieren. Dieser bereitet aber aufgrund des Verdampfungsvorgangs des Zinks erhebliche Probleme bei der Erzielung ausreichender Nahtqualitäten. Untersucht wurde deshalb der Einfluß definierter Fügespalte und verschiedener Zinkschichtdicken beim Strahlschweißen mit Hochleistungsfestkörperlasern und CO_2-Lasern der Multi kW-Klasse am 2 x 0,8 mm Überlappstoß (St 1403). Es konnte ein deutlicher, negativer Effekt des Zinks auf die maximal erreichbare Schweißgeschwindigkeit für eine vollständige Durchschweißung festgestellt werden.

1 Einleitung

Verzinkte Bleche werden insbesondere in der Automobilindustrie als wirkungsvolle Korrosionsschutzmaßnahme sehr häufig eingesetzt. Die dazu erforderlichen Schichtdicken betragen mindestens 5 bis 10 µm. Nur bei Schichtdicken dieser Größenordnung ist eine ausreichende kathodische Langzeit-Schutzwirkung erreichbar /1/. Zum Einsatz kommen ein- oder beidseitig elektrolytisch bzw. feuerverzinkte Bleche mit Schichtdicken von bis zu 25 µm. Diverse Mehrlagen Zinkbeschichtungen geringer Schichtdicke (< 5 µm) wie z. B. Zn/Ni, Zn/Al, etc. bieten nur kurzfristigen Korrosionsschutz und wurden daher bei dieser Untersuchung nicht berücksichtigt.

Die Zinkbeschichtung bereitet fertigungstechnische Probleme beim Laserstrahlschweißen, die das Fertigungsmittel und das Bauteil betreffen. Bedingt durch die niedrige Verdampfungstemperatur des Zinkes (T_{V-Zink} = 906 °C) im Vergleich zum Stahl (Schmelztemperatur von reinem Eisen ($T_{S-Eisen}$ = 1536 °C)) wird durch die Schweißwärme ein vehementer Verdampfungsprozeß verursacht /2-4/. Durch den hohen Dampfdruck des Zinks erfolgt eine rasche Ausbreitung des Zinkdampfs, wodurch die Prozeßstabilität beeinträchtigt werden kann. Die auftretende Zinkdampfmenge korreliert dabei mit der Beschichtungsdicke und der Wärmeeinwirkungsbreite aufgrund der Schweißung.

2 Problemstellung

Von essentieller Bedeutung hinsichtlich der Prozeßstabilität ist die Lage der Zinkschicht an der Stoßfuge. Befindet sich die Zinkschicht nur an der Werkstückober- und/oder -unterseite, so kann der entstehende Zinkdampf ungehindert entweichen und es ist keine Beeinträchtigung der Naht feststellbar /2, 5/. Ist in der Fügeanordnung die Zinkschicht zwischen den Blechen, so gestaltet sich der Entgasungsprozeß für den Zinkdampf schwieriger. Wird der Entgasungsvorgang stark behindert, so kommt es zu explosionsartigen Zinkeruptionen und einer damit verbundenen örtlichen Zerstörung der Schweißnaht infolge herausgeschleuderter Schmelzpartikel. Muß der Zinkdampf wegen des fehlenden Fügespalts ausschließlich über das Keyhole entweichen, so sind die zugehörigen Prozeßparameter sehr eingeschränkt. Wesentlich flexibler gestaltet sich die Prozeßparameterwahl bei vorhandenem Fügespalt. Ziel dieser Untersuchung war daher die Feststellung des Einflusses des Fügespalts in Abhängigkeit der Zinkschichtdicke im Spalt auf den Verfahrensparameter Schweißgeschwindigkeit am 2x0,8 mm Überlappstoß (Werkstoff St 1403), so daß eine vollständige Durchschweißung mit geschlossener parallelflankiger Nahtwurzel erzielt wird. Dieses Kriterium wurde aus Gründen praktikabler Qualitätskontrolle am Realbauteil gewählt, da im Gegensatz dazu eine definierte Einschweißung ins Unterblech aufwendig kontrollierbar ist.

3 Versuchsdurchführung

Die Untersuchung wurde mit 0,8 mm dickem, unverzinktem bzw. einseitig elektrolytisch verzinktem Tiefziehblech (St 14O3) durchgeführt. Die Schichtdicke betrug 10 μm. Damit ergab sich je nach Blechpaarung eine Zinkschichtdicke von 10 bzw. 20 μm im Fügespalt. Die Blechoberseite war stets

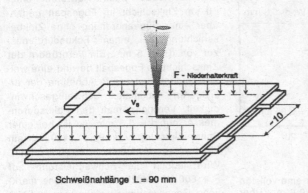

Schweißnahtlänge L = 90 mm

Bild 1: Schweißprobengeometrie

unbeschichtet. Der Fügespalt wurde durch Einlegen von Metallfolien entsprechender Dicke (0,05, 0,10 und 0,20 mm) parallel im Abstand von ca. 5 mm an beiden Seiten der Schweißnaht realisiert (Bild 1).

Die Schweißversuche wurden mit einem 2 kW Nd:YAG-Hochleistungsfestkörperlaser und einem 6 kW CO_2-Laser durchgeführt. Beim Festkörperlaser erfolgte die Strahlübertragung über eine Stufenindexfaser mit einem Kerndurchmesser von 0,6 mm bzw. 1,0 mm. Durch den Einsatz entsprechender Fokussieroptiken wurden die Fokusdurchmesser von 0,3, 0,5 und 0,8 mm erzielt. In ähnlicher Weise konnten beim CO_2-Laser durch Variation des

Betriebsmodus und der Fokussierspiegel Fokusdurchmesser von 0,25, 0,35 und 0,45 mm realisiert werden.

4 Ergebnisse

4.1 Festkörperlaser

Beim Überlappstoß kommt es bei einer Zinkschicht im Fügespalt zu einem interessanten Phänomen. Je größer die Zinkschichtdicke im Fügespalt ist, desto kleiner wird bei gleichen Prozeßparametern (Leistung, Fokusdurchmesser, Blechstärke und Fügespalt) die erreichbare maximale Schweißgeschwindigkeit für eine vollständige Durchschweißung. In Bild 2 ist dieser Zusammenhang dargestellt.

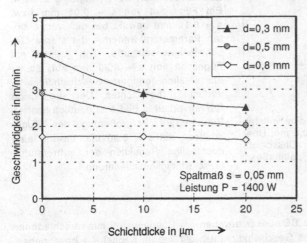

Während bei einem Fokusdurchmesser von d = 0,8 mm keine Änderung der maximal erzielbaren Schweißgeschwindigkeit auftritt, sinkt diese bei einem Fokusdurchmesser von d = 0,3 mm von v_s = 4,0 m/min auf v_s = 2,5 m/min ab. Durch die vergleichbaren Versuchsbedingungen ist dieser Effekt nur durch eine Veränderung im Metalldampf/Plasma aufgrund der unterschiedlichen Zinkschichtdicken erklärbar. Von Auswirkungen des Zinkdampfs auf das Schweißplasma wurde schon in /4/ berichtet, doch wurden diesbezüglich keine konkreten Aussagen gemacht.

Bild 2: Maximale Geschwindigkeit für eine vollständige Durchschweißung am 2x0,8 mm Überlappstoß in Abhängigkeit des Fokusdurchmessers d und der Zinkschichtdicke im Fügespalt

544

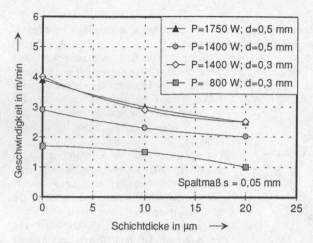

Bild 3: Maximale Geschwindigkeit für eine vollständige Durchschweißung am 2x0,8 mm Überlappstoß in Abhängigkeit der Zinkschichtdicke im Fügespalt, der Leistung P und des Fokusdurchmessers d

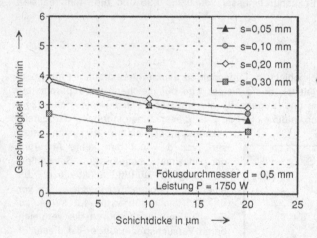

Bild 4: Maximale Geschwindigkeit für eine vollständige Durchschweißung am 2x0,8 mm Überlappstoß in Abhängigkeit der Zinkschichtdicke im Fügespalt und des Spaltmaßes s

Eine Beeinträchtigung des Metalldampfs/ Plasma durch den Zinkdampf wird durch folgendes Indiz untermauert. Im Bild 3 erkennt man eine drastische Geschwindigkeitseinbuße beim Auftreten einer 10 μm Zinkschicht im Fügespalt gegenüber einer Fügeanordnung ohne Zinkbeschichtung bei einem Fokusdurchmesser von d = 0,5 mm. Ein Vergrößern der Zinkschicht im Fügespalt bewirkt eine weitere, jedoch degressive Abnahme der erzielbaren maximalen Schweißgeschwindigkeit. Verringert man den Fokusdurchmesser auf d = 0,3 mm, so ist bei einer Schweißleistung von P = 1400 W ein identisches Verhalten zu beobachten. Wird jedoch die Schweißleistung auf P = 800 W reduziert, so tritt diese merkliche Geschwindigkeitseinbuße erst beim Übergang der Zinkschichtdicke von 10 μm auf 20 μm auf.

Das Ausmaß dieses Effektes ist somit von der Intensität am Werkstück und der auftretenden Zink-Konzentration im Metalldampf/Plasma abhängig. Letzteres wird mit Bild 4 untermauert. Die Zinkkonzentration ist durch Vergrößern des Fügespaltes auf einfache Weise reduzierbar, da die erzielten Nahtbreiten sich nur minimal und damit auch die verdampften Zinkmassen sich nur geringfügig ändern. Ein Fügespalt von s = 0,05 mm bzw. s = 0,10 mm bewirkt bei den verwendeten Parametern weiterhin die starke Geschwindigkeitseinbuße. Tritt jedoch ein Fügespalt von s = 0,20 mm auf, so ist ein merklich geringerer Geschwindigkeitsverlust zu beobachten. Der zugehörige Kurvenverlauf in Bild 4 ist deutlich flacher. Eine weitere Vergrößerung des Fügespaltes auf s = 0,30 mm bewirkt eine nochmalige Reduktion der auftretenden Geschwindigkeitsminderung.

4.2 CO₂-Laser

Schweißtechnische Untersuchungen mit dem CO_2-Laser ergaben am Überlappstoß mit verschiedenen Zinkschichtdicken im Fügespalt ein ähnliches Geschwindigkeitsprofil wie im Kapitel 4.1 beschrieben. Der Effekt tritt auch hier verstärkt bei kleinen Spaltmaßen, großen Schichtdicken und kleinen Fokusdurchmessern auf. Im Gegensatz zum Nd:YAG-Festkörperlaser ist die Verfahrensführung, insbesondere jene der Plasmakontrolle, zur effektiven Strahleinkopplung beim CO_2-Laser schwieriger. Zusätzlich tritt beim CO_2-Laser eine deutlicher ausgeprägte Schwellintensität auf, wodurch die Streuung der Versuchsergebnisse vergrößert wird. Dennoch ist auch hier dieses Phänomen zu beobachten.

5 Literatur

/1/ Drecker H., Meinhardt J., Wohnig W.: Auswahl und Verarbeitung vorbeschichteter Feinbleche im Karosseriebau, Blech Rohre Profile 39 (1992) 12, S. 1030-1035

/2/ Spies B.: Schmelzschweißen zinkbeschichteter Feinbleche mit CO_2-Laserstrahlen DVS-Bericht 127, 1989, S. 63-66

/3/ Akther R.; Steen W. M.: Laser welding of zinc coated steel, Proceedings of LIM 5, S. 195-206

/4/ Heyden J.; Nilsson K.; Magnusson C.: Laser welding of zinc coated steel, Proceedings of LIM 6, S. 93-104

/5/ Albrecht J. et al: Laserstrahlschweißen von Thyssen-Flachprodukten und großformatigen feuerverzinkten Feinblechen, Sonderdruck aus Thyssen Technische Berichte, Heft 2/86

Schweißtechnischer Vergleich zwischen Hochleistungsfestkörper- und CO$_2$-Laser

U. Bethke [1], F. Dausinger [2], D. Päthe [1], F. Trösken [3], P. Zopf [1]

1   INPRO Innovationsgesellschaft für fortgeschrittene Produktionssysteme in der Fahrzeugindustrie mbH
     Nürnberger Straße 68/69, D-1000 Berlin 30
2   Universität Stuttgart, Institut für Strahlwerkzeuge
3   Krupp Forschungsinstitut
     Münchner Straße 100, D-4300 Essen 1

Kurzfassung

Theoretisch ist beim Laserschweißen im Fall des Nd:YAG-Lasers bei vergleichbarer Strahlqualität und Leistung im Vergleich zum CO$_2$-Laser eine höhere Schweißeffizienz zu erwarten, da bei der Wellenlänge des Festkörperlasers metallische Werkstoffe eine höhere Grundabsorption aufweisen. Inwieweit hieraus allerdings auch beim Tiefschweißen ein höherer Wirkungsgrad resultiert, ist bisher fraglich. Zur Klärung dieser Problematik wurden mit einem Dauerstrichfestkörperlaser der 2 kW-Klasse sowie verschiedenen CO$_2$-Lasern Schweißversuche bei vergleichbaren Verfahrensparametern durchgeführt. Neben Blindschweißungen wurden auch Schweißungen an Überlappstößen (St 1403, 2x0,8 mm, unbeschichtet) ausgeführt. In beiden Fällen wurde für den Nd:YAG-Laser im Vergleich zum CO$_2$-Laser eine deutlich höhere Schweißeffizienz ermittelt.

1 Einleitung

Beim Schweißen an Dünnblechstrukturen in der industriellen Produktionstechnik dominieren derzeitig CO$_2$-Laser als Strahlwerkzeuge /1, 2/. Seit kurzem werden auf dem Markt in der Leistungsklasse bis 2 kW von verschiedenen Herstellern auch Nd:YAG-Laser angeboten /3/. Ihr Einsatz ist jedoch bisher auf den Laborbetrieb beschränkt. Während bei den etablierten CO$_2$-Lasern die Strahlführung ausschließlich durch Spiegelsysteme möglich ist, bietet der Festkörperlaser aufgrund seiner Wellenlänge die Möglichkeit der Leistungsübertragung durch Glasfasern. Flexibilität und Zugängigkeit werden besonders bei 3D-Applikationen in Verbindung mit Gelenkarmrobotern deutlich erhöht. Neben dem systemtechnischen Vorteil der Strahlübertragung durch Glasfasern müßten sich für den Festkörperlaser aufgrund seiner im Vergleich zum CO$_2$-Laser deutlich kürzeren Wellenlänge theoretisch auch bei der Energieeinkopplung in metallische Werkstoffe Vorteile ergeben /4/. Für das Laserschneiden wurde der im Vergleich zum CO$_2$-Laser höhere Prozeßwirkungsgrad des Festkörperlasers bereits nachgewiesen /5/. Inwieweit allerdings aus der besseren Energieeinkopplung beim Tiefschweißen möglicherweise eine höhere Effizienz resultiert, ist bisher noch nicht geklärt. Hierzu ist die Durchführung von Schweißversuchen mit beiden Lasertypen bei vergleichbaren Laser- und Verfahrensparametern notwendig. Besonders zu berücksichtigen ist in diesem Zusammenhang, daß mit CO$_2$-Lasern ein effektiver und reproduzierbarer Schweißprozeß in der Regel erst für Intensitäten ab 10^6 W/cm^2 realisiert werden kann.

2 Versuchsdurchführung

Grundlage für den Systemvergleich waren Schweißversuche, die mit einem 2 kW Nd:YAG-Dauerstrichlaser (Fa. HAAS) durchgeführt wurden. Die Strahlübertragung kann bei diesem System wahlweise über Stufenindexfasern mit einem Kerndurchmesser von 0,6 mm oder 1 mm erfolgen. Durch entsprechende Abbildung der Faserstirnfläche kann ein minimaler Fokusdurchmesser von d_F = 0,3 mm erreicht werden. Maximal stehen auf dem Werkstück nach der Strahlübertragung 2000 W zur Verfügung. Zum Vergleich mit dem CO$_2$-Laser wurden folgende Versuchsschweißungen durchgeführt:

- Schweißen unbeschichteten Karosserieblechs (St 1403) im Überlappstoß mit 2x0,8 mm
- Blindschweißungen in St 52-3 (t = 10 mm) mit sandgestrahlter Oberfläche.

Die Durchführung aller Schweißungen erfolgte in Wannenlage. Prozeß- bzw. Schutzgas wurde beim Nd:YAG-Laser nicht verwendet. Beim Überlappstoß wurde als Gütekriterium für eine ordnungsgemäße Durchschweißung die Ausbildung einer geschlossenen, gleichmäßigen Nahtwurzel (parallele Ränder) an der Blechunterseite herangezogen. Die quantitative Auswertung erfolgte anhand metallografischer Schliffe.

3 Schweißversuche mit Festkörperlaser

Der bei Blindschweißungen in St 52-3 für den Fokusdurchmesser d_F = 0,5 mm und die Laserleistung

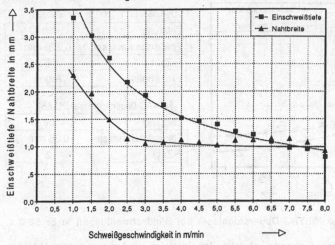

Laserleistung: 1,8 kW
Fokusdurchmesser: 0,5 mm
Blindschweißungen in St52-3, Oberfläche sandgestrahlt

Bild 1: Einfluß der Schweißgeschwindigkeit auf Nahtbreite und -tiefe bei Blindschweißungen in St 52-3 mit sandgestrahlter Oberfläche.
Die Laserleistung betrug 1800 W, der Fokusdurchmesser 0,5 mm

P = 1800 W ermittelte Zusammenhang zwischen Schweißgeschwindigkeit und Eindringtiefe ist in Bild 1 dargestellt. Tiefschweißungen mit einem Tiefe/Breite Verhältnis > 1 werden für Schweißgeschwindigkeiten bis v_s = 5,0 m/min erreicht. Die mit diesen Parametern bei Blindschweißungen ermittelten Einschweißtiefen ließen für das Schweißen von Blechen im Überlappstoß ebenfalls eine hohe Effizienz erwarten.

Für d_F = 0,5 mm und P = 1800 W konnten Durchschweißungen bis v_s = 4,5 m/min realisiert werden. Bei v_s = 3 m/min ergibt sich eine Verbindungsnahtbreite b_M = 0,75 mm, was knapp der einfachen Blechdicke und damit den üblichen Anforderungen entspricht (Bild 2).

4 Vergleich zwischen Nd:YAG- und CO$_2$-Laser

Zur Durchführung des schweißtechnischen Vergleichs zwischen beiden Lasertypen wurden Schweißungen an zwei CO$_2$-Laseranlagen durchgeführt. Hierbei kamen ein TLF 5000 (IFSW Stuttgart) des Herstellers Trumpf als auch ein RS 6000 (KFI Essen) von Rofin Sinar zur Anwendung. Die Fokusdurchmesser wurden an beiden Lasern mit dem Lasercope UFF 100 der Firma Prometec vermessen. Als fokussierende Elemente dienten Off-Axis-Parabolspiegel unterschiedlicher Brennweite. Beim Schweißen wurden Argon bzw. Helium als Prozeßgase zugeführt. Die Parameter für den CO$_2$-Laser wurden hierbei so gewählt, daß die Intensität oberhalb der kritischen Einkoppelschwelle lag und keine Einkoppelfluktuationen auftraten. Im Gegensatz zum Nd:YAG-Laser tritt hierbei beim

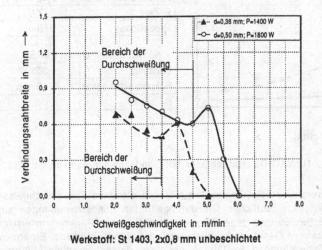

Werkstoff: St 1403, 2x0,8 mm unbeschichtet

Bild 2: Einfluß der Schweißgeschwindigkeit auf die Verbindungsnahtbreite beim Schweißen mit Nd:YAG Hochleistungsfestkörperlaser für unterschiedliche Fokusdurchmesser

548

Schweißen ein intensiv blau leuchtendes Plasma auf. Bild 3 zeigt die mit dem TLF 5000 bei 1400 W und einem Fokusdurchmesser d_F = 0,276 mm bei Blindschweißungen in St 52-3 erreichten Einschweißtiefen in Abhängigkeit von der Schweißgeschwindigkeit. Die mittlere Intensität lag mit

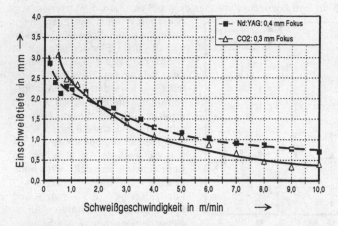

CO₂-Laser (TLF 5000, IFSW Stuttgart)	
Leistung	1400 W
Fokusdurchmesser	0,276 mm
F-Zahl	6
Prozeßgas	Argon

Nd:YAG Dauerstrichlaser (HAAS)	
Leistung	1400 W
Fokusdurchmesser	0,36 mm
F-Zahl	2

Bild 3:   Vergleich von CO₂- und Nd:YAG-Dauerstrichlaser bei Blindschweißungen in St 52-3

I = $2,3 \cdot 10^6$ W/cm² über der kritischen Einkoppelschwelle. Im Vergleich sind die mit dem Nd:YAG-Dauerstrichlaser erzielten Einschweißtiefen eingetragen. Obwohl die Intensität im Fall des Nd:YAG-Laser aufgrund des deutlich größeren Fokusdurchmessers mit I = $1,4 \cdot 10^6$ W/cm² geringer ist, erreicht er bei Geschwindigkeiten oberhalb von v_s = 2 m/min deutlich höhere Eindringtiefen. Unterhalb dieser Geschwindigkeit bewirkt offensichtlich die deutlich höhere Strahlqualität (die F-Zahl ist um den Faktor 3 höher) Vorteile für den CO₂-Laser. Der im Vergleich zum CO₂-Laser höhere

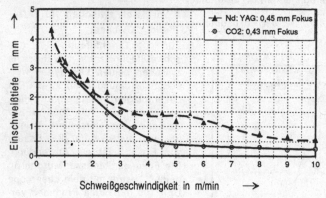

CO₂-Laser (RS 6000, KFI Essen)	
Leistung	2000 W
Fokusdurchmesser	0,43 mm
F-Zahl	7,6
Prozeßgas	Helium

Nd:YAG Dauerstrichlaser (HAAS)	
Leistung	2000 W
Fokusdurchmesser	0,45 mm
F-Zahl	3

Bild 4:   Vergleich von CO₂- und Nd:YAG-Dauerstrichlaser bei Blindschweißungen in St 52-3

Prozeßwirkungsgrad des Nd:YAG-Lasers wird beim Schweißen mit etwa gleichem Fokusdurchmesser von d_F = 0,43 mm bei P = 2000 W noch deutlicher. Die durchgeführten Blindschweißungen sind in Bild 4 den entsprechenden YAG-Schweißungen gegenübergestellt. Bei diesen Parametern erreicht der Festkörperlaser praktisch im gesamten Geschwindigkeitsbereich die höhere Effizienz. Ergänzend an Überlappstoßen durchgeführte Schweißungen bestätigen ebenfalls dieses Ergebnis. Während beim Schweißen mit einer Leistung von P = 1400 W und einem Fokusdurchmesser d_F = 0,3 mm mit dem Festkörperlaser bis v_s = 3,5 m/min eine einwandfreie Durchschweißung mit zusammenhängender Wurzel realisiert werden konnte, lag die Grenze für den CO₂-Laser für diese Fügeanordnung bei v_s = 2,5 m/min (Bild 5).

Zum weiteren Vergleich beider Lasertypen wurden unbeschichtete Überlappstöße mit einem Fokus-durchmesser d_F = 0,5 mm bei konstanter Geschwindigkeit v_s = 3 m/min geschweißt. Hierbei wurde am TLF 5000 (IFS-Stuttgart) die Laserleistung von P = 1940 W bis maximal P = 3970 W variiert. Ein stabiler Schweißprozeß ohne Plasmafluktuation wurde bei einer Leistung P = 2460 W erreicht, eine Durchschweißung mit P = 2550 W erzielt. Die Verbindungsnahtbreite lag hierbei im Bereich

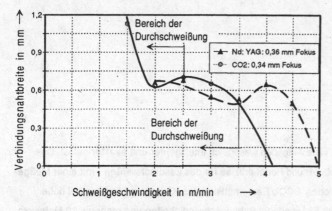

CO₂-Laser (RS 6000, KFI Essen)

Leistung	1400 W
Fokusdurchmesser	0,34 mm
F-Zahl	6
Prozeßgas	Helium

Nd:YAG Dauerstrichlaser

Leistung	1400 W
Fokusdurchmesser	0,36 mm
F-Zahl	2

Bild 5: Vergleich von CO₂- und Nd:YAG-Dauerstrichlaser beim Schweißen von 2x0,8 mm Überlappstößen (St 1403)

der einfachen Blechdicke. Der Nd:YAG-Laser bewältigt diese Aufgabe mit einer Leistung von P = 1800 W. Insgesamt zeigten die durchgeführten Versuche, daß der 2 kW Nd:YAG-Dauer-strichlaser hinsichtlich der erreichten Einschweißtiefen den Leistungen eines 3 kW CO₂-Lasers entspricht. Die Ursache für die höhere Effizienz des YAG dürfte die im Vergleich zum CO₂-Laser bessere Absorption an der Kapillarwand beim Tiefschweißen sein. Nachteilig für den Nd:YAG-Laser kann sich die im Vergleich zum CO₂-Laser deutlich schlechtere Strahlqualität bezüglich der zulässigen Toleranzen auswirken. Hinsichtlich des verfügbaren Leistungspotentials und der somit absolut erreichbaren Eindringtiefen und Schweißgeschwindigkeiten ist der CO₂-Laser dem Nd:YAG-Laser ebenfalls eindeutig überlegen.

Die in den Bildern 1 bis 3 dargestellten Schweißversuche mit Festkörperlaser wurden an einem Gerät der Firma Haas Laser GmbH D-7230 Schramberg, dessen Entwicklung unter dem Förder-kennzeichen 13 EU 00690 unterstützt wurde, durchgeführt. Wir möchten uns an dieser Stelle für die freundliche Unterstützung der Firma Haas Laser GmbH bedanken.

5 Literatur

/1/ Laserschweißen an den Astra-Motorhauben "Eine Maschine wie jede andere", Laser, November 1991, S. 328

/2/ Henn J.: Laserschweißen im Karosserie-Rohbau, VDI Berichte Nr.: 883, 1991

/3/ Göller E., Huber R., Iffländer R., Schäfer P., Wallmeroth K.: 2 kW-cw-Laser, Ergebnisdarstellung zu einem EUREKA-Projekt, Laser und Optoelektronik 25(2), 1993, S. 42

/4/ Dausinger F.: Lasers with different wave-length-Implications for varions applications, Proceedings of European Conference on Laser Treatment (ECLAT), Erlangen 1990, p. 1

/5/ Hack R., Faisst F., Meiners E., Dausinger F., Hügel H.: Schneiden mit fasergeführtem Nd:YAG-Hochleistungsfestkörperlaser - Festkörperlaser dringt in Bereiche der CO₂-Lasers vor, Laser und Optoelektronik 25(2), 1993, S. 62

Scout Nahtfolgesystem

Dr. K. Barthel, Fred Holick und Rudolf Pfefferle
Deutsche Aerospace AG, Laserstrahlführung
D-81663 München

EINLEITUNG

Mit Laser-Technologie werden zukünftig viele Aufgaben wesentlich intelligenter und exakter gelöst. Das Nahtfolgesystem SCOUT positioniert Roboter und Portal präzise für das Laserschweißen - mit einer Meßgenauigkeit von ± 0,05 mm in drei Dimensionen. SCOUT automatisiert das Kleben und Dichten bei hoher Geschwindigkeit. Die Bildverarbeitung mit 50 Hz erlaubt Bahngeschwindigkeiten von mehr als 20 Meter pro Minute.

Der herkömmliche, zeitaufwendige teach-Prozess zum definieren der Roboter-Bahn vor Beginn der Bearbeitung und auch bei jeder Änderung am Werkstück ist nun nicht mehr erforderlich.

Die Sensorentwicklung basiert auf der langjährigen Erfahrung auf dem Gebiet der Opto-Elektronik, insbesondere aber ist die Echtzeit Bildverarbeitung bereits seit 1984 ein Kernthema bei der DASA (MBB).

SYSTEMKOMPONENTEN

Das Sensor-System SCOUT zur dreidimensionalen Nahtverfolgung ist besonders auf die hohen Genauigkeitsanforderungen beim Laserschweißen zugeschnitten. Das System besteht aus den Hardwarekomponenten Sensorkopf und Sensorrechner.

Der Sensorkopf wird am Portal oder Roboter dem Werkzeug vorlaufend befestigt, ist also nahe der zu bearbeitenden Werkstückoberfläche und erzeugt das Videosignal für die Bildauswertung im Sensorrechner.

Der Sensorrechner besteht aus Bildverarbeitungshardware, Systemprozessor und - je nach Ausstattung - den Interface-Karten zur Maschinensteuerung und anderen peripheren Geräten, wie z.B. Arbeitslaser oder Schweißdrahtzuführung. Das Videobild aus dem Sensorkopf wird für die 50 Hz Echtzeit-Bildverarbeitung aufbereitet, bevor die Daten der schnellen Bildverarbeitungshardware übergeben werden. Diese besteht im Kern aus sechs Signalprozessoren in paralleler Architektur. Ein weiterer Prozessor steuert und kontrolliert den Datenfluß und führt die Kommunikation mit dem Systemprozessor.

Lage und Orientierung der Naht auf dem Werkstück werden von der Bildverarbeitungs-Software ausgewertet, das Werkzeug (TCP) an Roboter bzw. Portal folgt präzise dem dreidimensionalen Nahtverlauf.

MESSPRINZIP

Das Lichtschnittverfahren wird angewendet, wenn sich die zu verfolgende Naht durch einen Höhensprung (z.B. übereinanderliegende Bleche, Kehlnaht) gekennzeichnet ist. Bei einem Stumpfstoß (I-Stoß) ist dieses Verfahren nicht anwendbar, hier wird die Naht durch eine Kontrastbewertung im Graubild erkannt.

Beim Meßprinzip <u>Lichtschnittverfahren</u> wird ein Strichmuster unter einem Winkel auf das Werkstück projiziert und dieses Muster mit der CCD-Kamera senkrecht abgetastet. Das Video-Bild wird im Sensorrechner ausgewertet, wobei der Versatz in einer Linie einen Punkt der Kante definiert.

Vorteil dieses redundanten Verfahrens ist der Verzicht auf bewegte Teile im Sensorkopf (Scanner) und die gleichzeitige Auswertung von maximal 5 Kantenpunkten in einem Video-Bild.

So wird auch bei ungünstigen Meßbedingungen sichergestellt, daß alle 20 ms mehrere (maximal 5 und minimal 3) Kantenpunkte die Lage der Kante zeigen. Selbst der vollständige Ausfall einzelner Video-Bilder ist für die verringert die Systemgenauigkeit nicht. Das redundante Verfahren erzeugt eine sehr hohe Meßpunktdichte entlang der Naht. Erst bei einer Bahngeschwindigkeit von ca. 40 m/min überlappen die Video-Bilder nicht mehr. Aus der Lage der 5 Linien gewinnt die Bildverarbeitung zugleich die Information über alle drei Koordinaten (X, Y und Z) der Lage der Naht relativ zum Sensorkopf und die beiden Winkel der Flächennormale im Bildmittelpunkt.

Bei der reinen <u>Graubild</u> Kontrastbewertung wird der Nahtverlauf eines Stumpfstoßes erfaßt (X und Y). Es läßt sich jedoch keine Information über Höhe des Sensors über dem Werkstück oder die Flächenorientierung gewinnen.

Die <u>Kombination</u> beider Verfahren wird beim Stumpfstoß-Sensor angewendet. Die vollständige Information über Lage der Naht (X und Y), Abstand (Z) und Flächenorientierung wird mit diesem patentierten Verfahren meßtechnisch erfaßt. Mit diesem Sensor läßt ein Stumpfstoß oder sogar ein Anriß im Blech auf einer komplexen 3D-Oberfläche zu verfolgen.

SELBSTKALIBRIERUNG

Der Sensor kalibriert sich nach Anbau an die Roboterhand selbst, nachdem die menügeführte Kalibrationsprozedur gestartet wurde.

Anbau und Orientierung des Sensorkopfes sind unkritisch. Die Präzision der Messung liegt mit ± 0,05 mm in allen drei Achsen wesentlich höher als die Anforderung an die Einbaugenauigkeit.

Nach Umbau oder Wartung am Roboter erfolgt der Wiederanbau des SCOUT Sesorkopfes ohne besondere Anforderungen, lediglich eine einfache Neukalibrierung wird mit sehr geringem Zeitaufwand durchgeführt.

Die Lage des tool-center-point (TCP) muß <u>nicht</u> auf einer Symmetrie-Linie durch die Bildmitte liegen.

EICHDATEN

Sensorspezifische Eichdaten, die die Bauteilstreuungen und Fertigungstoleranzen, wie auch optische Parameter (FOV) beinhalten, sind im Sensorkopf gespeichert.

Auswechseln eines SCOUT-Sensorkopfes oder Betrieb mehrerer Köpfe an einem Roboter (Vor- und Rücklauf) sind daher ohne Zusatzaufwand, z.B. erneute Kalibration, direkt möglich.

SCHNITTSTELLEN GESAMTSYSTEM

In der Standard Konfiguration besteht das SCOUT-System aus Sensorkopf und Sensorrechner.

Der Datentransfer zur Steuerung erfolgt über eine schnelle Koppelkarten, deren Gegenstück steuerungs-seitig vom Anwender-System vorgesehen sein muß.

Mit einer üblichen seriellen Schnittstelle wird das SCOUT-System nicht in seiner vollen Leistung genutzt, kann aber die Anwender-Philosophie der Bahnkorrektur wirkungsvoll unterstützen. Die hohe Präzision der Meßdaten bleibt dem Anwender ebenfalls erhalten.

Weitere Schnittstellen (analog und digital) können optional geliefert werden, z.B. für die Leistungssteuerung des Arbeitslasers.

Es ist z.B. hierdurch möglich, das vom SCOUT erkannte Nahtende so für die Lasersteuerung zu verwenden, daß dieser exakt am Nahtende vom Sensorrechner ausgeschaltet wird. Ebenso kann bei der Wieder-aufnahme von Schweißnähten die Laserleistung "hochgefahren" werden.

Ein ebenfalls optional anschließbarer Monitor, der das Videobild des Sensors mit eingeblendeter Kantenlage zeigt, hilft das Funktionsprinzip des SCOUT zu verstehen.

Bei der Definition der Nahtparameter erfolgt auf dem Monitor die bildhafte Darstellung aller dem System bekannten Daten. Diese Option ist auch bei der Anwendung des SCOUT im Versuchsfeld sehr hilfreich und reduziert die möglichen Falscheingaben durch den Benutzer.

Ein Bediener-Terminal wird nur in der Integrationsphase erforderlich sein. Im Produktionseinsatz ist dieses jedoch in der Steuerung der Gesamtanlage integriert.

SCHNITTSTELLE SENSOR - SENSORRECHNER

Bis zu vier Sensorköpfe sind an einen SCOUT Sensorrechner ohne Zusatzausrüstung anschließbar und sequentiell zu betreiben. So kann z.B. an einer Roboterhand ein SCOUT-Sensorkopf in Vorlauf-Richtung und ein weiterer Kopf im Nachlauf angebaut werden; ohne Umorientierung des Roboters kann dann die Bewegungsrichtung reversiert werden, indem vorlaufender und nachlaufender Sensor-Kopf die Funktion tauschen.

Zwei im Bestückungswechseltakt arbeitende Roboterstationen können von einem SCOUT-Rechner bedient werden, indem während der Bestückung einer Station in der zweiten Station die Bearbeitung läuft. Hier werden beide Robotersteuerungen mit jeweils einer eigenen Steuerungskarte bedient.

SCHNITTSTELLE STEUERUNG - SENSORRECHNER

Der Datentransfer zwischen SCOUT-Rechner und der Anwender Steuerung erfolgt optimal über eine schnelle parallele Schnittstelle. Standardmäßig ist die Kopplung zu einer Maschinensteuerung vorbereitet und wird von der Betriebssoftware unterstützt.

Optional können mit zusätzlicher Hardware (Karte) und Software auch mehrere Steuerungen sequentiell bedient werden.

SPEZIELLE FEATURES

Das definierte Abbruchverhalten verhindert bei unkontrolliert auftretenden Situationen die möglichen Beschädigungen und Gefahren.

Fällt z.B. ein Gegenstand während der Nahtverfolgung auf die Bahn, so wird die Roboterbewegung sofort gestoppt, wenn der Gegenstand vom Sensor-Kopf erfaßt wird und die "falsche" Kontur nicht mit der gerade gültigen Bahnkonfiguration übereinstimmt.

So kann also auch nicht ein zufällig tangential auf der Naht liegendes Blech die Roboterbewegung ablenken, da die Blechdicke und der Höhensprung nicht der aktuellen Bahnkonfiguration entsprechen.

Der herkömmliche sehr zeitaufwendige teach-Prozess ist künftig überflüssig, da der Sensorkopf lediglich an den Nahtbeginn bewegt werden muß. Sobald die "passende" Bahnkonfiguration im Gesichtsfeld des Sensors liegt, kann die Nahtverfolgung beginnen, wobei SCOUT den Roboter führt.

Bei diesem speziellen teaching-Betriebsmodus liefert das SCOUT-System die Koordinaten-Tabelle während einer Nahtverfolgung im Handumdrehen.

Fracture Behavior of Silicon cut with High Brightness Nd:YAG-Lasers

J. Kalejs[1] ; C.C. Chao[1]; T. Franz[2] ; U. Dürr[3]; P. Verboven[3]

[1]Mobil Solar Energy Corp., 4 Suburban Park Drive, Billerica, MA 01821-3980, USA

[2]BIAS, Klagenfurter Straße 2, D-28359 Bremen, Germany

[3]LASAG AG, Mittlere Straße 52, CH-3600 Thun, Switzerland

Abstract

Thin silicon plates with thicknesses varying between s=0.3mm and s=0.5 mm were cut with a rod (v_c=2m/min) and a slab laser (v_c=4.8m/min). The samples were evaluated for microcracks. The microcrack configurations are observable by visual inspection with a microscope after etching. The fracture strength of laser cut material was measured in a fracture twist test, measured data were fit to a Weibull distribution.

Introduction

Laser beam cutting of silicon can become more important if it is possible to avoid microcracks which lead to a decrease of the mechanical strength. The combination of a brittle material and the thermal nature of the laser cutting process induce these microcracks during cutting. The influence of microcrack configuration must be evaluated to allow optimization of the cutting process.

The objective of this investigation is to compare thin silicon plates (s=0.3 to s=0.5mm) cut with high-brightness Nd:YAG-lasers (rod and slab) and to determine the microcrack configurations induced by the laser cutting process and its influence on the mechanical strength of the cut plates.

The intention to use a slab laser is to achieve higher cutting speeds than with the rod due to the rectangular beam's cross section.

Experiments

For the cutting experiments a LASAG KLS 322 (rod) and a LASAG SLAB R 200 were used. Both lasers were equipped with a focussing objective with a focal length of f=100mm. The optimized cutting parameters to minimize microcrack size and number as well as to maximize cutting speed were found. The pulse duration was t_i=0.1ms and pulse energy E_i=0.18J. The pulse frequency for the rod laser was f_i=375Hz the one for the slab laser was f_i=250Hz (limited by the maximum average power at the parameters m.a.). In both cases air at a pressure of p=0.3MPa was applied as cutting gas. These parameters led to a cutting speed of v_c=2m/min with the rod and v_c=4.8m/min with the slab laser. The slab laser spot has a larger dimension in cutting direction than the rod laser spot.

The material studied was polycrystalline edge defined film fed grown (EFG) silicon /1/ with a thickness varying between s=0.3mm and s=0.5 mm. The samples cut had the dimensions 50x50mm.

Evaluation Techniques

The applied evaluation techniques were focused on two goals: 1.) to study microcracks by visual inspection and 2.) to evaluate the influence of laser cutting on the mechanical strength of samples.

For optimization of the process, it is necessary to apply a method to detect the microcracks and microcrack configurations. These configurations are observable if the samples are prepared correctly. Therefore samples were etched in 9 parts of HNO3 and 1 part of HF to widen the microcracks. At a typical magnification of approx. 200, the microcracks appear as dark lines. This procedure can be performed in a very easy and fast way as compared to other methods. An example is given in figure 1. On the left the upper side of an unetched sample is shown. On the right a photograph of the same sample is given. After etching a microcrack becomes visible as a dark line:

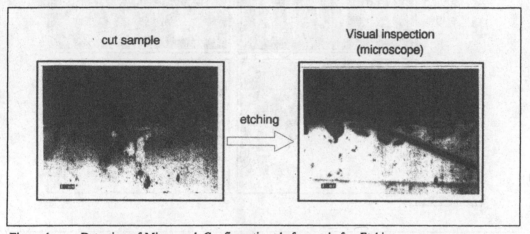

Figure 1:   Detection of Microcrack Configurations before and after Etching

In figure 2 a comparison of the detected microcrack configuration of samples cut with the rod and the slab laser are given. Samples cut with the rod (left side) have a very rough cutting edge caused by the circular shape of the rod beam. The cutting edges cut with the slab (rectangular beam) appear more smoother. With both lasers microcracks are induced by the cutting process. By applying the rod laser microcracks can be avoided at the lower side (laser exit) whereas the slab laser produces microcracks at the upper (laser entrance) and lower side as well as at the cutting edge.

The influence of these microcrack configurations on the mechanical strength of the samples was evaluated by fracture twist tests /1/. A sketch of the set up is given in figure 3.

The sample is stressed by four pins until it breaks. The fracture stress (fracture strength) can be calculated from the sample load. Measurements with sets of approx. 20 samples were done. The results were displayed in a cumulative probability plot and fitted to a Weibull distribution, as shown in figure 4 for silicon plates cut with the rod and slab laser, respectively.

Figure 4 shows a comparison of the fracture strength measured with the testing set up m.a. of samples cut with the KLS 322 (rod) and the SLAB R 200 (slab). The Weibull distribution of fracture stresses shows a lower fracture strength of the samples cut with the slab laser than of those cut with the rod laser.

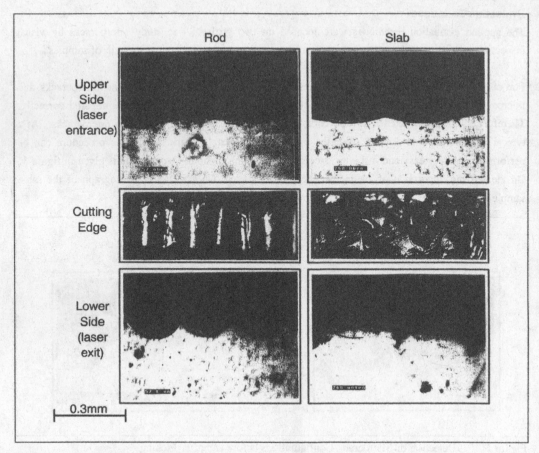

Figure 2: Comparison of Detected Microcrack Configurations

The graphs show a higher fracture strength for samples cut with the rod than those cut with the slab laser. This is attributed to an increase in the length of the microcracks for the latter.

Summary

Cutting of silicon with Nd:YAG lasers induces microcracks in the material. The microcrack configurations can be made visible by etching. Microcracks can be avoided at the lower side of the cut (laser exit) by using a rod laser. The cutting speed of the slab laser is more than two times higher than the one achievable with the rod laser.

From this results it is concluded that laser beam cutting of silicon wafers can be used for applications not requiring higher mechanical strength.

Literature

/1/   Chao, C.C.; et al.   Fracture Behavior of Silicon Cut with High Power Laser
                                 Mat. Res. Soc. Proc. Vol.226 1991

Acknowledgement

This work was supported in part by DOE/NREL through sub contract No. ZM-2-11040-3

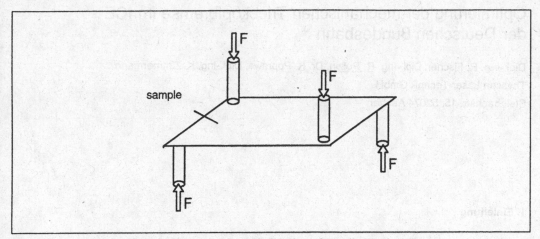

Figure 3:   Scheme of Testing Set Up for Determination of Tensile Fracture Stress

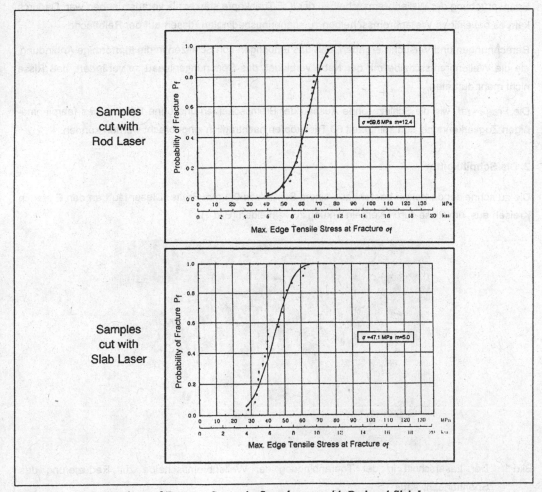

Figure 4:   Comparison of Fracture Strength; Samples cut with Rod and Slab Laser

Optimierung der mechanischen Triebkopfbremse im ICE der Deutschen Bundesbahn

Dipl.-Ing. R. Fischer, Dipl.-Ing. R. Polzin, Dr. R. Poprawe, Dipl.-Ing. K. Zimmermann
Thyssen Laser-Technik GmbH
Steinbachstr. 15, 52074 Aachen

1. Einleitung

Bereits in den ersten Monaten des regulären ICE-Betriebs im Jahre 1991 stellte sich heraus, daß die Beanspruchung der Wellenbremsscheiben der ICE-Triebköpfe stärker als vorauszusehen war. Dadurch kam es bei einigen Wellenbremsscheiben zu spannungsbedingten Rissen auf der Reibfläche.

Berechnungen und Versuche ergaben, daß das Einbringen von Schlitzen in die topfförmige Anbindung, die die Wellenbremsscheibe mit der Nabe verbindet, das Spannungsniveau so verändert, daß Risse nicht mehr auftreten.

Die Frage war, wie die Schlitze ohne Ausbau der Bremsscheiben und ohne Störung des fahrplanmäßigen Zugverkehrs bei den seinerzeit 60 Triebköpfen nachträglich eingebracht werden können.

2. Die Schnittkontur

Die zu schneidende Kontur besteht aus einem Schlitz von 0,1 mm Breite. Dieser läuft an den Enden in Kreisen aus, um Risse durch Kerbeinwirkung zu vermeiden (Bild 1).

Bild 1: Der Laserschnitt in der Topfanbindung der Wellenbremsscheibe zur Reduzierung des Spannungsniveaus

3. Anforderungen an die Schnittqualität

Die zu durchschneidende Materialdicke des Scheibentopfes beträgt ca. 4 mm. Die Schneidarbeiten waren so exakt auszuführen, daß eine Verletzung der in nur 1 mm Abstand hinter dem Topf befindlichen Nabe auszuschließen war. Die Schnittqualität hatte DIN-Anforderungen zu entsprechen. Bei der Suche nach einem geeigneten Schneidwerkzeug schieden alle konventionellen Verfahren aufgrund der gestellten Anforderungen und der Einbausituation im Triebkopf aus. Das einzig erfolgversprechende Verfahren war das Laserschneiden.

4. Auswahl des Lasersystems

Aufgrund der engen Einbaumaße der Wellenbremsscheiben zwischen Getrieben, Lagern und Motoren unter den Triebköpfen wurde ein langbrennweitiges CO_2-Lasersystem einem Nd:YAG-Laser vorgezogen (Bild 2).

Bild 2: Anordnung des Laserstrahlführungssystems unter dem Triebkopf des ICE beim Schneidvorgang

5. Unwucht der Wellenbremsscheiben

Die Hersteller der Antriebe, in denen jeweils 2 Wellenbremsscheiben auf einer Antriebswelle montiert sind, lassen bei einem Scheibengewicht von 93 kg eine Unwucht von 4,2 gm zu. Durch das Schlitzen änderte sich die Unwucht um ca. +/- 10%, also ca. 0,5 gm. Die beobachtete Veränderung ist durch die Entspannung der Scheibe, die die eingebrachten Schlitze hervorrufen, zu begründen.

6. Die mobile Laserschneidmaschine

Die mobile Laserschneidmaschine wurde speziell für diese Anwendung entwickelt, konstruiert und gebaut. Auf einem verfahrbaren Rahmen wurde ein Rofin Sinar 1700SM CO_2-Laser mit einem teilkartesischen, 3-dimensional arbeitenden GMF L-100 Roboter mit integrierter Strahlführung kombiniert.

Eine Dejustage der Laseranlage war durch eine Auflage des Rahmens auf drei Punkte auszuschließen. Die Höhe der Gesamtanlage durfte bei waagerechter Position der kartesischen Roboterachse 2,10 nicht überschreiten. Die Arbeitshöhe betrug 2,50 m.

7. Positionierung am Einsatzort

Ein spezieller, Positionierprozeß war erforderlich, um die Gesamtanlage der immer wieder veränderten Ausgangssituation anzupassen. Da die Achsen mit jedem eintreffenden Triebkopf an unterschiedlichen Positionen standen, mußte die mobile Laserschneidmaschine zur Bearbeitung von je zwei Wellenbremsscheiben auf einer Achse zu dieser neu positioniert werden. Die Positionierung gliederte sich in mehrere Teile. Zuerst erfolgte die Ausrichtung der mobilen Lasereinheit. Ein an der zu bearbeitenden Achse angebrachtes Lot bestimmt die Position der mobilen Lasereinheit zur Bearbeitung. Die Gesamtanlage wurde durch Parallelverschiebung entsprechend ausgerichtet. In einem zweiten Schritt erfolgte die Verlegung der programmierten Schnittkontur auf die zu schlitzende Wellenbremsscheibe. Anschließend wurde eine spezielle Vorrichtung auf die Wellenbremsscheibe aufgesetzt und der Roboter von einem Hilfspunkt aus hin zu einem definierten Punkt auf der Vorrichtung kartesisch verfahren. Diese Position diente als Ausgangspunkt für das Schneidprogramm.

8. Qualitätssicherung

Den Schneidvorgängen unter den ICE-Triebköpfen gingen vier Abnahmen durch die Deutsche Bundesbahn und BSI voraus. Durch die Laserbearbeitung wurde eine zeichnungsgerechte Schlitzkontur mit einer sehr guten Schnittqualität hergestellt. Aufhärtungen an der Schnittkante klingen nach 0,06 mm ab; Risse in der Schnittkante wurden in keinem einzigen Fall beobachtet.

Nach Freigabe des Verfahrens durch die Deutsche Bundesbahn wurde mit dem routinemäßigen Schneidbetrieb begonnen. Jede Schlitzung wurde durch eine abschließende Magnetrißprüfung im Schlitzbereich und die Messung der mittleren Unrundheit der Wellenbremsscheibe vor und nach der Bearbeitung begleitet. Die Ergebnisse wurden in einem Protokoll dokumentiert, das der Bremsenprüfdienst der Deutschen Bundesbahn bestätigte.

9. Die Bearbeitung

Die Bearbeitung der Triebkopfbremsscheiben im Betriebswerk I der Deutschen Bundesbahn in Hamburg-Eidelstedt erfolgte hauptsächlich nachts während der routinemäßigen Wartungsintervalle der ICE-Züge (Bild 3). Bei günstigen Bedingungen könnten in einer Nacht bis zu 10 Scheiben auf zwei Triebköpfen geschlitzt werden. Alle so bearbeiteten mehr als 350 Wellenbremsscheiben wurden nach einer Qualitätskontrolle durch das BSI-Personal und die Deutsche Bundesbahn freigegeben.

Bild 3:   Ansicht der mobilen Schneidmaschine unter einem ICE-Triebkopf bei der Bearbeitung

10. Auswirkungen auf den Fahrplan

Die Flexibilität der mobilen Laserschneidmaschine ermöglichte ein hohes Maß der Anpassung an die durch den Fahrplan vorgegebenen zeitlichen und örtlichen Gegebenheiten. Die Bearbeitungszeit war hinreichend kurz, um die Arbeiten innerhalb der nächtlichen, regulären Wartungszeiten des ICE abzuschließen. Dadurch kam es zu keiner Beeinträchtigung des fahrplanmäßigen Betriebes. Jeder Fehler während des Betriebs der Anlage hätte den Ausfall eines Triebkopfes und damit eines im Fahrplan fest eingeplanten Zuges sowie eine aufwendige, zeit- und kostenintensive Reparatur bedeutet. Das Gesamtprojekt wurde 9 Monate nach der ersten Anfrage erfolgreich abgeschlossen.

Die Rolle der exothermen Reaktion beim Laserstrahlbrennschneiden

J.Franke, W.Schulz, D. Petring, E. Beyer
Fraunhofer-Institut für Lasertechnik,
Steinbachstraße 15, D-5100 Aachen

1. Einleitung

Die exotherme Reaktion des Eisens mit dem Schneidsauerstoff hat einen bedeutenden Einfluß auf die Prozeßstabilität und die erreichbaren Schneidleistungen beim Laserstrahlbrennschneiden. Einerseits liefert die exotherme Reaktion zusätzliche thermische Schneidleistung, andererseits kann diese Reaktion unter bestimmten Bedingungen auch außerhalb des Laserstrahls zünden und ablaufen, so daß Prozeßinstabilitäten auftreten können.

Es wird ein neues Modell zum Laserstrahlbrennschneiden niedriglegierter Stähle vorgestellt, in dem gezeigt wird, daß die Reaktionsleistung den Schneidprozeß dominieren kann. Weiterhin wird eine Methode zur Stabilisierung des Schneidprozesses aufgezeigt, die nicht nur die Rauhtiefe der Schnittflächen insbesondere bei großen Blechdicken deutlich verringert, sondern auch höhere Vorschubgeschwindigkeiten und das Schneiden sehr großer Blechdicken bis zu 80 mm ermöglicht.

2. Leistungsbilanz des Laserstrahlbrennschneidens

Um zu klären, wie groß der Anteil der Leistung der Eisenverbrennung an der gesamten Schneidleistung ist, wird eine Leistungsbilanz aufgestellt. Die zum Trennen notwendige Schneidleistung P_{cut} beträgt /1/:

$$P_{cut} = 4 \cdot \lambda \cdot d \cdot \left[\left(T_p - T_o + \frac{H_m}{c_p} \right) \cdot \frac{b_s \cdot v_c}{4 \cdot \kappa} + (T_m - T_o) \cdot \left(\frac{b_s \cdot v_c}{4 \cdot \kappa} \right)^{0.36} \right] \qquad (1)$$

Mit λ = Wärmeleitfähigkeit des Stahls, d = Blechdicke, T_p = Prozeßtemperatur, d.h. die mittlere Temperatur der Schmelze an der Schneidfront, T_m = Schmelztemperatur, T_0 = Anfangstemperatur des Werkstücks. H_m = Schmelzenthalpie, c_p = spez. Wärmekapazität v_c = Vorschubgeschwindigkeit, b_s = Schnittfugenbreite, κ = Temperaturleitfähigkeit.

Dem steht die Verbrennungsleistung P_{reak} aus der Bildung des Eisenoxids FeO gegenüber:

$$P_{reak} = \rho_{Fe} \cdot b_s \cdot d \cdot v_c \cdot H_{FeO} \cdot X_{Abbr} \qquad (2)$$

mit H_{FeO} = Reaktionsenthalpie von FeO, r_{Fe} = Dichte von festem Eisen, X_{Abbr} = Massenanteil reagierten Eisens ("Abbrand") in der Schmelze. $X_{Abbr} = 1$ bedeutet vollständige Verbrennung des Eisens in der Fuge zu FeO. Höhere Oxide werden nach /3/ erst außerhalb der Fuge bei Abkühlung der Schlacke gebildet. Der Abbrand X_{Abbr} und damit die Reaktionsleistung P_{reak} wird durch Diffusionsvorgänge in der Gasgrenzschicht zur Schmelze sowie in der FeO-Schmelze begrenzt. Bei einer Sauerstoffreinheit unter 99,9% dominiert die Diffusion durch eine sogenannte Verunreinigungsgrenzschicht im Schneidgas (zuerst gefolgert von /2/). Aus einer eindimensionalen Diffusionsrechnung und der Annahme einer turbulenten Gasströmung wird folgende Abhängigkeit für den Abbrand aufgestellt (mit der Randbedingung, daß X_{Abbr} den Wert 1 nicht übersteigen kann):

$$X_{abbr} = 3,728 \cdot 10^{-4} \cdot \sqrt{\frac{b_s}{T_g \cdot d}} \cdot \frac{v_g}{v_c \cdot cos\phi} \cdot ln\left(\frac{1}{1-X_{O2}}\right) \qquad (3)$$

T_g = Temperatur der Gasgrenzschicht, v_g = mittlere Geschwindigkeit des Schneidgasstrahls, ϕ = Schneidfront-Neigungswinkel, X_{O2} = Sauerstoffreinheit.

In Abb.1 sind die notwendige Schneidleistung P_{cut}, die Verbrennungsleistung P_{FeO} sowie die beim Laserstrahlbrennschneiden mit einem 2500 W-CO_2-Laser nutzbare Leistung $P_{lcut} = P_{reak} + P_{labs}$ für eine Blechdicke von 10 mm, 1 mm Fugenbreite, 400 m/s Gasströmungsgeschwindigkeit und 80% Absorptionsgrad über der Vorschubgeschwindigkeit aufgetragen.

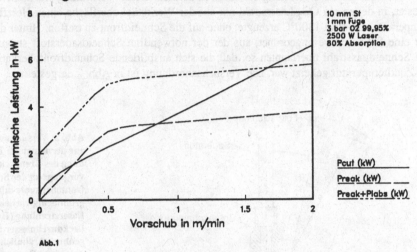

Leistungsbilanz beim Laserstrahlbrennschneiden

10 mm St
1 mm Fuge
3 bar O2 99,95%
2500 W Laser
80% Absorption

Pcut (kW)
Preak (kW)
Preak+Plabs (kW)

thermische Leistung in kW

Vorschub in m/min

Abb.1

Für oben angeführtes Beispiel mit einer Fugenbreite von 1 mm resultiert daraus eine Mindestvorschubgeschwindigkeit von 0,243 m/min, unterhalb derer sich der Brennschneidprozeß nicht mehr selbst tragen kann. Sollen daher dicke Bleche mit möglichst geringer Laserstrahlleistung laserstrahlbrenngeschnitten werden, müssen Fugenbreite und Vorschub genügend groß sein, was zunächst paradox erscheint.

3.Notwendigkeit einer zusätzlichen Heizquelle beim Brennschneiden

Beim Brennschneiden wird bekanntlich nicht nur zum Zünden beim Anschnitt, sondern permanent entweder einen Laserstrahl oder eine Brennerflamme genügender Leistung und Intensität benötigt, sonst erlischt die Eisenverbrennung. Der Grund liegt darin, daß der Verbrennungsprozeß nicht nur in Vorschubrichtung, sondern auch von der Schnitt-oberkante zur Schnittunterkante abläuft und die Oberkante dafür mindestens Zündtemperatur besitzen muß, die bei Baustahl ca. 1200°C beträgt /3/. Die Blechoberfläche wird jedoch durch den aus der Schneiddüse austretendenSchneidgasstrahl so stark gekühlt, daß die Zündtemperatur ohne zusätzliche Heizquelle an der Schneidfrontoberkante unterschritten wird. Außerdem können Verschmutzungen das zu verbrennende Eisen abdecken.

Das konventionelle Laserstrahlbrennschneiden niedriglegierter Stähle ist - je nach Laserleistung - bisher auf Blechdicken bis 20 mm, in Laborversuchen auch 30 mm, begrenzt. Eine Blechdicke von 30 mm beispielsweise ist mit einem 1500 W-CO_2-Laser auf die übliche Art nicht mehr schneidbar. Die Ursache dafür kann nur darin liegen, daß die Eisenverbrennung beim Laserstrahlbrennschneiden im Gegensatz zum autogenen Brennschneiden gestört abläuft. Beim Laserstrahlbrennschneiden kann es bei niedrigen Vorschubgeschwindigkeiten (<1 m/min) bekanntlich zu einer Zündung der Eisenverbrennung außerhalb des Laserstrahls kommen, da das umgebende Material durch Wärmeleitung auf Zündtemperatur geheizt wird. Das dadurch ausgelöste "Vorbrennen" der Schneidfront, auch "Self-Burning" genannt, das erstmals von Arata /4/ beschrieben wurde, ist instationär, da die Reaktion an der Schneidfront ohne Kontakt mit dem Laserstrahl erlischt, bis sie vom Laserstrahl durch den Vorschub wieder eingeholt und erneut gezündet wird. Eine großflächige Vorheizung der Schneidfrontoberkante auf Zündtemperatur wie beim autogenen Brennschneiden müßte die Eisenverbrennung stabilisieren können. Diese Vermutung wurde in Experimenten getestet, in denen ein CO_2-Laserstrahl nur auf der Werkstückoberfläche einen Heizfleck mit einer Temperatur von etwa 1300°C erzeugte, ohne auf die Schneidfront zu treffen. Hinter dem Heizfleck war eine Schneiddüse angeordnet, aus der der notwendige Schneidsauerstoff strömte. Heizfleck und Schneidgasstrahl überlappten so, daß die sich ausbildende Schneidfrontoberkante permanent auf Zündtemperatur geheizt war. Die Versuchsanordnung ist in Abb. 2 dargestellt:.

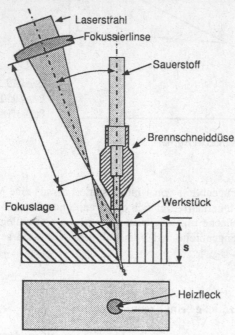

Laserstrahl

Fokussierlinse

Sauerstoff

Brennschneiddüse

Fokuslage

Werkstück

s

Heizfleck

Abb. 2: Versuchsaufbau zur permanenten Zündung des Brennschneidvorgangs an der Schneidfrontoberkante mittels großflächig vorheizender Laserstrahlung (Heizfleckdurchmesser > Fugenbreite), ähnlich dem autogenen Brennschneiden.

Mit dieser Anordnung konnten bei 1200 W Laserstrahlleistung selbst 80 mm dicke Stahlplatten noch mit 0,20 m/min bartfrei und mit guter Schnittqualität ($R_z < 50$ mm) getrennt werden (Abb. 3), wobei der Schneidleistungsbedarf rechnerisch bei 28,14 kW liegt. Die Versuche bestätigen, daß der Laserstrahl beim Laserstrahlbrennschneiden mit niedrigen Vorschubgeschwindigkeiten einen zu kleinen Durchmesser hat, um eine permanente Zündung der Verbrennung zu gewährleisten.

Abb. 3: Abbrandstabilisierter-Laserstrahlbrennschnitt von 80 mm St 52-3. Laserleistung nur 1200 W, Schneidgasdruck 9,5 bar O_2, Schneidgeschwindigkeit 0,20 m/min. Gemittelte Rauhtiefe R_z = 45 μm.

4.Übertragung der Versuchergebnisse auf das Laserstrahlbrennschneiden

Angesichts der niedrigen Investitionskosten des autogenen Brennschneidens gegenüber dem Laserstrahlbrennschneiden ist es natürlich nicht sinnvoll, die preiswerte Brennerflamme durch den teuren Laserstrahl zu ersetzen. Sinnvoll ist jedoch eine Hybridtechnologie, welche den Vorteil des autogenen Brennschneidens (stabilisierter, stationärer Verbrennungsprozeß an der Schneidfront) mit dem Vorteil des Laserstrahlbrennschneidens (Einkoppelung zusätzlicher thermischer Leistung in der Fuge) kombiniert. Ein solches Verfahren, das als abbrandstabilisiertes Laserstrahlbrennschneiden (AS-Laserstrahl-brennschneiden) bezeichnet wird, wurde inzwischen entwickelt und vom Fh-Institut für Lasertechnik zum Patent eingereicht /5/.

Im Gegensatz zum normalen Laserstrahlbrennschneiden wird beim AS-Làserstrahlbrennschneiden in Schneidrichtung vor der Schneiddüse oder alternativ rings um die Düse zusätzlich zum durch die Schneiddüse geführten Laserstrahl ein Heizfleck erzeugt, dessen Durchmesser größer ist als der des Schneidgasstrahls und der sich zumindest teilweise mit dem Schneidgasstrahl überlappt. Dieser Heizfleck kann durch einen geeignet fokussierten Laserstrahl, aber auch durch eine Induktionsschleife oder die Flamme eines Gasbrenners erzeugt werden. Die Temperatur im Heizfleck muß dabei oberhalb der Zündtemperatur liegen. Abb. 4 zeigt einen Versuchsaufbau mit Vorheizung durch eine Brennerflamme.

Ein Vergleich sowohl mit dem autogenen Brennschneiden als auch mit dem normalen Laserstrahlbrennschneiden zeigt, daß die von diesen Verfahren erzielten Schneidgeschwindigkeiten mit dem AS-Laserstrahlbrennschneiden erheblich übertroffen werden (Abb. 5). Die Abbildungen 6 und 7 zeigen AS-Laserstrahlbrennschnitte von 15 und 30 mm Dicke mit 2200 W bzw. 4900 W Laserstrahlleistung am Werkstück.

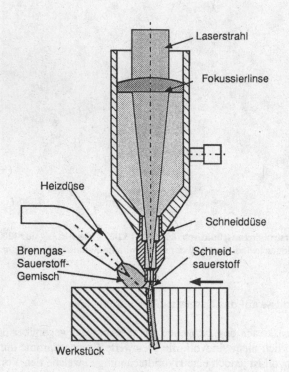

Abb. 4: Versuchsaufbau zum abbrandstabilisierten Laserstrahlbrennschneiden. Vorheizung durch Acetylenbrennerflamme.

Laserstrahl

Fokussierlinse

Heizdüse

Schneiddüse

Brenngas-
Sauerstoff-
Gemisch

Schneid-
sauerstoff

Werkstück

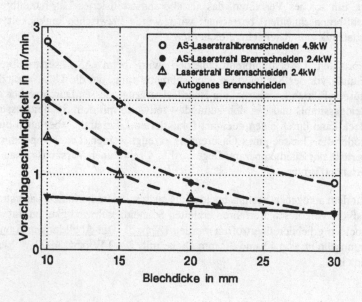

Abb. 5: Vorschubgeschwindigkeiten beim Laserstrahl- und autogenen Brennschneiden von Baustahlblechen für einen maximalen Riefennachlauf von 20% der Blechdicke (St 52-3, Sauerstoffreinheit 99.7%)

○ AS-Laserstrahlbrennschneiden 4.9kW
● AS-Laserstrahl Brennschneiden 2.4kW
△ Laserstrahl Brennschneiden 2.4kW
▽ Autogenes Brennschneiden

Vorschubgeschwindigkeit in m/min

Blechdicke in mm

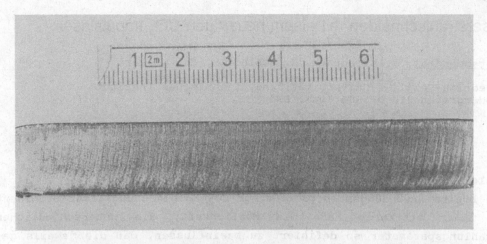

Abb. 6: Abbrandstabilisierter-Laserstrahlbrennschnitt von 15 mm St 52-3. Laserleistung 2200 W, Schneidgasdruck 3,0 bar O_2, Schneidgeschwindigkeit 1,2 m/min. Gemittelte Rauhtiefe R_z = 20 μm.

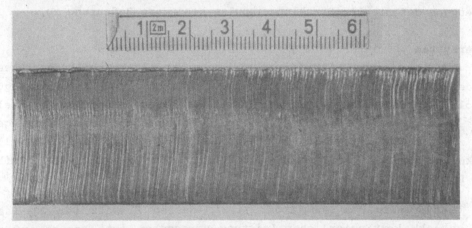

Abb. 7: Abbrandstabilisierter-Laserstrahlbrennschnitt von 30 mm St 52-3. Laserleistung 4900 W, Schneidgasdruck 4,0 bar O_2, Schneidgeschwindigkeit 1,0 m/min. Gemittelte Rauhtiefe R_z = 30 μm.

Schrifttum:

/1/ Schulz, W., D. Becker, J. Franke, R. Kemmerling, G. Herziger: Heat conduction losses in laser cutting of metals, eingereicht bei J.Phys.D: Appl. Phys., Aachen 1992.

/2/ Boschnakow, I.: Einige chemisch-metallurgische Teilvorgänge beim Brennschneiden Schweißtechnik, Berlin (Ost) 1973 (6), S.255-256.

/3/ Teske, K.: Über die Thermodynamik und Thermochemie des Brennschneidens, Dissertation, Berlin 1955.

/4/ Arata, Y. et al.: Dynamic behavior in laser gas cutting of mild steel, Transactions of JWRI 1979, S.15-25.

/5/ Franke, J., W. Schulz, G. Herziger: Verfahren zum Abtragen von Werkstoff von relativbe-wegten metallenen Werkstücken, Patentanmeldung P4215561.4 vom 12.05.92.

Laserstrahlschneiden mit einem neuartigen CO_2-Impulslaser

G.Staupendahl, J. Bliedtner, K. Schindler, Ch. Weikert

Friedrich-Schiller-Universität Jena, Technisches Institut
Löbdergraben 32, O-6900 Jena, BRD

Einleitung

Die breite Anwendungspalette moderner CO_2-Hochleistungslaser erfordert
in immer stärkerem Maße die Möglichkeit, die unterschiedlichen
Strahlungsparameter so definiert zu beeinflussen, daß die jeweils ge-
wünschte Wechselwirkung der Strahlung mit dem Material unter optimalen
Bedingungen abläuft. Für zahlreiche Anwendungen kommt dabei der
Erzeugung definierter Strahlungsimpulse besondere Bedeutung zu.

Lasersystem

Für die in diesem Beitrag beschriebenen Materialbearbeitungsprozesse
wurde ein neuartiger CO_2-Impulslaser eingesetzt. Die Impulserzeugung
erfolgt durch die Steuerung der Güte des Laserresonators. An Stelle
des üblicherweise eingesetzten teildurchlässigen Auskoppelspiegels
mit fester Reflektivität wird in diesem Lasertyp ein spezieller
Auskoppelmodulator eingesetzt. Abb.1 zeigt den schematischen Aufbau
des Lasers.

Das Grundprinzip des Modulators /1,2/, dessen Einsatz für CO_2-Laser
bis ca. 1kW kontinuierlicher Leistung vorgesehen ist, beruht auf dem
Prinzip des Fabry-Perot-Interferometers. Zwei ZnSe-Hochleistungsopti-

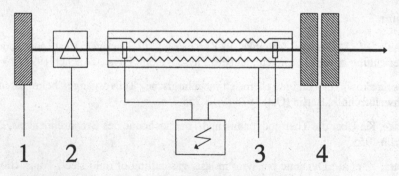

Abb. 1: Schematischer Aufbau des CO_2-Impulslasers
1 - Endspiegel   2 - wellenlängenselektives Element
3 - Wellrohr   4 - Auskoppelmodulator

ken (AR,R=0,5) werden schnell planparallel zueinander bewegt, damit die Reflektivität der Anordnung für eine feste Wellenlänge schnell durchgestimmt und somit die Güte des Resonators moduliert.

Die Umsetzung dieses Grundprinzips in eine technische Lösung erfolgte mit den in Abb. 2 und 3 dargestellten Modulatoren.

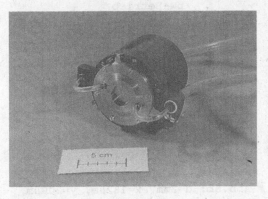

Abb.2:
Interferenzmodulator ILM-E

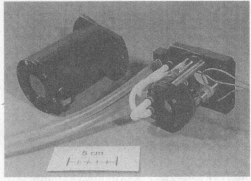

Abb.3:
Interferenzmodulator ILM-P

Im Interferenzmodulator ILM-E (Abb.2) erfolgt die schnelle Abstandsänderung der Interferometerplatten auf elektrodynamischem Wege, im Modulatortyp ILM-P (Abb.3) wird die Bewegung der Interferometerplatte durch drei Piezostellelemente realisiert. Dieser Modulator läßt Modulationsfrequenzen bis ca. 11kHz zu und besitzt ein vakuumdichtes Gehäuse, so daß er einfach an geeignete kommerzielle cw-CO_2-Laser angeblockt werden kann.

Eine typische Impulsform, die mit dem Modulatortyp ILM-P an einem CO_2-Laser SM 400 (Hersteller FEHA Halle GmbH) erzielt wurde, zeigt Abb.4.

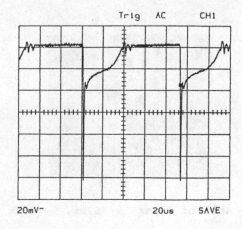

Abb.4: Typischer Impulsverlauf
Modulationsfrequenz : 11kHz
Mittlere Leistung   : 220W
Impulsspitzenleistung: 3.5kW

Die neuartigen Impulseigenschaften wurden zunächst an Bohrungen in speziellen Fe-Werkstoffen unterschiedlicher Dicke d erprobt. Die Abb. 5-7 zeigen Bohrlochgeometrien in Automatenstahl 9SMnPb2.3 mit geschliffener Oberfläche.

cw-Betrieb	pw-Betrieb

Abb.5a: d=2mm, x=403μm, y=450μm   Abb.5b: d=2mm, x=182μm, y=209μm

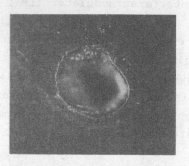 

Abb.6a: d=4mm, x=1649μm, y=1973μm   Abb.6b: d=4mm, x=243μm, y=362μm

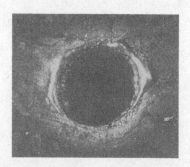 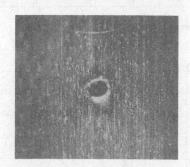

Abb.7a: d=6mm, x=1963μm, y=2217μm   Abb.7b: d=6mm, x=362μm, y=405μm

Um die Vergleichbarkeit der Bohrungen zu gewährleisten, wurde für beide Betriebsarten die gleiche mittlere Leistung von 220W gewählt. Im pw-Betrieb kamen die oben abgebildeten Impulsformen mit der Folgefrequenz von 11kHz zum Einsatz. Alle anderen Prozeßparameter blieben konstant.Im pw-Betrieb sind neben den auffallend kleinen Bohrlochdurchmessern auch kürzere Durchbohrungszeiten für größere Probendicken zu verzeichnen (Abb.8).

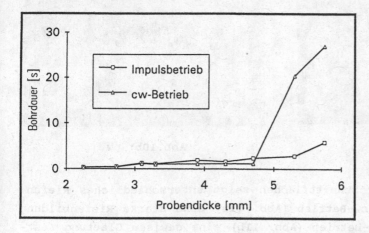

Abb.8:
Abhängigkeit der Bohrdauer von der Probendicke

Ein weiterer Vorteil des pw-Betriebes ist der geringere Wärmeeintrag in das Material, was eine kleinere Wärmeeinflußzone um das Bohrloch zur Folge hat. Dies wird in Abb. 9 am Beispiel des Stahlbleches ST 1203 sichtbar.

Bohrloch St 1203, d=1,5mm

Abb.9a: cw

Abb.9b: pw

Die erzielten Ergebnisse des Bohrens lassen sich beim Laserstrahlschneiden mit Einstechen auf der Kontur weiter verfolgen. Wie zu erwarten, zeigen die cw-Einstichlöcher bei vergleichbaren Prozeßparametern (Schneidgas: O_2 bei 2bar; Abstand Düse zum Werkstück: 2mm; Linsenbrennweite: 67mm) wesentlich größere Durchmesser.

Schneiden von St 1203 , d=1,5mm

Abb.10a: cw   Abb.10b: pw

Die Mikrostruktur der Schnittflächen zeigt unterschiedliches Riefen-verhalten. Während im cw-Betrieb (Abb. 11a) eine starke Riefenbildung vorherrscht, ist im pw-Betrieb (Abb. 11b) eine gewisse Glättung fest-zustellen.

Riefenstruktur St 1203
d=1,5mm, v=20mm/s

Abb.11a: cw   Abb.11b: pw

Abschließend sei bemerkt, daß die Bearbeitung von Fe-Werkstoffen mit dem vorgestellten Impulslaser wesentliche Vorteile in Bezug auf Effizienz und Qualität des Wechselwirkungsprozesses erbringt.

Ausblick

Die vorgestellten Resultate wurden unter experimentellen Bedingungen erhalten, die noch nicht optimal waren. Das betrifft insbesondere die relativ geringe Laserleistung, die zur Verfügung stand. Demnächst sind weiterführende Untersuchungen unter Einsatz eines CO_2-Laser-Verstärker Systems bzw. von Lasern höherer Leistungsklassen geplant.

Danksagung

Die diesem Beitrag zugrunde liegenden Ergebnisse ermöglichte ein Vorhaben, das aus Mitteln des BMFT unter dem Förderkennzeichen 13 N 5933 gefördert wurde.
Die Autoren möchten sich weiterhin bei den Mitarbeitern der Feinmechanischen Werke Halle GmbH für ihre freundliche Unterstützung bedanken.

Literatur

/1/ Staupendahl, G.; Pöhler, M. und Schindler, K. : Anordnung zur externen Modulation von IR-Laser-Strahlung hoher Leistung, DD-WP H 01 S/264 005 6, (1984)

/2/ Staupendahl, G. : Universeller Modulator für CO_2-Laserstrahlung und seine Einsatzmöglichkeiten, Laser und Optoelektronik **23** (3) (1991), S. 126-133

Laser Beam Fusion Cutting: Diagnostics and Modelling of Melt Drag and Ripple Formation

H. Zefferer[1], D. Petring[1], W.Schulz[2], F. Schneider[1], G Herziger[1, 2]

[1] Fraunhofer-Institut für Lasertechnik, Steinbachstr. 15, D-52074 Aachen
[2] Lehrstuhl für Lasertechnik, Steinbachstr. 15, D-52074 Aachen

1 Introduction

In laser beam fusion cutting the dynamics of the physical processes in the cutting front area and in the melt drag lead to ripple formation. In the case of high speed cutting the melt accelerated along the cut edge even includes a capillary [1].

Arata [2] suggested the edge cut as a means of visualizing the cutting front and the melt drag. In what follows, a modification of the edge cut is presented with which melt ejection comparable to that of a "normal" cut can be achieved with suitable forming and guiding of the gas jet. High speed photographs illustrate the geometry of the melt drag and of part of the cutting front during the cutting process.

In the case of oxygen cutting of steel with pulsed or modulated laser radiation a spectral analysis of the roughness profile shows that an externally exerted disturbance will only lead to ripple formation at frequencies of up to several hundred Hertz [3, 4]. New theoretical work by Schulz [5] concerning the time-dependent behaviour of the cutting front shows that the cutting front can be unstable. It then oscillates at frequencies which are typical for ripple formation. Here the dynamic behaviour of the cutting process is analysed using the methode of the frequency response of the cutting process to externally exerted time-variable laser beam power.

2 Edge cuts

Cuts are made along a pre-cut edge in order to visualize the zone of interaction. The laser beam is shifted in relation to this edge by less than one cut kerf width. The resulting cut ideally consists of a cut front with an azimuthal extension between 180° and 90° and a single cut edge.

In laser beam fusion cutting the cutting front is subjected to supersonic speeds of the gas jet [6]. In the form of edge cut suggested by Arata the expansion in the half-space left by the missing cutting flank leads to boundary layer separation in the area of cutting front and melt drag. Without additional measures this cutting process cannot be compared to that of a normal cut.

In the set-up shown in Fig. 1 the missing cutting flank is replaced by a transparent glass sheet. The gas jet expands without boundary layer separation because it is guided between the glass sheet and the cut edge. A gap is set between the edge to be cut and the glass sheet in order to prevent the destruction of the glass sheet by absorption of laser radiation.

A: distance of glass
sheet from the
edge to be cut

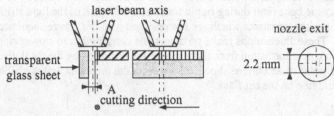

laser beam axis

nozzle exit

transparent
glass sheet

2.2 mm

A

cutting direction

Fig. 1   Experimental set-up of the edge cut procedure

The glowing zone of interaction is lit by a light source focussed on the relevant area to give contrast to the images from the high speed video camera. Fig. 2 shows such a video image. The cutting front area is bounded by the two light, vertical stripes on the left of the picture. The second stripe from the left is the result of direct reflection of the illumination beam path off the molten surface. It is at an azimuthal angle of around 70°-90° and indicates the transition from the cutting front to the melt drag. The cutting ripples, visible as light and dark stripes, continue to the right.

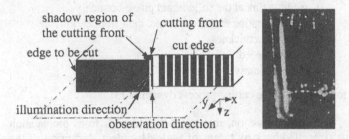

shadow region of
the cutting front

cutting front

edge to be cut

cut edge

illumination direction

observation direction

Fig. 2   Schematic of the recording arrangement and a corresponding video image of the cutting zone (CrNi-steel 2 mm, cutting speed: 2.0 m/min, laser beam power: 1200 W)

Fig. 3 shows a sequence of images. The width of the light stripe at the transition from cutting front to cut edge varies. During ripple formation this light stripe spreads from the front to the flank. As the ripple forms it separates itself from the transition area and thereafter undergoes no further changes in form: it is solidified. The ripples are formed in the area of transition from cutting front to cut flank.

Fig. 3   Sequence of an edge cut at an image sequence freqency of 1000/s
(CrNi-steel 2 mm, cutting speed: 2.0 m/min, laser beam power: 1200 W)

In this case ripple formation occurs with an average time-constant of T = 2 ms. In addition to this time behaviour during ripple formation the width of the light stripe varies locally according to time-constants which are not resolved with our image sequence frequency of f = 1000 Hz. These fluctuations make no detectable contribution to slower ripple formation. Here the even ripple formation from top to bottom is disturbed by the local discontinuity of the reflection point of an oblique shock. This effects the melt flux and is visible in the form of a horizontal line on the cut flank.

3 Dynamic process analysis

In [5] linear response theory is used to prove that the cutting front can be unstable. Fig. 4 shows a simplified two-dimensional geometry for modelling the cutting front dynamics.

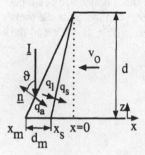

$\underline{I}$: vector of energy flux density of laser beam
ϑ: angle of incidence
$\underline{n}$: normal vector
q_a: heat flux absorbed
q_l, q_s: heat flux at the solid/liquid phase boundary
x_m, x_s: absorption front coordinate, melt front coordinate
d_m: melt film thickness
v_0: cutting speed
d: sheet thickness

Fig. 4   Geometry for the formulation of the cutting process dynamics [5]

The relaxing sub-systems, i.e. heat conduction, melt ejection and melt front propagation (solid/liquid phase boundary), are fed back by means of angle-dependent absorption. If the laser beam intensity is time-constant the amount of heat flux absorbed depends only on the position of the melt front and the melt film thickness:

$$q_a = \cos(\vartheta) \cdot A(\cos(\vartheta)) \cdot I \tag{3.1}$$

with $\quad \cos(\vartheta) = (1 + (d / x_m)^2)^{-1/2}$

and $\quad x_m = x_s - d_m$

Alterations to the amount of heat flux absorbed q_a determine variations in the heat flux q_l at the melt front by heat conduction in the melt film. A varying heat flux difference at the melt front leads to a change in its position x_s. The hydrodynamic reaction to this new flow condition (the speed of the melt front) is an adjustment of the melt film thickness d_m.

The extent to which the dynamics of the processes in the cutting front area effect ripple formation is analysed here using experiments with power-modulated laser radiation. The laser radiation in square wave signal form is modulated in such a way that the power varies between 80% and 100% of the maximum output power. Part of the secondary radiation emitted by the process is projected onto a photodiode with a scraper mirror arrangement [7] as shown in Fig. 5 for the same time-resolved recording as carried out for laser beam power.

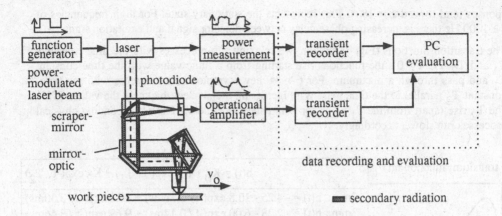

Fig. 5 Diagnostic set-up of the cutting experiments for dynamic process analysis

The measured signals are analysed according to the three following processing steps [8]:

1. Determination of the amplitude and phase of the complex frequency response $\underline{F}(\omega)$ for the selected excitation frequencies ω [9, p.177ff]. This describes the response behaviour of the detector signals to time-variable intensity.

2. Determination of an analytic approximation of the frequency response $\underline{F}(\omega)$.

3. Calculation of the transition function h(t) (step response) of the system [9, p.200ff].

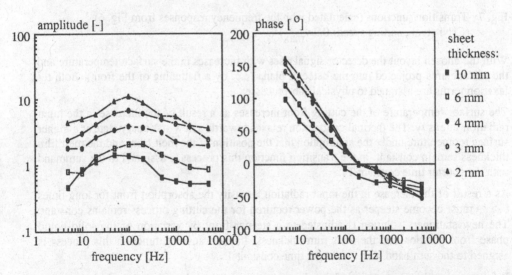

Fig. 6 Frequency response of the detected signal as a function of the excitation frequency (CrNi-steel, cutting speed: $0.9\ v_{max}$, laser beam power: 4.5 ± 0.5 kW)

The frequency responses $\underline{F}(\omega)$ (Fig. 6) show a resonant maximum at a frequency of around f = 100 Hz for all metal sheet thicknesses tested. The signals oscillate at a slightly higher frequency in phase with the excitation signal. For low frequencies of f -> 0 Hz the phase re-

578

sponses tend toward $\phi = 180°$. This represents the stationary state. For high frequencies of $f \geq 1000$ Hz there is increasing phase delay between detector signal and excitation signal.

The transition functions (Fig. 7) contain two relaxing sub-processes with the time-constants T_1 and T_2. For $t > 0$ s the functions rise starting from a finite value with the time-constant T_1 and pass through a maximum. For $t \to \infty$ they asymptotically converge with the time-constant T_2 parallel to the time axis. With increasing metal sheet thickness the values of T_1 and T_2 rise (apart from the T_1 value for 2 mm sheet thickness). The corresponding physical processes run slower accordingly.

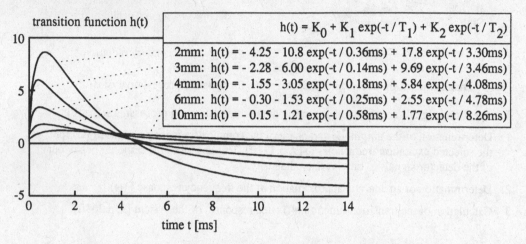

transition function h(t)

$$h(t) = K_0 + K_1 \exp(-t / T_1) + K_2 \exp(-t / T_2)$$

2mm: $h(t) = -4.25 - 10.8 \exp(-t / 0.36ms) + 17.8 \exp(-t / 3.30ms)$
3mm: $h(t) = -2.28 - 6.00 \exp(-t / 0.14ms) + 9.69 \exp(-t / 3.46ms)$
4mm: $h(t) = -1.55 - 3.05 \exp(-t / 0.18ms) + 5.84 \exp(-t / 4.08ms)$
6mm: $h(t) = -0.30 - 1.53 \exp(-t / 0.25ms) + 2.55 \exp(-t / 4.78ms)$
10mm: $h(t) = -0.15 - 1.21 \exp(-t / 0.58ms) + 1.77 \exp(-t / 8.26ms)$

time t [ms]

Fig. 7   Transition functions (calculated with the frequency responses from Fig. 6)
(CrNi-steel , cutting speed: $0.9 \, v_{max}$)

With the chosen layout the detector signal rises with increases in the surface temperature and the surface area projected into the detector plane (e.g. by a flattening of the front). Both relaxation terms are assigned to physical sub-processes.

The surface temperature of the cutting front increases as a result of the increase in the input radiation intensity. The thermal conduction relaxes towards a new stationary state at a higher surface temperature under the assumption that the position of the melt front and the melt film thickness remain constant. In the transition function this process is assigned to the summand with the shorter time constant T_1.

As a result of the increase in the input radiation intensity the absorption front for long times $t \to \infty$ must become steeper as the power required for the cutting process remains constant. The new stationary solution is adjusted to a steeper front by variation of the solid-liquid phase front position and the melt film thickness. In the transition function this process is assigned to the summand with the longer time constant T_2.

It follows that the propagation of the absorption front, which is assigned to the experimentally determined time-constant T_2, shows low-pass behaviour for exciting disturbances. Only disturbances with time-constants in the order of magnitude of this relaxation time or longer bring about changes in the position of the absorption front.

The processes in the cutting front area supply the boundary conditions for the melt flow and the solidification of the melt drag and thus for ripple formation. Spectral analysis of the ripple profiles (see Fig. 8) gives cut-off frequencies above which exciting frequencies are no longer to be found in the ripple profile. These rise with decreasing metal sheet thickness and have frequencies of f = 200-500 Hz. These cut-off frequencies are in the order of magnitude of the time-constants T_2 and show the same dependency on metal sheet thickness.

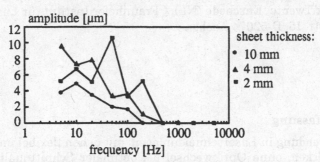

Fig. 8   Amplitudes of the Fourier transformed ripple profiles over the excitation frequency
(CrNi-steel, cutting speed: 0.9 v_{max}, laser beam power: 4.5 ± 0.5 kW)

This low-pass behaviour also corresponds with the findings from the high speed film images where the ripples are formed with a "natural" time-constant of $T_{ripples}$ = 2 ms. High frequency local disturbances do not effect this.

The dynamics of the physical processes in the absorption front area represent one cause of ripple formation. The application of the conventional methodes of process analysis leads as shown here to a sophisticated understanding of the process dynamics.

Acknowledgements

This work has been supported by Deutsche Forschungsgemeinschaft DFG (Az.: He 979/16).

Bibliography

[1] D. Petring, K.U. Preissig, H. Zefferer, E. Beyer: Plasma-Effects in Laser Beam Cutting. DVS-Berichte 135, Konferenz "Strahltechnik", Karlsruhe (1991), S. 12-15

[2] Y. Arata, H. Maruo, I. Miyamoto, S. Takeuchi: Dynamic Behaviour in Laser Gas Cutting of Mild Steel. Trans. JWRI (1979) Part 2, S. 15-26

[3] J. Powell, T.G. King: Cut Edge Quality Improvement by Laser Pulsing. Proc. 2nd Int. Conf. Lasers in Manufacturing, Birmingham (1985), S. 37-45

[4] I. Miyamoto, T. Ohie, H. Maruo: Fundamental Study of In-Process Monitoring in Laser Cutting. 4th Int. Col. on Welding and Melting by Electrons and Laser Beam, Cannes (1988)

[5] W. Schulz: Schmelzschneiden mit Laserstrahlung: Hydrodynamik und Stabilität des physikalischen Prozesses. Dissertation RWTH Aachen (Juni 1992)

[6] H. Zefferer, D. Petring, E. Beyer: Investigations of the Gas Flow in Laser Beam Cutting. DVS-Berichte 135, Konferenz "Strahltechnik", Karlsruhe (1991), S. 210-14

[7] E. Beyer, P. Abels, A. Drenker, D. Maischner, W. Sokolowski: New Devices for On-Line Process Diagnostics during Laser Machining. Proc. ICALEO, San Jose (1991), S. 133-38

[8] F. Schneider: Schmelzschneiden mit CO_2-Laserstrahlung: Identifikation des dynamischen System, Diplomarbeit, RWTH Aachen (Juli 1993)

[9] R. Isermann: Identifikation dynamischer Systeme Band II. Springer-Verlag Berlin, Heidelberg (1988)

Adaptives Linsensystem für Laserschneidanlagen

L. Beckmann[1] **, O. Märten**[2]

[1] Universität Twente, Enschede (NL), [2] Fraunhofer Institut für Lasertechnik, Steinbachstr. 15, D-52074 Aachen

Zusammenfassung

Für die Anwendung in Laserschneidanlagen, mit denen flexibel unterschiedliche Werkstückdicken, ohne Optikwechsel bei optimaler Schnittqualität bearbeitet werden sollen, wurde ein adaptives Linsensystem entwickelt. Mit diesem Linsensystem kann die Brennweite der Bearbeitungsoptik im Bereich von 63.5mm bis 170mm an die Werkstückdicke angepaßt werden. Dadurch läßt sich eine hohe Schnittqualität hinsichtlich Rauhtiefe und Flankensteilheit bei maximaler Schneidgeschwindigkeit erreichen. Es wurden zwei optische Entwürfe verglichen:

1. Ein Entwurf mit drei Linsen, wobei ein aus zwei positiven Linsen bestehendes Dublette gegenüber der negativen ersten Linse verschoben wird. Alle Linsen haben nur spährisch gekrümmte Flächen.

2. Ein Entwurf mit prinzipiell gleichem Aufbau, jeboch mit nur zwei Linsen, wobei eine Fläche der positiven Linse asphärisch gekrümmt ist

Die experimentelle Untersuchung der Fokussiereigenschaften beider Linsensysteme im Vergleich mit Einzellinsen mit festen Brennweiten zeigt, daß bei langer Brennweite gleiche Strahldurchmesser erzielt werden, während bei kurzen Brennweiten selbst etwas kleinere Fokusdurchmesser erzielt werden. Somit wird die bei der Auslegung des adaptiven Linsensystems erreichte hohe Korrektur des Öffnungsfehlers durch das Experiment bestätigt. Für den Anwender bedeutet dies, daß mit dem Linsensystem im Dünnblechbereich bei kurzen Brennweiten eine höhere Schnittleistung zu erwarten ist als mit Einzellinsen.

Einleitung

Das Schneiden mit Laserstrahlung ist das am weitesten in die Produktion eingeführte Laserbearbeitungsverfahren. Vornehmlich in der blechverarbeitenden Industrie werden mit dem Laserstrahl kleine bis mittlere Losgrößen wirtschaftlich gefertigt. Die Flexibilität des berührungslosen Werkzeugs Laserstrahl´ ermöglicht es dem Anwender, sich auf häufig wechselnde Bauteilgeometrien mit unterschiedlichen Werkstückdicken mit möglichst geringen Rüstzeiten der Laseranlage einzustellen.

Die Schnittqualität (Rauhtiefe und Kantensteilheit) und die Schneidleistung (Vorschubgeschwindigkeit) wird wesentlich von der Strahlgeometrie in der Schnittfuge beeinflußt. Hierbei läßt sich in Abhängigkeit von der Strahlleistung auf dem Werkstück und der Werkstückdicke eine Brennweite und eine Vorschubgeschwindigkeit angeben, mit der eine optimale Schnittqualität, d.h. minimale Rauhtiefe und senkrechte Schnittkanten, erzielt wird. Das bedeutet, daß bei einem Bauteilwechsel, bei dem sich die Werkstückdicke ändert, die Brennweite der Fokussieroptik für optimale Schnittqualität angepaßt werden sollte.

In der Praxis wird häufig das gesamte Werkstückdickenspektrum mit einer konstanten Brennweite der Fokussieroptik bearbeitet, da ein Optikwechsel mit anschließender Düsenjustage die Rüstzeiten erheblich verlängern würde. Als Konsequenz daraus wird eine entsprechend geringere Schnittqualität und nicht optimale Schnittgeschwindigkeit in Kauf genommen.

Die in diesem Beitrag beschriebene Entwicklung eines adaptiven Linsensystems zielt darauf ab, dem Anwender die Möglichkeit zu geben, die Brennweite der Fokussieroptik kontinuierlich an die Werkstückdicke anzupassen damit die Anlage immer eine optimale Schnittqualität und damit höchste Produktivität liefert. Im folgenden werden die Anforderungen an die Entwicklung eines Linsensystems mit einstellbarer Brennweite hinsichtlich der Abbildungseigenschaften erläutert. Weiterhin werden zwei optische Entwürfe vorgestellt. Abschließend werden die Eigenschaften der aufgebauten Prototypen sowie die Ergebnisse der Untersuchung der Fokussiereigenschaften diskutiert.

Optischer Entwurf des Linsensystems

Bei dem optischen Entwurf wird angestrebt, die Anforderungen an die Abbildungseigenschaften der Optik mit möglichst wenigen Linsenelementen voll zu erfüllen. Als Anforderungen an die Abbildungseigenschaften des Linsensystems sind zu nennen:

1. Es soll der für Schneidanwendungen typische Brennweitenbereich überdeckt werden (63.5mm bis 190mm) bei Strahldurchmesser bis 20mm.

2. Die Fokuslage sollte sich bei Brennweitenänderung nur geringfügig verschieben.

3. Geringe Aberrationen im gesamten Brennweitenbereich verglichen mit den Aberrationen von handelsüblichen Meniskuslinsen, die standardmäßig zum Schneiden eingesetzt werden.

4. Eine Erwärmung der Linsen durch absorbierte Leistung soll zu keiner Zunahme der Aberrationen bzw. Abnahme der Strahlqualität der fokussierten Laserstrahlung führen. Das Linsensystem soll bis Strahlleistungen von 2.5kW einsetzbar sein.

Aufgrund der beschriebenen Anforderungen wurden zwei Entwürfe für Linsensysteme entwickelt, wobei speziell die Änderung der Abbildungseigenschaften bei Erwärmung des Linsenmaterials berücksichtigt wurde /1,2/. Dabei wurde versucht, die Anzahl der optischen Komponenten zu minimieren um Leistungsverluste und optische Wirkung der absorbierten Leistung gering zu halten.

Beim ersten Entwurf wurden die geforderten Abbildungseigenschaften mit drei Linsen realisiert, wobei die Linsenoberflächen nur sphärische Krümmungen aufweisen. Abb. 1 zeigt schematisch die Anordung der einzelnen optischen Komponenten. Der einfallende Strahl wird durch eine erste Zerstreuungslinse aufgeweitet und mittels einer Linsendublette fokussiert. Der Abstand zwischen der

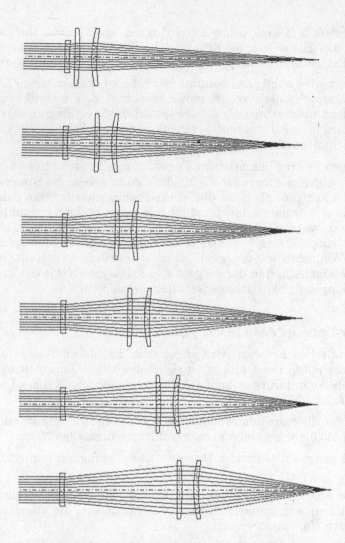

Abb. 1: Anordnung der Linsen im ersten optischen Entwurf (3-Linsen-System)

Zerstreuungslinse und der Linsendublette kann für die Einstellung der Brennweite verändert werden. Wie aus Abb. 1 zu erkennen ist, ergibt sich eine geringfügige Verschiebung der Fokuslage bei Änderung der Brennweite.

Beim zweiten Entwurf wurde im sammelnden Teil eine einzelne Linse mit einer asphärischen Oberfläche eingesetzt. Abb. 2 zeigt die Linsen-anordnung des zweiten Entwurfs für verschiedene Abstände zwischen der Zerstreuungslinse und der asphärischen Linse. Auch bei diesem Entwurf bleibt die Verschiebung bei Brennweitenänderung gering. Die Abbildungseigenschaften beider Entwürfe sind nahezu gleich. Bei kleinen Brennweiten liegen die berechneten Aberrationen bei beiden Entwürfen unterhalb der Werte für einzelne, handelsübliche Meniskuslinsen gleicher Brennweite.

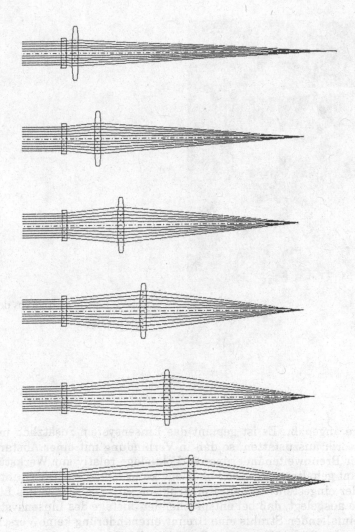

Abb. 2: Anordnung der Linsen im zweiten optischen Entwurf (Asphären-System)

Technische Realisierung des Linsensystems in einem Prototyp

Bei der technischen Umsetzung der optischen Entwürfe in einen Prototyp besteht die Aufgabe hauptsächlich darin, die Anforderungen an die Montage und Kühlung der Linsen, die notwendigen Bewegungseinrichtungen für Linsen und Schneiddüse sowie die notwendigen Justierfreiheitsgrade für das gesamte Linsensystem zu realisieren. Abb. 3 zeigt ein Foto des aufgebauten Prototypen. Die für die beiden optischen Entwürfe benötigten Linsen lassen sich im gleichen mechanischen Aufbau einbauen und testen. Der Prototyp besteht aus einem Führungsrohr, in dem die Einheiten für Linsen und Schneiddüse verstellt werden können. Die Verstellung von Linsenhalterung und Schneiddüse erfolgt über zwei Schrittmotoren. Dabei wird die Position der Schneiddüse automatisch an

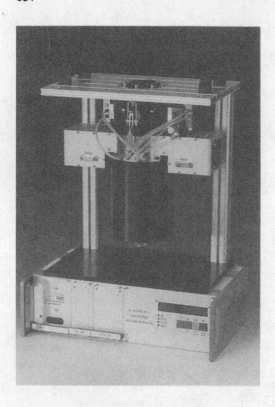

Abb. 3: Foto des Prototyps des Linsensystems

die Fokuslage angepaßt. Es ist geplant das Linsensystem zusätzlich mit einer Abstandssensorik auszustatten, so daß in Verbindung mit einer Abstandsregelung bei einer Brennweitenänderung die Fokuslage relativ zur Werkstückoberfläche konstant gehalten wird. Die Fokuslage kann relativ zur Düsenunterkante vom Anwender eingestellt werden. Konstruktion und Toleranzen des Linsensystems sind so ausgelegt, daß bei entsprechender Justage des Linsensystems bezüglich des einfallenden Strahls eine Brennweitenänderung keine Verschiebung des Strahls in der Düsenöffnung zur Folge hat. Die Linsen sind von ihrer Dicke her so dimensioniert, daß mit Schneidgasdrücken bis zu 10 bar gearbeitet werden kann. Tabelle 1 gibt einen Überblick über die mit dem Prototyp realisierten technischen Daten des Linsensystems.

Untersuchung der Fokussiereigenschaften des Linsensystems

Die Fokussiereigenschaften des Linsensystems wurden mit einem CO_2-Laser mit 1.5kW Nennleistung untersucht. Dazu wurde die Fokuskaustik bei verschiedenen Brennweiten mit einem Strahldiagnosesensor vom Typ UFF100 der Fa. Prometec vermessen. Zum Vergleich wurden Fokuskaustiken von drei Einzellinsen vermessen. Bei den Einzellinsen handelt es sich um handelsübliche Meniskuslinsen aus GaAs (63.5mm Brennweite) und ZnSe (127mm und 190mm Brennweite). Aus den Meßdaten wird jeweils der Strahlradius nach der 86%-Methode /3/ bestimmt und als Funktion des Strahlweges aufgetragen. Die

Tabelle 1: Technische Daten des Linsensystems

Bezeichnung	Entwurf 1	Entwurf 2
Einstellbereich der Brennweite	63.5mm - 170mm	65mm - 220mm
Eintrittsapertur	25 mm	25 mm
Verstellbereich der Schneiddüse	-1mm bis +5mm	-1mm bis +5mm
Durchmesser der Schneiddüse	1 mm	1mm
max. Schneidgasdruck	10 bar	10 bar
max. Einstellzeit der Brennweite	10 s	10 s

Strahlparameter wie Fokusdurchmesser, Rayleighlänge, Strahlqualitätskennzahl und F-Zahl werden durch Fit der Strahlausbreitungsgleichung /3/ an die Meßdaten nach der Methode der kleinsten Fehlerquadrate ermittelt. In den Abbildungen 4 bis 7 sind die genannten Strahlparameter als Funktion der Brennweite aufgetragen. Gegenübergestellt sind jeweils die Ergebnisse, die mit den beiden adaptiven Optiken und mit Einzellinsen ermittelt wurden. In Abb. 4 werden die Fokusdurchmesser der einzelnen Systeme verglichen. Mit dem 3-Linsen-System wurden die geringsten Fokusdurchmesser gemessen. Besonders bei kurzen Brennweiten ist zu erkennen, daß mit beiden Linsensystemen geringere Fokusdurchmesser erreicht werden als mit einer handelsüblichen Einzellinse. Bei großen Brennweiten (>150mm) werden mit allen drei Systemen im Rahmen der Meßgenauigkeit gleiche Fokusdurchmesser erreicht. Die Ursache für die deutlich unterschiedlichen Fokusdurchmesser bei kleinen Brennweiten liegt hauptsächlich darin, daß dort die Korrektur des Öffnungsfehlers bei bei einem Mehrlinsensystem besser ist als bei einer Meniskuslinse. Insgesamt lassen sich mit dem verwendeten Lasersystem Fokusdurchmesser im Bereich von ca 0.1mm bis 0.3mm einstellen. Abb. 5 zeigt die mit den Linsensystemen einstellbaren Rayleighlängen des Fokus. Hier wird ein Bereich von ca. 0.5mm bis ca. 4mm überdeckt. Zur Bewertung der Abbildungseigenschaften des Linsensystems bei den verschiedenen einstellbaren Brennweiten wird die aus der Fokuskaustik ermittelte Strahlqualitätskennzahl $K(f)$ im Verhältnis zum Maximalwert der mit einem Linsensystem ermittelten K-Zahl betrachtet. Diese Größe wird als optische Qualität bezeichnet und ist in Abb. 6 als Funktion der Brennweite für die drei betrachteten Systeme aufgetragen. Es zeigt sich, daß mit dem 3-Linsen-System die optische Qualität im Bereich von 0.9 bis 1.0 liegt, während das Asphären-System bei kleineren Brennweiten in der optischen Qualität auf ca. 0.75 abfällt. Wie sich inzwischen bei näherer Untersuchung herausgestellt hat, weicht die in diesem System eingebaute Zerstörungslinse von der berechneten optimalen Form ab. Bei den Einzellinsen wird bei der kürzesten Brennweite lediglich eine optische Qualität von ca. 0.5 erreicht. Aus den Fokuskaustiken wird die F-Zahl der Abbildung bestimmt und in Abb. 7 dargestellt. Es zeigt sich, daß mit dem Linsensystem ein F-Zahlbereich von ca. 3 bis 12 überdeckt werden kann. Tabelle 2 gibt einen Überblick über die mit den Linsensystemen und den Einzellinsen ermittelten Strahleigenschaften.

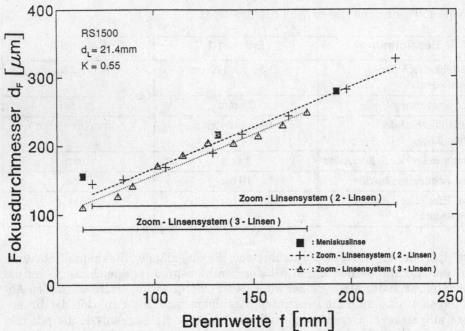

Abb. 4: Mit dem Linsensystem und Einzellinsen gemessene Fokusdurchmesser als Funktion der Brennweite. Verglichen werden die Fokussiereigenschaften beider optischer Entwürfe mit denen von Einzellinsen.

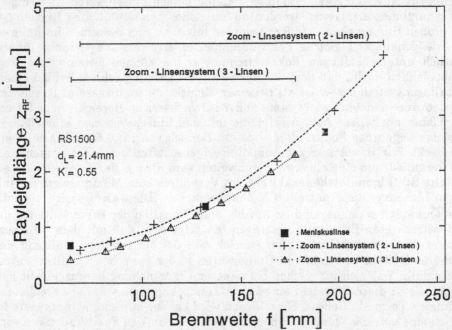

Abb. 5: Mit dem Linsensystem und Einzellinsen gemessene Rayleighlängen des Fokus als Funktion der Brennweite. Verglichen werden die Fokussiereigenschaften beider optischer Entwürfe mit denen von Einzellinsen.

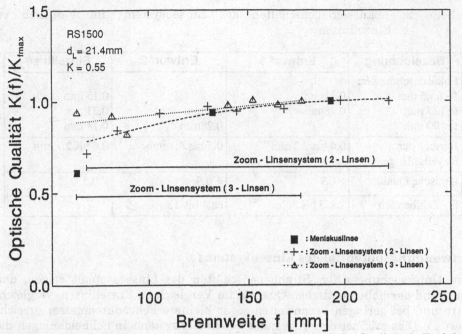

Abb. 6: Mit dem Linsensystem und Einzellinsen gemessene optische Qualität als Funktion der Brennweite. Verglichen werden die Fokussiereigenschaften beider optischer Entwürfe mit denen von Einzellinsen.

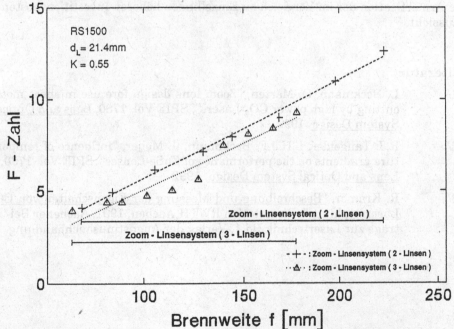

Abb. 7: Mit dem Linsensystem gemessene F-Zahlen der Fokuskaustik als Funktion der Brennweite.

Tabelle 2: Fokussiereigenschaften des Linsensystems im Vergleich zu Einzellinsen

Bezeichnung	Entwurf 1	Entwurf 2	Einzellinse
Fokusdurchmesser: f= 63,5 mm f=127 mm f=190 mm	0.11mm 0.18mm	0.2 mm 0.28mm	0.15 mm 0.21 mm 0.28 mm
Bereich der Rayleighlänge	0.4 bis 2.2 mm	0.5 bis 4.1 mm	0.6 bis 2.7 mm
Optische Qualität	> 0.9	> 0.8	> 0.5
F - Zahlbereich	ca. 3 bis 8	ca 4 bis 13	

Anwendungspotential des Linsensystems

Die Untersuchungen der Strahleigenschaften des Linsensystems zeigen, daß aufgrund der hohen optischen Qualität im Vergleich zu Einzellinsen, vergleichbare und bei geringen Brennweiten sogar kleinere Fokusdurchmesser erreicht werden. Dies eröffnet dem Anwender des Linsensystems in Schneidanlagen die Möglichkeit, die Brennweite optimal und kontinuierlich an die Blechdicke anzupassen und im Dünnblechbereich sogar die Schneidleistung der Anlage im Vergleich zur Verwendung einer Einzellinse zu erhöhen. Dies stellt eine kurzfristige Armortisation der im Vergleich zu Einzellinsen höheren Investitionskosten in Aussicht.

Literatur

/1/   L. Beckmann, O. Märten, "Zoom lens design fore use in sheet metal cutting by high power CO_2-Lasers", SPIE Vol. 1780, Lens and Optical System Design, 1992

/2/   R.J. Tangelder, L.H.J.F. Beckmann, J. Meijer, "Influence of temperature gradients on the performance of ZnSe-Lenses", SPIE Vol. 1780, Lens and Optical System Design, 1992

/3/   R. Kramer, "Beschreibung und Messung der Eigenschaften von CO_2-Laserstrahlung", Dissertation RWTH Aachen, 1991, Aachener Beiträge zur Lasertechnik Bd.1, Verlag der Augustinusbuchhandlung

Anforderungen an Laser-Materialbearbeitungssysteme für den Einsatz in mittelständischen Betrieben

Von Johannes Gartzen, Andreas Gebhardt und Hans Lingens

Der LASER wird gerne als hochflexibles, weitgehend losgrössen-
unabhängiges Werkzeug angepriesen, dessen schier unübersehbare
Vielfalt an Applikationen ihn zum prädestinierten Werkzeug für
mittelständische Strukturen macht. Entsprechend optimistisch
waren die Prognosen.
Der Grad der Marktdurchdringung bleibt aber tatsächlich hinter
den Erwartungen zurück.

Andererseits beobachten wir den LASER-Einsatz in der Materialbearbeitung überall dort, wo die Applikationen Grossserien-
Charakter haben. Das Schweissen von Rasierklingen und Tassen-
stösseln, das Beschichen von Ventilen und das Härten von Zy-
linderrohren und Nockenwellen sind bekannte Beispiele.
Die LASER-Strahlquellen sind in der Regel zwar der technologische Kristalisationspunkt der Produktionslinie, machen bezüglich des technischen und finanziellen Aufwandes für die Gesamtanlage aber meistens nur einen kleinen Prozentsatz aus.
Sie behaupten sich excellent in komplizierten Produktionsli-
nien mit hoher Gesamtverfügbarkeit, die bekanntlich das Produkt aus den Einzel-Verfügbarkeiten ist.

Prof.Dr.rer.nat. Johannes Gartzen,
Fachhochschule Aachen, Lehrgebiet Schweisstechnik/Lasertechnik,
Sprecher des Forschungsverbundes Lasertechnik der Fachhochschulen NRW.

Dr.-Ing. Andreas Gebhardt,
LASER Bearbeitungs- und Beratungszentrum NRW (LBBZ-NRW), Aachen

Dipl.-Ing. Hans Lingens,
FH-Aachen, Kontaktperson Forschungsverbund Lasertechnik - LBBZ-NRW.

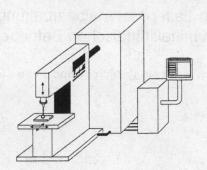

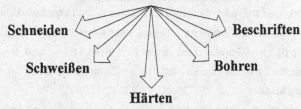

Der Laser kann

Schneiden   Beschriften

Schweißen   Bohren

Härten

Bild 1. Hersteller-Versprechen I

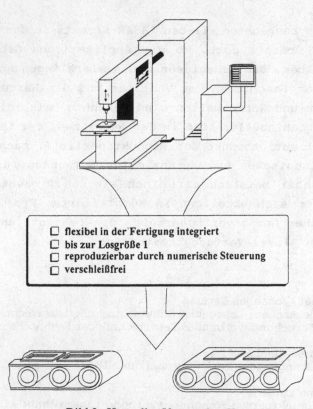

☐ flexibel in der Fertigung integriert
☐ bis zur Losgröße 1
☐ reproduzierbar durch numerische Steuerung
☐ verschleißfrei

Bild 2. Hersteller-Versprechen II

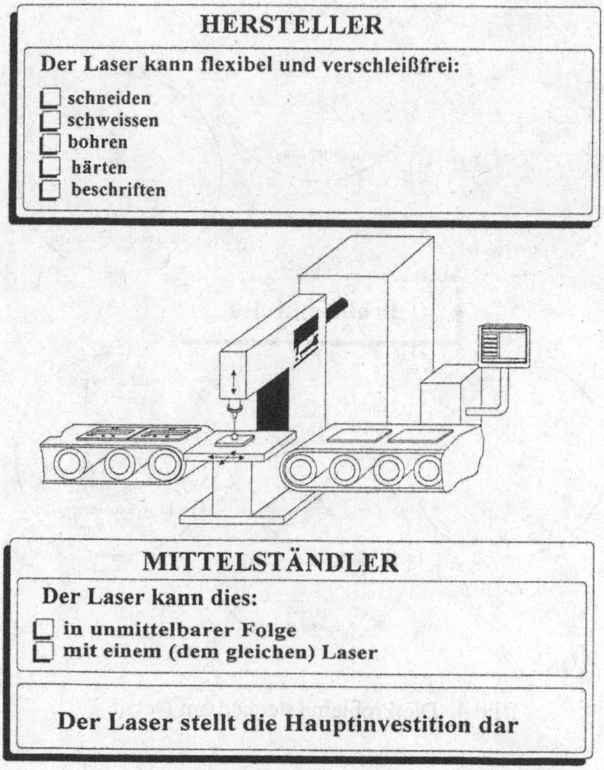

HERSTELLER

Der Laser kann flexibel und verschleißfrei:

- [ ] schneiden
- [ ] schweissen
- [ ] bohren
- [ ] härten
- [ ] beschriften

MITTELSTÄNDLER

Der Laser kann dies:

- [ ] in unmittelbarer Folge
- [ ] mit einem (dem gleichen) Laser

Der Laser stellt die Hauptinvestition dar

Bild 3. Mittelständler-Vorstellung vom "LASER"

Bleibt dem Mittelständler nur die Erkenntnis, dass auch diese neue Schlüsseltechnologie vorwiegend der Grossindustrie nutzt und geeignet ist, den Vorsprung der Grossen noch grösser werden zu lassen ?

Fügt sich der Hersteller darein, dass der Markt doch bei den Grossen liegt ?

Konzentriert sich der Verfahrensentwickler auf Anwendungen, die nur vor dem Hintergrund grosser Serien in industrielle Massstäbe umzusetzen sind ?

Um diesen grossen und weitgehend noch neuen Markt zu erschliessen, müssen LASER-Strahlquellen, -Systeme und -Verfahren angeboten werden, die den speziellen Anforderungen mittelständischer Unternehmen gerecht werden.

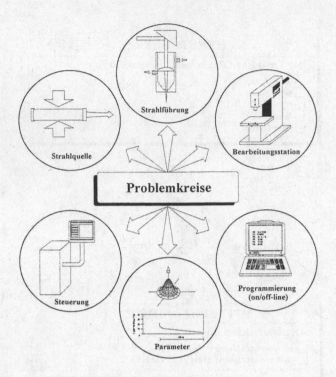

Bild 4. Die Probleme stecken om Detail

Der Beitrag zeigt die besondere Situation mittelständischer
Betriebe auf. Er setzt an bei den Missverständnissen zwischen
Herstellern und Mittelständlern, die aus dem unterschiedlichen
Wissen um diese neue Technik resultieren (Bilder 1-3).

Aus der detaillierten Information über die unzähligen Möglichkeiten, eine geeignete LASER-Anlage aufzubauen (Bild 4) ,
wächst Unsicherheit und Misstrauen anstelle von Vertrauen
(Bild 5). Abwarten mit der Folge zu geringer Marktdurchdringung ist die Folge.

Aus diesem Spannungsfeld leiten sich Forderungen LASER-Mate-
rialbearbeitungssysteme ab, die dem taktischen Charakter mittelständischer Investitionsentscheidungen Rechung tragen.
Daraus ergeben sich Wünsche an die Verfahrensentwickler, die
Hersteller der Strahlquellen und -führungssysteme, die Steuerungen, die Prozesskontrolle, den Service und die qualifizie-
renden Einrichtungen, aber auch an die Organisation der Betriebe (Bild 6).

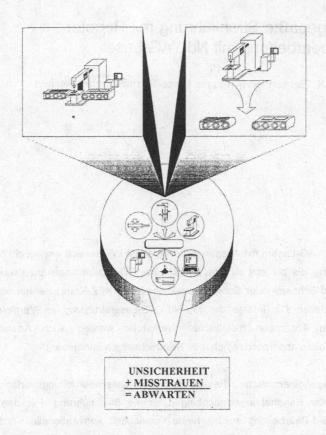

Bild 5. Konfliktentstehung zwischen Mittelständler und Hersteller

☐ **Anwendungsfall genau analysieren**

☐ **Fertigungsprozeß berücksichtigen**

☐ **Komplettangebote, keine
Komponenten - Baukästen**

☐ **Kurze Lieferzeiten**

☐ **Praxisgerechte Schulung**

☐ **Leichte Bedienbarkeit**

☐ **Einfacher Service, Tele-Service**

☐ **Klare Kalkulation**

Bilkd 6. Schlußfolgerungen

Anwendungsangepaßte Strahlführung mit Roboter für die Materialbearbeitung mit Nd:YAG-Laser

R. Klein, G. Neumann, K. Zimmermann, Thyssen Laser-Technik GmbH, Aachen

Einleitung

Die Entwicklung von Nd:YAG-Lasern mit Ausgangsleistungen im kW-Bereich ermöglicht Anwendungen in der Materialbearbeitung, die bis jetzt als eine Domäne der CO_2-Laser angesehen werden konnten. Als Beispiele stehen 3-d-Schneid- oder Schweißapplikationen an KFZ-Komponenten wie Karosserie- oder Innenausstattungsteilen /1/. Infolge der für Nd:YAG-Laserstrahlung im Vergleich zur CO_2-Laserstrahlung höheren Absorption metallischer Werkstoffe werden auch Anwendungen zur Oberflächenveredelung ohne absorptionserhöhende Beschichtungen interessant.

Wichtig für die fertigungstechnische Realisation von Laserbearbeitungsverfahren ist das Vorhandensein geeigneter Handhabungseinrichtungen für die Strahlführung. Für den Einsatz von Nd:YAG-Laser in der 3-d-Bearbeitung werden heute vorwiegend konventionelle Knickarm-Roboter eingesetzt /2,3/. Die Strahlführung erfolgt über Lichtleitkabel (LLK), die mit der Fokussiereinrichtung an der Roboterhand angeflanscht werden. Über eine weiterführende Roboterentwicklung für einen praxisgerechten Einsatz in der industriellen Fertigung mit Nd:YAG-Laser wird in diesem Beitrag berichtet.

Konventionelle Faserankopplung an Knickarm-Roboter

Die Anbindung von Nd:YAG-Lichtleitkabel an Knickarm-Roboter erfolgt üblicherweise durch Anflanschung an das Handstück des Roboters /4/. Dabei wird ein Bearbeitungskopf angebaut, der neben der Fokussieroptik mit Schutzgasdüse die Strahlkollimationsoptik und die Steckverbindung für das Lichtleitkabel umfasst (Abb.1). Im Regelfall erfolgt die Anbringung mit einer Orientierung der Strahlachse senkrecht oder mit einem großen Neigungswinkel zur Drehachse der letzten Roboterhandachse.

Zur Kompensation von Bahnungenauigkeiten und Werkstücktoleranzen wird beim 3-d-Schneiden zusätzlich am Bearbeitungskopf eine Abstandssensorik integriert, die kapazitiv oder taktil die Ist-Position der Fokuslage relativ zur Werkstückoberfläche erfasst und eine Stellgröße für die Nachregelung bereitstellt.

Während der Bearbeitung erfolgt die Nachregelung der Fokuslage durch eine Korrektur der berechneten Bahnkoordinaten in der Robotersteuerung oder durch eine in den Bearbeitungskopf integrierte hochdynamische Zusatzachse mit eigener Steuerung /5/. Die letzt genannte Möglichkeit stellt dabei die i.A. realisierte Variante dar .

Die Nachführung der Fokuslage über die Korrektur der Bahnkoordinten durch die Robotersteuerung stellt hohe Anforderungen an die Regeltaktzeiten, da hierbei alle Roboterachsen während der Bewegung des Roboters fortlaufend korrigiert werden müssen. Stimmt zudem die Fokusposition nicht exakt mit dem Werkzeugpunkt (TCP) des Roboters überein, ergeben sich zusätzliche Ungenauigkeiten in der Bahnbewegung, die größer werden, je weiter der TCP und die Fokusposition von der Roboterhand und deren Drehachse entfernt liegen.

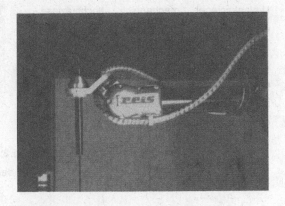

Abb.1: Konventionelle Anbindung der Strahlfokussierung einschließlich Lichtleitkabel an eine Roboterhand

Ein weiterer Nachteil der konventionellen Anbindung der Strahlführung an Knickarm-Roboter stellt die Zuführung des Lichtleitkabels dar. Die Roboterhandachsen sind i.A. als hochdynamische Bewegungsachsen ausgeführt, die entlang der Bahnbewegung bei ungünstigen Stellungen der Roboterachsen geschwindigkeitsabhängig schnelle Umorientierungen ausführen. Diese können das Lichtleitkabel durch Biegen oder Verdrillen stark belasten und beschädigen. Zur Vermeidung solcher Überbeanspruchungen ist eine besondere Sorgfalt bei der Bahnprogramierung bezüglich der Anstellung der Roboterachsen sowie ein vorsichtiges Heranführen der Bahngeschwindigkeit an die programmierte Sollgeschwindigkeit im Testbetrieb notwendig.

Integration der Strahlführung in die Roboterhand

Die in der industriellen Praxis gestellten Anforderungen an die Anlagensicherheit und die Bedienerfreundlichkeit einer Kombination von Nd:YAG-Laser mit Knickarm-Roboter führten in Zusammenarbeit mit dem Roboterhersteller Reis zur Entwicklung einer neuen Roboterhandachse. Dabei wird die Strahlführung über Spiegelelemente in die Roboterhand integriert. Die Anbindung des Lichtleitkabels erfolgt seitlich parallel zur vierten Roboterachse in einem Bereich, bei dem das Lichtleitkabel gegen plötzliche Richtungsänderungen und Verdrillungen geschützt ist **(Abb.2)**.

Die Einkopplung der Laserstrahlung erfolgt über einen seitlich an der Handachse angeflanschten justierbaren Umlenkspiegel. Innerhalb der Handachse ist ein weiterer Umlenkspiegel mechanisch hochpräziese eingebaut, der die Drehbewegung der Handachse mit ausführt und die Laserstrahlung senkrecht zur Drehbewegung umlenkt. An der Austrittseite der Handachse ist die Fokussieroptik mit koaxialer Schutz- oder Prozeßgasdüse angebaut (vgl. Abb.2).

Abb.2: Roboterhand mit integrierter Strahlführung und seitlicher Anbindung des Lichtleitkabels

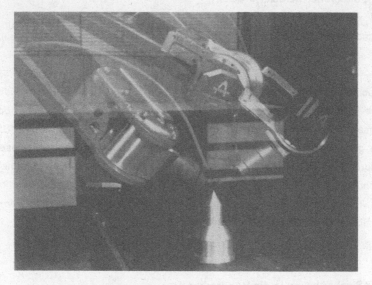

Abb.3: Demonstration der Beweglichkeit der entwickelten Roboterhand

Neben dem Schutz des Lichtleitkabels gegen mechanische Überbeanspruchung ist ein weiterer Vorteil der neuentwickelten Roboterhand die volle Ausnutzung von Arbeitsraum und Bewegungsmöglichkeit des Knickarm-Roboters für die Laseranwendung **(Abb.3)**. Durch den kompakten Aufbau der Roboterhand können selbst schwierig zugängliche Bauteilgeometrien bearbeitet werden, bei denen konventionelle Lösungen nur mit erhöhtem Programmieraufwand oder überhaupt nicht eingesetzt werden können. Der TCP rückt sehr nahe an den Drehpunkt der Hand heran, wodurch schnelle Umorientierung begünstigt werden.

Für 3-d-Schneidanwendungen im industriellen Einsatz kann ein kapazitiv oder taktil arbeitender Abstandssensor ohne großen technischen Aufwand in die Bearbeitungsoptik integriert werden. Für Geschwindigkeiten im Bereich einiger m/min kann das Abstandssignal nach elektronischer Aufbereitung direkt über die Sensorschnittstelle der Robotersteuerung für die Korrektur der Bahnbewegung über die Roboterachsen eingesetzt werden. Eine hochdynamische Zusatzachse, die eine Nachstellung der Bearbeitungsoptik in Strahlrichtung durchführt, wird somit in diesem Geschwindigkeitsbereich ersetzt. Erst bei höheren Bearbeitungsgeschwindigkeiten wird der Einsatz einer Zusatzachse sinnvoll, da hier der Lageregeltakt der Robotersteuerung nicht mehr ausreichend schnell ist, um alle Roboterachsen synchron während der Bewegung zu korrigieren.

Weitere Ergänzungen, die mit geringem technischen Aufwand in die Roboterhand integriert werden können sind Prozeßdiagnostiken wie Plasmaüberwachung für das Laserstrahlschweißen oder eine CCD-Kamera als Justierhilfe für die Eingabe der Bahnkoordinaten über Teach-In-Programmierung.

Beispiele für die Anwendung

Das neue Konzept für die Handhabung der Strahlführung hat sich in Anwendungen bewähren können, von denen zwei im näheren beschrieben werden. Ein Beispiel ist das Verschweißen von Behälterkanten aus Edelstahl (**Abb.4**). Die Kanten des Bauteils wurden zur Fixierung vorher durch einzelne Schweißpunkte verbunden. Anschließend erfolgte das Dichtschweißen der drei aufeinander folgenden Kanten an jeder Schmalseite des Behälters. Dabei mußte durch den Roboter neben der exakten Bahnverfolgung auch der Anstellwinkel der Fokussierung zur Stoßkante eingehalten werden, um Störungen der Schweißnaht zu vermeiden. Dies konnte durch die integrierte Strahlführung ohne Umorientierungen der Handachsen während der kontinuierlichen Bewegung des Strahlfokusses auf der Kante durchgeführt werden.

Abb.4: Verschweißung eines Edelstahlbehälters entlang der Stoßkante

Ein weiteres Anwendungsbeispiel der neu entwickelten Handachse zeigt **Abb.5**. Hier muß eine

598

Edelstahlmembran auf eine Edelstahlplatte aufgeschweißt werden. Die Verschweißung erfolgt in einer kontinuierlichen Bewegung um das gesamte Bauteil. Aus Zugänglichkeitsgründen zum Fügebereich, die durch die spezielle Form der Membran gegeben sind, muß beim Schweißen ein konstanter Anstellwinkel eingehalten werden. Mit einer konventionellen Faseranbindung an den Roboter kann diese Aufgabe nicht ohne Unterbrechung des Schweißprozesses zur Umorientierung der Handachse ausgeführt werden, da sonst die Lichtleitfaser verdrillt oder sogar geknickt werden würde. Durch die seitliche Anbringung der Faser an die Handachse mit integrierter Strahlführung kann die geschilderte Problematik umgangen werden.

Abb.5: Positionierung der Roboterhandachse zum kontinuierlichen Umfangsverschweißen von Edelstahlmembran und Edelstahlplatte

Zusammenfassung

Die fertigungstechnische Realisation von Laserbearbeitungsverfahren mit Hoch-leistungs-Nd:YAG-Lasern erfordert neue Konzepte bei der Anbindung von Lichtleitfasern an Handhabungsroboter. Die aus der industriellen Praxis gestellten Anforderungen führten zur Entwicklung einer Roboterhandachse mit integrierter Strahlführung, die eine verbesserte Bedienerfreundlichkeit und Anlagensicherheit aufweist. Die Anbindung der Glasfaserstrahlführung erfolgt seitlich am Roboterarm. Die weitere Strahlführung in der Handachse wird mit zwei jeweils koaxialen Spiegeln durchgeführt. An den Strahlaustritt an der Handachse wird direkt die Fokussieroptik angeflanscht. Anwendungsbeispiele zum Schweißen zeigen die Tauglichkeit des Systems bei der Bearbeitung komplexer Bahnkonturen.

Literatur
/1/ C. Olianeck; ND:YAG-LASER UND ROBOTER ALS TEAM; Laser-Praxis, Mai 1992,
 C. Hanser Verlag
/2/ D. Cheval, J. Vogt, YAG-LASERANWENDUNGEN IN DER FRANZÖSISCHEN
 AUTOMOBILINDUSTRIE; Proceedings European Laser Marketplace 92, Hannover 1992
/3/ N.N.; DIE ZEIT IST REIF, LASERROBOTER SCHWEIßT ALUMINIUM; Laser, Okt. 1992,
 IVA-Verlag, München
/4/ R. Grundmüller; LICHTLEITER-KONZEPT; 3-D-LASERSCHNITT MIT ND:YAG LASER UND
 ROBOTER; Laser, Juni 1991, IVA-Verlag, München
/5/ F. Wild; JOB-SHOP-ERFAHRUNG FÜR SYSTEMNUTZER; Laser-Praxis, Okt 1991,
 C. Hanser Verlag

Investigations on the Generation and Propagation of Hazardous Substances During the Laser Treatment of Ceramics

R. Dierken[1], E. Schubert[1], H.W. Bergmann[1], W. Zschiesche[2]

[1] *ATZ-EVUS*, Außenstelle Vilseck

   Rinostr.1, 92249 Vilseck

[2] Friedrich-Alexander-Universität Erlangen-Nürnberg, Institut für Arbeits- und Sozialmedizin

   Schillerstr. 25 - 29, 91030 Erlangen

Introduction

In the last years the laser treatment of ceramics entered production subsidiary to the well-established treatments of metals. Especially scratching of alumina substrates is state-of-the-art in production right now. This treatment comprises a thermal process of melting and evaporating material. As a consequence of this an airborne emission (fumes, dust, aerosols etc.) occurs, which might be hazardous.

The targets of the following investigations are the characterization of the emission in quality (Laser-light-sheet-technique (LLST), Laser Doppler anemometry (LDA), Scanning electron microscopy (SEM)) and quantity (gravimetric, Atomabsorption spectroscopy (AAS)). Thereout, the optimization of the process parameters itself, and the necessity and design of an optimized exhaust system should be concluded.

Experimental Setup

The scratching of alumina substrates was carried out using a 1 kW HF-(High frequency)-CO_2-Laser at production near parameter setting (P_L = 50 W, f = 250 Hz, v = 50 mm/s) and work capacity. In addition simulation of this process without laser was performed for the experiments concerning the flow visualization and Laser Doppler anemometry. An Ar^+-Laser visualized the gas flow around the cutting head (LLST). A cylindrical lense spread the laser beam into a light sheet enabling to detect a two dimensional air flow by a CCD-camera (see **Fig. 1**). The LDA-measurements were executed with a PC-controlled fibre optic LDA including a diode laser (λ = 830 nm) (see **Fig. 2**). Here a seeding of the air (addition of smoke to the air) is necessary for achieving a sufficient high data rate. Particles were

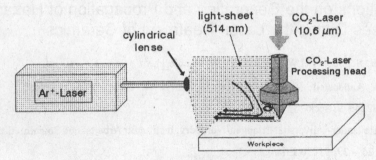

Fig. 1: Drawing of the setup for flow visualization (LLST)

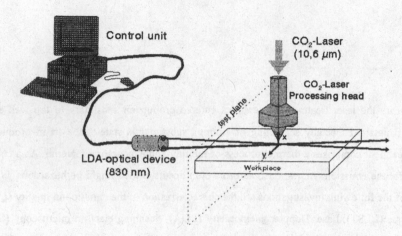

Fig. 2: Experimental Setup for LDA-measurements

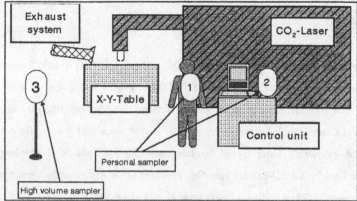

Fig. 3: Schematic drawing of the sampling positions during the measurements of workplace exposure

characterized in size and morphology by SEM. Fume sampling of the emission (related to person) lead to data of the workplace exposure. Therefore, the particles were sampled on celluloseacetate-filter with a constant flow velocity of 1.25 m/s. The sampling positions are presented in **Fig. 3**. In Pos. 1 and 2 the collected volume corresponded to human (personal) capacity, whereas Pos. 3 was a high volume sampler. The content of Aluminum was obtained by AAS and recalculated to the sampled emission for comparison with the threshold limit value (TLV).

Results

Figure 4 shows the air flow, as it was found with LLST during the simulated process. The main stream occured as a fast laminar flow radial to the surface of the workpiece, as it results from an impinging jet. Through the high velocity of this stream under-pressure arise and ambient air was taken into the main stream (**Fig. 4** and **5**). It is obvious that the mass transport of emitted particles occured mainly within this air stream.

From these results it can be concluded that an optimized position of an exhaust system would be a ring system around the processing head or the area around it. A height of several centimetres would be sufficient for the inlet of such a exhaust ring system.

The emitted particles have a globular morphology and their size range was inbetween less than 1 upto 50 μm (**Fig. 6a**). The coarse part of the dust found on the surface of the working area has a mean size diameter of appr. 40 μm (**Fig. 6b**). For the health surveillance the particles of interest are in a size range from 0.1 to 5 μm, which are sampled on the fine filtering area. The SEM-micrograph shows the collected particles. The AAS evaluation of the sampled emission are given in **Table 1**.

The two columns provide the results wtih and without an exhaust and filter system. The data clearly show that the measured values do not exceed the given TLV (Threshold Limit Value), even if no exhaust system was connected.

	Experimental data without Exhaust system	Experimental data with Exhaust system
Emission (gravim.)	0,73 mg/s	1,06 mg/s
Emission / exposure time	90,91 %	92,7 %
Air load	0,75 mg/m³	0,007 mg/m³
TLV (air)	6 mg/m³ (fine part)	
Evaluation	secure below TLV	secure below TLV

Table 1:   Experimental data of emission at the workplace

Fig. 4:   Videoprint of the visualized flow
(p = 0,5 bar; d_{nozzle} = 1 mm)

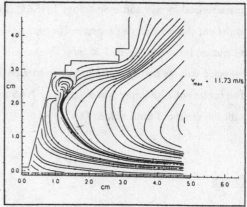

Fig. 5:   LDA-measured flow-field of the
process; parameter as in Fig. 4

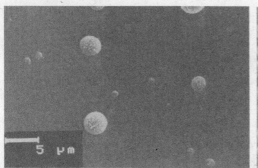

Fig. 6:   SEM-micrographs of a) fine sample of emitted particles and b) coarse sample

From another point of view there was a remarkable reduction of the total exposure due to the use of an exhaust and filter system. This appears to be convenient for reasons of occupational health. A biomonitoring of the workers urine took place, too. Here the results showed a significant increase of Al-concentration during the exposure time. Further work yields on statistical clearance of the data and on the research of other materials.

Conclusions

o   Emission follow the main air stream, which flow radially from the processing head as it is expected by an impinged jet perpendicular to the surface.

o   Emission of hazardous substances due to the scratching of alumina substrates with a 1 kW HF-CO_2-Laser does not exceed the TLV.

o   Exhaust system minimizes the workplace exposure.

o   Optimized exhaust system would be a ring exhaust system around the processing area

Acknowledgement

The authors like to thank the BMFT for their support within the EUREKA-Project "Laser Safety INDAL" EU 643 (Proj.-Nr. 13 EU 0117). Also Messer-Griesheim, Puchheim, and Hoechst CeramTec, Selb, are to be thanked for their support and test materials for the experiments.

Rotationsoptik zur Lasermaterialbearbeitung für einen Gelenkarmroboter

Gerhard Kröhnert, Axel Laas, Günter Lensch
NUTECH GmbH, Institut für Lasertechnik, Optische Schichten
und Materialprüfung, Ilsahl 5, 24536 Neumünster

Gerhard Kröhnert, Axel Laas, Günter Lensch

Sind in der Lasermaterialbearbeitung kleinere rotationssymmetrische Konturen anzubringen, ergeben sich mit konventionellen Bewegungssystemen oftmals Probleme.

Qualitäts- und Zeitvorgaben sind schwer einzuhalten, wenn es sich um große bzw. unhandliche Werkstücke handelt, die oft noch unterschiedlich geneigte Bearbeitungsebenen aufweisen.

Zum Teil können die Beabeitungsorte nur mit solchen Robotern erreicht werden, die aufgrund ihrer Bauart nur bedingt zum präzisen Konturenfahren geeignet sind. Auch lassen sie nur begrenzte Werkzeuglasten zu.

Daraus entstand die Idee, eine leichte kompakte Rotationsoptik für Roboteranwendungen zu entwickeln.

Bereits mit Beginn der Entwicklungsarbeiten wurde berücksichtigt, daß es möglich sein sollte, dem Anwender auf seine Bedürfnisse angepaßte Ausbaustufen anzubieten. Auch sollte die Möglichkeit bestehen, die Rotationsoptik wie eine Standardoptik für lineare Konturen einzusetzen.

Durch Einsatz unterschiedlicher Fasern sind Schneid- und Schweißanwendungen möglich.

Grundsätzlich werden zwei Prinzipien zur Radiusverstellung vorgesehen, die je nach erforderlichem Durchmesserbereich ausgewählt werden:

1. Linsenverschiebung   für kleinere Durchmesser und hohe Drehzahlen
2. Spiegelverschiebung   für mittlere und große Durchmesser

Ausbaustufen der Rotationsoptik

- mit festeingestelltem Radius, als kostengünstige Variante für Serienfertigung. Kleinste Abmessungen und geringstes Gewicht.

- Manuelle Radiusverstellung und Skalenablesung bei geringem Automationsbedarf und ausreichender Flexibilität z. B. für Klein- und Vorserienfertigung.

- motorische Radiusverstellung mit Radiuskontrolle und CNC Anbindungsmöglichkeit. Elektronische Übertragung erfolgt über Schleifkontakte. Version geeignet für flexible Automation.

- Der Antrieb erfolgt grundsätzlich über einen Servomotor mit Drehzahlregelung und als Option mit einer integrierten Drehpositionserfassung.

- Die Integration einer Abstandssensorik befindet sich in der Planung. Die Signale müssen auch hier über Schleifkontakte übertragen werden.

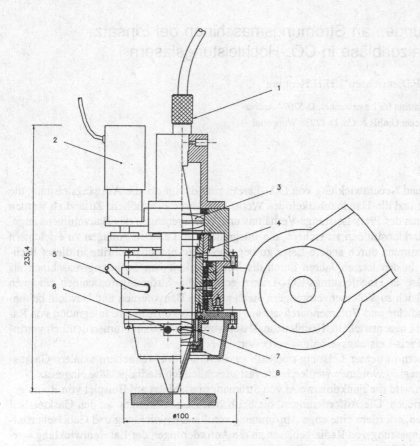

Baugruppenbeschreibung

1 Faseraufnahme - Anschluß verschiedener gängiger Fasern durch Anpassung an das jeweilige System des
Laserherstellers

2 Antriebsmotor mit Planetengetriebe
- Drehzahlbereiche werden durch festgelegte
Getriebeübersetzungen angepaßt

3 Kollimatoroptik - Erzeugung eines parallelen Strahlenbündels
von max. 20 mm Durchmesser

4 Drehdurchführung - Einspeisung des Prozeßgases in rotierende
Komponenten der Rotationsoptik

5 Gaszuführung - für Prozeßgase

6 Verstellmechanik - für Linsenverschiebung bei Durchmessern
von 0 bis 12 mm
- für Spiegelverschiebung bei Durchmessern
von 12 bis 100 mm (graphisch nicht dargestellt)

7 Fokussieroptik - Brennweiten 50 oder 100 mm je nach Anwendung

8 Schwenkgehäuse - trägt gemeinsame Linsen- und Düsenaufnahme

Anforderungen an Strömungsmaschinen bei Einsatz als Umwälzgebläse in CO_2-Hochleistungslasern

U.Jarosch[1], F.Diedrichsen[2], H.H.Henning[2]

[1] Fraunhofer Institut für Lasertechnik, D-52074 Aachen
[2] Fa. Gebr. Becker GmbH & Co, D-42238 Wuppertal

1. Einleitung

Die Weiter- und Neuentwicklung von CO_2-Lasern zielt darauf ab, die Ausgangsleistung, die Strahlqualität und die Handhabbarkeit des Werkzeugs Laser zu erhöhen. Zusätzlich werden Verbesserungen des Preis-Leistungs-Verhältnis und Verringerungen des Bauvolumens angestrebt, um Marktpositionen zu sichern, die Integration der Laser in Anlagen zu erleichtern und der Konkurrenz durch andere Laser zu begegnen. Wesentliche Schritte in dieser Richtung wurden in den letzten Jahren durch die Verfügbarkeit von Strömungsmaschinen als Umwälzgebläse in Hochleistungs-CO_2-Lasern ermöglicht. Strömungsmaschinen zeichnen sich im Vergleich zu Volumenverdrängern durch geringes Bauvolumen und Gewicht bei hoher Leistungsdichte und Volumendurchsatz aus [1]. Durch konstruktive Integration von Radialgebläsen in neuentwickelte Hochleistungslaser wurde deren Bauvolumen deutlich verringert und das Preis-Leistungsverhältnis verbessert.

Bei CO_2-Lasern mit einer Leistung oberhalb ca. 6kW und mit schnellem axialem Gasaustausch im Anregungsvolumen werden heute fast ausschließlich Radialgebläse eingesetzt.

Im Folgenden wird die Funktionsweise von Strömungsmaschinen am Beispiel von Radialgebläsen beschrieben. Die Anforderungen, die bei hohen Laserleistungen an den Gaskreislauf gestellt werden, erfordern eine enge Abstimmung von Laserentwicklung und Gebläsehersteller. Für die Anpassung von Radialgebläsen an die Anforderungen der Laserentwicklung werden geeignete Maßnahmen vorgestellt.

2. Grundlagen energiezuführender Strömungsmaschinen

Die Leistungsumsetzung in einer Strömungsmaschine vollzieht sich in zwei Schritten.

Im ersten Schritt wird ein Fluid mit niedriger Geschwindigkeit in einem Laufrad beschleunigt, wodurch es einen Drall in Richtung der Umfangsgeschwindigkeit erhält.

Im zweiten Schritt wird der Drall in einem Leitschaufelgitter in Druck umgewandelt, indem die Umfangskomponente des Geschwindigkeit verringert wird. Im Idealfall ist die Strömung am Austritt aus dem Leitapparat drallfrei.

Die Leistungsübertragung stellt dabei eine Funktion der Laufradform, speziell des Schaufelabströmwinkel β_2, der Umfangsgeschwindigkeit u_2 und des Volumenstrom $\dot{V}$ dar. Die Absolutgeschwindigkeit des Fluids c kann in die meridionale Komponente c_m und die Umfangskomponente c_u zerlegt werden. c_m ist proportional zum Volumendurchsatz, c_u ist proportional zu Drehzahl n und Durchmesser des Laufrades.

In Bild 1 ist am Beispiel eines dreistufigen Radialgebläses die Aufeinanderfolge von Laufrad und Leitapparat dargestellt. Der linke Bildteil zeigt die Geschwindigkeiten in Umfangs- und Meridiankomponente zerlegt bei Eintritt und Austritt aus dem Laufrad, sowie bei Austritt aus dem Leitrad. Die Austrittsgeschwindigkeiten aus dem Laufrad entsprechen in diesem Fall näherungsweise den Eintrittsgeschwindigkeiten in das Leitrad. Die Umfangskomponenten der Geschwindigkeiten $c_{i\,u}$ der drei aufeinanderfolgenden Stufen sind nur drehzahlabhängig und im Vergleich der Stufen zueinander gleich. Die Meridionalkomponenten $c_{i\,m}$ verringern sich bei kompressiblen Fluiden mit der Erhöhung des statischen Drucks p_{stat}, sofern keine Anpassung der Kanalquerschnitte an die Kompression erfolgt ist. Im unteren Bildteil ist der Verlauf von statischem Druck p_{stat} und Ruhedruck p_0 in den drei Stufen aufgetragen. Die Differenz zwischen p_0 und p_{stat} ergibt sich aus der absoluten Geschwindigkeit des Fluids c_i. Der Unterschied der Ruhedrücke an den Stellen 3 und 4 kommt durch die Strömungsverluste im Leitapparat zustande. Das Druckverhältnis Π zwischen saugseitigem und druckseitigem Druck kann aus dem Verhältnis der jeweiligen Ruhedrücke berechnet werden, wenn die Geschwindigkeiten im Eintritt und Austritt des Gebläses klein sind.

Bei einem Abströmwinkel aus dem Laufrad von $\beta_3 \sim 90^0$ und der Annahme drallfreier Zuströmung zum Laufrad beträgt die spezifische Schaufelarbeit Y_{sch}

$$Y_{sch} = u_2\,c_{3u} - u_1\,c_{1u} = u_2^2$$

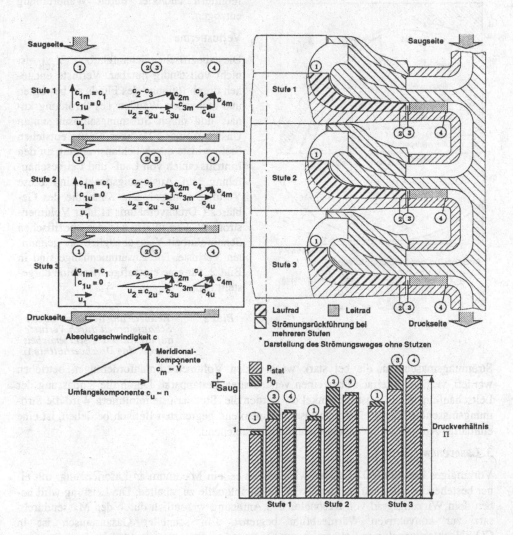

Bild 1   *Schnitt durch Lauf- und Leiträder eines dreistufigen Radialgebläses mit Darstellung des Geschwindigkeits- und Druckverlaufs*

608

Dabei bezeichnet u die Umfangsgeschwindigkeit des Laufrades, die Indizes bezeichnen die Stelle in einer Stufe und bei der Geschwindigkeit c des Fluids die Komponente (vergl. Bild1).

Auf das Leitrad trifft das Fluid mit der Umfangsgeschwindigkeit c_{3u}, sowie der Meridiankomponente c_{3m}. Die Meridiankomponente ergibt sich aus dem Volumenstrom $\dot{V}_3$ und der Eintrittsfläche in das Leitrad. Der Winkel zwischen den beiden Geschwindigkeitskomponenten α_3 gibt an, wie das Fluid auf die Leitschaufeln trifft (vergl. Bild 2).
Der Leitapparat kann aus einem Leitschaufelgitter bestehen oder als unbeschaufelter Kanal ausgeführt sein. Bei mehrstufiger Ausführung ist zusätzlich eine Rückführung in die nächste

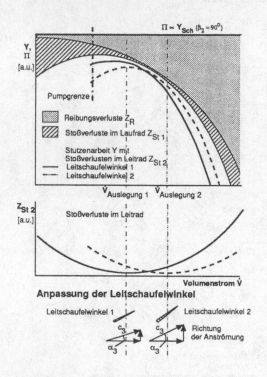

Stufe erforderlich. Der Drall wird dem Fluid durch die Umlenkung in den Schaufelgittern und/oder durch Wandreibung entzogen.

Verlustterme

Die spezifische Schaufelarbeit Y_{sch} ist nicht vollständig nutzbar. Verluste entstehen durch Reibung des Fluids an bewegten und stehenden Wänden des Strömungskanals und durch Strömungsablösungen an Unstetigkeiten. Darüber hinaus entstehen Stoßverluste durch Fehlanströmung an den Eintrittskanten von Lauf- und Leitbeschaufelung. Im Auslegungspunkt sind diese Verluste minimal. Die Kennlinie des Gebläses (Druckverhältnis Π über Volumenstrom $\dot{V}$) resultiert aus der spezifischen Schaufelarbeit Y_{sch} abzüglich der genannten Verluste. Die Zusammenhänge sind in Bild 2 für eine einstufige Maschine dargestellt [2].

Bild 2   *Reduzierung der spezifischen Schaufelarbeit durch Verluste auf die nutzbare Stutzenarbeit Y, bzw. das Druckverhältnis Π*

Strömungsmaschinen, die bei stark wechselnden Volumenstromanforderungen betrieben werden, verfügen vielfach über einen verstellbaren Leitapparat. Durch die Anpassung der Leitschaufeln an den Anströmwinkel α_3 werden die Stoßverluste minimiert. Wird die Strömungsmaschine wie in einem Gaslaser in einem eng begrenzten Bereich betrieben, ist eine einmalige Anpassung der Leitbeschaufelung ausreichend.

3. Laserentwicklung

Vorrangiges Ziel der CO_2-Laserentwicklung ist es, ein Maximum an Laserleistung mit einer bestehenden oder neu zu entwickelnden Strahlquelle zu erhalten. Die Leistung wird neben dem Wirkungsgrad von Resonator und Anregung wesentlich durch den Massendurchsatz zur konvektiven Wärmeabfuhr begrenzt. Ein schneller Gasaustausch ist in CO_2-Hochleistungslasern notwendig, um die unvermeidlich entstehenden Verlustleistungen im Bereich der Anregung des aktiven Mediums in ausreichendem Maß abführen zu können.

Grundsätzlich besteht die Möglichkeit, die Strömung senkrecht (transversal) zum oder axial mit dem Laserstrahl zu führen.

Bei transversaler Strömung erfolgt der Gasaustausch im Anregungsbereich im Vergleich zur axialen Strömung durch große Querschnitte mit geringer Geschwindigkeit. Bei der Gasumwälzung entstehen daher nur geringe Verluste, die mit Axial- oder Tangentialgebläsen bei Druckverhältnissen $\Pi < 1,2$ überwunden werden. Derartige Laser werden mit Laser-Leistungen bis ca. 25kW cw betrieben. Sie sind i.d.R. gleichstromangeregt (DC). Wesentliche Nachteile entstehen bei dieser Bauweise durch die Ausbildung transversaler Temperaturunterschiede aufgrund der Erwärmung des Gases beim Durchlauf durch den Entladungsbereich. Durch Ausbildung eines leistungsabhängigen Brechungsindexprofils senkrecht zum Laserstrahl wird eine optische Keilwirkung erzeugt, die zu einer leistungsabhängigen Dejustage des Resonators führt. Dies bewirkt, daß der Einsatz derartiger Laser auf Aufgaben mit konstanter Leistungsabgabe begrenzt ist.

Bei CO_2-Lasern mit axialer Gasströmung erfolgt der Gasaustausch durch die Entladungsrohre des Lasers. Die minimale Querschnittsfläche wird durch den Durchmesser der Strahlung multipliziert mit der Anzahl der Entladungsstrecken vorgegeben. Da die Erwärmung des Gasstroms axial mit der Laserstrahlung innerhalb des Resonators erfolgt, beeinflussen die leistungsabhängigen Änderungen des Brechungsindex den Laserstrahl nicht. Bei vergleichbarem Massendurchsatz bedingen die kleineren Querschnitte der axialen Gasströmung höhere Geschwindigkeiten und damit höhere Strömungsverluste. Daher liegen die Druckverhältnisse axial schnellgeströmter CO_2-Laser i.d.R. bei Werten $\Pi > 1,5$. Strömungsführungen, wie sie bei DC-angeregten Lasern und/oder bei Lasern mit Roots-Gebläsen zu finden sind, weisen in ungünstigen Fällen Druckverhältnissen $\Pi > 2$ auf. Bei Lasern mit optimiertem Strömungssystem treten Druckverhältnisse Π zwischen 1,2 und 1,7 auf. Diese Laser sind in den meisten Fällen auf den Einsatz von Strömungsmaschinen zur Gasumwälzung ausgelegt oder lassen sich mit Radialgebläsen betreiben.

Unabhängig von Strömungsführung und Anregungsart werden bei CO_2-Lasern Gasgemische mit 70 bis 85 Volumen% Helium, 10 bis 20 Volumen% Stickstoff und 3 bis 10 Volumen% CO_2 eingesetzt. Die Gasdrücke in der Entladung liegen je nach Art der Entladung und Abmessung des Entladungsvolumens zwischen 40 und 200 hPa.

Die Austauschrate in der Entladung wird bei vorgegebenem Gasgemisch einerseits durch die Effizienz der Entladung $\eta_{Entladung}$ (aus dem aktiven Medium extrahierte optische Leistung $P_{optisch}$ durch eingebrachte elektrische Leistung $P_{elektrisch}$) und anderseits durch die zulässige Temperaturdifferenz im Gas bestimmt. Die Eintrittstemperatur ist nach unten durch die Taupunkttemperatur und die Grenzen des Kühlaggregats begrenzt. Sie liegt zwischen 290 und $320^{\circ}K$. Die Austrittstemperatur wird durch Minderung der Effizienz $\eta_{Entladung}$ und die Dauerbelastbarkeit von Dichtungswerkstoffen auf 500 bis $520^{\circ}K$ begrenzt. Die Anforderung an den Massenstrom ergibt sich damit zu

$$\dot{m} = \frac{P_{elektrisch}(1 - \eta_{Entladung})}{c_p \Delta T}$$

Die Effizienz der Entladung $\eta_{Entladung}$ entspricht in 1. Näherung dem Laserwirkungsgrad η_{Laser} (Verhältnis von Laserleistung zu eingekoppelter elektrischer Leistung). Mit Laserwirkungsgraden von 15 bis 20 %, spezifischen Wärmekapazitäten zwischen 2000 und 2500J/kg/K und Temperaturdifferenzen ΔT von 180 bis 230 $^{\circ}K$ ergeben sich Massenströme von 7 bis 16 g / s / $kW_{Laserleistung}$. Bei einem Saugdruck des Gebläses von 100 hPa und einer Eintrittstemperatur von $300^{\circ}K$ entspricht dies einem Saug-Volumenstrom $\dot{V}$ von 700 bis 1200 m^3/h /$kW_{Laserleistung}$. Die z.Zt. verfügbaren Strahlquellenkonzepte erreichen mit einem Gebläse maximal ca. 6kW Laserleistung.

Durch die Verwendung hochfrequenter Wechselspannung zur kapazitiven Leistungseinkopplung und /oder durch kleinere Querabmessungen des Entladungsvolumens, können vergleichbare Laserleistungen aus kleineren Anregungsvolumina bei höherer Leistungsdichte extrahiert werden. Aufgabe der Laserentwicklung ist es, ein Optimum zwischen den Vorgaben der Resonatorgeometrie (Fresnelzahl, Strahlqualität), der Anregungstechnik (Leistungsdichte, Entladungslänge, Homogenität der Anregung) und der Strömungstechnik (Druckverluste, Strömungshomogenisierung, Kühlung) zu finden. Als Tendenz läßt sich abschätzen, daß bei sonst identischen Randbedingungen (Strömungsgeometrie, Massenstrom, Leistungseinkopplung) eine Erhöhung des Saugdrucks am Gebläse und damit eine Erhöhung des Gesamtdruckniveaus im Laser zu einer näherungsweise quadratischen Senkung der Druckverluste im System führt. Daraus resultiert eine Senkung der Druckverhältnis Π und eine Steigerung des Volumenstroms. Besteht der wesentliche Unterschied zwischen zwei Lasern in ihrer Strömungsführung, so ist es möglich, die Gebläse auf die speziellen Anforderungen anzupassen (vergl. Bild 3).

Das Widerstandsverhalten (Druckverhältnis Π über Volumenstrom V) wird näherungsweise durch Parabeln durch den Ursprung beschrieben. Mit Erhöhung der Laserleistung wird der Widerstand im Strömungssystem erhöht (Beschleunigung des Gases durch Aufheizung, Widerstandszunahme durch höhere Geschwindigkeiten hinter der Entladung). Die Widerstandsparabel verlagert sich nach links, d.h. das gleiche Druckverhältnis stellt sich bei kleineren Volumenströmen ein. Die Arbeitspunkte bei unterschiedlichen Laserleistungen werden durch die Schnittpunkte der Widerstandsparabeln mit der Gebläsekennlinie ermittelt.

Bei Lasern, die schon ohne elektrische Anregung eine steile Kennlinie, also hohe Widerstände aufweisen, ist der Unterschied der Arbeitspunkte klein. Der überwiegende Anteil der Druckverluste ist unabhängig von der Laserleistung. Die verfügbaren Volumenströme unterscheiden sich nur gering und liegen i.d.R. im Bereich der höchsten Druckverhältnissen auf der Gebläsekennlinie.

Bei strömungstechnisch optimierten System sinkt der Anteil der leistungsunabhängigen Widerstände. Die Widerstandsparabeln verschieben sich zu höheren Volumenströmen und rücken für unterschiedliche Laserleistungen auseinander. Die Schnittpunkte der Widerstandskurven mit der Gebläsekennlinie liegen dadurch ebenfalls bei höheren Volumenströmen. Setzt man gleiche Saugdrücke voraus, so bedeutet dies unmittelbar eine Steigerung des verfügbaren Massenstroms. Der Laser kann bei höheren Leistungen betrieben werden.

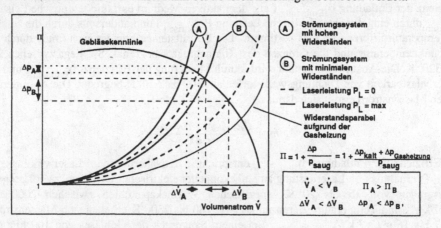

Bild 3   *Widerstandsverhalten von Strömungssystemen in axial schnellgeströmten Hochleistungs-Lasern und Kennlinie der Gebläse (schematische Darstellung)*

Dem Vorteil des höheren Volumenstroms steht gegenüber, daß sich die Betriebspunkte in den Bereich ungünstigerer Wirkungsgrade des Gebläses verlagern. Soweit eine entsprechende Auswahl möglich ist, muß geprüft werden, ob Gebläse höherem Volumenstrom geeigneter sind. Eine Erhöhung des Volumenstroms ist i.d.R. mit einer flacheren Gebläsekennlinie verbunden.

Die richtige Auswahl eines Gebläses und die Anpassung der Gebläsekennlinie beeinflußt den Gesamtwirkungsgrad des CO_2-Lasers deutlich, da bei allen Lasern das zwei bis vierfache der Laser-Ausgangsleistung als Antriebsleistung der Gasumwälzung benötigt wird. Mit Blick auf die Betriebskosten der Strahlquellen ist eine optimale Abstimmung der Gebläse auf die strömungstechnischen Anforderungen erforderlich. Ohne Strömungsoptimierung kann die Gasumwälzung zu kritischen Komponente bei der Strahlquellenentwicklung bei hohen Laserleistungen werden.

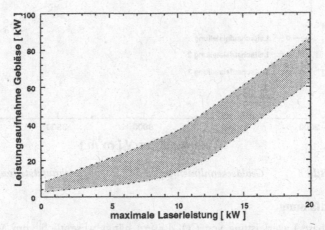

Bild 4   Leistungsaufnahme für die Gasumwälzung verschiedener axial-schnellgeströmter CO_2-Laser über deren Laserleistung

In Bild 4 ist die maximale Leistungsaufnahme der verwendeten Gebläse über der Ausgangsleistung axial-schnellgeströmter Laser aufgetragen. Unterschiede ergeben sich zum einen durch die Art der verwendeten Gebläse (Wirkungsgrad) und die Druckdifferenzen innerhalb der Strömungssystem, zum anderen durch die Wirkungsgrade der Laserresonatoren, die die erforderlichen Massenströme bestimmen. Bei Laserleistungen bis ca. 6 kW sind deutliche niedrigere Gebläseleistungen durch die Verwendung von Radialgebläsen anstelle von Rootsgebläsen zu erreichen. Bei sehr hohen Laserleistungen $P_L \gg 10kW$ (derartige Geräte befinden sich derzeit in der Entwicklung) werden zumeist instabile Resonatoren mit relativ großen Querschnittsflächen der Entladungsvolumina verwendet. Deren niedrigerer Wirkungsgrad und die mit den Entladungsquerschnitten sinkenden Saugdrücke erhöhen den Leistungsbedarf der Umwälzgebläse überproportional.

4. Anpassung von Radialgebläsen

Wie in Abschnitt 2 gezeigt, ist es möglich, den Bereich des optimalen Gebläsewirkungsgrades und die Steigung der Gebläsekennlinie im Arbeitsbereich zu beeinflussen, indem die Leitschaufelstellung dem Volumenstrom $\dot{V}$ angepaßt wird. Bild 5 zeigt den Einfluß unterschiedlicher Leitschaufeleinstellungen auf die Kennlinie eines dreistufigen Radialgebläses.

612

Die dargestellten Gebläsekennlinien sind alle bei Leitschaufeleinstellungen für den Bereich hoher Volumenströme aufgenommen. Anpassungen an die Forderung nach hohen Druckverhältnissen führt zu steileren Kennlinien bei niedrigeren maximalen Volumenströmen. Bei konstantem Druckverhältnis sind Veränderungen des Volumenstroms um bis zu 20 % zu erreichen.

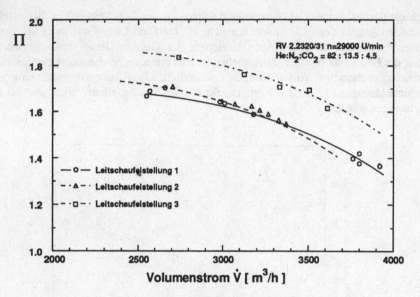

Bild 5   Gebläsekennlinien bei veränderten Leitschaufelstellungen

5. Zusammenfassung

Die extrahierbare Laserleistung von CO_2-Lasern hängt wesentlich vom Massenstrom zur konvektiven Kühlung des laseraktiven Mediums ab. Dieser Massenstrom läßt sich steigern , wenn die Widerstände des Strömungssystems verkleinert werden und darüber hinaus der vom Gebläse umgewälzte Volumenstrom vergrößert wird. Radialgebläse sind für Volumenströme bis ca. 4000 m^3/h und maximale Druckverhältnissen Π_{max} = 1,7 bis 2,0 je nach Laser-Gasgemisch verfügbar, Gebläse mit Volumenströmen von ca. 6000 m^3/h bei Druckverhältnissen Π_{max} < 1,5 stehen vor der Markteinführung.

Bei der Auswahl des Gebläses stehen ein-, zwei- oder dreistufige Radialgebläse je nach Anforderung an das Druckverhältnis und den Volumenstrom zur Auswahl. Ein ausgewähltes Gebläse kann durch die Anpassung der Leitschaufelstellung an den Arbeitsbereich des Lasers feinabgestimmt werden. Verschiebungen der Kennlinie sind um bis zu 20 % des Volumenstroms bei konstantem Druckverhältnis möglich. Damit verbunden ist auch eine Verbesserung des Gebläsewirkungsgrades im Arbeitsbereich.

6. Literatur

[1]   U.Jarosch, F. Diedrichsen, H.H. Henning, M.Franke
         Verhalten von Radialverdichtern als Umwälzgebläse in CO_2-Lasern
         bei Variation von Gasgemisch und Druck, Proceedings Laser 91, Springer

[2]   C. Pfleiderer, H. Petermann
         Strömungsmaschinen, 5. Auflage , Springer 1986

Bestimmung der Abmessungen von Laserstrahlen gemäß des Normenvorschlags ISO 172

B. Küppers[1], H. Staubach[2], M. Scholl[2], O.Märten[1]

[1] Fraunhofer Institut für Lasertechnik,

[2] Lehrstuhl für Lasertechnik der RWTH Aachen

Steinbachstr. 15, D-52074 Aachen

Kurzfassung:

Der Normenvorschlag ISO 172 enthält Richtlinien zur Messung und Charakterisierung des Ausbreitungsverhaltens von Laserstrahlung aus stabilen Resonatoren /1/. Die dort definierte Beschreibungsform der Laserstrahlung erfordert die Messung von Leistungsdichteverteilungen (LDV) entlang einer Strahlkaustik. Aus den LDV werden die Strahlabmessungen bestimmt und daraus weitere Strahlparameter.

In dem Normenvorschlag werden die Strahlabmessungen über die 2.Momente der LDV definiert, es werden allerdings weitere Definitionen zugelassen. In der vorliegenden Arbeit wurden zwei verschiedene Definitionen anhand gemessener Leistungsdichteverteilungen eines 10KW-CO_2-Laserstrahls entlang einer Fokuskaustik miteinander verglichen.

Die Definition der Strahlabmessungen über die 2.Momente der LDV in Verbindung mit einer Gauß-Filterung der LDV erscheint geeignet zur Charakterisierung von Laserstrahlung sowohl aus stabilen, als auch aus instabilen Resonatoren.

Abstract:

The proposal for international standardization ISO 172 deals with measurement and characterization of the propagation of laser beams from stable resonators /1/. The definitions made require the measurement of power density distributions (PDD) along the beam path. From these the beam dimensions and further beam parameters can be derived.

The beam dimensions are defined as the second moments of PDD, but some other definitions are also discussed. Two different definitions have been examined with measured PDD along the beam path of a focussed 10KW-CO_2-laser beam.

The definition of beam dimensions as the second moments of PDD in connection with Gauß-filtering of PDD turn out to be applicable to laser beams from stable resonators as well as from unstable resonators.

Einleitung:

Bislang gibt es eine Vielzahl von Meßmethoden und uneinheitlichen mathematischen Beschreibungsformen für Laserstrahlung. Aus diesem Grund wird in der Norm ISO 172 die Vermessung und Charakterisierung von Laserstrahlung international standardisiert.

Als Meßmethode wird die Messung der Leistungsdichteverteilung (LDV) gefordert, da diese die höchstwertige Informationsform darstellt, aus der alle zur Charakterisierung notwendigen Strahlparameter abgeleitet werden können. Die Bestimmung der Abmessungen von Laserstrahlung aus den LDV erfolgt standardmäßig über die 2.Momente der LDV. Diese haben gegenüber anderen Definitionen der Strahlabmessungen den Vorteil, daß sie entlang einer Strahlkaustik mathematisch exakt beschreibbar propagieren. Bei Laserstrahlung aus instabilen Resonatoren liefert diese Definition allerdings zu große Abmessungen, so daß die Norm vorläufig auf Laserstrahlung aus stabilen Resonatoren beschränkt ist.

Die Anwendbarkeit der Norm auf einen industriellen 10KW-CO_2-Laserstrahl aus instabilem Resonator wird im folgenden überprüft. Hierzu werden zwei verschiedene Verfahren zur Bestimmung der Strahlabmessungen miteinander verglichen. Da die gemessenen LDV gestört sind, werden weiterhin zwei verschiedene Datenaufbereitungsverfahren untersucht. Es soll festgestellt werden, ob durch eine Filterung der LDV mittels eines Gaußfilters auch eine Anwendbarkeit der Norm auf Laserstrahlung aus instabilen Resonatoren erreicht werden kann.

Strahlradiusdefinitionen:

Der Normenvorschlag enthält neben der Definition der Strahlabmessungen über die 2.Momente der LDV noch die Option, die Strahlabmessungen über Strahlblenden mit variabler Apertur zu bestimmen, durch die 86.5% der Strahlleistung hindurchtreten. Die mathematische Beschreibung der beiden Definitionen ist:

86%-Definition

$$0{,}865 * P_{Ges} = \int_G I(x,y)\,dx\,dy$$

G: Gebiet, welches durch eine Ellipse mit den Hauptachsen r_{lx} und r_{ly} um den Schwerpunkt beschrieben wird.

86%-Definition mit Gaußfilter

$$K(x,y) = I(x,y) * e^{\left(-\frac{x^2}{A r_{Gx}^2} - \frac{y^2}{A r_{Gy}^2}\right)}$$

$$0{,}865 * P_{Ges} = \int_{-\infty}^{\infty} K(x,y)\,dx\,dy$$

r_{Gx} und r_{Gy} sind zu so zu bestimmen, daß obige Gleichungen gelten.

$A = 1{,}78732$

Momenten-Definition

$$N_{xx} = \frac{1}{P_{Ges}} \int_{-\infty}^{\infty} I(x,y) * (x - x_s)^2\,dx\,dy$$

$$r_{Mx} = 2 * \sqrt{N_{xx}}$$

Momenten-Definiton mit Gaußfilter

$$N_{Gxx} = \frac{1}{P_{GGes} * 0.86} \int_{-\infty}^{\infty} K(x,y) * (x - x_{Gs})^2\,dx\,dy$$

$$r_{GMx} = 2 * \sqrt{N_{Gxx}}$$

Abbildung 1: Strahlradiusdefinitionen

Die Methode der variablen Apertur wird im vorliegenden Fall softwaretechnisch realisiert, indem in gemessenen LDV die Gebiete G so bestimmt werden, daß die obige Definition erfüllt ist. Hierzu wird der Schwerpunkt, die Hauptachsen einer Ellipse und ihr Drehwinkel gegenüber den Koordinatenachsen bestimmt. Der Leistungsinhalt der Ellipse beträgt 86.5% der Gesamtleistung der LDV.

Datenaufbereitungsverfahren:

Untersucht wird ein industrieller 10 KW-Laser mit instabilem Resonator, der mit seiner Nennleistung betrieben wird. Die LDV wird mit einem Gerät des Typs UFF100 der Fa. PROMETEC gemessen /2/. In unmittelbarer Fokusnähe können keine Meßdaten aufgenommen werden, weil dort die Leistungsdichte über 10^7 W/cm^2 liegt. Da Hochleistungslaserstrahlung aus physikalischen und technischen Gründen starke Leistungsfluktuationen aufweist, werden die LDV aus Mittelwerten mehrerer Messungen bestimmt.

Die gemessenen LDV sind von farbigem Rauschen /3/ und systematischen Meßfehlern überwiegend im Randbereich der Verteilungen überlagert (siehe Abbildung 2). Dies führt zu großen Fehlern bei der Bestimmung der Strahlabmessungen.

Aus diesem Grunde werden verschiedene Datenaufbereitungsverfahren getestet, die Störungen aus den Verteilungen beseitigen sollen, ohne dabei den Informationsgehalt der LDV nennenswert zu beeinflussen.

Das erste Verfahren ist ein Gauß-Filter, das die Eigenschaft besitzt, die LDV in dem Bereich um ihren Schwerpunkt kaum zu beeinflussen. Im Randbereich der Verteilung werden Störungen und Beugungsanteile stark gedämpft. Die Dämpfung von Beugungsanteilen im Randbereich der LDV ist bei Laserstrahlung aus instabilen Resonatoren wichtig, falls die Strahlabmessungen aus den 2.Momenten der Verteilung bestimmt werden.

Das zweite untersuchte Verfahren ähnelt einem Median-Filter /4/. Die Filterung wird dadurch bewirkt, daß die Umgebung eines LDV-Punktes nach bestimmten Kriterien untersucht wird, und der LDV-Punkt entsprechend modifiziert wird.

Strahlkaustiken:

In der Abbildung 3 sind die Strahldurchmesser nach Momenten-Definition und 86.5 %-Definition entlang der beschriebenen Fokuskaustik aufgetragen. Parabeln 2.Ordnung sind an die Meßpunkte in beiden Hauptachsenrichtungen angenähert. Dargestellt sind sowohl die Diagramme für die Originaldaten, als auch für die beiden beschriebenen Datenaufbereitungsverfahren. An der rechten Seite der Diagramme kann jeweils der innerhalb der Strahlabmessungen vorhandene Leistungsanteil abgelesen werden.

Ergebnisse:

Die Bestimmung der Strahlabmessungen aus unbearbeiteten Daten ist nur näherungsweise möglich, da die Verteilungen im Randbereich stark gestört sind. Eine Aufbereitung der Daten ist notwendig. Die Gauß-Filterung der Daten ist gegenüber der digitalen Filterung mittels modifizierten Median-Filters vorzuziehen, da bei der letzteren die Filterparameter jeweils individuell an die LDV angepaßt werden müssen.

Die Gauß-Filterung hat neben der Beseitigung von Störungen noch die positive Eigenschaft, daß hohe Beugungsanteile der Verteilungen stark gedämpft werden, und somit eine Berechnung der Strahlabmessungen mittels Momenten-Definition auch bei Laserstrahlung aus instabilen Resonatoren sinnvolle Ergebnisse liefert.

Die Momenten-Definition hat gegenüber der 86.5%-Definition den Vorteil, daß die Strahlabmessungen mathematisch exakt in einer Parabel 2.Ordnung propagieren.

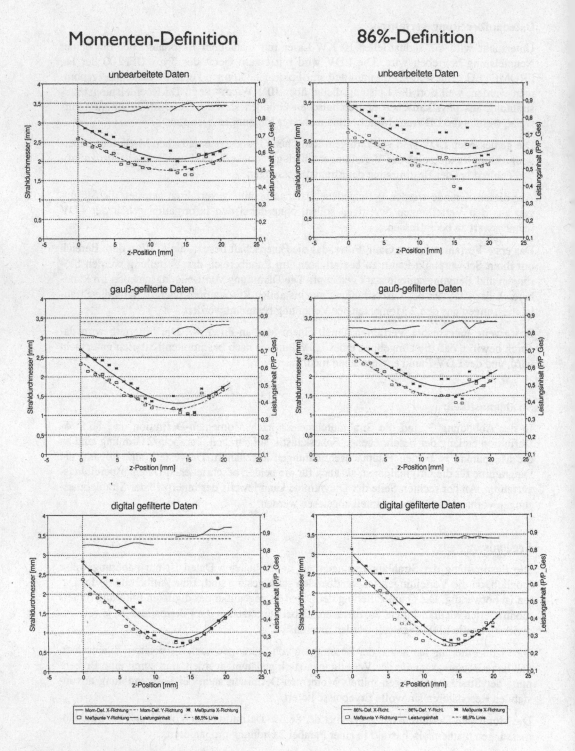

Abbildung 2: Leistungsdichteverteilungen

Meßebene außerhalb Fokus
(z=0mm)

Meßebene in Fokusnähe
(z=13mm)

unbearbeitete Daten
- 10 fache Mittelung
  von Einzelmessungen.

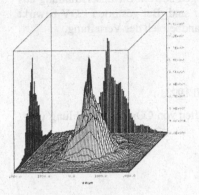

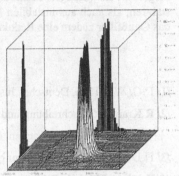

gauß-gefilterte Daten
- Definition des Gauß-
  filters siehe links.

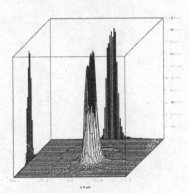

digital gefilterte Daten
- Filterung der unbear-
  beiteten Daten mit
  modifiziertem Median-
  filter.

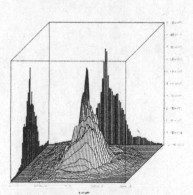

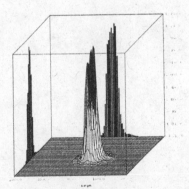

Abbildung 3: Strahlkaustiken

618

Der Vorteil der 86.5%-Definition gegenüber der Momenten-Definition besteht darin, daß der Leistungsinhalt innerhalb der Strahlabmessungen konstant ist, was für viele Materialbearbeitungsprozesse von Bedeutung ist.

Die Ermittlung der Strahlabmessungen über Momenten-Definition mit Gaußfilter erscheint insgesamt als günstigste Methode, da die Strahlabmessungen sowohl für Laserstrahlung aus stabilen, als auch aus instabilen Resonatoren berechenbar sind. Bei gestörten LDV bewirkt das Gaußfilter zudem eine Reduktion der Störungen im Randbereich der Verteilung.

/1/ ISO/ WI 11146, Deutsches Institut für Normung, (April 1993)

/2/ R.Kramer, "Beschreibung und Messung der Eigenschaften von CO_2-Laserstrahlung",

Dissertation RWTH Aachen, (1991)

/3/ H.D.Lüke, "Signalübertragung", Springer Verlag

/4/ B.Jähne, "Digitale Bildverarbeitung", Springer Verlag

Richten von Blechformteilen mittels laserinduzierten thermischen Spannungen

P. Hoffmann; J. Kraus; M. Geiger

Bericht aus dem Anwenderlabor Lasermaterialbearbeitung Erlangen

Haberstraße 2, D-91058 Erlangen, Tel.: 09131/85-8341; FAX: 09131/36403

1 Einleitung

Die Fertigung komplexer Blechformteile aus Feinblech ist gekennzeichnet durch zahlreiche Einflußgrößen. Halbzeuge aus verschiedenen Chargen werden formgebend bearbeitet und anschließend formschlüssig, häufig thermisch gefügt. Hierbei werden in vorausgegangenen Fertigungsschritten eingebrachte Eigenspannungen freigesetzt sowie zusätzliche Spannungen erzeugt. Die Folge sind Verzugserscheinungen am Fertigteil. Der Stand der Technik sind hier mechanische, manuell durchgeführte und damit kostenintensive Richtoperationen. Mit Blick auf humane Fertigungsmethoden und Wirtschaftlichkeit des Fertigungsprozesses wird nach flexiblen, automatisierbaren Verfahren gesucht. Das Biegen von Blechteilen mittels laserinduzierter Spannungen bietet hierfür die notwendigen Voraussetzungen [1,2,3]. Die Umformung wird berührungsfrei durchgeführt, d.h., es muß keine äußere mechanische Kraft aufgebracht werden, und ein aufwendiges Spannsystem für das Bauteil ist somit nicht erforderlich. Dadurch ist eine zusätzliche Beschädigung der Werkstückoberfläche durch den Richtvorgang von vornherein ausgeschlossen. Da aber auch größere Umformgrade mit diesem Verfahren erzeugt werden können, sind neben dem Richten auch andere Anwendungsgebiete wie z.B. in der Prototypen- und Kleinserienfertigung denkbar.

Im weiteren soll eine Modellvorstellung zum Umformmechanismus dieses Verfahrens näher erläutert werden.

2 Theoretisches Modell zum Biegen von Blechformteilen

Für das Laserstrahlbiegen von ebenen Blechen ist bereits ein theoretisches Modell [1,2] erarbeitet worden. Im folgenden soll aufbauend auf diese Theorie der Umformmechanismus an räumlichen Bauteilen vorgestellt werden. Zu Beginn des Prozesses führt der von der eingebrachten Strahlungsleistung absorbierte Anteil zur Aufheizung der Blechoberfläche. Die thermische Expansion dieser

620

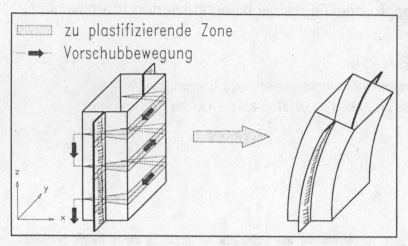

Zone bewirkt einen Volumenzuwachs, der vom Wärmeausdehnungskoeffizienten α abhängig ist. Da die benachbarten, nicht bestrahlten Gebiete relativ dazu kalt bleiben und somit nicht expandieren, wird die Ausdehnung behindert, was zur Entstehung von Druckspannungen führt. Erreichen diese Druckspannungen die jeweilige Warmfließgrenze, wird der Werkstoff örtlich gestaucht. Die Ausdehnung dieser Stauchzone ist von einer ganzen Reihe von Parametern abhängig [2,4], von denen hier nur Laserleistung P, Fokuslage z, Vorschubgeschwindigkeit v, Blechdicke s und Wärmeleitfähigkeit λ genannt seien. Diese zu Beginn der Bestrahlung auftretende thermische Expansion wird als Gegenbiegung bezeichnet. Nachdem der Laser die erwärmte Zone passiert hat, kommt es überwiegend durch Wärmeleitung zur Abkühlung der gestauchten, vorwiegend oberflächennahen Bereiche. Die Gegenbiegung verschwindet und kehrt sich in die eigentliche Biegerichtung um. Da die Blechunterseite bei Feinblech - und nur dieses wird im weiteren behandelt - durch den Wärmefluß ebenfalls beträchtlich aufgeheizt wird, wird die Ausbildung der Biegeendform durch die herabgesetzte Fließgrenze zusätzlich erleichtert. Die Endform ist dann erreicht, wenn das gleichmäßig erwärmte Bauteil auf nahezu Umgebungstemperatur abgekühlt ist. Prinzipiell gilt, daß das Blech zur Generierung eines möglichst großen Biegewinkels nur an der Oberfläche gestaucht werden darf. Erfolgt die Plastifizierung über die ganze Blechdicke, so wird das Bauteil nur verkürzt, jedoch nicht gebogen. Für geschlossene Profile mit hohem Flächenträgheitsmoment muß für eine effiziente Prozeßführung ähnlich wie beim Flammrichten [5] das Bauteil durch Einbringung einer keilförmigen Stauchzone (Bild 1) verkürzt und somit die Biegung bewirkt werden. Hier ist es vorteilhaft, die Stauchzone nicht nur oberflächennah zu gestalten, sondern möglichst über die gesamte Blechdicke auszuweiten. Eine keilförmige Bestrahlungszone mit konstanter Intensität, wie in Bild 1 gezeigt, kann durch Veränderung der Fokuslage und Einsatz von Laserleistungssteuerung erzielt werden.

3 Zusammenfassung und Ausblick

Es wurde ein Modell für eine neue Technologie vorgestellt, die Synergien zwischen der Umformtechnik und der Lasertechnik aufzeigt. Anhand der Bearbeitung eines Realbauteils konnte inzwischen auch nachgewiesen werden, daß hiermit eine präzise Umformung selbst an komplexen Blechkonstruktionen möglich ist, so daß die geforderten Toleranzen eingehalten werden konnten.

Ein Einsatz dieser Technologie erscheint sowohl als Richtwerkzeug als auch für das Rapid Prototyping sinnvoll. Hier sei nur beispielhaft das Space-Frame-Konzept genannt, das in neueren Entwicklungstendenzen der Automobilindustrie angeführt wird. Die Formgebung an den Strangpreßprofilen, die für die Rahmenteile erforderlich sind, könnte durchaus durch Laserstrahlbiegen erfolgen.

In weiteren Arbeitsschritten sollen Konzepte für ein automatisiertes System erstellt und realisiert werden. Durch die Integration in einem geschlossenen Regelkreis mit Online-Kontrolle des Umformergebnisses kann ein sicherer und adaptiver industrieller Prozeß aufgebaut werden. Die Erprobung eines weiteren Maschinenkonzeptes mit einem Nd:YAG-Laser als Strahlquelle soll Aufschluß darüber geben, ob wegen der besseren Absorption bei dieser Wellenlänge auf ein zusätzliches Coating verzichtet werden kann.

Literatur

[1]   Geiger, M.; Vollertsen,F.; Amon,S.
        Flexible Blechumformung mit Laserstrahlung - Laserstrahlbiegen
        BLECH ROHRE PROFILE 38 (1991) 11 856-861
[2]   Frackiewicz, H.; Kalita, W.; Mucha,Z.; Trampczynski, W.
        Laserformgebung der Bleche
        VDI-Berichte 867 (1990) 317-328
[3]   Geiger, M.; Hoffmann, P.; Hutfless, J.
        Lasertechnik in Synergie zur Umformtechnik
        BLECH ROHRE PROFILE 40 (1993) 4 324-330
[4]   Namba,Y.
        Laser forming in space
        International conference on lasers
        Procedings 1985/86 403-407
[5]   Pfeiffer, R.
        Richten und Umformen mit der Flamme
        DVS-Verlag. Düsseldorf 1989

Laserbeschriftung durch Gradientenindex - Glasfasern mit weitgehender Beibehaltung der Strahlqualität

S. Geiger

ROFIN-SINAR LASER GmbH, Bereich Markieren

Neufeldstr. 16, D-85232 Bergkirchen

1. Einleitung

Festkörperlaser, eingesetzt zur Materialbearbeitung, sind in der Industrie schon alltäglich. Sie erfüllen Aufgaben, wie z.B. Trennen, Fügen, Bohren, für welche hohe Leistungen benötigt werden, wie auch Arbeiten im Bereich der Strukturierung, Skalierung, Mikrobearbeitung und Beschriftung, wo hohe Genauigkeiten und hohe Arbeitsgeschwindigkeiten erreicht werden.

Insbesondere die Beschriftungstechnologie mit Nd-YAG-Festkörperlasern erfordert eine hohe Strahlqualität am Ort der Galvanometerspiegel-Ablenkeinheit, begründet durch den Wunsch nach hoher Leistungsdichte, Auflösung, Markiergeschwindigkeit, großem Bearbeitungsfeld und Arbeitsabstand sowie mechanischer Flexibilität zwischen Laser und Bearbeitungsoptik durch die Verbindung mittels Lichtleitfasern. Dabei kann jedoch das Öffnungsverhältnis der Fokussieroptik nicht beliebig gewählt werden, da eine große Eingangsapertur entsprechend große Ablenkspiegel mit hohem Massenträgheitsmoment bedingen, welches die maximale Markiergeschwindigkeit herabsetzt.

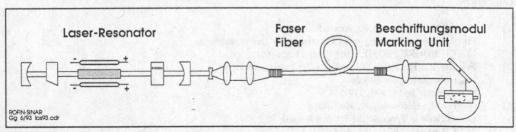

Bild1: Optische Anordnung des Aufbaus

2. Stand der Technik

Glas wurde erstmals vor 3000 Jahren in Ägypten für Dekorationsgegenstände verwendet, wobei eine Zugabe von Metalloxiden verschiedene Farben erzeugte. Seither wurde die Lichtdurchlässigkeit mehrmals um Größenordnungen verbessert bis es in den 60er Jahren gelang, erste Glasfasern herzustellen [1]. Durch weitere Entwicklungen wurde die Transmission bei 1064 nm, der Wellenlänge für Nd-YAG-Festkörperlaser, auf bis zu 90 % und mehr gesteigert [2, 3]. Dabei werden im industriellen Einsatz Durchschnittsleistungen bis in den kW-Bereich übertragen [4, 5]. Zur Anwendung kommen hierbei überwiegend Stufenindexfasern, welche den Einfallswinkel des Laserstrahls am Fasereingang in der Faser nahezu konstant halten. Strahlqualitätserhaltend arbeitet diese Art der Übertragung folglich nur dann, wenn der gesamte Faserkerndurchmesser ausgeleuchtet wird.

Die aufgrund unvermeidlicher Systemtoleranzen und Fehljustagen entstehende Problematik der Strahlungsein-kopplung in den Fasermantel kann durch geeignete Faserkonfektionierung ausgeschlossen werden. Ein sehr gutes System, welches durch ein patentiertes Mode-Stripping-Verfahren sogar extreme Fehljustagen ohne Schaden übersteht, bietet die Fa. ROFIN-SINAR in Zusammenarbeit mit dem schwedischen Hersteller Permanova an [4, 6]. Durch ein Abbildungsverhältnis von 1:1 bzw. 1:0,66 lassen sich für Schneid- bzw. Schweißanwendungen Foki zwischen 150 µm und 1 mm erzielen.

Für Beschriftungsapplikationen sind jedoch Fokusdurchmesser um 100 - 150 µm bei typischerweise 160 mm Objektivbrennweite und bedingt durch die Galvanometerspiegel 14 mm Eingangsapertur gefordert. Die Kollimierungsbrennweite nach einer 400 µm Stufenindexfaser würde daher 2,6 bzw. 4 mal 160 mm betragen, also ca. 400 bis 600 mm. Dies ist für die Praxis ein unakzeptabel hoher Wert, da die durch die Glasfaser erst gewonnene Flexibilität mechanisch erheblich beeinträchtigt würde.

Die Übertragung der Laserstrahlung für Beschriftungsapplikationen mit Galvanometerablenkeinheit über Stufen-indexfasern scheidet folglich aus.

3. Anpassung Laser - Glasfaser

Bei der Anpassung des Lasers an die Glasfaser bzw. der Glasfaser an den Laser ist zu gewährleisten, daß das Strahlparameterprodukt des Lasers BP_{Laser} kleiner gleich dem Strahlparameterprodukt der Stufen- bzw. Gradientenindexfaser $BP_{SI-Faser}$ $BP_{GI-Faser}$ bleibt. Gleichzeitig darf weder der Faserkerndurchmesser noch die numerische Apertur überschritten werden. Berechnungsgrundlagen sind in Bild 2 zusammengestellt [7, 8]. Es gilt:

$BP_{Laser} \leq BP_{SI-Faser}$ bzw. $BP_{GI-Faser}$ Fokusradius ≤ Faserkernradius

$M^2_{Laser} \leq M^2_{SI-Faser}$ bzw. $M^2_{GI-Faser}$ Einkoppelwinkel ≤ Akzeptanzwinkel

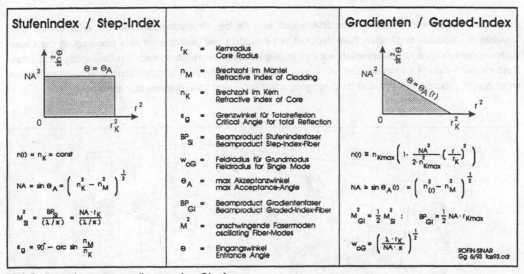

Bild 2: Berechnungsgrundlagen der Glasfaser

Obige Bedingungen gelten sowohl für Stufen-, als auch für Gradientenindexfasern. Bei ersteren ist der Strahl-qualitätsverlust nach der Übertragung umso geringer, je näher das Strahlparameterprodukt des Lasers bei dem der Faser liegt. Ab ca. 5 - 10 m ungerader Faserlänge füllt die Ausgangsstrahlung den vollen Kerndurchmesser und den vollen Akzeptanzwinkel der Stufenindexfaser aus (Bild 3l).

624

Für den industriellen Einsatz und die damit verbundenen hohen Anforderungen an Reproduzierbarkeit und Zuverlässigkeit wurde bisher der Faserdurchmesser und der Akzeptanzwinkel nur zu jeweils maximal 70 % ausgenutzt. Daraus resultiert eine Verschlechterung der Strahlqualität um $1/0,7 \cdot 0,7$ entsprechend dem Faktor 2 im günstigsten Fall der Anpassung (maximales Strahlparameterprodukt, entsprechend maximale Modenzahl des Lasers). Bei geringerer Pumpeingangsleistung bzw. dem Einsatz von Modenblenden im Resonator verbessert sich das Strahlparameterprodukt des Lasers zum Teil erheblich. Nach der Stufenindexfaser erscheint das gleiche Strahlbild wie im obigen Fall, und der Anwender erfährt eine unakzeptable Verschlechterung der Laserstrahlqualität um bis zu einem Faktor 5, welche die Folge eines Modenkopplungsprozesses innerhalb der Faser ist [9, 10].

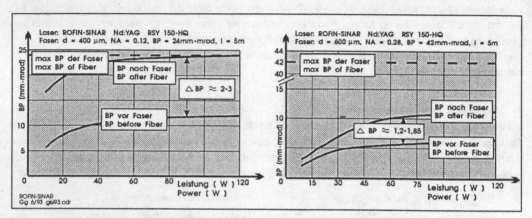

Bild 3l: Strahlparameterprodukt BP vor und nach einer Stufenindexfaser
3r: Strahlparameterprodukt BP vor und nach einer Gradientenindexfaser

Durch den Einsatz von Gradientenindexfasern wird eine flexible Beschriftungstechnologie eröffnet, die dem Standard der Industrie hinsichtlich Zuverlässigkeit voll entspricht und zudem noch eine Übertragung der Laserstrahlung nahezu ohne Qualitätsverluste ermöglicht. Bei korrekter Einkopplung nach der Bedingung Laserfokusdurchmesser = Faser-Felddurchmesser, $w_{t2} = w_{OG}$, sollte eine Übertragung gänzlich ohne Strahlqualitätsverluste möglich sein. Beachtenswert ist, daß die Bedingung unabhängig vom Faserdurchmesser bleibt.

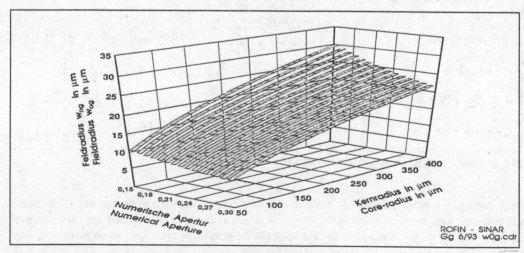

Bild 4: Feldradius für Grundmodus einer Gradientenindexfaser

Bedingt durch das Wandern der Strahltaille und deren Durchmesseränderung, sowie die Abweichung des Brechzahlprofils der Faser von der Idealparabel als auch Abbildungsfehler des optischen Systems, kann völlig strahlqualitätsverlustfrei in der Praxis nicht gearbeitet werden. Bei einer Abweichung $w_{t2} \neq w_{OG}$ pendelt der Strahlradius im Faserinneren zwischen den Werten w_{t2} und w_{OG}^2/w_{t2}. Bild 3r zeigt die Strahlqualität vor und nach einer Gradientenindexfaser mit 600 µm Durchmesser und einer maximalen numerischen Apertur von 0.28 in der Kernmitte. Die Strahlqualitätseinbuße lag zwischen Faktor 1,2 und 1,85. Diese Werte wurden erreicht, obwohl das Strahlparameter-Produkt der Faser zu nur ca. 25 % ausgenutzt wurde und ermöglicht erstmals eine Beschriftung von Metallen über ein flexibles Lichtleitersystem mit hoher Geschwindigkeit und akzeptabler Bearbeitungsfeldgröße.

4. Zusammenfassung

Für den industriellen Einsatz der Laserbeschriftung stehen heute Glasfaserübertragungssysteme mit ausgezeichneter Standfestigkeit zur Verfügung.

Die flexible Kopplung ermöglicht ein getrenntes Aufstellen von Laser und Beschriftungskopf, mit einer Reihe sich daraus ergebender Vorteile. Durch die Montage des Beschriftungskopfes auf einen Roboter kann das Beschriftungsfeld beliebig auf der Oberfläche eines bzw. verschiedener Körper positioniert werden.

Bei optimierter Auslegung des Systems Laser / Glasfaser wird eine Leistungsübertragung von 90 % erreicht. Durch den Einsatz eines Mode-Stripping-Verfahrens konnten die Strahleingangsparameter einer Stufenindex-Glasfaser voll ausgenutzt und somit die erreichte Strahlqualität nach Austritt der Faser gegenüber herkömmlichen Systemen verbessert werden.

Durch den Einsatz einer Gradientenindex-Glasfaser war es möglich, um einen weiteren Schritt zu optimieren und Strahlqualitätsübertragungsverluste von Faktor 1,2 bis 1,85 unabhängig von der Anzahl der Lasermoden zu erreichen. Dabei bestätigte sich die Theorie, daß bei geeigneter Anpassung der Laserstrahltaille an den Faser-Feldradius die Strahlqualitätsübertragung unabhängig vom Faserdurchmesser bleibt. Erhebliche Vorteile hinsichtlich Standfestigkeit und Justagefreundlichkeit des Systems werden erzielt.

5. Literatur

[1] Drexhage, M., Moynihan, C., Infrarot-Lichtleiter, Spektrum der Wissenschaft (1/1989), S. 104 ff

[2] Prause, C., Hering, P., Lichtleiter für gepulste Laser: Transmissionsverhalten, Dämpfung und Zerstörschwellen, Laser und Optoelektronik (1/1987), S. 25-31

[3] Geiger, S., Hefter, U., Robotertaugliche Laserbeschriftung durch den Einsatz von Glasfasern, Laser und Optoelektronik (4/1988), S. 70 ff

[4] Fa. Rofin-Sinar Laser GmbH, Produktinformation

[5] Göller, E., Huber, R., Iffländer, R., Schäfer, P., Wallmeroth, K., 2kW-CW-Laser, Ergebnisdarstellung zu einem Eureka-Projekt, Laser und Optoelektronik (2/1993), S. 42 ff

[6] Fa. Permanova, High Power Optical Fiber with Improved Covering, US-Patent Nr. 4, 678, 273, 1987

[7] Geckeler, S., Lichtwellenleiter für die optische Nachrichtenübertragung, Springer-Verlag, 1990

[8] Geiger, S., Laserbeschriftung durch Glasfasern mit Nd:YAG Q-Switch Laser, Laser und Optoelektronik (3/1992), S. 58 ff

[9] Lundgren, L., Vilhelmsson, K., Mode Excitation in Grade-Index Optical Fibers, Journal of Lightwave Technology, Vol. LT-2, No. 4, 1984

[10] Schildbach, K., Mode coupling in optical Fibers used for Laser Spot Welding, Information Philips Eindhofen

Nd:YAG Laser Oscillator with Unstable Resonator

J.Marczak, R.Ostrowski,J.Owsik, A.Rycyk and A.Sarzyński
Institute of Optoelectronics, Military University od Technology
01-489 Warsaw, 2 Kaliski str, Poland

Unstable resonator configuration has been extensively investigated and it appeared in result to be successfully applied to high gain gas and solid state lasers, delivering much greater output energy, better energy extraction and lower beam divergence than the stable resonator in TEM_{oo} - mode operation. Unfortunately, the output beam quality of the most unstable configuration in the near field suffers from strong intensity modulations within the beam. Therefore, generation of diffraction limited lasers beam with the very good transverse distribution is the principal problem in laser engineering.

These shortcomings can be partially avoided by using self-filtering n-branch and p-branch unstable resonators described first by G.C.Reali et al. and K.Vogler et al. Another method is using output mirror whose reflectivity decreases radially from a suitable value at the center to a sufficiently low value at the radial position corresponding to the edge of the laser rod. Unstable resonators with variable reflectivity mirrors (VRM) as an output-coupler offer a great number of advantages. The most promising technique demonstrated first by Lavigne et al. was based on the deposition of thin electric films of shaped thicknesses on a transparent substrate. Recently, for a greater mode overlap efficiency with the active medium instead of gaussian reflectivity profile, DeSilvestri et al. have proposed a novel supergaussian reflectivity profile. Such a profile substantially increases the mode volume in the active material and preserves good diffraction and propagation properties of the output beam.

We designed and operated a positive branch Nd:YAG unstable resonator oscillator fig.1. The laser head contains a 4 mm x 50 mm Nd:YAG rod 6° wedged with antireflection coated end faces, which is pumped by one linear xenon flashlamp and they are situated together in elliptically formed reflector.

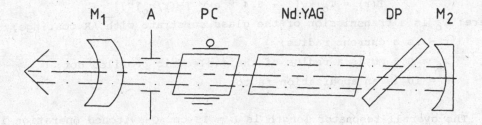

Fig.1. Optical scheme of the Nd:YAG oscillator with VRM as the
output coupler. M_1 - VRM, P.C.- Pockels cell, D.P.-dielectric
polarizer, A-aperture, M_2 - 100% reflectivity concave mirror.

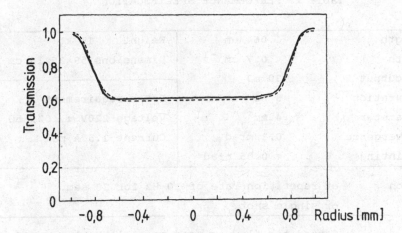

Fig.2. Comparison of experimentally (contineous line) measured
transmission of the mirror M_1 with approximated profile (dotted
line).

In p-branch configuration, the resonator is limited by two mirrors
with radii of curvature R_1=-500 mm and R_2=1033 mm. The concave mirror
M_2 with 100% reflectivity at operation wavelength is located near one
end of the laser. The mirror M_1 has partially reflecting (fig.2) and
profiled dielectric layer in form of dot with diameter d=2 mm deposited
on a convex-concave lens with the thickness of 4.5 mm. Both surfaces of
this mirror were first antireflection coated. Then, on the convex
surface of this lens a profiled layer was deposited. As a result,
transmission of the mirror is:

$$T(r) = T_o \times [1 - 0.4 \times \exp [-(r/r_o)^N]]$$

where: T_o is a transmission of the glass substrate with AR coatings;

r - is a current radius;

$r_o = 0,9$ mm is a radius of the dielectric profiled dot;

N = 10 is an apodization factor.

The overall resonator length is L = 18 cm. Q-switched operation is achieved using Pockels cell on $LiNbO_3$ crystal with 6° wedged AR coated end faces. The Pockels cell driver is based on avalanche transistors stacked in series.

Table 1. PERFORMANCE SPECIFICATION

Wavelength	1.064 μm	Weight 15 kg	
Linewidth	< 0.7 cm $^{-1}$	Dimensions 59×36×16 cm	
Energy output	30 mJ		
Pulse duration	5 ns	Power requirement	
Beam diameter	4 mm	Voltage 220V ± 10%, 50 Hz	
Beam divergence	0.3 mrad	Current 1.5 A max.	
Beam pointing	< 0.05 mrad		
Operation	at repetition rate of 10 Hz for 30 sec. or single shot		
Cooling	forced air flow is used to cool the laser rod and flashlamp		

NOTES ON SPECIFICATION

* high power
* variable reflectivity mirror as an output coupler
* p-branch unstable resonator
* possibility to generate second or higher harmonic of light

APPLICATIONS

* nonlinear optics
* laser spectroscopy
* material processing
* industry

Preparation of Ultrafine Oxide Particles by Solid State Evaporation with XeCL-Laser Radiation

I.Bothe, W.Rath and F.Bachmann
Rofin-Sinar Laser GmbH,
Berzeliusstr. 87, D-22093 Hamburg, Germany

W.Riehemann and B.L.Mordike,
Technische Universität Clausthal,
Agricolastr. 6, D-38678 Clausthal-Zellerfeld, Germany

1) Introduction

Among other well known methods /1/ laser ablation from solid targets has advantages for the production of extremely fine powders. The results of initial investigations using CO_2 and Nd:YAG lasers /2,3,4/ are described in the literature. The advantages of the laser ablation method compared to more conventional techniques such as gas atomisation, electron beam evaporation, sol-gel techniques are:

- the sublimation process is contactless without crucible, therefore a potential source for contamination is eliminated
- the process is applicable to virtually all materials
- different gas atmospheres (reactive or non-reactive) can be used during sublimation process
- powder specifications can be controlled by laser parameters
- the powder produced is very small (average diameter 5-20 nm)

A disadvantage of the laser ablation method is, that the production rate is low compared to other methods which produce larger particles.

As is shown here, production of fine powder is also possible using excimer laser light. The advantage of the excimer laser compared to other laser sources derives from its short wavelength and high energy pulses. One consequence of the short wavelength is the extremely high absorption in almost any material. Very shallow layers ($<< 1$ µm) can be removed by a single shot. Furthermore, with the short pulse length (20-60 nsec) the ablation takes place on such a short time scale, that the heat load to the substrate is minimized. The particles, molecules or clusters are ejected from the thin surface layer with a high kinetic energy.

In the following the first experiments using an high pulse energy XeCl excimer laser for powder formation from an Al_2O_3 target are described.

2) The mechanism of powder formation

If the target material is exposed to intensive laser radiation, immediate sublimation of the target material leads to formation of a supersaturated plume. Out of this plume the material can be solidified and extremely fine particles with diameters in the nanometer range are formed. They can be collected easily on the target surface and on the walls of the surrounding containment. The process is shown schematically in fig.1.

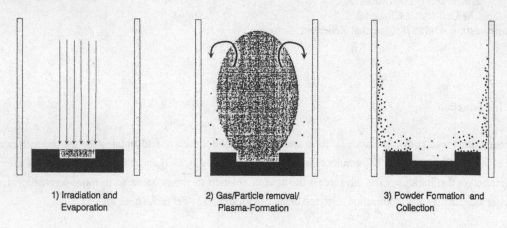

| 1) Irradiation and | 2) Gas/Particle removal/ | 3) Powder Formation and |
| Evaporation | Plasma-Formation | Collection |

Fig. 1: Schematic presentation of the powder formation process

3) Experimental

An x-ray preionized XeCl excimer laser (Siemens XP 2020) was used. The laser delivers pulses of 2J puls energy and a pulse length of 45 nsec (FWHM) at a maximum repetition rate of 20 Hz; thus an average power of 40 W is available. The beam size is approximately 5,5 cm x 4,5 cm and shows a top hat profile. A rectangular aperture is used to trim the raw beam; the resultant beam is concentrated by a field lens and is transformed by two quartz lenses (f_{ges} = 102,56 mm), so that energy densities up to 25 J/cm^2 can be delivered to the substrate.

The powder is produced in a stainless steel vacuum chamber. This chamber can be evacuated by a roughing pump and flushed with gases from a gas inlet tube. The powder is collected on the inner side of a glass tube, which is surrounds the laser beam. Fig. 2 shows the experimental setup.

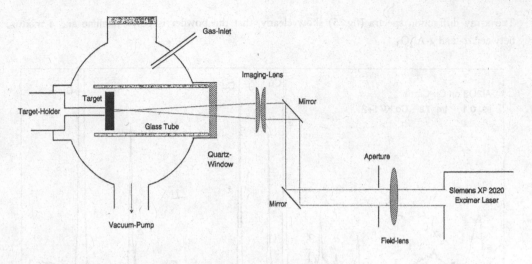

Fig. 2: Experimental setup for powder formation by excimer laser

4) Results and Discussion

First experiments on using the XeCl excimer laser and the vacuum chamber were made with α-Al_2O_3-bulk materials. The ultrafine powder has been characterized by x-ray diffraction (XRD) and transmission electron microscopy (TEM). Copper grids, which are mounted in the particle stream are used as specimen holders for the TEM investigation. The TEM photograph in fig. 3 shows the size and the form of the particles. The particle size was obtained by measuring 150 - 200 particles assuming, them to be spherical. Fig. 4 shows the average particle size and the size distribution, which can be approximated by logarithmic standard distribution. It should be pointed out, that nearly 95% of the powder lies between 3 nm and 15 nm.

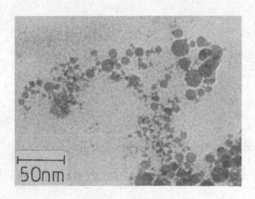

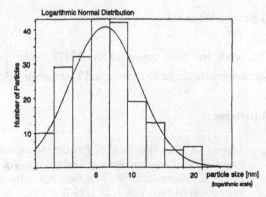

Fig.3: TEM photograph of powder
Fig.4: Size distribution of the powder

632

The x-ray diffraction spectra (fig. 5) show clearly, that the powder is polycrystalline and a mixture between α- and γ-Al_2O_3.

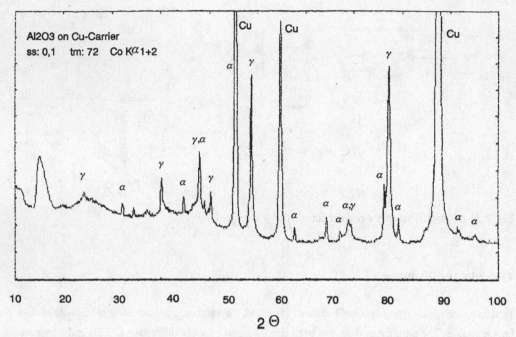

Fig. 5: x-ray diffraction pattern for the nanocrystalline powder

5) Conclusion

Nanocrystalline powders with an average diameter of 8 nm have been produced from α-Al_2O_3 targets by excimer laser ablation under air of normal pressure. The powder formed consisted of α- as well as γ-Al_2O_3 as has been demonstrated by the analysis of x-ray diffraction spectra.

6) Acknowledgment

This work has been supported by BRITE under project number 5147/91 and by the Deutsches Bundesministerium für Forschung und Technologie, FRG.

7) References

/1/ Lawley, A.: An Overview of Powder Atomisation Processes and Fundamentals, Int. Journ. of Powder Metallurgy and Powder Technology, Vol. 13, No.3 (1977)
/2/ Fritze, L.; Riehemann, W.; Mordike, B. L.: Int. Conference of Powder Metallurgy PM `90, The Institute of Metals, Vol. 3, 23 (1990)
/3/ Lee, H.-Y.; Riehemann,W.; Mordike, B. L.: Z. Metallkunde, 84, 2 (1993)
/4/ Riehemann, W.: Kleines Korn, Maschinenmarkt, 96, 20 (1992)

Strahlüberwachung als Teilbereich der Qualitätssicherung (ISO 9001)

Reinhard Kramer und Joachim Franek
PRIMES GmbH
Adolf Damaschke Str. 14, 64319 Pfungstadt

Einleitung

Im Jahr 1987 sind mit den ISO-Normen 9000 verschiedene Modelle der Qualitätssicherung genormt worden. Inzwischen wird in vielen Firmen nach ISO 9000 produziert und die Qualitätssicherung ist zu einem wichtigen Wettbewerbsfaktor geworden.

Bei der Fertigung mit Lasern werden zur Zeit nur in geringem Umfang Verfahren zur Qualitätssicherung eingesetzt. Dabei kann es in weiterer Zukunft ein gravierender Nachteil für die lasergestützten Fertigungsverfahren sein, wenn keine Fertigung nach ISO 9000 möglich ist. Im folgenden wird deshalb ein Konzept zur Überwachung des Laserstrahls im Rahmen eines Qualitätssicherungssystems (QSS) vorgestellt.

Vielfach werden die Begriffe der Qualitätssicherung nicht definitionskonform verwendet, was zu Mißverständnissen mit den Fachleuten für Qualitätssicherung führt. Beispielsweise wird der Begriff „Qualitätssicherung" oft synonym für Prozeßstabilisierung und die Optimierung von Prozessen gebraucht. Im Sinne der Definition kann allerdings nur bei der Prozeßstabilisierung von einer qualitätssichernden Maßnahme gesprochen werden, da die Prozeßoptimierung auch eine zeitliche Optimierung des Prozeßablaufs beinhalten kann, die aber keine qualitätssichernde Maßnahme darstellt. Für eine exakte Beschreibung der Probleme sollte deshalb auf eine normgerechte Verwendung der Begriffe Wert gelegt werden.

Laser und Qualitätssicherung

Nach ISO 9000 umfaßt ein Qualitätssicherungssystem alle Produktphasen von Design, Fertigung, Instandhaltung bis zur Entsorgung. Die wichtigsten Bestandteile eines Qualitätssicherungssystems im Produktzyklus sind schematisch im Qualitätskreis dargestellt. Viele Bestandteile des Qualitätskreises sind nicht laserspezifisch und können deshalb von anderen Produktionsverfahren übernommen werden. Bei der Materialbearbeitung mit Lasern ist neben dem Entwurf („Lasergerechtes Konstruieren") die Fertigung der zentrale Bereich, in dem die laserspezifische Qualitätssicherung ansetzen muß.

Die Norm ISO 9001 schreibt für die Qualitätssicherung im Bereich Fertigung keine Verfahren vor, sondern gibt in Abhängigkeit der Anforderungen an das Produkt und den Fertigungsverfahren nur Richtlinien an die Hand, mit denen ein System geplant und installiert werden soll.

Der Umfang und der Aufwand für ein Qualitätssicherungssystem hängt sehr stark vom jeweiligen Produkt und von den Anforderungen an dieses Produkt ab. Besonders hohe Anforderungen werden zum Beispiel an die sicherheitsrelevanten Bauteile in Kraftfahrzeugen gestellt.

Spezieller Prozeß

Der Begriff der „speziellen Prozesse" (vgl. Kap. 4.9.2 ISO 9001) ist für die Qualitätssicherung von besonderer Wichtigkeit, weil ein großer Teil der lasergestützten Verfahren spezielle Prozesse sind. Im folgenden wird eine an die Lasertechnik angepaßte Definition gegeben:

> *Spezielle Prozesse sind Verfahren, bei denen wichtige Qualitätsmerkmale des Produktes festgelegt oder beeinflußt werden, die nach der Bearbeitung nicht oder nur schwer meßbar sind. Solche Merkmale können sein: Dauerfestigkeit, Spannung, Zugfestigkeit, Härte, Korrosionsfestigkeit, etc.*

Das Laserstrahlschweißen und -härten fällt im großen Umfang in diese Kategorie, das Schneiden nur in einzelnen Fällen. Für die speziellen Prozesse fordert die Norm, daß bei Bedarf die Auswirkung der einzelnen Prozeßparameter auf das Bearbeitungsergebnis bekannt sein muß und die wichtigen Prozeßparameter durch Messung zu überwachen sind. Daraus ergibt sich für die Strahlüberwachung folgende Zielsetzung:

> *Die Strahlüberwachung zum Zweck der Qualitätssicherung muß alle Veränderungen des Strahls erfassen, die das Bearbeitungsergebnis negativ beeinflussen.*

Für die gerätetechnische Realisierung enthält die Norm keine Vorschriften, weil die notwendigen Überwachungssysteme nach den speziellen Erfordernissen des Prozesses ausgewählt werden müssen. In der Regel müssen neben den Strahlparametern weitere Prozeßparameter, wie Prozeß-gasdruck und -menge, überwacht und protokolliert werden.

Planung einer Strahlüberwachung

Für eine optimale Anpassung des Qualitätssicherungssystems an das Bearbeitungsproblem muß eine Reihe von Punkten bearbeitet werden, die im folgenden ausgeführt sind:

Schritt 1: Bestimmung kritischer Produkteigenschaften

In diesem Schritt sollen die kritischen Produkteigenschaften benannt werden, die von dem Bearbeitungsprozeß festgelegt oder beeinflußt werden und für die Qualität des Produktes entscheidend sind. Besonders wichtig sind die Produkteigenschaften, die im weiteren Fertigungsprozeß nicht mehr meßtechnisch verifiziert werden können.

Schritt 2: Parameteridentifikation

Aus dem Prozeßverlauf müssen mit denen aus Schritt 1 bekannten kritischen Eigenschaften die zugehörigen kritischen Strahleigenschaften ermittelt werden. Für diesen Zweck ist eine genaue Kenntnis des Bearbeitungsprozesses und seiner Abhängigkeiten notwendig.

Relevante Strahleigenschaften können sein:

- Geometrie
  - Position
  - Durchmesser
  - Divergenz

- Leistung
- Polarisation
- Symmetrie
- Fokussierkennzahl

Schritt 3: Parameterreduktion

Für jeden zu messenden Strahlparameter muß in der Regel ein eigenes Meßsystem installiert werden. Unter Kostengesichtspunkten ist es deshalb sinnvoll, die Anzahl der zu messenden Strahlparameter möglichst weit zu reduzieren. Dies kann durch die Untersuchung der folgenden Fragestellungen geschehen:

- Ist der Strahlparameter zeitlich stabil, weil er beispielsweise nur durch einen Umbau der Anlage geändert werden kann?

- Korreliert die Veränderung des Strahlparameters mit der Änderung anderer Strahlparameter?

Im ersten Fall kann auf eine Überwachung verzichtet werden, weil der Parameter zwar einen Einfluß auf das Bearbeitungsergebnis hat, sich aber nicht verändert. Im zweiten Fall kann der Parameter durch die Überwachung des korrelierenden Parameters indirekt überwacht werden.

Ein gutes Beispiel ist die Korrelation der Strahlleistung mit der Symmetrie der Intensitätsverteilung. Bei konstanten Betriebsparametern des Lasers hängt die Leistung empfindlich von der Intensitätsverteilung ab. Schon geringe Asymmetrien führen zu einer leicht zu beobachtenden Abnahme der Strahlleistung. Durch eine genaue Messung der Strahlleistung kann auf eine konstante Intensitätsverteilung zurückgeschlossen werden.

Als Ergebnis des Schrittes 4 stehen nun die Strahlparameter zur Verfügung, die bei dem gegebenen Prozeß überwacht werden müssen.

Schritt 4: Allgemeine Anforderungen an das Meßsystem:

Unabhängig von dem jeweiligen Fertigungsverfahren und denen in Schritt 3 zur Überwachung festgelegten Strahlparametern muß das auszuwählende Meßsystem eine Reihe äußerer Randbedingungen genügen, beispielsweise:

- keine Veränderung des Laserstrahls während der Materialbearbeitung

- vollständige Integration in die Bearbeitungsanlage und den Fertigungsprozeß

- Protokollierung der Meßergebnisse

- genaue und reproduzierbare Meßergebnisse

- an die Qualifikation des Personals angepaßte Bedienung

- Kosten-Nutzen-Relation.

Besonders der letzte Punkt kann erhebliche Bedeutung erlangen, weil die Kosten eines Fertigungsverfahrens auch von den Kosten der Qualitätssicherung beeinflußt werden. Die Auswahl der Fertigungsverfahren wird deshalb in Zukunft auch unter dem Aspekt geschehen, wie einfach die Qualitätssicherung realisiert werden kann.

Schritt 5: Auswahl des Überwachungssystems

Entsprechend den im Pflichtenheft festgelegten Aufgaben und Randbedingungen wird das am besten geeignete System ausgewählt.

Schritt 6: Installation und Inbetriebnahme

Bei der Inbetriebnahme des Überwachungssystems müssen die vorhandenen Informationen über den Prozeß und das Bearbeitungssystem dem Qualitätssicherungssystem in Form von Grenz- und Schwellwerten eingegeben werden. Die Einstellungen sind sorgfältig vorzunehmen, damit die kritischen Parameter empfindlich überwacht und Fehlalarme vermieden werden.

Zusammenfassung

Der zunehmende Einsatz von Qualitätssicherungssystemen stellt an die lasergestützten Fertigungsverfahren neue Anforderungen hinsichtlich der Prozeßsicherheit und der Überwachung. Damit der Laser gegenüber anderen Fertigungsverfahren nicht in einen Nachteil gerät, müssen die zur Qualitätssicherung notwendigen Verfahren und Meßgeräte verfügbar sein. Eine den Qualitätsanforderungen angepaßte Überwachung des Werkzeug „Laserstrahl" ist dabei unverzichtbar.

Literatur:

DIN/ISO 9000-9003,   Begriffe und Definitionen der Qualitätssicherung,
Qualitätssicherungssysteme

Measurement of Industry Laser Polarization

Joachim Franek, Reinhard Kramer, Primes GmbH
Adolf-Damaschke-Str. 14, 64319 Pfungstadt, Germany

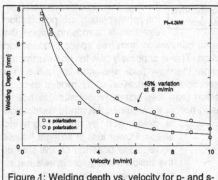

Figure 1: Welding depth vs. velocity for p- and s-polarization /1/

1. Introduction

Polarization dependent effects during material processing with CO_2-lasers are well known. An example for this is shown in figure 1: the dependence of welding depth as function of the welding velocity. Two curves are seen for p- and s-polarization (plane of polarization parallel (p) and perpendicular (s) to the plane of incidence). As can be seen, especially for high welding velocities (> 3 m/min) there is a considerable difference between the welding depths for s- and p-polarization.

2. Influence of Beam Quality

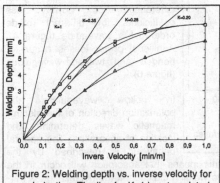

Figure 2: Welding depth vs. inverse velocity for p-polarization. The line for K=1 is extrapolated from the other beam qualites and is the theoretical limit /1/.

Figure 3: Welding cross sections for different beam qualities
left: K=0.28, right K=0.35
Heat affected zones are shown also

If energy losses due to heat conduction can be neglected - this can usually be done for high velocities - there is a simple formula for welding depth calculation. Welding Depth depends in a linear way on laser power and is inversly related to focus radius and welding velocity /1/. Drawing welding depth against inverse velocity shows lines with slopes that are proportional to beam quality. From three known beam qualities the theoretical limit for a beam quality of 1.0 (TEM_{00}-Mode) is extrapolated.

3. Theoretical Background for Polarization Dependent Welding Effects

The following reflections explain the influence of polarization during laser beam welding. A laser beam is an electromagnetic wave which popagates in free space as TEM wave. The keyhole that is formed during laser beam welding changes the boundary conditions for the electromagnetic wave. For conductive boundaries, the waves will spread as TE-, TM- and hybrid-waves. This has been sudied extensively for hollow waveguides /2,3/.

638

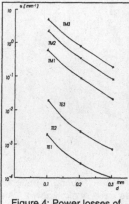

Figure 4: Power losses of straight hollow waveguides for different waveguided modes /from 2/

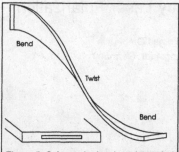

Figure 5: Schematic drawing of bended and twisted hollow waveguides /from 3/

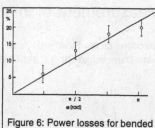

Figure 6: Power losses for bended hollow waveguides /from 3/

Taking this into consideration, the preservation of polarization direction is surprising, since the modes are not preserved during conversion from free space propagation to hollow waveguide propagation. This is especially true for non symmetric hollow waveguides. For example the waveguide shown in figure 5 with a rectangle cross section preserves the polarization of its modes through bends and twists. The transition from free space propagation to hollow waveguide propagation is explained by antennae theories. In hollow waveguides the mode distribution is determined by their geometric shape. Power loss in straight hollow waveguides depends on the size of the hollow waveguide in relation to the wavelength. This results in the curves shown in figure 4. When drawn on log-log paper the graph will show straight lines /4/. Remarkable are the great differences of power loss coefficients for different mode orders of a waveguide (figure 4). Additional power losses occur for bended and twisted waveguides (figure 6).

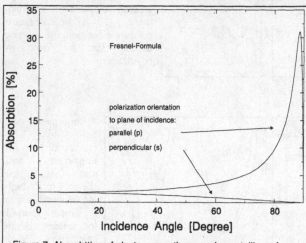

Figure 7: Absorbttion of electromagnetic power by metallic surfaces vs incident angle for p- and s-polarizattion

In hollow waveguides the polarization direction of the electro-magnetic waves determines the location of absorption. The general behaviour is described by the fresnel formulas. With regards to laser beam welding this means that absorption will occur in the anterior part of the keyhole if the polarization is parallel to welding direction; if the polarization is per-pendicular, absorption will occur at the side-wall of the keyhole. If welding velocity is low, the diameter of the welding capillary will be low against the dimension of the melting pool. In this case the heat conduction from the wall of the keyhole to the border of the melting pool does not depend much on the polarization direction since the heat conductivity differs little between the area of laser power absorption and the border of the melting pool. The situation is very different for high speed welding (v < 3 m/min). Now the thickness of molten material is small compared to keyhole diameter. Heat transfer from anterior part of the keyhole to the border of molten / solid material is high for parallel polarization. In the case of perpendicular polarization the side-walls of the keyhole are heated. The thermal transfer resistence from this heat source to the anterior part of the border between molten and solid material is high, whereas to the side there is a small heat conductivity resistance. This results in wider and less deep welding cross sections.

From the viewpoint of electromagnetic theory of waveguides the welding process shows as a method to analyse waveguide modes. High order waveguide modes originated during transformation from free space propagation to propagation with conductive boundary conditions have high power losses in the upper part of the keyhole. The power loss for low order modes is low for this case (see figure 4). This results in deep penetration of low order modes and explains the usually obtained V-shapes.

At high velocities the polarization dependent welding depth varies widely with polarization direction. As can be seen in figure 1, at 6 m/min welding velocity the relative relation between p- and s-polarization comes close to 50% of the mean value between the two. In this case the polarization properties of the laser and the beam path are relevant for the welding performance. To date, the polarization properties of industrial CO_2-Lasers are not well known. It is assumed that the beam from folded resonators is polarized linearly, and that this is not time dependent. This linear polarization is changed to circular by phase retarders, which must be hit at an angle of 45°. Other (bending) mirrors must not introduce phase shifts, so the circular polarization is preserved.

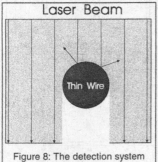

Figure 8: The detection system consists of thin wires

4. Detection System

For quality control of material processing machines we developed a measurement device to monitor the polarization of industrial CO_2-lasers. The detection device consists of thin wires. In figure 8, a cross section of a thin wire is shown. A CO_2-laser beam is directed onto these wires. Laser power absorbed by the wires is locally dependent on incident angle. Especially on the sides of the wires (high incident angles), the absorbed power depends strongly on the polarization direction as shown by Fresnel-formulas (figure 7). By changing the orientation between the wire and the laser beam (figure 10) power absorbed by the wire changes. Through the temperature dependent resistence change this absorbed power is detected by the

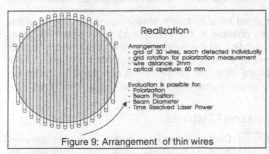

Figure 9: Arrangement of thin wires

measurement electronics. In our device we have arranged 30 wires with a distance of 2 mm between each. This gives an optical aperture of 60 mm (figure 9). To detect the CO_2-laser beam polarization, the grid of wire rotates and obtains a sine curve in case of non circular time independent polarization (figure 11). The zero crossing point of this sine curve indicates the polarization direction of the laser beam, whereas the amplitude of the sine curve indicates the polarization depth (100%=linear, 0%=circular). This measurement principle is especially useful to check circular polarization. For the case of rotational beam symmetry no sine curve (amplitude=0) is expected.

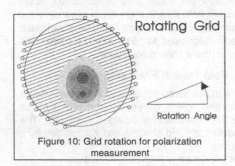

Figure 10: Grid rotation for polarization measurement

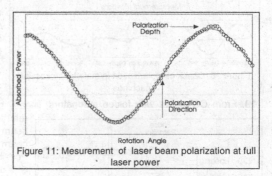

Figure 11: Mesurement of laser beam polarization at full laser power

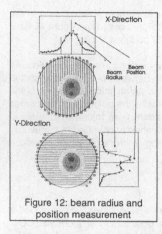

Figure 12: beam radius and position measurement

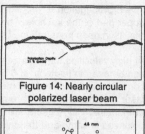

Figure 13: Nearly linear polarized laser beam

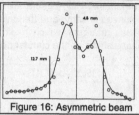

Figure 14: Nearly circular polarized laser beam

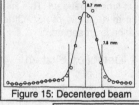

Figure 15: Decentered beam

Figure 16: Asymmetric beam

5. Typical measurement results

Typical mesurement results are shown in figures 13 -17. A nearly linearly polarized beam is indicated by the measurement in figure 13, a nearly circular in figure 14. As can be seen in figure 14 -17, there are additional measurement data collected for beam diagnostic. Each wire is read out individually. The measured data show beam cross sections for x- and y-directions

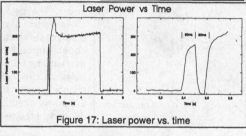

Laser Power vs Time

Figure 17: Laser power vs. time

(figure 12). Beam position and radius are calculated by a microprocessor. Figures 15 and 16 show decentered and asymmetric beam profiles. Also possible is the detection of laser power vs. time. Figure 17 shows a laser pulse of approx. 3,5 s duration. At the beginning of this pulse there is a drop out of about 60ms. After this, the laser power rises up to 20 % over the cw level. After about 1 s, stable laser power output for welding purposes is obtained. With this behaviour it is convenient to switch on the discharge for about 2 s and then to open the beam shutter without disruption of the discharge. The time resolution for this picture is 10 ms.

6. Check of Polarization of Material Processing Machines

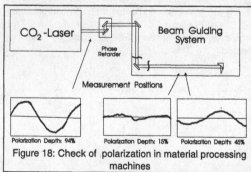

Figure 18: Check of polarization in material processing machines

During material processing of metals direction dependent effects have been observed. To check the polarization of this machine we first measured in front of the focussing optics: 45% polarization depth were detected. Second we checked the linear polarization of the laser source: 95% polarazation depth were ok. After changing mirror by mirror suddenly polarization depth decreased to 15%, and the mirror with faulty phase shifting was found.

7. Polarization Fluctuations

The experimental setup used is shown in figure 19. From CO_2-laser with folded resonators, linearly polarized light is generated, and with a phase retarder the polarization is changed to circular. The beam passes the grid of wires and is absorbed by a beam dump.

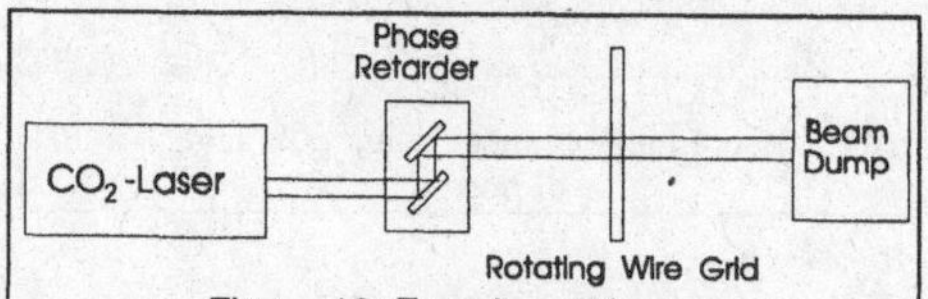

Possible causes for polarization fluctuations are:
- unstable plane of polarization
- transversal modes with different polarization
- longitudinal modes with different polarization
- optical feedback from material processing zone

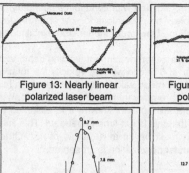

Figure 19: Experimental setup

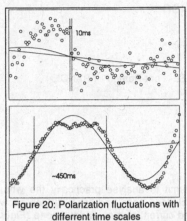

Figure 20: Polarization fluctuations with differrent time scales

The obtained different time scales for polarization fluctuations are seen in figures 20 and 21. The time between the two peaks in figure 21 is about 200 ms, in figure 20: 10 ms and 450 ms. The polarization fluctuation amplitude depends on the laser power. Figure 21 shows a nearly linear increase of polarization fluctuation amplitude with laser power for a CO_2-

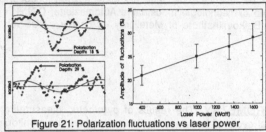

Figure 21: Polarization fluctuations vs laser power

laser with a nominal output power of 1750 Watt. This is explained by us in the following way: if the discharge is heated by the input power, turbulence of laser gas flow increases and with this the fluctuations of the polarization plane of the laser beam. The condition of 45° incidence for changing linear to circular polarization on the phase retarder is violated and elliptical polarization with amplitudes > 0 is measured. These fluctuations are not small effects with no relevance for material processing. Usually obtained values are 20% - 30% polarization depth. With this values the variation in welding depth can be calculated, assuming a welding velocity of 6 m/min with a difference of about 50% in welding depth between s- and p-polarization, and assuming circular polarization welding depth is half between s- and p-polarization. With fluctuations of about 30%, welding depth fluctuations up to 10% can be estimated. For assessment of welding depth variations, polarization fluctuations must be taken into consideration together with laser power variations, focus postion, shielding gas flow, etc.

8. Summary

Polarization determines performance of material processing. In this paper we have discussed polarization dependent beam coupling into the workpiece for welding of metals (waveguide model). For quality control purposes we have developed a measurement divice called "MultiMonitor" for CO_2-laser beam diagnostics. This device can check polarization porperties at laser power up to 5 kW as well as beam position, radius, shape and laser power vs time. Typical measurements of linear, circular and elliptical polarized beams are shown in figures. In industrial material processing machines, polarization is usually achieved linear by folded resonators and changed to circular by phase retarders.

Origins for perturbations are:
- fluctuations of polarization plane (probably power dependent)
- phase shift of phase retarders differs usually +/-6%
- phase shift of coated mirrors

Polarization measurement of processing machines shows two important fields:
- if direction dependent effects occur, check polarization properties (linear, circular)
- if irregular results are obtained, check polarization fluctuations (time scale, amplitude)

/1/ Joachim Franek, Keming Du, Silke Pflüger, Ralf Imhoff, Peter Loosen
   Comparison of Welding Results with Stable and Unstable Resonators
   Proceedings Gas Flow and Chemical Lasers, 1990,Madrid, Spain, SPIE Volume 1397, 791-795
/2/ Hans Opower
   Die physikalischen Grundlagen des Schneidens von Metallen mit CO_2-Hochleistungslasern
   Laser und Optoelektronik 1982, Nr. 2, 47ff
/3/ Elsa Garmire, T. McMahon, M. Bass
   Flexible Infrared Waveguides for High-Power Transmission
   IEEE Journal of Quantum Optics, Vol. QE-16, NO. 1, January 1980, 23ff
/4/ E. A. J. Margatili, R.A. Schmeltzer
   Hollow Metallic and Dielectric Waveguides for Long Distance Optical Transmission and Lasers
   The Bell System Technical Journal, July 1961, 1783ff

Heat and Mass Transfer in Pulse Laser Action

I. Smurov, G. Flamant *, L. Covelli **, A. Lashin **, M. Ignatiev***

ENISE - Saint-Etienne - FRANCE
*Institut de Science et de Génie des Matériaux et Procédés, C.N.R.S. - Font-Romeu - FRANCE
**Istituto di Tecnologie Industriali e Automazione, C.N.R. - Milano - ITALY
***A.A.Baikov Institute of Metallurgy - Moscow - RUSSIA

PYROMETRY. The results of pyrometric measurements permit to analyse practically the whole thermal cycle of laser action. On-line temperature monitoring of cladded bead and base surface allows to determine maximum temperature, heating and cooling rates and current size of cladded zone (Figure 1a). The expulsed metal particles (droplets) and plume can cause a high degree "noise" of pyrometer output signal in laser welding process (Figure 1b). These disturbances may be eliminated by signal filtering technique and by the use of average temperature values, as a result weld seam size may be controlled. The developed pyrometer can be used for process control on line in continuous, pulse and pulse-periodic laser machining.

MELTING/SOLIDIFICATION. In 2D model of melting/solidification surface evaporation is assumed to be week enough to neglect the movement of irradiated surface but significant enough to be taken into account as a heat losses [1]. The marked difference between the results obtained on the basis of the models taking into account radiation heat losses only, and both: evaporation and radiation losses is presented in Figure 2. The influence of the thickness of irradiated slab on melting/solidification dynamics is demonstrated on the example of 1 mm and 0.27 mm slabs. For all the above mentioned cases the parameters of laser action are as follows: energy density flux $q_0=1.5 \cdot 10^5$ W/cm^2, coefficient of Gaussian distribution $k=10^2$ cm^{-2}, pulse duration $t=7$ ms, material - Titanium.

CONVECTION. The main influencing factors on convective mass transfer (including spatial distribution of surface temperature, ambient gas atmosphere, angle of incidence of laser beam, etc) are analysed. The multi-cells flow pattern for melt convection is shown and explained on the basis of surface tension phenomenon. A 2-D model of transient convective mass transfer in laser alloying is developed [2,3]. As an example typical stages of convective mass transfer in case of positive value of surface tension coefficient $d\sigma/dT$ are presented in Figure 3.

COMBINED ACTION. The peculiarities of melt hydrodynamics in combined continuous CO_2 + pulse-periodic Nd: YAG laser action are illustrated by alloying of Ti in the flow of nitrogen (Figure 4). The corresponding parameters are as follows: power 1750 W, velocity 0.2 m/min (CO_2 laser); pulse energy 25 J, pulse duration 10 ms, frequency 4 Hz (Nd: YAG laser). High power continuous action produce the melted layer and high energy density flux in pulse-periodic action intensifies its remixing.

REFERENCES

1. Smurov I., Lashin A., Poli M. "Peculiarities of pulse laser melting: Influence of surface evaporation". to be published in the Proceedings of International Congress on Applications of Lasers and Electro-Optics, October 1992 (ICALEO'92), Florida, USA.
2. Smurov I., Covelli L. "Pulse laser alloying: Theory, experiment". Ibid.
3. Smurov I., Covelli L., Tagirov K., Aksenov L. "Peculiarities of pulse laser alloying: Influence of the beam spatial distribution". J. Appl. Phys., Vol. 71, No. 7, (1992), pp. 3147-2158.

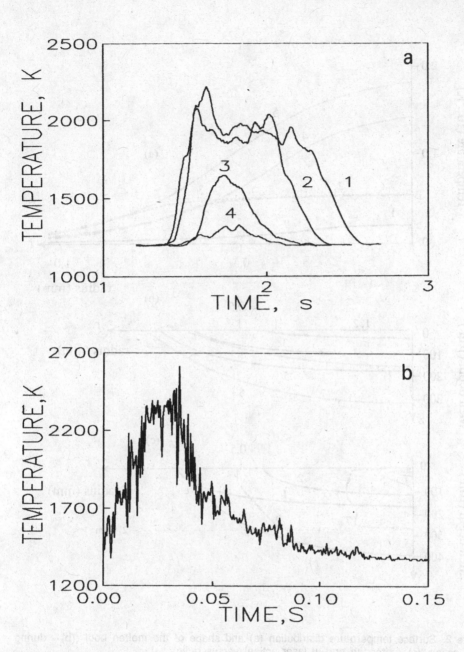

Figure 1. The thermocycles for laser: (a) cladding, bead size -3.0 mm; (b) welding of mild steel. Laser power: (a) 1.9 kW; (b) 2.3 kW. Beam traverse speed: (a) 300 mm/min; (b) 1000 mm/min. Distance from the trace centre: (a) curve 1-0 mm, 2-1.4 mm, 3-1.6 mm, 4-1.8 mm; (b) 0.9 mm.

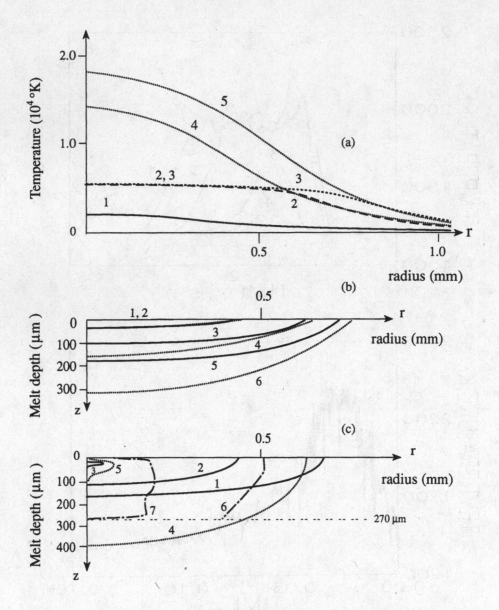

Figure 2. Surface temperature distribution (a) and shape of the molten pool ((b) - during laser action, (c) - after the end of laser action) versus radius.

(a) curve 1 - $t=3 \cdot 10^{-2}$ ms, curves 2,4 - 2 ms, curves 3,5 - 7 ms; for the curves 1,3,4 evaporation is taken into account, curves 1,4,5 - evaporation is neglected.

(b) curves 1,2 - $t=0.2$ ms, curves 3,4 - 2 ms, curves 5,6 - 7 ms; for the curves 1,3,5 evaporation is taken into account, curves 2,4,6 - evaporation is neglected.

(c) curve 1 - 8 ms, 2 - 14 ms, 3 - 19 ms, 4 - 14 ms, 5 - 44 ms, 6 - 20 ms, 7 - 40 ms; curves 1,2,3,6,7 - surface evaporation is taken into account; curves 4,5 - evaporation is neglected; curves 1-5 - thickness of the slab equals 1 mm; 6,7 - thickness equals 270 μm.

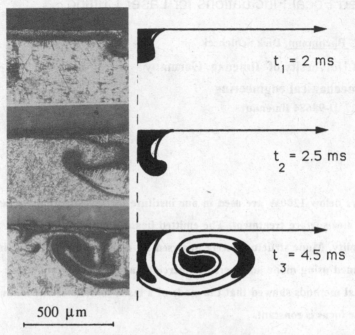

500 μm

Figure 3. The propagation of admixture in pulse laser alloying from pre- deposited coatings for positive value of surface tension coefficient. Numerical simulation: $q_0=5\ 10^4$ W/cm^2, $k=50$ cm^{-2}, the corresponding moments of time: 2, 2.5, 4.5 ms. Experiment: graphite coating 10 μm thickness on Ti, pulse Nd:YAG laser.

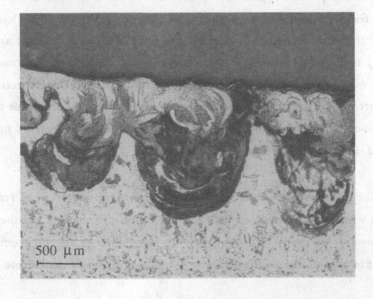

500 μm

Figure 4. Melt hydrodynamics in combined continuous CO_2 + pulse-periodic Nd: YAG laser action. Illustration by alloying of Ti in the flow of nitrogen.

Minimized Focal Fluctuations for Laser Cutting

Matthias E. Bachmann, Dirk Schlebeck

Technical University of Ilmenau, Germany
Dept. of mechanical engineering
Postfach 327, D-98684 Ilmenau

Abstract

CO_2- Lasers below 1200W are used in our institute for investigations in laser cutting, hardening and surface treatment. The emitted beam quality was analysed to improve cutting quality. Mode structures were analysed by using screen burnings in plexiglass and confirmed using mode analysers commercially available. Investigations using these conventional methods showed that the mode of a laser consists of one main mode. The shape of the focus is constant.

Newest experiments with fast mode analysers showed that the real focus geometry is a result of interference and superposition of the main mode and various spurious modes of weaker intensity oscillating in the resonator. The vincinity conditions in the focus are changing frequently and therefore the mode interferences are fluctuating. That is the reason why the focus diameter is oscillating with some Megahertz. It is not as stable as commonly regarded. The integral of the intensity distribution over time measured in the focus may be represented by a Gaussian distribution, too. These effects have not been detected unil now. With an improved resonator geometry it is possible to prevent spurious modes from oscillating, interferences are reduced and we reached nearly doubeled feed rates with reduced peak- to- valley- depth.

To describe the quality of a resonator the K-factor was introduced. K->1 represents a resonator with only one mode oscillating. Interferences are not able and the focus remains constant. We propose to use not only the laser power to describe treatment processes but to expand it to an effective laser power including laser quality and resonator output.

K-factor

Hence it's easily seen that the cutting or welding quality should be affected. A K-factor has been introduced for simplified description of beam quality. *Muys* et al. introduced it as the relation between theoretical expected and experimental found laser foci of given systems. We prefer to consider not only the beam waist but also the relative aperture. The K-factor then may be described as $K = \frac{w_{0,theor} * \Theta_{theor}}{w_{0,ex} * \Theta_{ex}}$. It is inverse proportional to the mode number and will be exactly 1.0 in case of an ideal stable TEM_{00} without interferences of spurious modes. The picture gives an overview on K-factors of various lasers for material processing in dependence on their output. Altough the manufacturers state that their lasers oscillate in TEM_{00} their K-factors vary between 0.2 and 0.6. This can be explained, with modes of higher orders ($TEM_{5\,0}$ to $TEM_{20\,0}$) able to oscillate because of reduced resonator quality and interfere in the focus.

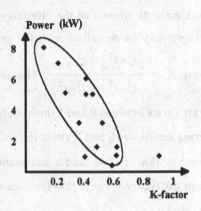

When introducing the K-factor into the analysis to describe the cutting process it correlates with caustic surface and Rayleigh- length. So it is of a very strong influence on treatment qualtiy and feed rate taking into account that the diameter of a bundle of rays observed will be doubled with the distance $z = z_R$ and that intensity will be reduced to half of the value measured in the focus. In case of a K-factor near 1 only one mode is able to oscillate in the resonator. The focal fluctuations caused are weaker and the treatment becomes more stable. By changing the properties of a resonator the diffraction losses for high order modes (i,j) $K_{diff.} = \frac{4\pi(8\pi N)^{2i+j+1}}{p!(i+j+1)!}e^{-4\pi N}$ are influenced. The Fresnel-number $N = \frac{a^2}{\lambda L}$ was changed by changing aperture diameter between 0.3 and 2 a_0. It was possible to change the number of modes in the resonator over a wide range, to induce interferences between the modes and to vary the K-factor between 0.1 and 0.9 in this way.

Mode interferences

Modes oscillating in the resonator are coherent and able to interfere.

If two modes (i,j) represented by $Y_{i,j} = A_{i,j}\cos(\omega t + \alpha_{i,j})$ interfere the resulting wave will satisfy the equation $Y = A\cos(\omega t + \alpha)$. The generated maxima and minima of the interference fringes may be described with the contrast $K = \dfrac{\left(A_i^2 + A_j^2\right) - \left(A_i^2 - A_j^2\right)}{\left(A_i^2 + A_j^2\right) + \left(A_i^2 - A_j^2\right)}$. The con-

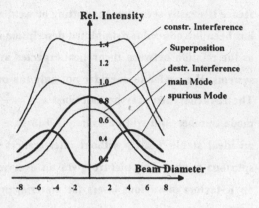

trast varies between 0 and 1 theoretically. The resulting intensity generated by the interfering intensities I_i and I_j yields to $I = I_i^2 + I_j^2 + 2\sqrt{I_i I_j}\ \cos$. In case of a path difference $\delta = 2k\pi$ $\cos\delta = 1$ and a maximum is generated with $I = I_i^2 + I_j^2 + 2\sqrt{I_i I_j}$. If the path difference is equal $(2k+1)\pi$ because of $\cos\delta = 0$ a minimum with $I = I_i^2 + I_j^2$ is generated.

Even if we assume the intensity of the spuriuos mode to be one fifth of the main mode's intensity (assumed as 1) the resulting interference fringes show a maximum with a relative intensity of 1.93 while the minimum is 1.04. The intensity resulting on non- coherent superposition is 1.2.

Taking the temporal behaviour of these interferences into consideration the result will be a fluctuation of focal intensity depending on the time. First investigations led to the result that the analysed mode interferences cause fluctuations of the beam diameter between 0.5 and $2w_0$ with a frequency of some megahertz.

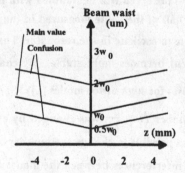

Treatment results

Following our theory feed rate and peak- to- valley- depth depend strongly on the effective power defined as $P_{wirk} = P_{Laser} * K$. Therefore the behaviour of v and Rz is proportional to K and P_{Laser}. The treatment results described illustrate the possibility of improving cutting parameters by increasing the K-factor while laser power remains constant. In case of 600W cw laser power and a K of 0.5 it is possible to expect an effective power of 300W, resulting a

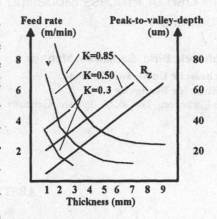

cutting velocity of 7m/min. Increasing K to 0.85 led to 510W effective power and a cutting velocity of 12m/min with diminished peak- to- valley- depth.

References

[1] *Berger, M. et al.*   "Optics for high- power CO_2- lasers"
     SPIE 1983, Proc.,52-58

[2] *Brunner, W.; Junge, K.*   "Lasertechnik"
     Fachbuchverlag Leipzig 1987

[3] *Dickmann, K. et al.*   "Fein- und Mikrobohren mit Nd:YAG- Laser hoher
     Strahlqualität"
     Laser und Optoelekronik 23(6)1991, 56-62

[4] *Muys, P.; Sona, P.*   "Beam quality evaluation of high- power- lasers"
     Laser Magazin

[5] *Nuss, R; Geiger, M.*   "Verfahrensgrundlagen zum Laserstrahlschneiden"
     Laser und Optoelektronik 21(2)1988, 88-93

The Use of Process Modelling of Laser Hardening in Close Limit Production

Schlebeck, Dirk; Bachmann, Matthias
Technische Universität Ilmenau,
Institut für Fertigung
Am Ehrenberg, D- O6300 Ilmenau, Germany

ABSTRACT

Though laser hardening moved to production about twenty years ago that technology is still of subordinated importance. Reasons for that lack of acceptance are not only high costs for investments but also missing knowledge about that technology. Computer aided inquiry of technological data are state of the art in the laser centres and universities. Most of that programs have been made only for internal use. With the help of our program the designer shall get suitable advise in material selection and designing for close limit production.

CLASSIFICATION

A profund knowledge about the thermal field induced by laser beam in the workpiece is essentially to a computer aided optimization of parameters. We developed fast analytical procedures to solve the heat diffusion equations to find out the relation between process parameters and geometry of hardened tracks. The power density distribution often may assumed as a Gaussian or rectangular one. So it is possible to reduce computing time.
More than 80% of the pieces we have hardened may be covered with a classification valid for semi- infinitive, quarter or other parts of simply shaped geometries. Using that classification key we can seperate our orders into groups of repetition or similar parts. Repetition parts are in the same class, have identical geometries and are made of materials with similar behaviour while laser treated. In that case same parameter set lead to comparable machining results. The field of parameters is limited by the following conditions- based on the setup of the laser machine and material`s properties:

- maximum power; power density distribution and beam diameter
- lowest scanning velocity limit to realise the quenching of the material
- ratio between austeniting and melting temperature
- hardening depth limited by thermal diffusion and part`s geometry

We developed the simulation program during the last two years. It allows to predict the shape of the hardened tracks in dependence of process parameters, various material properties and the simplified geometry of the hardened areas fast and convenient. An interactive material data bench including temperature dependent properties and different methods calculating the important austentiting temperature is part of our solution. In practical work it is helpful to save time and money because minimizing test runs.

SCHEME FOR COMPUTING

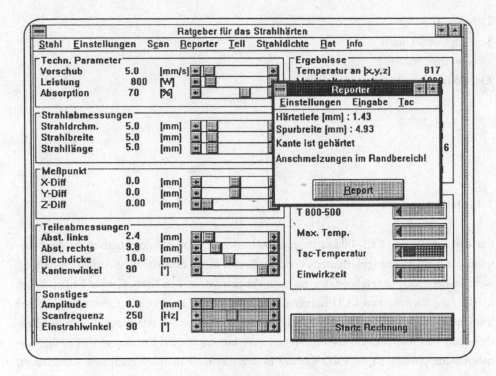

Ratgeber für das Strahlhärten
Stahl Einstellungen Scan Reporter Tell Strahldichte Rat Info

Techn. Parameter
Vorschub   5.0   [mm/s]
Leistung   800   [W]
Absorption   70   [%]

Strahlabmessungen
Strahldrchm.   5.0   [mm]
Strahlbreite   5.0   [mm]
Strahllänge   5.0   [mm]

Meßpunkt
X-Diff   0.0   [mm]
Y-Diff   0.0   [mm]
Z-Diff   0.00   [mm]

Teileabmessungen
Abst. links   2.4   [mm]
Abst. rechts   9.8   [mm]
Blechdicke   10.0   [mm]
Kantenwinkel   90   [°]

Sonstiges
Amplitude   0.0   [mm]
Scanfrequenz   250   [Hz]
Einstrahlwinkel 90   [°]

Ergebnisse
Temperatur an [x,y,z]   817

Reporter
Einstellungen Eingabe Tac
Härtetiefe [mm] : 1.43
Spurbreite [mm] : 4.93
Kante ist gehärtet
Anschmelzungen im Randbereich!

Report

T 800-500
Max. Temp.
Tac-Temperatur
Einwirkzeit

Starte Rechnung

RESULTS

Target was refing the analytical and empirical required knowledge for making it exploitable for not specially in laser technology educated personal too. The designer receives now with the beginning of his work informations about choice of materials, design suitable for laser treatment and technological realizability of his ideas. So it is possible to include requirements of laser technology at an very early state in the developing process for new products to save costs.

Although the laser treatment of functional areas leads to a considerable improvement of their properties, marketing activities at the customer`s markets are to be monitored hardly. Competetive advertising and sales promotion should emphasize laser technology as promoting factor to separate high quality tools from those of standard degree. The figure on the right illustrates the average price difference between normally and addional at the functional areas hardened tools.

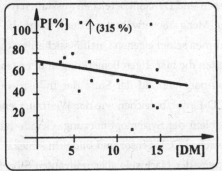

Systemanalyse an CCD-Bildaufnahmeanordnungen

U. Schmidt
Jena-Optronik GmbH
Göschwitzer Straße 33
O-6905 Jena-Göschwitz

1. Einleitung

Für den Einsatz von CCD-Bildaufnahmesystemen im Hochtechnologiebereich ist die Beurteilung des Übertragungsverhaltens von Sensor und Kanal eine wichtige Aufgabenstellung, welcher der CCD-Kamerahersteller bisher nicht gerecht werden kann. Gerade heute, da am Markt eine Vielzahl von CCD-Kameras angeboten wird, kommt einer Bewertung dieser Systeme auf Eignung für den jeweiligen Anwendungsfall große Bedeutung zu. Für den Produzenten von wissenschaftlich-technischen oder kommerziellen Geräten mit CCD-Bildaufnahmekomponenten ist die CCD-Kamera in den meisten Fällen lediglich eine Kaufeinheit, welche bestimmten Forderungen genügen muß. Dabei zeigt die Erfahrung, daß es für CCD-Anwender im beschriebenen Sinne äußerst schwierig ist, ein geeignetes Kamerasystem auszuwählen, bzw. dieses System in seiner Leistungsfähigkeit einzuschätzen. Der Anwender von CCD-Komponenten kann sich damit nur anhand von Herstellerangaben orientieren und ist auf deren Richtigkeit angewiesen. Allein die Tatsache, daß CCD-Kamerahersteller ihre Produktkataloge mit großzügigen und vor allem werbewirksamen Signal/Rausch-Verhältnissen und Dynamikbereichen ausstatten (wobei häufig weder Meßmethode noch Definition der Meßgröße erwähnt werden), zeigt die Notwendigkeit, daß der CCD-Anwender im Rahmen seiner eigenen Qualitätssicherung diese Parameter nachvollziehen muß.

Gelten die bisherigen Bemerkungen in erster Linie für den Einsatz von vorgefertigten CCD-Komponenten auf der Stufe der Industrieautomatisierung, so stellt sich in so exklusiven CCD-Einsatzbereichen wie der Weltraumsensorik und der messenden Bildverarbeitung das Problem der Systemoptimierung. Gerade hier kommt es auf eine gelungene Abstimmung zwischen CCD-Sensor und Bildaufnahmekanal an. Wesentliche Voraussetzungen sind dabei erstens der Nachweis aller relevanten Störprozesse im System und zweitens die Kenntnis der Übertragungseigenschaften der am Bildaufnahmeprozeß beteiligten Teilsysteme.

2. Der CCD-Bildaufnahmekanal aus signal- und systemtheoretischer Sicht

CCD-Bildaufnahmesysteme mit digitaler Schnittstelle wandeln das orts- und wertkontinu-ierliche 2d-Leuchtdichtesignal der realen Bildszene im optoelektronischen Kanal in ein dis-kretes 2d-Zahlenfeld um. Auf dieses Zahlenfeld im Speicher eines Digitalsystems (Framegrabber o.ä.) setzt die Algorithmik der digitalen Bildverarbeitung auf. Gerade im Bereich der messenden Bildverarbeitung (Radiometrie, Photogrammetrie) stellt sich die Frage nach dem Informationsgehalt der digitalisierten Bilddaten. Aus signaltheoretischer Sicht sind damit Parameter wie Signal/Rauschverhältnis, Dynamik, Linearität und me-trische Genauigkeit des CCD-Bildaufnahmesystems angesprochen. Für eine diesbezügliche Systemanalyse als unmittelbare Voraussetzung für die Optimierung der Bildaufnahmean-ordnung ist die Definition eines signal- und systemtheoretischen Modells des Gesamt-systems erforderlich. Abbildung 1 verdeutlicht die Bildsignalwandlung vom orts- und wert-kontinuierlichen 2d-Leuchtdichtesignal zum diskreten 2d-Zahlenfeld im Digitalspeicher.

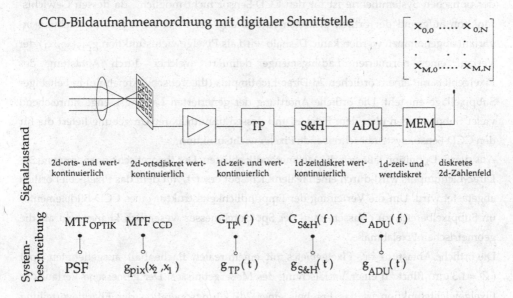

Abb. 1 Blockschaltbild eines CCD-Bildaufnahmesystems

Die höchste Stufe der Abstraktion zur Beschreibung von komplexen Systemen wird durch systemtheoretisches Herangehen erreicht. Dabei rückt die Physik der Signale und Systeme in den Hintergrund. Baugruppen wie Optik, Sensor und Kanal werden gleichermaßen als Systeme mit bestimmten Übertragungseigenschaften aufgefaßt. In Abbildung 1 sind die Gewichts- und Übertragungsfunktionen der Teilsysteme der CCD-Bildaufnahmeanordnung aufgeführt.

Da gerade im Bereich der visuellen Weltraumsensorik (star tracker, remote sensing) ein optimiertes CCD-Systemdesign notwendig ist, wurde bei der Jena-Optronik GmbH ein optoelektronisches modulares Meßsystem entwickelt [SCZ92], welches die Vermessung aller Teilkomponenten des CCD-Bildaufnahmekanals ermöglicht. Mit dem theoretischen Hintergrund der zwei- und eindimensionalen Systemtheorie und diesem Meßsystem kann eine radiometrische, photogrammetrische sowie dynamische Verifikation optoelektronischer CCD-Sensorsysteme durchgeführt werden.

3. Meßtechnische Erfassung der Pixelgewichtsfunktion

Die Gewichtsfunktion spielt eine zentrale Rolle bei der Beschreibung von signalverarbeitenden Systemen. Im systemtheoretischen Sinn stellt sie die Systemantwort auf einen Einheitsstoß zum Zeitpunkt t_0 bzw. am Ort (x_{02}, x_{01}) dar. In optischen Systemen findet sie ihre Entsprechung in der PSF (point spread function). Eine formale Übernahme dieser Definition aus der optischen Systemtheorie ist für den CCD-Sensor nicht möglich, da dessen Gewichtsfunktion aufgrund der örtlich diskreten Anordnung der Pixel nicht als verschiebungsinvariant angenommen werden kann. Deshalb wird als Pixelgewichtsfunktion $g_{pix}(x_2, x_1)$ der Verlauf einer normierten Ladungsmenge definiert, welcher durch Abtastung des Pixelgebiets mit einem örtlichen 2d-Dirac-Lichtimpuls (theoretische Annahme) in beliebiger Subpixellage entsteht. Die örtliche Anordung der generierten Ladungsmenge hinreichend vieler Subpixellagen über dem Pixelort und dessen unmittelbarer Umgebung liefert die für den CCD-Sensor systembeschreibende Pixelgewichtsfunktion.

Aus dieser Definition läßt sich die Meßmethodik ableiten. Der theoretisch angenommene 2d-Dirac-Lichtimpuls wird durch einen realen Lichtspot ersetzt, mit dem das Pixelgebiet örtlich abgetastet wird. Um die Verteilung der Empfindlichkeitsstruktur eines CCD-Bildelementes im Subpixelbereich zu erfassen, muß der Spotdurchmesser wesentlich kleiner sein als die geometrischen Pixelabmaße.

Die örtliche Abtastung des Pixelgebietes mit einem realen flächenhaft ausgedehnten Spot ($\varnothing = 1.3\,\mu m$) führt zu einer Verfälschung des Meßergebnisses. Der gemessene Verlauf der Pixelgewichtsfunktion ist das Ergebnis einer 2d-Faltungsoperation der Energieverteilung der Spotabbildung mit der tatsächlichen Pixelgewichtsfunktion. Demzufolge muß der gemessene 2d-Datensatz mit der 2d-Energieverteilung der Spotabbildung (durch Optikrechner und Messung verifiziert) entfaltet werden. Abbildung 2 zeigt das Resultat dieser Operation für die CCD-Matrix TH7895A in perspektivischer 3d-Darstellung.

Parameter der Messung:

Pixelgröße	: $19\,\mu m \cdot 19\,\mu m$
Spotdurchmesser	: $1.3\,\mu m$
Spot-Positionierschrittweite	: $1.3\,\mu m$
monochromatisches Licht	: $\lambda = 546\,nm$

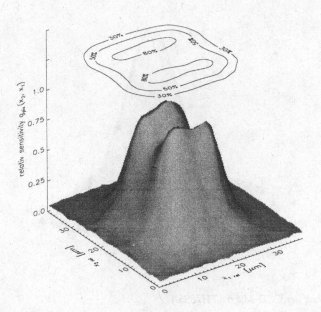

Abb. 2 gemessener Verlauf einer 2d-Pixelgewichtsfunktion

Bemerkenswert ist, daß das Pixel der Thomson CCD TH7895A zwei empfindliche Gebiete enthält. Dies wird durch die 80%-igen Höhenlinien in Abbildung 2 verdeutlicht. Der gemessene Verlauf der Subpixelempfindlichkeit korrespondiert direkt mit dem Halbleiterlayout der Pixelstruktur. Der Empfindlichkeitseinbruch in Pixelmitte ist auf das überlagerte Polysilizium-Gate der Φ_{V1}-Vertikaltransportphase zurückzuführen.

Die meßtechnische Ermittlung der Pixelgewichtsfunktion ist eine wesentliche Voraussetzung für den Einsatz modellgesteuerter Subpixelalgorithmen in der messenden Bildverarbeitung (Photogrammetrie).

Durch das Ausführen der 2d-Fouriertransformation auf die Pixelgewichtsfunktion läßt sich die MTF (modulation transfer function) des Pixels und damit des CCD-Bauelementes ermitteln. Abbildung 3 zeigt den errechneten 3d-MTF-Verlauf der CCD-Matrix TH7895A. Gegenüber der herkömmlichen Darstellung der MTF im 1d-Schnitt für beide unabhängigen Ortsfrequenzen kann jetzt auch das Isotropie-Verhalten des örtlich abtastenden CCD-Sensors in beliebigen Ortsfrequenzrichtungen erfaßt werden. Mit Kenntnis der MTF des CCD-Bauelementes läßt sich ein optimiertes und vor allem kostengünstiges Optikdesign durchführen. Die Optik kann auf diese Weise dem Auflösungsvermögen des CCD-Sensors angepaßt werden.

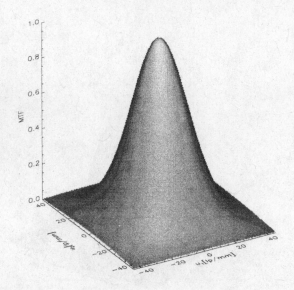

Abb. 3 MTF-Verlauf der CCD-Matrix TH7895A

4. Zusammenfassung

In diesem Beitrag wurde gezeigt, wie das sehr komplexe CCD-Bauelement auf der Abstraktionsstufe der Systemtheorie als Teilsystem einer Bildsignalverarbeitungskette erfaßt werden kann.

Unter Beachtung der Definitions- und Wertebereiche während der Bildsignalverarbeitung im CCD-Kanal (siehe Abb. 1) und der typischen Stör- und Rauschsignale der Teilsysteme ist eine signaldynamische Bewertung des Gesamtsystems möglich. Diese erfolgt durch eine spektrale Rauschanalyse im 2d-Fourierspektrum [SCH92] auf der Basis eines geeigneten digitalisierten Bildsignals. Damit sind die Voraussetzungen für ein optimales Design des CCD-Bildaufnahmesystems gegeben.

[SCZ92]   Sczepan, A., Brandt, M., Linhart, B., Ehrich, J.
            Modulares Meßsystem für optische und optoelektronische Bildeinzugskanäle und deren Komponenten
            37. Internationales Wissenschaftliches Kolloquium der TU Ilmenau, Vortragsreihen Band 2, S. 590-595, 1992

[SCH92]   Schmidt, U.
            Rauschanalyse in CCD-Bildaufnahmesystemen
            37. Internationales Wissenschaftliches Kolloquium der TU Ilmenau, Vortragsreihen Band 2, S. 584-589, 1992

Congress D

Congress-Chairman: F. Lanzl

Optoelektronische Komponenten und Systeme
Optoelectronic Components and Systems

Investigation of Optical Components Under Irradiation with High-Power, cw CO_2 Lasers

A. Angenent[*], C.-R. Haas[**], J. Koch[*], E. W. Kreutz[*], and D. A. Wesner[*]

[*]Lehrstuhl für Lasertechnik, RWTH Aachen
Steinbachstraße 15, D-52074 Aachen, FRG
[**]Fraunhofer Institut für Lasertechnik
Steinbachstraße 15, D-52074 Aachen, FRG

1. Introduction

Optical components used in high-power, cw-lasers must withstand power densities of up to several tens of kW/cm^2 over long periods of time without appreciable changes in their properties [BERGER & REEDY, GUENTHER]. Most important for applications in resonator cavities and in beam guiding and delivery systems are such characteristics as optical absorptivity, thermal conductivity and capacity, chemical stability, and surface morphology. Changes induced in one of these by high-power laser radiation might also affect the others, resulting in a relatively complex physical process, possibly leading to long-term degradation or catastrophic failure of the components. A basic understanding of such processes and the connection to damage is a goal of the present work.

Here we present results on the thermal, optical, morphological, and chemical properties of Cu mirrors exposed to controlled irradiation with cw, CO_2 laser radiation [ANGENENT et al.]. Surface morphology is monitored by interferometry, which can also be used in-situ, during irradiation. Both the mean mirror temperature and the temperature distribution in the surface during irradiation are monitored. The latter gives information on the thermal gradients in the surface due to absorbed optical energy. After irradiation the surface chemical state is analyzed ex-situ by means of x-ray photoelectron spectroscopy (XPS) and scanning Auger electron spectroscopy (AES, SAM), giving detailed chemical information (e.g., stoichiometry, nature of the bonding) on the near-surface region, within a sampling depth of about 10 nm. Surface morphology is measured by scanning electron microscopy (SEM). The SEM apparatus is also equipped to provide a site-specific chemical analysis by means of energy-dispersive x-ray analysis (EDX). Atomic force microscopy (AFM) and microscopic interferometry are also used in order to get information on the surface microstructure. Finally, photothermal deflection spectroscopy (PDS) is used to measure the relative optical absorptivity at $\lambda = 10.6$ μm across the surface.

2. Experimental details

A 2.5 kW cw-CO_2 laser (Spectra Physics 975) is incident at a 45° angle to the mirror surface. By varying the focus size between about (3x2) mm^2 and ~1 mm^2, power densities between 40 and 250 kW/cm^2 can be achieved. Commercially available, diamond-turned plane mirrors made of OFHC (oxygen-free high-conductivity) copper are studied. They are 25 mm in diameter and 6 mm in thickness, and are polished on one side to an accuracy of $\lambda/20$ (at $\lambda=10.6$ μm). Both uncoated and coated mirrors are studied, for which the reflectivity at $\lambda=10.6$ μm is ~99% and ~99.5% (according to manufacturer's specifications). Coated mirrors have a Au film deposited on a thin Ni buffer layer atop the Cu substrate. A protective dielectric overlayer of Y_2O_3 atop the Au completes the film system. The total overlayer system (comprising dielectric, Au, and Ni buffer layer) is about 1 μm in thickness.

Two types of mirror holder are used during irradiation (Fig. 1), which was in every case done in air under normal laboratory conditions. A three-point holder, in which the mirror is held at its periphery by three stainless steel cylinders, provides relatively little cooling and an essentially thermally isolated mirror. In the cooled holder the mirror is secured to a large thermal mass (a water-cooled Cu block), ensuring efficient removal of the absorbed optical power.

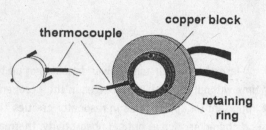

thermocouple · copper block · retaining ring

Fig. 1: Sketches of the two types of mirror holder used during irradiation: uncooled (left) and cooled (right).

In the PDS setup, the deflection of a HeNe laser probe beam specularly reflected from the mirror surface is monitored while a focused, chopped CO_2 laser pump beam impinges near it. Absorbed optical energy causes localized heating and expansion of the surface near the probe beam, deflecting it. The sample is scanned in two directions, giving a two-dimensional mapping of the relative optical absorptivity, A.

3. Results and discussion: uncooled mirror holder

Generally, the mean temperature of mirrors irradiated with a power density of 40 kW/cm^2 in the uncooled holder increased monotonically with irradiation time, reaching values as high as 300 °C after several minutes. At this temperature there is visible degradation of the mirror surfaces. Uncoated mirrors acquire a relatively dull finish over the whole surface. The changes in coated mirrors are less pronounced and consist of a darkening of the surface which is difficult to specify. Inspection in an optical microscope reveals relatively diffuse structures, possibly fissures, distributed over the whole surface. These visible effects for both coated and uncoated components first appear for irradiation times greater than ~60 s and mean temperatures above ~150 °C. At about the same time the irradiation-induced thermal gradients in the surfaces of both types of mirror increase markedly. The laser focus in thermographic images, difficult to discern for short irradiation times, becomes clearly delineated by temperature contour lines; maximum gradients are of order 50-80 °C across the mirror surface. These changes occurring for irradiation times >60 s are irreversible and suggest an increased thermal loading of the component, probably due to increased optical absorption.

Irradiation immediately distorts the mirror surface mechanically. In interferographs of uncoated and coated mirrors taken before and 6 s after irradiation at a power density of 40 kW/cm^2, the laser focus is in both cases visible at the mirror center as a distortion of the interference bands. The effect, however, is much more pronounced for the coated mirror (surface deformation ~1 μm vs ~0.1 μm). One explanation for this is a differential thermal expansion of the various layers of the coating, although increased optical absorption cannot be ruled out. The large distortion of the coated mirror relaxes somewhat with increasing irradiation time as the temperature distribution becomes more uniform.

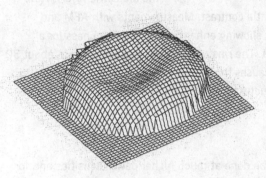

Fig. 2: Surace morphology of a permanently distorted, uncoated mirror

In addition to these reversible distortions, a *permanent* distortion of both mirror types sets in near mirror temperatures of 150 °C. Figure 2 gives an example of a mirror that was irradiated for 210 s at a power density of 40 kW/cm^2. The surface distortion amounts to about 1 μm. Although the laser was focused 7 mm off-center, the pattern is symmetric, indicating a connection with thermomechanical stresses induced by the high temperature, rather than with a localized process at the laser focus.

Chemical analysis of uncoated mirrors with XPS and AES shows little change after an irradiation for 60 s at a power density of 40 kW/cm^2. Longer irradiation times produce first an oxidation to Cu_2O and finally to CuO over the entire mirror surface [SCHÖN]. The degree of oxidation is only slightly higher in the laser focus, indicating that the process is mainly thermally driven by the high mean mirror temperature. The oxidized surface in SEM micrographs is rough, accounting for the dull surface appearance. This probably also increases the optical absorptivity, contributing to the large thermal gradients observed during irradiation. Coated mirrors are chemically much stabler. No change in the Y_2O_3 overlayer (the other, deeper layers are invisible to XPS or AES) other than a thermally induced desorption of an hydroxide contaminant layer is found.

In spite of this, coated mirrors are degraded for irradiation times of 100 s or more; i.e., they have permanent mechanical distortions and large temperature gradients similar to those of uncoated mirrors. The optical absorptivity as measured in PDS also increases by 50% after irradiation (Fig. 3). This is most likely due to a structural change in the coating system, as implied by the optical microscopy results mentioned above. XPS results on such mirrors also suggest

662

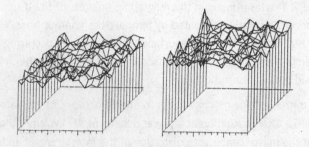

Fig.3: PDS absorptivity distributions over a 10 x 10 mm^2 region of a coated mirror before (left) and after (right) irradiation for 100 s at a power density of 50 kW/cm^2 (uncooled)

structural damage. As an electron spectroscopy, XPS is sensitive to charging and the electrical conductivity near the surface. With increasing irradiation time, XPS spectral features shift to higher binding energies, as expected for a decreasing electrical conductivity (caused, e.g., by structural damage to the dielectric layer). However, SEM-micrographs of such mirrors show little contrast. Measurements with AFM and microscopic interferometry are more revealing, showing enhanced surface roughness for a damaged mirror compared to unirradiated ones. The rms roughness is relatively small (about 30 nm), so that it seems unlikely that this alone causes the increased optical absorptivity at λ=10.6 µm. It indicates, rather, an overall structural degradation of the film.

4. Results and discussion: cooled mirror holder

Irradiation of mirrors in the cooled holder can be done at much higher power densities and for longer times (e.g., 150 kW/cm^2, 2 h). For both coated and uncoated mirrors the maximum mean temperature of ~40 °C is reached shortly after beginning the irradiation, thereafter remaining constant. Thermography measurements show that in the laser focus the local temperature is only a few degrees higher. Reversible surface distortions are seen for both types of mirror, similar to the uncooled case, but no irreversible ones. Similarly, no large-scale chemical changes are detected in XPS or AES, the only hint of damage being a slight increase in charging effects localized at the laser focus on coated mirrors. Without a strong thermal loading of the mirror (as, e.g., in the uncooled configuration), there is little degradation until, eventually, at very high power densities (~0.2 MW/cm^2), catastrophic failure may occur.

For the long exposures with the cooled mirror holder "extrinsic" damage also appears. An example is the adsorption within the laser focus of dust particles, which effectively absorb the optical energy, creating a localized heat source. SEM micrographs show such particles broken apart under the thermal stress during irradiation of an uncoated mirror, resulting in a localized heating and melting of the surface. SAM and EDX analysis show that they are composed of $CaCO_3$, Al_2O_3, or SiO_2. On a coated mirror such particles are found near damaged positions on the surface involving removal of the Y_2O_3 protective overlayer, exposing the Au layer beneath.

5. Conclusions

In summary, the irradiation conditions used here are extreme in that in the uncooled configuration there is an overall thermal loading of the mirror that would occur in a cooled configuration only at higher power densities. Nonetheless, the damage mechanisms in the two cases are probably similar. A threshold temperature of about 150 °C marks the onset of oxidation (for uncoated mirrors) or for structural changes in the overlayers (for coated mirrors), both of which lead to similar thermal loading and morphological effects under irradiation. The thermal loading, due to an increased optical absorptivity (and/or to a decreased thermal conductivity), leads to large thermal gradients in the surface, as the increasing amount of absorbed optical energy cannot be efficiently conducted away from the mirror.

6. Acknowledgments

The authors are grateful to the German Research and Technology Ministry (BMFT) for their financial support of this work under FKZ 13 N 5862.

7. References

A. Angenent, C. R. Haas, J. Koch, E. W. Kreutz, and D. A. Wesner, SPIE Proc. Ser., in press.
M. R. Berger and H. E. Reedy, SPIE Proc. Ser. 455, 52 (1983).
K. H. Guenther, SPIE Proc. Ser. 801, 200 (1987).
G. Schön, Surface Sci. 35, 96 (1973).

Non-Toxic Coatings for CO_2-Laser Optical Systems

H. Hagedorn, R. Anton, G. Lensch
NU-TECH GmbH
Ilsahl 5, D-24536 Neumünster

Introduction

The standard antireflection coating of transmitting ZnSe optics for high power CO_2 laser radiation (10.6μm) consists of a ThF_4/ZnSe double layer. The best available coatings exhibit an absorption coefficient of approximately 10 cm^{-1} , corresponding to a total absorption of less than 0.2%, and a damage threshold of more than 10 J/cm^2. Since Th is a radioactive α-emitter, it would be advantageous to replace it by a less hazardous compound of similar optical and mechanical properties. Thus, the utilization of other metal fluorides as substitute low index materials has been the focus of many material studies [PELLICORI].

The crucial issue of producing low absorption films is the minimization of water vapour incorporation. Due to its high absorption coefficient ($\approx 850\,cm^{-1}$) at 10.6 μm [PALIK], even small amounts of water markedly raise the absorption of the coating. Fluoride films often show a porous structure of low packing density and a related tendency to adsorb water vapour [PULKER]. This is especially true for rare earth fluorides [BARYSHNIKOV].
Ion Assisted Deposition (IAD) has been applied to produce films of enhanced density with closed microstructure and high mechanical performance.

This paper documents the results of an investigation into the influence of ion energy and ion current on the optical and mechanical properties of single layer YbF_3 and YF_3 IAD films.

Experiments

The films were prepared on ZnSe substrates at a temperature of 380 K in a cryopumped box-coater by electron beam evaporation at an angle of impingement of 36° with respect to the surface normal. The Ar^+-beam of a 3 cm rf ion source (RIM4) was incident at 53° . The substrate holder allows transmission measurement at normal incident by an optical monitoring system (OMS 3000) on the rotating substrate . The evaporation rate was kept constant by an oscillating quartz monitor. The partial pressure of H_2O was controlled by a quadrupole mass spectrometer before and after deposition, and was maintained below $1*10^{-7}$ mbar. The ion current density and average energy at the substrate position was determined by a Faraday cup with retarding grid.

The ion to molecule fluxes were set between 1:20 and 4:3 by variation of the evaporation rate (0.4 - 2 nm/s) and the ion current density (20 - 150 μA/cm²) . Average ion energies of 60, 140 and 330 eV were used, resulting in average energies per incident molecule (EpM) of up to 450 eV . The transmission between 2.5 and 25 μm of the coated substrates was measured by FTIR spectroscopy after approximately 2 weeks exposure to air. The water absorption lines at 2.9 and 6.1 μm were used to estimate the water content of the films. The absorption at 10.6 μm was measured by a laser calorimeter. The 1on1 50% laser induced damage threshold (LIDT) was examined with an 100 ns (4.5 μ tail) puls of 1.4 mm diameter.

The film density was computed from the mass of the material deposited, as determined by weighing, and from film thickness data, as derived from optical measurements via the envelope method [SWANEPOEL]. All films were λ/4 layers for 10.6 μm with a thickness of about 2 μm.

Results

YbF_3-films prepared at an evaporation rate of 2nm/s without ion assistance (EpM $\approx$ 0.11 eV) showed a density of 75% of the bulk value. With increasing EpM, the film density increased up to 90% of the bulk value (Fig. 1).

In the region of 10-20 eV EpM, a distinct change in the water content of the coatings occurred (Fig. 2). While films prepared with lower EpM showed high absorption indicative of incorporated water, no water absorption could be detected with films produced at higher EpM.

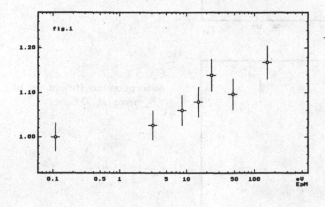

Fig. 1
Rel. densities of YbF_3-films compared with a coating without ion assistance.

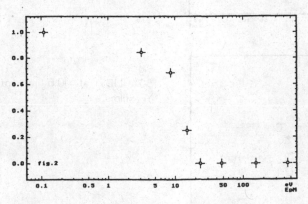

Fig. 2
Rel. water absorption at 2.9 and 6.1 μm compared with an coating without ion assistance.

666

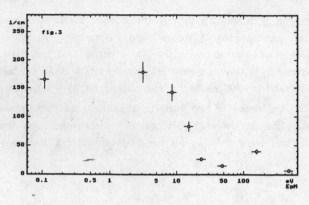

Fig. 3
Absorption coefficient α of YbF$_3$ -films at 10.6μm.

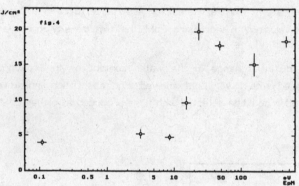

Fig. 4
50% LIDT at 10.6 μm of YbF$_3$-films.

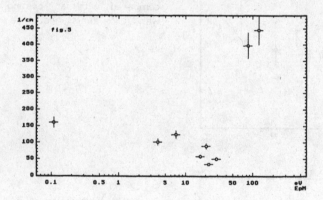

Fig. 5
Absorption coefficient α of YF$_3$ -films at 10.6μm.

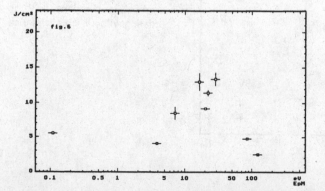

Fig. 6
50% LIDT at 10.6 μm of YF$_3$-films.

A similar dependence of the EpM was observed for the absorption coefficient at 10.6 μm and the damage threshold. Low values of about 5 J/cm^2 for the damage threshold and high values of the absorption coefficient are obtained with EpM below 10 eV. The application of high EpM resulted in damage thresholds of up to 20 J/cm^2 and absorption coefficients of less than 10 cm^{-1} (Fig. 3 and 4).

YF$_3$-films prepared at low EpM showed similar behaviour as observed with YbF$_3$. High EpM yielded films without any detectable water absorption, however, exhibiting a very high absorption coefficient at 10.6 μm. The best films that could be prepared had a LIDT between 10 and 15 J/cm^2 and an absorption coefficient below 50 cm^{-1} (Fig. 5 and 6).

Conclusion

To predict and control the film properties of YbF$_3$ and YF$_3$, the average energy per incident molecule was found to be an appropriate parameter. Films with enhanced density and low absorption coefficient at moderate substrate temperature can be obtained when EpM of more than 10 to 20 eV are applied. YbF$_3$ seems to be a promising candidate for a non radioactive substitute for ThF$_4$ in Laser optics coatings.

Acknowledgement

This work was supported by the Bundesministerium für Forschung und Technologie, Förderkennzeichen 13 N5686 9.

References

BARYSHNIKOV, KARPOV and GUSHCHUINA ; Sources of Oxigen in Rare Earth Metal Fluorides Izvestiya Akademii Nauk SSSR, Neorganicheskie Materialy,Vol. 4, No. 4, pp. 532–536, 4/1968

PALIK ; Handbook of Optical Constants of Solids II, Academic Press, 1991

PELLICORI, COLTON ; Flouride compounds for IR coatings, Thin Solid Films, 209(1992) 109–115

PULKER ; Coatings on Glass, Thin Films Science and Technology, 6, Elsevier, 1984

SWANEPOEL ; Determination of the thickness and optical constants of amorphous silicon, J. Phys. R. Sci. Instrum., Vol. 16, 1983

Chiral Mesomorphic Materials as Optical Isolators for High Power Laser Systems

J.Owsik, E.Szwajczak[*], A.Szymański[*]
Military University of Technology, Institute of Optoelectronics,
Warsaw, Poland
[*] Chair of Physics, Technical University of Rzeszów

INTRODUCTION

Liquid crystals constitute a unique state of matter. They can flow over surfaces as easily as liquids and yet they possess a long range structural order characteristics of crystalline solid. These substances have found a wide range of applications and are commonly used as numerical displays in calculators and watches as well as in thermometers, computer information displays, electronic games and even as an artistic medium for paintings.

This paper describes how cholesteric liquid crystal might be used as large aperture optical isolators (or one way light valves) in the laser systems.

All laser systems used in experiments, for investigation of high power laser interaction with matter, must focus their optical radiation onto small targets.

The laser optical components can be damaged if infrared radiation reflected off a target surface is allowed to propagate back onto the laser system. Optical isolators are therefore placed at the output of a laser beamline. These devices permit light propagation only in a forward (to the target) going direction, acting as optical light valves to stop the propagation of any back reflected radiation.

Recently, in the experiment of laser - plasma interaction, the Faraday isolators are used to protect high power laser system. They require a high voltage supply and additional, expensive dielectric polarizers. They are not convenient to use and they occupy a lot of space in a laser system. It is very difficult to achieve a large diameter, good quality Faraday isolator. Liquid crystal isolators may be used in optical communication, too. Cholesteric liquid crystal ability of selective reflection of light permits to use the layers of this material with a pitch and a refractive index, matched to a wavelength of laser radiation as an optical isolator.

The CLC pitch p and an avarage refractive index n must satisfy the equation:

$$\lambda_L = n \ p$$

Where - λ_L is a laser radiation wavelength.

PRINCIPLE OF OPERATION OF A LIQUID CRYSTAL ISOLATOR.

The liquid crystal isolator (LCI) is a passive device that requires no electronics, electric or magnetic fields as well as no synchronization with the firing of the laser system.

Models of the isolator, based on a liquid crystal, of a clear aperture diameter of 50 mm and 100 mm respectively, have been designed and built (see figure 1).

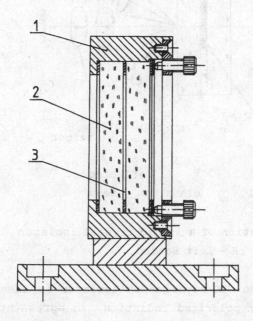

Fig.1. The structure of a liquid crystal cell
1. Housing;
2. Glass plates;
3. Liquid crystal layer.

A cholesteric liquid crystal layer is located between two windows of BK 7 glass, 10 mm thick. The thickness of CLC layer has been determined by the thickness of a separator, inserted between the two windows pressed together by the special bolts. The structure of such a liquid crystal cell is shown in Fig.1. This cell was used to study the influence of a cholesteric liquid crystal (CLC) layer on the distribution of an energy in the cross - section of the transmitted laser beam.

Figure 2 shows schematically how the liquid crystal isolator (LCI) works. Right - hand circularly (RHC) polarized infrared radiation, generated from a laser system, passes unattenuated through the LCI and propagates to the target. A portion of the light not absorbed by the target is specularly reflected and undergoes a 180° shift in its phase of vibration. The polarization sense of this retroreflected light is now left - hand circular (LHC) and the liquid crystal cell will not permit this state of polarization to pass.

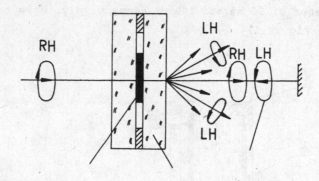

circularly polarized   target
laser radiation

                                                    reflected
        LH liquid crystal   glass   beam

Fig.2. Principle of operation of a liquid crystal isolator
      RH - Right Handed; LH - Left Handed.

An application of the CLC layer as an isolator is based on the fact that the circularly polarized radiation, in agreement with the sense of the helix is dispersed on the CLC layer. The spectral range of a selective reflection of the CLC must be equal to the wavelength of the radiation.

Advantages of liquid crystal optical isolators.
- they don't require any voltage supply;
- they don't require additional polarizers;
- high contrast ratio;
- good optical quality at large apertures;
- low optical loss for light propagating in the forward direction;
- simplicity of fabrication;
- simplicity of installation and operation;
- low costs.

The beam progagated in a laser system is circularly polarized, what minimizes the stress birefringence, induced by laser rods.

EXPERIMENTAL RESULTS

Fig.3 illustrates a transmission of the CK mixture as a function of a wavelength for various temperatures. At room temperature the wavelength of selective reflection varies insignificantly and the value of the rate of variation is 2.3 [μm/°C].

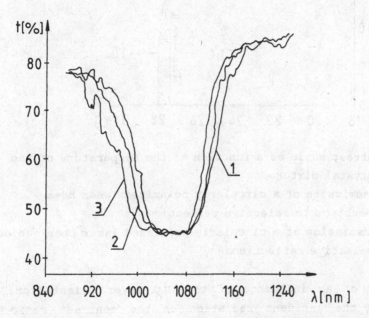

Fig.3. Transmission of a liquid crystal mixture as a function of wave length;
   Temperature: 1 – T=19.2 C; 2 – T=22.0 C; 3 – T=25.0 C.

The spectral half-width of the interval of selective reflection reaches the value of 140 [μm], within the range of room temperatures and increases insignificantly with the increase of a layer temperature.

Fig 4 shows the results of the contrast study of the CK mixture as a function of temperature. Within the region of the temperatures in which the CLC layer was investigated, there is a stability of a contrast ratio within the measurement error. Above 27° the contrast ratio decreases rapidly due to the temperature of phase change from the CLC to an isotropic liquid. The transition temperature is 27.5°C. The contrast reaches a level of K =100.

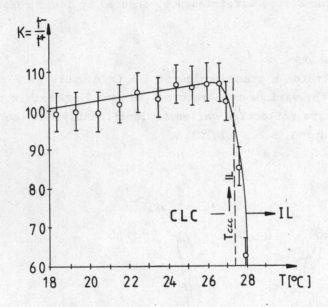

Fig.4. The contrast ratio as a function of the temperature of the liquid crystal mixture
   t⁻ - transmission of a circularly polarized laser beam non-subject to selective reflection;
   t⁺ - transmission of a circularly polarized laser beam subject to selective reflection.

The study of an influence of the CLC layer orientation, in refer-ence to the incident radiation on the contrast ratio, is illustrated in Fig 5. It follows from the diagram that the location of the CLC layer, with an accuracy to ±10°, in reference to the axis of the laser beam, does not change the value of the contrast ratio for the transmitted laser radiation.

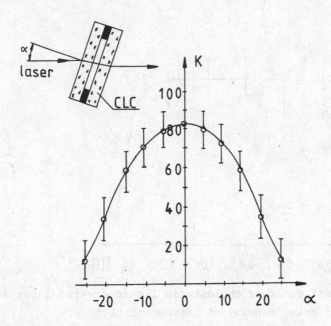

Fig.5. Dependence of the contrast ratio of a liquid crystal cell on the
incidence angle of laser radiation
K - contrast ratio;
α - angle of incidence of laser radiation on liquid crystal.

An application of the CLC layers in isolators depends on the power
density of radiation, which produces irreversible changes in the liquid
crystal. The results of investigations of the contrast ratio of the CLC
layer as a function of power density of laser radiation inciding the
CLC layer are illustrated in Fig 6. For a power density of more than
1.2 GW/cm^2 the CLC layer is damaged, some opacity is the first symptom
which can be observed. It may be assumed that the admissible power density of radiation which does not damage the CLC layer is 1.2 GW/cm^2.
For the nanosecond pulses the value of the contrast ratio was:

$$K = 100 \pm 10$$

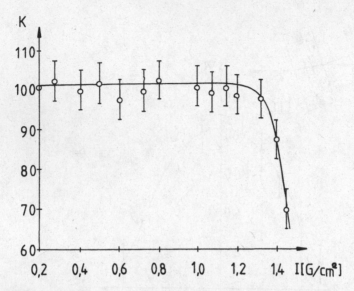

Fig.6. The contrast ratio of cholesteric liquid crystal layer as a function of power density of laser radiation

TABLE.1. COMPARISON OF PROPERTIES OF ISOLATORS

CHARACTERISTICS	ISOLATOR	
	FARADAY'S	CLC
NUMBER OF SURFACES	6	2
OPERATION	ACTIVE	PASSIVE
OPTICAL ALIGNMENT	DIFFICULT	EASY
APERTURE	LIMITED	NOT LIMITED
DAMPING dB	20	20
WEIGHT	VERY HIGH	INSIGNIFICANT
COST	VERY HIGH	LOW

CONCLUSIONS

The results of the CLC test, disscused in this paper make it possible to design and produce the liquid crystal isolators. Table 1 presents in a comparative manner the experimental results obtained for the parameters of isolators based on liquid crystals with the ones obtained for the conventional devices.

It is found that the elements based on liquid crystals are competitive in relation to the conventional solutions.

Design and Experiments of Hundred Joule-Level KrF Laser Pumped by Intense Electron Beam

Shan Yusheng Wang Naiyan Zeng Naigong Zhou Chuangzhi Yang Dawei
Ma Weiyi Huang Debao Wang XiaoJun Gao Junsi Wang Ganchang
China Institute of Atomic Energy, P.O.Box 275-7,Beijing 102413, China

Introduction

In the inertial confinement fusion (ICF) resarch, the KrF laser is considered to be a promising candidate for the reactor driver due to its high efficiency, scalability[1] and advantages of UV laser in laser-plasma coupling[2]. One hundred-joule KrF laser[3] pumped by electron beam has been constructed for the foundamental reserch of ICF and with a view to addressing some of the key technologies of pulsed power and laser generation at CIAE. In order to upgrad the KrF laser energy to 100J level our original 80 GW REB accelerator[4] (1MV, 80kA, 80ns) was modified by adding a prepulse switch following the Blumlein transmission line and a taper transmission line which changes the impedance of Blumlein line from 6 Ω to 2.5 Ω and a large area diode and a laser cavity of large volume are constructed. The KrF laser output energy of 100J and 100ns in duration have been achieved. The design and experiments of 100J-level KrF laser are briefly introduced in this paper.

Design and Design Parameters

The physical design was performed by combining the analyses and kinetic simulation methods. According to our zero-dimensional kinetic simulation of KrF laser[5] which include 56 reaction channels and 15 species of particles, the dependences of laser intrinsic efficiency (η_{in}) on the electron beam density transsimited into cavity (J_{in}) for three different condition of laser osillators are shown in Fig 1. From this figure, we can see that the intrinsic efficiency $\eta_{in} \geqslant 8\%$ may be expected when the pressure of mix is 0.25MPa, $J_{in}. \geqslant 100A/cm^2$, which is equivalent to the energy deposition rate is about 1.5MW/cm³,under this pressure values and pumping condition the fractional laser energy output is about 10J/l, so total laser volume can be determined to be 20 l so as to obtain a desirable output energy which is large than 100J. A diamiter of active volume (D) can be chosen to be 20cm so as to deposite 40% of the injected electron energy at the gas pressure of 0.25MPa and at the chosen e-beam voltage 500kV.

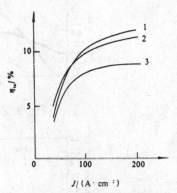

$J/(\text{A} \cdot \text{cm}^{-2})$

Fig. 1 Theoretical calculation of the dependence of the intrinsic efficiency η_{in} on the beam density
1. $\varphi 70mm, L=38cm, p=0.3MPa, R_s=30\%$
2. $\varphi 110mm, L=60cm, p=0.3MPa, R_s=30\%$
3. $\varphi 200mm, L=75cm, p=0.25MPa, R_s=30\%$

On the basis of consideration mentioned above the preliminary parameters of laser cavity are: the KrF laser gas mixture: $Ar/Kr/F_2$=89.6%/10%/0.4%; the pumping parameters: P_{in}=1.5MW/cm³, p=0.25MPa, J_{in}=100A/cm²: the cavity parameters: V=23l, D=20cm, L=75cm, reflectivity of output mirror~30%; the laser output parameters: E_L≥100J, ε =10J/l. The small signal gain g_0, the nonsaturable absorption α and the saturation intensity I_s are determined by using various pumping parameters and the zero–dimensional kinetic simulation model, as shown in Table 1.

Table 1

$J_{in}/(A \cdot cm^{-2})$	$P_{in}/(MW \cdot cm^{-3})$	g_0/cm^{-1}	α/cm^{-1}	$I_s/(MW \cdot cm^{-2})$	$\eta_{in}/\%$	$I_{out}/(MW \cdot cm^{-2})$
50	0.79	0.046	0.004	3.3	5.8	4.5
75	1.2	0.063	0.004	3.5	7.2	7.6
100	1.6	0.080	0.004	3.8	7.9	11
125	2.0	0.093	0.004	4.0	8.3	14
150	2.4	0.11	0.004	4.2	8.5	17
200	3.2	0.12	0.004	4.7	8.8	22

As seen, the laser energy output E_L>100J can be obtained by using parameters metioned above and the chosen operation conditions of the total pressure of 0.25MPa and pumping e–beam energy deposition rate of 1.5MW/cm³.

The one–dimentional kinetic simulation of KrF laser pumped by e–beam[6] was also performed with above parameters, taking into account for 24 species of particles and 119 reaction channels. The main results are shown in Fig 2 and Fig.3, g_0=6%, α =0.55% are obtained by average of pulse duration and peak value P_{in}=2MW/cm³, I_s=2MW/cm², η_{in}=7.5%, E_L=95J.

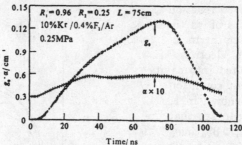

Fig 2 g, α as a function of time

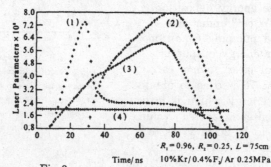

Fig.3 Laser parameters as a function of time

(1) side light × 300/ $(J \cdot s^{-1} \cdot cm^{-1})$;   (2) laser output power/ $(J \cdot s^{-1} \cdot cm^{-1})$;

(3) pump rate × 3.0/ $(W \cdot cm^{-3})$;   (4) saturation flux/ $(W \cdot cm^{-2})$

According to the requirements of pumped rate the cathode area of 12cm×75cm for diode design are chosen then the total e-beam current transmited into cavity I_{in}=220kA are required, provided the efficiency of e-beam transmited through the anode foil and Hibachi and pressure foil is 50% and diode voltage $V_a\approx$550kV; that is, diode impedence $Z_a\approx$2.5Ω and the anode-cathode spacing of diode d=2cm are required.

Apparatus and Experiments

On the basis of mentioned above design requirements, we calculated and designed and constructed the all of elements of KrF laser device including e-beam pumped source and e-beam transmission system (foils, hibachi) and laser cavity. An 0.14 TW intense electron beam accelerator has been built, tested based on original 80 GW accelerator, and now is operated at CIAE. The designed parameters of it are 550-600kV, 220-240kA, 100ns for e-beam in diode. The total energy and peak power in diode are 10kJ and 0.14TW respectively. This accelerator consists of an oil-immersed Marx generator, a main switch, a Blumlein transmission line (BTL) followed by prepulse switch and a taper transmission line (TTL) and a field emission large area diode. A side view of BTL followed by prepulse switch and TTL is shown in Fig.4.

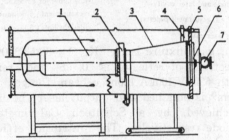

1. Blumlein transmission line ; 2. prepulse switch ; 3. taper transmission line ; 4. resistance divider
5. current shunt ; 6. diode ; 7. laser oscillator

Fig.4. The side view of the Blumlein transmission line followed by a taper transmission line

The Marx generator consists of 20 capacitors of 0.7μF, 100kV with 10 gas park gaps. It stores 34kJ energy when capacitors are charged to 70kV. By using TTL , the impedance of BTL can be changed from 6.5 Ω to 2.5 Ω. The voltage efficiency of TTL has been studied theoretically and experimentally to be 0.62. Two advantages have been obtained by adding multichannel prepulse switch between BTL and TTL. The first is the amplitude of prepulse could be reduced by 100 times to be less than 1kV and the second one is the risetime of the voltage and current becomes short as gas pressure increases in the switch.

The large area diode with carbon felt as a cathode was constructed and tested, as shown in Fig. 5. Typical impedance of diode is 2.5Ω matched to TTL and emission area is 12cm×75cm which can provide 500-600kV, 200-240kA, 100ns (FWHM) electron beams. In our experiments, carbon felts is proved to be superior to other field emitters (such as multipin brass, graphite brush-type). From the point of view of the anode foil surrivability, 25μm Ti foil or anode made of eleven copper wirers of 1.3mm in diameter are chosen. The anode foil pressure main foil (30μm Ti foil and its support structure called hibachi are proved to be able to provide a transmission efficiency of 50% for e-beam into laser cavity, as shown in Fig.6. The laser cavity and active region length with 20cm of diamiter are 130cm and 75cm respectively.

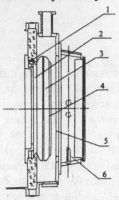 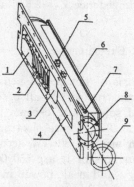

Fig. 6 The schematic diagram of the anode foil,

hibachi and pressure foil of the cavity

1. anode foil ; 2. hibachi ; 3. main foil ;

4. window for e−beam ; 5. hole for pressure

measurement ; 6. window for meas urement ;

7. sidelight hole ; 8.laser window ; 9.laser reflector

Fig. 5 The side view of the diode and laser cavity

1. clamping ring ; 2. cone base ; 3. adjustment pedestal

4. cathod ; 5. anode ; 6. laser cavity

For KrF laser(248nm), gas mixture is Ar: Kr: F_2=89.6%:10%:0.4% and total pressure is 0.25MPa, reflectivity of reflector is 0. 96 and reflectivity of output coupler is 0.75 with 620kV,160kA,120ns e-beam for gas excitation, the pumping rete of 1.5MW/cm³ is obtained and output laser beam 106J energy and 100ns in duration were achieved. by a Scientech Calorimeter with 20cm in diameter and by aphotodiode respectively. The intrinsic efficiency of 7% is obtained by using the transient pressure rise methode.

The values of the small signal gain g_o, the nonsaturable absorption coefficient α and saturation intensity I_s are determined to be 7%/cm, 0.6%/cm, 2MW/cm², respectively, by using self-injection proble methode.

Reference
[1] Sullivan, J. A. Fusion Technology 11,684(1987)
[2] Owadano, Y. et al, Plasma Physics and controlled Nuclear Fusion 1982, 1, 125(1983 IAEA Vienna)
[3] Wang Neiyan, Shan Yusheng, High Power :aser and Particle Beam Vol.3, No.3, P411(1991)
[4] Wang Naiyan, et at, Proceeding of 5th International Topical Conf. on High Power Electron and Ion Beam Research and Tech. P60, 1983, San Francisco
[5] Kang Xiangdong, Shan Yusheng, Wang Naiyan, High Power Laser and Particle Beams, Vol.3, No.4, p457(1991)
[6] Feng Guogang and Wang Naiyan, High Power Laser and Particle Beams, Vol.5, No.1, p29(1993)

Diffraktive Elemente für die Lasermaterialbearbeitung

M. Pahlke, C. Haupt, H.-J. Tiziani

Universität Stuttgart, Institut für Technische Optik, Pfaffenwaldring 9, D - 7000 Stuttgart 80

1. Einleitung

In der Materialbearbeitung werden immer häufiger CO_2-Hochleistungslaser eingesetzt. Beim Schneiden, Schweißen und Bohren sind Energiedichten von $\sim 10^7 W/cm^2$ erforderlich, die durch Fokussierung des Laserstrahls erreicht werden, wobei als optische Elemente Linsen bzw. Hohlspiegel eingesetzt werden.

Geringere Energiedichten sind zum Härten erforderlich. Dabei muß die Temperatur einer zu härtenden Fläche für eine bestimmte Zeit innerhalb eines bestimmten Temperaturintervalls liegen. In der Praxis wird ein Laserstrahl über die zu härtende Fläche geführt. Ein Aufschmelzen der Oberfläche muß dabei vermieden werden. Die so erzeugte Intensitätsverteilung auf dem Werkstück führt nicht zu idealen Ergebnissen, weil das Temperaturfeld im Werkstück den Härtungsprozeß nicht im erforderlichen Maß unterstützt. Ein optimiertes Intensitätsprofil zum Härten wurde von Burger /1/ vorgeschlagen. Dieses Intensitätsprofil kann mit klassischen optischen Komponenten nicht erzeugt werden. Diffraktive Elemente (DE) können grundsätzlich derartige Intensitätsverteilungen erzeugen, deren Struktur jedoch nur mit aufwendigen numerischen Verfahren berechnet werden können.

Das Härten von runden Öffnungen, beispielsweise bei Ventilen oder die Lauffläche in einem Konuslager, könnte dadurch erleichtert werden, daß die gesamte zu härtende Flache mit einer ringförmigen Intensitätsverteilung bearbeitet wird. Die Erzeugung dieser Intensitätsverteilung soll hier beschrieben werden. Dazu wird ein analytisches Verfahren zur Berechnung der Beugungsstrukturen und der lithographische Herstellungsprozeß eines DE beschrieben. Ein diffraktives Element wurde hergestellt und untersucht.

DE zur Materialbearbeitung werden in Form von computergenerierten Phasenhologrammen realisiert. Sie sollen einen möglichst hohen Beugungswirkungsgrad und eine möglichst geringe Absorption besitzen. Die Funktion der Elemente beruht auf der Beugung von Lichtwellen an einer Oberflächenstruktur auf einem ebenen Substrat. Für den Einsatz mit Hochleistungslasern bieten diffraktive Reflexionselemente die gleichen Vorteile wie konventionelle Reflexionsoptiken, insbesondere im Hinblick auf Absorption und Kühlung.

2. Berechnung der Oberflächenstruktur diffraktiver Reflexionsoptiken

Die Oberflächenstruktur wird mit der skalaren Beugungstheorie berechnet. Das Beugungsintegral beschreibt das komplexe Lichtfeld in einer Ebene B, das von einer Apertur in Ebene A erzeugt wird (Fig. 1). In Ebene A liegt die erwünschte Intensitätsverteilung. Das Beugungsintegral wird in der Fraunhofer-Näherung verwendet (Gleichung 1).

$$h(x', y') \propto \int_{-\infty}^{\infty} a(x, y) e^{-i \frac{2\pi(xx' + yy')}{\lambda z}} \, dx\,dy \qquad (1)$$

Das hat in der Praxis zur Folge, daß zur Rekonstruktion ein zusätzliches fokussierendes Element benötigt wird, wodurch die Beugungsstrukturen gröber werden und damit leichter herstellbar sind. Für die Berechnung bedeutet dies, daß das komplexe Lichtfeld der Apertur A in der Ebene B mit einer Fouriertransformation ermittelt werden kann. Die Fraunhofer-Näherung ist jedoch beschränkt auf den Fall, daß die Apertur A und die Ebene B parallel zueinander sind und die Lichtwelle senkrecht zur Apertur einfällt.

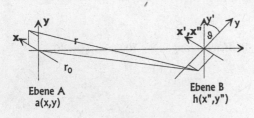

Fig. 1: Die geometrischen Verhältnisse zur Berechnung von DEs; Ebene A: Rekonstruktionsebene; Ebene B: Ebene des DE

Die diffraktiven Reflexionselementen werden jedoch schräg beleuchtet, was zur Folge hat, daß die Fraunhofer-Näherung erweitert werden muß /2/,/3/. Berücksichtigt man die geometrischen Bedingungen des Strahlengangs, so kann das Beugungsintegral derart angenähert werden, daß der schräge Einfall des Laserstrahls auf das diffraktive Element berücksichtigt wird. Man erhält als Ergebnis

$$h(x'', y'') \propto \int_{-\infty}^{\infty} a(x, y) e^{-i\frac{2\pi(xx'' + yy''\cos\vartheta)}{\lambda r_0}} \, dxdy \qquad (2)$$

mit der Größe r_0

$$r_0 = \sqrt{z^2 + x''^2 + y''^2 + 2zy'' \sin\vartheta} \qquad (3)$$

Diese modifizierte Näherung des Beugungsintegrals führt dazu, daß die Beugungsstrukturen des DE in y-Richtung mit dem Faktor $1/\cos\vartheta$ gestreckt werden. Im Vergleich zur Fraunhofer-Näherung werden die Raumfrequenzen für $\psi < 0$ größer und für die $\psi > 0$ kleiner.

Dieses Ergebnis dient als Grundlage für das folgende Berechnungsverfahren, das sich insbesondere zur Berechnung von diffraktiven Strukturen zur Erzeugung von Fokuskurven eignet /4/,/5/. Die gewünschte Fokuskurve soll sich durch eine Funktion $y = f(x)$ beschreiben lassen. Die Intensitätsverteilung $a(x,y)$ wird dann mit Hilfe der Dirac'schen Deltafunktion dargestellt, und man erhält $a(x,y) = \delta(y - f(x))$. Setzt man $a(x,y)$ in Gleichung (2) ein, dann läßt sie sich über eine Variable direkt integrieren. Zur Lösung des verbleibenden Integrals wird die Methode der Stationären Phase verwendet. Da die meiste Information über eine Wellenfront in der Phasenfunktion /6/ enthalten ist, wird nur diese von dem

Fig. 2: Beugungsstruktur für Kreisring

berechneten komplexen Feld zur Erzeugung der Beugungsstruktur verwendet. Mit der Berücksichtigung der reellen Amplitude wären zusätzliche Energieverluste verbunden.

Die Intensitätsverteilung eines Kreisrings mit dem Radius ρ ist gegeben mit

$$a(x, y) = \delta(y - \sqrt{\rho^2 - x^2}) \qquad (4).$$

Die daraus resultierende Phasenfunktion lautet

$$\Phi(\xi, \psi) = \frac{2\pi\rho}{\lambda r_0} \sqrt{\xi^2 + \psi^2 \cos^2 \vartheta} \qquad (5).$$

3. Herstellung der Diffraktiven Elemente

Aus herstellungstechnischen Gründen ist es notwendig den Wertebereich der Phasenfunktion auf einen diskreten, endlichen Wertebereich zu begrenzen (quantisieren). Die Phasenfunktion kann nun auf Grund der Periodizität trigonometrischer Funktionen in den Wertebereich zwischen Null und 2π abgetragen und in äquidistante Intervalle unterteilt werden. Die Werte eines Intervalls werden schließlich auf einen Wert festgelegt, was einer Annäherung der berechneten Funktion durch eine Treppenfunktion entspricht. Die Effizienz eines diffraktiven Elements steigt mit der Anzahl der Stufen, da die berechnete Funktion mit zunehmender Zahl der Stufen immer besser angenähert wird. Begrenzt man den Wertebereich der obigen Phasenfunktion auf zwei Werte, dann erhält man die Oberflächenstruktur in Fig. 2. Bei einem Einfallswinkel $\vartheta = 45°$ und einer Wellenlänge von 10.6µm beträgt die Tiefe der Struktur 3,75µm.

Als Substratmaterial der diffraktiven Strukturen wurde Silizium ausgewählt /7/, weil es im Hinblick auf die thermischen und verfahrenstechnischen Eigenschaften für die Anwendung geeignet ist. Darüber hinaus sind die erforderlichen Prozeßparameter für Silizium bei der Strukturierung mit reaktivem Ionenstrahlätzen aus der Mikroelektronik bekannt.

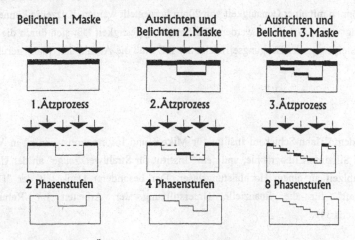

Fig. 3: dreistufiger Ätzprozeß zur Erzeugung einer 8-Stufigen Beugungsstruktur

Im ersten Schritt werden mit einem Laserscanner die berechneten Muster in den Photoresist von Chrommaskenblanks geschrieben. Die Anzahl der Stufen wächst als Zweierpotenz mit der Anzahl der eingesetzten Masken. Dazu wird im ersten Schritt die erste Maskenstruktur photolithographisch auf das Substratmaterial durch Ätzung übertragen, wobei die Strukturen eine bestimmte Tiefe bekommen. Danach wird die zweite Maske zu der bereits geätzten Struktur positioniert und wieder photolithographisch übertragen. Durch den zweiten Ätzprozeß erhält man schließlich vier Stufen. In Fig. 3 ist der Prozeß schematisch dargestellt. Da Silizium für elektromagnetische Strahlung der Wellenlänge $\lambda = 10{,}6µm$ transparent ist muß nach der Strukturierung eine harte, hochreflektierende Schicht aufgebracht werden.

682

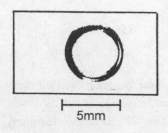

|⊢—————— 5mm ——————⊣|

Fig. 4: Optische Rekonstruktion der
Beugungsstruktur in Fig. 2

Es wurde ein diffraktives Element mit vier Stufen hergestellt. Als reflektierende Schicht wurde Gold aufgedampft. Die optische Rekonstruktion zeigt, daß beim Ätzprozeß die notwendige Tiefe von 5,6μm der Struktur mit großer Genauigkeit hergestellt wurde, was sich dadurch äußert, daß der Gleichlichtanteil verschwindet. Der rekonstruierte Kreisring (Fig. 4) weist keine meßbaren Abweichungen von der Kreisform auf. Die Beugungseffizienz beträgt 73% bei der Rekonstruktion mit einem CO_2-Laborlaser (Ausgangsleistung: 20W). Die Erwärmung eines ungekühlten diffraktiven Elements war dabei gering. Im Hochleistungslaserstrahl wurde das ungekühlte Element mehrmals kurzzeitig mit 1kW Strahlleistung belastet, ohne daß eine Beschädigung der Beugungsstrukturen auftrat.

4. Zusammenfassung und Ausblick

Es wird ein Verfahren zur Berechnung von schräg beleuchteten Beugungsstrukturen vorgestellt. Das Verfahren ist für die Erzeugung von Fokuskurven geeignet. Die berechnete Beugungsstruktur wurde mit einem Mehrmaskenprozess durch reaktives Ionenstrahlätzen in Silizium hergestellt. Mit einem vierstufigen diffraktiven Element wird eine Effizienz von 73% erreicht. Die erforderlichen Tiefen der Beugungsstruktur von 5,6μm können mit einer Genauigkeit von 20nm hergestellt werden. Kuzzeitig können die DE mit einer Strahlleistung von 1kW belastet werden. Die Strahlungsfestigkeit läßt sich durch die Kühlung der DE erhöhen. Zur Steigerung der Beugungseffizienz der DE muß die Anzahl der Quantisierungsstufen erhöht werden.

5. Danksagung

Die Autoren danken dem "Hahn-Schickard Institut für Mikro- und Informationstechnik" in Villingen-Schwenningen für die Silizium Lithographie, und dem "Institut für Strahlwerkzeuge" an der Universität Stuttgart für die Strahlzeit an einem Hochleistungslaser. Der besonderer Dank gilt der "Deutschen Forschungsgemeinschaft" für die finanzielle Unterstützung der Arbeiten im Rahmen des Sonderforschungsbereichs 349.

6. Literatur

/1/ Burger, D., Optimierung der Strahlqualität beim Laserhärten, in LASER Optoelektronik in der Technik, Springer-Verlag Berlin, 1990.
/2/ Patorski, K., Fraunhofer diffraction patterns of tilted planar objects, Optica Acta 30, 673 (1982).
/3/ Leseberg, D.,Frere, C., Computer-generated holograms of 3D-Objects composed of tilted planar segment, Appl. Opt., 27(14), 3020 (1988).
/4/ Jaroszewicz, Z., Kolodziejczyk, A., Mouriz, D., Bara, S., Analytic design of computer-generated Fourier-transform holograms for plane curves reconstruction, J. Opt. Soc. Am. A, 8 ,559 (1991).
/5/ Pahlke, M.; Optimierung von Beugungsstrukturen für hohen Beugungswirkungsgrad, Diplomarbeit Universität Stuttgart, Institut für Technische Optik, (1992).
/6/ Kermisch, D., Immage Reconstruction from Phase Information Only, J. Opt. Soc. Am., 60, 15(1970).
/7/ Haupt, C., Pahlke, M., Jäger, E., Tiziani, H. J., Design of diffractive optical elements for CO2-laser material processing, SPIE Vol.1718 Workshop on Digital Holography (1992), p. 175-180.

Miniaturized Polychromator with Holographic Optical Elements

Günter Lensch, Peter Lippert, Dirk Probian,
NU TECH GmbH, Ilsahl 5, 24536 Neumünster,
Tel. 04321-30620, fax 04321-30631

A report on a compact polychromator will be given.
Miniaturisation is achieved through an optimized arrangement of HOE-diffraction grating, fibre optics and diode array.
The polychromator is suitable for use from the near IR into the UV region.

The polychromator is to be used for spectral analysis. The number of optical components has been minimized through a design employing diffractive holographic imaging optics (gratings).
The spectral energy distribution is almost concurrently recorded by a CCD-array and transformed into an electrical signal via an electronic processor. A schematic diagram of the setup is depicted in **Figure 1**.
 The imaging characteristics of the diffractive optics and the spectral width of the polychromator, as determined from these, was calculated (see **Figure 2**) and optimized be means of raytracing computer programs.

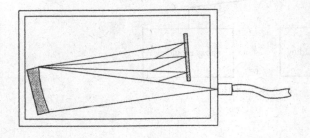

Figure 1: Schematic diagram of the
polychromator

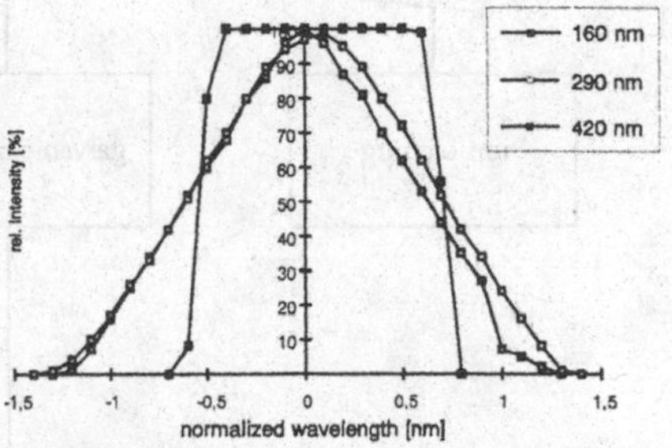

Figure 2: Spectral width of the
polychromator

The steps in the process of manufacturing diffractive optical elements is shown in **Figure 3**.

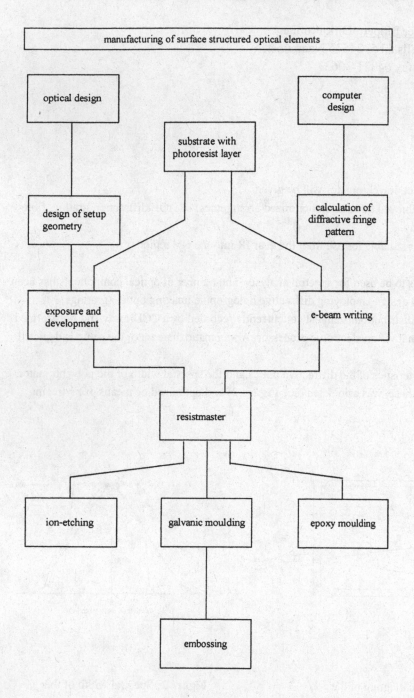

The DOE was interferometrically produced by means of the holographic setup shown in **Figure 4** using resist technology.

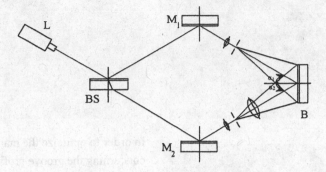

L : Laser
BS : beam splitter
M1,2 : mirrors
B : holographic plate

Figure 4: Holographic setup for the photoresist exposure of polychromator gratings

Due to the spatial frequency of the DOE's resulting from the optical design, it was necessary to optimize the groove profile in order to increase the diffraction efficiency.
Using rigorous diffraction theory, the dependence of the diffraction efficiency on the form of the groove profile was calculated (see **Figure 5**).

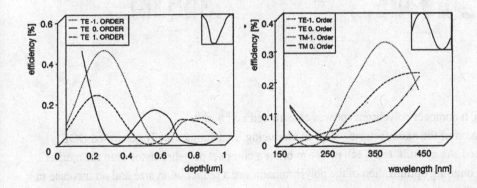

Figure 5: Calculation of diffraction efficiecy using the model of infinite conductivity

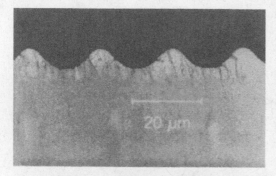

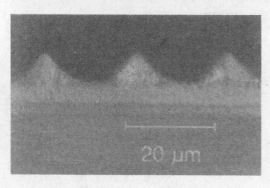

In order to optimize the manufacturing process concerning the groove profile, studies of the resist parameters were performed. Through variation of the manufacturing and processing parameters, various profile forms could be produced in a controlled manner (see **Figure 6**). Controlled post-processing via ion-etching led to a further improvement of the profile. The grating masters produced in this manner were used in a special copying process to produce duplicates in epoxy technology.

Figure 7 shows the diffraction efficiency of the grating master compared to the efficiency of the duplicated grating.

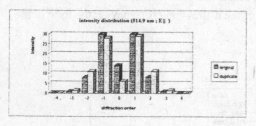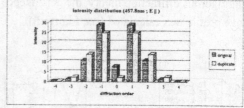

Summary:

At NU-Tech, a compact polychromator was constructed and built.

An optimization of the manufacturing steps in producing interferometrically generated DOE's was performed. As a result, NU-Tech is able to bring a compact polychromator onto the market. Goals of the ongoing optimization of the polychromator are a reduction in size and an increase in diffraction efficiency.

Diffractive Optical Components with Increased Radiation Resistance

G. Lensch, W. Rudolph, P. Lippert, Ch. Budzinski*,
NU TECH GmbH, Ilsahl 5, 24536 Neumünster,
phone. 04321-30620, fax 04321-38435
* Berliner Institut für Optik GmbH, Rudower Chaussee 5,
12489 Berlin, phone 030-63923453, fax 030-63923452

The production and the properties of radiation resistant holographic gratings are reported.
Diffractive optical components (DOC's) open new possibilities for splitting, steering and forming
(focusing) of laser beams [1].

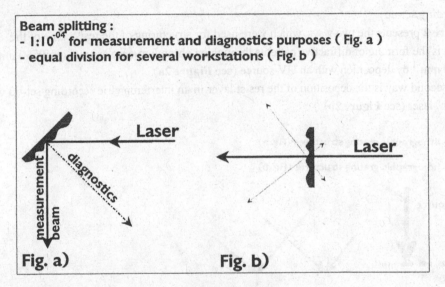

Figure 1 Applications examples of beam splitting

A DOC for splitting laser beams (see **Figure 1a**) in the ratio of $1:10^{-4}$ for measurement and
diagnostics purposes has been developed. **Figure 1b** shows an application for several
workstations. DOC's also were used for wavelength selection or tuning of lasers. DOC may be
produced using resist techniques. However, they are not able to withstand high laser powers, i. e.
they are destroyed by an irradiation of a few Watts per cm^2.
To overcome this disadvantage we transferred the diffractive structures onto substrates such as

glass (for transmission) or copper (for reflection) by ion-etching.

In particularly the following steps are necessary for fabrication of DOC:

1. Preparation of substrate
- Required surface flatness $\lambda/10$
- Scratch and micro-roughness $\lambda/100$ less than 5 nm
- Glass substrate: customary technology
- Copper substrate: ultra high precision machined, oxygen-free copper
- Cleaning, preparation

2. Resist coating
- Resist type S 1400 is spin coated on substrates
- layer thickness and resist treatment are optimization parameters

3. Recording set-up

Figure 2 presents the two ways, which were used for structuring of the resist. One of the ways is the forming of diffractive patterns through an chromium-on-glass mask with 50 mm^{-1} or 20 mm^{-1} by deposition with an UV-source (see **Figure 2a**).

The second way is the deposition of the resist layer in an interferometric recording set-up with an Ar$^+$ laser (see **Figure 2b**)

- lithographic grating structures (Fig. a)

- holographic grating structures (Fig. b)

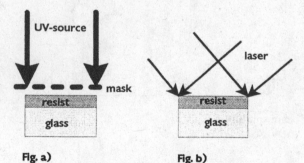

Fig. a)   Fig. b)

Figure 2 Diffractive structuring of the resist

Profile forms and plateaus are functions of deposition and development parameters.
Figure 3 depicts the dependence of the layer thickness on application and the removal as a function of intensity for various development and exposure parameters [2].

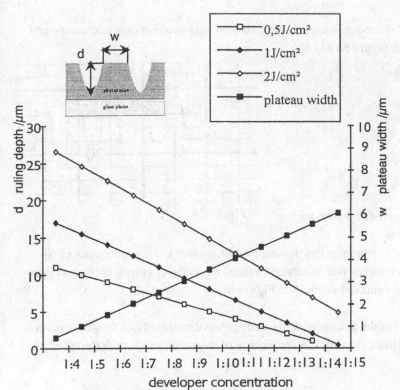

Figure 3
Structuring parameters

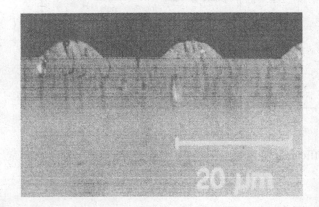

Figure 4 shows the interferometrically fabricated etch-masks of resist (S 1400) on a copper surface.

This structures were transfered onto the surface by plasma etching. The etch rate of copper was found by 340 nm/h and for glass 47 nm/h (400 W, 0.35 Pa).

The depths of the etched structures in copper and glass were measured via light microscope and with a micro-scanner (see **Figure 5a** and **5b**).

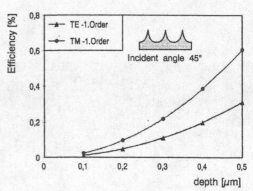

Figure 5a Etch-mask of resist on a glass surface   Figure 5b

The diffraction efficiency is determined by the shape of the grooves and the reflectance of the surface. The diffraction efficiency was calculated by using the model of infinite conductivity for copper for various profile forms and depths (see **Figure 6**).

The diffraction efficiency of the first order of the gratings was determined in a CO_2-laser beam (150 W). The results are listed in **table 1**. A consistency of measuring values with the calculated has been found.

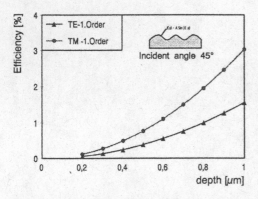

Figure 6 Calculation of the diffraction efficiency

No.	grating	depth	diffractionefficiency in the ratio of	
	l/mm	nm	measured	theoretical
5	50	ca 200	0.50%	0.4 - 0.7 %
8	20	260	< 0.3 %	< 0.1 %

Table 1

Summary
Lithographic and holographic fabricated profiles were transfered onto copper and glass surfaces by plasma etching. The diffraction efficiency was calculated by using the model of infinite conductivity for various profile forms and depths.
The results were presented in comparison to theoretical values.
A minimum of ratio of diffraction efficiencies of $3:10^{-4}$ has been reached.
This work was sponsored by the Bundesministerium für Wirtschaft, Außenstelle Berlin, KZ. 828/2, 1992

References:
[1] Lensch G., et. al., " Abstandssensor mit holografisch optischen Elementen ", Jahrestagung der DGaO 1993, Wetzlar
[2] Lensch G. , et. al.," Optimized replikation of interferometrically generated deep diffractive structures by embossing into thermoplastics ", SPIE Paper No. 1780-27

Singlemode 1,3/1,3 Duplexer for FTTH-FTTC Networks

V,Stappers, J.J.Mailard, M.Poulain, R. Puille, A.Richard
Direct Communications GmbH, St Jean de la Ruelle/F

This paper reports the development and tests made on a singlemode duplexer.

In order to answer to the need of components for FTTH-FTTC networks taking place in the future ISDN network, Compagnie DEUTSCH has developed a singlemode duplexer. Two versions exist, one connectorised, one pigtailed. The connectorised version is associated to a singlemode jumper fully compatible with the standard DIN 41612 connector and works like an "active" connection.

A duplexer allows simultaneous emission and reception of optical signals into and out of a singlemode line. It includes a laser diode, a photodiode and a splitter in order to seperate the beams. The technology is based on zhe concept of expanded beam for the connection and coupling. Different ratios allow to answere to several performances (80/20, 20/80, 50/50).
This product is based on a low cost concept with molded pieces in order to reach with accuracy in the requirement of the markets.

The main characteristics for a 1,3 μm version and 50/50 ratio, are more than -7dBM (200 um) launches power, more than 0,3 A/W responsivity, less than 110 dB/Hz RIN, less than 10 EE-11 BER at -19 dBm at 622,08 MBit/S operation, less thann -25 dB for crosstalk and backreflection. The operating temperature range is from -40°C to +70°C,

Different versions havew been developed and tested within RACE LoCo consortium (Research and develoment in Advanced Communications Technologies in Europe).

Performance Properties of Polymer Optical Fibers

A. HOFFMANN[1], W. DAUM[2], J. KRAUSER[3]

[1] A. Hoffmann, Deutsche Bundespost Telekom FH Berlin now Bundesanstalt für Materialforschung und -prüfung (BAM), Unter den Eichen 87, 12200 Berlin, Tel. (030) 8104-3633

[2] W. Daum, Bundesanstalt für Materialforschung und -prüfung (BAM), Unter den Eichen 87, 12200 Berlin, Tel. (030) 8104-1613

[3] J. Krauser, Deutsche Bundespost Telekom FH Berlin, Ringbahnstr. 130, 12103 Berlin, Tel. (030) 7574-4516

Abstract

Polymer optical fibers (POF) are of increasing importance for optical short-distance signal transmissions especially in industrial applications. Here they are exposed to a multitude of environmental stress factors that can affect the functionality of the optical fibers. This article describes investigations of cyclic bending with different bending radii at combined temperature and humidity changes on transmission loss of POF. To this end the POF were exposed to the combined environmental stress factors in a special experimental setup and the effects on the transmission measured. The introduction into the group of subjects dealt with, where, among others, the environmental stress factors affecting the optical fibers are shown, is followed by a description of the test equipment and measuring philosophy. Above and beyond this typical transmission curves of the test results are shown which are then discussed.

1. Introduction

POF may be used in many short-distance transmission applications. They offers the advantages of easy handling and flexibility and the high numerical aperture allows to use lower cost components. The advanced durability make POF more and more attractive for optical data communication systems in industrial applications such as short data links in robot and machine control, automotive systems and fiber optical sensors.

During operation under industrial conditions, environmental stress has an important influence on the durability and functionality of a POF data transmission system /1/. Environment stress factors for POF in industrial applications can be characterised by mechanical stress factors (e.g. static and cyclic bending, vibration, torsion), climatic stress factors (e.g. temperature, humidity, temperature and humidity changes) and chemical stress factors (e.g. solvents, fuel and types of oil). The intensity and frequency of these stress factors vary over a wide range, so that an exact characterisation can only be given for specific industrial applications.

1.1 POF-Structure

In the investigations, POF are used with a core material of polymethylmethacrylat (PMMA), a cladding of a special fluorpolymer and a coating of polyethylen (PE).

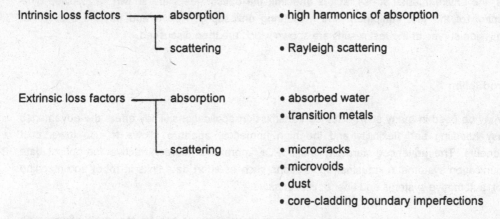

Structure	stepindex
Core diameter	980µm
Cladding diameter	1000µm
Total fiber diameter	2.2 mm
Acceptance angle	56°
Core refractive index	1.492
Cladding refractive index	1.415
Attenuation at 650nm	150 dB/km

Fig. 1. Structure and properties of the POF

1.2 Transmission loss due to environmental stressing

The most interesting POF property is the transmission loss and its variance with environmental stressing. Loss mechanisms and loss factors in POF are well known /2/. Loss factors are divided into two categories: intrinsic and extrinsic. Both consist of scattering and absorption.

Intrinsic loss factors ⎯⎯⎯┌── absorption • high harmonics of absorption

                                  └── scattering • Rayleigh scattering

Extrinsic loss factors ⎯⎯⎯┌── absorption • absorbed water
                                             • transition metals

                                  └── scattering • microcracks
                                             • microvoids
                                             • dust
                                             • core-cladding boundary imperfections

With regard to transmission loss caused by environmental stress factors, extrinsic loss factors have the major influence. Due to climatic stressing the absorption of water is most interesting effect. Investigations have shown that the influence of absorbed water becomes a significant constituent for the loss of POF /3/. Due to mechanical stressing new impurities and imperfections can arise. Furthermore existing defects can increase. This can lead to an early failure of a POF data communication system.

2. Methods

2.1 Setup of the test equipment

Figure 2. shows the principal setup for the simulation of cyclic bending under different climatic conditions in industrial applications. The cyclic bending equipment is integrated into the environmental chamber. Simulations with different bending radii in a range of 5mm to 40mm are possible.

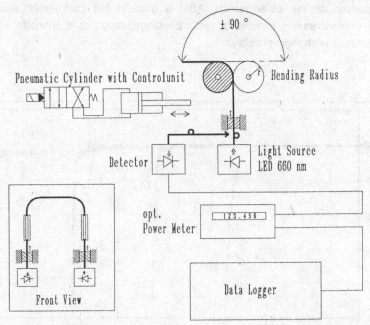

Fig. 2. Principal setup for cyclic bending under climatic conditions

The optical transmittance was measured in the unbended state of the POF before start of the simulation, and during the simulation at constant time periods in the unbended state after a relaxion time of 1 minute. In order to study the influence of different climate conditions on the bending behaviour of the POF, tests were carried out at room temperature, -40°C and +85°C/85% RH.

3. Results

In Figure 3. results from cyclic bending tests of a POF with a PE coating at room temperature and bending radii of 5, 10, 20 and 40mm are presented. All graphs show nearly the same characteristic of change in transmittance. Depending on the bending radius the POF endures the cyclic bending procedure without any change in transmittance up to a characteristic number of cycles. Then a dramatic change in transmittance appears. The durability (transmittance ≤50%)

varies between 300 cycles (R=5mm) and 12300 cycles (R=20mm). For a bending radius of R=40mm the POF endures 100.000 cycles without any significant change in transmittance.

In order to study the influence of combined environmental stress factors, the cyclic bending test was reproduced under two different climatic conditions at -40°C and +85°C/85% RH. Figure 4. shows a typical result from these tests. For the same bending radius of R=10mm durability is changing with the climatic conditions. In this case the durability for -40°C is approximately half of the durability at room temperature. At 85°C/85% RH optical transmittance decrease in a different way in comparison to the other results. After a gradual but permanent decrease of the transmittance with increasing number of cyclic bendings, again at a characteristic number of cycles transmittance is changing rapidly.

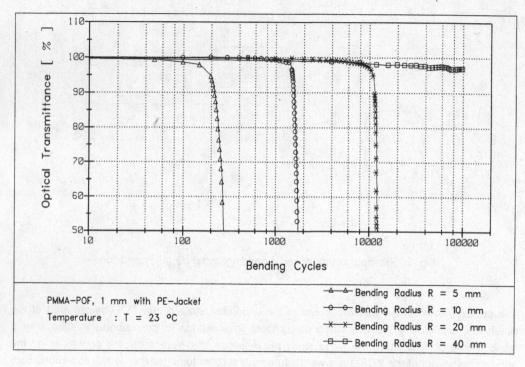

Fig. 3. Transmittance vs. cyclic bending with different bending radii

Characteristic test results for a POF with a PE coating from cyclic bending under different climatic conditions and different bending radii are summarised in Fig. 5. It shows the durability of the POF as a function of four bending radii and three different climatic conditions. The durability of a POF at a bending radius of R=5mm is less than the durability of bigger bending radii. At low temperature and under a bending radius of R=5mm the durability is less than at other climatic conditions, but for bending radii in a range of R=20mm it becomes nearly the same durability.

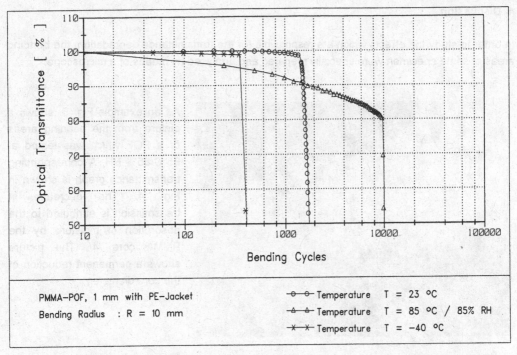

Fig. 4. Transmittance vs. cyclic bending under room temperature and extreme climate conditions

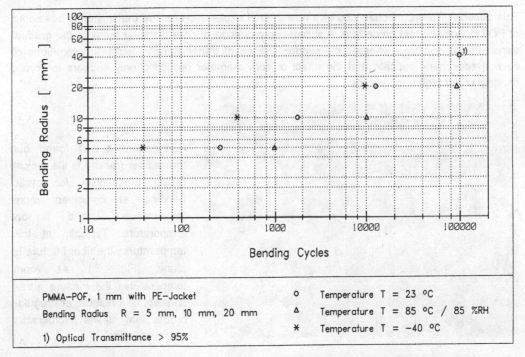

Fig. 5. Durability for cyclic bending as a function of bending radii and climatic conditions

698

4. Discussion

In order to study the effects which are responsible for the transmittance degradation, the bending areas of some specimen were specially prepared and analysed by means of a microscope.

As an example Fig. 6. shows a picture from the bending areas of a POF which was tested at 85°C/85% RH. A corresponding transmittance graph is shown in Fig. 4. The decrease in transmission is attributed to the absorption of moisture by the PMMA core /4/. The picture shows a permanent reduction of the core diameter.

Fig. 6. Failures in the bending area due to cyclic bending under 85°C/ 85% RH

This observed failure mechanisms give now a good explanation for the transmittance behaviour of POF during the test procedure. A successive reduction of core diameter leads to the gradual transmittance decrease whereas the sudden change of transmittance results. Investigations of several specimens indicate that the effect of core diameter reduction only appears for cyclic bending under high temperature.

Furthermore it was found out that a fiber fracture is the typical failure mechanism for cyclic bending at combined room temperature and low temperature. Typically at low temperatures the fiber fracture is plane (Fig. 7.). At room temperatures the fracture areas consists of a number of irregular cracks that means microcracks and microvoids.

Fig. 7. Failure in a bending area due to cyclic bending under -40°C

5. Conclusions

As a result of the systematic investigations the following conclusions can be made: The durability of a POF for cyclic bending under high temperature and relative humidity is higher in comparison to durability at room temperature. For small bending radii and low temperature the durability concerning cyclic bending is less than the durability at normal conditions, but for bending radii in a range of R=20mm it becomes nearly the same. This demonstrates that extreme climatic conditions have no negative influence on the excellent flexibility of POF.

6. Acknowledgement

The authors wish to thank the staff of BAM-Laboratory 6.13 "Optical measuring techniques; experimental stress analysis"; especially Dipl.-Ing. A. Rahvali, cand.-ing. S. Zedler, D. Kadoke for their invaluable assistance.

7. References

/1/   W. Daum, A. Brockmeyer and L. Goehlich:
        Environmental qualification of polymer optical fibers for industrial applications
        Proc. of the Plastic Optical Fibres and Applications Conference, Paris June 22-23, 1992,
        pp. 91-95

/2/   T. Kaino: Absorption Losses of Low Plastic Optical Fibres
        Japanese Journal of Applied Physics, Vol. 24, 12(1985), pp. 1661-1665

/3/   T. Kaino: Influence of water absorption on plastic optical fibres
        Applied Optics Vol. 24, No. 23 (1985), pp. 4192-4195

/4/   A. Brockmeyer, R. Kaps, J. Theis: Polymer optical fibers for industrial applications
        Proc. of the Plastic Optical Fibres and Applications Conference, Paris June 22-23, 1992,
        pp. 119-122

Fiber Positioner with a new Drive Concept

R.Glöß, H.Marth
Physik Instrumente (PI) GmbH
7517 Waldbronn

Requirements for a Fiber Positioner

Handling and packaging of monomode fibers, waveguides, coupling to laserdiodes and the arrangement of miniature optical components, require high precision and stability in the nanometer range.

Often, scan functions for millimeter range length are useful. Manual mechanical handling or touching of the positioner during alignment in the nanometer range is not acceptable. This means that all moving functions must be controlled by a computer.

Positioner Concept

The LightLine combines:

- 3-axis moving system for a range of 6x6x6 millimeters
- Integrated Piezo Drives
- Integrated absolute sensors
- Remote control electronics for all functions
- An optional 3 angle tilt unit

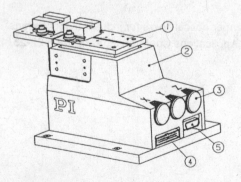

1 Moving Platform
2 Chassis
3 "Optical Micrometer" for manual control
4 Position Sensor Connector
5 PIEZO WALK connector

3-axis LightLine Positioner

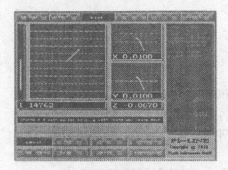 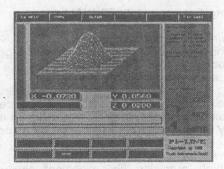

Graphic software tool for the control function

The mechanical Leading Concept:

To reach the required accuracy, a high precision ball-bearing system was chosen. It creates much less stress into the chassis than spring bending systems, gives more stiffness and permits long linear tracking without crosstalk. However, it requires a stiff, sensor controlled driving system because of the variations in the roller resisting force.

Four-joint leading   **Roberts-driver**

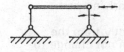   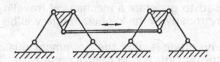

Disadvantages: *Disadvantage:*
Circular moving,   Instability for large
Crosstalk   travel ranges

Sliding bearing   **Ball bearing**

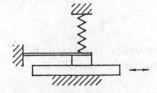   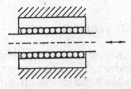

Disadvantage: *Disadvantage:*
High friction coefficient   Small variations of
                    roller resisting force

Driving Concept of PIEZO WALK

Requirements for the drive:
- highest resolution in the nanometer range
.- velocity up to a few mm/sec
- force of more than 5N
- long travel range, only restricted by bearings
- smooth movement, no vibration influence
- small dimensions, easy to integrate into a stage
- only remote controlled operation by a computer

Known Precision Drives:
For high precision fiber alignment always a drive-sensor concept is necessary.

DC-motor and spindle:
For small size and high force a lossless gear is necessary. The drive gives vibration influence. The length of the drive does not match with small stages, so it is difficult to integrate it into small stages.

DC-motor, spindle and hybrid piezo stack:
They have the same disadvantages as mentioned before. Fast scans can only be performed by piezos within very small ranges.

Stepper motor and spindle:
Vibration influence might be very high. High resolution can only be achieved with gearheads.

Piezo ultrasonic motors:
There are many different constructions. Generally a resonance frequency of one part is used to generate a mechanical traveling wave or vibration. This vibration frequency mostly in the kiloherz range will be applied to the driving part.

The disadvantage is, that the movement quality depends strictly from the friction coefficient- a very critical parameter. The velocity can not change to extremly small values, because it depends on the amplitude of wave. For small amplitudes the moving stability and smoothness is not given.

Piezo drives of Inchworm type:
This type has the advantage that the velocity can be controlled over a wide range. The critical parts are the clamps. The clamps induce spikes into the motion. Either the clamps are closed with preload and can be opened, or they are opened and can be closed. Because the piezo stroke is only a few micrometers, mechanical tolerances are very small. The construction must be optimized to compensate thermal influence. Mostly high voltage piezo systems are used.

Principle of PIEZO WALK Operation

The PIEZO WALK works like a walker, standing on a movable band with high load on his shoulders. He is not strong enough to spring. He holds tight at a stake. The movable band is the driving part.

What occurs?
If he walks, the movable band is driven forward or backward. Always the walker is pressed to the band. He can only raise on one foot. If he has thin or thick shoes, it has no influence. If the band has not stable thickness - it has no influence. If it is warm or cold - it has no influence.

The PIEZO WALK

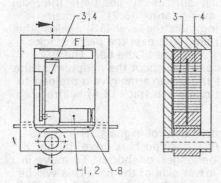

1,2 Piezo Pusher
3,4 Piezo Stacks
8 Steel Bar

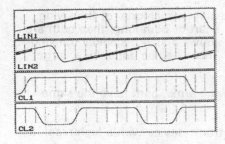

Control Sequence

The Miniaturized System
The system was miniaturized and optimized to small dimensions, especially in the moving direction. All piezoceramic stacks should be preloaded.
Manufactoring tolerances should be as large as possible.

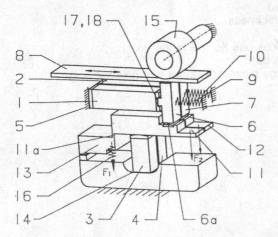

Miniaturized PIEZO WALK

The driving part is a steel bar (8) with parallel surfaces in line with the roller. This bar is clamped between roller (15) and swing arms (6,7). The clamp force comes from one spring (16). This spring pulls the lever (13) against the case, so that the two double levers (11,12) are tilted over the piezo-stacks (3,4) and the swing arms are pressed to the lower side of the bar. The two piezo-pushers (1,2) are placed between the case and about the midpoint of the swing arms. Two springs on the right side of the swing arms give a preload to the piezo-pushers (1,2). Preload, in case that one piezo-stack (3,4) is lifted, comes from two springs described by (F2).

The moving sequence results from the electronic control signals. The piezo-pushers are driven by signals LIN1 and LIN2. These pushers drive the bar forward or backward even though they may be longer or shorter. To move in any way, the swing arms must lift from the lower side of the bar. This will be done by control of the piezo-stacks (3,4). The control signals are CL1 and CL2.

Only the length difference of these stacks decides which one of the swing arms are pressed against the bar. If both are short or both long, than both are pressed against the bar. The signal sequence shows a smoot sinoidal transition between low and high level of the signals. These can be done in a large range.

During contact of one and lift of the other swing arms, both have the same velocity.

The drive will be digital controlled by an electronic device with a high resolution linear sensor, mounted inside the stage.

Advantages of the PIEZO WALK

- small dimension, very short length
- moving range is not limited
  bars or bands as driven parts can be used
  nanometer resolution is available at any time and position
- the drive is always clamped, even it is not energized
  by error control no opening function is possible
- very low influence of mechanical tolerances
  several materials can be combined
- the moving part can change its thickness
- no influence of thermal effects
- the abrasion at the contact surfaces has no influence on
  operation as long as the same on both
- low voltage as well as high voltage piezo actuators can be used

Two-Dimensional Offset Microsensing by Fibre-Fibre Coupling

R.Güther, M.Becker, R.Staske

Ferdinand-Braun-Institut für Höchstfrequenztechnik im Forschungsverbund Berlin

Rudower Chaussee 5, D-12489 Berlin-Adlershof, FRG

and H.Richter

Forschungsinstitut der Deutschen Bundespost TELEKOM

Am Kavalleriesand 3, D-64295 Darmstadt, FRG

Abstract: A weakly multimode fibre is incoherently excited by a single mode fibre. The change of the farfield pattern emitted by the multimode fibre is used to determine the two-dimensional offset between the two butt end coupled fibres.

Introduction: One of the main problems in laser-fibre-coupling and fibre-fibre-coupling is the offset of the fibre relative to the laser in transverse and lateral direction. Offset measurements can be done by measuring the position of submounts of the elements or by observing light transmission. If applicable, the ladder case would be the the better and more precise possibility. By FAUGERAS et al. a butt end fibre coupling was used for a detection of one offset co-ordinate. Here we determine the two transverse offset co-ordinates by butt-end fibre-fibre coupling. This combination can simulate a chip-fibre-pair for testing a chip-fibre coupling method. Another possibilitiy ist the integration of this offset measuring pair into to a microoptical connection to be tested.

   Setup and simulation model: Setup and steps of theoretical simulation are shown in figure 1. Light of a superluminescence laser diode (λ = 820 nm, $\Delta\lambda$ > 10 nm) propagates as single mode in fibre 1 (λ_{cutoff} = 750 nm). Fibre 2 is a weakly guiding birefringend multimode fibre (λ_{cutoff} = 1250 nm) the modes of which were modeled by elliptical Hermite-Gaussian modes as given in SNYDER et al.. Fibre 2 was a York fibre in which the modes [0,0],[0,1],[0,2],[1,0],[1,1] were possible with our light source. The theoretical treatment starts with a (generally elliptical, here rotational symmetrical) Gaussian 0,0-mode, which is matched to the above mentioned mode combination in fibre 2 with inclusion of the 2-dimensional offset co-ordinates ξ and η (GRAU et al. , KRYVOSHLYKOV et al.). The farfield after the exit face of fibre 2 we get by Fourier transform. Because of high line width of the light source and the low beat distances between the modes an incoherent mean procedure is used in the camera receiving plane. Control calculation have shown that the mode patterns are very weakly influenced by spectral mean procedures in 10nm- or 100nm-ranges. Theoretical mode patterns in dependence on ξ and η are shown in figure 2. Experimental patterns are very similar with figure 2. The main deatures of figure 2 are different width and different height. Therefore we construct two recognition parameters p_x and p_y by the following procedure: 1. Normalization of the integral pattern intensities to the value 1. Then the shape of patterns is essentially. 2. p_x = integral along the x-axis, x-ing the centre of gravity of pattern. 3. p_y = integral along the y-axis, x-ing the centre of gravity. The analytical formulae for px and py can be approximated by a lowest order expansion in ξ and η : $p_x \sim (1 - ... \eta^2)$ and $p_y \sim (1 - ... \xi^2)$. If this rough approximation is assumed to be exact, then ξ and η can be calculated from p_x and p_y in a simple way, but with uniqueness in a quadrant of the $\xi - \eta$-plane only. Programming the analytical formulae we get figure 3 for p_x and p_y. The paraboloidal shape near to $\xi = \eta = 0$ is obviously. Furthermore, we see that p_x does not disturb the uniqueness, but the uniqueness range is limited by the p_y-curves because of twofold solution in some inversion cases. This results in figure 4.

Experiment: The camera in figure 1 was followed by a PC with image processing. Fiber 2 was carried by piezidriven slides and scale steps down to 0.01 μm. The scanned ranges were different fibre combinations between 2 x 2 μm^2 and 6 x 6 μm^2. The parameters p_x and p_y were determined from experimental patterns. We made an approximation ansatz

$$p_x = a_{00} + a_{20} \xi^2 + a_{02} \eta^2 + a_{22} \xi^2\eta^2$$

$$p_y = b_{00} + b_{20} \xi^2 + b_{02} \eta^2 + b_{22} \xi^2\eta^2$$

(or other appropriate functions) and fitted at four points of the $\xi - \eta$-plane. The we can compute the offset ξ and η from the measured p_x and p_y. This ξ and η can be compared with that ξ and η shown by the piezodrivers. The relative errors in relation to the ξ-range or η-range are given in figure 5. The maximum error is 15%. In other cases approximations by planes resulted in 10% error within a 2 x 2 μm^2 - range. In summary we obtained for 3 or 4-point fittings 8 - 20 % error. With increasing fitting number the error decreases.

In figure 5 we show 9 patterns of a 200 x 200 nm - surroundings and the differences versus the central pattern by experiment and calculation. The signs of differences determine the position uniquely. Figure 6 explains that in the vicinity of a given pattern two signals from three receivers are sufficient for determination of ξ and η.

A **relative sensitivity** we could see in the equipower patterns down to 10 nm. Two patterns with an offset-difference 15 nm are shown in figure 7. We guess the possibility of better sensitivities.

Applications we see in the connection techniques for microptical components, two-dimensional microsensing and testing of precision sligdes in the sub-μm-range (typical values for turning $\sim$ 50 - 100 nm).

We thank to Lutz Olivier and his group, FI-TELEKOM, Berlin, for interesting discussions and technical support.

References:

[1] FAUGERAS,P., PAGNOUX,D., DI BIN,F. and BLONDY,J.-M., 'Measurements of fiber offsets in multimode connectors using single-mode excitation', *Opt. Commun.*,1992,**90**,pp. 35-38

[2] GRAU,G., LEMINGER,O. and SAUTER,E., 'Mode Excitation in Parabolic Index Fibres', *Arch. Elektr. Übertr.techn.*,1980,**34**,pp. 259-265

[3] KRIVOSHLYKOV,S.G., and SAUTER,E.G.,'Excitation of Gauss-Laguerre modes in parabolic index fibers by astigmatic Gauss-Hermite beams or by Gauss-Laguerre beams with offset, tilt and gap', *Arch. Elektr. Übertr.techn.*,1990, **44**,pp. 462-469

[4] SNYDER,A.W., and LOVE,J.D., 'Optical Waveguide Theory' (Chapman and Hall, London, 1983), Ch.16.3

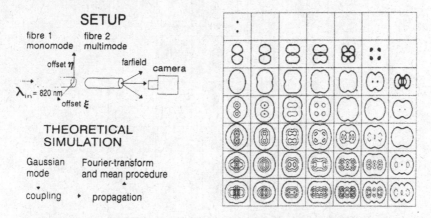

Fig. 1 Setup and method of simulation

Fig. 2 Farfield patterns for variation of ξ (abscissa) and η (ordinate)

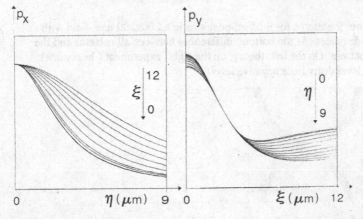

Fig. 3 Recognition parameters p_x and p_y on the $\xi - \eta$ -plane.

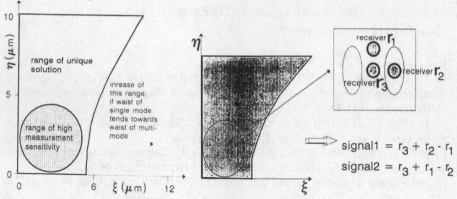

Fig. 4 Range of unique solution

Fig. 6 3 receivers generate 2 signals

$$\text{signal1} = r_3 + r_2 - r_1$$
$$\text{signal2} = r_3 + r_1 - r_2$$

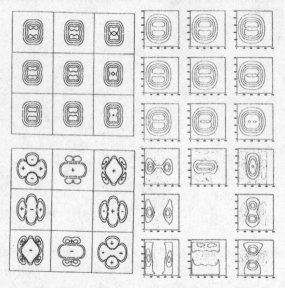

Fig. 5 On the top: 9 patterns for 9 offset-positions in a 200x200 nm^2-field with 100 nm-distances. At the bottom: differences between all patterns and the central pattern. On the left: theory, on the right: experiment (heavy lines: positive level,dotty lines:negative level)

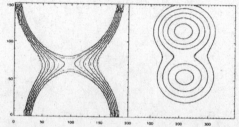

Fig. 7 Sensitivity for a 15 nm- offset-change: On the right: survey of pattern, on the left: many level detail: heavy lines: with offset change,light lines: without offset change. The sharp levels result from image processing (smoothing).

The Description of an Ideal Cylindrical Multimode Step Index Waveguide by the Means of a Transfer Matrix

M. Eckerle, A. Chakari, P. Meyrueis

Laboratoire des Systèmes Photoniques

Ecole Nationale Supérieure des Physique de Strasbourg

Université Louis Pasteur

7, rue de l'Université

F-67084 Strasbourg cedex

1. Introduction

In our laboratory a computer controlled analyser for research, development and testing of sensors and optical components made of step index wave guides is currently under development. The analyser characterises and analyses mode coupling in all known kinds of optical multimode step index wave guides (msiw) like fibres and optical components like couplers or modulators based on such wave guides.

For the correct interpretation of the automatically obtained data, a model has to be set up for each msiw geometry (elliptical, planar, cylindrical...). These models describe the distribution of the light intensity in the far field of the wave guide depending on the injection angle of the laser light. Hereby, the msiws are thought to be ideally uniform.

In the following chapters a model for the probably most common case of an optical fibre i.e. cylindrical msiw geometry will be presented.

Moreover, remarks will have to be made about the setup of the transfer matrix, whose coefficients represent a relative intensity. The coefficients are based on the relative output light power $P_{out,rel}(\Theta_i)$ as function of the injection angle Θ_i . However, most important are the geometrical limits, which have to be respected, to obtain useful results.

2. Assumptions

1. For the modelisation, it is assumed to have an uniform fibre with a cylindrical core of a refractive index n_{co} , a cylindrical clad of a refractive index n_{cl} and a perfectly light absorbing, cylindrical jacket (figure1).

2. A HeNe laser with a Gaussian distribution of light intensity and a wave length λ serves as light source.

3. The core radius $r_{co} = a/2 \gg \lambda$ which is the wave length of the laser light, so that the ray theory can be applied.

4. The laser beam is adjusted in the way that the symmetry line of the fibre and the laser beam intersect (figure1).

5. The acute angle of the light cone is twice as large as the injection angle Θ_i.

6. The light in the fibre's far field is distributed uniformly within 2 concentric circles ($\Theta_{min} < \Theta_i \leq \Theta_{max}$); Θ_{max} can be chosen within the limits $0° \leq \Theta_{max} < 90°$ having always the same distance. The distance between the circles equals the diameter a of the fibre core (figure1).

7. Due to the small difference between the refractive index of the core and the clad, the transition between reflection and transmission at the boundary core cladding is thought to be abrupt. An example of the idealised and the real reflection is given in graph 1. Therein the refractive indices are taken from a widespread multimode step index fibre. Furthermore, graph 1 underlines that it makes no difference whether the light is polarised parallel or perpendicular to the plane of incidence, so that the chosen approximation is justified.

8. Within the fibre leaky rays are assumed to propagate whereas refracted rays assumed to dissipate in the cladding and the jacket. This assumption can made because of assumption 7 and because even for $\Theta_c < \Theta_i < 90°$ light can be detected at the output, especially when short fibres are taken. $\Theta_c = \arcsin(NA)$ is the critical angle and NA the numeric aperture of the fibre.

9. Modes do not couple in the fibre.

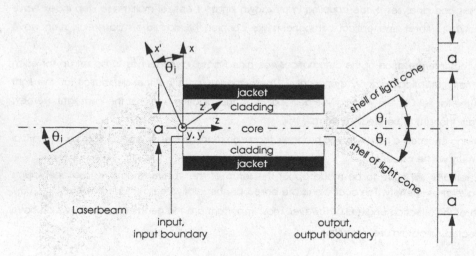

Figure 1
The Illustration of the model

3. The relative output intensity matrix

The exact determination of the relative output power $P_{out,rel}$ and of the intensity transfer matrix $\underline{I}$ would be to long to be explained in detail. Therefore, only the basic ideas are presented:

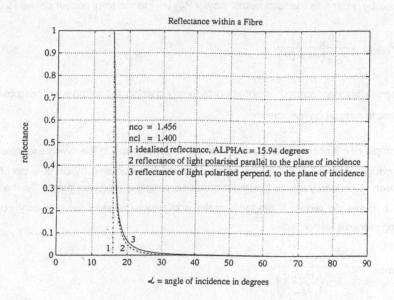

Reflectance within a Fibre

reflectance

nco = 1.456
ncl = 1.400
1 idealised reflectance, ALPHAc = 15.94 degrees
2 reflectance of light polarised parallel to the plane of incidence
3 reflectance of light polarised perpend. to the plane of incidence

$\measuredangle$ = angle of incidence in degrees

Graph 1:
Real and idealised reflection within a fibre

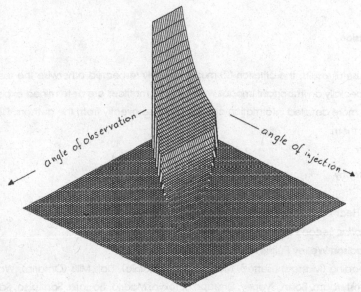

angle of observation

angle of injection

Graph 2
A typical intensity transfer matrix where criteria (2) is not respected

In a first step the laser power on the input boundary (core surface) is determined. Then the total light power P_{out} in the light cone is derived while taking account of transmission at the input and output boundary [1]. The infinite number of reflections between both boundaries and the transmission at every instance of reflection at the output boundary counts as well for the power P_{out}. Finally, $P_{out,rel}$ is obtained by relating to the total output power $P_{out}(\Theta_i)$ to the total output power $P_{out}(\Theta_i = 0°)$.

$$(1) \quad P_{out,rel} = \frac{P_{out}(\Theta_i)}{P_{out}(\Theta_i = 0°)}$$

The intensity transfer matrix $\underline{I}$ is obtained by simply dividing the relative output power $P_{out,rel}$ by the appropriate surface on the screen.

An example of it is of a typical intensity matrix $\underline{I}$ in given graph 2. The given example seems to suggest mode coupling although there is no mode coupling. This is caused by an overlap of the light rings on the screen. An intensity matrix $\underline{I}$ without mode coupling is a diagonal matrix. So the result obtained in graph 2 makes no sense. A diagonal matrix and hence a useful result is only obtained when the following criteria is respected

$$(2) \quad r_{max} > \frac{a}{2} \cdot \frac{\tan\Theta_{max}}{\tan\left(\frac{\Theta_{max}}{N-1}\right)}.$$

Therein r_{max} is the maximum radius of the base of the light cone obtained with the injection angle Θ_{max}.

4. Conclusion

To obtain useful results, the criterion (2) must absolutely respected otherwise the results are of no value. This has especially an important impact when transfer matrices are determined experimentally.
More, and more detailed information can be obtained directly from the authors. Please do not hesitate to contact them.

5. Bibliography

[1]   E. Hecht and Alfred Zajac
        Optics, Second Edition
        Addison-Wesley Publishing Company 1987
        Reading (Massachusettes), Menlo Park (California), Don Mills (Ontario), Wokingham (England),
        Amsterdam, Bonn, Sydney, Singapore, Tokyo, Madrid, Bogota, Santiago, San Juan.
        chapter 4.2, pages 92 - 113.

Congress E

Congress-Chairman: E. Wagner

Mikrosensorik und Faseroptik
Micro-Sensors and Fiber Optics

Micro Engineered Displacement Sensor

Günter Lensch, Peter Lippert, Dirk Probian, Horst Kreitlow*,
NU TECH GmbH, Ilsahl 5, 24536 Neumünster,
Tel. 04321-30620, fax 04321-38435
* Fachhochschule Ostfriesland, Constantiaplatz 4
26723 Emden, Tel. 04921-807345, fax 04921-807201

A sensor is presented, produced using micro-system technology, which is suitable for use in measuring small distances in manufacture. The miniaturised sensor head comprises specially developed holographic optical elements for beam splitting, combining and forming, a laser diode and external signal processing electronics. The sensor is capable of measurements of distances of several centimeters with a travel on the order of millimeters.

The distance sensor functions using the astigmatic-error-method [1]. The principle set-up and beam geometry for a macroscopic design of this apparatus is shown in **Figure 1**. Dependent upon the distance between the sensor and the object being measured, one obtains, between the beginning of the measurement range (greatest distance to be measured) and the end of the measurement range (smallest distance to be measured), different, characteristic intensity distributions in the plane of the detector. **Figure 2** depicts this dependence for the case of a steel surface with $Rm = 0.6\mu m$. Via a measurement of the major and minor axes of the elliptical intensity distribution, e.g. with a 4-quadrant diode and suitable processing electronics or a CCD-camera with an image processing system, unambiguous determinations of distances are possible. An important advantage of holographic optical elements (HOE) is that they may be produced as diffractive structures on flat, very thin substrates. This allows the construction of compact optical measurement systems. Furthermore, the ability to generate diffractive structures with the necessary spatial frequency in small areas (e.g. 1 mm²), allows a further spatial miniaturisation of optical measurement systems. The distance sensor depicted in **Figure 1** may be constructed as a miniature measurement head using three HOE´s

→ polarising beam splitter
→ collecting lens
→ cylindrical lens

combined with a quarter wave plate, a semiconductor laserdiode and a 4-quadrant diode. The three HOE´s are combined on the three surfaces of a prism to a single compact element as depicted in **Figure 3**. The optical quality requirements placed on the HOE´s are listed in table 1.

For the production of these optical elements in photoresist, holographic exposures as well as computer generated electron beam scribing techniques were studied and applied. The resist masters produced in this manner were then directly replicated on the surface of the prism via epoxy moulding. This process is detailed in **Figure 4**, and in **Figure 5** a schematic diagram of the set-up used for the holographic exposures is depicted.

Variation of the groove profile served to optimize the optical characteristics of the HOE's. To this end, parameters concerning the photoresist such as:

→ layer thickness
→ baking duration
→ energy density during exposure
  ● power density of the object wave and the reference wave
  ● duration of exposure
→ duration of development
→ concentration of developer

were studied experimentally, the results of which are detailed in **Figure 6** and in [2] . Computer simulations following a model due to Bartolini [3] were also carried out, the results being displayed in **Figure 7** . In order to optimize the diffraction efficiency with respect to its dependence upon the groove profile, computer simulations following Petit's model [4] were performed (see **Figure 8**) .

The development of the individual components of the distance sensor was successful enough to allow the production of a functional prototype for the complete optical component comprising all three HOE`s using the replication techniques described here. For the HOE`s, the following manufacturing techniques were employed:

1. polarising beam splitter
   Holographic exposure of photoresist

2. collecting lens
   Electron beam scribing in photoresist using computer generated data

3. cylindrical lens
   Exposure in photoresist using computer generated masks

Tests of those components gave good results and lead us to expect the miniaturized version of the sensor to function at least as well as the macroscopic version. Following coupling of this combined optical element with light source (laser diode), the 4-quadrant sensor and previously developed signal processing electronics, testing of the complete system will be undertaken.

References

[1] **Kimiyuhi Mitsui**, et. al., Development of high resolution sensor for surface roughness, Optical Engineering, 6(1988), pp. 498-502

[2] **Lensch, G.**, et. al., Optimized replication of interferometrically generated deep diffractive structures by embossing into thermoplastics, SPIE Paper No.1780-27

[3] **Bartolini, R. A.**, Holographic Recording Materials (edited by Smith, H. M.), Topics in Applied Physics, Vol. 20, Springer Verlag, 1977 pp. 209-227

[4] **Petit, R.**, et. al., Electromagnetic Theory of Gratings (edited by Petit, R.), Topics in Current Physics, Springer Verlag, pp. 181ff

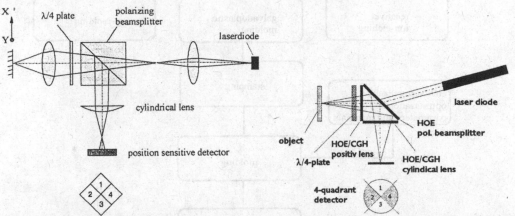

Fig 1 : Macroscopic set-up of the distance sensor (principle)

Fig 3: Miniaturized distance sensor (principle)

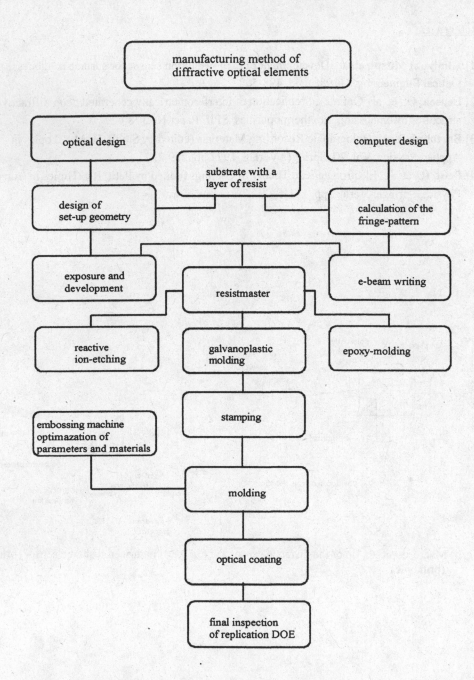

Fig 4: Manufacture of surface-structured optics

end of measuring range

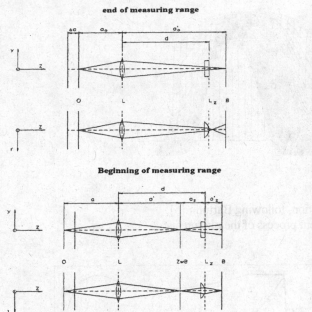

Beginning of measuring range

Fig 2: Distribution of intensity for various distances
between the sensor and the object

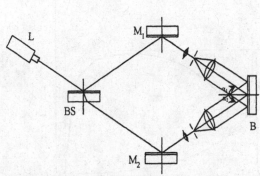

Fig 5: Schematic setup for the holographic resist
exposure

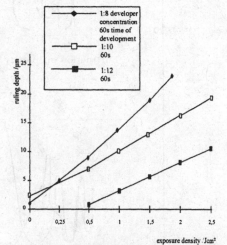

Fig 6: Dependence of the groove (ruling)
depth in the photoresist upon the
exposure energy density

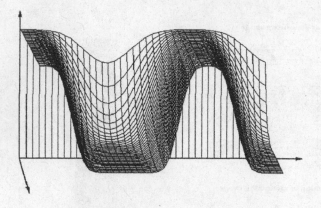

Fig 7: Computer simulation, following Bartolini[5],
of the development process of the photoresist

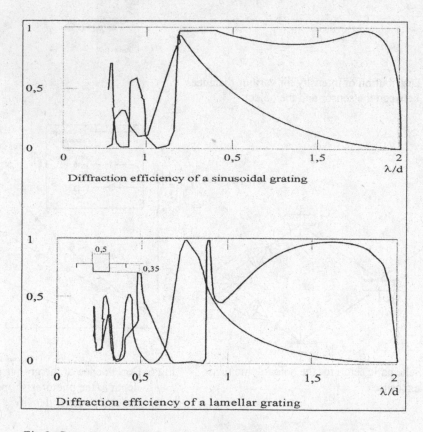

Fig 8: Computer simulation, following Petit[4], of the diffraction
efficiency of HOE`s

Acoustically Tunable Integrated Opticl Filters for the 0.8 μm Wavelength Range

I.Hinkov and V. Hinkov

Fraunhofer Institut für Physikalische Meßtechnik

Heidenhofstr. 8, 7800 Freiburg i.Br.

1. Introduction To date, most research on integrated optical acoustically tunable wavelength filters (ATOF) is concentrated in the wavelength range 1.3 μm to 1.5 μm, which is of interest for the communication systems [HEFFNER et al.]. However, systems operating near 0.8 μm in the first telecomunication window may also be expected to find applications for short distance distribution and local area networks. Several sensor applications can also be envisaged. The main advantage of systems operating in this wavelength region is the availability of a large choice of low cost AlGaAs thansmitters and silicon receivers. Recently, thulium-doped fiber amplifiers operating at around 0.8 μm have also been presented [SMART et al.].

In a recent investigation, we have demonstrated, that the collinear acousto-optic interaction near 0.8 μm allows the construction of devices with significantly lower RF-drive power, smaller bandwidth, increased channel density etc. [I.HINKOV and V.HINKOV]. Now we have developed and investigated for the first time very low power, narrowband integrated ATOFs operating in the first communication window. They are composed of overlapping stripe guides for the optical and the surface acoustic waves (SAW) and a new type efficient, easy to fabricate integrated polariser.

2. Fabrication of the filter devices The structure of the ATOF we have developed is schematically shown in Fig.1. It essentially consisits of an integrated, acoustically driven TE-TM mode converter and an integrated polariser.

2.1. Polarisers for the filter In order to design an ATOF with good filter characteristics an efficient, low loss polariser is indispensible. The well known dielectric/metal or proton exchanged segment polarisers have some drawbacks: they require precise control and time consuming fabrication; it is difficult to obtain low insertion loss.

We have developed a new type polariser [I.HINKOV and V.HINKOV] consisting of a proton exchanged and annealed region fabricated directly on the Ti-waveguide (see Fig.1). In the

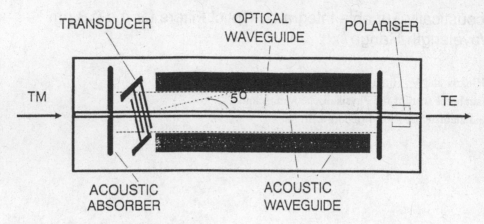

Fig1. Schematic diagramm of the ATOF

proton exchanged region the positive ordinary index change from the Ti-diffusion is compensated by the negative change from the exchange and only the extraordinary(TE) mode is further guided. The length and width of the exchanged regions were 2.5 to 3.5 mm and 50 μm respectively. To test the sensitivity of the polariser performance on the fabrication tolerances we have used a variety of fabrication conditions: pure and up to 1 mol% diluted (with lithium benzoate) benzoic acid; diffusion temperatures between 180°C and 220°C; different diffusion times; annealing at temperatures between 350°C and 400°C for several hours.

The polariser fabrication tolerances were rather uncritical and we easily obtained polarisation extinction ratios (TM-mode) of >40 dB and excess insertion losses (TE-mode) of better 0.1 dB. Additionally, it was relatively insensitive to changes in the operating wavelength, which is very important for filter applications.

2.2. Integration of the filter components The TE-TM mode converters were fabricated on 10x30 mm^2 X-cut LiNbO$_3$ substrates. In a first step, 130 nm to 160 nm thick Ti-stripes, 60 μm apart were indiffused to form the acoustic waveguides. The diffusion was performed at 1050°C in dry oxygene for 24 hours. The optical waveguides were fabricated by indiffusing photolitho-graphically formed 2.5-4.5 μm wide and 50-65 nm thick vacuum evaporated Ti-stripes. The same diffusion conditions were used, only the diffusion time was reduced to 4-6 hours. In a next technological step the integrated polariser was fabricated as already desribed above.

For the excitation of the SAWs an interdigital transducer consisting of six pairs of 150 nm thick Al-fingers with 2.6 μm width, and spacing between the fingers was photolithographically patterned on the sample surface. The transducer fingers were tilted at 5° away from the Z-axis to ensure a SAW generation parallel to the Y-axis [SMITH and JOHNSON]. The aperture of the transducer was equal to the width of the acoustic waveguides (60 μm).

3. Acousto-optic measurements and discussion The measurements of the filters were performed on a set-up schematically shown in Fig.2. The beam of a superlumineszenz diode

TE-TM MODE CONVERTER

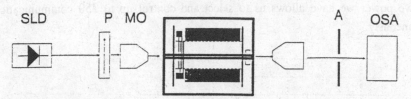

Fig.2 Measurement set-up.

3. Acousto-optic measurements and discussion The measurements of the filters were performed on a set-up schematically shown in Fig.2. The beam of a superlumineszenz diode (SLD) with wavelength near 0.8 μm was passed through a polariser (P) and coupled into the channel waveguide by means of a microscope objective (MO). After the interaction the outcoupled beam was passed through an aperture (A) to block out the substrate light and was detected with a photodiode (PD) or an optical spectrum analyser (OSA).

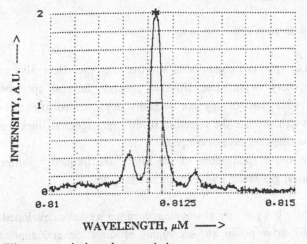

Fig.3 Filter transmission characteristic.

In Fig.3 the transmission characteristic of a filter is shown. The TM-polarised light source was coupled into the optical waveguide. A fixed frequency (365 MHz) SAW was excited and the filter output was scanned with an optical spectrum analyser. The ATOF transmission line, with nearly 100% mode conversion efficiency has a bandwidth of only 0.25 nm, obtained with RF-power as low as 1.2 mW (interaction length 25 mm). Both characteristics are the best reported yet for such filters. The filter tuning rate was -2 nm/MHz and the interdigital transducer allows a tuning range of 120 nm near the central wavelength of 0.8 μm.

The acoustically tunable filters we have developed can be used in communication systems e.g. as controllable demultiplexers. The narrow bandwidth allows a very high channel density, in the whole tuning range approx. 500 channels can be selected. An unique property of this type

724

of filter is the possibility to select several channels simultaneously [CHEUNG et al.]. The low RF-drive power we have allows us to select and control up to 250 communication channels simultaneously.

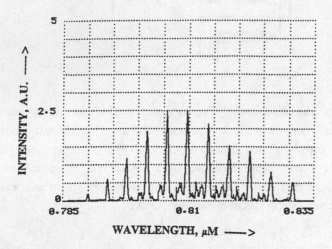

Fig.4 Selection of several lines with an ATOF.

The filter can be applied as a fast (μs-rate) controllable spectrometer too. Here also several lines can be selected simultaneously to simulate different spectra. Several spectroscopic methods like differential or derivative spectroscopy can be easily implemented to detect gases or other chemical species. Fig.4 demonstrate the selection of different spectral lines from an SLD spectrum performed with the filter we have developed.

We are also working on further sensor applications of the filter, like frequency shifters for heterodyne detection, acousto-optical gas sensors with sensitive films etc..

4. Conclusions The ATOFs for the 0.8 μm optical wavelength range we have developed have a narrow bandwidth, very low RF-drive power and are highly efficient. Several applications can be envisaged in the optical communication or sensor systems.

References

B.L.HEFFNER, D.A.SMITH, J.E.BARAN, A.YI-YAN and K.W.CHEUNG, Electron.Lett., 24 (1988) pp.1562-1563

R.G.SMART, J.N.CARTER, A.C.TROPPER, D.C.HANNA, S.F.CARTER and D.SZEBESTA, Electron.Lett., 27 (1991) pp.1123-1124

I.HINKOV and V.HINKOV, Electron.Lett., 27 (1991) pp.1211-1213

I.HINKOV and V.HINKOV, "Integrated Optical Polariser", pending patent

D.A.SMITH and J.J.JOHNSON, IEEE Photon.Technol.Lett., 3 (1991) pp.923-925

K.W.CHEUNG, D.A.SMITH, J.E.BARAN and B.L.HEFFNER, Electron.Lett. 25 (1989) pp. 375-376

Integrated Acousto-Optical Heterodyne Interferometers in LiNbO$_3$

F. Tian, R. Ricken, St. Schmid, and W. Sohler

Angewandte Physik, Universität-GH Paderborn, D-4790 Paderborn, Germany

Abstract -- Integrated optical heterodyne interferometers in X-cut LiNbO$_3$ have been developed for applications in metrology and vibration analysis. They combine 7 or 9 devices, respectively, on a single chip. 75 dB signal-to-noise ratio has been achieved with 3 kHz bandwidth by using a commercial DFB laser diode (λ=1.5μm) and a PIN-FET balanced receiver.

I. Introduction

In the last few years there is a growing interest in integrated optical (heterodyne) interferometers for applications such as Doppler-velocimetry, frequency analysis of vibrating surfaces and contact-free distance measurements. Different versions of interferometers have been developed in glass [1-2], on silicon [3] and in LiNbO$_3$ [4-6]; the first optical systems with integrated optical interferometric sensors on silicon are now commercially available.

Contrary to glass and silicon, LiNbO$_3$ allows to take advantage of its excellent electro-optical and acousto-optical properties; both can be exploited to develop heterodyne interferometers of ultimate sensitivity using integrated frequency shifters, beam splitters and polarizing optics. Recently, we have presented the design of an integrated acousto-optical heterodyne interferometer in LiNbO$_3$ and have reported experimental results of all discrete components developed for the integrated optical circuit (IOC) [7].

In this contribution we present and discuss the performance of this interferometer in X-cut, Y-propagating LiNbO$_3$ (see Fig.1); and describe the current developments towards more complex structures, including polarization diversity at the detector side and a waveguide laser at the input (see Fig.3).

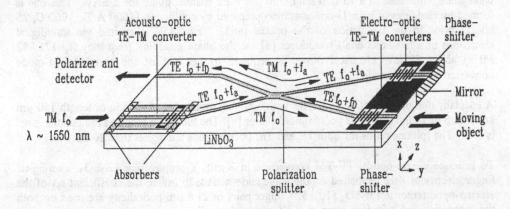

Fig.1 Schematic diagram of the integrated acousto-optical heterodyne interferometer in LiNbO$_3$.

II. Principle of Operation

TM-polarized light of frequency f_0 (λ=1545 nm) is fed into the integrated optical chip. It passes the acousto-optical TE-TM mode converter first, converting half of the power into TE-polarized light. As the mode conversion is induced by a running surface acoustic wave (SAW), the frequency of the generated TE-mode is simultaneously shifted by the acoustic frequency f_a (about 171 MHz in our experiment).

Both TE- and TM-polarized waves are separated by the subsequent passive polarization splitter and fed into the reference and measuring arms of the (Michelson) interferometer. In both arms electro-optical TE-TM mode converters and phase shifters are used to rotate the polarizations of the back-reflected waves by 90° without an additional frequency shift. The result is that both waves are recombined by the polarization splitter and fed into the output arm without any principle loss and without feedback to the optical source. The reference arm is terminated by a metallic end face mirror. The measuring arm is extended via an external collimating optics which will collect the back-reflected, phase-/frequency-modulated light from the moving object to be measured.

In the output guide, the polarization of the reference wave is orthogonal to that of the measuring wave. Therefore, an external polarizer with 45°-orientation with respect to the main axes of the waveguide has to be inserted in front of an optical receiver to "mix" both polarizations and to generate the phase-/frequency-modulated heterodyne signal. Electronic signal processing can then be used to determine e.g. the velocity or the vibration frequency and amplitude of the moving object.

III. Integrated Waveguide Devices

The IOC has been fabricated in a 42 mm long X-cut, Y-propagating LiNbO$_3$ substrate using Ti-diffused (1050 Å, 1030°C, 9 hours), 7 µm wide, single-mode optical channel waveguides; it was designed for an optical wavelength λ~1550 nm. Waveguide losses are ~ 0.2 dB/cm and ~ 0.5 dB/cm for TM- and TE-modes, respectively.

The acousto-optical TE-TM mode converter/frequency shifter consists of an optical waveguide, embedded in a 13 mm long, 100 µm wide channel guide for a SAW. This one is formed by claddings of high Ti-concentration prepared by indiffusion (1600 Å Ti, 1060°C, 25 hours) prior to the fabrication of the optical guide. The SAW is excited via interdigital electrodes of a (unidirectional) transducer [8]. At the phase matching frequency (f_a=171.083 MHz) about 58 mW electrical power is required to get half of the coupled TM-mode converted to the TE-mode.

A specially designed directional coupler with a double mode central section of length 150 µm and width of 14 µm is used as polarization splitter [9]. The full angle between the coupler arms is 0.5°. The splitting ratios for both TE and TM polarizations are higher than 20 dB.

To produce electro-optical TE-TM conversion in X-cut, Y-propagating LiNbO$_3$, interdigital finger electrodes were deposited on the waveguide surface to utilize the coefficient r_{51} of the electro-optic tensor in LiNbO$_3$ [10]. 324 finger pairs of 21.6 µm periodicity are used on both the measuring and reference arms. A 50% conversion efficiency is achieved with an applied voltage of 8.5 V; the device temperature is adjusted to 42.3°C.

The phase shifters are 4 mm long Al-electrodes on both sides of the waveguides. They allow to control, via the electro-optical coefficients r_{31} and r_{33}, the relative phase of TE- and TM-components. In this way a complete polarization conversion can be achieved in boths arms after a double-pass of the electro-optical mode converters. The gap between the electrodes of the phase shifters is 15 μm. Their half-wave voltage is 18 V.

IV. Interferometer Performance

The characteristics of our packaged integrated heterodyne interferometer with single-mode fiber pigtails was measured by using a commercial DFB laser diode of 1545 nm wavelength with isolator as optical source, its emmission bandwidth is ~ 20 MHz. About 400 μW was coupled into the input channel of the IOC leading to the acousto-optical mode converter. The total insertion loss of the IOC is about -14 dB for both optical paths through the reference and measuring arms. A vibrating mirror acts as the object to be measured, which is ~10 cm apart. Two PIN-FET detector/preamplifier modules are used as the balanced receiver; and a fiber optic polarization rotator and a fiber polarization splitter are inserted in front of the receiver in order to "mix" the measuring and reference waves.

The heterodyne signal was investigated by using a RF-spectrum analyzer. If operated with a static mirror, the intermediate frequency of 171.083 MHz was measured with 75 dB signal-to-noise ratio at 3 kHz bandwidth (see Fig.2, left). If operated with a vibrating mirror of sinusoidal oscillation $A\sin(2\pi f_v t)$, different sidebands arise in the spectrum at multiples of f_v (see Fig.2, right; $f_v=50$ kHz). The amplitude of the nth sideband is determined by the square of the Bessel function J_n with the argument $2A(2\pi/\lambda)$. Therefore, the amplitude A of the vibration can be determined by comparing different sidebands. The minimum detectable vibration amplitude A_{min} of our present system is less than 45 pm, derived from the 75 dB signal-to-noise ratio. The noise floor in our experiment was mainly attributed to the laser phase noise and the receiver noise. Therefore, there is still a great potential to improve these results up to shot-noise limited operation by using a laser of narrower linewidth and higher optical power.

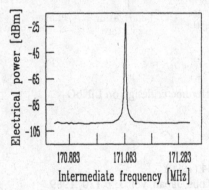 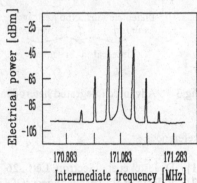

Fig.2   Measured spectrum of the heterodyne signal for operation with a static (left) and vibrating (right) external mirror terminating the measuring arm.

V. Discussions and Conclusions

We have demonstrated a new integrated acousto-optical heterodyne interferometer in X-cut, Y-propagating $LiNbO_3$ for sensor applications. An acousto-optical TE-TM mode converter, a

728

polarization beam splitter, and two electro-optical TE-TM converters and phase shifters have been integrated on a common substrate. By using a hybride construction with fiber optic polarization rotator and polarization splitter and balanced detection, the packaged integrated heterodyne interferometer with single-mode fiber pigtails shows the promising result of 75 dB signal-to-noise ratio. The sensitivity of detectable amplitude is 45 pm.

The simple structure, large frequency bandwidth, high sensitivity, and potential low insertion loss make the integrated heterodyne interferometer very attractive for a range of possible applications; vibration analysis of the amplitude of the oscillating cantilever of an atomic force microscope (AFM) is just one example.

To further improve the IOC, we are working towards the fully integrated version of the heterodyne interferometer as shown in Fig.3. On the right of dashed line is our present version. The fourth TE-TM mode converter and the second polarization splitter for polarization diversity detection have been integrated onto a common chip, which is under investigation. An Erbium-doped waveguide laser, pumped by a laser diode (λ_p~1480 nm), is planned as integrated coherent light source with ~1530 nm emission wavelength.

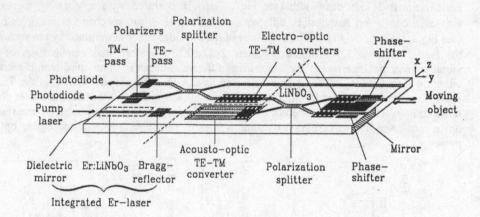

Fig.3   Advanced integrated heterodyne interferometer design on LiNbO$_3$

References

[1]   D. Jestel et al., Electron. Lett., 26, p.1144 (1990)
[2]   P. Roth et al., Techn. Digest IOOC'89, Kobe, Japan, vol. 3, p.120, 1989
[3]   C. Erbeia, in "Optical fiber sensors", Springer Proc. in Physics, vol.44, pp. 234, (1989)
[4]   H. Toda et al., J. Lightwave Technol., LT-5, p. 901, (1987)
[5]   H. Toda et al., Techn. Digest IOOC'89, Kobe, Japan, vol. 2, p. 168, 1989
[6]   P. G. Suchoski et al., IEEE Photonics Technol. Lett., 2, p. 81, (1990)
[7]   H.Herrmann et al., Techn. Digest IOOC'91, Paris, France, 1991, vol. 1, p. 537
[8]   J. Frangen et al., Electron. Lett., 25, p.1583, (1989)
[9]   L. Bersiner et al., J. Opt. Soc. Am. B, 8, p. 422, (1991)
[10]   R. C. Alferness et al., Electron. Lett., 19, p.40, (1983)

Reflected Mode Operation of Grating Couplers for Chemical Sensing

A. Brandenburg
Fraunhofer-Institut für Physikalische Meßtechnik
Heidenhofstraße 8, 79110 Freiburg, Germany

Introduction

The use of grating couplers as sensor elements was first published
by W. Lukosz and K. Tiefenthaler in 1983 [1]. This type of sensor
makes use of the well defined angle at which light couples into a
waveguide by means of a grating coupler. The effective refractive
index of the waveguide mode can be measured very accurately by
determining this coupling angle. The dependence of coupling angle
α on effective refractive index n_{eff} of the waveguide modus is
given by:

$$n_a \cdot \sin\alpha = n_{eff} - k \cdot \lambda_0 / \Lambda ,$$

where n_a, λ_0 and Λ denote the refractive index of the ambient,
vacuum wavelength and grating constant, respectively. k is the
diffraction order.

Chemical sensing

The change of effective refractive index may be caused by
different mechanisms: The refractive index of the waveguiding film
may change by chemical absorption of molecules from the ambient
medium. A change of the index of the surrounding medium also
changes the effective refractive index via the evanescent field of
the guided mode. The medium surrounding the waveguide may be e.g.

a thin film, that is specifically sensitive towards the substance
to be measured. An adsorption of molecules on the surface of the
waveguide can be considered as an increase of thickness of the
waveguiding layer that affects the effective refractive index,
too. This mechanism has been successfully used for the detection
of antigen-antibody reactions [2].

Reflected mode operation

Sensitive detection requires a very high resolution of the
coupling angle. In [3], two optical configurations are proposed:
The input- and the output grating coupler. Here, reflected mode
operation is proposed. Instead of a parallel light beam,
convergent or divergent beams may be irridiated onto the grating
(fig. 1). The convergent beam is realized by using a focussing
lens, while the divergent rays may easily be produced by
positioning a fiber end or a semiconductor laser close to the
grating. In both cases, at the same time not a discrete angle of
incidence, but an angular spectrum is given. The range of angles
is chosen in such a way that input coupling always occurs under
one of these angles. In order to detect the coupling angle, the
reflected light is analysed (zeroth diffraction order in
reflection). As the coupling into the waveguide is much more
efficient than the diffraction into radiating modes, the intensity
of reflected light is reduced considerably at the angle under
which coupling occurs. Hence, a dark line can be observed in the
reflection which changes its angular position with the effective
refractive index. With the aid of a CCD array this position can be
detected very accurately.

A sensor element was fabricated in order to test the measuring
principle. Using a monomode optical fiber the light of a
semiconductor laser emitting at a wavelength of 790 nm is guided
to the grating sample. Without any further optical elements the
intensity distribution of the reflected light is detected by a CCD
array, having 1728 pixels, each being 10 μm wide. The distance
between grating coupler and detector array is 80 mm. The range of

detectable angles is therefore ca. 12°, which is sufficient for
most applications. The width of the dark line has been in the
range of 0.02° ... 0.4°, depending on the quality of the grating.
The resolution of this setup in terms of n_{eff} is about $5 \cdot 10^{-4}$.
Higher resolutions are achieved with improved electronics, smaller
width of the dark line and larger distances between grating
coupler and CCD array.

Waveguiding films have been made of evaporated SiO_2 and TiO_2. In
the surfaces of these films, gratings with a 0.7 μm-period were
etched by ion milling. Water absorption of these films from the
surrounding atmosphere results in a sensitivity towards humidity.
Relative humidity was changed from 0 % to approximately 80 % at
room temperature, resulting in a change of coupling angle of e.g.
2.2° for the 1 μm thick TiO_2-film. An example for the output sinal
of the CCD array is given in fig. 2 for dry and humid atmosphere.

Conclusions

An optical setup for the evaluation of the coupling angle of
integrated optical grating couplers is proposed that may be
interesting for sensor applications due to its simplicity.
Especially no moving parts are required. Critical adjustments,
which form an essential obstacle for commercialization of
integrated optical devices are not necessary. The sensor can be
highly sensitive. It may serve as a transducer for numerous sensor
applications like gas detection, detection of components in
liquids and monitoring antibody-antigen reactions.

References

1. K. Tiefenthaler and W. Lukosz: Integrated optical switches
    and gas sensors, Opt. Lett. 10 (1984) 137
2. Ph.M. Nellen, K. Tiefenthaler and W. Lukosz: Integrated
    optical input grating couplers as biochemical sensors,
    Sensors and Actuators 15 (1988) 285

732

3. W. Lukosz, Ph. Nellen, Ch. Stamm, P. Weiss: Output grating
    couplers on planar waveguides as integrated optical
    chemical sensors, Sensors and Actuators B1 (1990) 585

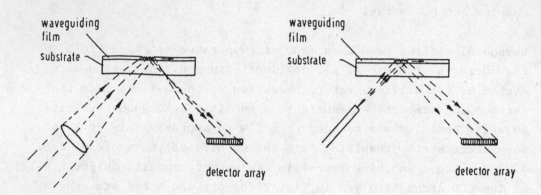

Fig. 1:   Reflected mode operation of grating coupler: Convergent
input beam, Divergent input beam.

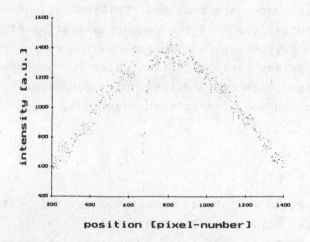

Fig. 2:   Signal of the CCD array.

Optical Gas Detection with Lightguides Using Fluorinated Polymers

M.Osterfeld, R.P.Podgorsek, H.Franke, A.Brandenburg*, C.Feger**

FB Physik, Univ. Duisburg, Lotharstr. 1, 4100 Duisburg 1, FRG

* Fraunhofer Institut für Physikalische Meßtechnik,

D-7800 Freiburg, Heidenhofstr.8

** T.J. Watson Research Center, Yorktown Heights, NY 10598

1 Introduction

Although the use of solvents may be further reduced there is still a growing need for the detection, measurement or control of organic vapors.

For the protection of employees the law provides maximum alowable concentrations of solvents. For many substances the maximum concentrations are in the range of a few parts per million. This limits the use of many common sensor systems. Therfore it is necessary to develop new sensor concepts especially for gas detection. Nowadays optical methods find increasing interest. Most of them based on the methods of waveguide- and fiber optics. The lack of any electric contacts in the sensor head is an advantage considering safety in many chemical processes.

The optical sensor principles presented in this paper make use of optical waveguides. The light is guided in a thin optically clear polymer film. The chosen polymers do not react with the particular gas phase. The sensor effects are phase effects and they are completely reversible.

2 Material

Fluorinated polymers such as polyimides [1] or Polytetrafluoroethylen (PTFE) and its copolymers are chemically and thermally stable coating materials. Furthermore polyimides show an reversible absorption of moisture and other polar molecules. The hydroscopic effect is due to the formation of hydrogene bonds of water molecules with the imide group. The stiff aromatic monomer units in the polyimide are highly anisotropip with respect to their polarizability. Thin film processing with polyimides generally leads to a negative birefringence of up to $\Delta n = 0.1$. The optical axis is found to be normal to the film plane. The anisotropy of polyimide films is explained by a preferential in-plane-orientation of the chain units.

Other versatile polymers are the recently developed amorphous fluorocarbons, which are comercially available under the tradename Teflon AF. They are copolymers of tetrafluoroethylene and 2,2-bis(trifluoromethyl)-4,5- difluoro- 1,3-dioxole [2]. The polymers are highly transparent, isotropic materials. They are inert materials and only soluble in fluorinated solvents. The net dipolar moment of the monomer units is rather poor. A remarkable low refractive index of $n_{Teflon} = 1.3$ is provided by the weak polarizability of the monomer units. Teflon AF films are very efficient diffusion barriers against moisture.

Recently the use of Teflon AF for optical sensing was shown [3]. An in-diffusion of nonpolar vapor molecules into the polymer matrix leads to an increase of the refractive index.

3 Sensor Arrangements and their Characteristics

3.1 Index matched Polyimide Lightguides

Fig.1a shows a waveguide set-up placed between two crossed polarizers. The polyimide is spincoated on a glass substrate. With respect to the anisotropy the thickness has to be adjusted in order to get a phase matched propagation of both the TE- and TM-modes in the polyimide film [4]. The incident He-Ne-laser radiation has to be polarized at 45° to the plane of incidence.

The polyimide changes its optical properties with the vapor pick up. Similar to a retardation plate the intensity $I(t)$ at the detector oscillates according to :

$$I(\Gamma) = I_0 \sin^2\left(\frac{\Gamma}{2}\right) \qquad with \quad \Gamma = \frac{2\pi}{\lambda}(N_{TE} - N_{TM})l = \frac{2\pi}{\lambda}\Delta N l \qquad (1)$$

The change of the effective index ΔN is calculated from the number of oscillations.

As sketched in Fig.1a the waveguide is exposed to a defined atmosphere by a flow cell. The humidity is adjusted by mixing dry and humid air and it is measured with a psychrometer set-up. The response times and the saturation values strongly depend on the concentration of the vapor content (Fig.1b). Within several seconds the saturation is achieved. The response follows a stretched exponential law. It indicates a superposition of several pure exponential processes. We found that the waveguide method allows the determination of single polar components (H_2O, CH_3OH, ...) in a mixed atmosphere in a dynamic measurement. Each type of molecule shows a characteristic diffusion.

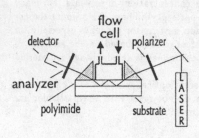

Fig.1 : a)The waveguide setup.

b) Change of the optical anisotropy versus as a function of time for different humidities . The solid line represent the best fit to an stretched exponential law.

3.2   Metal enhanced Leaky Mode Spectroscopy

If a mode is launched by prism coupling technique a dark line appears in the reflection spectrum. Contributing to their resonance nature the lines exhibit a half width in the angular spectrum. In order to achieve lines of improved sharpness the air gap between the prism and the guiding dielectric is replaced by a thin (50nm) silver layer. Such a metal enhanced reflection spectrum versus the angle of incidence is shown in Fig.2 for both polarizations. On the right side of the TM-spectrum a wide resonance line can be seen. It corresponds to the excitation of a surface plasma wave at the metal-dielectric boundary.

The angular position and the distance of the leaky modes depend quite sensitively on the optical properties and the film thickness. If an undisturbed guide is fully characterized it is sufficient to follow the shift of one mode in the angular spectrum. As indicated in eq.2 variations of the film thickness and the refractive index contribute to the shift of the coupling angle Θ_m of the m-th mode.

$$\Delta\Theta_m = \frac{\partial\Theta_m}{\partial n}\frac{\partial n}{\partial T}\Delta T + \frac{\partial\Theta_m}{\partial t}\alpha\, t\, \Delta T \qquad (2)$$

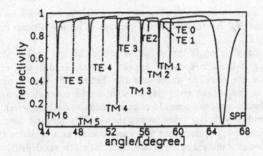

Fig.2: Metal enhanced leaky mode spectrum for both polarizations. The TM 0 mode is degenerated to a surface plasmon (SPP) excitation.

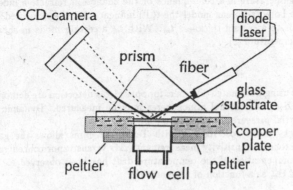

Fig.3: Optical sensor. The shift of a dark line in the reflection spectrum is monitored with a CCD-camera.

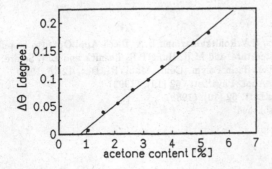

Fig.4 : The mode shift $\Delta\Theta$ versus the acetone concentration in dry nitrogen.

Fig.3 shows a metal enhanced leaky mode sensor. The thin films are deposited on a glass substrate, which is brought into an optical contact with the coupling prism by an immersion fluid. A laser diode (790nm) coupled to a polarisation maintaining fiber acts as a divergent light source. The angular distribution is detected with a CCD-camera.

As an intensity minimum in the gaussian shaped reflection spot the excitation of a mode is apparent. The CCD-Camera allows an in-situ measurement of the modes angular shift.

The guide is exposed e.g. to an acetone-N_2 atmosphere, which is adjusted by mixing a saturated Acetone-N_2 atmosphere with pure N_2. Acetone is chosen because of its low dipolar moment and to show the chemical stability of Teflon AF.

From the mode shift of the coupling angle the refractive index change can be calculated. Additional experiments show that a swelling of the film does not appear with the in-diffusion of vapor. A linearity between the acetone content and the coupling angel into the Teflon AF -film is obvious (Fig.5). The extrapolation of the fitted line to the abzissa gives rise to a threshold concentration above which an in-diffusion takes place. Again we found response times to concentration variations which are below 15s. The optical changes are totally reversibel with the vapor atmosphere. This method does not depend on the kind of the sensing material. Hence every transparent film can be used as a sensitive layer.

One major problem of especially optical sensors are their large thermal drift. This drift is often suppressed by tempering the optics. A sensor with a low thermal cross sensitvity is desirable. We conducted some experiments where the coupling optic is heated. We found that a particular mode shows a low thermal drift. Other modes shift to the right or left in the angular spectrum. The drift of the resonance lines in the angular spectrum with temperature is a consequence of the change in refractive index n=n(T) and film thickness t(T). Using the Lorentz-Lorenz model the t(T) influence may be expressed by n(T) (eq.2). Setting $\Delta\Theta_m = 0$ in eq.2 leads to a certain thickness t_m. With t_m a certain mode m should not drift in the angular spectrum with temperature [5].

Conclusions

Two different arrangements using fluorinated polymers for optical gas detection are demonstrated . With fluorinated polyimide guides H_2O and alcohol atmospheres may be measured. Dynamic measurements allow the determination of partial pressures in mixed atmospheres.

Metal film enhanced leaky mode spectroscopy with Teflon AF films allows the gas detection of apolar vapors like CCl_4 or Acetone. A sensitivity measurement ($\Delta\Theta$ versus vapor concentration) has been performed for acetone. The compensation of the temperature drift has been observed for one particular mode by measuring the shift $\Delta\Theta_m$ as a function of temperature.

Acknowledgement

The authors wish to thank the BMFT for financial support.

References

[1] H.Franke, D.Wagner, T.Kleckers, H.V.Rohitkumar and B.A. Blech Appl.Opt. (to be published)
[2] Teflon$^{(R)}$AF, DuPont product literature, and M.H. Hung P.R. Resnick and E.N. Squire, Cyclic Fluorinated Polymers, presented at 1st Pac. Basin Polym. Conf., Maui, HI, Dec. 12-15, 1989
[3] M.Osterfeld, H.Franke, C.Feger, Appl. Phys.Lett. 62 (19), (1993)
[4] R.Reuter, H.Franke, Appl. Phys.Lett. 52 (10), (1988)
[5] M.Osterfeld, H.Franke, to be published

Micro-Sensors and Fibre Optics

Optical fibre - channel waveguide butt coupling : modelling of the system, new manufacture of a channel waveguide on a glass substrate and experimental verification of the theoretical model at 1.3 µm.

M. Réglat, I. Verrier, M. Ramos, J.P. Goure, P. Sass
Laboratoire TSI - Faculté des Sciences 23 rue du Dr Paul Michelon 42023 Saint-Etienne Cedex 2
G. Clauss, A. Kévorkian, F. Rehouma
GeeO - LEMO 46 Av. Felix Viallet 38031 Grenoble Cedex

1- Introduction

It is becoming essential to master the technology for connecting optical fibres to integrated optical waveguides. The need is large in the field of optical communications and network components as well as in the field of sensors.

To date these connections that have been developed and produced industrially are fixed and permanent (for instance connectors for 1 -> n splitters). Such connectors are sometimes called "pig-tail" connections by analogy with the laser diode to fibre devices.

Yet, the circuits to be made in integrated optics may need connectors that can be unpluged (for technological or economic reasons).

This is the approach that we propose to discuss here. Within a project called CIOS (Connectorisation of Integrated Optical Sensors) we have studied a singlemode connector which could be dismantled. This paper presents a feasibility study of such a connector.

In the first part, the aim is to model a direct coupling system and to determine its characteristics in order to obtain the best coupling between a singlemode fibre and a singlemode waveguide.

Then, the second part deals with the manufacture of such a waveguide, designed to optimize the coupling, using integrated optics technology on glass (IOG).

Finally, the last part validates the theoretical model by measurements carried out on these new waveguides.

2 - Modelling

One important characteristic of a junction is its coupling factor [that expresses how much light can be coupled from one side (fibre or waveguide) to the other (waveguide or fibre)]. We first of all establish how this factor relates the various parameters of a connection. The coupling factor between two structures (fig.1) is expressed by :

$$\eta = \frac{\left| \iint_{-\infty}^{+\infty} \psi_{f0} \psi_{g0}^* \, dS \right|^2}{\left(\iint_{-\infty}^{+\infty} \psi_{f0} \psi_{f0}^* dS \right) \times \left(\iint_{-\infty}^{+\infty} \psi_{g0} \psi_{g0}^* dS \right)} \tag{1}$$

ψf0 is the field at the output of the fibre, expressed in an arbitrary plane of integration, ψg0 is the field at the output of the waveguide expressed in the same plane, (*) is for complex conjugate and "dS" is the elementary surface of integration.

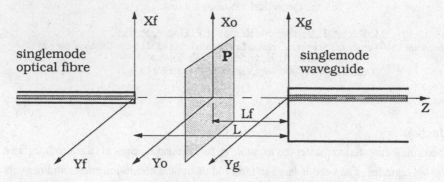

Fig 1 : coupling between the two structures.

2.1 Conditions that maximize η .

In order to optimize coupling, the coupling factor η has to be maximized. Two solutions are possible :

- direct coupling (ref [1] [2] [3]) by fabrication of a waveguide with a field ψg0 close to the field ψf0 of the fibre.

- indirect coupling by use of an optical system between the waveguide and the fibre to accommodate the fields as they are different.

As fabrication technique of glass substrate waveguides is evolving very fast, we have chosen the first solution (ie a direct coupling system.)

In order to maximize η , the shape of the field at the output of the waveguide and the fibre have to be known in a plane at a distance Z. We consider (fig 1) the following coupling system : the waveguide and the fibre are separated by a distance "l", the plane P of integration is placed at a distance l_f from the fibre and $l-l_f$ from the waveguide. In the (x_f, y_f) plane at the output of the singlemode fibre, the field $\psi_f (x_f, y_f)$ can be satisfactorily modellized by a gaussian curve having the same parameters for x_f and y_f (assuming that the gaussian curves are centered).

$$\psi_f(x_f, y_f) = \sqrt{2/\pi} \left(\frac{1}{w_f} \right) \exp\left(-\frac{x_f^2}{w_f^2} - \frac{y_f^2}{w_f^2} \right) \tag{2}$$

w_f is the modal radius of the field at the output of the fibre.

Between the output plane $(x_f, y_f, 0)$ and the P plane of coordinates (x_0, y_0, l_f), the gaussian field $\psi_f (x_f, y_f)$ is transformed after free propagation into $\psi_{f0} (x_0, y_0)$, which is also gaussian. (ref [4])

$$\psi_{f0}(x_0, y_0) = \sqrt{2/\pi} \frac{1}{W_f} A(l_f) \exp\left[-\left(\frac{x_0^2 + y_0^2}{W_f^2} \right) - ik \frac{x_0^2 + y_0^2}{2 R_f} \right] \tag{3}$$

with $\quad R_f = l_f\left[1+\left(k\,w_f^2/2l_f\right)^2\right]$ (4) and $W_f = w_f\left[1+\left(2l_f/kw_f^2\right)^2\right]^{1/2}$ (4)

$$A(l_f) = \frac{2i\pi\,w_f^2}{\lambda\left(2l_f + ik\,w_f^2\right)}e^{-ikl_f}$$ (5)

The shape of the field at the output of the waveguide, in a plane P, has now to be determined. By examining the formula (1) that gives the coupling coefficient η, we can notice that η will be maximum ($\eta = 1$) when $\psi_{g0} = \psi_{f0}$ in the P plane.

The field ψ_{g0} has then to be gaussian after propagation until the plane P, and consequently gaussian at the output of the waveguide. The expression of the field at the output of the waveguide is then :

$$\psi_g(x_g, y_g) = \sqrt{2/\pi}\left(\frac{1}{w_g}\right)\exp\left(-\frac{x_g^2}{w_g^2} - \frac{y_g^2}{w_g^2}\right)$$ (6)

w_g is the modal radius of the field.

Backpropagating the field $\psi_g(x_g, y_g)$ over the distance $1 - l_f$, we obtain the field in the P plane : (ref 4)

$$\psi_{g0}(x_0, y_0) = \sqrt{2/\pi}\,\frac{1}{W_g}A(1 - l_f)\exp\left[-\frac{x_0^2 + y_0^2}{W_g^2} - ik\frac{x_0^2 + y_0^2}{2R_g}\right]$$ (7)

with $\quad R_g = (l_f - 1)\,[1 + (k\,w_g^2/2(l_f - 1))^2]$ (8) and $W_g = w_g\left[1+\left(2\frac{(l_f-1)}{kw_g^2}\right)^2\right]^{1/2}$ (8)

$$A(1 - l_f) = \frac{2i\pi\,w_g^2}{\lambda\left(2(l_f - 1) + ik\,w_g^2\right)}e^{+ik(1 - l_f)}$$

2.2 Relations between waveguide parameters and fibre parameters.

In order to maximize the coupling coefficient, we have shown that ψ_{f0} has to be equal to ψ_{g0}.

As we know the analytical expressions of each field, we can write the following identity:

$$\sqrt{2/\pi}\,\frac{1}{W_f}A(l_f)\exp\left[-\frac{x_0^2 + y_0^2}{W_f^2} - ik\frac{x_0^2 + y_0^2}{2R_f}\right] = \sqrt{2/\pi}\,\frac{1}{W_g}A(1 - l_f)\exp\left[-\frac{x_0^2 + y_0^2}{W_g^2} - ik\frac{x_0^2 + y_0^2}{2R_g}\right]$$

and then $\quad \dfrac{W_g}{W_f}\dfrac{A(l_f)}{A(1 - l_f)} = \exp\left[\left[x_0^2 + y_0^2\right]\left[\left(\frac{1}{W_f^2} - \frac{1}{W_g^2}\right) - \frac{ik}{2R_g} + \frac{ik}{2R_f}\right]\right]$

that can be expressed in the following manner :

$$\text{Log}\left[\frac{W_g}{W_f}\frac{A(l_f)}{A(1-l_f)}\right]=\left[x_0^2+y_0^2\right]\times\left[\frac{1}{W_f^2}-\frac{1}{W_g^2}-ik\left(\frac{1}{2R_g}-\frac{1}{2R_f}\right)\right] \tag{10}$$

To optimize the field overlapping, this expression has to be valid for any point M (x_0, y_0) of a plane at a distance l_f of the fibre, we must then have :

$$\frac{1}{W_f^2}-\frac{1}{W_g^2}+ik\left(\frac{1}{2R_g}-\frac{1}{2R_f}\right)=0 \tag{11}$$

and $\quad \text{Log}\left[\frac{W_g}{W_f}\frac{A(l_f)}{A(1-l_f)}\right]=0 \Rightarrow \frac{W_g}{W_f}\frac{A(l_f)}{A(1'-l_f)}=1 \tag{12}$

By separating real and imaginary parts from (11) and (12), we get the following conditions

$(11) \Rightarrow \qquad \frac{1}{W_f^2}=\frac{1}{W_g^2} \text{ et } \frac{1}{2R_f}=\frac{1}{2R_g} \tag{11'}$

$(12) \quad \text{real part } \left[\frac{W_g}{W_f}\frac{A(l_f)}{A(1-l_f)}\right]=1 \text{ and imaginary part } \left[\frac{W_g}{W_f}\frac{A(l_f)}{A(1-l_f)}\right]=0$

the conditions (11') give then:

$$8l_{1f}-\frac{4l_f^2\left(w_f^2-w_g^2\right)}{w_f^2}=4l^2-k^2w_g^2\left(w_f^2-w_g^2\right)$$

and $\quad 4l_f^2 \, 1 + l_f \, (k^2 w_f^4 - k^2 \, w_g^4 - 4l^2) - 1 \, k^2 \, w_f^4 = 0$

This latter conditions are true in any plane of integration and particulary for $l_f = 0$, ie when $w_g k \, (w_f^2 - w_g^2)^{1/2} = 2l$ and for $l = 0$ which implies $w_f = w_g$.
The identities (12) are true under the conditions (11') ie for $l_f = 0$ and $w_f = w_g$.

As we have guessed, to get the maximum coupling coefficient η, the distance between the waveguide and the fibre has to tend to zero and the size of the two gaussian fields have to be as close as possible.

The waveguide also have to be perfectly aligned in the calculation of gaussian field propagation. Waveguide fabrication could allow the realization of the condition $w_f = w_g$.

2.3 *Misalignment influences on η.*

As both fibre and waveguide fields have gaussian distributions that are independent of x or y, the double overlap integrals in the expression of η can be simplified into the product of simple integrals in x_0 et y_0. The coupling coefficient η can then be expressed as follows (ref [5]) :

$$\eta = \frac{4}{\left[\left(\dfrac{w_f}{w_g}+\dfrac{w_g}{w_f}\right)^2+\left(\dfrac{2}{kw_fw_g}\right)^2 l^2\right]} = \eta_{x0}\eta_{y0} \tag{13}$$

with $$\eta_{x0} = \frac{2}{\left[\left(\dfrac{w_f}{w_g}+\dfrac{w_g}{w_f}\right)^2+\left(\dfrac{2}{kw_fw_g}\right)^2 l^2\right]^{1/2}} \tag{14}$$

and $$\eta_{y0} = \frac{2}{\left[\left(\dfrac{w_f}{w_g}+\dfrac{w_g}{w_f}\right)^2+\left(\dfrac{2}{kw_fw_g}\right)^2 l^2\right]^{1/2}} \tag{15}$$

where η_{x0} is the coupling coefficient in the Oxz plane and η_{y0} is the coupling coefficient in the Oyz plane.

with $k = 2\pi / \lambda$

For a transversal misalignment "d" of the fibre in the "x" or "y" direction, we can show that:

$$\eta_d = \eta \, \exp\left[-\left(\frac{d}{D}\right)^2\right] \tag{16}$$

with $$D = \frac{2^{1/2}}{\eta_{x0}\left[\dfrac{1}{w_g^2}+\dfrac{1}{w_f^2}\right]^{1/2}} \quad \text{for a misalignment in the } Ox_0z \text{ plane} \tag{17}$$

and $$D = \frac{2^{1/2}}{\eta_{y0}\left[\dfrac{1}{w_g^2}+\dfrac{1}{w_f^2}\right]^{1/2}} \quad \text{for a misalignment in the } Oy_0z \text{ plane}$$

We then get the analitic expressions for the coefficients η et η_d.

3- Realization of well-matched waveguides.

The way to obtain a good modal adaptation is to fabricate waveguides with characteristics as close as possible to those of the fibre. Thanks to its flexibility, integrated optics technology on glass is a good candidate (fig 3)(ref [7] [8] [10]). As the waveguides obtained by a simple thermal ion-exchange present non homogeneous field shapes, waveguides buried using a second electric-field-assisted exchange exhibit a better homogeneity (ref [9] [11] [12] [13]).

By an appropriate choice of exchange parameters, it is possible to achieve circular waveguides with modal field very close to that of optical fibres. (fig 2)

PROFIL D'INTENSITE FIBRE A 1.3 µm (Y)
GUIDE (Y)
INTENSITE (U.A)
DISTANCE (µm)
0.00
20.00
40.00
60.00
80.00
100.00
120.00
5.00
10.00
15.00
20.00
PROFIL D'INTENSITE FIBRE A 1.3 µm (X)
GUIDE (X)
INTENSITE (U.A)
DISTANCE (µm)

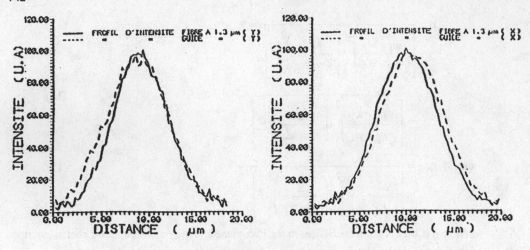

fig 2. Comparaison of intensity mode profiles of (—) a fibre and (---) a waveguide.

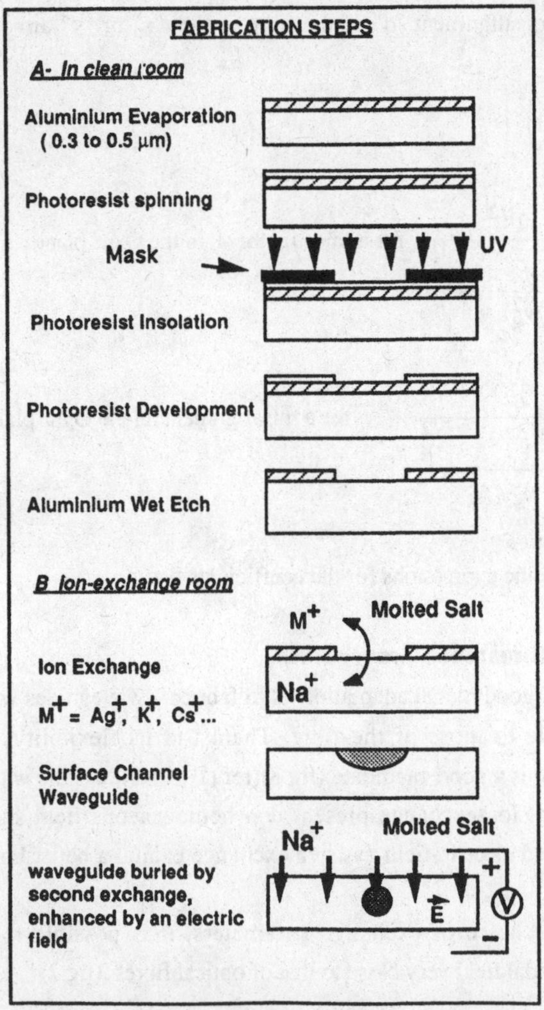

FABRICATION STEPS
A- In clean room
Aluminium Evaporation
(0.3 to 0.5 µm)
Photoresist spinning
Mask
UV
Photoresist Insolation
Photoresist Development
Aluminium Wet Etch
B Ion-exchange room
M+
Molted Salt
Ion Exchange
Na+
M+ = Ag+, K+, Cs+...
Surface Channel
Waveguide
Na+
Molted Salt
waveguide buried by
second exchange,
enhanced by an electric
field
E
V
+
-

fig 3. Integrated optics technology on glass

4- Experimental verification

The mode width of the buried waveguide on glass substrate measured at 1.3 μm is $w_{gx}=5.0\mu m$ and $w_{gy}=5.25$ μm, the mode width of the fibre is $w_f = 4.5$ μm.

All the theoretical calculations were carried out at the wavelength of 1.3 μm.

With the two values w_f and w_g, and the expressions for η and η_d, we can calculate the maximum coupling coefficient and the variation of the coupling coefficient as a function of the various misalignments.

In order to check our theoretical values, we used the following experimental set up (fig. 4)

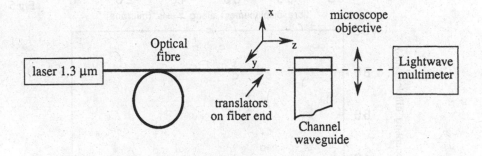

Fig 4 : experimental set up

The fibre is connected to a 1.3 μm source and positionned in front of the waveguide by an x y z stage controlled by a step-by-step motor. The waveguide is fixed and a microscope objective is positionned in order to focuse the light on the detector. Maximum coupling is obtained when the faces of the fibre and the waveguide are almost in contact and can reach 85%. The theoretical model gives 97,6 %.

We show coupling efficiency curves :
- as a function of the longitudinal misalignment of the fibre (z axis) (fig. 5)

- as a function of transversal misalignment (x axis) (fig. 6) and (y axis) (fig. 7).

744

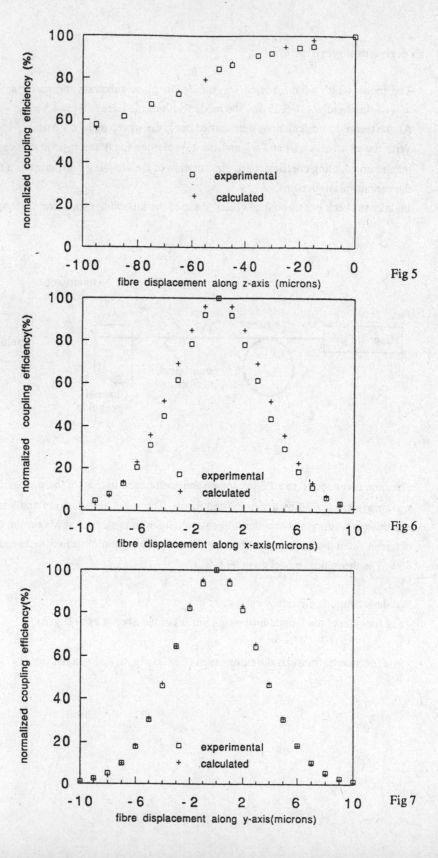

Fig 5

Fig 6

Fig 7

A good agreement is found between the theoretical model and the experimental results.
The experimental maximum efficiency is found lower than the theoretical one because waveguide propagation losses, Fresnel losses at the interfaces as well as losses originated by angular tilt between the guides were neglected in the theoretical model (ref[6]).

5 - Conclusion.

The manufactured waveguides, made by IOG, exhibit modal fields well matched with those of standart optical fibres .

For the future, following on from the CIOS project, an acceptable coupling performance is within reach. In fact misalignment of a few microns for the x and y axes, and a few dozen microns for the z axis, result in losses less than 1dB, tolerances acceptable to modern micromechanics.

With other technologies, using for instance LiNbO3, it is much more difficult to adapt the waveguide size and consequently it is necessary to find a different coupling system (from this direct coupling) in order to match both fields. e.g. using one or several lenses.

* CIOS (Connectorisation of Integrated Optical Sensors) is a pluriannual international Research Programme "Guided Optics" : Region RHONE-ALPES-FRANCE and LAND BADEN-WÜRTTEMBERG - FEDERAL REPUBLIC OF GERMANY.

References

[1]   O. G. Ramer : "Single-mode fibre-to-channel waveguide coupling" ; J. of Opt. Commun. 2 (1981) 4, 122-127
[2]   E. J. Murphy : "Fibre attachment for guided wave devices" ; J. of Lightwave Technol. 6 (1988) 6, 862- 871
[3]   G. Grand, H. Denis, S. Valette : "New method for low cost and efficient optical connections between singlemode fibres and silica guides" ; Electron. Lett. 27 (1991) 1, 16-18
[4]   J. W. Goodman : "Introduction à l'optique de Fourier et à l'holographie" ; Masson et Cie, Paris (1972)
[5]   W. B. Joyce, B. C. DeLoach : "Alignment of Gaussian beams" ; Appl. Opt. 23 (1984) 23, 4187- 4196
[6]   V. Ramaswamy, R. C. Alferness, M. Divino : "High efficiency single-mode fibre to Ti:LiNbO3 waveguide coupling" ; Electron. Lett. 18 (1982) 1, 30-31
[7]   T. Findakly " Glass waveguides by ion exchange : a review," Optical Engeneering 24(2), 244-250 (Mars/April 1985).
[8]   R.V. Ramaswamy, "ion-exchanged glass waveguides : a review," J. of LIGHTWAVE TECHNOLOGY vol. 6, No. 6, June 1988.
[9]   F. Rehouma, D. Persegol, G. Clauss, " Fibre compatible waveguides made on glass substrates developed for Ag+ <-->Na+ ion-exchange, " , 2nd French-German workshop on optical measurement techniques, fibres optics and instrumentation, Saint-Etienne, october 13 and 14, 1992.
[10] S. Iraj Najafi " Introduction to glass integrated optics ", The Artech House optoelectronics library Boston . London
[11] J. Albert and J. M. Y. Lit, " Full modeling of field-assisted ion exchange for graded index buried channel optical waveguides," Applied Optics vol. 29, No. 18 June 90.
[12] R. V. Ramaswamy, H. C. Cheng and R. Srivastava, " Process optimization of buried Ag+-Na+ ion-exchanged waveguides : theory and experiment," APPLIED OPTICS /Vol. 27, No. 9 / 1 May 1988.
[13] A. Tervonen, P. Pöyhönen, S. Honkanen, M. Tahkokorpri, and S. Tammela, "Examination of two-step fabrication methods for single-mode fibre compatible ion-exchanged glass waveguides,"APPLIED OPTICS / Vol. 30, No. 3 / 20 January 1991.

Polarimetric Sensor Using Multimode Optical Fibre

A. CHAKARI, N. MANCIER, L.F. MASSOUMU, P. MEYRUEIS

LSP - ENSPS - ULP ; 7, rue de l'université, 67084 Strasbourg Cedex, France

Single-mode fibres, and especially polarisation-maintaining fibres are used in polarimetric optical fibres sensors. A stress (pressure[1], bending[2], twisting[3], etc.) applied to the fibre leads to a modification of the birefringence which can result either in the rotation of the polarisation plan or in a change of the state of polarisation. Both can be detected by a variation in the light intensity at the end of the fibre. In a single mode optical fibres the fundamental mode HE_{11} is degenerated in two orthogonal modes $HE_{11}(x)$ and $HE_{11}(y)$, linear polarised in the Ox and Oy directions and with propagation constants being respectively $\beta(x)$ and $\beta(y)$. The birefringence Δn can be defined as :

$$\Delta n = (\beta(x) - \beta(y))/K \qquad (1) \qquad \text{with } K = 2\pi/\lambda$$

In the case of multimode optical fibres, the analysis of propagation based on ray theory can be applied to uniform-core multimode fibres with a good accuracy. The considered fibre is composed of a uniform core with radius r_c and refractive index n_1, and a cladding with refractive index n_2, where $n_2 < n_1$. The rays that propagate in this fibre may be classified in two kinds: meridian rays, which travel in a plane that includes the fibre axis and skew rays, which never cross the fibre axis.

1- Meridian rays

We consider an incident ray arriving on the centre of a uniform-core fibre with an angle θ_0 as shown on Fig.1. That ray propagates by total reflection at the core-cladding boundary, if the incident angle θ_0 satisfies:

$$\theta < \theta_1 \text{ with } \theta_1 = \text{Arccos}(n_2/n_1) \text{ for } n_1 > n_2, \quad \text{Sin}(\theta_0) = (n_1^2 - n_2^2)^{1/2} / n_a = NA \quad (2)$$

With $n_a = 1$, and NA is called the numerical aperture of the optical fibre.

2- Skew ray

We consider the case shown in Fig.2, in which an incident ray is defined by his input position on the fibre face $P_0 = i.X_0 + j.Y_0$ with a direction vector $S_0 = i.L_0 + j.M_0 + k.N_0$ where i, j and k are unit vectors, with $N_1 = N_0/n_1$, $M_1 = M_0/n_1$, $L_1 = L_0/n_1$ and $r_c^2 = x_1^2 + y_1^2$.

If the mth reflection point is represented by a vector a_m and the ray direction just before the mth reflection by a direction vector $S_{m,}$ the following equation can be presented with the following conditions:

- the vector a_m, the incident and reflected rays are in the same plane:

$$(S_m - S_{m+1}) \, \Lambda a_m = 0 \qquad (3)$$

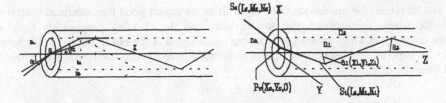

Fig. 1:
A meridian ray in a uniform core fibre.

Fig. 2:
A skew ray in a uniform core fibre

- the incident and reflection angles are equal to each other:

$$(S_m + S_{m+1}) \bullet a_m = 0 \qquad (4)$$

- for the case of total reflection:

$$S_m \bullet (a_m / |a_m|) \leq (n_1^2 - n_2^2)^{1/2} / n_1 \qquad (5)$$

The equation (5) can be rewritten as a function of the parameters L and M:

$$\left[(L_0^2 + M_0^2) - \left[(X_0.M_0 - Y_0.L_0)/r_c \right]^2 \right] \leq N A' \qquad (6)$$

The equation (6) gives an important property of skew rays : When $X_0=r_c$ and $Y_0=0$ the equation (6) becomes $L_0 \leq NA$ and when $Y_0=r_c$ and $X_0=0$ equation (6) becomes $M_0 \leq NA$. This means that even those incident rays, almost parallel to the y or x axis, may be confined in the core. Such a ray will travel along an helical path on the core-cladding boundary, but with a low axial velocity. Hence, if such helical rays are excited, the uniform-core multimode fibre may exhibit an important dispersion. According to wave theory, total dispersion consists in three factors: multimode dispersion, wave guide dispersion, and material dispersion. In multimode fibres, *multimode dispersion predominates*; the other two factors may usually be neglected.

3 - Application to polarimetric sensor

Multimode fibres can be used to design polarimetric sensors[4] provided that the cladding propagation and skew rays are minimised, so that the polarisation dispersion is reduced. The meridian propagation creates a polarisation dispersion more important than in monomode

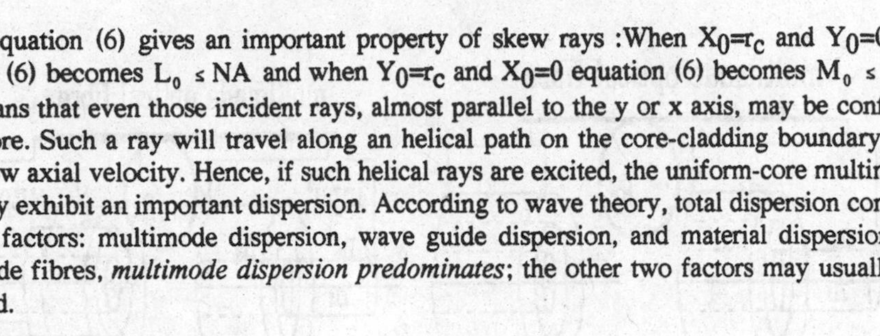

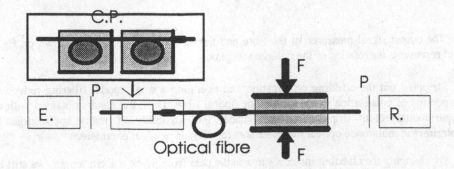

Fig. 3: Schematic set-up of the sensor.

fibres and therefore, the multimode fibre capabilities are not so good than obtained with the monomode fibres. Furthermore, the non-meridian and the cladding propagation have a different polarisation which causes perturbations in the signal at the end of the fibre. We have used for an experimental set up a step index multimode fibre FP-200-LMT with and without modal filtering (Fig.3).

- Without modal filtering

The optical fibre was tested without modal filtering by using 4 metre length of multimode optical fibre. A coil with a radius varying from 5 mm to the infinite (no coil) was made to the fibre (fig.4); a -5,9 to -20 dB extinction ratio was then measured.

- With modal filtering

In this case with a weak modal filter, the extinction ratio raised to -36,90 dB. This was obtained by stripping out the cladding propagation at the input and changing the cladding index in the output of the fibre (fig.5). In this way, the non-meridian and principally cladding propagation, were minimised.

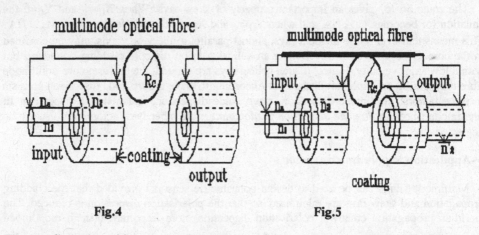

Fig.4

Fig.5

Fibre configuration without modal filtering   Fibre configuration with modal filtering

The output signal measured at the fibre end through a Polaroid filter is shown on fig. 6, and represents the rotation of the polarisation plan.

Stripping out the cladding propagation and operating a weak modal filtering reduce the dispersion of polarisation in the multimode optical fibre. Thus, the results obtained with our experimental set-up (fig. 3) let us consider the possibility of many applications for polarimetric multimode optical fibres sensors for various physical parameters.

By changing the cladding index n'_2 to a value near from n_1 on a 1 cm length, we still can enhance the system capability. An other possibility is to create a local thinning near the fibre output.

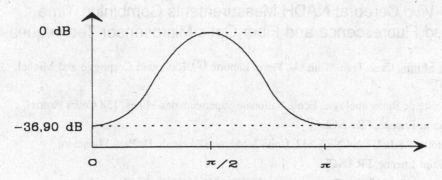

0 dB

−36,90 dB

0   π/2   π

Fig.6 : Normalised output power at the fibre end as a function
of the rotation of the polarisation plan.

References:

1- K. S. Chiang, J. Light. Technol., Vol. 8, n° 12, p. 1850-55 (1990).

2 -R.Ulrich, S.C.Rashleigh, and W.Eickhoff, Opt. Lett., Vol. 5, N° 6, p. 273 (1980).

3- R. Ulrich, M. Johnson, App. Opt., Vol. 18, N° 11, p. 1857-61 (1979).

4- A. Chakari, *Effet de certaines perturbations sur la propagation guidee de la lumière; application à la réalisation de quelques capteurs à fibres optiques*, Thèse de doctorat d'état à l'Université Louis Pasteur, Strasbourg, France (1989).

Fast *In-Vivo* Cerebral NADH Measurements Combining Time Resolved Fluorescence and Fiber Optic Microsensor Techniques

Stéphane Mottin, Canh Tran-Minh [1], Pierre Laporte [2], Raymond Cespuglio and Michel Jouvet [3].

[1]Laboratoire de Biotechnologie, Ecole Nationale Supérieure des Mines, 158 Cours Fauriel, F-42023 Saint-Etienne, FRANCE.

[2]Laboratoire T.S.I., URA CNRS 842, Univ. J. Monnet, 23 rue du Dr Paul Michelon, F-42023 Saint-Etienne, FRANCE.

[3]Département de Médecine Expérimentale, INSERM U 52, Univ. Cl. Bernard, F-69373 Lyon Cedex 08, FRANCE.

Abstract—A nanosecond time resolved fluorescence (TRF) from the nucleus raphe dorsalis of rats in chronic conditions, is investigated. The same single optical fiber (Ø=200 µm), ten meters long, transmits a subnanosecond nitrogen laser pulse (337 nm) and collects the brain autofluorescence from 390 nm to 550 nm. The fluorescence emission spectrum collected through the fiber, exhibits two maxima at 400 nm and 460 nm which principally comes from NAD(P)H (Nicotinamide-Adenine-Dinucleotide). The advantages of TRF is discussed. Preliminary data of the variation of NADH observed immediately after death of the animal are presented.

I. INTRODUCTION

The first detailed study of surface microfluorometry of brain *in situ* was reported in 1962 by B. Chance et al. [1]. A. Mayevsky and B. Chance [2] have since used Y-shaped optical fibers (Ø=2000µm) for monitoring surface NADH fluorescence in brains of anesthetized rats. We report the feasibility of NADH brain measurement in deep tissues through a fiber optic chemical sensor (FOCS) in freely moving unanesthetized rats.

To achieve this approach, the first determinant choice resides in the use of Laser Induced Fluorescence technique (LIF) which induces a rather high fluorescence level and therefore compensates the rather poor collection efficiency of small diameter optical fiber.

The second important technical choice consists in the use of time-resolved fluorescence (TRF) technique in a subnanosecond scale, widely used in fields as diverse as molecular biology, polymer science and solid-state physics [3]. Such methodology presents different advantages:

      —TRF allows a nice temporal separation between the optical fiber residual autofluorescence and the signal obtained with different lengths of the fiber;

      —the TRF can give more accurate characterizations of observed phenomena (comparison and assignment of fluorescence decay times).

The third choice is represented by a single fiber configuration used. With it the same multimode fiber transmits the excitation pulses and collects the fluorescence. This design, easy to build, limits the volume of the sample while permitting a good fluorescence collection.

Finally, the average optical power injected into the living medium is less important than in the case of steady state fluorescence, thus preventing the possibility of photophysical and photochemical injuries.

II. EXPERIMENTAL SECTION

A. *In vivo experimental procedure.* OFA male rats (IFFA.CREDO, n=4) weighting 260-300 g were anesthetized (chloral hydrat 400 mg/kg) and implanted with a guide cannula in both the n.RD (Raphe Dorsalis nucleus) and the frontal cortex. After ten days of recovery, the optical fiber was placed in the cannula (2mm out of the inner edge of the cannula). Measurements were performed during several sessions of six hours each without technical impairments.

B. *Time-resolved Fluorescence Measurements.*
The experimental setup has been previously described [4]. Briefly, the source is a nitrogen laser which delivers pulses of 300 ps (FWHM); the photomultiplier (PMT) of R3810 type (Hamamatsu) is selected in gain in order to detect single events. The 50Ω output of the PMT is coupled to a transient digitizer (Tektronix Model 7912 AD mainframe, Model 7A19 amplifier unit, Model 7B90P time base unit). The fiber used is a step index multimode, model PCS 200 from Quartz et Silice ($\emptyset_{core}$=200 µm).

III. RESULTS AND DISCUSSION

Figure 1 presents emission spectra established from the integrated intensity of pulsed fluorescence. The four spectra derived from n. RD show the temporal evolution of the signal versus time. When the fiber is pushed into the cannula, time is set to zero. The realization of each emission spectrum takes 18 min. The four spectra are recorded on a total period of 90 min. A good stability of the signal without normalization is found overtime. Two peaks are recorded: the first is near 400nm and the second near 460nm. In all experiments the first peak decreases consistently with time during a period which does not exceed half an hour after the zero set. For the wavelengths below 415 nm, the filter effect of the dichroic mirror must be taken into account in order to define more precisely the position of the first peak. Data of the litterature [5] gives normalized static spectra obtained from *post mortem* rat brain tissue with an excitation at 337 nm. The above two peaks are also observed in the same spectral region.

In order to identify more clearly the chemical species contributing to these signals, we first studied the emission spectrum of NADH in a buffer at pH7 (Fig. 1). A good correlation between this spectrum with those registered *in vivo* is obtained at the 460 nm peak and its red edge up to 530 nm. Knowing the strong absorption at 337 nm of NAD(P)H, the 460 nm *in vivo* signal, in agreement with surface microfluorometry, results most probably from this compound [1, 2]. The

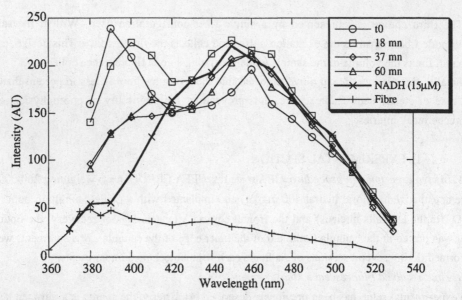

Fig 1. Emission spectra obtained *in vivo* in the raphe nucleus, of *in vitro* NADH solution and buffer pH7. Emission spectra of the raphe nucleus are monitored versus time, with time zero taken as the entry of the optical fiber into the cannula. The last curve is the spectrum without fluorescent matter at the tip (fibre alone in distilled water).

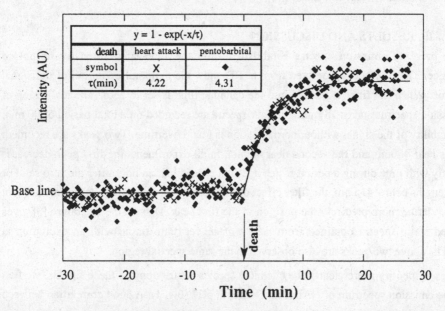

Fig 2. Normalized integrated fluorescence signal of n. RD versus time are registered at 460nm. The death occurs at time zero. The fitted functions $1-\exp(-x/\tau)$ and their parameters are shown. Time intervals between two successive experimental points are 30s (X) and 11s ($\blacklozenge$).

signal observed *in vitro* with a NADH concentration of 15µM is equal to that measured *in vivo*.

The normalized fluorescence signal following the death is presented on Figure 2.

This variation reported (40 to 60 %) is corrected from the fluorescence noise of the fiber i.e. the residual signal given when the fibre is uncoupled (last curve of Fig. 1).

The quickness of the death effect is clearly shown by the fitted function 1-exp(-x/τ). τ is close to 4.3 mn. The increase of NAD(P)H fluorescence signal is correlated with the increase of its endogenous concentration resulting from the rapid decrease of oxygen that follows the death. Oxygen is the main biological parameter affecting the NAD(P)H / NAD$^+$ ratio.

IV. CONCLUSION

The specificity of the measurement is based on three parameters : (1) pulsed excitation, (2) emission wavelengths, (3) decay times. The use of the time resolved LIF method together with a FOCS seems to be a feasible, powerful and promising tool adapted to brain measurement in unanesthetized and freely moving rats.

REFERENCES

[1] B. Chance, P. Cohen, F. Jobsis and B. Schoener, "Intracellular Oxidation-Reduction states in vivo", *Science*, 1962, vol. 137, pp. 499-508.
[2] A. Mayevsky and B. Chance, "Intracellular Oxidation-Reduction state measured in situ by a multichannel fiber-optic surface fluorometer", *Science*, 1982, vol. 217, pp. 537-540.
[3] V. Brückner, K.H. Feller and U.W. Grummt,. *Application of Time-resolved Opitcal Spectroscopy;* Studies in Physical and Theoretical Chemistry vol. 66. Amsterdam: Elsevier, 1990.; pp. 61-181.
[4] S. Mottin, C. Tran-Minh, P. Laporte, R. Cespuglio and M. Jouvet "Fiber-optic time-resolved fluorescence sensor for in vitro serotonin determination", *Appl. Spectrosc.*, 1993, vol. 47, 5, pp. 590-597.
[5] S. Andersson, A. Brun, E. Kjellen, L.G. Salford, L. Stromblad, K. Svandberg, S. Svandberg, "Identification of brain tumours in rats using LIF and haematoporphyrin derivative", *Lasers. Med. Sci.* , 1989, vol. 4, pp. 241-249.

Ein effizientes Verfahren zur Einkopplung von Laserstrahlung in Faserbündel

H. Gerhardt[1], U. Gladbach[2], G. Hillrichs[1], W. Neu[1]

[1]Laser-Laboratorium Göttingen e.V., D-37077 Göttingen
[2]CeramOptec GmbH, D-53121 Bonn

Zusammenfassung:

Messungen der Transmission von XeCl-Excimerlaserstrahlung ($\lambda = 308$ nm) durch Quarzfaserbündel zeigten, daß mit Hilfe von Linsenarrays die Einkoppeleffizienz erheblich gesteigert werden kann. Herkömmliche Beleuchtung eines Faserbündel mit identischen Fasern ergibt typisch etwa 33%, während unter Verwendung von Linsenarrays bis zu 70% Transmission erreicht wurden. Die verwendeten Quarzfaserbündel bestehen aus sechs bis zehn Fasern gleichen Typs mit Kerndurchmessern von 400 μm, 200 μm und 100 μm. Der Laserstrahl wird durch ein System aus gekreuzt angeordneten Zylinderlinsen oder durch aneinandergesetzte sphärische Linsen mit quadratischem Rand in konvergente Teilstrahlen aufgespalten, die dann in die einzelnen Fasern des Bündels eingekoppelt werden. In Kombination mit einer weiteren Einzellinse werden Fokusdurchmesser unter 90 μm erzielt. Die Faserhalterung der Einkoppeleinheit ermöglicht sowohl die Fixierung als auch die Justierung der Faserenden des Bündels, um die Einkoppelflächen der Fasern im Fokusraster des Linsenarrays zu positionieren. Als präziseste und zuverlässigste Methode zur Herstellung der Aufnahmebohrungen für die Fasern erwies sich die Markierung der Fokuspositionen mit dem XeCl-Excimerlaser und anschließende Mikromaterialbearbeitung mit Hilfe eines ArF-Lasers ($\lambda = 193$ nm). Durch die erzielte Verbesserung der Einkopplung ist es möglich, Laserstrahlung höherer Energie zu übertragen, ohne dabei die Quarzfasern zu zerstören und somit eine effektive Laseranwendung zu gewährleisten. Dieses Einkoppelverfahren kann ohne Schwierigkeiten für das gesamte Spektrum von Lasern hoher Leistung verwendet werden.

1. Einleitung

Die Übertragung von Laserpulsen hoher Leistung wird im wesentlich von der Zerstörschwelle des Fasermaterials an der Einkoppelfläche limitiert. Die Leistungs- bzw. Energiedichte läßt sich relativ einfach durch Verwendung entsprechend dickerer Fasern oder durch spezielle Einkoppelgeometrien, wie z.B. Taperfasern reduzieren, wobei jedoch zunehmend die Flexibilität eingeschränkt wird [1]. Die Biegekräfte wachsen um das Vierfache mit dem Faserdurchmesser und quadratisch mit dem Biegeradius. Darüber hinaus ist zu beobachten, daß die Transmissionsverluste nicht mit kleiner werdendem Biegeradius konstant ansteigen, sondern unterhalb eines bestimmten Biegeradius (ca. 100facher Faserdurchmesser) stark zunehmen [2]. Die Verwendung von Bündeln aus Lichtleitfasern zur Übertragung von Laserstrahlung verbindet mechanische Flexibilität mit hohen übertragbaren Leistungen bzw. Pulsenergien. Konventionelle Methoden zur Einkopplung des Laserlichtes in Faserbündel, wie sie insbesondere im medizinischen Bereich angewendet werden, sind allerdings mit erheblichen Verlusten behaftet. Die Faserendflächen, die alle bündig in einer Ebene abschließen, werden mit paralleler oder leicht fokussierter Strahlung bestrahlt. Dabei wird der Akzeptanzwinkel der Faser jedoch nicht völlig ausgeschöpft, so daß durch unzureichende Ausnutzung der schwingungsfähigen Moden störende Intensitätsüberhöhungen auftreten können.

In der Regel werden kreisförmig angeordnete Fasern verwendet, wobei die dichteste Packung bei hexagonaler Anordnung etwa 77% der Stirnfläche ausgefüllt. Bei einem typischen Kern/Mantel-Verhältnis von 1,1 beträgt der nutzbare Flächenanteil 60% und bei einem Verhältnis von 1,2 nur 47%. In der Praxis liegt der Flächenanteil der Kernflächen in der Regel unter 50%, da die ideale hexagonale Anordnung in der Herstellung nur bedingt möglich ist. Die Transmissionen, die mit dieser Art von Faserbündeln erreicht werden liegen zwischen 20% und 45%. Von Nachteil ist weiterhin, daß die Faserenden, die frei im Anschlußstecker in Richtung des Lasers herausragen, durch Kleber fixiert sind, der teilweise bei Einstrahlung des Lasers in Mitleidenschaft gezogen wird. Sofern sich ablatierte Partikel des Klebers oder des Steckers auf die Fasereinkoppelfläche ablagern, kann das zu einer Zerstörung der Faserfläche an dieser Stelle führen.

Die im folgenden dargestellte Einkopplungsmethode unterscheidet sich von den üblichen Systemen dadurch, daß der Laserstrahl mit Linsenanordnungen in Teilstrahlen aufgespalten wird, so daß eine gezielte Einkopplung in jede einzelne Faser des Faserbündels möglich wird. Auf der Basis dieses Einkoppelverfahrens lassen sich flexible Laserstrahlführungssysteme für viele Anwendungsgebiete realisieren.

2. Experimenteller Aufbau
2.1. Linsensysteme

Abbildung 1 zeigt ein System von gekreuzt angeordneten Zylinderlinsen das zur Aufspaltung des Laserstrahls verwendet wurde. Die Zylinderlinsen des verwendeten Arrays haben eine Brennweite von f=100 mm. Um den rechteckigen Abmessungen des für die Transmissionsmessungen verwendeten Excimerlaserstrahls zu entsprechen, werden in einer Ebene drei, in der anderen sieben Linsen

angeordnet, so daß das Array von 12mm x 28mm eine Aufspaltung in 21 Teilstrahlen ergibt.

Ein zweites Linsensystem wurde aus plankonvexen, auf quadratische Abmessungen abgeschliffenen Einzellinsen aufgebaut. Dazu wurden 10 Linsen mit einer Brennweite von f=60 mm, Abmessungen von 5 mm x 5 mm auf eine polierte Quarzplatte aufgesprengt. Da Quarzplatte und Linsen aus gleichem Material und optisch kontaktiert sind, lasse sich Reflektionsverluste vermeiden. Die sphärischen Linsen sind in zwei Reihen zu je fünf Linsen angebracht; so daß ein Array von 10 mm x 25 mm mit 10

Einzelstrahlen entsteht. Bei diesem Linsenarray entsteht eine exakte Brennebene, in der die zehn Fokuspunkte vorliegen. Um die Fokiausdehungen darzustellen, die auf die Fasereinkoppelflächen abgebildet werden, wird der vom LLG entwickelte UV-Laser Beam Profiler eingesetzt. Bei Verwendung des Arrays (f=60 mm) beträgt die Ausdehnung ca. 190 µm x 50 µm (hor. x vert.), so daß eine Einkopplung in 200 µm-Fasern realisierbar wird. Mit einer zusätzlichen Linse mit f=100 mm verringert sich die Ausdehnung auf ca. 140 µm x 40 µm. Die Linsenanordnung von sphärischem Array mit der Brennweite f=60 mm und zusätzlicher Linse mit f=100 mm ergibt eine Gesamtbrennweite von etwa 38 mm, wenn man die Linsen als hintereinander angeordnete dünne Linsen betrachtet. Verwendet man eine Linse mit der Brennweite f=50 mm, so erreicht man eine Fokusausdehnung von etwa 75 µm x 40 µm, die zur Einkopplung in 100 µm-Fasern notwendig ist.

2.2. Faserhalterung

Um das Fokiraster in der Brennebene, welche durch das starre Linsenarray vorgegeben ist, auf die Fasereinkoppelflächen genau abzubilden, muß die Positionierung der Fasern in einer Faserhalterung präzise ausgelegt sein. Die Genauigkeit der Positionierung ergibt sich dabei aus dem Durchmesser der verwendeten Fasern und der Fokusgröße bei dem jeweils genutzten Linsenarray.

Mit Hilfe einer Mikrometerverschiebevorrichtung kann die Frontseite der Faserhalterung in die Brennebene verfahren werden, wobei durch Laserablation die Foki markiert werden. Je nach Faserdurchmesser werden die Führungslöcher mechanisch oder durch Lasermaterialbearbeitung gebohrt. In die mechanisch gebohrten Löcher wurden teilweise Metallkanülen zur Faserführung eingeschoben. Die Fasern ragen dabei in Richtung des Linsenarrays etwa 2 - 3 mm aus den Kanülen heraus, um eine Ablation der Faserhalterung zu vermeiden. Für hochpräzise Bohrungen der Führungslöcher wurde eine im Laser-Laboratorium Göttingen entwickelte Mikromaterialbearbeitungsapparatur verwendet [3]. In diesem Fall wird auf die Führungskanülen verzichtet, denn die Faserenden werden hierbei in die gebohrten Führungen einer PMMA-Halterung eingeklebt. Nach Markierung der Halterung werden durch Ablation mit einem ArF-Excimerlaser (λ=193 nm) Löcher mit einem Durchmesser von 230 μm bei 200/220 μm-Fasern und von 130 μm bei 105/125 μm-Fasern gebohrt. Das Einfädeln der Faserenden erfolgte unter einem Lichtmikroskop.

2.2. Transmissionsmessungen

Der experimentelle Aufbau zur Messung der Absoluttransmission von Laserstrahlung durch ein Bündel optischer Fasern ist in Abb. 2 dargestellt. Ein XeCl-Excimerlaser (λ=308 nm, Lambda Physik LPX 210iCC) mit einer Pulsdauer von $\Delta\tau$=28 ns wurde für die Transmissionsmessungen verwendet. Der Laserstrahl weist in vertikaler Richtung eine Divergenz von 1 mrad, in horizontaler Richtung von 3 mrad auf; die unterschiedliche Divergenz wird durch ein Zylinderlinsenteleskop angeglichen. Ein variabler, schrittmotor-gesteuerter dielektrischer Abschwächer erlaubt eine quasi-kontinuierliche Einstellung der Pulsenergie. Zwischen Linsenarray und Faserhalterung koppelt eine Quarzglasplatte etwa 8% der Strahlung zur Kalibrierung der eingekoppelten Pulsenergie aus. Die eingekoppelte und transmittierte Energie wird mit je einem pyroelektrischen Detektor (RjP-735, Laser-Precision) gemessen, wobei sich durch Quotientenbildung unmittelbar die Absoluttransmission ergibt. Die folgende Tabelle stellt die Ergebnisse der Transmissions-

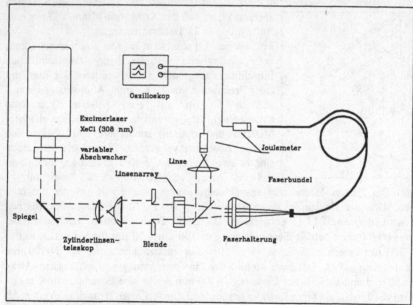

messungen mit den unterschiedlichen Parametern und Linsensystemen dar.

Verwendetes Linsensystem	Fasern ($\varnothing_{Kern}$ in μm)	Anzahl	Transmission in %
Zylinderlinsenarray (f=100 mm)	400	9	72
sphärisches Linsenarray (f=60 mm)	200	10	73
sphärisches Linsenarray (f=60 mm)	105	6	64
mit zusätzlicher Linse (f= 50 mm)	115	6	70

Zum Vergleich wurde ebenfalls die Transmission eines Faserbündels aus zehn Einzelfasern mit 200/220 μm bei konventioneller Einkopplung gemessen. Die Einkoppelfläche des Faserbündels weist einen Durchmesser von etwa 1 mm auf und wird möglichst homogen durch Abbildung einer Blende von 10 mm Durchmesser beleuchtet. Dabei konnte maximal eine Transmission von 33% erreicht werden.

3. Diskussion

Die Messungen zeigen, daß mit Hilfe der verwendeten Linsenarrays die Transmission durch Quarzfaserbündel von Excimerlaserstrahlung der Wellenlänge $\lambda = 308$ nm etwa um den Faktor zwei gesteigert werden kann [4]. Bei Einkopplung mit dem Zylinderlinsenarray (f = 100 mm) kann problemlos eine Transmission von 70% durch neun 400 μm-Fasern realisiert werden. In Relation dazu ist eine Transmission von bis zu 90% durch Einzelfasern und etwa 35% durch Faserbündel üblich. Die Fokusabmessungen von etwa 260 μm x 100 μm, die mit dem Zylinderlinsenarray erreicht werden, sind für eine Einkopplung in 200 μm-Fasern aber nicht geeignet. Mit Hilfe der kürzeren Brennweite der sphärischen Linsen des zweiten Linsenarrays (f = 60 mm) kann die Einkopplung in zehn 200 μm-Fasern erfolgen, wobei der erreichte Transmissionwert von 70% ein zufriedenstellendes Ergebnis darstellt. Die Vergleichsmessung mit Hilfe der herkömmlichen Einkoppelmethode ergab eine typische Transmission durch zehn 200 μm-Fasern von ca. 33%. Diese direkte Gegenüberstellung bei identischen Fasern zeigt die Überlegenheit der Einkopplung mittels eines Linsenarrays. Durch Kombination des sphärischen Linsenarrays mit einer zusätzlichen Linse (f = 50 mm) wird die Gesamtbrennweite auf etwa 27 mm verringert mit Fokusabmessungen von ca. 75 μm x 40 μm. Allerdings mußte damit aufgrund der völligen Ausleuchtung der Zusatzlinse eine erhebliche Zunahme der Größe der Randfoki durch sphärische Aberration in Kauf genommen werden. Zur Auswertung kamen daher nur die zentral gelegenen Fasern mit Transmissionwerten von 64% bzw. 70%. Bei der Herstellung der Faserhalterung erweist sich die Methode des mechanischen Bohrens und die Verwendung von Führungskanülen als zu ungenau und insgesamt zu aufwendig. Bei der Fertigung der Halterung für die 200 μm-Fasern ist das Bohren mit Hilfe des Lasers, nach Markierung, zuverlässig und ausreichend genau. Obwohl die erreichte Verbesserung der Einkopplung mit daraus resultierender höherer Transmission durch Faserbündel aufgezeigt werden konnte, wäre als nächster Schritt eine Erhöhung der Anzahl der Einzelfasern des Bündels zu überlegen.

Will man in Zukunft die Einkopplung mit Hilfe von Linsenarrays weiterführen, so müssen reproduzierbare Linsenarrays konzipiert werden, die den Einsatz von universell gefertigten "Serienfaserhalterungen" erlauben. Eine Möglichkeit, ein Linsenarray mit definierten Fokusabständen zu fertigen, wäre die Herstellung durch holographische Verfahren. Mittles Ätztechniken wird ein holographisches Array hergestellt, welches den Vorteil einer idealen Abbildung der Faserenden und eine vollständige Ausnutzung der numerischen Apertur beinhaltet. Die Anzahl der Linsensegmente bei diesem Verfahren kann fast nach Belieben gewählt werden.

Das oben beschriebene Verfahren für Einkopplung von Laserstrahlung läßt sich ohne weiteres auf den sichtbaren wie den nahen infraroten Spektralbereich ausdehnen.

Literatur:

[1] K.F. Klein, G. Hillrichs, W. Neu, H. Fabian, U. Grzesik: UV-Laserlichtübertragung mit Quarzglasfasern - Stand der Technik. In: Laser. Technologie und Anwendungen. Jahrbuch 1993. Hrsg. H. Kohler. Essen: Vulkan-Verlag 1993 (im Druck)

[2] K.-H. Schönberg, H.Ch. Bader: Advatages of using thin fibers. Optical fibers in Medicine III, SPIE **960**, 238 (1988)

[3] J. Ihlemann, H. Schmidt, B. Wolff-Rottke: Excimer laser micromachining. Adv. Mat. Opt. Electr. **2**, 87 (1993)

[4] U. Gladbach: Aufbau und Test einer Zylinderlinsenanordnung zur Einkopplung von Excimerlaser-. strahlung in Faserbündel. Fachhochschule Aachen. Diplomarbeit 1992 (unveröffentlicht)

Congress F

Congress-Chairman: J. Franz

Optische Kommunikationstechnik
Optical Communications Technology

The Communication Subsystem for Silex

Christoph Nöldeke
ANT Nachrichtentechnik GmbH
Gerberstrasse 33, D-7150 Backnang

Abstract

The communication subsystem for the ESA project SILEX (Semiconductor Intersatellite Link Experiment) features direct detection and NRZ modulation with remote driving of 60mW diode lasers transmitters, and ultra-low noise SLIK APD optical receivers (better than 100 photons/bit at BER=10^{-6}). Dense packaging of service modules including hybrid DC/DC converters allows the concentration of the functional electronics into few boxes. Direct interfaces with the payload primary power and On-Board Data Handling buses are provided.

1. Introduction

The SILEX project was originally planned in the mid 80s comprising both an Inter-orbit link (IOL) between a low earth orbit satellite (LEO) link and an Inter-satellite link (ISL) between two geostationary satellites, with features such as 4 optical channels of 120 Mbps Quaternary Pulse Position Modulation (QPPM) each. Today's scenario is depicted in Fig.1.1.

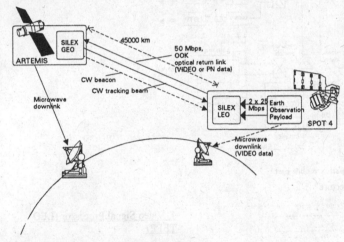

Fig.1 The SILEX Scenario

2. Communication subsystem requirements

The main requirements for the communication subsystem are:

- 45000 km inter-orbit link distance with up to 76 dB of signal loss,
- Transmission of 2x25 Mbps digital video data streams from the imaging payload of SPOT4,
- pseudo-random data generation and transmission for on-ground bit error rate measurements at 2x25 Mbit/s,
- one of three redundancy for 60 mW, 850 nm laser diodes
- commandability of laser levels via On-Board Data Handling (OBDH) bus,
- receive sensitivity of -58.3 dBm with 400 pW of detector straylight.

The forward link GEO-LEO is operated in CW and acts as a tracking beacon. However, the GEO terminal has the option to transmit data in order to be compatible for second use in a forthcoming ISL experiment.

3. Communication subsystem architecture

In order as not to overload the telescope platform of the LEO and the GEO terminals, part of the communication electronics was confined to the "fixed part", namely the video signal processing and PN pattern generation functions and the regenerator function. This leads to the denominations TEMP (Telecom Electronics Mobile Part) and TEFP (Telecom Electronics Fixed Part). The laser drivers and their service functions (DC/DC Converter and OBDH interface) had to be kept in proximity to the lasers and were integrated into a single housing (the TEMP). This device is implemented to both LEO and GEO payloads. For GEO, it supplies as well the optical Receiver Front End (RFE). The architectures for LEO and GEO are shown in Fig.2 and Fig.3, respectively.

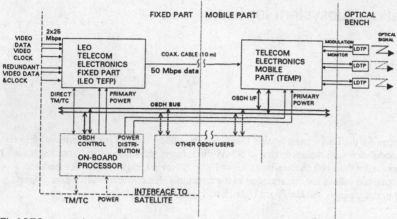

Fig.2 LEO communication subsystem architecture. LDTP: Laser Diode Transmitter Package, not part of communication subsystem.

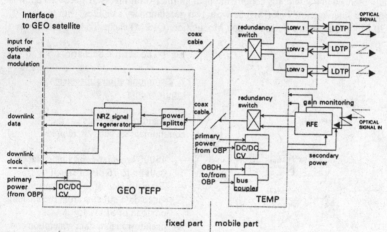

Fig.3 GEO communication subsystem architecture

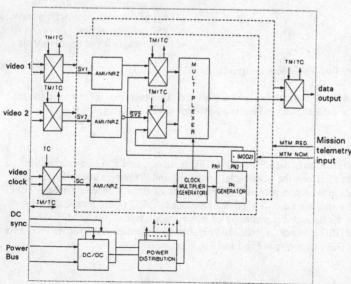

4. Video Signal Processor (LEO TEFP)

This unit is internally redundant. It contains DC/DC converters (DC/DC CV) which are supplied from the satellite. The block diagram of the unit is shown in Fig.4.

Fig.4 LEO TEFP (Video Signal Processor, VSP) block diagram.

Both video channels are converted from AMI to NRZ and multiplexed. For PN transmission, the VCXO normally used for video data reclocking is running as master clock oscillator at 25 MHz. A PN 15 sequence is generated and supplied to the multiplexer, once as original, and once as a maximum length shifted version. This second version is used as a carrier for low (500 kbps) sensor data from the SILEX payload, called Mission Telemetry (MTM).

5. Telecom Electronics Mobile Part (TEMP)

For the block diagram, refer to Fig.3. This device performs the following functions:

- laser diode remote driving via low impedance coaxial cable,
- up to 250 mA , 3ns rise time pulses,
- laser diode monitoring,
- laser diode protection,
- data signal redundancy switching for receive and transmit functions,
- OBDH interface,
- secondary voltage generation,
- RFE supply and monitoring (GEO only).

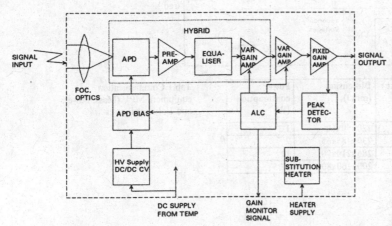

DC SUPPLY FROM TEMP   GAIN MONITOR SIGNAL   HEATER SUPPLY

6. Receiver frontend (RFE)

The RFE is a 40 MHz bandwidth direct detection receiver equipped with a super-low noise SLIK APD ($k_{eff} = 0.005$). This part has already been used in the benchmark 39 photons/bit QPPM optical receiver developed in an earlier phase of SILEX [1].

Fig.5 Receiver Frontend block diagram

The device is equipped with its own focusing lens and with a 450 V APD supply. The block diagram is given in Fig.3.3. The stringent in-orbit stability requirements (100 µrad) are realized by minimizing thermoelastical mismatch between parts of the housing.

A constant output level of 500 mV over the dynamical range from 250pW to 16 nW of input power is obtained from the amplifier stages using an automatic level control (ALC) system which acts on RF gain stages as well as on the APD voltage.

7. Regenerator (GEO TEFP)

The function of this device is to recover clock data from the noisy RFE output signal. The device is internally redundant and comprises its own DC/DC converters. The block diagram is shown in Fig.6.

8. System Performance

The system figure of merit in terms of bit error rate versus received power can be calculated using the standard Gaussian model as described previously [2]. The results are displayed in Fig.7. Mechanical outlines, mass, and power consumption of the communication units are summarized in Tab.1.

764

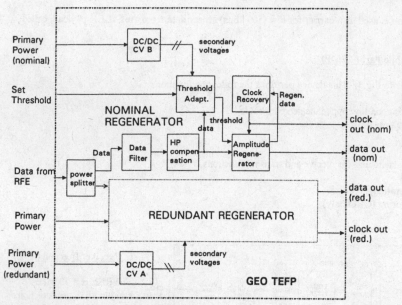

Fig.6 GEO TEFP (regenerator) block diagram. The PLL pull-in range is +/- 150ppm. HP: High Pass

Device	Mass (kg)	Dimension (mm3)	Power consumption (W)
RFE	0.44	120x50x60	1.3
GEO TEFP	1.9	230x185x65	7
TEMP	3.6	286x210x90	12
LEO TEFP	2.1	200x160x80	12.5

Tab.1 Communication equipment MPV (Mass, Power, and Volume) figures

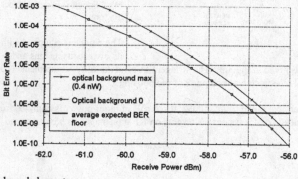

Fig.7 System figure of merit in terms of received power versus bit error rate (Gaussian model, simulation for SLIK APD, keff=0.005, transimpedance amplifier). The expected BER floor is due to radiation in the GEO orbit.

Acknowledgments

This work was funded by ESA, under prime contractorship of MMS, France. The responsibility for the contents lies with the author. The author recognizes substantial contributions from D.Appel, M.Alberty, M.Clostermann, H.Fried, P.Greulich, B.Günther, B.Hespeler, G.Hüther, U.Hildebrand, B.Smutny (ANT), B.Dion (EG&G Optoelectronics, Canada) and M.Kowatsch (Schrack Aerospace, Austria).

References

[1]   A.MACGREGOR, B.DION, CH.NÖLDEKE, O.DUCHMANN: "39 photons/bit direct detection receiver at 810 nm, BER=1x10^{-6}, 60 Mb/s QPPM SPIE Proceedings 1417 , 374 (1991).
[2]   CH.NÖLDEKE: "Transmission schemes for optical telecommunications in space" SPIE Proceedings 1131 , 24 (1989)

Solacos YAG Communication System-YKS Implementation of the Syncbit-Concept

K. Pribil[1], Ch. Serbe[1], B. Wandernoth[2], Ch. Rapp[2]

[1]Dornier GmbH, PO BOX 1420, D-7990 Friedrichshafen
[2]DLR Institut für Nachrichtentechnik, D-8031 Wessling

ABSTRACT

The syncbit system is a new type of an optical homodyne transmission system. In the syncbit system the phase synchronisation information to maintain the receivers phase lock is time -multiplexed into the transmitter data stream. Compared to the Costas-Loop-system this concept avoids the 90° hybrid and the Q-channel for the expense of increased complexity in the digital signal processing unit. During the acquisition process two subsequent acquisition steps have to be carried out: In the first step the receiver is tuned to a low-IF and held there fore the subsequent acquisition phase. In the second step the position of the syncbits in the incoming datastream are located. Sampling the syncbits yields the error signal to pull the optical PLL into homodyne operation mode. To achieve reliable operation of such a system a very robust syncbit detection algorithm must be implemented. In the present work schemes for channel and syncbit coding are investigated. First results from computer simulations are presented.

1. INTRODUCTION

The SOLACOS program is a German national program for the development of laser communication systems for space applications. The aim of the present phase C is the development of an experimental communication system for technology verification in a laboratory based experiment. The phase C experimental hardware consists of two communication terminals and the necessary command and control infrastructure. The system implements an asymmetric IOL link in which data between the terminals are transmitted on a 650 MBit/s link in one direction and on a 10 MBit/s link in the opposite direction. For the high data rate (HDR) path the coherent PSK-transmission scheme will be implemented to achieve optimum performance.

2. PSK HOMODYNE TRANSMISSION CONCEPTS

In a homodyne receiver the local oscillator laser must be locked to the phase of the transmitter laser. There are several methods to provide the synchronisation information:

- transmission of a residual carrier (DC-coupled receiver)

- generation of a phantom-carrier in the receiver (Costas-Loop-receiver, AC-coupled receiver)

- transmission of synchronisation information multiplexed into the data stream (syncbit-receiver, AC-coupled receiver)

For the HDR channel of the SOLACOS system we chose binary PSK modulation with a Costas-Loop- and syncbit-receiver as the two most promising alternatives. The Costas-Loop-receiver generates a phantom carrier by means of multiplying the signals in an I- and Q-channel. The received and the LO-signal are optically superimposed into I and Q with a relative 90° phase shift of the LO-signal in the Q-channel. The key component to split the signals is an optical 90° hybrid which has been fabricated in bulk optics [1] and in fibre optics [2]. The alignment of the hybrid is critical and additional loss compared to a 180° beamsplitter is encountered. Another drawbacks of the Costas-Loop-receiver is the additional Q-channel path which doubles the number of hardware components. Several papers were published on the realisation of Costas-Loop-receivers and the behaviour of such a system is well known.

The syncbit-receiver uses a simpler analogue receiver with a 180° hybrid and only one receiver signal path. But additional hardware is required in the transmitter and in the receiver for syncbit processing. Therefore optical setup and analogue electrical circuitry is simplified. On the other hand certain ammount of digital signal processing is introduced to provide the syncbits. In future implementations this additional digital circuitry can be integrated into the ASIC, that is needed for channel coding and I/O data multiplexing. Therefore we expect the syncbit system to be less expensive than the Costas-Loop system.

3. OPERATION OF THE SYNCBIT SYSTEM

Figure 1 shows how syncbits are merged into the datastream in the transmitter:

- transmitter frame filled with data TX_D (a),

- transmitter frame with the syncbit TX_S (b),

- transmitter frame TX (c) assembled from TX_D and TX_S, according to:

$$TX = TX_D + \tfrac{1}{2}TX_S \qquad {}^{*}$$

(1)

The assembled transmitter frame is a three-level analogue signal which is fed to the modulator amplifier of the laser transmitter. Fig. 2 shows the phase state diagram of the optical signal on the channel. The receiver must synchronise with the syncbit information in the Q-axis and detects data in the I-axis. Therefore in the photo diode the received field is multiplied with the LO-field giving an electrical input signal U_{RX} to the receiver:

$$U_{RX} \propto \cos\varphi$$

(2)

This is amplified, filtered and amplitude stabilised in the subsequent stages. Then the received signal is fed into a 3-level amplitude discriminator with two outputs for data and syncbit. A databit is detected within -1 and +1 by an ordinary comparator with the decision level set to 0. A syncbit is detected if the signal is within -s and +s range by means of a window comparator.

In the design phase we have considered two methods to obtain the syncbit from the received datastream:

- syncbit extracted by the channel decoder (requires complicated coder/decoder circuits which can handle the syncbit induced code violations)

- syncbit extracted with a 3-level comparator. The channel decoder is subsequent to the comparator. (simple circuit, channel decoder can be synchronised by the syncbits)

For the high data rates we chose the latter because this method leads to a simpler hardware realisation (less power, mass and volume).

4. IMPLEMENTATION IN THE PRESENCE OF NOISE

To implement the syncbit algorithm in a practical system the behaviour in a noisy environment has to be evaluated. An appropriate noise model is Additional White Gaussian Noise (AWGN). AWGN in the receiver adds to the desired input signal as shown in fig. 3. The error rates for databits and syncbits can be calculated to:

$$P_{DB} = Q\left(\sqrt{\frac{2 \cdot E_b}{N_0}} \right)$$

(3)

$$P_{SB} = Q\left(\sqrt{2\left(\frac{E_b}{N_0} - 6dB \right)} \right) \qquad \text{where } Q(x) = \frac{1}{\sqrt{2\pi}} \cdot \int_x^\infty e^{\frac{-u^2}{2}} \, du$$

(4)

The bit error rate as a function of signal to noise ratio (SNR) is shown in fig. 4. From this graph one can see that the error rate of the syncbit P_{SB} is significantly higher than the bit error rate P_B. This is due to the reduced noise margin of a multilevel signal. The exact knowledge of the position of the syncbit is essential both for phase synchronisation of the optical PLL (OPLL) and for the synchronisation of the data recovery circuit. To overcome the poor syncbit-error rate and obtain reliable information of the syncbit position stochastic means must be applied. This means that a set of several syncbits has to be evaluated.

For the system in operation two different state transitions have to be considered:

- lock-in-process: The receiver is out of phase synchronisation and searches for syncbits to enter phase lock. Provided the bit clock recovery is already operating the lock-in-probability for a single data frame as a function of SNR is shown in fig. 5. From this diagram the number of dataframes that must be observed to achieve lock with can be calculated, see fig. 6 (for 99% lock-in-probability)

- lock-out-process due to AWGN: The receiver is in phase lock. If n subsequent syncbits are corrupted the receiver erroneously interprets this as a loss of synchronisation, see fig. 7.

Currently we are investigating two different implementations of the syncbit finder which differ in the way they recognise and process lock-in and lock-out transitions:

- System 1, shown in fig. 8, looks for one fully uncorrupted dataframe of 12 databits and one syncbit which must immediately follow the DBs to switch to lock-in. Lock-out is set if n subsequent false syncbits are detected. n has to be chosen high enough to avoid too early lock-out by noise but small enough to quickly detect real loss of synchronisation.

- System 2, shown in fig. 9, looks for 3 subsequent syncbits to enter lock. Lock-out is set if 4 subsequent false syncbits are detected.

To obtain a more detailed insight into the operation characteristics of the two systems a simulation model was built and both systems were evaluated. For system 1 fig. 10 shows the number of symbols (databits plus syncbits) required to enter lock, the number of symbols necessary to recognise loss of synchronisation and the number of symbols to reenter lock after loss. The results of the respective simulation for system 2 are shown in fig. 11.

The following conclusions can be drawn from these results:

- lock-in performance for both systems is very good for high SNR, system 1 is faster than system 2 while for bad SNR system 1 requires more time to lock.

- lock-out performance is mainly affected by the length of the shift-register. In the simulation system 1 uses a 7 bit counter while system 2 uses a 4 bit shift-register which set lock-out after 3 frames.

- System 1 works best with a high threshold of $s=0.6$, while system 2 works best around $s=0.4$ (obtained by simultions with varied thresholds)

Both algorithms are currently implemented in a low data rate breadboard at DLR to experimentally verify these results.

5. SYNCBIT IMPLEMENTATION IN THE SOLACOS SYSTEM

For SOLACOS one concept will be selected and implemented. The preliminary parameters for the SOLACOS implementation are listed in fig. 12. Channel coding is performed by a 10B12B coder [5].

The block diagram of the SOLACOS transmitter is shown in fig. 13: The incoming datastream is processed according to the syncbit-scheme and fed to optical modulator in which the cw light of a Nd:YAG laser is phase modulated. Via a fibre the transmitter signal is guided to the optical bench of the terminal from where it is coupled into the free space transmission channel.

The block diagram of the SOLACOS receiver is shown in fig. 14: In the receiver terminal the incoming light from the optical bench is guided to the receiver via a fibre. In the receiver the LO-signal is superimposed in the 180° hybrid and fed to the balanced receiver. After postamplification, filtering and gain control the signal enters the data-clock-syncbit recovery stage and subsequently the channel decoder. In the S/H stage the analogue signal is sampled and fed to the OPLL controller. Additional hardware is required for the frequency acquisition process and to derive signals for the fibre nutator beam position sensor.

In frequency acquisition the frequency acquisition system of the receiver tunes the LO towards the OPLL lock-in range. At a beat frequency of approx. 100 kHz the syncbit finding process starts. If the syncbits are found the analogue error signal can be sampled and fed to the OPLL. Figure 15 shows that the time available to find the syncbits is limited by the beat signal envelope. This shows clearly that lock-in time is a major selection criterion for the syncbit finding algorithm.

6. CONCLUSIONS

With the concepts described it will be possible to reliably enter synchronisation within sufficiently short time and to keep lock even under bad SNR conditions. We feel that the syncbit-system is a reliable and very attractive solution for optical PSK transmission.

7. ACKNOWLEDGEMENTS

This work is partly supported under DARA contract 50 YH 9207/8.

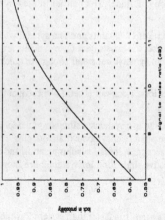

Figure 3: Amplitude distribution with AWGN and decision windows for data and syncbits.

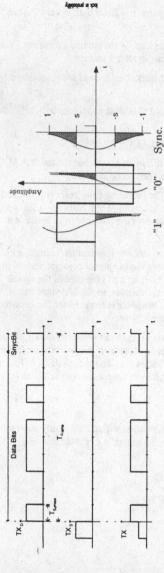

Figure 1: Data TX_D (a), syncbits TX_S (b), and assembled transmitter frame TX(c).

Figure 2: Phase state diagram of the optical carrier on the transmission channel.

Figure 5: Lock-in-probability for a single data frame as a function of SNR

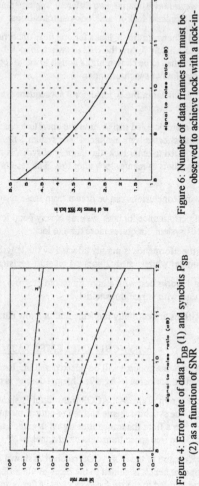

Figure 4: Error rate of data P_{DB} (1) and syncbits P_{SB} (2) as a function of SNR

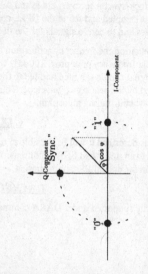

Figure 6: Number of data frames that must be observed to achieve lock with a lock-in-probability of 99%.

769

Figure 11: System 2: number of symbols required to enter lock, number of symbols necessary to recognise loss of sync., and number of symbols to reenter lock after loss of sync.

Synchronisation :	Sync. Bits (12DB1SB)	
Channel Coding :	10B12B	
Bit Rate :	650 MBit/s	
Symbol Rate :	845 MBit/s	
t_{Symbol} :	1.18 ns	
t_{Frame} :	15.38 ns	
f_{Beat} :	100 kHz	
$t_{Lock In}$:	2.05 μs	$\approx 130\ t_{Frame}$

Figure 12: Preliminary parameters for the SOLACOS implementation

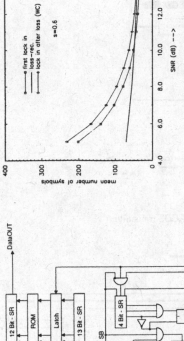

Figure 9: Syncbit processing system 2

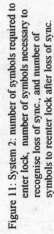

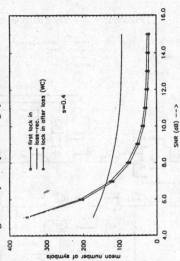

Figure 10: System 1: number of symbols required to enter lock, number of symbols necessary to recognise loss of sync., and number of symbols to reenter lock after loss of sync.

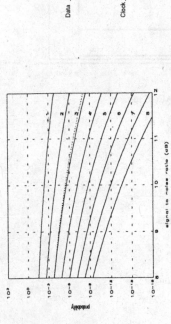

Figure 7: Lock-out-probability for n subsequent false syncbits as a function of SNR.

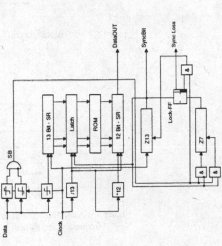

Figure 8: Syncbit processing system 1

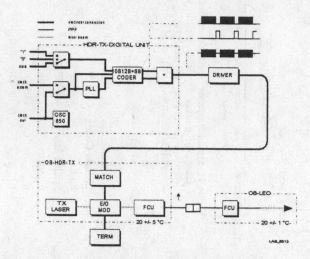

Figure 13:   Block diagram of the SOLACOS transmitter

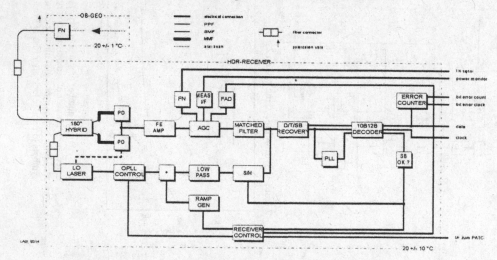

Figure 14:   Block diagram of the SOLACOS receiver

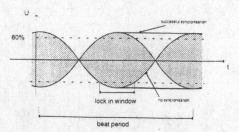

Figure 15: Available time to find syncbits during
frequency acquisition.

8. REFERENCES

[1] Garreis R., "90° optical hybrid for coherent receivers", SPIE Vol. 1522, 1991, pp 210

[2] Schöpflin K., et al. "PSK Optical Homodyne System with Nonlinear Phase-Locked-Loop", Electronics Letters, Vol. 26, No. 6, 1990, pp 395

[3] Bopp M., et al. "BPSK Homodyne and DPSK Heterodyne Receivers for Free-Space Communication with Nd:Host Lasers", SPIE Vol. 1522, 1991, pp 199

[4] Patent: P4110138.3

[5] Morgenstern G., "Die Leistungsdichtespektren einiger nB(n+1)B-blockcodierter Signale", Techn. Ber. d. FI der DBP, Nr. 44 TBr 98, März 1985

Activities of the European Space Agency in the Field of Optical Space Communication

B.Furch, M.Wittig, A.F.Popescu
European Space Agency

ABSTRACT

The European Space Agency ESA is developing an optical inter-orbit communication system enabling a link between a low-Earth orbiting (LEO) and a geostationary spacecraft (GEO). This so-called SILEX system allows a transmission of 50 Mbps from LEO to GEO in an experimental and pre-operational mode. The system uses GaAlAs laser diodes operating at 60 mW average power and direct detection. Both terminals are in the main development phase C/D. The terminals will be launched on board the ESA satellite ARTEMIS and the French satellite SPOT 4 with nominal launch dates in 1996 and 1995, respectively. A fully operational data relay satellite DRS 1 will be launched in 1999 to complement the European data relay infrastructure. Apart from the SILEX project the Agency has investigated advanced system concepts to optimize the terminals and to fulfil the data rate requirements of future Earth observation instruments. Results of system studies and ongoing R&D activities in the areas of high-power NdYAG lasers and coherent receivers are presented.

1. INTRODUCTION

Recognizing the potential performance edge of optical communications over RF technologies in terms of size, weight and power, which are important parameters for space applications, the European Space Agency ESA started in 1977 a research and development programme for the exploitation of the optical spectrum for intersatellite crosslinks. A great number of study contracts and preparatory hardware developments were undertaken in the frame of the Basic Technology Research Programme (TRP), which is a mandatory programme funded by all ESA member states, and the Advanced Systems Technology Programme (ASTP), which is an optional programme funded by specific member states. In the late 80's an ambitious step was taken by embarking on a programme aimed at the experimental demonstration of an optical link between two satellites: SILEX (Semiconductor laser Intersatellite Link Experiment). The objective of this project is the development of all elements of a space-based optical communication system and the demonstration of its capabilities in an experimental and pre-operational mode. SILEX will play a key role in ESA's data relay scenario, which is described in Chapter 2. The characteristics and performances of the SILEX terminals and its development status are presented in Chapter 3.
The successful operation of SILEX in the second half of this decade will be a major achievement in (civilian) space communications. Needless to say that - due to the complexity and the financial volume - SILEX currently forms the key element in the optical communications activities at ESA leaving not much financial room for additional research. Nevertheless, R&D activities have to go on. Within the ASTP programme and the Advanced Research in Telecommunications Systems programme (ARTES), and other technology supporting programmes like the General Support Technology Programme (GSTP) the Agency continues to investigate and develop attractive second-generation systems and related technology. Advanced system concepts are presented in Chapter 4. Results of recently completed breadboard contracts and objectives of running contracts are summarized in Chapter 5.

2. THE EUROPEAN DATA RELAY SCENARIO

Towards the end of the century two ESA satellites in geostationary orbit (GEO) will serve as data relay satellites and provide a communications infrastructure for low-Earth orbiting (LEO) spacecraft. A very peculiar characteristic of this communications scenario is the asymmetry of the links: high-rate data generated by Earth-observation payloads or scientific instruments on board the LEO satellite are sent down to a ground station via the optical interorbit link and the RF feeder link (return link), whereas low-rate telecommand data is sent from ground to the LEO spacecraft (forward link). This link asymmetry has a considerable impact on the design of the terminals. ESA's first element of the European data relay infrastructure is ARTEMIS (Advanced Relay and Technology Mission Satellite), which is a geostationary satellite carrying the optical data relay payload SILEX, a combined S/Ka-band data relay payload, and an L-band land mobile payload. The delivery of the SILEX GEO-terminal to the spacecraft prime is planned for mid 1995. ARTEMIS will be launched by Ariane-5 in mid 1996. The SILEX LEO-terminal will be carried by the French Earth observation satellite SPOT 4. The SPOT 4 satellite has to be ready for launch in May 1995 with a terminal delivery date of end 1994. The actual launch date will depend on the performance of SPOT 3 to be launched in September 1993. The link between SPOT 4 and ARTEMIS will transmit a data stream of 50 Mbps. In addition to the primary interorbit link mission, an experimental optical GEO-to-ground link will be established between ARTEMIS and an optical ground station. The optical ground station will be located in an observatory on the Canary Islands. Preparatory atmospheric link experiments have been carried out in the frame of IMOCE (Intermountain optical communication experiment)/1/. An international cooperation with NASDA in Japan is now seriously considered. Management and technical level negotiations have been initiated to explore the possibility of an experimental link between ARTEMIS and a dedicated Japanese LEO satellite OICETS (Optical Intersatellite Communications Experimental Test Satellite). ESA's first operational data relay satellite is DRS 1 with a nominal launch date in 1999. It is worth mentioning that ARTEMIS is conceived as an experimental satellite with pre-operational capabilities. Thus, if everything goes fine, ESA will have available a two-satellite data relay system at the turn of the millenium. As far as the optical terminal is concerned the current baseline is that DRS 1 will carry a terminal similar to SILEX on

ARTEMIS. For spare policy reasons the procurement of equipment for DRS 1 will be initiated in mid 1993. In any case the optical terminal on DRS 1 has to be fully compatible with SPOT 4.

3. SILEX CHARACTERISTICS AND PERFORMANCES

SILEX is a free-space optical communications system consisting of two optical payloads, called optical terminals, to be carried on the French Earth-observation satellite SPOT 4 and on the ESA technology mission satellite ARTEMIS. The detailed design of the SILEX system and its equipments and experimental results obtained on breadboards in phase B have been presented in a number of papers in the past /2, 3, 4/. The prime contractor for SILEX is Matra Marconi Space in Toulouse, France.

Despite the strong asymmetry of the link the two terminals on GEO and LEO, respectively, are very similar. For cost and schedule reasons a high degree of hardware commonality was desired. Both terminals are equipped with a 25 cm telescope mounted on a coarse-pointing mechanism with more than hemispherical pointing range. The differences between the two terminals concern the acquisition beacon and the optical data receiver, which are on GEO only. Also the interface structures for spacecraft accommodation are different. Surprisingly enough, they contribute substantially to the overall mass.

Table 1: Main characteristics of the SILEX system

link range	45.000 km , worst case of SPOT4/ARTEMIS space configuration
link capacity	return link: 50 Mbps, SPOT 4 useful data rate forward link: unmodulated, but 2 Mbps modulation capability on GEO
wavelength range	797 - 853 nm
optical power	communications laser diodes: 60 mW average beacon laser diodes: 500 mW CW
beam divergence	communications beam: 8 μrad (telescope diameter 250 mm) beacon (on GEO): 750 μrad
terminal pointing	angular mobility: quasi hemispherical initial uncertainty on GEO angular position: ∅ 8500 μrad communications pointing accuracy: probability less than 10^{-3} for > 2 μrad
link performances	acquisition: < 4 minutes, 0.95 probability of success data transmission: BER better than 10^{-6} availability: better than 95% of the LEO-GEO annual visibility
operations	GEO multi-user capability (operation of 2 LEO's in an interleaved way) LEO multi-relay capability (operation of 2 GEO's, 1 per communication session)
terminal mass	~150 kg
terminal volume	~1600 x 900 x 800 mm^3
power consumption	~160 W (in communications mode)

4. ADVANCED LASER COMMUNICATIONS SYSTEMS

In an Inter-Orbit Link (IOL) scenario, where two terminals in geostationary orbit have to relay data from several LEO spacecraft to ground stations, the mass, the cost and the robustness of the LEO user terminals are characteristics of paramount importance for the competitiveness of laser communication technology versus RF-technology. For this scenario the present mass of 150 kg per user terminal is definitely too high. Although it is not at all obvious, the timing requirement of the acquisition phase has a considerable impact on the complexity and the mass of the user terminal. This applies especially to the maximum duration of the time slot within which the LEO terminal has to detect the beacon and to point its communications beam towards GEO. In addition to the reduction of the mass, the requirements for the interface between the laser communication terminal and the host spacecraft have to become more user friendly (sensitivity towards spacecraft vibrations, dynamic disturbances induced onto the spacecraft). Finally, the data rate capacity has to be increased beyond the 50-Mbps capacity of SILEX in order to satisfy the needs of future missions (Earth observation, Columbus Space Station). Taking these future needs into due account the European Space Agency has initiated in 1989 system studies for advanced laser communication systems emphasizing the following requirements:

* The design shall provide acquisition and tracking interoperability with the SILEX GEO terminal on

ARTEMIS. This includes a compatible wavelength/polarization plan and compatible photometric and timing requirements.
* The terminal design shall be optimized for asymmetric IOL applications.
* The LEO-to-GEO data rate shall cover the range of 200 to 500 Mbps, and the user terminal mass shall not exceed 50 kg.
* For low-data-rate users (2 to 10 Mbps), the terminal mass shall be less than 30 kg.
* The GEO-to-LEO data rate shall be typically 2 Mbps.
* The design of the user terminal shall be modular in order to fulfil different data rate and spacecraft accommodation requirements.
In order to achieve these goals several conceptual and technological guidelines were given:
* The (high-data-rate) design shall be based on high-power transmitters of the 1-W class.
* The telescope diameter of the user terminal shall be reduced to 10-15 cm. The rationale of this guide-line was to allow for the simplification of the pointing, acquisition and tracking system and to reduce, at the same time, the system sensitivity with respect to host spacecraft microvibrations.
* The mobile part shall be minimized.
* The design shall make use of fiber technology and micro-optics in order to take heat generating equipments away from the optical bench and to allow for a miniaturization of the opto-mechanical part. Modularity and easiness of redundancy implementation are additional benefits.
The results of these studies are best illustrated by two representative designs: The Advanced Laser Communication Terminal (ALCT) for high-data-rate applications (200 to 500 Mbps) and the Small Optical User Terminal (SOUT) for low-data-rate applications (2 to 10 Mbps) which is subject of /5/.

4.1. Advanced Laser Communication Terminal (ALCT) design

The overall ALCT design is shown in Fig. 1. Optimized for a LEO-to-GEO data rate of 250 Mbps this design results in a terminal with an estimated mass of 58 kg. The design is based on a 10 cm telescope diameter, diode-pumped Nd:YAG lasers and coherent reception on the GEO-side. Since the mobile part consists only of the telescope, which is coupled to the optical bench via a Coudé path, the torques to the platform are kept to a minimum. The small telescope diameter contributes not only to the compactness of the mobile part, but allows also for the relaxation of the pointing accuracy requirements (tracking and point ahead), which has a major impact on the mass, the complexity, and the cost of the user terminal. The fiber-coupled optical bench allows to keep the highly dissipative equipments (e.g. 1-W transmitter) away from the bench, and to reduce the size of the optical bench to a minimum. Both aspects are very beneficial to the thermo-mechanical and long-term stability of the opto-mechanical system. The high stiffness of the optical bench contributes also to the simplification of the tracking loop design. The fiber-coupling gives flexibility in the implementation of redundancy and in the accommodation of different transmitters and receivers without changing the opto-mechanical design. For acquisition and tracking the wavelength and polarization plan of the ALCT is compatible with the SILEX one. For the 1.064 μm return channel a dedicated "add-on" receiver has to be put on GEO.

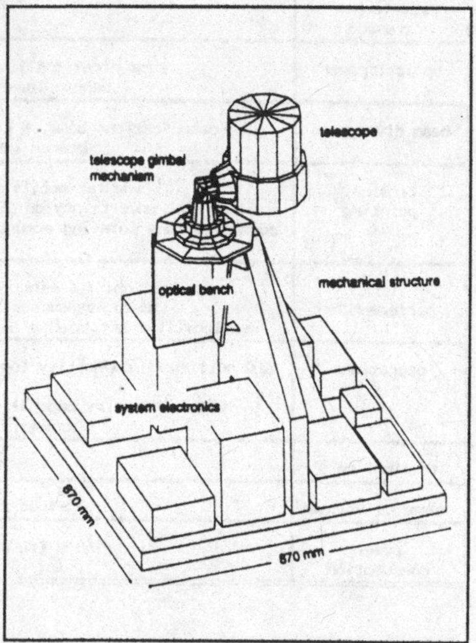

Figure 1: The design of the Advanced Laser Communication Terminal (ALCT).

5. R&D ACTIVITIES

In the following, results of recently completed contracts and objectives of running or planned contracts are summarized:

5.1 Telescope Array Communication Terminal.

Volume, mass and cost of a terminal increase rapidly with the telescope diameter. That is a reason why we have investigated the possibility of synthesizing a large self-phased diffraction limited optical antenna by using small-diameter telescopes /7/. More details of this concept are presented in /8/.

5.2 High-power transmitter laser

A diode-pumped NdYAG laser consisting of a discrete twisted-mode linear resonator oscillator and a 3-stage amplifier has been developed and breadboarded . 1 W single-frequency output power has been achieved in CW mode. The technical requirements and performances of this transmitter laser are described in /9/.

In a follow-on experiment the two 1-W pump diodes of the oscillator were replaced by 3-W pump diodes (SDL 2480). Without using the amplifier 1.15 W single-frequency output power has been achieved. Recently a contract has been placed for the development of a high-brightness pump source of sufficient reliability (redundancy) to guarantee a 10-year lifetime of the transmitter laser.

5.3 External modulator for NdYAG radiation

An engineering model of an electrooptic intensity modulator for NdYAG radiation has been developed and tested. Two LiNbO crystals of 0.5 x 0.5 x 40 mm each were cascaded to constitute a traveling-wave bulk modulator. The crystals are embedded in a beryllia holding structure. More details can be found in /10/. The optical transmittance is 86.9% over the whole operating temperature range (20°-50°C). The modulator has passed thermal cycling between -10°C and +50°C and sinusoidal and random vibration tests without degradation of its performance.

Based on the development of this intensity modulator a detailed design study of a phase modulator and its drive electronics was performed. The main design features of the phase modulator are an optical insertion loss of < 0.5 dB and an RF power consumption of 2.6 W for +/- 90° phase shift at 650 Mbps.

5.4 Advanced Coherent Optical Receiver

The development of a 500 Mbps coherent optical receiver for Nd:YAG systems was started in 1992. This receiver is implemented in single-mode fiber technology and has a sensitivity requirement of better than 50 photons/s for a BER<10^{-6}. This heterodyne receiver will also offer the possibility of receiving up to 7 closely spaced 150 Mbps data channels.

5.5 Single-mirror two axis fine pointing assembly

This activity is concerned with the development of a small (volume <50 cm^3) single-mirror fine pointing assembly (FPA) with an angular range of 100 mrad and a bandwidth of 1 kHz. The position sensors and the driving electronics shall be integrated into the FPA.

5.6 Combined Acquisition and Tracking Sensor (ATS)

This activity aims at the development of a single integrated acquisition and tracking sensor with a frame rate in excess of 10 kHz. The CCD sequencing, the pointing error signal processing, and the analog-to-digital conversion shall be integrated within the ATS.

6. CONCLUSIONS

The design of SILEX, the first European laser intersatellite communication system, has been completed. In the frame of this project, which is now in phase C/D, the engineering and flight models of two terminals are in manufacturing. The nominal launch dates for the LEO and GEO terminal are 1995 and 1996, respectively. The data rate of the SILEX system is 50 Mbps, the terminal mass is approximately 150 kg. ESA's first operational data relay satellite DRS 1 (nominal launch date is 1999) will carry a terminal very similar to SILEX. In parallel to SILEX, the European Space Agency is preparing the development of advanced systems with data rate capacities up to 500 Mbps for future Earth observation missions. In this development emphasis is put on low mass, complexity and cost as well as on full compatibility with SILEX. Key equipments like the 1-W NdYAG transmitter laser, the electrooptic modulator, and the coherent optical receiver have already been breadboarded, while equipment like the integrated acquisition and tracking sensor, the single mirror fine pointing assembly, and the self-phased receiving telescope array are under development.

7. REFERENCES
1. E. Fernández, P. Menéndez-Valdés, M.E. Wittig, A. Comerón, "Link budget model and applications for laser communications through the atmosphere", Proc. SPIE vol.1866 (these proceedings, paper 1866-15), Los Angeles, 1993.
2. G. Oppenhäuser, M.E. Wittig, A.F. Popescu, "The European SILEX project and other advanced concepts for optical space communications", Proc. SPIE vol.1522, 2 - 11, Munich, 1991.
3. J.L. Perbos, O. Duchmann, "The first European optical space communication system", First European Conference on Satellite Communications, Munich, November 1989.
4. various papers in Proc. SPIE vol.1218 (Los Angeles 1990) and Proc. SPIE vol.1522 (Munich 1991).
5. G. Baister, P. Gatenby, J.Lewis, M.Wittig, "Small Optical User Terminal for Intersatellite Communications", to be presented at LASER '93, Munich, 21.-23. June 1993 .this volume,
6. H. Sontag, U. Johann, K. Pribil, "Second generation high data-rate inter-orbit link based on diode-pumped Nd:YAG laser technology", Proc. SPIE vol.1522, 61 - 69, Munich, 1991.
7. ESA patent application ESA/PAT/241 "Optical communications terminal", 1991.
8. K. Kudielka, W.M. Neubert, "Design of an experimental optical multi-aperture receive antenna for laser satellite communications", to be presented at LASER '93, Munich, 21.-23. June 1993 .
9. U. Johann, W. Seelert, "1 W CW diode-pumped Nd:YAG laser for coherent space communication systems", Proc. SPIE vol.1522, 158 - 168, Munich, 1991.
10. T. Petsch, "Electrooptic modulator for high speed Nd:YAG laser communication", Proc. SPIE vol.1522, 72 - 82, Munich, 1991.

DLR Experimental Systems for free Space Optical Communications

Ch. Rapp, B. Wandernoth, G. Steudel and A. Schex
German Aerospace Research Establishment (DLR)
Institute for Communications Technology
D - 8031 Oberpfaffenhofen

ABSTRACT

The DLR experimental systems for coherent free space optical communications are presented. These systems have been designed for experimental studies in the laboratory and in a stationary free space testbed. The aim is to demonstrate the suitability of high bit rate coherent optical data transmission schemes for future space applications. A very high sensitive breadboard system uses PSK modulation with homodyne reception. A new synchronization technique made it feasible to build up a low complexity receiver allowing a power efficient transmission 3.5 dB above the shot noise limit at a data rate of 565 MBit/s. A second PSK homodyne system, which is now in construction, uses a smaller bit rate of about 1-5 MBit/s and allows further optimization of the time and frequency synchronization methods with smaller hardware effords. This work is sustained by a simulation program, which has been designed to simulate the whole optical communication system, enabling BER and signal analysis with different noise sources. In parallel , an earlier constructed DPSK heterodyne system has been embedded in a stationary test facility, which allows free space communication and PAT experiments between two buildings (720m).

1. INTRODUCTION

The institute for communications technology at DLR is investigating optical transmission schemes with regards to future application in space. Due to their superior performance in power efficiency for the uncoded case, the work is concentrated at the coherent transmission schemes (in the sense of hetero/homodyning shemes [1]) at the time. The new high quality and tunable monomode lasers (diode pumped Nd:YAG) and the invention of a new phase synchronization scheme (SYNCbit receiver) made it possible to construct a low complexity PSK-homodyne system, which is the highest sensitive coherent transmission scheme at the time [2]. This system is now considered for the implementation in a future high data optical inter orbit link (IOL) in the frame of the German national SOLACOS program [3]. After a short introduction of the already realized 565 MBit/s homodyne system, the actual works on a low bit rate PSK homodyne prototype and in the field of system simulation are presented. Additionally, the DLR testbed for coherent free space experiments (COFEX) and a concept for an alternative PAT (pointing, acquisition and tracking) experiment (PATEX) is shortly presented.

2. PSK HOMODYNE TRANSMISSION SYSTEM

2.1 High Bitrate PSK System

Phase control of the local oscillator in an optical PSK homodyne system is usually the most critical task. This can be achieved by the classical methods using a Costas Loop or a pilot carrier. Both methods have significant drawbacks, e.g. the necessity of an optical 90 degree hybrid in the Costas Loop receiver and the technical problems in the pilot carrier receiver (DC coupling needed !). A new synchronization technique, called SYNCbit method, uses a time multiplexed pilot carrier, which represents a transmitted "0" degree phase, while data is transmitted in the "+/-90" degree phases. This modulation states can easily be achieved by driving the optical phase modulator with a tenery signal. The 0° SYNCbits are transmitted every N bits and must be sampled in the receiver at the right time to build up an error signal for the local oscillator. The receiver frontend is quite simple and the most part of complexity is now shifted into the digital signal processing part of the receiver, which has to extract the position of the SYNCbits in the data stream for sampling. The first breadboard system with a 565 MBit/s SYNCbit receiver has been successfully demonstrated [2] and the characteristics of this remarkable system are shown in table 1.

2.2 Low Rate PSK Homodyne System

Since further optimization of the receiver became difficult to realize in hardware at 565 MBit/s, a second, low rate (1-5 MBit/s) PSK homodyne system has been designed.This system provides fully autonomous operation of the receiver including a clock and SYNCbit synchronizer. For better timing characteristics and spectral shaping, a self designed 10B12B line code has been implemented. The SYNCbit frame is N=13 in this case. All digital logic functions are implemented in self programmable EPLD chips. First tests of the low rate system performed in these days showed good performance of the optical PLL and of the transmitter with the line code.

Future versions of this system will also make use of channel coding, e.g. with a convolutional code and Viterbi detection. This coded system should then allow coherent optical communication below the shot noise limit of 9 Photons/ Bit for PSK.

3. SIMULATION TOOL

Accompanying the optimization of the SYNCbit concept, a new simulation program has been written, which allows a variety of analysis of optical communication systems. Figure 1 shows the simulation-block diagram for the homodyne system described above, of course, it is for use with also a variety of other modulation schemes. The program uses the complex baseband representation of the signal with a variable oversampling factor of 4...64. Nearly all standard filters and pulses are available, also a variety of data generators. All signals can be viewed graphically in the time or frequency domain. In most cases, a very fast half analytical simulation method can be applied, where a Monte Carlo simulation of the gaussian noise is performed only in the control loop for clock, SYNCbit recovery and OPLL.

This simulation model is well suited for the analysis of the bit error rate (BER) in the presence of shot noise, thermal noise, laser phase noise and other perturbations e.g. like amplitude modulation by the fiber nutator. Also the acquisition and tracking performance of the optical PLL, the SYNCbit-finder algorithm and the clock recovery can be simulated. First investigations concerned the optimization of the SYNCbit-finder algorithm. Two similar approaches have been implemented and some of the results can be seen in [3] in this proceedings.

4. FREE SPACE EXPERIMENT

In order to have a testbed for a coherent free space experiment (COFEX) a stationary test facility has been built up on the flat roof of the institute building. The communication link is established by sending the transmitted light to a mirror (retro) at another building and receiving the reflected beam. The length of this link is ~720 m (fig. 2). In a future experiment, the whole transmitter will be separated from the receiver, allowing communication without retro. The transmitter will then be remote controlled by a FM link.

The testbed is equipped with an older DPSK heterodyne system achieving a sensitivity of ~150 Photons/Bit [4]. Figure 3 shows the block diagram of the COFEX arrangement. The retro between transmitter and receiver has been already left out. For adjustments, a red HeNe-Laser can be superimposed to the transmitting beam.

Besides the possibility of testing the components under the harder conditions and to experiment under nearly far field conditions, one of the most interesting problems in COFEX is caused by the atmospheric turbulences. The main disturbances by atmospheric turbulence are intensity fluctuations (scintillation), beam wandering (angle of arrival fluctuations) and beam broadening. The strongest effect in this experiment is caused by beam wander (angle of arrival fluctuations). Since we employ a coherent receiver with

Transmitter

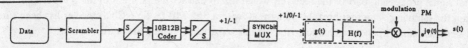

Receiver

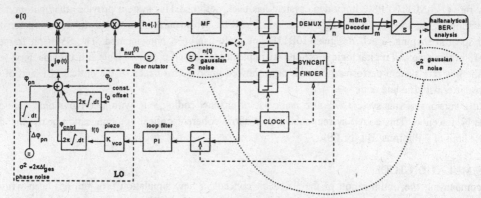

Fig. 1: Block diagram of the PSK homodyne system simulation model (same as in the low rate system will be realized)

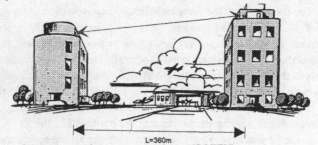

Fig. 2: Site view of the coherent free space experiment (COFEX)

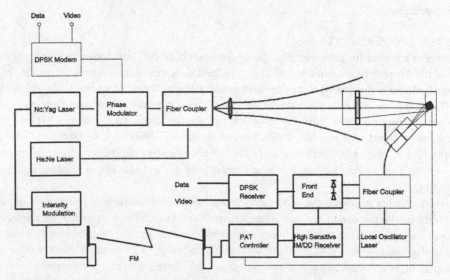

Fig. 3: Block diagram of the COFEX experiment with a simple PAT subsystem.

diffraction limited field of view, angle of arrival fluctuations lead to enormous power fluctuations in the receiving fiber. First experiments with a provisional 6m atmospheric link have shown already power fluctuations in the fiber in the order of ~20 dB. In the experiments with the 720 m link, huge power fluctuations didn't allow any communication. For further atmospheric experiments, some kind of a PAT system for COFEX was needed.

It is known, that the atmospheric turbulences have frequencies in the range from 0.1 ...100 Hz (1 kHz) and the angle of arrival fluctuations can vary from some urad to ~ 20 urad. The same orders of magnitude are expected for the disturbances due to microaccelerations on a satellite system [4]. Therefore, it is obvious to check the possibility of using the free space testbed for a realistic PAT experiment. Some work is now in progress, to achieve a PAT system for COFEX in special and to investigate alternative PAT methods for space applications in general.

5. CONCLUSIONS
The high sensitive PSK homodyne systems with SYNCbit synchronization have a big potential for further studies in the next jears. Improvements of the receiver algorithms and the combination with line and channel coding let us expect performances below the shot noise limit.

6. REFERENCES

[1] Leeb, W.R., "Coherent Optical Space Communications", in "Advanced Methods for Satellite and Deep Space Communications", Springer 1992, Lecture Notes in Control and Information Sciences 182

[2] Wandernoth, B., "20 Photon/Bit, 565 MBit/s PSK Homodyne Receiver Using Synchronization Bits", Electronic Letters 28 (1992)

[3] Pribil, K. et.al., "SOLACOS YAG Communication System-YKS, Implementation of the Syncbit- Concept", this issue

[4] Wittig, M. et. al., "In-Orbit Measurements of Microaccelerations of ESA's Communication Satellite OLYMPUS", SPIE Vol. 1218 (1990)

Wavelength	$\lambda = 1064$ nm (Nd:YAG)
Information bit rate	$f_B = 565$ Mbit/s
Channel bit rate	$f_C = 635.625$ Mbit/s
Bit error rate	$BER = 10^{-9}$
Data (pseudo random)	$L_p = 2^{10} - 1$

Received light power required for an ideal system (shot noise limit)	-60.3 dBm	≙ 9 Photons/Bit

Losses:

Increase of channel bit rate by 9/8	0.5 dB
Quantum efficiency of photodiode ($\eta = 0.75$)	1.3 dB
Other Implementation losses	1.7 dB

Received light power required in the realized system	Σ − 56.8 dBm	≙ 20 Photons/Bit

Table 1: Characteristics of the realized homodyne system

Small Optical User Terminal for Intersatellite Communications

G.Baister, P.Gatenby, J.Lewis, and M.Wittig[*]

British Aerospace Space Systems, Gunnels Wood Rd, Stevenage, UK,

[*] ESTEC, Noordwijk, The Netherlands

1. Introduction

The European Space Agency (ESA) has programmes underway to place data relay satellites (DRS) in geostationary orbit (GEO) within the next decade. The first satellite to fly will be Artemis which will carry a SILEX optical data relay terminal with a single 50 Mbps optical channel, followed by the operational European Data Relay System (EDRS) which will also carry SILEX terminals. Once these elements of Europe's DRS infra-structure are in place, there will be a need for optical communications terminals on low Earth orbiting (LEO) satellites (user terminals) capable of transmitting data to the GEO DRS terminals. ESA initiated a development programme in 1992 for a LEO optical terminal with a 2 Mbps data rate. This Small Optical User Terminal (SOUT) has distinctive features of low mass, small size, and compatibility with SILEX. The prime contractor is British Aerospace, with the Canadian companies CAL and Spar, and the Belgian company Spacebel as major subcontractors.

2. Differentiating features of the SOUT

The equipment consists of two main parts - a terminal head unit and a remote electronics module (REM). The REM contains the digital processing electronics for the pointing, acquisition, and tracking (PAT) and terminal control functions together with the communications electronics. The REM will take advantage of advanced packaging, ASIC and hybrid technologies to obtain a compact low mass design.

The terminal head unit performs the critical functions of generating and pointing the transmit laser beam and acquiring and tracking the received beacon and tracking beams. The receive/transmit telescope aperture has a baseline value of 70 mm to satisfy acquisition, tracking and communications link budget requirements. The telescope line-of-sight has to be pointed over a large angular range (360° azimuth, ±120° elevation) to meet the LEO to GEO look angle constraints. Rather than gimballing the complete head unit, a periscopic coarse pointing assembly (CPA) is located outside the telescope. This employs a pair of elliptical mirrors which rotate about azimuth and elevation axes. It has the advantage of a low moving mass around 4 kg, thus minimizing torque transfer to the host spacecraft during slewing manoeuvres.

The telescope is a refractive afocal design which does not have the obscuration loss associated with reflective systems. All other optical components are mounted on a double sided optical bench. The lasers and communication receivers are coupled to the optical system by single-mode and multi-mode fibres respectively. This approach offers advantages In component layout and redundancy switching implementation.

The terminal head is mounted on a three point anti-vibration mount (softmount). This acts as a low-pass filter for the micro-vibrations generated by the spacecraft and allows the 3 dB bandwidth of the fine-pointing tracking loop to be 550 Hz. The acquisition and tracking functions can consequently be carried out by a single CCD detector array, enabling a compact sensor subsystem to be engineered.

3. Structural configuration

There are three key drivers for the structural configuration of the head unit, namely, the optical bench pointing stability, softmount constraints, and base-bending moments associated with the telescope-CPA. Detailed modelling and analysis has led to the design shown in Figures 1 and 2. Figure 1 shows the truss frame assembly, while Figure 2 includes the softmount support frame.

There has to be a high degree of co-alignment between the transmit and receive beam paths on the optical bench in order that the transmit beam can be pointed towards the GEO terminal with an acceptably small pointing loss. Detailed analysis of the terminal pointing performance led to an allocation of 10 µrads to the maximum transmit/receive beam misalignment on the bench. To maintain the required stability over periods up to 24 hours in the presence of orbital variations in thermal environment, the bench has to be structurally and thermally de-coupled from the rest of the system as far as possible. This is achieved by means of an isostatic interface which minimizes heat flow to the bench from the terminal structure and prevents stresses in the bench due to differential expansion effects. The interface consists of three flexural titanium blades attached to the top side of the bench. All components which have critical pointing requirements are referenced to the under-side of the bench. The bench is square with a maximum area of 28 by 28 cm. The need to have at least 100 fixing points on the bench, together with excellent thermal and mechanical properties led to the choice of beryllium for the bench material.

Effective operation of the softmounts requires that the centre of mass (CoM) of the terminal head unit is located in or very close to the plane containing them. The three softmounts are located at the corners of a triangle. The CPA mass together with the mass of the telescope objective lens lies some distance above the softmount plane and hence a counter-balance has to be introduced below the softmount plane. This comprises the optical bench and a close proximity electronics module (CPEM) suspended below the softmounts. The CPEM contains the mechanism electronics and other circuits required for local control.

The cantilevered telescope with the CPA attached would result in excessive base bending moments at launch. During this phase, the softmounts are launch-locked and the CPA must be off-loaded to these by an outer support structure. The SOUT design employs a tripod truss structure to off-load the CPA and objective lens to three points aligned with the softmount launch-locks to give common load paths.

4. Thermal configuration

The SOUT thermal configuration was designed assuming a worst case of the SOUT mounted in the centre of the anti-Earth face of a large spacecraft in a low, circular, eclipsing orbit. The optical bench is the element requiring the tightest control of temperature levels and gradients. Several of the units mounted on the optical bench dissipate heat. To keep the bench temperature low, as many of these units as possible have been moved on to the lower truss frame leaving only 3.6 W dissipation on the bench. This heat is allowed to radiate upwards to the lower truss frame and all surfaces are painted

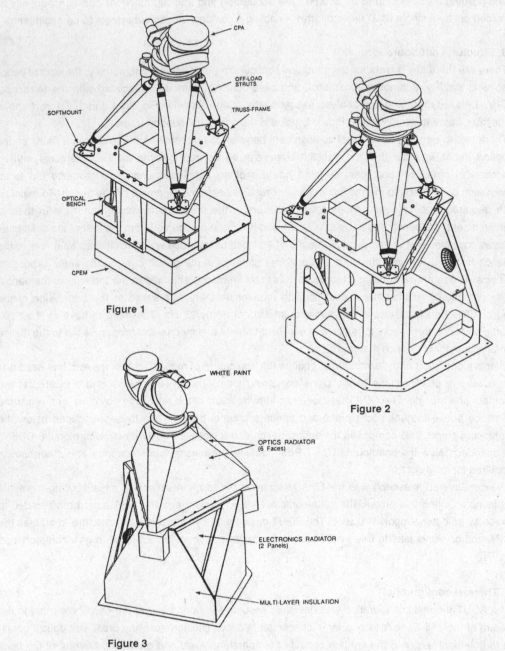

Figure 1

CPA

OFF-LOAD
STRUTS

SOFTMOUNT

TRUSS-FRAME

OPTICAL
BENCH

CPEM

Figure 2

WHITE PAINT

Figure 3

OPTICS RADIATOR
(6 Faces)

ELECTRONICS RADIATOR
(2 Panels)

MULTI-LAYER INSULATION

783

black to facilitate this transfer. The beryllium bench has a high thermal conductivity and a high specific heat capacity. This ensures that temperature gradients are minimized and short term temperature variations are strongly damped. To keep the optical bench at a constant temperature despite degradation of thermal control surfaces, it is necessary to introduce an active thermal control element. The lower truss ring is heated immediately above the bench using a thermostatically controlled circuit. The CPEM package is supported by three aluminium brackets below the lower truss ring. Two of these brackets form upwards sloping surfaces and have been used as thermal control radiators for the CPEM due to their limited view factor to the spacecraft blanket, and are covered with silvered teflon. Radiators comprising silvered teflon film (painted black on the inwards facing surface) are also arranged around the tripod truss structure in a quasi-tetrahedral configuration.

A thermal model of the SOUT has been generated using THERMICA and SINDA software. The optical bench is kept within a temperature band of 20 to 25 °C at all times, while short term temperature variations due to the spacecraft orbit are less than 0.5 °C. Temperature gradients through the bench are only about 0.1 °C, lateral gradients are less than 0.5 °C, and these do not vary significantly around the orbit. The external appearance of the SOUT with the thermal radiators in place is indicated in Figure 3.

5. Mass, power and size

The baseline SOUT as shown in Figure 2 has a total mass (including the remote electronics module) of 25.8 kg and a dynamic mass of 3.7 kg. The maximum power dissipation is around 65 W. The overall terminal height above the spacecraft must be compatible with accommodation within the launch vehicle fairing at launch. The height of the terminal above the spacecraft depends upon the mounting interface; options include mounting through a hole in the sidewall of the spacecraft (suitable for large platforms), external mounting on a support frame as shown in Figure 2, or mounting on a deployment mechanism. The package dimensions can be tuned to suit accommodation on a wide range of LEO spacecraft; height is nominally 87 cm.

6. Conclusions

Detailed design and modelling of the SOUT has been carried out under an ESA programme and has provided a high confidence level that the terminal can be built and qualified with a total mass around 25 kg. The next phase of the programme will be to integrate and test a breadboard terminal and this is scheduled to be complete around the end of 1994.

The SOUT has considerable stretch potential with respect to data rate. Advances in laser diode technology are expected to produce space-qualified devices with an order of magnitude increase in output power over the next few years. This will allow major increases in bit rate (or alternatively simplifications in terminal design) to be achieved in the future with little mass penalty.

7. Acknowledgements
The work reported here was carried out under ESA contract 9728/91/NL/PM.

VSOUT (Very Small Optical User Terminal) for Intersatellite Communications

M.Wittig & R.Czichy
ESTEC, Noordwijk/NL

1. History of Optical Space Communication

With the demonstration of the first operating laser in 1960, optical communication systems became a growing area of research and development activity between 1960 and 1970.
From 1970 onwards, major efforts were directed towards the guided propagation of optical signals through optical fibres. Unguided optical communication systems are influenced by adverse atmospheric conditions which may even interrupt the link. In free space such atmospheric problems do not exist. Free space optical communication systems were investigated starting in 1970, leading to the first terminal designs at the end of the 7th decade. Two terminals with different laser technology were developed: One used a flashlamp pumped Nd:YAG laser operating at 1064 nm wavelength in conjunction with direct detection /1/ and the other terminal was implemented with CO_2 lasers operating at 10.6 μm wavelength and heterodyne detection /2/. Neither terminal was ever put into orbit. A review of the history of space laser communication is given in /3/.
In the mid 80s the European Space Agency initiated work on free space laser communication systems. A terminal, based on CO_2 lasers, was designed /4/ and a laboratory model realised and tested /5/.
Around 1984 a European industrial consortium, led by Matra, defined a free space laser communication system for the European Data Relay System DRS /6/. This terminal is based on semiconductor lasers, operating at around 820 nm wavelength, and using direct detection with Avalanche Photodiodes (APD's) as very sensitive receivers. This terminal became later known as SILEX (Semiconductor Laser Intersatellite Link Experiment) and entered the main development phase (Phase C/D) in 1991 for scheduled launches in 1994 and 1996 on a low orbiting (LEO) and a geostationary satellite (GEO) respectively.
The SILEX terminal is able to transmit and receive a data rate of 50 Mbit/s. The mass of one terminal is about 140 kg. To attract future users to apply an optical terminal for data-relay applications instead of a microwave terminal, the mass, power consumption and accommodation on the host spacecraft must be improved.

2. Review of Laser Sources

Several laser sources might be attractive for optical space communication systems, such as:
 1: Semiconductor lasers in the 800 nm wavelength region;
 2: Semiconductor lasers in the 680 nm wavelength region;
 3: Semiconductor lasers in the 1300 to 1600 nm wavelength region;
 4: Semiconductor laser diode pumped Nd:YAG lasers operating around 1064 nm wavelength;
 5: Semiconductor laser diode pumped frequency doubled Nd:YAG lasers operating around 532 nm wavelength;
The most promising laser sources are the recently developed 800 nm laser diodes delivering 1 W cw in a single longitudinal and spatial mode. This is designed as a monolithic MOPA (Master Oscillator Power Amplifier) configuration, developed by SDL /7/.
The laser diodes operating around 680 nm wavelength are not mature enough for a space communication system. The long wavelength laser diodes require heterodyne receivers to achieve a sensitivity comparable with 800 nm direct detection systems.
Nd:YAG lasers operating at 1064 nm wavelength have a long development history. Several years ago semiconductor pumped Nd:YAG lasers were developed. The Agency developed a 1 W CW Diode Pumped Nd:YAG Laser /8/. This laser type is one strong candidate for a heterodyne system. For a direct detection system the lack of very sensitive photodetectors is not in favour for a system using this laser.
A frequency doubled Nd:YAG laser operating at 532 nm wavelength has the advantage that the link budget shows an improvement of more than 3 dB {=$(800/532)^2$ = 3.5 dB } compared with the 800 nm system. In addition, a direct detection system can be realized with good sensitivity.

3. Optical Communication System Concepts

The first choice is to consider a communication system operating in the 800 nm wavelength region. Optical heterodyne systems have a higher sensitivity compared with direct detection systems but are more complex. Considering the fact that very sensitive photodetectors are available based on silicon technology, which show the highest sensitivity around 800 nm, a direct detection system operating in the 800 nm region is still the most attractive candidate for space applications.
Direct detection systems at 1064 nm are out of any question due to the low quantum efficiency of the photodetector (typically below 10 %). Further it is interesting to look at shorter wavelength, e.g. the frequency doubled ND:YAG laser emitting at 532 nm. Here also high laser output powers can be expected, however such a laser needs to be developed for space applications. In addition the quantum efficiency of the silicon photodetector is not as good as at 800 nm.
For a direct detection system using semiconductor laser diode MOPA operating at 830 nm with intensity modulation via the injection current the achievable data rate for a link distance of 45 000 km is plotted in Figure 1.
With a system operating at 532 nm the achievable data rate is approximately doubled if transmit power, receiver sensitivity and system losses are identical.

4. Link Scenarios

All possible free space link scenarios around the planet Earth are shown in Figure 2.
The "classical" Data Relay Applications are the
- LEO to GEO Link (called Inter Orbit Link, abbreviated IOL)
- GEO to GEO Link (called Inter Satellite Link, abbreviated ISL)
and the more and more emerging
- LEO to LEO Links.
In addition, the GEO to Ground link to suitable locations, like Izana at Teneriffe, should not be forgotten.

5. The Very Small Optical User Terminal (VSOUT) Design

Usually data relay applications call for an unsymmetrical link in terms of capacity. Consequently, the transmitter telescope should be as small as possible, to reduce the effect of beam pointing errors, and the receiver telescope should be as large as possible.
The scenario considered further is based on an unsymmetrical link with a high capacity demand from the user spacecraft to a geostationary data relay satellite.
The availability of single longitudinal mode high power laser diodes allows to design compact optical communication terminals with telescope diameters ranging from 2.5 cm up to 7 cm, depending on the data rate to be transmitted.

5.1 Very Small Optical User Terminal Design Concept

The terminal optical head consists of an optics block and an attached pointing assembly. The required radiating/ receiving aperture of 25 mm diameter calls for a design concept with high miniaturization potential. Such concept was elaborated making use of hybrid optics, a novel type of optics combining diffractive optical elements (DOE) and refractive/reflective components in one single element. This gives to the optics designer an additional degree of freedom for design optimization, since any type of effective surface shape can be realized on the diffractive component and the total number of required lens elements can be minimized, thus contributing to an overall optimization in terms of mass, volume and reliability. Hybrid elements and DOE's are fabricated by methods adopted from semiconductor manufacturing processes (pattern/mask generation, photolithographic replication, etching technology etc), which have proven to be highly efficient and reproducible.
Basic aspects of diffractive optics, the design methods for hybrid optics and the manufacturing technology involved were developed in frame of ESA research activities /9,10/.
Figure 3 shows the complete optical head including pointing assembly. The transmitter part consists of two laser diode packages coupled via polarizing beam splitters in cold redundancy. High power laser diodes are assumed. The beam is expanded with an afocal telescope to 25 mm diameter and transmitted via a two mirror pointing assembly. Separated from the communications beam a beacon beam can be transmitted in the acquisition phase. The beacon transmit optics is mounted on top of the telescope front system, the beam direction is controlled via the same mirrors as the communications beam. The receive path uses only the telescope front lens and is then deflected into the acquisition sensor path resp. the tracking
sensor/receiver path. A CCD serves as acquisition sensor. A special arrangement combines the receiver and the tracking sensor in one common axis, which allows for a very compact design of the receive part.
Hybrid optics is used within this optical system in the laser transmitter package, the telescope eyepiece, the rear group of the acquisition path and in the tracking sensor/receiver combination.
The optical elements and the necessary electro-optical components are mounted into a single block of glass ceramic or similar material, which fully compensates for thermal expansion effects.
The optical head is therefore thermally fully stabilized by passive means, not requiring power consuming thermal control provisions for proper operation. This structural optics block, the core of the monolithic optical system, can be manufactured with highly accurate mounting interfaces which makes obsolete implementation of complex/heavy adjustment provisions, isostatic mounts etc., and allows for cost efficient fabrication/assembly.
In front of the optics block the pointing assembly is mounted, an advanced two mirror coelostat pointing system with combined coarse/fine pointing capability. The axes of the system can be driven by either direct drive DC torque motors with incremental encoders or stepper motors. The basics of such configuration were studied in the frame of an ESA contract. However, the initial concept is modified here by replacing one of the mirrors by a single element two axis fine pointing assembly, which was also developed under ESA contract. The advanced pointing assembly implemented here provides full hemispherical pointing capability with a fine pointing range of $\pm 2^{\circ}$ and a bandwidth in the order of 500 Hz.
The terminal design as presented up to now does not include a Point Ahead Assembly (PAA). If the required point ahead angular range is lower than the emitted beam divergence, no additional PAA is required. This might be the case for several LEO-LEO links. However, most of the missions would require a point ahead angular range which exceeds the emitted beam divergence. In that case the foreseen two laser diodes mounted on the glass block are replaced by an integrated laser diode transmitter, redundancy switching and point ahead assembly, which is under development by an ESA contract.
To process all the signals obtained from the different detectors into commands to control the different pointing mechanism and to execute the different modes of operation, a special Pointing Acquisition and Tracking-controller is required. The obvious solution is to design a VLSI containing an application specific digital signal processor tailored to these tasks.
Every optical terminal has to fulfil the optical isolation requirement of > 110 dB. The proposed solution to operate the system is the following: Each communication channel may use one out of two carrier frequencies which are phase modulated (BPSK, QPSK, COPSK,...). With this scheme only two subcarrier operating at one and the same laser wavelength are required to operate the system and allowing in addition the operation of many more optical terminals without any interference problems in a multiple access mode. The required optical isolation is low because the transmit and receive carrier are always different, i.e. the optical isolation problem is transferred into the rf-spectrum range.
The tracking channel may also use two different (low-frequency) carriers similar to the communication channel.

5.2 Terminal Mass, Power, Volume

The total terminal mass is approximately 10 kg. The estimated volume is 200 x 140 x 100 mm^3. The estimated power consumption is around 20 W. For an effective radiating aperture of 2.5 cm the terminal design is comparable in size and volume to a standard hand-held video camera.

6. Conclusion

A very small optical user terminal concept is presented which is suitable to transmit up to 150 Mbit/s over a distance of more than 40 000 km. The optical subsystem of the terminal is realized around a single glass ceramic block carrying all optical and electro-optical elements of the terminal (monolithic design). An incomparable miniaturization can be achieved due to consequent use of advanced micromechanic design concepts, a new type of optical elements (hybrid optics) in connection with a novel optical system design approach, a new frequency allocation scheme for optical transmitters/receivers, operating in multiple access, and implementation of ASIC technology for most of the electronic functions.

7. References

1: M.Ross, et al : Space Optical Communications with the Nd:YAG laser. Proceedings IEEE, Vol. 66, No.3, March 1978

2: J.McElroy, et al : CO_2 laser Communications Systems for Near-Earth Space. Applications. Proceedings IEEE, Vol. 65, No.2, Feb. 1977

3: M.Ross : The History of Space Laser Communication. SPIE Proceedings, Vol.885, Free-Space Laser Communication Technologies, January 1988

4: W.Englisch, M.Endemann, W.Diehl, W.Wiesemann: CO_2 Laser Transceiver Package with Gbit/s Capability for Space-to-Space Communications. Proceedings of an ESA Workshop on Space Laser Applications and Technology, SPLAT, ESA SP-202, March 1984

5: P.Huber, W.Reiland, V.Klein, A.Popescu: Full-scale laboratory breadboard model of a free space laser transceiver package. SPIE Proceedings, Vol.1218, Free-Space Laser Communications Technology II, January 1990

6: J.L.Perbos, B.Laurent : Laser diodes communications for the European Data Relay. System. SPIE Proceedings, Vol.756, Optical technologies for Space Communication Systems. January 1987

7: D.F.Welch: Coherent Lasers Turn Up the Power. Circuits and Devices, September 1992, p.17...23

8: U.Johann, W.Seelert: 1 W CW Diode-Pumped Nd:YAG Laser for Coherent Space Communication Systems. SPIE Proceedings, Vol.1522, Optical Space Communication II, 10-11 June 1991, p.158...168

9: H.Buczek, J.M.Mayor, P.Regnault, J.M.Teijido, H.P.Herzig: Holographic Optics. ESA contract (1990)

10: R. Czichy: Hybrid Optics for Space Applications, PhD Thesis (1991)

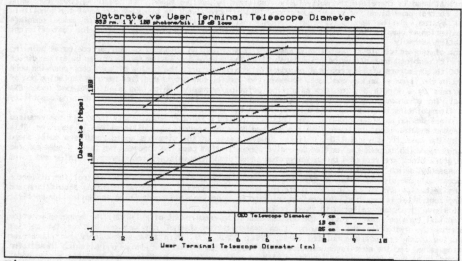

Figure 1 Datarate as function of Telescope Diameter.

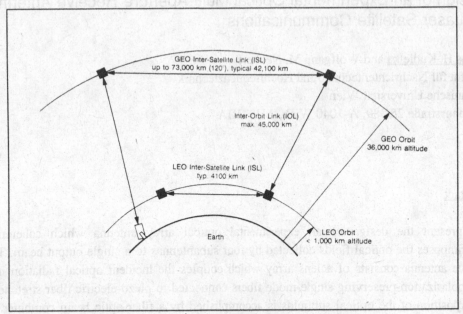

Figure 2 Free Space Communication Link Scenarios.

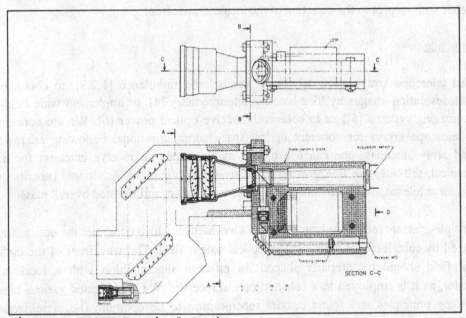

Figure 3 VSOUT Terminal Design.

Design of an Experimental Optical Multi-Aperture Receive Antenna for Laser Satellite Communications

Klaus H. Kudielka and Wolfgang M. Neubert
Institut für Nachrichtentechnik und Hochfrequenztechnik
Technische Universität Wien
Gußhausstraße 25/389, A-1040 Wien, AUSTRIA

Abstract

We present the design of an experimental optical array antenna which coherently superimposes the optical fields collected by four subantennas to a single output beam. The receive antenna consists of a lens array which couples the incident optical radiation into four polarization-preserving single-mode fibers connected to piezo-electric fiber stretchers. Superposition of the optical subfields is accomplished by a fiber-optic beam combiner. A control unit automatically adapts the optical phase differences between the four subantenna branches with respect to the direction of the incident wavefront.

Introduction

Phased telescope arrays allow to correct atmospheric turbulence [1,2,3], to obtain high angular resolution images by long-baseline interferometry [4], to implement wide field-of-view imaging systems [5], or to coherently receive optical power [6]. We are concerned with telescope arrays for coherent optical space communications. Following microwave phased array antennas, one can use optical multiple-aperture receive antennas for non-mechanical, self-adaptive fine steering of the main lobe direction. Additional benefits over single, large telescopes are modularity, implicit redundancy, and reduced overall mass.

A multiple-aperture receive antenna requires a mechanism which combines the optical fields collected by subtelescopes into a single optical output field. The wavefront of the optical output field should be properly shaped for efficient superposition with a local laser oscillator, as it is employed in a coherent optical receiver. We investigated various beam-combining principles and found coaxial superposition to be optimum (i.e. a subsequent optical heterodyne receiver achieves maximum electrical intermediate frequency power).

Subbeam superposition can be achieved either in the optical regime (by beam splitters or fiber directional couplers) or in the electrical intermediate frequency domain. The electrical approach, already demonstrated by MERCER [6], is very complex. High-data-rate receiver electronics are required for every single subaperture branch. We decided in favour of optical subfield superposition, which results in a much simpler, transparent array antenna system and which operates independently of the subsequent optical receiver.

Design of experiment

Figure 1 shows the block diagram of the experimental optical four-aperture receive antenna. By four lenses the optical input wave ($\lambda=1.06\mu m$) is coupled into polarization-maintaining single-mode fibers. Phase actuators shift the varying optical subfield phases and set constant phase relationships. The beam combiner is realized by a cascade of polarization-maintaining 3dB fiber directional couplers. If the correct phase relationships are set, the optical subfields are coherently collected and directed to output fiber 4. In this case output fibers 1, 2, and 3 carry no optical power. The optical power sensors attached to these fibers drive a control unit which in turn drives the piston actuators. Proper operation of the control loop thus minimizes the optical powers in fibers 1, 2, and 3 and hence maximizes the optical power available for the receiver (output fiber 4). This system automatically adapts itself to the direction of the incident wavefront.

To find the optical power minima at fibers 1, 2, and 3, the control unit employs the so-called perturbation method, outlined in Figure 2. If two optical subfields of equal powers $P_A=P_B=P$ and of phases φ_A and φ_B are applied to a symmetric fiber directional coupler, the optical power sensor detects

$$P_2 = P\left(1 - cos(\varphi_A - \varphi_B)\right) . \tag{1}$$

The desired phase difference is $\varphi_A-\varphi_B=0$, resulting in $P_1=2P$ and $P_2=0$. An additional sinusoidal disturbance of amplitude $\Delta\varphi$ and frequency f_p is applied to the phase actuator. Hence the power sensor detects oscillations at f_p and its multiples. It is sufficient to synchronously demodulate the line at f_p into the baseband. The resulting signal is proportional to

$$PJ_1(\Delta\varphi)sin(\varphi_A - \varphi_B) , \tag{2}$$

where J_1 denotes the first-order Bessel function. In combination with a loop filter and an integrator this signal is used to form an optical phase-locked loop (OPLL). The perturbation frequency f_p must be significantly higher than the bandwidth of the phase-locked loop, and the perturbation amplitude $\Delta\varphi$ must be very small compared to 2π. The OPLL locks at $\varphi_A-\varphi_B=0$ and directs the total optical input power $P_A+P_B=2P$ to output fiber 1.

The perturbation method was successfully implemented in various adaptive optical phased arrray systems. BRIDGES and PEARSON [1,2] employed different dither frequencies for each subaperture to control the phases of multiple-aperture arrays for atmospheric turbulence compensation and adaptive glint tracking.

790

For the experimental four-aperture receive antenna, we use a single perturbation signal at $f_p=10kHz$. By perturbing only input fibers A and D with amplitudes $\Delta\varphi_A$=0.1rad and $\Delta\varphi_D$=-0.1rad, respectively, all phase differences can be measured. Figure 3 depicts the resulting block diagram of the control unit. To decouple the three OPLLs, the integrator outputs are linearly combined according to a simple algorithm. The resulting four control signals are applied to the phase actuators shown in Figure 1.

First experiments with a setup according to Figure 2 confirm the design. With an optical input power of only 1nW per subaperture it was possible to control the optical phases with a residual error of 0.04rad. Within 400µs, the control loop adapted the phase difference with respect to a step-shaped change.

References

[1]   W. B. BRIDGES, "Coherent optical adaptive techniques", Appl. Opt. 13, 291 (1974)
[2]   J. E. PEARSON, "Atmospheric turbulence compensation using coherent optical adaptive techniques", Appl. Opt. 15, 622 (1976)
[3]   C. L. HAYES et al., "Experimental test of an infrared phase conjugation adaptive array", J. Opt. Soc. Am. 67, 269 (1977)
[4]   S. SHAKLAN, "Fiber optic beam combiner for multiple-telescope interferometry", Opt. Eng. 29, 684 (1990)
[5]   C. R. DeHainaut et al., "Wide field of view phased array telescope", Proc. SPIE 1236, 456 (1990)
[6]   L. B. MERCER, "Adaptive coherent optical receiver array", Electron. Lett. 26, 1518 (1990)

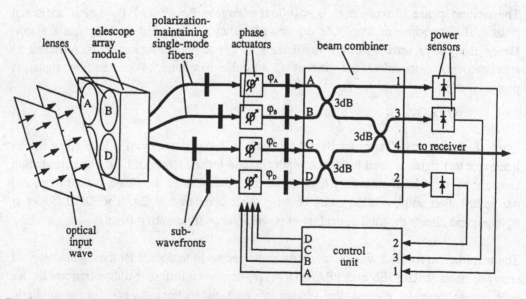

Figure 1: Design of the experimental optical four-aperture receive antenna

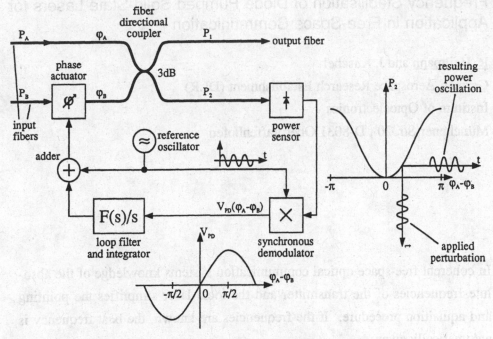

Figure 2: An optical phase-locked loop employing the perturbation method for phase sensing

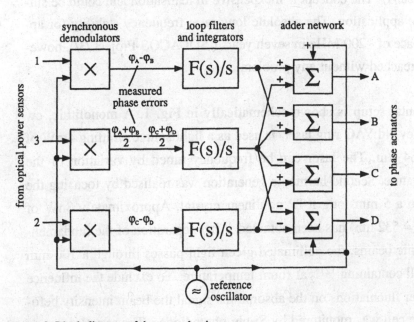

Figure 3: Block diagram of the control unit

Frequency Stabilisation of Diode Pumped Solid-State Lasers for Application in Free-Space Communication

R. Heilmann and J. Kuschel

German Aerospace Research Establishment (DLR)

Institute of Optoelectronics

Münchener Str. 20 , D-8031 Oberpfaffenhofen

In coherent free-space optical communication systems knowledge of the absolute frequencies of the transmitter and the local laser simplifies the pointing and aquisition procedure. If the frequencies are known, the beat frequency is automatically given.

In this paper a simple method of frequency locking to an iodine absorption line is presented /1/. The concept is inexpensive in realisation and could be suitable for space application. The absolute long term frequency stability for application in space of ~200 MHz in seven years (SOLACOS-Project /3/), however, could be reached without any problem.

The experimental setup is shown schematically in Fig. 1. A monolithic, cw single frequency Nd:YAG ring laser is used as a light source emitting 4mW of power at 1.064 μm. The laser can be frequency tuned by variation of the crystal temperature. Second-harmonic generation was realised by focusing the laser light into a 5 mm long KTP nonlinear crystal. Approximately 3nW of green light ($\lambda \approx 532$ nm) has been obtained. After separation of the fundamental and harmonic beams, the collimated green light passes through a 100 mm long quartz cell containing $^{127}I_2$ at room temperature. To exclude the influence of output power fluctuations on the absorptions signal the beam intensity before and after the cell was monitored by Si-pin-photodiodes. The servoloop

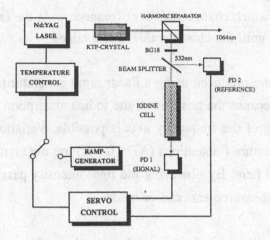

Fig. 1 Block diagram of experimental arrangement for frequency stabilisation

stabilises the laser frequency in the following manner: On a flank of a selected absorption line, each measured intensity I is directly related to a definite frequency f of the laser light (see Fig 2). Thus, within a frequency interval, scanned approximately by the line flank, a laser frequency calibration with respect to the measured intensity is possible. To maintain the laser at a constant frequency, the light intensity traversing the iodine cell must also be kept constant. If there is a frequency deviation from a chosen value f_{lock}, the servocontrol

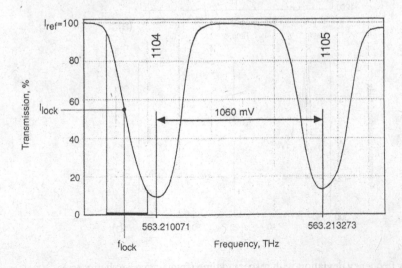

Fig.2 Absorption spectrum of molecular iodine near frequency of 563.21 THz

produces a voltage, which changes the laser frequency, via the external temperature control input, until the chosen value is reobtained.

Absorption spectra are produced using a linear ramp generator to create a frequency variation. Because the position of the iodine absorption lines are well
known /4/, a scaling of the frequency axis is possible. Variations in the line
shapes due to temperature fluctuations ($\Delta T \leq 0.1$ K) are not significant for the
conditions presented here. By observing the light intensity passing the iodine
cell frequency drift measurements can be made.

Fig 3a illustrates the operational mode of the servoloop after switching on the
laser. The frequency deviation is shown as a function of time. The initial frequency was set about 1 GHz below where the frequency of the laser should be

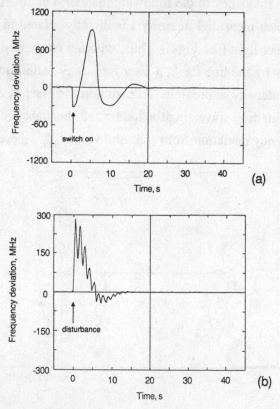

Fig. 3 Laser frequency deviation with respect to time (frequency search process)
(a) After switch on the laser
(b) After an externally induced frequency disturbance

locked. About 20 s after switch on, the frequency is then locked. The remaining frequency jitter was measured to be less than 17 MHz, i. e. a relative frequency stability $\Delta f/f < 6 \cdot 10^{-8}$ has been achieved. Fig. 3b shows the frequency response of a locked laser after a disturbance. A frequency deviation of ~300 MHz is compensated within 15 s. The superposed oscillation, having a frequency of 0.8 Hz, arises from the internal temperature control of the laser.

Considering calibration problems, inaccuracy of the measurement instruments, external electrical disturbances etc., the laser frequency can be determined and reproduced with an absolute accuracy of $\Delta f/f < 10^{-7}$.

Applying frequency offset locking the absolute frequency stability of the ring laser can be transferred to an twisted-mode-cavity (TMC) laser with an output power of about 1 W. The frequency drift of the TMC-laser has been compensated by light absorption, which induces rapid changes in the temperature inside the gain medium and the laser frequency, respectively /2/.

Because the concept of the absolute frequency stabilisation can be realised in a relatively robust, compact, light and inexpensive arrangement, application to coherent free-space communication or lidar systems should be possible.

/1/ R. HEILMANN and J. KUSCHEL, Electron. Lett. **29**, 810 (1993)

/2/ R. HEILMANN and B. WANDERNOTH, Electron. Lett. **28**, 1367 (1992)

/3/ K. PRIBIL, U. JOHANN, and H. SONTAG, Proc. SPIE **1522**, 36 (1991)

/4/ S. GERSTENKORN and P. LUC, Laboratoire Aimé Cotton, Orsay, France, 1980

Characteristics of an Integrated Optical Phase Modulator at High Optical Energy Densities

M. Fickenscher* and A. Rasch**

*DLR, Institute of Optoelectronics, D-82230 Oberpfaffenhofen

**Friedrich-Schiller-University, D-07743 Jena

1 Introduction

For coherent optical communication in space based on ND:YAG (1.064 µm) a transmitter with about 1 Watt output power is needed. To reach this aim, two different transmitter concepts were extensively discussed in the nationally funded German technology program SO-LACOS [1].

Concept 1 consists of a high power laser oscillator and a subsequent bulk-pulse modulator; Concept 2 consists of a medium power oscillator, an Integrated Optical phase Modulator (IOM) and (a) laser amplifier(s) [1].

The bulk modulator was not available for us up to now. In [1] is stated, that Concept 2 shows the better overall efficiency of the transmitter subsystem, "if the IOM can handle optical power levels of some 100 mW".The Friedrich-Schiller University Jena presented a IOM for 1300 nm at the congress at the fair "Laser '91" in Munich two years ago [2]. Meanwhile they have developed a device for the wavelength 1064 nm, which is described below.

2 Integrated Optical phase Modulator (IOM)

Device Fabrication

A large variety of devices has been demonstrated and waveguide devices in lithium niobat have shown improved performance as external modulators in high-bit-rate systems, sensors and coherent communications, where pure phase modulation is required. Due to their large index change, high optical damage threshold and polarisation selectivity, proton exchanged (PE) optical waveguides in $LiNbO_3$ have been regraded for some years as one possible alternative to their titanium indiffused counterparts commonly used for certain device applications. The large extraordinary index change Δn_e has been found to be related to relatively high attenuation values, long term index instability and, especially, the degradation of electrooptic effect [3]. However, by an annealing procedure following the proton exchange, the electrooptic effect in the PE region can be restored [4].Low-loss wave guides and efficient electrooptic devices have been fabricated.

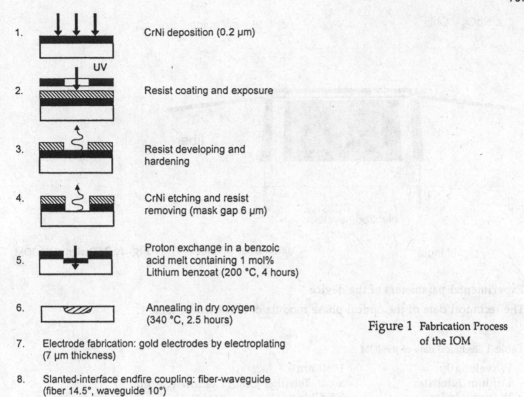

1. CrNi deposition (0.2 µm)

2. Resist coating and exposure

3. Resist developing and hardening

4. CrNi etching and resist removing (mask gap 6 µm)

5. Proton exchange in a benzoic acid melt containing 1 mol% Lithium benzoat (200 °C, 4 hours)

6. Annealing in dry oxygen (340 °C, 2.5 hours)

7. Electrode fabrication: gold electrodes by electroplating (7 µm thickness)

8. Slanted-interface endfire coupling: fiber-waveguide (fiber 14.5°, waveguide 10°)

9. Packaging

Figure 1 Fabrication Process of the IOM

In Figure 1 the fabrication technology of the device is shown. In order to define the PE-waveguides a 0.2 µm thick chromium layer and a positive resist layer is first deposited on a x-cut LiNbO$_3$ substrate. The waveguide pattern is then transferred photolithographically from an electron beam written mask down to the resist and then by chemical etching into the chromium layer. The waveguide width (width of the opening in the mask) is typically 6 µm. Proton exchange is carried out in a benzoic acid melt containing 1 mol% lithium benzoat at 200 °C for 4 hours. The PE waveguides annealed at 340 °C for 2.5 hours in dry oxygen atmosphere.

For the high-speed electrooptic devices, the electrodes are made e.g. by deposition of a thin metal layer (e. G. Ni), patterning the layer, again using electron-beam-lithographically produced mask, increasing the electrode pattern thickness by electroplating with copper/gold up to 7 µm and etching the thin metal layer outside the electrodes. A 200 nm thick SiO$_2$ buffer between the metallic electrodes and dielectric waveguides is arranged in order to avoid optical loss by absorption in the metal. In Figure 2 the design of the integrated optical phase modulator is shown.

The optical fibers are adjusted precisely with respect to the waveguides in order to achieve maximum overlap of the light intensity distributions of both light guides. Then the fiber is connected to the waveguide by UV-cured epoxy.

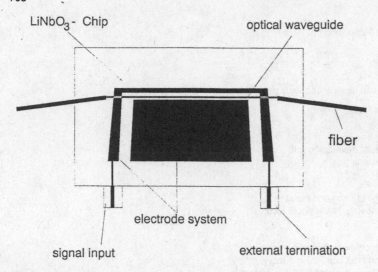

optical waveguide

fiber

electrode system

signal input

external termination

Figure 2 Design of the IOM

Experimental parameters of the device

The technical data of the optical phase modulator is shown in Table 1.

Table 1 Technical data of the IOM

Wavelength:	1060 nm
Lithium Niobate:	x-cut, Telefilter GmbH Teltow (Germany)
Waveguide loss:	0.8 dB/cm
Coupling loss:	0.7 dB per coupling place
Total insertion loss:	4.0 dB
Input optical power:	≤ 500 mW
Reflection suppression:	54 dB
Modulation voltage:	5V/π (at 1 GHz)
Bandwidth:	2 GHz
Input termination.	50 Ω
Connector:	SMA
Fibers:	HIBI-YORK HB 1000
package dimensions:	(105 x 22 x 15) mm^3

3 Investigation of the IOM at high optical intensities

In Figure 3 the experimental set-up for testing the IOM in a heterodyne-experiment at high optical intensities is shown.

Part of the beam of a SLM-laser [5] is frequency-shifted by an Acusto-Optic Modulator (AOM). The unshifted main part is coupled into the IOM where a sinewave with a frequency of 1 MHz and variable amplitude for phase modulation is applied. The light leaving the IOM and the frequency shifted beam from the AOM are heterodyned. Without phase modulation only one line can be seen at the spectrum analyzer at the frequency of the AOM at 70 MHz. With small phase modulation applied, sidebands of this line with 1 MHz distance appear, and grow with increasing phase modulation, whereas the original line is decreasing. At a certain modulation depth the original line at 70 MHz is disappearing.

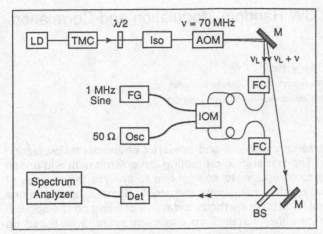

Figure 3 Experimental set-up for testing the IOM at high optical intensities

With the given sensitivity of the experiment, this point (amplitude of the modulation voltage) can be measured with an accuracy of 10^{-3}. No change has been seen up to an optical input power of 280 mW, that is a transmitted power of 100 mW of the IOM (total insertion loss 4.5 dB). The optical power applied to the IOM was only limited by the output power of the laser and the losses in the devices between laser and IOM. As seen in table 1, an improved version of the IOM (total insertion loss 4.0 dB) was tested at FSU Jena in a transmission experiment. Up to an optical input power of 500 mW (limited by the laser and losses) no change could be seen.

Conclusion: For discussion of the problem regarding the overall efficiency of the transmitter subsystem more measurements at even higher intensities are necessary, to find the maximum optical power the IOM can handle.

Acknowledgement: The project of developing the IOM was supported by DARA-BMFT.

[1]   Pribil, K.; Johann, U., Sontag, H.;"SOCACOS: a diode pumped Nd:YAG laser breadboard for coherent space communication system verification", Opt. Space Communication II, J. Franz, Editor, Proc. SPIE 1522 (1991), pp. 36 - 47

[2]   Rasch, A.; Buß, W.; Göring, R; Steinberg, S.; Karthe, W.: "Optical carrier modulation by integrated optical devices in lithium miobate" Opt. Space Communication II, J. Franz, Editor, Proc. SPIE 1522 (1991), pp. 83 -92

[3]   Minakata, M.; Kumagai, K.; Kawakami, S.: "Lattice Constant and Electro Optic Effects in Proton Exchanged LiNbO$_3$ Optical Waveguides", Appl. Phys. 49 (1986), pp. 992-994

[4]   Rottschalk, M.; Rasch, A.; Karthe. W.: "Behaviour of Proton Excanged LiNbO$_3$ Optical Waveguides"; J.of Opt. Commun. 9 (1988) pp. 19-23

[5]   Wallmeroth, K., Peuser, P., "High power, cw single frequency, TEM00, diode-laser-pumped Nd:YAG laser", Electron. Lett. 24 (1988), pp. 1086 - 1088

Improvement of LIDAR by CW Random Modulation and Correlation Detection

Petra Bisle[1], Jürgen Franz[2], Daniel Mengistab[2]

[1]Deutsche Forschungsanstalt für Luft- und Raumfahrt, 8031Oberpfaffenhofen
[2]FH Düsseldorf, Fachgebiet Optische Nachrichtentechnik, 4000 Düsseldorf 30

1 Introduction:

Today environmental pollution is a serious problem and powerful enviromental protection becomes more and more important. The first step in protecting the environment and understanding the reasons of pollution is to recognize, to control and to analyze the process of pollution. This basic task requires an efficient measuring technique. Remote sensing is one of the most promising environmental measuring methods today. Depending on the application, remote sensing systems can be either installed on earth, on aircrafts or based on satellites. During recent years the advantage of using lasers in environmental remote sensing has been demonstrated in many different applications, experiments and papers [i.e. 1-5]. Optical remote sensing allows the monotoring and analyzing of most of todays relevant environmental pollutions: traffic-induced air pollution, uncontrolled gas emission in chemical plants, the ozone hole, the greenhouse effect, vegetations stress and water pollution (oil). One of the most important optical measurement methods is LIDAR, which stands for Light Detection and Ranging. In the frame of environmental remote sensing LIDAR plays a key role today. LIDAR is similar to Radar (Radio detection and ranging). However, whereas Radar is based on radio waves, LIDAR uses UV, visible or IR light. Most of today´s LIDAR systems are based on high-power-pulse lasers which send short pulses with high energy into the atmosphere. The light is scattered in all directions by aerosols and molecules. A part of the light is scattered back to the source, where the detector is also placed. The back-scattered and detected light intensity enables the registration of the absorption characteristic as a function of range (distance) for a given gas.

Fig. 1 shows the principle of LIDAR and the simplified blockdiagram of its communication representation. Whereas in an actual LIDAR system the transmitter and receiver are located at the same place, in the communication representation the transmitter and receiver are usually located at both ends of the system. The basic function, however, is exactly the same. One disadvantage of conventional LIDAR is that high-power light pulses are required for most of its applications. Therefore conventional LIDAR systems are usualy not eye-safe. However, eye-safety is a strong demand if LIDAR is to be used to monitor the environment from aircrafts or satellites. Moreover, nowadays LIDAR show a somewhat limited mobility. Developing small, flexible and eye-safe LIDAR systems is

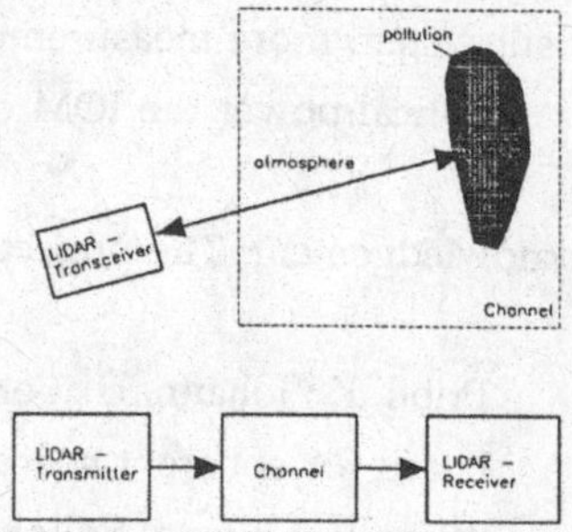

Fig.1 Principle of LIDAR and its simplified communication representation

a challenge of today. The most simple solution is to change power and time. To overcome the inherent quality loss if transmitting light power is to be decreased, the measurement time must be increased appropriately. This idea can be realized by modulating a low-power continuous wave laser beam with a binary pseudo random signal and comparing transmitted signal sequences and back-scattered, received signals by a conventional correlation receiver. By using modulation techniques and correlation receivers LIDAR now takes advantage of typical telecommunication methods which are well known and well established in communication engineering since a long time. Regarding low-power continuous wave lasers, semiconductor lasers would be a very attractive solution. The following chapters focus on the communication aspects of random-modulated continuous wave LIDAR techniques (RM-CW-LIDAR).

2 Communication Aspects of RM-CW-LIDAR:

Fig. 2 shows the communication related blockdiagram of a RM-CW-LIDAR system, including the CW-laser transmitter, modulated by a binary pseudo random signal a(t), the channel and the correlation receiver. The periodical random signal a(t), usually obtained by a simple binary pseudo noise (PN) generator, is defined by its impulse rate $1/\Delta t$ (often denoted as bitrate) and by the number M of bits per

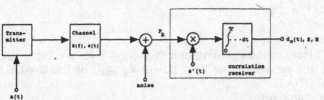

Fig. 2 Simplified communication blockdiagram of RM-CW-LIDAR

sequence with a periode of $T = M \Delta t$ (fig. 3). Each sequence includes the same configuration of M impulses; (M+1)/2 are "1" and (M-1)/2 are "0". The channel, described by its frequency response S(f) or impulse response s(t), includes the characteristics of the atmosphere as well as those of the measurement object, i.e. of the pollution. Morever, the channel also takes into account twice the range to the pollution object. The periodical pseudo random sequence a'(t) in the correlation receiver is of the same configuration as a(t) in the transmitter. Regarding the system performance, the receiver output signal, here denoted as the detection signal $d_o(t)$, signal power S and noise power N are of interest.

Similar to conventional LIDAR, based on high-power pulse lasers, the detection signal $d_o(t)$ of a RM-CW-LIDAR receiver also shows small impulses (fig. 3). The impulse width is $2 \cdot \Delta t$. In the case of a fixed and plane target as a channel (i.e. a mirror) all impulses are equal in shape and height. In the case of pollution, the measurements shape and height now depend on the range and kind of pollution and enables to observe the pollution process. The impulse destination in the detection signal $d_o(t)$ is defined by the period T of the PN sequence. The impulse peaks, approximately given by

$$d_o = R \, P_R \, \frac{M+1}{2} \, s(0) \, , \qquad (1)$$

depend on received light power P_R, sensitivity R of the photodiode, channel characteristics, here described by s(0), and on the number M of impulses per period T. Provided there is a fixed period T,

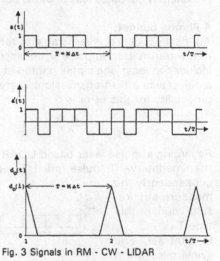

Fig. 3 Signals in RM - CW - LIDAR

amplitude d_o as well as the signal power $S = (d_o)^2$ increases with M; whereas the impulse width decreases with M. Both, increased signal power and decreased impulse width improve system performance and measurement accuracy. To discuss the system performance in more detail and to compare a RM-CW-LIDAR system with a conventional pulse laser based LIDAR system, either the signal to noise ratio or the photon budget can be taken into account.

3 Signal to noise ratio:

System performance and measurement accuracy of RM-CW- and pulse laser based LIDAR systems can be regarded to be the same if the signal to noise ratios are the same:

$$(S/N)_{Pulse} = (S/N)_{RM-CW}. \qquad (2)$$

In a correlation receiver of a RM-CW-LIDAR system the signal power S increases approximately with M^2 whereas the noise power N merely increases with M. For M >> 1 equ. (2) yields the following result [6]:

$$\frac{P_{R,pulse}}{P_{R,CW-RM}} \left(\frac{\Delta t_{CW-RM}}{\Delta t_{pulse}}\right)^{\frac{3}{2}} \approx \frac{1}{2}\sqrt{M}. \tag{3}$$

The higher the number M of binary impulses during one PN-period T (or the higher the PN-generator bitrate), the lower the required input light power $P_{R,\,RM\text{-}CW}$ of RM-CW-LIDAR in contrast to pulse laser based LIDAR.

Example: Assuming a PN-generator with M = 4095 and $\Delta t_{RM\text{-}CW} = \Delta t_{puls} = 1ns$ a RM-CW LIDAR systems require a 15 dB less back-scattered light power than an appropriate pulse laser LIDAR system. This gain in receiver sensitivity is, of course, also valid if the ratio of required transmitter light powers are taken into account. Moreover equ. (3) also gives the ratio of the required number of photons per bit. Here, RM-CW-LIDAR requires approximately 32 times less photons per bit.

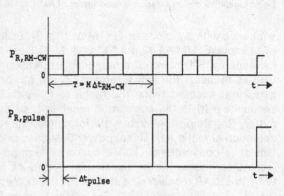

Fig. 4 RM-CW-LIDAR and pulse laser based LIDAR

4 Photon budget:

An ideal optical receiver is able to recognize a transmitted impulse if the impulse includes at least one single photon at the receiver side after transmission. Therefore an error first occurs if no photon is received. The probability for this error is

$$p = e^{-\frac{P_R \,\Delta t}{h\,f}}. \tag{4}$$

Regarding a pulse laser based LIDAR system, which is based on one single impulse during the time interval T (pulse rate 1/T), equ. (4) gives the probability of no received signal and consequently no measurment results during this time T .

In contrast, one single missed impulse (no received photon) is not critical in a RM - CW - LIDAR with a total number of M impulses during the same time interval T. As shown in fig. 5a the peaks in the detection signal $d_o(t)$ can still be easily identi-

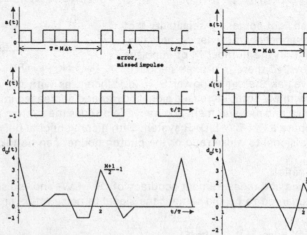

Fig. 5a Example 1: M = 7 and one missed impulse (no received photon)

Fig 5b Example 2 : M = 7 and two missed impulses (no received photon)

fied and no break in obtaining measurement results occurs. However, a small loss in signal power and therefore a somewhat decreased measurement accuracy are unavoidable. Each single impulse, which cannot be detected yields a signal power degradation of approximately 2/M %. This means, that the loss in power degradation is small if M is high.

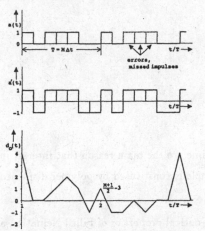

The higher the number M of transmitted impulses during the time intervall T the more missed impulses are tolerable. However, if as much as 50 % or more of all impulses are lost during transmission an errorless idendification of the detection signal peaks is no longer possible (fig. 5c). From this point of time on RM-CW-LIDAR can be regarded as worse as a pulse laser based LIDAR system with one missed impulse. The probability of exactly r missed impulses in a RM-CW-LIDAR system is given by

$$
p_r = \begin{pmatrix} \dfrac{M+1}{2} \\ r \end{pmatrix} \left(e^{\frac{-P_{R,\,CW-RM}\,\Delta t}{h\,f}} \right)^r \left(1 - e^{\frac{-P_{R,\,CW-RM}\,\Delta t}{h\,f}} \right)^{\frac{M+1}{2}-r} \tag{5}
$$

Fig 5c Example 3 : M = 7 and three missed impulses (no received photon)

If the probability of one non-detectable pulse-laser LIDAR impulse equals the probalility of 50 % or more non-detectable RM-CW-LIDAR impulses then the system performance of both systems can be regarded the same:

$$
e^{-\frac{P_{R,\,pulse}\,\Delta t}{h\,f}} = \sum_{r=(M+1)/4}^{r=(M+1)/2} \begin{pmatrix} \dfrac{M+1}{2} \\ r \end{pmatrix} \left(e^{\frac{-P_{R,\,CW-RM}\,\Delta t}{h\,f}} \right)^r \left(1 - e^{\frac{-P_{R,\,CW-RM}\,\Delta t}{h\,f}} \right)^{\frac{M+1}{2}} \tag{6}
$$

Equ. (6) provides there is the same impulse width Δt and same period T. The lower bound of the sum denotes 50% of all non-zero impulses, the upper bound 100%. To solve equ. (6) means to determine the ratio $P_{R,\,pulse}/P_{R,\,RM\text{-}CW}$. Because an analytical solution is non-existent, appropriate approximations or computer solutions have to be taken into account [7]. The result equals the result of the signal to noise consideration, cosidered above.

5 Conclusion:

By the use of conventional communication methods, such as modulation and correlation, LIDAR systems can be improved in respect of the receiver sensitivity: a lower required reveiver input power or a lower number of photons per bit (per impulse), respectively. This result is a consequence of changing power with time and enables the replacement of conventional high-power pulse-laser LIDAR systems by low-power RM-CW-LIDAR systems.

References

[1] Hinkley, E.D.: Laser Monotoring of the Atmosphere. Topics in Applied Physics, Vol. 14,Springer- Verlag 1976.

[2] Takeuchi, N.; Sugimoto, N.; Baba, H.; Sakurai, K.: Random modulation cw-lidar. Applied Optics 22(1983)1, 1382-1386.

[3] Takeuchi, N.; Baba, H.; Sakurai, K.; Ueno, T.: Diode-laser random-modulation cw-lidar. Applied Optics 25(1986)1, 63-67.

[4] Ehret, G.; Franz, J.; Günther, K.; Klingenberg, H.; Werner, Ch.: Laser in der Umweltmeßtechnik. DLR-Nachrichten, Heft 63 (1991), 19-23.

[5] Mesures, R. M.: Laser Remote Sensing. Wiley-Interscience Publication (1984).

[6] Bisle, P.: Anwendung kohärent-optischer Modulationsverfahren in der Umweltmeßtechnik. Diplomarbeit, Lehrstuhl für Nachrichtentechnik, TU München und DLR-Oberpfaffenhofen, Feb. 1992.

[7] Mengistab, D.: Bewertung von Signalrauschverhältnis und Photonenbilanz bei RM-CW-LIDAR und Puls-LIDAR, Diplomarbeit, FH Düsseldorf, FG Optische Nachrichtentechnik, Jan. 1993.

New Materials for Optical Data Storage and High Information Content Displays

M. Kreuzer, T. Tschudi

Institut für Angewandte Physik, TH Darmstadt,

Hochschulstr. 6, D–64298 Darmstadt

Introduction

High contrast ratio and high lumen efficiency at the same time are the main reason that interest in scattering displays has been revived. The best–known example is constituted by polymer dispersed liquid crystals (PDLC) [1].

In this paper we want to discuss the electro–optical and all–optical properties of Filled Nematics, a new liquid crystal display using static light scattering. Bistable operation allows laser addressing for high resolution graphic displays [2].

With polymerized Filled Nematics (PFN) we introduce a new concept for a full electrically addressed simple and low cost scattering display. No need of index matching, no dependence on viewing angle and high contrast are the characteristics of the material.

Material

Filled Nematics consist of small solid particles of pyrogenic silica dispersed in a nematic liquid crystal [3, 4]. Typically, the diameter of the particles is chosen in the range 7 nm to 40 nm. Inside the dispersion the particles form agglomerates and aggregates with a large specific surface [5] (typ. 50 – 380 m^2/g) deviding the liquid crystals into small nematics domains (50 -200 nm in diameter). Due to the low bulk density of the dispersed pyrogenic silica, we can achieve stable dispersions with high volume ratio for the nematic phase (typically 97 – 99 Vol%). Therefore, the dispersed particles have no influence on the optical properties and matching between the refractive indices of liquid crystal and dispersed material is not required. The material is placed between untreated glassplates with transparent ITO electrodes in a conventional display configuration.

Laser addressed bistable display

After preparation, the liquid crystal display is switched to a homeotropically aligned tranparent state by applying a low frequency electrical field. On removal of the electric field, the display remain in the transparent homeotropic state. Illumination of the aligned Filled Nematic with a focussed beam of a

Table 1: Summarizing of the properties of a laser addressed Filled Nematic display

Mechanism	: Thermal	Linearity	: Good
Thickness d	: $2 - 20\,\mu m$ (typ. $10\,\mu m$)	Global erase	: $\succeq 70\,$V @ $500\,$Hz
Sensitivity	: $< 1\,$nJ$/\mu m^2$	Selective erase	: $\approx 25\,$V @ $500\,$Hz
Contrast ratio	: $> 30 : 1$	Liquid crystals	: ZLI 1132 (Merck),
Resolution	: $> 500\,$lines/mm		: E48, E44 (BDH)
Grey–levels	: Yes	Aerosil (Degussa)	: R 812/R 709 (1:1),
Dye	: SC 1515 (BASF)		: R 974

semiconductor laser switches the display to a strongly sacttering state. The laser writing, based on a thermal shock mechanism, causes a new configuration of the framework [6] of the particles and leads to a random alignment of the nematic domains resulting in strong scattering of visible light. The written information can be erased by strong electric fields (global) or in a selective way by applying a weak external field in combination with the laser beam again.

Table 1 shows the display properties used in a first prototype of a laser addressed projection system using a semiconductor laser ($\lambda = 785\,$nm,, $I_m$ax $= 25\,$mW).

Since contrast depends on laser energy grey levels can be realized. Using a conventional laser scanning system high resolution imaging with 16 grey levels (4096×4096 pixels on an area of 2×2 cm^2) has been demonstrated [6].

Electrically addressed polymerized Filled Nematic

To realize a monostable electrically addressed scattering display we have added a small amount ($\approx$ 1.5%) of prepolymer and a small amount of photo–initiator to the dispersion. After preparation the display was switched to a strong scattering state (e.g. by ultrasound or laser writing) and photocured in the scattering state. The polymer network stabilizes the the solid frame work of the agglomerates forming a monostable display. Depending on preparation parameters and on concentration of the dispersed particles switching times ($50\ \mu s$ to 10 ms (switching on), 2 – 10 ms (switching off)) and switching voltages (between 35 V and 70 V, AC) can be varied in a wide range. A typical switching behaviour is shown in figure 1.

The transmittance was measured using a HeNe laser (collection angle $\approx 2°$). A remarkable contrast of 1:330 was observed using a liquid crystal with high Δn (E44, BDH). Since the free configuration of the agglomerates of Aerosil particles is responsible for the physical mechanism of Filled Nematic displays [6] any nematic liquid crystal can be used also in polymerized Filled Nematics. On the other side no dependence on viewing angle allows direct view applications.

806

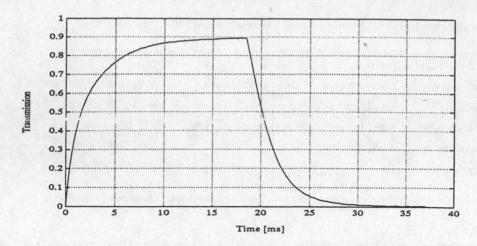

Figure 1: Switching behaviour of a polymerized Filled Nematic (response to a pulse of 18 ms width, 50 V)

References

[1] See, for example, J.W. Doane, N.A. Vaz, B.-G. Wu and S. Zumar, Appl. Phys. Lett. 48, 269 (1986)

[2] M. Kreuzer, T. Tschudi, R. Eidenschink, High resolution projection displays by use of Filled Nematics. Proceedings of the 12th Int. Display research conference, SID (Japan Display '92), 1992

[3] M. Kreuzer, T. Tschudi, R. Eidenschink, Erasable Optical Data Storage in Bistable Liquid Crystal Cells, Mol.Cryst.Liq.Cryst. 223 pp. 219–227 (1992)

[4] R. Eidenschink, W.H. de Jeu, Static Scattering in Filled Nematic: New Liquid Crystal Display Technique, Electronic Letters, 27, 1195, 1991.

[5] Degussa AG, Firmenzeitschrift 'Aerosil'

[6] M. Kreuzer, T. Tschudi, W.H. de Jeu, R. Eidenschink, A new Liquid Crystal Display with Bistability and Selective Erasure using Scattering in Filled Nematics, Appl. Phys. Lett., 12 April 1993, in print

Congress G

Congress-Chairmen: F.P. Schäfer and M. Stuke

Laser in der Forschung
Laser in Research

Experimental Study of 248 nm and 308 nm Ablation in Dependence on Optical Illumination Parameters

B. Burghardt*, U. Sarbach, B. Klimt, H.-J. Kahlert
MicroLas Lasersystem GmbH, O-6900 Jena
*Lambda Physik GmbH, W-3400 Göttingen

Abstract

Optical illumination parameters such as principal ray tilt and numerical aperture of illumination (NA: divergence or light cone angle) play a dominant role in the generation of well defined microstructure wall angles. High power (up to 200 W, Lambda 4000) Excimer Lasers with pulse energy up to 700 mJ are applied to ablate polyimide at 248 nm and 308 nm using a new 5x large field lens. As zoom lens homogenizer set up is used to arrange a numerical aperture of 0.13. Results of a systematic study with beam delivery systems of different NA are presented. The variation of the numerical aperture clearly results in steeper walls at higher NA. The impact of the applied energy density is investigated over a range of 100 mJ/cm^2 to 11 J/cm^2.

Introduction

Due to the unique properties of pulsed ultraviolet laser radiation, Excimer Lasers are widely used for micromachining of all kinds of materials. Typical feature size of 100 μm down to 0.2 μm can be processed, which is not possible with other lasersystems like Nd:YAG of CO_2. The ablation rate is in the range of 0.1 to 1.0 μm/laser pulse. Therefore, the depth of the ablated structure can be controlled with high precision by the number of superimposed laser pulses.

Excimer Lasers have been applied in research laboratories since 1977. About ten years later, they were introduced into industrial processing and manufacturing lines. Meanwhile Excimer Lasers with an average power of up to 200 W and pulse energy of up to 700 mJ (Lambda 4000) have become a reliable tool for industrial microprocessing.

One of the most interesting industrial application is the well known ablation of polymeres by UV Excimer Laser light (1, 2). Laser ablation of polyimide and other polymeres is used in manufacturing of MCM`s, TAB-substrats, FPCB`s and in processing of micro-nozzles for bubble jet printer heads (3).

Fig.1: Micro-Structure in Polyimide processed by Excimer Laser
(248 nm, 1 J/cm^2, mask imaging technique)

High resolution microstructuring with Excimer Lasers uses mask imaging
techniques. Fig. 1 shows a microstructure in polyimide which had been
processed by an Excimer Laser running at 248 nm. The walls of the
generated pattern are in general not parallel but show a conical shape
with a wall angle of 5 to 10 degrees (defined as the angle between the
surface normal and the direction of the wall). The specific wall angle
originates in the lateral energy density distribution because the ablation
depth is function of the local energy density on the substrate. The energy
density distribution at the edges of the structures is determined by the
interference pattern and imaging quality of the optical system.

By controlling optical parameters such as principle ray tilt and numerical
aperture and the energy density on the substrate surface as a process
parameter, microstructures with well defined wall angles can be generated.
The energy density, necessary for processing polymeres is in the range of
100mJ/cm^2 up to several Joules/cm^2. To investigate the resulting wall angle,
the numericl aperture of the imaging system and the energy density at the
polymere surface have been varied over a wide range.

Experimental Set-Up

Three different optical illumination and mask imaging systems have been
used for experimental work.

A simple one (fig.2) consists of a mechanically drilled or chemically etched
metal mask which is illuminated by the homogenious central part of an
LPX 300 Excimer Laser beam. A variable beam attenuator is used to adjust

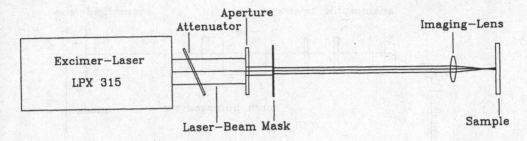

Fig.2: Simple Excimer Laser Mask Projection System

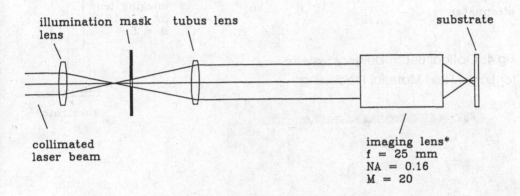

*Leica Mikroskopie and Systeme GmbH
Wetzlar, Germany

Fig.3: Experimental Optical Set-Up for high Energy Density
and High Numerical Aperture

the pulse energy at the mask position. The mask itself is imaged onto the polymere surface by a single lens element. To avoid optical distortion and aberations the numerical aperture was limited to 0.01. By varying the demagnification factor and the pulse energy the energy density at the substrate could be set between 0.1 and 4.0 J/cm^2.

To get higher energy densities at a high numerical aperture a specially designed eight elements short focal length imaging lens was used (4). The optical set-up of the complete system is shown in fig. 3.

The imaging lens with a focal length of 25 mm and a numerical aperture of 0.16 is corrected for infinity at 248 nm. In combination with a 500 mm tubus lens the demagnification factor of the system is set to 20. The metal mask is illuminated with a diverting beam, so that the pupile of the imaging lens is filled correctly. Only the central part of the Excimer Laser beam is used. In this way energy densities of up to 11 J/cm^2 at the sample position could easily be achieved. The processing field is restricted by the lens to 0.25 mm diameter.

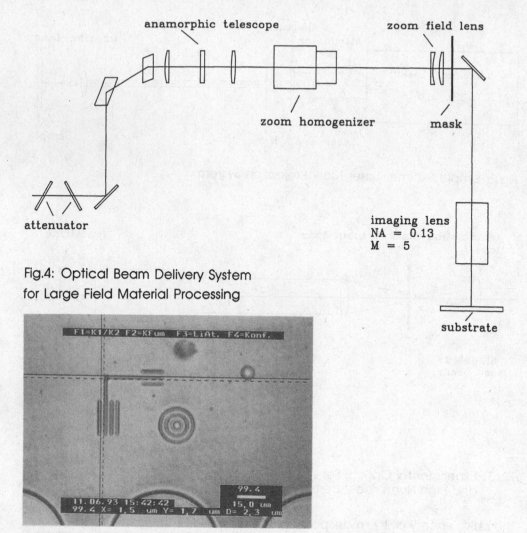

Fig.4: Optical Beam Delivery System
for Large Field Material Processing

Fig.5: Microstructure processed in Polyimide (10 μm thick)
using a 5x Large Field Imaging Lens

For large field high resolution micromachining a more complex beam delivery system is needed (fig. 4). It allows to create an energy density of up to 1 J/cm^2 at the substrate plane within a large image field of 18 mm diameter.

The system consists of a beam steering optics (beam displacement compensation), an illumination module (anamorphic beam forming telescope with a zoom homogenizer), the mask (chromium on fused silica substrate) and the 5x imaging optics (telecentric combination of a zoom field lens with the high resolution imaging lens). Again a variable beam attenuator is used to adjust the energy density.

The specifications of the imaging lens are summarized in table 1. With the numerical aperture of 0.13 the lens provides a theoretical resolution of 1.44 µm at 308 or 1.16 µm at 248 nm (5). The distrotion is less than +/- 0.5 µm across the image field. To get the optimum performance of the imaging lens, the numerical aperture of the illumination optics has to match the object side numerical aperture of the imaging lens.

The video print of a microscope image (fig. 5) shows the ablation result in polyimide at 308 nm. The test pattern is processed in a 10 µm thick layer of polyimide on a silicon wafer with 400 mJ/cm^2. It gives a practical resolution of better than 1.7 µm, close to the theoretical limit.

Table 1   **Optical Parameters of the Large Field 5x Lens**

demagnification	$5x \pm 1 \times 10^{-5}$
wavelenth	248 nm
transfer length SS`	(800 ± 1) mm
image field diameter	18 mm
work distance	50 mm
pupil position / object	577
pupil position image (telecentric)	∞
max. NA	0.13
distortion	+/- 0.5 µm
max energy density	2 J/cm^2
max. energy transfer per pulse	200 mJ

Experiments and Results

With the different beam delivery systems vias have been processed into polyimide and polysulfone films of 30 and 250 µm thickness. The samples were positioned, that the plane of best image was just on top of the films. The energy density at the substrate surface was set between 0.1 and 11 J/cm^2. The diameter of the holes at the entrance and the exit side have been measured by a microscope. From that, the corresponding wall angle was calculated. As an example the entrance and exit of a via in 250 µm thick polysulfone are shown in fig. 6a and 6b.

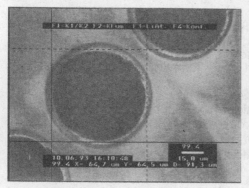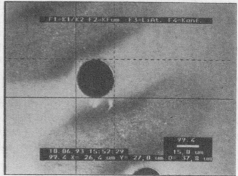

Fig.6a and b:
Microscope Image of an Excimer Laser Processed
Via in Polysulfone of 250 μm Thickness
(a: front side, b: rear side)

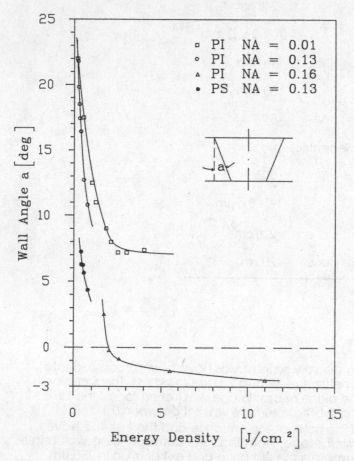

Fig.7: Influence of the Energy Density and the Numerical Aperture
on the Wall Angle of Excimer Laser processed Micro Structures

The experimental results of the processed wall angles are summarized in fig. 7. It shows the dependence of the generated wall angle on the energy density with the different realized numerical apertures. The wall angle is decreasing with increasing energy density and can be controlled by this process parameter. For low energy density near ablation threshould the wall angle is mainly determined by the ablation process itself whereas for higher densities an influence of the optical system besomes obvious. For a given numerical aperture the angle goes to a fixed value at high energy densities.

With a numerical aperture of 0.16 and at a fluence of more than 2.0 J/cm^2, holes with even negative wall angle can be processed. In this region the optical parameters clearly dominate the ablation process and determine the 3-dimensional ablation pattern.

Summary

The design and layout of an Excimer Laser based optical system is of great importance for reliable microstructuring. The illumination optics the mask and the imaging optics determine the 3-dimensional ablation pattern. By controlling the optical and the processing parameters microstructures with well defined wall angles can be processed. Thus for optimum performance the beam delivery system has to be designed and adapted to the special needs.

(1) **R. Srinivasan** in "Laser Processing and Diagnostics", D. Bäuerle, ed., Springer Verlag 1984

(2) **D.P. Brunco, M.O. Thompson, C.E. Otis and P.M. Goodwin,** J. Appl. Phys. Vol. 72, No.9, Nov. 1992

(3) **Multi-Chip-Modul,** Siemens Magazin COM, 2/89, p42

(4) **Leica** Mikroskopie and Systeme GmbH, Wetzlar

(5) **W.J. Smith,** Modern Optical Engineering", McGraw Hill, New York 1966, p140

Change of the Ablation Rates with Ablation Structure Size

B. Wolff-Rottke, J. Ihlemann, H. Schmidt, A. Scholl

Laser-Laboratorium Göttingen e.V.

Im Hassel 21, 37077 Göttingen

Introduction

Material processing with excimer lasers has been studied in great detail during the last years [1,2]. The short wavelength of this laser type allows the ablation of nearly every material. A further advantage is the increase in resolution of the structures producible with decreasing wavelength. With UV laser radiation a resolution down to one micrometer or even less can be achieved. Therefore the excimer laser is a useful tool for generating microstructures in various materials.

An important parameter for the ablation process is the amount of material which is removed by a single laser pulse. This ablation rate (ablation depth per pulse) depends on the material, the laser parameters (wavelength, pulse duration, energy density reaching the sample surface (J/cm^2)), and the surrounding (atmosphere, pressure). For very small laser spot sizes on the sample surface also a dependence on the size of the ablated area is observed. This has been reported by M. Eyett et al. [3] and Th. Beuermann et al. [4] for $LiNbO_3$ at the wavelengths 248 nm and 308 nm. They found an increase of the ablation rate if the spot diameter is decreased below about 200 µm.

In this work we investigated the size dependence of the ablation rate at the laser wavelengths 193 nm, 248 nm, and 308 nm. We determined the ablation rate as a function of fluence and laser spot diameter for various materials (polymeres, glasses and ceramics). The attenuation of the laser radiation by the ablation plume was measured.

Experimental Setup

Ablation experiments are performed by mask imaging using the Lambda Physik excimer lasers EMG 301 MSC at 248 nm or 193 nm and EMG 203 MSC at 308 nm. Pulse durations are about 20-30 nsec FHWM, and pulse energies used reach up to 1 J. The fluence is adjusted by tilting a dielectrically coated fused silica plate placed directly in front of the excimer laser. Pulse energies are measured by pyroelectric detectors (Gentec ED-200). The experimental setup is shown in figure 1. A mask is homogeneously illuminated by the laser beam, which is either unfocused or slightly focused by a 1500 mm focal length fused silica lens. This mask is imaged onto the sample surface using a fused silica lens (f=100 mm) or a reflective objective (Fig. 1). Part of the experiments was performed in a vacuum chamber, which can be evacuated down to a pressure of about $4 \cdot 10^{-3}$ mbar.

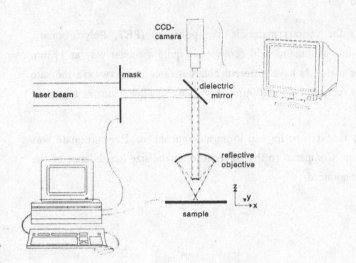

Ablation depth and hole diameters are measured by a Dektak 3030 Auto II stylus profilometer and a confocal laser-scan microscope (Leica). The morphology of the laser drilled holes is investigated with an optical microscope and a scanning electron microscope (Zeiss DSM 962).

Fig.1: Experimental setup

To determine the attenuation of the laser radiation by the ablation plume, first a tiny hole is drilled through the samples. Then the area around this hole is ablated and the pulse energy is measured in front and , through the hole, behind the sample.

Experimental Results and Discussion

Ablation starts if the fluence of the laser radiation is higher than a certain threshold fluence specific for the different materials. Then the ablation rate increases with increasing fluence until a saturation value is reached. For higher fluences the ablation still increases but only very slowly. This saturation is supposed to be mainly caused by the attenuation of the incoming laser radiation by the material already ablated. The fluence values where ablation starts and saturates and the maximum ablation rate depend on the material and laser wavelength.

Figure 2 shows the ablation rate of Polyethylenterephthalate (PET) as a function of the fluence at 193 nm for different laser spot diameters on the sample surface. The size of the ablated area significantly influences the ablation rate at high fluences if the diameter becomes smaller than about 300 µm. With decreasing spot diameter the ablation rate in the saturation range increases.

In Figure 3 the dependence of the ablation rate on the spot diameter is given as a ratio between the ablation rate for a certain diameter (AR(x)) and the ablation rate for a diameter of 300 µm

(AR(300μm)). Shown is the change of the ablation rate for PET at the wavelengths 193 nm, 248 nm, and 308 nm. For the ablation at 193 nm and 248 nm a similar size dependence of the rate is observed. At 308 nm the ablation rate remained constant down to a spot diameter of 30 μm. By further reducing the diameter (20 μm) the rate suddenly rises by a factor of about 1.9.

This size dependence appears for various materials like polymers (PET, Polycarbonate, Polyamide), borosilicate glass, and aluminium oxid ceramics in quite a similar way at 193nm and 248nm. Although all these materials have different ablation rates in every case the rate increases if the spot diameter is reduced below 200μm, identifying this behaviour as a general effect.

All measurements described so far were performed in an air atmoshpere. Polycarbonate was also ablated in vacuum at 248nm. Compared to the ablation in air the size dependence of the ablation rate is slightly more pronounced.

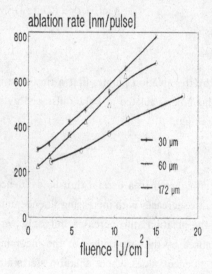
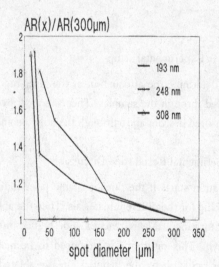

Fig. 2: Ablation rate of PET at 193 nm as a function of the fluence for different spot diameters.

Fig. 3: Spot diameter dependence of the ablation rate of PET at different wavelengths (15 J/cm^2).

Measurements of the pulse energy in front of the sample and behind it, after passage through the ablation plume, show a strong attenuation of the radiation at high fluences. For example for PET at 248 nm and 20 J/cm^2 only 35% of the pulse energy reaches the sample surface. This investigation was restricted to large ablation areas to achieve a detectable signal. But it shows the important role of the interaction between the laser radiation and the ablation plume in the range of saturation.

The ablated material is ejected mainly perpendicular to the sample surface [5]. For large ablation areas the portion of material which leaves the laser beam during the laser pulse is therefore negligible. In contrast for small ablation areas the material loss becomes important. If the density of the ablation plume is reduced significantly already during the laser pulse the fluence on the sample surface is increased. This effect is supposed to be the main reason for the increase of the ablation rate with decreasing laser spot diameter.

The decrease in density within the ablation plume over small ablation areas is an exclusively geometric effect and should therefore appear for all ablated materials. In the case of PET ablated at 308 nm the size dependence is observed only for very small spot sizes. This may be due to the high ablation rate at this wavelength (3 μm/pulse at 308 nm, 15 J/cm^2; 1 μm/pulse at 248 nm, 15 J/cm^2). The large amount of ablated material must be diluted to a higher degree before the laser radiation can penetrate through the ablation plume.

If the ablation is performed in vacuum the plume can expand unhindered. Therefore the increase of the ablation rate for decreasing spot diameters is slightly more pronounced.

In conclusion, we observed a size dependence of the ablation rate for various materials and wavelengths. This seems to be a general effect caused by the faster thining out of the ablation plume over small ablation areas.

References

1) R.Srinivasan, B.Braren, Chem. Rev. **89** (1989) m1303

2) S.Lazare and V. Granier, Laser Chem. **10** (1989) 25

3) M.Eyett, D.Bäuerle, Appl. Phys. Lett. **51** (1987) 2054

4) Th.Beuermann, H.J.Brinkmann, T.Damm, M.Stuke, Mat. Res. Soc. Symp. Proc. Vol. **191** (1990) 37, Material Research Society

5) R.Srinivasan, B.Braren and R.W.Dreyfus, J. Appl. Phys. **61** (1987) 372

Laser Induced Etching of Silicon with Fluorine and Chlorine and with Mixtures of Both Gases

U. Köhler, A. Guber, W. Bier

Kernforschungszentrum und Universität Karlsruhe
Institut für Mikrostrukturtechnik
Postfach 3640, D-76021 Karlsruhe

1.   Introduction

Thanks to its high photon energy the excimer laser does not only allow direct photolytical ablation to be achieved but also indirect laser induced etching of materials with gases. The source gases, most of them halogenated, can be dissociated with the laser so that the atoms and radicals formed enter into reactions with the substrate to be etched [1]. The Cl_2 and F_2 halogens have great practical importance in silicon etching. Due to its absorption band, F_2 can e.g. be excited near the KrF and XeCl lines whereas Cl_2 is excited solely by the XeCl wavelength. Most of the gaseous reaction products can be traced in the infrared by FTIR spectrometry due to their characteristic absorption bands. The quantitative evaluation of the individual bands allows conclusions to be drawn with respect to the reaction mechanisms which are effective in the etching process.

2.   Experimental

For specimen conditioning the silicon wafers are thermally oxidized. The approximately 1 µm thick SiO_2 layer is provided with a number of test patterns in a photolithography step and broken into specimen pieces about 15 mm x 20 mm in size. The specimens are positioned approx. 1 cm behind a quartz glass window in a purpose-built stainless steel cuvette which, at the same time, serves as IR measuring cell. The IR measurements are made with an FTIR spectrometer, type IFS 88, supplied by BRUKER, in a range of 4500 cm^{-1} to 400 cm^{-1}, with a resolution of 0.5 cm^{-1}. By means of a homogenizer the silicon specimens are irradiated with an excimer laser EMG 103 MSC, supplied by LAMBDA PHYSIK. The energy density is about 110 mJ/cm^2 with approx. 90 mm^2 beam cross-section. After placement of the specimen the system is carefully evacuated and the gas to be studied is introduced. It is excited and dissociated by the incoming laser light in the volume between the specimen and the quartz window. First,

3.3. Chlorine/Fluorine Mixtures

When the two gases, F_2 and Cl_2, are mixed, traces of ClF are generated during some hours in the absence of laser irradiation. In the IR spectrum, between 810 cm^{-1} and 730 cm^{-1}, the ClF typical band spectrum can be recognized. The formation of SiF_4 is deferred by the competing Cl_2/F_2 reaction and the adsorption of Cl_2 on the silicon surface is delayed depending on the fraction of Cl_2. After switching on of the laser the formation of ClF is greatly enhanced and after some delay ClF_3 is formed. This is accompanied by an increase in SiF_4 formation (Fig. 6). A similar production of $SiCl_4$ could not be detected in the spectrum. The IR main band for ClF_3 occurs at 710 cm^{-1} (Fig. 7). The ClF_3 formed reacts spontaneously with silicon to become SiF_4 and ClF. As long as F_2 is present in excess, the ClF formed is always fluorinated into ClF_3. The ClF_3 partial pressure on the average remains constant and decreases only when most of the fluorine content in the mixture has reacted and been converted into SiF_4.

4. Summary

The investigations have shown that by means of FTIR spectroscopy the elemental etching reactions can be represented and the etching rates measured. In laser induced etching using molecular chlorine e.g. $SiCl_4$ can be detected. By etching with fluorine the etching rate can be increased by one order of magnitude due to laser irradiation. If chlorine/fluorine mixtures are used the mixed compounds ClF and ClF_3 are formed in addition. Scanning electron micrographs (SEM) show an anisotropic etching profile for the chlorine involving process whereas in all experiments involving fluorinated gases isotropically etched patterns are produced.

References

[1] Bäuerle: Chemical Processing with Lasers. Berlin: Springer Verlag 1986
[2] Winters, Coburn: Surface Science Aspects of Etching Reactions. North-Holland: Surface Science Report 14 (1992) 161-169
[3] Horiike, Nayasaka et al. Excimer Laser Etching on Silicon. Appl. Physics A 44, 313-322 (1987)
[4] Guber, Köhler: Some Investigations into Plasmaless Etching of Silicon and its Oxide with Molecular Fluorine. J. Fluorine Chemistry 54,4 (1991)

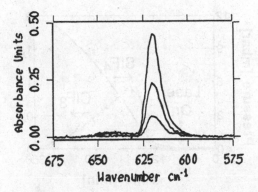

Fig. 1: Development of the $SiCl_4$ absorption band at 619 cm^{-1} versus time.

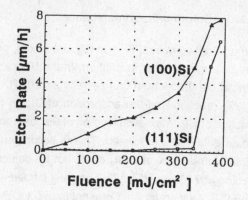

Fig. 2: SEM of a (100) silicon structure etched with 40 mbar Cl_2 at 308 nm and SiO_2 mask left.

Fig. 3: Intercomparison of the etching rates in (100) and (111) silicon.

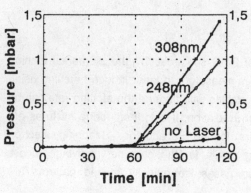

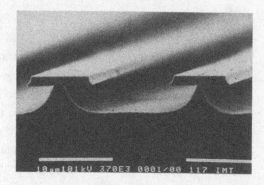

Fig. 4: Development of the SiF_4 partial pressure in an etching experiment with and without laser.

Fig. 5: SEM of a silicon structure etched with 20 mbar F_2 at 308 nm and SiO_2 mask left.

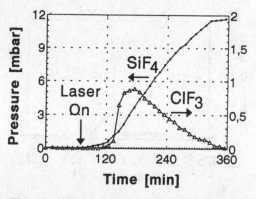

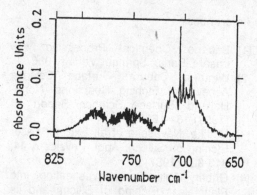

Fig. 6: Plots of SiF_4 and ClF_3 partial pressures in an experiment performed with 5/45 mbar Cl_2/F_2.

Fig. 7: FTIR spectrum of a Cl_2/F_2 mixture during laser irradiation.

PLD of Thin Films for Applications

A. Voss, W. Pfleging, M. Alunovic, E.W. Kreutz

Lehrstuhl für Lasertechnik, RWTH Aachen, Steinbachstraße 15, D-52074 Aachen, Germany

1. Introduction

Thin films are widely used in a variety of applications either as structural overcoats or as functional coatings /1/. Pulsed laser deposition (PLD) shows various advantageous properties for the deposition of thin films for applications. Firstly, evaporation by laser radiation is not limited by the electrical charging of the target or the substrate. Secondly, very short laser pulses allow transfer of the target material with stoichiometric deposition at the substrate /2/. Also, deposition of materials without a stable melting phase (e.g. SiC) is possible.

Ceramic films were deposited with PLD on metallic substrates by a variation of laser parameters and processing variables. The structure, morphology and composition, as well as the mechanical properties, such as wear, corrosion resistance and adhesion of the deposited films were examined. The broad variety of film structures obtained by varying the deposition conditions are discussed in view of applications.

2. Experimental setup

The PLD process was performed in a high vacuum chamber at a base pressure of 6×10^{-5} mbar. The laser radiation was guided through a window onto a rotating target. The substrate surface was parallel to the target surface allowing a distance variation in the range of 0.5-4.5 cm. The experiments were carried out with different laser radiation sources (Table 1). The fluence on the target was adjusted by variation of the spot diameter, while the energy per pulse remained constant. The substrate holder was electrically insulated from the vacuum chamber and connected with an rf source either for sputter cleaning of the substrates before deposition or for use as bias during film deposition.

Thin films were prepared either for matched (alternating phases of electrical and optical field) or for unmatched (phase of superposed electrical and optical field followed by a field-free phase) conditions of the additional rf field /3/. The films were deposited onto stainless steel or refractory metal plates. The morphology, structure and composition of the films were investigated by SEM, TEM, XRD, EDX and AES. The wear, adhesion and corrosion resistance of the films were studied by the ball grinding method, scratch test and current density potential measurements. Further details of the PLD setup have been described elsewhere /4/.

laser system	wavelength /μm	average power /W	repetition rate /Hz	mode of operation	pulse length /ns
TEA CO$_2$	10.6	26	10	pw	200 + 3500
CO$_2$	10.6	1000	- -	cw	- -
Nd:YAG	1.06	20.5	5000	pw	20
Excimer	0.308	35	10	pw	50

Table 1   Laser Parameters (pw pulsed wave, cw continuous wave)

3. Results

SEM images of ceramic films (Fig. 1) deposited with different laser wavelength show the variety of film structures which are achieved by PLD. Al$_2$O$_3$ films deposited with pulsed CO$_2$ laser radiation have a dense and glassy structure in the bulk and droplets at the surface. These films are amorphous or consist of nanocrystallites in an amorphous matrix, as there is no indication for a crystalline structure in XRD /5/. SiC films deposited with Nd:YAG laser radiation exhibit a columnar structure and a smoother surface than the films deposited with CO$_2$ laser radiation. There are large outgrowths extending from the substrate surface throughout the film bulk, which are only partially connected to the film. The ZrO$_2$ films deposited with Excimer laser radiation have a slightly textured film structure in combination with a flat and nearly defect free surface. TEM analysis shows that these films are composed of nanocrystallites.

a) Al$_2$O$_3$ (TEA CO$_2$ laser, t_p = 3.5 μs, ν = 10 Hz, ε = 27 J/cm^2, p_{Ar} = 2 x 10^{-2} mbar)

b) SiC (Nd:YAG laser, t_p = 20 ns, ν = 5 kHz, ε = 23 J/cm^2, p_{Ar} = 2 x 10^{-3} mbar)

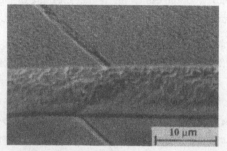

Fig. 1 Morphology and structure of films deposited with different laser wavelength

c) ZrO$_2$ (Excimer laser, t_p = 50 ns, ν = 10 Hz, ε = 9.8 J/cm^2, p_{Ar} = 2 x 10^{-2} mbar)

The influence of the rf bias was examined with respect to the film structure /3/ and the mechanical properties. The bias mode induces very dense and glassy Al_2O_3 films. The ball grinding test revealed that the adhesion of Al_2O_3 films on stainless steel or refractory metal is generally poor, but improves with the use of a bias and a processing gas mixture of Ar and O_2 (Fig. 2a). The volume loss determined by ball grinding of an Al_2O_3 film deposited under matched conditions is 10 % less compared to a polycrystalline target. Fig. 2b shows the current density-potential curves of a non-coated substrate and of substrates coated with Al_2O_3 under matched bias conditions and without bias. The film deposited without bias shows a stronger increase in current density at higher positive potentials than the film deposited with rf bias. The latter shows no indication of dielectric breakdown up to a potential of 2500 mV.

a) Ball grinding of Al_2O_3 film with good adhesion (film thickness 36 μm, processing gas Ar/O_2, rf bias matched conditions)

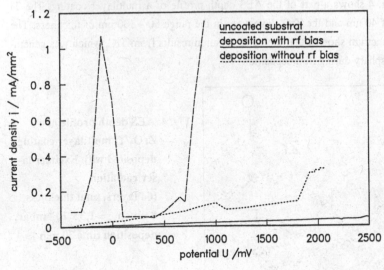

b) current density-potential measurements of Al_2O_3 films deposited with and without rf bias

Fig. 2 Mechanical properties of amorphous Al_2O_3 films

ZrO_2 films are deposited with pulsed CO_2 laser radiation at short distances from the target (1 - 2 cm) with contact of the luminous plasma plume with the substrate. They show fine grained or columnar structures at deposition rates of 1 μm/min on non-heated substrates (Fig. 3), /3/. Similar film structures are achieved with cw CO_2 laser radiation and electron beam evaporation if the films are deposited at substrate temperatures up to 950 - 1100 °C with deposition rates of 5 - 10 μm/min.

a) TEA CO_2 laser, $\varepsilon = 27$ J/cm², $p = 2 \times 10^{-2}$ mbar (Ar/O₂), non-heated substrate, deposition rate 1.0 μm/min

b) CO_2 laser, $I = 1600$ W/cm², $p = 4 \times 10^{-2}$ mbar (Ar/O₂), substrate temperature 950 °C, deposition rate 10 μm/min

Fig. 3 ZrO_2 films with columnar structure

Coating systems such as multilayers, graded or homogeneously doped layers are achieved with Excimer laser radiation with low technical effort by the use of segmented targets resulting in flat and nearly defect-free surfaces /6/. Fig. 4 shows a part of the AES depth profile of a multilayer coating. The Ti layers exhibit a thickness of 40 nm and the ZrO_2 layers are in the range 80 - 100 nm of thickness. The shape of the carbon AES spectrum showed that the contamination results from TiC, which was genera-ted in the surrounding atmosphary during film deposition.

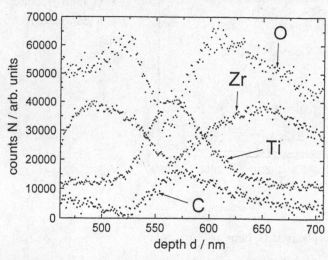

Fig. 4 AES depth profile of a ZrO_2/Ti multilayer coating deposited with Excimer laser radiation
(64 layers, total thickness 8 μm, $p_{Ar} = 1,3 \times 10^{-2}$ mbar, deposition time 30 min)

4. Discussion

The deposition of thin films with shorter laser wavelength (Fig. 1) generally leads to smoother surfaces of the films /7/. At shorter laser wavelength the penetration depth of the laser radiation into the target and the resulting geometry of the molten pool is smaller for most ceramic materials, which reduces the emission of particulates from the target. This advantage was used for the deposition of coating systems from segmented targets by excimer laser radiation.

To a large extend the mechanical properties of the deposited films depend on their structure. The resistivity against corrosion is improved by depositing the film under matched bias conditions, since glassy structures have no grain boundaries, which are favoured points for the start of the chemical attack. The corrosive medium may diffuse along the grain boundaries allowing diffusion at the film-substrate interface. The wear of Al_2O_3 against diamond is also lower for amorphous than for crystalline material because of the higher ductility of the amorphous material.

The ZrO_2 film (Fig. 3a) deposition at a short distance from the target (1 cm) results in high energies of the arriving particles at the substrate surface /8/. Thus, the film deposition occurred under non-equilibrium conditions before the cooling by expansion of the plasma was finished. Consequently the deposited particles exhibit a high mobility, which is comparable to that occurring for film deposition at high substrate temperatures (Fig. 3b). As ZrO_2 has a low thermal conductivity the films are used as thermal barrier coatings. The columnar structure of the films also allows for stress compensation during thermal shock treatments. This prevents the total delamination of the film from the substrate. Due to the low substrate temperatures during the deposition of ZrO_2 films with pulsed CO_2 laser radiation the application of the columnar films as thermal shock protection for steel or other temperature-sensitive materials is obvious.

5. Conclusion

Thin films of ceramic materials (Al_2O_3, ZrO_2, SiC, TiN, BN) and composed coatings (ZrO_2,Ti) for technical applications were deposited by the PLD at different laser parameters (wavelength, fluence, mode of operation) and processing variables (processing gas pressure and composition, rf bias, target-substrate distance). Dense and glassy Al_2O_3 films deposited on non-heated substrates with pulsed CO_2 and Excimer laser radiation show high wear and corrosion resistivity. The application as protective coatings against corrosion and erosion in wet tribological systems is promising, especially if the thermal treatment of the substrate must be below 500 K. ZrO_2 films deposited with CO_2 laser radiation show columnar structures which are applicable for heat protection. The application as thermal shock protection coatings for steel or other temperature sensitive materials is obvious. For such applications short deposition times or deposition at low substrate temperatures are required, which are both fulfilled by the PLD. Multilayer systems (Ti/ZrO_2) with 3000 double layers were produced with excimer laser radiation because this method yields films with nearly flat and defect-free surfaces. Graded layers (Ti -> ZrO_2) deposited with excimer laser radiation show a linear variation of the composition from Ti to ZrO_2 in Auger depth profile analysis.

References
/1/ H. Frey and G. Kienel, Dünnschichttechnologie, VDI-Verlag, Düsseldorf, 1987
/2/ P. E. Dyer, A. Issa and P. H. Key, Appl. Surf. Sci., 46 (1990) 89
/3/ H. Sung, G. Erkems, J. Funken, A. Voss, O. Lemmer and E. W. Kreutz, Surf. Coat. Technol., 54/55 (1992) 541
/4/ E. W. Kreutz, M. Krösche, H. Sung, A. Voss, A. Jürgens and T. Leyendecker, Surf. Coat. Technol., 53 (1992) 57
/5/ J. Funken, E. W. Kreutz, M. Krösche, H. Sung, A. Voss, G. Erkens, O. Lemmer and T. Leyendecker, Surf. Coat. Technol., 52 (1992) 221
/6/ H. Sung, W. Pfleging, A. Voss, E. W. Kreutz and G. Erkens, Dünne Schichten, 4 (1992) 16
/7/ W. Kautek, B. Roas and L. Schultz, Thin Solid Films, 191 (1990) 317
/8/ A. Voss, E. W. Kreutz, J. Funken, M. Alunovic and H. Sung, Appl. Surf. Sci., 69 (1993) 174

Tunable Subpicosecond Light Pulses in the Mid Infrared Produced by Difference Frequency Generation

C. Lauterwasser, P. Hamm, M. Zurek and W. Zinth

Institut für medizinische Optik

Universität München

Barbarastr. 16, D-80797 München

With the advent of Ti:Al$_2$O$_3$ laser / amplifier systems that produce highly energetic light pulses in the near infrared [1,2] there exits a powerful light source to generate femtosecond light pulses further in the infrared (1 µm to 12 µm) by nonlinear processes. Tunable mid-infrared pulses can be obtained by difference frequency mixing of light pulses of two different frequencies in a nonlinear medium like LiIO$_3$ or AgGaS$_2$ [3-5]. The excellent stability and high repetition rate of Ti: sapphire systems make it possible to produce light pulses well suited for sensitive spectroscopic investigations with a subpicosecond time resolution. Here we present a new parametric system generating subpicosecond pulses tunable from 6 to 11.5 µm with a repetition rate of 1 kHz and intensity fluctuations of less than 10 %.

Our system is based on a Ti:Al$_2$O$_3$ oscillator and regenerative amplifier. The oscillator delivers stable pulses at a central wavelength of 815 nm with pulse durations of 60 fs. In a chirped pulse amplification scheme similar to that of Salin et al. [2] these pulses are amplified to energies of 700 µJ at a repetition rate of 1 kHz. After recompression pulse durations of 120 fs are obtained.

In order to get a second wavelength for the difference frequency process a small part of the energy of these pulses (20-40 µJ) is used to pump a traveling wave dye laser (TWDL). For the generation of tunable IR-pulses between 6 and 11 µm, the central wavelength of the TWDL should be tunable between 870 and 930 nm.

A single stage TWDL emits femtosecond pulses at the long wavelength absorption edge of the laser dye used with a bandwidth of approximately 30 nm FWHM [for a description of the setup see 6]. It is tunable only by the choice of the laser dye. Since there are only few laser dyes available in the region between 870-1000 nm, it is necessary to increase the tunability of the system by other means. To this end we use a two stage traveling wave dye laser-amplifier configuration. Approximately 15 μJ of the amplified Ti:Al₂O₃-pulses are focused by a cylindrical lens (6 cm focal length) onto a first TWDL-cell. A spectrally narrow part (15 nm) of the emitted light is selected by a dispersion compensated double pass grating spectrometer. These pulses are then focused onto a second TWDL stage which is pumped by approximately 20 μJ and which acts as an amplifier for the seed pulses.

By this arrangement approximately 70% of the total energy from the TWDL amplifier (0.5 - 1.5 μJ) is emitted at the wavelength of the seed pulse. A tuning range of 60 nm is realized for one single laser dye. This makes it possible to cover the whole wavelength region between 870 nm and 1000 nm by the two laser dyes IR 140 (Lambda Physics) and IR 143 (Radiant Dye).

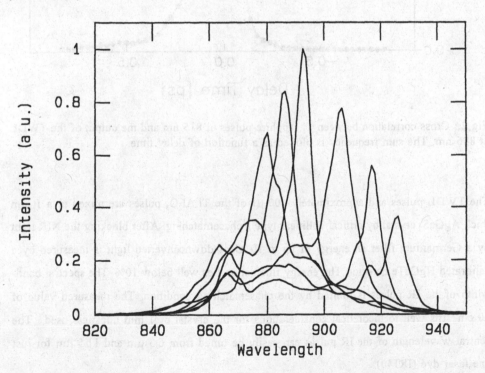

Fig 1.: Tunability of the two stage TWDL using the laser dye: IR140.

The pulse duration of the TWDL pulses lies between 100 fs and 200 fs. In general, they are somewhat longer than those of a single stage TWDL. Fig.1 shows a cross correlation between the output of the Ti:Al₂O₃ amplifier and a TWDL pulse at λ = 886 nm illustrating the short duration of these pulses as well as their small mutual temporal jitter.

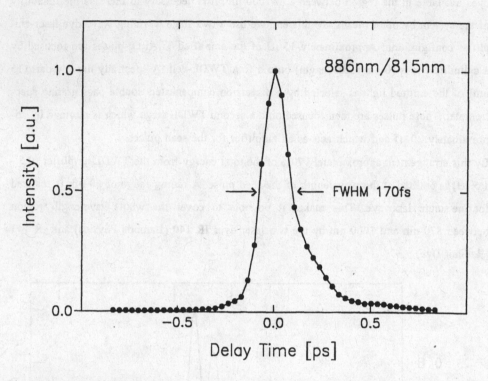

Fig 2.: Cross correlation between Ti:sapphire pulses at 815 nm and the output of the TWDL at 886 nm. The sum frequency is plotted as a function of delay time.

The TWDL pulses and approximately 100 μJ of the Ti:Al₂O₃ pulses are mixed in a 1 mm thick AgGaS₂ crystal by critical collinear type I phasematching. After blocking the NIR light by a Germanium filter an energy of up to 10 nJ of downconverted light is measured by a calibrated HgCdTe detector. The energy fluctuation lies well below 10%. The spectral bandwidth of the IR pulses is limited by the phasematching condition. The measured value of 60 cm⁻¹ fits well to theoretical considerations for the crystal of 1 mm thickness used. The central wavelength of the IR pulses can easily be tuned from 6.3 μm and 11.5 μm for just one laser dye (IR140).

The pulse duration of the infrared light pulses at a wavelength of 8 μm was measured via the rise of the reflectivity of a thin Silicon plate after generation of free carriers by exciting the

Silicon plate with a intense Ti:Al$_2$O$_3$-pulse. From the rise time of the signal the cross-correlation width between IR-pulses and Ti:Al$_2$O$_3$ pulses can be deduced. The best fit gives a FHWM time of 450 fs corresponding to a pulse duration of approximately 400 fs of the infrared pulses.

The pulse duration of the IR pulses is limited mainly by the group velocity dispersion between the NIR pulses and the mid IR pulse in the generating nonlinear crystal which is calculated to be approximately 300 fs for the present case. A time bandwidth product of $\Delta v \Delta t = 0.8$ is calculated which is approximately two times the bandwidth limit of pulses with gaussian pulse shape.

In conclusion, we have presented a scheme for the generation of intense and stable subpicosecond pulses in the spectral region of 6 to 11 μm at a high repetition rate. To our knowledge these are the shortest pulses ever produced in this spectral region.

1 D.E.Spence, P.N.Kean, W.Sibbett; Optics Letters 16 (1991) 42
2 F.Salin, J.Squier, G.Mourou, G.Vaillancourt, Optics Letters 16 (1991) 1964
3 T.Elsässer, H.Lobentanzer, A.Seilmeier; Optics Comm. 52 (1985) 355
4 D.S.Moore, S.C.Schmidt; Optics Letters 12 (1987) 480
5 T.Elsässer, M.C.Nuss; Optics Letters 16 (1991) 411
6 J.Hebling, J.Kuhl; Optics Letters 14 (1989) 278

Self-Phase Modulation of Ultrashort Light Pulses in the Second-Harmonic Nonlinear Mirror

K. A. Stankov

Laser Laboratory, Im Hassel 21, D-37077 Göttingen, Germany

V. P. Tzolov

Institute of Electronics, Bulgarian Academy of Sciences, 72 Trakia blvd, Sofia, Bulgaria,

Abstract: The second-harmonic nonlinear mirror arrangement can produce (self)-phase modulation of ultrashort light pulses and correspondingly chirp of arbitrary sign. The sign and the magnitude of the chirp can be easily varied by changing the amount of the dispersion in the space between the nonlinear crystal and the dichroic mirror. Potential applications in modelocked laser are discussed.

Self-phase modulation and chirping of light pulses has been usually associated with the third-order nonlinearity /1/. Although it has been known for a relatively long time that an effective third-order nonlinearity can be generated by cascading of second order nonlinearities /2/, the phenomenon was only recently demonstrated in a cascaded second-harmonic generation arrangement /3/. The phenomena may have interesting applications since the second-order nonlinearity manifests itself generally at considerably lower light intensities and an intensity-dependent refractive index n_2 of the order 10^{-10} to 10^{-12} can be realized /3/. However, certain limitations arising from the phase-mismatched condition or from the losses at the second harmonic wave are imposed.

It was also shown by Piskarskas et al /4/ that a quadratic nonlinearity can produce chirp reversal and enhancement by using parametric amplification. These effects, however, are intensity-independent and rely only on the specific crystal dispersion properties .

We show in this paper that the second-harmonic (frequency-doubling) nonlinear mirror /5-8/ can be used to produce phase modulation and chirping of light pulses in a way similar to the cascaded second-order nonlinearity, but in a perfectly phase-matched condition.

The analysis of the phase modulation and chirping refers to the second-harmonic nonlinear mirror arrangement as shown in Fig. 1.

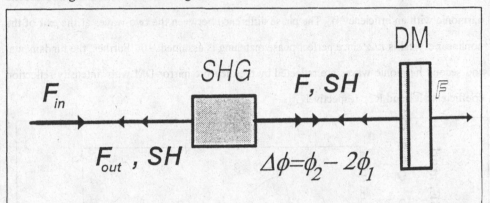

Fig. 1. The basic configuration of the frequency-doubling nonlinear mirror. SHG - second harmonic crystal; DM - dichroic mirror; F, SH - fundamental and second-harmonic radiation.

It consists of a phase-matched frequency-doubling crystal SHG and a dichroic mirror DM with arbitrary (but differing from zero) intensity reflection coefficient at the fundamental wavelength, R1, and the second harmonic, R2. Its operation is described as a two-step process: first, generation of second-harmonic in the first pass through the nonlinear crystal, and second, the interaction between the fundamental and the second harmonic in the second pass through the crystal, after reflection from the mirror. The present treatment refers to the phases of the interacting fundamental and second harmonic waves. The coupled equations for the fundamental and second harmonic phases in the case of perfect phase matching and plane waves are given by /10/:

$$d\phi_1/dz = -\sigma A_2 \cos(\Delta\phi) \tag{1a}$$

$$d\phi_2/dz = -\sigma(A_1^2/A_2)\cos(\Delta\phi) \tag{1b}$$

A_1 , A_2 denote the amplitudes of the interacting wavelengths. The nonlinear interaction is governed by the coupling constant σ, and the phase difference $\Delta\phi=\phi_2-2\phi_1$.

The above equations describe interaction for arbitrary conversion efficiency. The methodology in analysing the nonlinear mirror replicates that of the earlier papers /7, 8/, with the exception that the phase difference does not acquire the two particular values $\pi/2\pm2m\pi$, or $\pi/2\pm(2m+1)\pi$, but has an arbitrary value.

After the first pass through the crystal, the fundamental intensity is converted into second harmonic with an efficiency η. The phase difference between the two waves at the exit of the nonlinear crystal is $\pi/2$, since perfect phase matching is assumed /10/. Further, the fundamental and second harmonic waves are reflected by the dichroic mirror DM with intensity reflection coefficients R1 and R2, respectively.

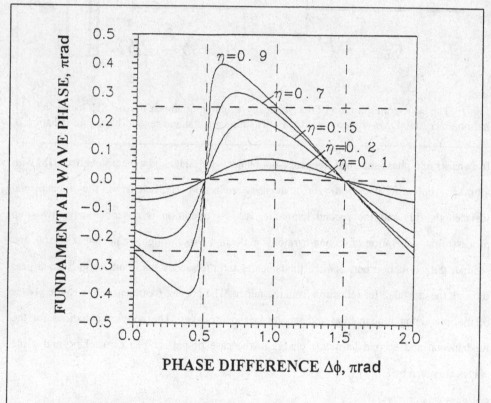

Fig. 2. The fundamental phase variation calculated as a function of the phase difference $\Delta\phi$ for several values of the conversion efficiency η.

We have calculated the fundamental phase variation as a function of the phase difference for several values of the conversion efficiency. The results are shown in Fig.2.

Positive and negative phase modulation can be easily achieved by changing the initial phase difference of the two waves reflected by the dichroic mirror. Experimentally this has been performed by utilizing the dispersion in air by varying the distance between the nonlinear crystal and the dichroic mirror /5/ or by tilting a parallel glass plate /11/

From equation (1b) it is evident that the second harmonic will also suffer phase modulation. One can introduce an effective nonlinear refraction index assuming that the phase change arises from an equivalent Kerr medium. Thus, from the data in Fig.2, for 50% conversion efficiency and optimum phase difference the effective nonlinear reflection coefficient is calculated for a nonlinear crystal of 4 mm length taking into account that the phase change takes place in a double pass through the crystal. Then for z=8 mm, $\Delta\phi$ =0.18π rad, λ=1000 nm, the refractive index change $\Delta n_2 = \Delta\phi\lambda/2\pi z = 11.25\times10^{-6}$. This large nonlinear refraction coefficient can be

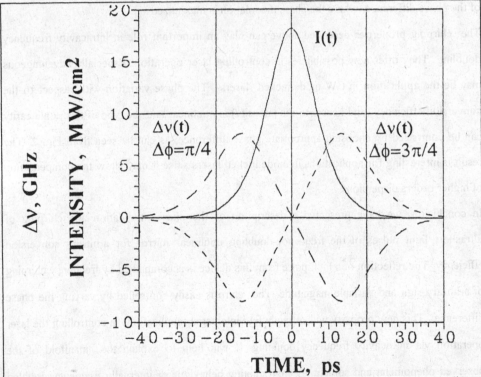

Fig. 3. Positive and negative chirping of a gaussian pulse after reflection from the nonlinear mirror for two optimized values of the phase difference. The pulse intensity and the instantaneous frequency deviation are plotted as a function of time.

realized for efficient nonlinear crystal like KTP which would provide such conversion efficiency at an input intensity of 200 MW/cm^2 /12/. The value of n_2= 5.6x10^{-11} cm^2/kW is to be compared with the most efficient (electronic) Kerr media like CS_2 and nitrobenzene, $\approx$8x10^{-11} cm^2/kW /1/.

The phase modulation as demonstrated in Fig.2 will result in a chirp for a light pulse reflected by the nonlinear mirror. As an example, we analyse the pulse chirping $\Delta v(t)= -\partial\phi/\partial\omega$ of a gaussian pulse having full width at half maximum (FWHM) 20 ps, with peak conversion efficiency of 50%. We assume that R1=R2=1, but the results are valid for any value for R1 differing from zero. Thus, the pulse form is given by $I_{in}(t)=I_0\exp([-4\ln 2t^2/\tau_p^2]$ where I_0 is the peak amplitude, and τ_p is the pulse duration.

In Fig.3. the time dependence of the instantaneous frequency deviation $\Delta v(t)$ is shown together with the pulse form of the gaussian pulse. The two curves for the chirp refer to the two values of the phase difference, where the phase modulation is maximized.

The chirping properties described above can play an important role in intracavity frequency doubling. They offer new possibilities in controlling laser operation. Especially advantageous may be the application in CW mode-locked lasers. The phase variation with respect to the conversion efficiency (and hence, to the intensity) is nonlinear and the type of this nonlinearity can be controlled by choosing appropriate phase difference, as can be seen from Fig. 2. The result is interesting for applications in mode locked lasers since it may allow the compensation of higher orders dispersions.

In conclusion, we have presented a description of the phase modulation and chirping of ultrashort light pulses in the frequency-doubling nonlinear mirror for arbitrary conversion efficiency. The reflection of a light pulse from this device is accompanied by frequency chirping of arbitrary sign and variable magnitude. The chirp is easily controlled by varying the phase difference. This new property will certainly find interesting applications in controlling the laser operation via intracavity frequency doubling. It will help to explain the manifold of the observed phenomena and sometimes contradicting behaviour of internally frequency doubled lasers.

References:

1. S. A. Akhmanov, V. Vysloukh and A. Chirkin, in *Optics of Femtosecond Laser Pulses*, Chapter 2, (AIP, New York, 1992)

2. C. Flytzanis, Theory of Nonlinear Optical Susceptibilities, in Quantum Electronics, vol. 1 Nonlinear Optics, Part 1, Ed. H. Rabin and C.L.Tang, (1975).

3. R. DeSalvo, D.J.Hagan, M.Sheik-Bahae, G. Stegeman, and E. W. Van Stryland, Opt. Lett. **17,** 28 (1992).

4. A. Piskarskas, A. Stabinis, and A. Yankauskas, Sov. Phys. Usp. **29,** 869 (1986)

5. K. A. Stankov and J. Jethwa, Opt. Commun. **66,** 41 (1988).

6. K. A. Stankov, Opt. Lett. **14** (1989) p. 51

7. K. A. Stankov, Appl. Phys. B **52** (1991) p. 158

8. K. A. Stankov, V. P. Tzolov, and M. G. Mirkov, Appl. Opt. **31** N. 24, 5003 (1992).

9. K. A. Stankov, V. P. Tzolov, I. Y. Milev, and M. G. Mirkov, "A Novel Light Modulator Based on the Second-Harmonic Nonlinear Mirror", submitted to IEEE J. Quant Electr, 1992.

10. J. A. Armstrong, N. Bloembergen, J. Ducuing and P. S. Pershan, Phys.Rev. **127,** 1918 (1962)

11. K. A. Stankov, K. Hamal, H. Jelinkova and I. Prohazka, Opt. Commun. **95,** 85 (1993).

12. C. L. Tang, W. R. Bosenberg, T. Ukachi, R. J. Lane, and L. K. Cheng, Laser focus World, **Sept.** , 87 (1990).

Generation and Amplification of Subpicosecond Pulses in the VUV - Towards Soft X-Rays with Nonlinear Optics

B. Wellegehausen

Institut für Quantenoptik, Universität Hannover

Welfengarten 1, W-3000 Hannover 1, Germany

Due to the development of short-pulse powerful laser systems, interesting perspectives for the generation of coherent radiation at short wavelengths by nonlinear optical processes exist.

So, high order harmonic generation up to the 109th or 135th order with wavelengths of 7.4 nm (168 eV) and 7.8 nm (159 eV), respectively, has been demonstrated, starting from a Ti:sapphire laser at about 806 nm /1/ or a high intensity Nd:glass laser system at 1.05 μm /2/. Harmonic generation has also been demonstrated using a subpicosecond KrF excimer laser at 248 nm, with the observation of shortest wavelengths at 10 nm (25th order) /3/. In these and further experiments /4,5/ it has been observed, that after an initial strong decrease of the intensity with the harmonic order, finally a plateau is reached with only a relatively slow decrease of the intensity up to a cut-off frequency. It seems to be interesting to extend the KrF laser conversion experiments and consider harmonic generation by even shorter wavelength lasers, with the possibility to generate soft x-rays by high efficient low order processes. Lasers that may be used for this concept are the discharged pumped excimers like ArF (193 nm), and F_2 (157 nm) or even the electron beam pumped excimers like Xe_2 (172 nm), Kr_2 (146 nm), and Ar_2 (126 nm).

Unfortunately, these systems as well as the KrF laser do not directly deliver the required intense short-pulses, and therefore these materials have to be used as amplifiers. Consequently, first the generation of picosecond or femtosecond pulses at these amplifier wavelengths has to be considered. Second, the short-pulse amplification characteristics of these materials have to be explored in order to realize high peak intensity laser systems. Finally, frequency tripling or higher harmonic generation in suitable nonlinear materials has to be investigated. In the contribution, results of first steps along this concept will be reported.

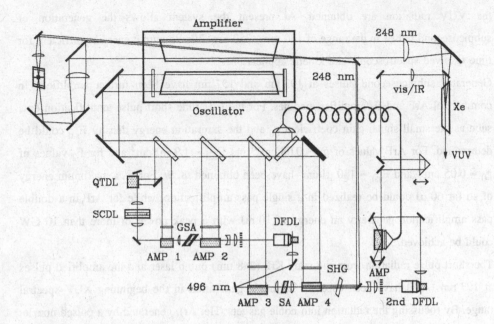

Fig 1. Experimental setup with fs-KrF-excimer laser system (248 nm) and additional distributed feedback dye laser (vis / IR) for the generation of tunable subpicosecond VUV pulses

For the generation of tunable subpicosecond pulses in the VUV, a four wave difference frequency mixing process in xenon is used, based upon a near resonant two photon excitation with KrF (248 nm) pulses /6/. The used KrF pump laser system consists of a dye laser chain to generate 400 fs pulses at 496 nm, which are then frequency doubled and finally amplified in a KrF amplifier to pulses with an energy of up to 25 mJ and a pulse duration of about 400 fs (Fig.1). The VUV radiation with frequency ω_{VUV} is obtained according to $\omega_{VUV} = 2*\omega_{248} - \omega_{vis}$, where ω_{vis} is a laser field in the visible or near infrared spectral range. We used for ω_{vis} either ns pulses from an independent laser system or could also directly generate tunable pico- or femtosecond pulses from our main pump laser system, by exciting a travelling wave or a distributed feedback dye laser as indicated in Fig. 1. Both, the pump and the mixing radiation were focussed into a cell (length about 40 cm) containing xenon at pressures between 20 - 600 mbar. Phase matching is accomplished in a non-collinear matching geometry with matching angles of a few degrees, which can simply be adjusted by a proper choice of the focussing lenses. At typical pump energies of 5 mJ for the KrF laser radiation and 50 μJ (ps pulses) for the mixing radiation ω_{vis}, output energies of several μJ for the VUV radiation are obtained. At present the system allows the generation of subpicosecond pulses in the range of 130 nm - 350 nm /7/ with output energies sufficient for time resolved spectroscopic and kinetic applications.

Generated subpicosecond pulses at 193 nm and 157 nm have been further amplified in commercial ArF and F_2 amplifier modules. For the first time short pulse amplification data such as the small signal gain coefficient g_0 and the saturation energy density E_{sat} could be determined. For ArF values of $g_0 = 0.08$ cm^{-1} and $E_{sat} = 1.9$ mJ/cm^2 and for F_2 values of $g_0 = 0.05$ cm^{-1} and $E_{sat} = 180$ µJ/cm^2 have been obtained /8, 9/. For F_2 a maximum energy of so far 60 µJ could be realized in a single pass amplification, while for ArF in a double pass amplification geometry an energy of 10 mJ with a peak power of more than 10 GW could be achieved.

The short pulse radiation from the main KrF (248 nm) pump laser and the amplified pulses at 193 nm (ArF) have been used for harmonic generation in the beginning XUV spectral range. By focussing the radiation into noble gas jets (He, Ar), generated by a pulsed nozzle, so far the 9th harmonic of KrF in He at 28 nm and the 5th harmonic of ArF in Ar at 38 nm have been obtained.

References

/1/ J. J. Macklin, J. D. Kmetec, C. L. Gordon III; Phys. Rev. Lett. 70, 766 (1993)

/2/ A. L'Huillier, P. Balcou; Phys. Rev. Lett. 70, 774 (1993)

/3/ N. Sarukura, K. Hata, T. Adachi, R. Nodomi, M. Watanabe, S. Watanabe; Phys. Rev. A 43, 1669 (1991)

/4/ A. Huillier, P. Balcou, S. Candel, K. J. Schafer, K. C. Kulander; Phys. Rev. A 46, 2778 (1992)

/5/ L. A. Lompré, A. Huillier, M. Ferray, P. Monot, G. Mainfray, C. Manus; J. Opt. Soc. Am. B 7, 754 (1990)

/6/ A. Tünnermann, K. Mossavi, B. Wellegehausen; Phys. Rev. A 46, 2707 (1992)

/7/ A. Tünnermann, C. Momma, K. Mossavi, C. Windolph, B. Wellegehausen; IEEE QE-29 (1993)

/8/ C. Momma, H. Eichmann, H. Jacobs, A. Tünnermann, H. Welling, B. Wellegehausen; Opt. Lett. 18, 516 (1993)

/9/ C. Momma, H. Eichmann, A. Tünnermann, P. Simon, G. Marowsky, B. Wellegehausen; Opt. Lett. 18, 1180 (1993)

Parametric Picosecond Laser System with very Broad Tunability for Nonlinear Optical Interface Spectroscopy

H.-J. Krause, U. Reichel, and W. Daum

Institut für Grenzflächenforschung und Vakuumphysik, Forschungszentrum Jülich, 52425 Jülich, D

Abstract:

Nonlinear optical spectroscopy by second-harmonic generation (SHG) or sum-frequency generation (SFG) is a powerful technique for probing interfaces. To be able to excite electronic transitions as well as vibrational modes at interfaces, a high-power laser with broad tunability from the visible to the medium infrared is required. We report on a picosecond laser system developed for this purpose. An optical parametric generator and amplifier with two LiB_3O_5 crystals and a diffraction grating is pumped by the third harmonic of a 25 ps Nd:YAG laser. Spectrally narrow high power pulses in the visible and near-infrared range with continous tunability from 0.41 µm to 2.4 µm are generated. A subsequent optical parametric amplifier with two $AgGaS_2$ crystals, pumped by the fundamental of the Nd:YAG laser, extends the spectral range to 12.9 µm. The duration of the tunable pulses is between 14 and 19 ps, the bandwidth ranges from 3 cm^{-1} in the infrared to 30 cm^{-1} in the violet. Experimental results on clean, oxidized and hydrogen-terminated silicon (100) and (111) surfaces demonstrate the versatility of this laser system. The SHG and SFG spectra presented are discussed.

1. Introduction

In addition to established probes of surfaces and interfaces based on the scattering of electrons, atoms, and ions, optical methods are of interest because they allow the investigation of any interface accessible by light. A major drawback of linear optical spectroscopies is their lack of interface sensitivity requiring the elimination of a large signal from the bulk in order to detect relatively small interface signals. Nonlinear optical techniques such as *second-harmonic generation* (SHG) and *sum-frequency generation* (SFG), however, possess inherent surface sensitivity[1,2]. Provided that the adjacent media are centrosymmetric (all gases, liquids, amorphous solids, and many crystals), the only electric-dipole allowed contribution to the second-harmonic or the sum-frequency signal is generated at the interface which breaks the symmetry.

Most of the previous work has been concentrated on SHG measurements with fixed excitation wavelength. Structural symmetries of interfaces and adsorption of molecules on surfaces have been examined[3,4]. More information on interface properties can be obtained by varying the excitation wavelength. Transitions between electronic states of the interface can be identified from SHG and SFG spectra. So far, only very few spectroscopic nonlinear-optical investigations on semiconductor interfaces have been reported[5,6].

One reason that this powerful technique has not become more widespread is the lack of suitable high-power laser systems with large tunability. The wavelength of dye lasers can be varied easily only in a small range, and dye changes are tedious. Tunable solid state lasers like the Ti-sapphire laser are promising alternatives, but even with frequency mixing techniques wavelength gaps remain. This disadvantage has been overcome by optical parametric systems[7,8]. They combine a very large tunability with easy handling and comparatively weak dependence of the output energy on frequency. The new nonlinear crystals beta-barium borate (β-BaB_2O_4)[9] and lithium borate (LiB_3O_5)[10] have allowed the design of high-power laser systems in the visible and near-infrared range without damage threshold problems[11,12,13,14]. Thus tunable parametric systems are ideal sources for spectroscopic applications.

2. Tunable parametric laser system

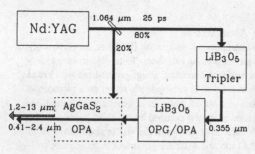

Fig. 1. Schematic diagram of the tunable laser system

The experimental setup of our system[14] is schematically sketched in Fig. 1. An actively and passively mode-locked Nd:YAG laser pumps the system with 25 ps pulses at a repetition rate of 10 Hz. After propagation the beam has a pure Gaussian spatial profile, and a pulse energy of 22 mJ. 4 mJ are split off to pump the $AgGaS_2$ crystals, the rest is used for third harmonic generation.

The third harmonic of the Nd:YAG laser frequency is generated efficiently in two uncoated LiB_3O_5 crystals, a type-I doubler and a type-II mixer. The external 1.064 µm-to-0.355 µm conversion efficiency is about 45%, corresponding to a IR-to-UV photon conversion of 56%.

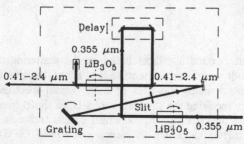

Fig. 2. Schematic diagram of the LiB_3O_5 optical parametric generator and amplifier (OPG/OPA)

The setup of the LiB_3O_5 parametric generator and amplifier is shown in Fig. 2. It consists of two angle-tuned LiB_3O_5 crystals cut for type-I phase matching, each 15 mm long and uncoated. The 0.355 µm pump beam enters the first crystal with an intensity of 5-6 GW/cm². At this intensity, parametric superfluorescence is generated in a broadband, divergent beam. A standard dielectric 0.355 µm mirror separates the virtually undepleted pump beam from the parametrically generated components which are used for seeding the subsequent parametric amplification in the second crystal. Either the visible signal or the infrared idler component is spectrally narrowed by a diffraction grating in Littrow geometry with high dispersion and high reflectivity. The pump beam merges with the seed beam after traveling an adjustable delay line. In the second LiB_3O_5 crystal the seed beam is parametrically amplified. When seeded with the narrow-band signal, the idler is efficiently generated also with narrow bandwidth, and vice versa. Simple rotation of the crystals and the grating allows a continous tuning from 0.41 to 2.4 µm.

The tuning range is extended to the medium IR region by subsequent optical parametric amplification in of two AgGaS$_2$ crystals (Fig. 1). The crystals are 18 mm long, antireflection-coated for 1.064 µm, and cut for type-I phase matching. The near infrared output from the LiB$_3$O$_5$ parametric amplifier between 1.16 and 2.13 µm is parametrically amplified in the AgGaS$_2$ crystals with 4 mJ, 0.7 GW/cm^2 pump pulses at 1.064 µm from the Nd:YAG laser. Of course, the amplification of the injected infrared pulses is accompanied with the generation of a corresponding idler component in the wavelength range from 2.13 to 12.9 µm.

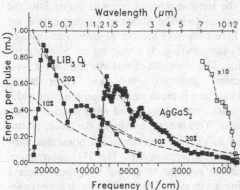

Fig. 3. Output energy per pulse as a function of frequency

The system yields high power pulses continuously tunable in the wavelength range from 0.41 to 12.9 µm. Fig. 3 shows the output energy per pulse. The conversion efficiency of the system lies between 10 and 20% for almost the whole spectral range. The pulses from the LiB$_3$O$_5$ parametric amplifier have a spectral width between 7 and 30 cm^{-1}, the medium infrared pulses from the AgGaS$_2$ amplifier between 3 and 8 cm^{-1}. The pulse durations are 14 and 19 ps, respectively.

3. SHG and SFG spectroscopy at silicon interfaces

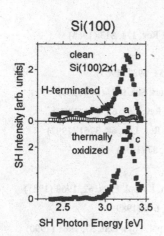

Fig. 4. SHG spectra of H-terminated, clean, and oxidized Si(100) samples

To give an insight into the variety of possible spectroscopic applications feasible with this laser system, we present some results of experiments on silicon interfaces.

The intensity of the second-harmonic photons generated at an interface can be resonantly enhanced if the fundamental or the second harmonic frequency is tuned into an electronic transition of the interface. In the following, we give an example of such a resonant enhancement. Fig. 4 a displays the SHG Spectrum of a hydrogen-covered, bulk-like terminated Si(100) surface prepared by wet-chemical treatment and introduced into ultrahigh vacuum. Then the hydrogen was desorbed by annealing the surface and a clean, 2×1-reconstructed surface was obtained. The SHG spectrum of the clean Si surface exhibits a strong resonance band at 3.3 eV SH photon energy (Fig. 4, b). Likewise, a band at 3.3 eV is also observed for oxidized samples (Fig. 4, c). Its energetic position is close to the well-known E$_1$ peak at 3.4 eV which is caused by direct interband transitions in bulk silicon. The band in our spectra, however, is caused by interband transitions localized at the interface because it is absent for the hydrogen-terminated surface. The red shift in energy indicates strained and weakened Si-Si bonds within a few silicon monolayers at the interface[15].

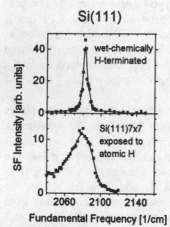

Si(111)

Fig. 5.   SFG spectra of H-covered Si(111) samples.

A big advantage of our laser system is that vibrational SFG spectroscopy is feasible without any changes of the experimental setup. With this technique, surface vibrations can be detected by resonant enhancement of the sum-frequency signal when the tunable infrared beam is tuned into the frequency of a surface vibration. As an example, we show here two SFG spectra of hydrogen-covered Si(111) samples (Fig. 5) displaying the Si-H stretch vibration. The linewidth of this vibration for a wet-chemically prepared, bulk-like H-terminated Si(111) surface is much smaller than the bandwidth of the laser system because all adsorption sites are equivalent. This is not the case for the clean, reconstructed Si(111)7×7 surface which possesses various different adsorption sites. The SFG spectrum of this surface exposed to atomic hydrogen displays therefore a vibrational band which is inhomogeneously broadened.

4. References

1.   N. Bloembergen, R. K. Chang, S. S. Jha, and C. H. Lee, Phys. Rev. **174**, 813 (1968)

1.   Y. R. Shen, Nature **337**, 519 (1989)

3.   H. W. K. Tom, T. F. Heinz, and Y. R. Shen, Phys. Rev. Lett. **51**, 1983 (1983)

4.   H. W. K. Tom, X. D. Zhu, Y. R. Shen, and G. A. Somorjai, Surf. Sci. **167**, 167 (1986)

5.   T. F. Heinz, F. J. Himpsel, E. Palange, and E. Burstein, Phys. Rev. Lett. **63**, 644 (1989)

6.   M. S. Yeganeh, J. Qi, A. G. Yodh, and M. C. Tamargo, Phys. Rev. Lett. **68**, 3761 (1992)

7.   A. Laubereau, L. Greiter, and W. Kaiser, Appl. Phys. Lett. **25**, 87 (1974)

8.   A. H. Kung, Appl. Phys. Lett. **25**, 653 (1974)

9.   D. Eimerl, L. Davies, S. Velsko, E. K. Graham, and A. Zalkin, J. Appl. Phys. **62**, 1968 (1987)

10. Ch. Chen, Y. Wu, A. Jiang, and G. You, J. Opt. Soc. Am. **B6**, 616 (1989)

11. J. Y. Huang, J. Y. Zhang, Y. R. Shen, C. Chen, and B. Wu, Appl. Phys. Lett. **57**, 1961 (1990)

12. A. Fix, T. Schröder, and R. Wallenstein, Laser und Optoelektronik **23**, 106 (1991)

13. H.-J. Krause and W. Daum, Appl. Phys. Lett. **60**, 2180 (1992)

14. H.-J. Krause and W. Daum, Appl. Phys. **B56**, 8 (1993)

15. W. Daum, H.-J. Krause, U. Reichel, and H. Ibach, to be published

Ultrashort Laser Pulses in Surface Mass Spectrometry

M. Schütze, C. Trappe, M. Raff, H. Kurz

Institut für Halbleitertechnik II, RWTH Aachen

D–52056 Aachen

Introduction

Pulsed laser beams have been applied to mass spectrometry since many years [1]. Laser pulses as excitation for the emission of surface particles provide some decisive advantages compared to techniques using particle beams: 1. There is no charge transfer due to the excitation conditions. 2. A wide range of input parameters of the laser beam can be varied, e.g. wavelength, input power– and energy–density. 3. It is a "clean" technique, since no particles are deposited on the surface.

Using ultrashort (picosecond or even femtosecond) laser pulses for the analysis of surfaces may provide two main advantages compared to the commonly used ns–pulses: The irradiation of the surface with ultrashort pulses can deposit comparably little energy at high power densities on the solid. In consequence, the emission process may not longer be dominated by thermal processes. Instead, electronically induced emission may play a role. On the other hand, ultrashort laser pulses can provide the feasibility of non–resonant multi–photon ionization (NR–MPI). In consequence, the detection of emitted neutral particles can be used for quantification in mass spectrometry. The present contribution briefly addresses these two points and shows possible applications of ultrashort laser pulses in surface mass spectrometry.

Experimental

Figure 1 shows the set–up of the Picosecond Laser–Desorption Mass–Spectrometer (ps–LDMS) which was used in our experiments. An active/passive mode–locked Nd:YAG laser (Quantel, France) providing ps–pulses was used for both desorption of surface particles and post–ionization of emitted neutrals. Due to frequency doubling and quadrupling IR, visible and UV radiation could be used. UV radiation was always applied to post–ionization, because of the high photon energy of 4.67 eV. A Time–of–Flight Mass–Spectrometer (TOF–MS) performed the detection of the ions. More details of our system are described elsewhere [2].

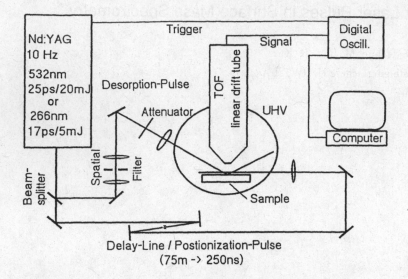

Figure 1: Set–up of the ps–LDMS system

To determine the desorption rates we measured the depths of craters formed by a sufficient number of laser pulses. Even several thousand pulses formed craters of extremely shallow sizes. Therefore, we used a laser–interference microscope to obtain the depth values. To determine ionization yields, we calculated the amount of detected ions by integration of the mass–peaks.

Desorption

Figure 2 shows crater depths each time formed by 200.000 single laser pulses on InP for green excitation ($\lambda = 532$ nm). The solid line shows calculations according to a theoretical model proposed by Strekalov [3]. Its main ansatz is that the desorption process is stimulated by recombining electron-hole pairs within the semiconductor.

Under appropriate excitation conditions, the ultrashort laser pulse can produce a dense electron–hole plasma while not exceeding a surface temperature which would lead to melting. The recombination energy (e.g. the gap energy) of electron–hole pairs can be transferred directly to single surface atoms. Their binding energy therefore is lowered by the amount of the gap energy. This leads to a high flux of desorbates compared to thermal desorption at temperatures below the melting threshold. Hence, in contrast to the application of longer pulse durations, the emission process is mainly due to electronical processes instead of thermal effects. Nevertheless, the particle distribution of the desorbates above the sample is of the Maxwell–Boltzmann type. The temperature drawn from this distribution is the same as the surface temperature. Our experiments showed good agreement with the theory for different materials like InP and GaAs as well as for different excitation conditions ($\lambda = 532$ nm and $\lambda = 266$ nm).

An important fact of the use of ultrashort laser pulses is the extremely shallow form of the craters: Even 200.000 pulses produce craters with a depth in the nm-regime. Calculating the amount of emitted particles per single laser pulse we found a desorption rate of well below one monolayer being detracted per pulse. Therefore, the information depth of the system is one monolayer because there occurs no layer alteration induced by melting.

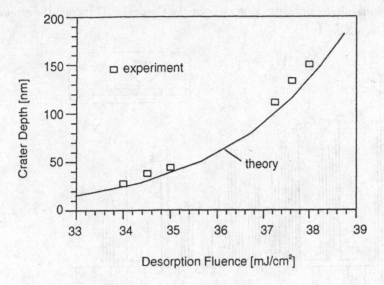

Figure 2: Crater depths on InP formed by 200.000 single laser pulses at $\lambda = 532$ nm. The solid line is a theoretical calculation (see text).

Ionization

As known from other sputtering techniques (e.g. SIMS), the dominant part of emitted particles consists of neutrals. This is true also in ps-LDMS. Therefore, post-ionization has been established to the desorption system in order to improve the sensitivity. Due to the use of ultrashort laser pulses, power densities of about 10^{13} W/cm^2 were reached within the focus of the post-ionization pulse. This power density leads to saturation of non-resonant multi-photon ionization processes. We have proven the saturation for two-photon processes (e.g. Ga and In) as well as for three-photon processes (e.g. P and As). Therefore, we are able to determine the amount of desorbed particles from the detected ones, since the particle distribution above the sample is known. In consequence, quantification in the ps-LDMS system seems to be feasible.

The useful yield (i.e. the ratio between detected to emitted particles) of our system has proven to be about 10^{-3}. Therefore, despite the low amount of desorbed particles, mass spectrometric analysis of surfaces can be performed with ps-LDMS. From the values given above we are able to estimate the sensitivity of the system. With material consumptions of well below a single monolayer and a desorption beam diameter of 50 μm the detection limit should be about 10^{10} atoms/cm^2.

In order to test this prediction, we applied the ps–LDMS system to a Copper contaminated Silicon surface. The contamination was applied to the surface by a solution. The resulting Cu–concentration was 2×10^{10} atoms/cm^2, which was determined by total reflection X–ray fluorescence spectrometry (TXRF). The spectrum in Fig. 3 clearly shows the mass peaks of Copper. The logarithmic plot shows two orders of magnitude signal–to–noise ratio for the Cu–peaks. Therefore, the estimate for the detection limit is confirmed.

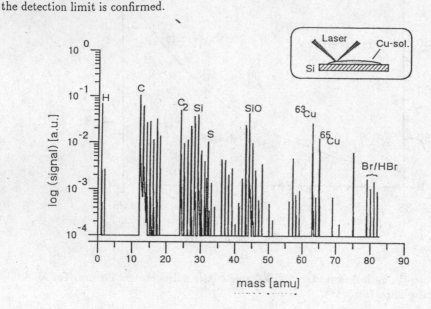

Figure 3: Mass spectrum of a copper contaminated Si–surface. The Cu–concentration on the surface had been calibrated by TXRF and was found to be 2×10^{12} atoms/cm^2.

Conclusion

In conclusion, we have presented an all–optical mass spectrometer capable to analyze the first few layers of solids. Due to the use of ultrashort laser pulses, particle emission occurs without melting of the surface. The system has high sensitivity because of saturation in the non–resonant post–ionization process. Therefore, standard free quantification of LDMS mass spectra is feasible.

We are grateful to Dr. R. Girisch from Philips, Nijmegen for prividing the Si-sample and the TXRF measurement.

References

[1] F. Hillenkamp, E. Unsöld, R. Kaufmann, and R. Nitsche, Appl. Phys. **8**, 341 (1975).

[2] C. Trappe et al., Fres. J. Anal. Chem. **346**, 368 (1993).

[3] V. Strekalov, Sov. Phys. Semicond. **20**, 1218 (1986).

Second Harmonic Generation on Chemically Treated Surfaces of Vicinal Si(111) Wafers

U. Emmerichs, C. Meyer, K. Leo, H. Kurz,

Institute of Semiconductor Electronics II, Rheinisch-Westfälische Technische Hochschule,

D-52056 Aachen, Germany,

C.H. Bjorkman, C.E. Shearon Jr., Y. Ma, T. Yasuda, G. Lucovsky,

Departments of Physics, and Material Science and Engineering, North Carolina State University,

Raleigh, NC 27695, USA

INTRODUCTION

Surface Second Harmonic Generation (SSHG) is a sensitive probe for surface studies in centrosymmetric materials [1, 2]. Recent investigations of the Si/SiO_2 system have shown that optical harmonic generation can be used to study this technologically important material [3, 4, 5, 6]. The electric field generated at the second harmonic frequency $E(2\omega)$ is due to a nonlinear polarization $P(2\omega)$ that is determined by the susceptibility tensor $\chi^{(2)}$:

$$E_l(2\omega) \propto P_l(2\omega) \approx \chi_{lmn}^{(2)} E_m(\omega) E_n(\omega) \tag{1}$$

For silicon, a centrosymmetric crystal, $\chi^{(2)}$ in the electric dipole approximation is only non-zero at the surface where the symmetry is broken. However, additional higher order processes from the bulk can also contribute to the signal. Using surface second harmonic generation (SSHG) at 1053nm, we study the influence of off–axis orientation and surface structure of silicon (111) surfaces. We examine wafers cut at an off–axis angle θ between $0°$ and $5°$ in the $[11\bar{2}]$ direction. The surface structure is varied by thermal oxidation at $850°C$, annealing and removing of the oxide in a HF solution resulting in a H–terminated surface. We find that these chemical modifications of the interface result in dramatic changes of the nonlinear optical response. The results are discussed in comparison with a microscopic model of the oxidized misoriented surface.

EXPERIMENTAL PROCEDURE

Experiments are performed on Si(111) samples which are misoriented with a tilt angle between $0°$ and $5°$ in $[11\bar{2}]$ direction. On these wafers thermal oxides (thickness $\approx$ 60nm) are grown at $850°C$ in a dry oxygen ambient. Some of these samples are annealed by Rapid Thermal Annealing (RTA) at $1100°C$ in a dry argon atmosphere, containing sufficient O_2 to prevent reduction of the oxide layer on the Si surface. Parts of the wafers are etched back to oxide thicknesses of $\approx$ 5nm and $\approx$ 1.5nm in a HF solution. Finally, all oxide is removed and the freshly etched, H–terminated surface is investigated.

The SHG experiments are performed using the 1053nm emission of a Nd:YLF regenerative amplifier operating at 1kHz seeded by a cw–mode locked Nd:YLF oscillator. The 40ps laser pulses are filtered by a glass filter and a polarizer before striking the sample at an incident angle of 45 degrees. The spot size on the sample is 2mm. The reflected fundamental laser light and the third harmonic are suppressed by a combination of dichroic mirrors, glass filters and a polarizer. The energy per laser pulse used is kept at approximately 1mJ for crossed polarizers of laser light and second harmonic signal and 0.5mJ for equal polarizations. The generated second harmonic signal is focused into a single photon counting photo multiplier. The counting electronics are "gated" by the laser with a gate width of $\approx 1\mu s$ to suppress background signal. A schematic of the experimental set-up and the sample geometry are shown in Figs. 1 and 2, respectively.

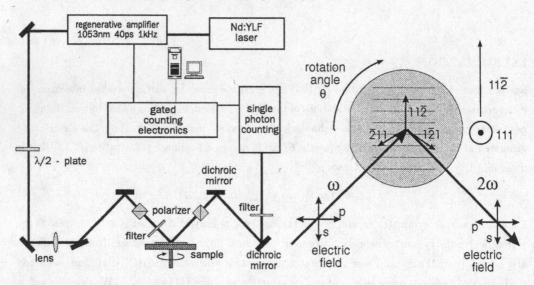

Fig. 1 *Schematic diagram of the setup*   Fig. 2 *Sample geometry*

EXPERIMENTAL RESULTS

In all figures presented in this paper, the intensity of SHG signal is plotted versus the rotation angle of the sample about its surface normal. $\phi = 0°$ means that the $[11\bar{2}]$ crystal axis lies in the plane of polarization of the incident laser light ($E \parallel [11\bar{2}]$).

The notation "ps" means that the laser is p-polarized (parallel to the plane of incidence) and the s-polarized (perpendicular to plane of incidence) of the SHG is measured. The counting rates are corrected assuming a Poisson distribution for the probability that more than one photon is generated per laser pulse [7].

$$N = -\ln(1 - n) \tag{2}$$

where n are the measured counts and N are the corrected counts.

The experimentally obtained signal intensities are fitted with the following equation:

$$I(2\omega) = \text{const} * |E(2\omega)|^2 = |\sum_{k=0}^{3} a_k e^{i\psi_k} \cos(k\phi)|^2 \qquad (3)$$

where ϕ is the angle of rotation, ω is the frequency of the incident laser light, ϕ_k are the fitted phase angles, a_k are the fitted amplitudes, and k is the index of summation. The $a_k e^{i\psi_k}$-terms give the amplitude and phase of the k-fold symmetry contribution to the measured signal. Introducing vicinal steps on the surface will change the symmetry of the surface introducing a $\cos(\psi)$ component to the measured signal. The investigations show that these different contributions react in different ways to surface treatments allowing the bonding characteristics at atomic surface steps to be studied. As already reported [3, 8, 6, 9, 10] the k=1 contribution depends on the off–axis angle θ. While the $\cos(3\phi)$ component remains nearly constant, the $\cos(\phi)$ component scales with off–axis angle θ.

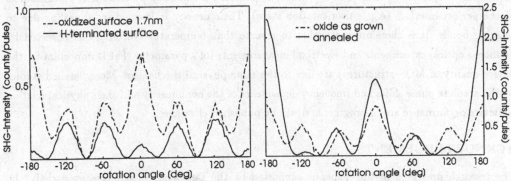

Fig. 3 Comparison between oxidized (1.7nm) and H-terminated (HF-etched) (ss-pol., off-axis angle $\theta = 3°$)

Fig. 4 Comparison between annealed and not annealed oxide films (pp-pol., off-axis angle $\theta = 5°$)

A freshly etched surface can be ideally H–terminated [11]. We observe that the nonlinear response of this surface strongly differs from that of an oxidized surface (Fig. 3). The peaks at $\phi = 0°$ or $180°$ decrease by a factor of 5 when the oxide is removed, while the height of the other peaks changes similarly to the change for an unstepped surface (not shown). This clearly demonstrates that the change of the signal for oxidized surfaces by introducing a tilt angle θ cannot simply be explained by a tilt of the silicon crystal with respect to the laser beam: parts of the k=1 contribution originate from surface/interface bonds at the steps. We tentatively assign the phase shifts to changes in the resonance energy of particular bond configurations at the steps.

Further evidence for this interpreation is given by the dependence of the nonlinear response on a thermal treatment of the wafer. When the samples are annealed, the measured intensity strongly increases for the pump polarization in the [11$\bar{2}$]-direction (Fig. 4). The phase analysis shows that this enhancement in intensity is not due to an enhancement in the electric field, but again due to a change in phase between $\cos(3\phi)$ and $\cos(\phi)$ contributions, while the corresponding amplitudes stay approximately the same. An annealing series shows that the observed shift of the phase ψ_1 of the electromagnetic field occurs around a transition temperature of appr. 900°. This is the same

852

temperature, where the density of interface traps determined by electrical capacitance–voltage (C–V) measurements on the same Si/SiO$_2$–structures decreases [12].

DISCUSSION

We observe that the generated SHG signal strongly depends on the interface structure. The difference in signal between etched Si-surfaces and oxidized surfaces clearly indicates that part of the signal originates from the bonds on the surface. The thermal treatment changes the nonlinear optical response of the oxide film. We tend to explain this behavior as a rearrangement of the Si/SiO$_2$ bonds into their thermodynamically most favorable arrangement. If we model the bond by an anharmonic oscillator, the rearrangement will shift the resonance of the oscillator and therefore shift the phase of the nonlinear optical response. Interface traps are generally assumed to be located at perturbations of the perfect interface (e.g., misorientation steps). Their density is lowered by annealing due to relaxing bonds. It is therefore reasonable to assume that temperature threshold as observed by nonlinear optical experiments and electrixal measurements (of a parameter that is important for the device quality of MOS–structures) are due to the same physical mechanism. More detailed studies on the absolute phase shifts and frequency dependence of the nonlinearity and their physical relation to device performance are in progress and will be published elsewhere.

ACKNOWLEDGEMENTS

The research presented in this paper is supported by the Deutsche Forschungsgemeinschaft, the Otto-Junker-Stiftung, the Alfried Krupp–Stiftung, the National Science Foundation and the Office of Naval research, and by a joint collaboration between the states of Nordrhein-Westfalen and North Carolina. We would like to thank Dr. W. Zulehner of Wacker Chemitronic for the supply of some of the wafers used.

[1] N. Bloembergen, R. Chang, S. Jha, and C. Lee, Phys. Rev. **174**, 813 (1968).

[2] H. Tom, T. Heinz, and Y. Shen, Phys. Rev. Lett. **51**, 1983 (1983).

[3] C. van Hasselt, M. Verheijen, and T. Rasing, Phys. Rev. B **42**, 9263 (1990).

[4] L. Kulyuk, D. Shutnov, E. Strumban, and O. Aktsipetrov, J. Opt. Soc. Am. B **8**, 1766 (1991).

[5] R. Hollering and M. Barmentlo, Opt. Comm. **88**, 141 (1992).

[6] M. Verheijen, C. van Hasselt, and T. Rasing, Surf. Sci. **251/252**, 467 (1991).

[7] T. Heinz, Ph.D. thesis, University of California, 1982.

[8] R. Hollering, D. Dijkkamp, H. Lindelauf, and P. van der Heide, J. Vac. Sci. Technol. A **8**, 3997 (1990).

[9] U. Emmerichs et al., in MRS Symp. Proc. (1992), Vol. 281.

[10] C. Bjorkman et al., J. Vac. Sci. Techn. A **11**, (1993), in press.

[11] G. Higashi, Y. Chabal, G. Trucks, and K. Raghavachari, Appl. Phys. Lett **56**, 656 (1990).

[12] C. Bjorkman et al., J. Vac. Sci. Techn. B (1993), in press.

Application of an Intracavity Laser Spectroscopy Based on Ar+ Laser for the Detection of Unstable Molecules in Liquid Phase

S.A.Mulenko, A.I.Chaus
Institute of Metal Physics AS Ukraine
252680, Kiev-142, 36 Vernadsky str., Ukraine.

An intracavity laser spectroscopy (ILS) method is one of the most high-sensitive and high-speed method in optical spectroscopy for the molecule detection in gas and liquid phases[I]. Mainly molecules in aqueous solutions have broad-band absorption spectra in visible and UV ranges. These spectra are sufficiently broader than the laser bandwidth. While placing such broad-band absorbers inside the laser cavity all laser modes are being quenched equally. And there is no mode competition as it occurs between those modes which are affected by selective losses and other modes of the laser while having been detected of narrow-band absorbers (molecules in the gas phase). So the intracavity sensitivity for the detection of molecules in aqueous solutions is not as high as for the detection of molecules in the gas phase. But however the intracavity detection of broad-band absorptions in aqueous solutions is more sensitive than a single pass detection - an ordinary spectroscopy.

In the present work experimental data on the detection of hydrogen peroxide molecules (H_2O_2) in aqueous solution with ILS method based on Ar^+ laser are presented. To detect H_2O_2 molecules in aqueous solution is very actual as many processes in living tissues are taking place in this solution. Furthermore H_2O_2 molecules are quite unstable owing to their dissociation under the action UV range of spectrum resulting to the formation of oxygen atoms[2]. The absorption coefficient of H_2O_2 molecules in aqueous solution is very low in the visible range and is very high in UV range of spectrum (at $\lambda < 3000$ Å, $K > 10$ cm^{-I}) [3]. So to detect H_2O_2 molecules in aqueous solution is suitable with ILS method based on Ar^+ laser using 4880 Å and 5145 Å lines. As absorption spectrum of H_2O_2 molecules in aqueous solution in the visible range is more broader with respect to Ar^+ laser bandwidth ($\Delta\lambda_L \approx 0.I$ Å) to place the absorption cell inside the laser cavity provides quenching all laser modes and may quench the laser generation entirely even when absorption cell is filled with distilled water. To prevent laser quenching one can use the adjacent cavity (three mirror cavity) for the intracavity measurements of broad-band absorptions.

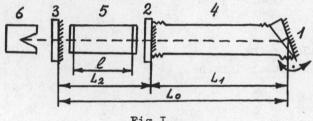

Fig.I.

The optical scheme of the experimental arrangement.

I-dispersive element formed with Littrov's prism covered by I00% reflactive layer;2-output mirror of Ar^+ laser;3-output mirror of the adjacent cavity formed by 2 and 3 mirror; 4-active medium of Ar^+ laser;5-absorption cell with H_2O_2 molecules in aqueous solution;6-laser power meter. The length of Ar^+ laser cavity (L_1) was 96 cm and the length of the adjacent cavity (L_2) was 32 cm. Argon laser was being tuned by Littrov's prism on two lines: 4880 $\overset{o}{A}$ and 5I45 $\overset{o}{A}$. The output power of Ar^+ laser with the cell inside the cavity was about 0.3 W on the line 4880 $\overset{o}{A}$ and 0.2 W on the line 5I45 $\overset{o}{A}$. Experiments were being carried on with the absorption cell as inside and as outside of the laser cavity. To eliminate of influence the laser reflection and absorption of cell windows on the measuring passed laser power there was experiment with two absorption cells with different optical lengthes. The absorption coefficients $K(\lambda_L)$ were calculated from measuring the laser power to have been passed through the cell outside the laser cavity with H_2O_2 molecules in aqueous solutions of different concentrations.

$$K(\lambda_L) = \frac{1}{l_{10}-l_2} ln \frac{P_2}{P_{10}} \qquad (I)$$

where P_2, P_{10}-laser power to have been passed through the absorption cell with 2 and I0 cm solution length.

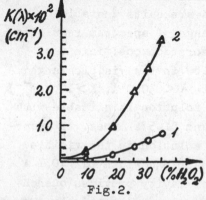

Fig.2.

The absorption coefficients of H_2O_2 molecules in aqueous solution.

I-absorption coefficients to be calculated for the line 4880 A;2-for the line 5I45 $\overset{o}{A}$. After that we can evaluate an intracavity enhancement on two Ar^+ laser lines with different output mirrors (3) of the adjacent cavity. This evaluation was made from the absorbed laser power while having been placed the absorption cell as inside and as outside of laser cavity. The absorbed laser power in solution of 8 cm length is a difference of the passed laser power through two absorption cells of 2 and I0 cm length:

$$P_{abs.} = P_{t_2} - P_{t_{10}} = P_o(I - e^{-KNl}) \qquad (2)$$

where P_{t_2}, $P_{t_{10}}$ - output laser power with absorption cells of 2 and 10 cm length inside and outside the laser cavity; P_o-output laser power without the absorption cell; N-number of photon passes inside the laser cavity; l-optical length of absorbing solution (l=8 cm). When KNl ≤ 0.1, the ratio $P_{abs.}/P_o \simeq$ KNl. The number of photon passes through the absorption cell with absorbing solution is an intracavity enhancement.

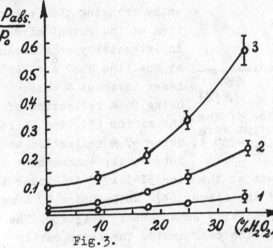

Fig.3.

The dependences of $P_{abs.}/P_o$ as a function of H_2O_2 molecules concentration o in aqueous solution for the line 4880 A.

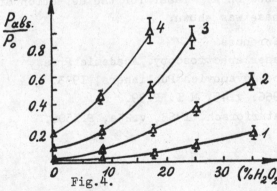

Fig.4.

The dependences of $P_{abs.}/P_o$ as a function of H_2O_2 molecules concentration o in aqueous solution for the line 5145 A.

1-reflection of the output mirror (3) is 0 - outside the laser cavity measurement; 2-reflection of the output mirror (3) is 70 %;3-reflection of the output mirror (3) is 97 %.
It is possible to find out the number of photon passes while placing the absorption cell with the solution inside the adjacent cavity - an intracavity enhancement from data to be presented in Fig.3 and Fig.4.

1-reflection of the output mirror (3) is 0 - outside the laser cavity measurement; 2-reflection of the output mirror (3) is 70 %;3-reflection of the output mirror (3) is 80 %;4-reflection of the output mirror (3) is 99 %.
The more reflection of the output mirror (3) we have the more of photon passes are in the laser cavity. The number of photon passes is the same as for distilled water and as for H_2O_2 molecules aqueous solution of dif-

ferent concentrations. Generaly an intracavity enhancement in this ex-

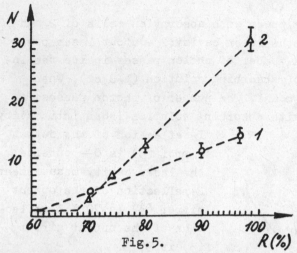

Fig.5.

The dependences of N as a function of the reflection coefficient of the output mirror (3) for two laser lines: $I - \lambda_L = 4880$ Å, $2 - \lambda_L = 5145$ Å.

periment is equal to the number of photon passes through the absorption cell. While using two Ar^+ laser lines it is possible to detect H_2O_2 molecules in aqueous solution in the broad range of concentration by changing the reflaction of the output mirror. An intracavity enhancement at the line 4880 Å of Ar^+ laser is about 4 while using 70 % reflection of the mirror (3). And while using 97 % reflection an intracavity enhancement is (13±1). An intracavity enhancement at the line 5145 Å of Ar^+ laser is 3 while using 70 % reflection of the mirror (3). While using 99 % reflection of the mirror (3) an intracavity enhancement is (30±3). The application of more longer absorption cell inside the laser cavity gives the possibility to increase the sensitivity of an intracavity method. In this paper the perspectivity of the application of an intracavity laser spectroscopy based on Ar^+ laser for the detection of unstable molecules in liquid phase was shown.

References

I.Klinger D.S.Ultrasensitive laser spectroscopy. Academic Press, A Subsidiary of Harcourt Brace Jovanovich Publishers, 1983.

2.Greiner N.R. J.Chem.Phys., 1966, v.45, N I, P.99.

3.Schürgers M., Welge K.H. Z.Naturforsch, 1968, v.23a, P.1508.

Deposition of Diamond-Like-Carbon Films by Laser PVD Technique Using Nanosecond and Femtosecond Excimer Lasers

K. Mann and F. Müller
Laser-Laboratorium Göttingen e.V.
Im Hassel 21, W-3400 Göttingen

1. INTRODUCTION

The inherent simplicity of laser induced physical vapor deposition (LPVD) as well as its large success in growth of high T_c superconducting films[1] have led to much interest in this process, although the fundamental mechanisms are still far beyond a comprehensive understanding. In many cases target ablation is provided by an excimer laser which delivers ultraviolet laser pulses of nominally 15-30ns duration. Consequently, much of what is currently understood concerning ablation and deposition mechanisms is based on this tens of nanosecond timescale. However, advances in short pulse high energy KrF excimer amplifiers have provided a means to study ablation and deposition processes on much shorter ($> 10^5$) timescales. In this work we have investigated the influence of the pulse duration on the plasma plume formation and deposition characteristics of laser deposited amorphous diamond-like-carbon (DLC) films[2-3], using 30ns as well as 500fs KrF excimer laser pulses.

2. EXPERIMENTAL

The experimental arrangement used for laser induced deposition of DLC films has been described in detail elsewhere[5,9]. Graphite targets are ablated with excimer laser pulses at 248nm and either 30ns (Lambda Physik EMG 202), or ~500fs duration. The subpicosecond 248nm photons are obtained from a system[4] developed at Laser-Laboratorium Göttingen. A commercially available version has recently been introduced (Lambda Physik LPDfs).

3. PLASMA ANALYSIS

Fig. 1 shows the results obtained from time of flight (TOF) spectroscopy during laser ablation of a graphite target as used for DLC deposition. The spectra results from positive carbon ions detected by an electron multiplier placed at the end of a field free drift tube. For 30ns laser pulses and an energy density of $20J/cm^2$ two distinct regions are observed: a broad continuum at high flight times ($> 250\mu s$), which is due to large carbon cluster formation, as well as sharp peaks at early flight times ($< 100\mu s$), which can be attributed to small molecular ions. The earliest arriving species have been identified as C_3^+ and C_4^+ for 30ns laser pulses at energy densities above $9J/cm^2$.[5] At lower energy densities cluster formation is reduced. Particle energies fall off linearly as the energy density is decreased. Kinetic energy values ranging from 220eV at $23J/cm^2$ to $< 5eV$ at $0.5J/cm^2$ are observed.[7]

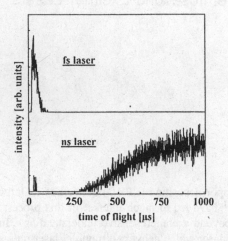

Fig. 1: TOF spectrum resulting from ablation of graphite with 500fs (top) and 30ns (bottom) KrF laser pulses.

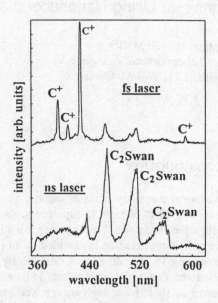

Fig. 2: Plasma emission spectrum of graphite obtained with 500fs and 30ns pulses

The results obtained for target ablation with 500fs laser pulses are strinkingly different as seen from the upper part of Fig. 1. Although the spectrum has been measured at a high energy density of $30J/cm^2$, no evidence of large cluster formation is observed in contrast to the nanosecond results. All of the ablated ions arrive at the detector in under $100\mu s$, indicating high kinetic energies. Assuming C^+ as the major ablation product (see below), kinetic energies of about 2.5keV can be estimated.

In addition to the TOF measurements, the emission from the ablation plume was characterized by optical spectroscopy using an OMA spectrometer (EG&G OMAII). The OMA results obtained over the spectral region from 360nm to 620nm are presented in Fig. 2. The ablation plume induced by 30ns laser pulses and an energy density of $5J/cm^2$ ($\sim 1.7 \times 10^8 W/cm^2$) shows strong emission characteristic of the C_2 Swan band system, in agreement with data of Chen et al.[6] The emission peaks lie on top of a very broad background with little structure, which can be attributed to larger molecules and cluster formation as seen in the TOF measurements.

The emission from the femtosecond laser plasma (cf. Fig. 2 below) shows strong deviations from the ns results. The spectra measured at power density of $1.2 \times 10^{13} W/cm^2$ is dominated by C^+ ion emission near 425nm and 385nm, justifying the assumption made for the evaluation of the corresponding TOF spectra. A smaller contribution due to C_2 Swan emission is also observed. No evidence is found for the existence of neutral carbon in the ablation plume over the region from 260nm to 650nm.

4. FILM CHARACTERIZATION

The structural nature of the 30ns carbon films has been studied previously, utilizing uv-vis-nir transmission, Raman spectroscopy and x-ray diffraction.[5] It was shown that films deposited with energy densities in excess of $8J/cm^2$, corresponding to kinetic particle energies of >80-100eV (cf. Sect. 3), had an amorphous structure with a very small crystalline component. These films exhibit physical properties characteristic for diamond-like carbon (DLC), indicating a considerable degree of sp^3 type bonding. Optical band gap values of 0.2 to 1.5eV depending on laser energy density have been determined. A similar analysis is applied to films prepared by 500fs LPVD. These films also display an amorphous structure and exhibit DLC properties as well. However, they are improved over the 30ns films with respect to adhesion and surface smoothness. In contrast to the ns films, essentially <u>no particulate formation</u> can be observed[9], which is in accordance with the absence of large clusters in the corresponding TOF spectra.

The influence of the laser energy density on DLC film thickness, as measured by a mechanical profilometer, is compiled in Fig. 4 for ns and fs laser deposition on quartz substrates. The data points have been taken at the zero angle peak of the ablation plume and normalized for equal spot size on the target and constant number of pulses (10000). In case of the fs laser the deposition rate increases drastically above a certain ablation threshold and rapidly reaches a saturation value. Due to the much higher pulse power density, this <u>ablation threshold is nearly 2 orders of magnitude smaller</u> than in the case of the ns laser, for which no saturation of the deposition rate can be observed in the investigated energy density range. Küper and Stuke[8] have obtained similar results for 248nm ablation of PMMA.

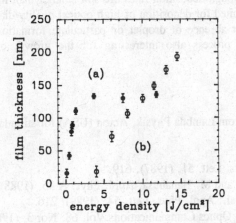

Fig. 3: Normalized DLC film thickness on quartz substrates as a function of energy density for 10000 KrF laser pulses of 500fs (a) and 30ns (b).

5. DISCUSSION

We have studied laser ablation of graphite and subsequent film deposition using 248nm laser pulses with temporal durations of 30ns and ~500fs. In both cases DLC films could be grown, although ns and fs ablation are each dominated by different interaction mechanisms:

i) Due to the extremely short pulse of the fs laser, the plasma plume formation is effectively decoupled from the energy transfer to the target, i.e., other than for the ns laser, there is virtually no interaction of the pulse with the plasma cloud.

ii) The investigation of the ablation products indicates much higher kinetic energies and a higher degree of ionization as well as atomization for the fs pulses; especially, there are no large particle clusters present in the fs plasma.

iii) From the morphology of the ablation targets a strongly reduced thermal component of the ablation can be derived.

The two latter observations can be explained by multiphoton effects, which surely play an important role at pulse power densities in the range of 10^{13}W/cm^2. Obviously, the amount of energy transfered to a single ejected target particle is much higher for the 500fs pulse. This may also be the reason for the observed saturation behaviour in Fig. 3: Since the fs pulse is effectively absorbed in a thinner surface layer of the target than for the ns laser, target ablation is limited by this reduced absorption depth, and, in contrast to the ns laser, an increase in energy density does not lead to a further removal of material. The excess pulse energy will either lead to a further excitation and increase of kinetic energy of the emitted particles, or, which is also conceivable, reflected by the dense and hot electron gas in the interaction zone[10]. Clarifying experiments in this direction are in progress.

In summary we can say that, although deposition rates are still smaller than for ns lasers, 500fs excimer laser pulses seem well suited for deposition of high optical quality diamond-like carbon films. Furthermore, the apparent absence of droplet or particulate formation makes the sub-picosecond excimer laser PVD process also interesting for the growth of other kinds of functional coatings.

ACKNOWLEDGEMENT
We like to thank J.S. Bernstein from Lambda Physik, Acton (USA) for stimulating discussions.

REFERENCES
1. D. Dijkkamp et al.; Appl. Phys. Lett. 51, (1987), 619.
2. T. Sato, S. Furuno, S. Iguchi, and M. Hanabusa; Appl. Phys. A 45, (1988), 355.
3. C. B. Collins, F. Davanloo et al.; Appl. Phys. Lett. 54, (1989), 216.
4. S. Szatmári and F. P. Schäfer; Optics Communications Vol. 68, No. 3, (1988), 196.
5. F. Müller and K. Mann; Diamond and related Materials 2 (1992), 233.
6. X. Chen, J. Mazumder, and A. Purohit; Appl. Phys. A52, (1991), 328.
7. R. C. Dye et al.; Proc. of OPTCON'92, Boston, Nov. 1992 (to be published).
8. S. Küper and M. Stuke; Appl. Phys. B 44, (1987), 199.
9. F. Müller, K. Mann, P. Simon, J. S. Bernstein and G. J. Zaal; in: Proc. of OE LASE '93, Los Angeles, Jan. 1993, SPIE Vol. 1858 (to be published).
10. M. M. Murnane et al.; Phys. Rev. Lett. 62 (2), (1989), 155.

Ein Modell für die Optimierung der PLD

G. Granse, H. Mai, B. Schultrich, S. Völlmar

Fraunhofer-Einrichtung für Werkstoffphysik und Schichttechnologie (IWS)

Helmholtzstraße 20, 01171 Dresden

1. Zielstellung

In den letzten Jahren hat sich mit der Pulse Laser Deposition (PLD) eine neue Methode der Dünnschichtabscheidung erfolgreich eingeführt, die sich in Ergänzung anderer Abscheideverfahren durch hohe Energien und Dichten des kurzzeitigen Teilchenstromes auszeichnet und sowohl auf Reinstelemente als auch auf stöchiometrisch exakt einzuhaltende Verbindungen nahezu unbeschränkt angewendet werden kann /POMPE/. Der Energieeintrag in das Targetmaterial durch den Laserpuls löst je nach Impulsdauer Prozesse von einem schnellen Verdampfen bis zum explosionsartigen Ablösen einer Oberflächenschicht aus. Durch die Wechselwirkung der Strahlung mit dem Festkörper und der sich ausbreitenden Materialwolke wird in Abhängigkeit von der Wellenlänge ein mehr oder weniger ionisiertes Plasma mit Teilchenenergien von typischerweise 0.5 eV bis 1 keV gebildet. Die letztlich die Schichtqualität mitbestimmenden Parameter der lateralen und zeitlichen Verteilung der Teilchendichte und Teilchenenergie werden durch die Wechselwirkung der Laserstrahlung festgelegt. Für diesen Prozeßbereich wird ein anwendungsorientiertes Modell als erster Schritt einer Gesamtmodellierung der PLD vorgestellt.

2. Phasen der Modellierung des PLD Prozesses

Die Laserenergie wird im Festkörper an der Oberfläche bzw. in einem oberflächennahen Gebiet absorbiert und führt zum Aufheizen und Verdampfen des Festkörpers. Die Dynamik der Phasenübergänge (fest - flüssig - gasförmig) wird durch Wärmeleitungsvorgänge bestimmt.

Beim Überschreiten einer kritischen Leistungsdichte wird der sich rasch ausbildende Neutralteilchendampf ionisiert.

Im relevanten Temperaturbereich kann das entstehende Plasma als quasineutrale Zweikomponentenflüssigkeit, bestehend aus Elektronen und Ionen, beschrieben werden. Die bei der PLD typischen Laserleistungsdichten, Pulscharakteristiken und Wellenlängen bedingen die Absorption der Laserenergie im Plasma hauptsächlich durch inverse Bremsstrahlung.

Die Vorgänge sind solange in einem eindimensionalen Modell beschreibbar, wie die charakteristischen Längen kleiner als der Durchmesser des Laserspots bleiben.

3. Wärmeleitung — Aufheizen und Verdampfen des Festkörpers

Das eindimensionale Wärmeleitungsproblem mit frei beweglichen Phasengrenzen wird mit dem Programmpaket LASTEC-2 /MASHUKIN/ gelöst. Dabei sind temperaturabhängige Materialeigenschaften berücksichtigt. Für den Nichtgleichgewichtsvorgang Verdampfen wird die Näherung der Knudsenschicht zur Bestimmung von T_v und ρ_v angewendet.

Ergebnisse der Wärmeleitungsrechnungen für Nickel bei Einwirkung eines gaussförmigen Laserpulses (Leistungsdichte 10^9 Wcm^{-2}) und Annahme von Oberflächenabsorption (Reflexionsvermögen R_{solid}=0.8, R_{liquid}=0.2) sind zum Beispiel die Dynamik der Phasengrenzen (Bild 1), sowie der zeitlicher Verlauf der Oberflächentemperatur (Bild 2) .

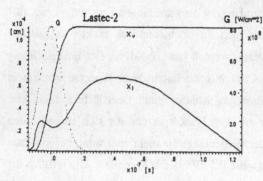

Bild 1: Absorbierte Laserleistungsdichte G
Dicke der Schmelzzone X_l und
abgetragene Materialdicke X_v

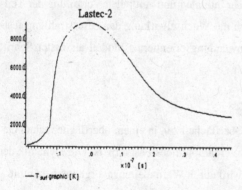

Bild 2: Oberflächentemperatur im Zeitverlauf

4. Plasmamodell und Wechselwirkung Laser - Plasma

Elektronen und Ionen bilden eine quasineutrale Flüssigkeit und befinden sich im lokalen thermodynamischen Gleichgewicht (LTE). Das die Hydrodynamik beschreibende Gleichungssystem beinhaltet die Teilchen-, die Impuls- und die Energiebilanz, wobei die Zustandsgleichung für das ideale Gas verwendet wird. Die Viskositäten und Wärmeleitungsbeiträge werden proportional zum negativen Gradienten von Geschwindigkeit bzw. Temperatur angenommen (klassisches stoßdominiertes Plasma).

Zur Beschreibung der Elektronen-Ionen Wechselwirkung wird der Quellterm der Boltzmanngleichung für die Elektronenverteilungsfunktion in Relaxationszeitnäherung berücksichtigt und der den Coulomb-Streuquerschnitt mitbestimmende Coulomblogarithmus aus physikalisch anschaulichen cut-off Parametern b_{max}(Abschirmung der Elektronen in dünnen Plasmen, Debyelänge) und b_{min}(Ionenkorrelation in dichten Plasmen, größte Annäherung im Coulombfeld) berechnet. Die Elektronentemperatur wird in Grundzustandsnähe durch die Fermitemperatur korrigiert.

Die Bestimmung der absorbierten Laserenergie beruht auf klassischer Elektrodynamik.

Aus der Dispersionsrelation folgt der komplexe Brechungsindex für (LASER-) Licht im Medium:
$$\underline{n} = n - i\kappa = \sqrt{1 - \frac{\omega_p^2}{\omega^2(1 + \frac{i}{\omega \tau_{ei}})}}$$

wobei die Elektronen-Ionen Relaxationszeit τ_{ei} hier ihren Ursprung im verallgemeinerten Ohmschen Gesetz (Schlueter Diffusionsgleichung) bzw. der mittleren Driftgeschwindigkeit der Elektronen zwischen zwei (Coulomb-) Stößen hat.

Die Lösung des hydrodynamischen Gleichungssystems in Lagrange-Koordinaten (Massezellen) ergibt im Zeitablauf ein stufenartiges Dichteprofil im Simulationsgebiet. In jeder Massenzelle gibt es Transmission, Reflexion und Absorption ($T_j + R_j + A_j = 1$). Speziell Matrixmethoden /BERNIG/, wie sie zur Berechnung der optischen Eigenschaften von Multischichten eingesetzt werden, erlauben die Ermittlung der absorbierten Energie pro Schicht (aus der Änderung des Pointingvektors) unter Berücksichtigung von Einfallswinkel Θ und Polarisation.

Der aktuelle Ionisationszustand Z wird, ausgehend von einer Startionisation Z_0 (z.B. 10^{-2}), durch Minimierung der Freien Energie des Systems bestimmt.

5. Ergebnisse der Plasmasimulationsrechnungen und Ausblick:

Die folgenden zwei Bilder zeigen Ergebnisse von Simualtionsrechnungen für Nickel bei Laserleistungsdichten von 10^9 Wcm^{-2}.

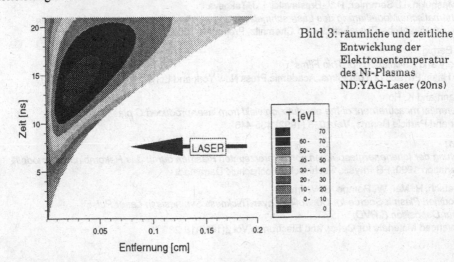

Bild 3: räumliche und zeitliche Entwicklung der Elektronentemperatur des Ni-Plasmas ND:YAG-Laser (20ns)

864

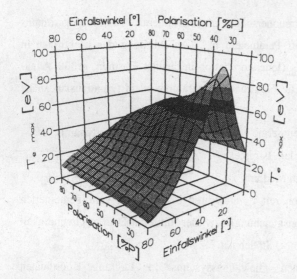

Bild 4: maximal erreichbare
Elektronentemperatur in
Abhängigkeit von
Einfallswinkel und
Anteil P-Polarisation
ND:YAG-Laser (20ns)

Zur Vervollständigung des Modells sind anstelle der Minimierung der Freien Energie die Energieumverteilungen (Anregung, Ionisation, Rekombination) im expandierenden Plasma entsprechend experimentellen Ergebnissen (Ionenenergiespektren) /MANN/ durch Simulation verschiedener Stoßprozesse (Rekombinationsdynamik) /KUNZ/ sowie die Modellierung der sich anschließenden stoßfreien Freiflugphase (Skalierungsgesetze) zu berücksichtigen.

Damit sind verbesserte Modellrechnungen zur Schichtdickenhomogenität bei der PLD bei bewegter Plasmafackel /DIETSCH/ möglich.

Quellenverzeichnis:

W. Pompe, H. Mai, H.-J. Scheibe, S. Völlmar
"Laser-Impuls-Gasphasenabscheidung dünner Schichten und Vielfachschichtsysteme"
in H. Kohler: Laser-Jahrbuch Technologie und Anwendungen, Vulkan-Verlag Essen, 2(1990)300-306

V.I. Mashukin, U. Semmler, P.V. Breslavskij. L.J. Takoeva
"Mathematische Modellierung des Laserschmelzens und -verdampfens homogener Materialien"
in Wissenschaftliche Zeitschrift der TU Chemnitz, Preprint Nr. 208/5.Jg./1991

P.H. Bernig
"Theory and Calculations of Thin Films"
in G. Hass: Physics of Thin Films, Academic Press New York and London, Vol.1(1963)69-120

K. Mann and K. Rohr
"Differential measurement of the absolute ion yield from laser-produced C plasmas"
Laser and Particle Beams, Vol.10, 3 (1992) 435-446

I. Kunz
"Deutung der Ionenenergiespektren von lasererzeugten Plasmen durch das Rekombinationsmodell"
Dissertation 1990, FB Physik, Technische Hochschule Darmstadt

R. Dietsch, H. Mai, W. Pompe, S. Völlmar
"A Modified Plasma Source for Controlled Layer Thickness Synthesis in Laser Pulse Vapour Deposition (LPVD)"
in Advanced Materials for Optics and Electronics Vol.2(1993)19-23

Stimulierte Brillouin Streuung durch breitbandige XeCl Laserstrahlung: Untersuchungen zur Formung des Strahlprofils

H.-J. Pätzold, R. König

Max-Born-Institut für Nichtlineare Optik und Kurzzeitspektroskopie
Rudower Chaussee 6, 12489 Berlin
(gefördert durch das BMFT, Kennz. 13 N 5935)

Einführung

Störungen einer Wellenfront, die z.B. beim Durchlauf durch einen Verstärker auftreten, können durch einen phasenkonjugierenden Spiegel (PC) korrigiert werden. Eine effektive Methode der Erzeugung von PC Spiegeln ist die Stimulierte Brillouin Streuung (SBS), die Laserstrahlung hoher longitudinaler Kohärenz benötigt. Bisherige Untersuchungen zur Korrektur einer Wellenfront erfolgten vorwiegend an schmalbandigen Oszillator-Verstärker Kombinationen sowohl im sichtbaren, als auch mit Excimerlasern im UV-Spektralbereich. Excimerlaser ohne Bandbreitenreduktion emittieren Linienbreiten, die einem Vielfachen der Brillouin Linienbreite des SBS Mediums entsprechen, so daß nur geringe Reflexionsgrade des PC Spiegels und geringe phasenkonjugierende Eigenschaften zu erwarten sind.

Im Folgenden soll gezeigt werden, wie mit einem nur gering modifizierten XeCl Excimerlaser hohe Reflektivitäten und bei starker Fokussierung (f = 20 mm) Änderungen des Strahlprofils erreicht werden.

Experiment

Der experimentelle Aufbau ist prinzipiell in /1/ beschrieben, wobei zusätzlich für einige Untersuchungen mit einem Gitter (21oo Linien/mm) der Resonator modifiziert und die Emission auf ca. 60 GHz eingeengt wurde. Verglichen mit der Brillouin Linienbreite (4 GHz) ist breitbandige Anregung gegeben. Für die Messung der transversalen Kohärenzlänge wurde die Methode des Doppelspaltes nach Young /2/ benutzt und die Modulation der entstehenden Interferenzstrukturen ausgewertet. Als eine Möglichkeit der Charakterisierung der phasenkonjugierenden Eigenschaften des SBS Strahles kann das Verhältnis der Divergenz des Laserstrahles zur Divergenz des SBS Strahles dienen. In der Brennebene einer f = 4,4 m Quarzlinse wurden aus den Strahlradien die jeweiligen Divergenzen bestimmt. Die Strahlprofile wurden durch die CCD-Kamera eines Beamprofiler Systems (Laserlaboratorium Göttingen) vermessen.

Ergebnisse

Für verschiedene Anwendungen des Excimerlasers, z.B. Photolithographie ist eine geringe transversale Kohärenzlänge wünschenswert. Die transversale Kohärenzlänge des breitbandigen Anregungslasers lag bei 120 μm und vergrößerte sich auf ca. 180 μm bei Bandbreitenreduktion (60 GHz).

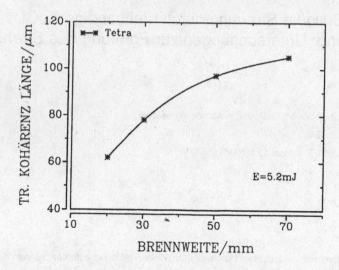

Abb. 1 Transversale Kohärenzlänge der SBS bei breitbandiger Anregung

Die transversale Kohärenz der SBS entsprach der des Pumplasers bei Fokussierungen $f \geq 100mm$, wohingegen eine starke Abnahme bei Verringerung der Brennweite auch bei Bandbreitenreduktion festzustellen war. Die Abhängigkeit von der Pumpenergie war, bei Änderung um den Faktor 5, gering, wogegen sich bei den Untersuchungen in Abb. 1 die Leistungsdichte um zwei Größenordnungen änderte.

Durch die unterschiedlichen Divergenzen in horizontaler und vertikaler Richtung im Strahlquerschnitt ergaben sich auch unterschiedliche Grade der Rekonstruktion der Wellenfront durch die SBS (sog. Fidelity).

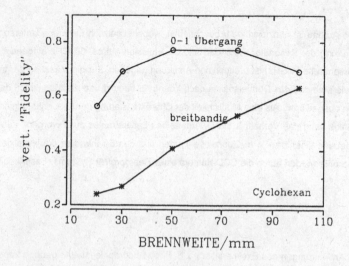

Abb. 2 Abhängigkeit der Phasenkonjugation von der Bandbreite des XeCl Lasers (vertikale Richtung)

Bei breitbandiger Anregung wird das reflektierende Phasengitter von den zwei intensiven Emissionslinien (0-1 und 0-2) des XeCl Lasers erzeugt, die einen geringen longitudinalen Kohärenzgrad repräsentieren. Eine Bandbreitenreduktion auf eine Emissionslinie führt durch Erhöhung der Kohärenzlänge zu einer Erhöhung der Reflektivität und zur Verbesserung der Phasenkonjugation. Bisherige Untersuchungen z.B. /3/, /4/ verwendeten keine Fokussierungen, bei denen die Brennweite $f < 100mm$ war, da einsetzende nichtlineare Effekte befürchtet wurden. Bei den hier gezeigten Experimenten ergeben sich Leistungsdichten von mehr als 100 GW/ cm^2, so daß das reflektierende Phasengitter bei starker Fokussierung von anderen NLO Effekten, wie z.B. der Wirkung des nichtlinearen Brechungsindex beeinflußt wird.

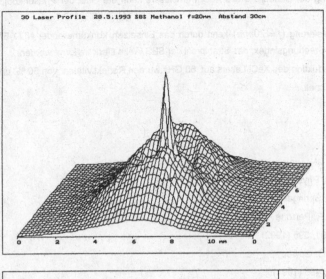

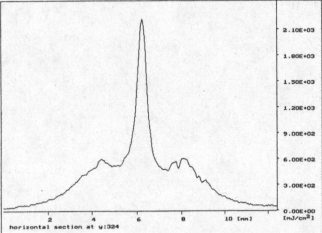

Abb. 3 Formung des SBS Strahlprofils bei starker Fokussierung (f = 20 mm)

Das entstehende Strahlprofil stellt im äußeren Randbereich einen vorwiegend durch SBS reflektierten Teil dar, während im inneren Bereich die Wirkung des Brechungsindex $n_2(E^2)$ in Wechselwirkung mit dem Phasengitter vorherrscht. Das Entstehen von Plasmafunken oder Dampfblasen in der Flüssigkeit konnte nicht mit der Änderung des Strahlprofils und mit der Reflektivität /1/ bzw. der "Fidelity" in Übereinstimmung gebracht werden.

Zusammenfassung

Die transversale Kohärenz der SBS Welle kann durch starke Fokussierung reduziert werden.

Durch Verringerung der Bandbreite des XeCl Pumplasers kann die Güte der Phasenkonjugation stark verbessert werden.

Bei starker Fokussierung ($f = 20 mm$) kann durch das Einsetzen konkurrierender NLO Effekte, z.B. die Wirkung des nichtlinearen Brechungsindex, das Strahlprofil der SBS Welle stark verformt werden.

Bei Bandbreitenreduktion des XeCl Lasers auf 60 GHz wurden Reflektivitäten von 60 % und ohne Reduktion von 40 % (n-Hexan) erzielt.

Literatur

/1/ H.J. Eichler et. al
   J.Phys. D. Appl. Phys. 25, 1161 (1992)
/2/ M. Born, Optik, Springer Verl. 1985, 3. Ausg. S. 114
/3/ A.A. Filippo, M.R. Perrone
   Opt. Commun. 91, 395 (1992)
/4/ G.M. Davis, M.C. Gower
   IEEE, J. QE-27, 496 (1991)

Zwei-Strahl-PLD für Metalle

A. Lenk, B. Schultrich, G. Granse und S. Völlmar

Fraunhofer-Einrichtung für Werkstoffphysik und Schichttechnologie, FhE IWS

Helmholtzstraße 20, O-8027 Dresden, D

1. Problemstellung

Die Bestrahlung von Targets mit Lasern hoher Impulsleistungen ($> 10^5$ W) ist hervorragend geeignet, um ein breites Spektrum angeregter Teilchen zu erzeugen, das für die Abscheidung qualitativ hochwertiger Schichten genutzt werden kann. Dafür muß auf der Targetoberfläche ein Grenzwert der Leistungsdichte $> 10^7$ W/cm² überschritten werden. Dieser Grenzwert hängt von Reflexionkoeffizient, Extinktionskoeffizient, Wärmeleitfähigkeit und Bindungsenergien des Targetmaterials sowie von der Impulsdauer und der Größe der bestrahlten Fläche ab. Bei der Abscheidung von Keramiken und Halbleitern wird diese Methode der **P**ulse **L**aser **D**eposition (PLD) erfolgreich angewendet. Die mittleren Abscheideraten liegen allerdings deutlich unter den mit Elektronenstrahlverdampfen oder Magnetron-Sputtern erreichbaren. Um der PLD ein breiteres Anwendungsfeld zu verschaffen, ist es notwendig sowohl die mittlere Abscheiderate wesentlich zu erhöhen, als auch den Bereich der verwendeten Targetmaterialen auf die Metalle auszudehnen. Hinsichtlich der verfügbaren mittleren Leistung (bis 10 kW) und des energetischen Wirkungsgrades (> 10 %) sind Hochleistungs-CO_2-Laser, wie sie gegenwärtig bei der Oberflächenbehandlung eingesetzt werden, ohne Konkurrenz. Deshalb kommen nur diese Laser für eine PLD-Hochrateabscheidung in Betracht. Das hat allerdings eine Reihe schwerwiegender Probleme zur Folge. Einerseits liefern diese Laser Impulsleistungen von nur einigen 10^3 W und andererseits reflektieren Metalle bei einer eingestrahlten Wellenlänge von 10,6 μm besonders gut.

2. Lösungsmöglichkeiten

Für die Realisierung einer PLD-Hochrateabscheidung ist es notwendig, dem Target eine hohe Leistungsdichte "anzubieten" und diese auch effektiv einzukoppeln. Die einzige Möglichkeit zur Erreichung einer hohen Leistungsdichte bei gegebener Pulsleistung besteht in der maximalen Fokussierung des Laserstrahls. Der theoretisch erreichbare minimale Fokusdurchmesser entspricht dem Produkt aus Divergenz des Laserstrahls und Brennweite der fokussierenden Optik. Typische Werte sind 1,5 mrad für die Divergenz und 100 mm für die Brennweite. Eine Reduzierung der Divergenz durch Strahlaufweitung verbietet sich, da der Durchmesser des Fensters in der Vakuumkammer, durch das der Laserstrahl eingekoppelt wird, aus Kostengründen beschränkt ist. Eine Verringerung der Brennweite ist aus geometrischen Gründen nicht möglich. Der theoretisch erreichbare minimale Fokusdurchmesser ist somit 150 μm. Bei einer Impulsleistung von beispielsweise $5*10^3$ W ergibt das eine maximale Leistungsdichte von $> 3*10^7$ W/cm^2. Unter bestimmten Vorraussetzungen läßt sich dieser Wert mit Hilfe einer Superpulseinrichtung um den Faktor 3 erhöhen.

Diese Leistungsdichte reicht vor allem auf Grund des hohen optischen Reflektionsvermögens im allgemeinen nicht aus, um auf Metallen ein Plasma zu zünden und einen effektiven Materialabtrag zu erzeugen. Bekannte Methoden zur Erhöhung der einkoppelten Leistung (> 50%) sind die Bestrahlung unter dem Brewsterwinkel sowie die vorherige Modifizierung der zu bestrahlenden Oberfläche. Für Metalle läßt sich der nahe 90° liegende Brewsterwinkels technisch nicht realisierbar. Die zweite Methode läßt sich ebenfalls nicht anwenden, da schon nach dem ersten Laserimpuls die verbleibende Metalloberfläche durch Aufschmelzen und Wiedererstarren sich in der Regel so verändert, das sie wieder hochreflektierend wird. Die gesuchte Methode zur Erhöhung der einkoppelten Leistung muß also eine "insitu"-Modifizierung der Targetoberfläche erlauben. Diese Anforderung erfüllt die im folgenden vorgestellte Zwei-Strahl-Methode.

3. Zwei-Strahl-Methode

Diese Methode beruht auf der Einkopplung der Laserleistung durch ein bereits auf der Oberfläche des Targets vorhandenes Plasma . Diese Plasma wird durch einen zweiten sogenannten Zünd-Laser synchron zum Beginn des CO_2-Laserimpulses erzeugt. Für diesen Zweck wird sinnvollerweise ein gütegeschalteter cw-angeregter Nd-YAG-Laser (eventuell mit SHG) verwendet. Dieser Lasertyp stellt bei geeigneter Fokussierungsoptik die notwendige Leistungsdichte von $> 10^8$ W/cm^2 zur Verfügung und erlaubt eine Wiederholfrequenz im kHz-Bereich. Außerdem ermöglicht der im allgemeinen geringere Reflexionskoeffizient eines Metalltargets bei kürzeren Wellenlängen eine bessere Einkopplung des Zündlasers (Interband- Übergänge).

Theoretische Modellrechnungen zeigen die Möglichkeit auf, das durch den Zündlaser (mit einigen mJ Pulsenergie) erzeugte Plasma mit dem CO_2-Laserpuls (einige 100 mJ Pulsenergie) soweit aufzuheizen (inverse Bremsstrahlung), daß eine Energieübertragung bis zum Target stattfindet. Dies hätte ein entscheidende Erhöhung der ins Target eingekoppelten Laserleistung zur Folge.

4. Experimentelle Realisierung

Der Kern der experimentellen Anlage ist eine modular aufgebaute HV-Anlage (Enddruck 10^{-4} Pa). Als CO_2-Hochleistungslaser kommt ein hochfrequenzangeregter 6 kW-Laser und als Zündlaser ein cw-angeregter Nd-YAG-Laser mit Güteschaltung und optionaler SHG zum Einsatz. Durch geeignete Fenster bzw. Optiken werden die beiden Laserstrahlen in die Vakuumkammer eingekoppelt und unter einem Winkel von 45 ° auf ein und dieselbe Stelle des Targets fokussiert. Die rechnergesteuerte Synchronisation der Laserpulse wird mittels schneller Sensoren kontrolliert (siehe Tabelle und Abb. 1). Zur Beurteilung der Targeterosion sowie für die Untersuchung der Plasmabildung stehen eine Reihe leistungsfähiger insitu-Meßsysteme zur Verfügung (Mikrowaage, 15 kV REM, Spektrometer und verschiedene Sonden).

Parameter	CO_2-Laser	Nd-YAG-Laser
Puls-Leistung	4 000 W	50 000 W
Puls-Folgefrequenz	0..25 kHz	0..25 kHz
Puls-Dauer	0.1 ms	120 ns
Puls-Energie	400 mJ	6 mJ
min. Fokusdurchmesser	200 µm (f = 100 mm)	120 mm (f = 300 mm)
max. Leistungsdichte	$1,3*10^7$ W/cm²	$3,5*10^8$ W/cm²

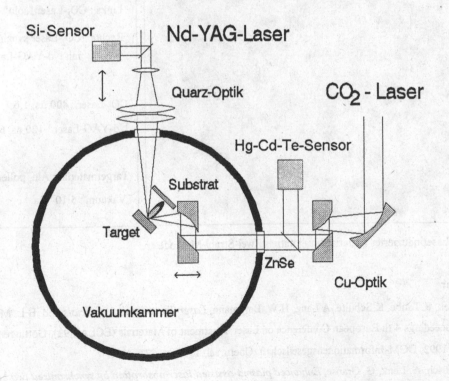

Abb. 1 Schematische Darstellung der Laserstrahlführung in der Zwei-Strahl-Anlage

5. Erste Ergebnisse

Mit der an der Kammer installierten Mikrowaage ist es möglich die Masse wie auch die mittlere Geschwindigkeit des durch den Laser ablatierten Materials zu messen. Erwartungsgemäß wird bei der Bestrahlung von Kohlenstoff mit einem CO_2-Laserimpuls (1 ms) weit mehr Energie absorbiert und teilweise in kinetische Energie umgewandelt als bei der Bestrahlung von Aluminium :

Target	ablatierte Masse	mittlere Geschwindigkeit	kinetische Energie
Aluminium	565 µg	30 m/s	0,25 mJ
Kohlenstoff	43 µg	1100 m/s	52 mJ

Setzt man nun die Zwei-Strahl-Methode bei der Ablation von Aluminium ein, so wird die Erosion des Targets erheblich verstärkt (siehe Abb. 2). Die Quantifizierung und Optimierung dieses Effektes sowie der Vergleich verschiedener Targetmaterialien stehen im Mittelpunkt unserer weiteren Arbeit.

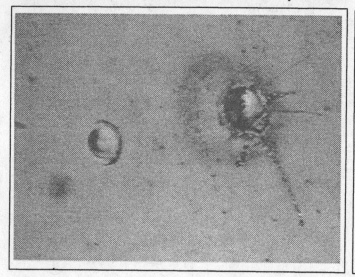

Links : CO_2-Laser "solo"

Rechts : CO_2-Laser synchron
mit Nd-YAG-Laser

CO_2-Laser : 400 µs, 1,6 J

Nd-YAG-Laser : 120 ns, 6 mJ

Targetmaterial : Alu, poliert

Vakuum : $5 \cdot 10^{-4}$ Pa

Abb. 2 Laserinduzierter Materialabtrag mittels Zwei-Strahl-Methode

Literatur

R. Jaschek, R.Taube, K.Schulte, A.Lang, H.W. Bergmann, *Laser Treatment of Materials* (Ed. B.L. Mordike, Proceedings 4 th European Conference on Laser Treatment of Materials (ECLAT'92), Göttingen October 1992, DGM-Informationensgesellschaft, Oberursel, 1992, p. 673

B. Schultrich, A. Lenk, G. Granse, *Enhanced plasma-assisted laser absorption by synchronized two beam irradiation for laser pulse evaporation and deposition of metals*, Third Int. Conf. on Plasma Surface Engineering, Garmich-Partenkirchen, 25.-29. 10 1

A Frequency Tunable High Power Mid- and Far-Infrared Laser System

W. Schatz[1], _K.F. Renk_[1], _and P.T. Lang_[2]

[1] Institut für Angewandte Physik, Universität Regensburg, 93053 Regensburg, FRG
[2] MPI für Plasmaphysik, 85748 Garching, FRG

Abstract A laser system composed of a grating tuned high-pressure CO_2-laser pumping a single-pass molecular gas laser is demonstrated to deliver high intense coherent light pulses covering the spectral range of the mid- and far-infrared. The transversal electrically excited CO_2-laser emitts radiation pulses of up to 300 ns length and powers up to several MW; using mode-locking techniques single nanosecond radiation pulses can be obtained. Using the CO_2-laser radiation for optical pumping of appropriate molecular gases, frequency tunability can be transferred into the far-infrared via Raman tuning. Highest efficiencies up to quantum conversion rates exceeding 0.1 have been found for several isotopes of the methylhalides (CH_3X; $X = F$, Cl, Br, J), heavy water (D_2O) and ammonia (NH_3). For these gases, more than a thousand Raman tuning regions are available ranging from the spectral region of millimeter waves to wavelength of 40 μm. Pulse durations below 100 ps can be achieved when mode-locking techniques are applied at the pump laser. The laser system has already been used for numerous spectroscopic applications.

The High-Pressure CO_2-Laser

The resonator of our CO_2-laser consists of a concave germanium output coupler (reflectivity ≈ 0.7; radius ≈ 25 m) and a 150 l/mm grating for frequency tuning; to enhance the resolution and to avoid radiation damage an off-axis Galilean telescope is inserted to expand the beam diameter at the grating surface. Optical access is given to the high pressure chamber by two NaCl Brewster's angle windows. With this configuration, frequency tuning is possible in steps of 1 GHz over ranges of about 500 GHz, the spectral width of the emitted radiation is 5 GHz (FWHM) [1]. In Fig. 1 the four tuning ranges of our CO_2-laser are illustrated where the energy density of the discharge was 21 $\frac{J}{l \cdot atm}$ at the R-branch and 26 $\frac{J}{l \cdot atm}$ at the P-branch; an increase of the tuning regions was obtained with an energy density of 30 $\frac{J}{l \cdot atm}$ (dashed lines). The emitted pulses have a total duration of about 300 ns and a maximum energy of 250 mJ, limited by the damage threshold of the NaCl Brewster windows.

For generation of short infrared-pulses several methods can be used. Operating the CO_2-laser near threshold deliveres strong spiking in the temporal structure of the laser pulses due to self-mode-locking of different longitudinal modes. Another more reliable methode to generate short pulses is passive mode-locking by use of p-Germanium at 77 K. In Fig. 4a a passive mode-locked CO_2-laser pulse train with single pulse durations of about 2 ns is shown.

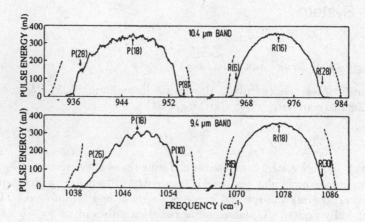

Fig.1: *Tuning ranges of the high pressure CO_2-laser*

The Optically Pumped Far-Infrared Laser

The CO_2-laser radiation is focussed by a molybdenum mirror (radius $\approx$ 5 m to obtain maximum efficiency) into the fused quartz waveguide (diameter 7 mm) of the single-pass Raman laser [1]. This laser is equipped with a BaF_2 entrance and a TPX (polymeric 4-methyl-1-pencene) output window. Since TPX does not completely absorb the CO_2-laser radiation transmitted through the FIR laser tube, a crystalline quartz plate is used to block remaining CO_2-laser radiation.

Fig.2: *Survey on all measured FIR frequencies*

In our experiments we investigated the istopic molecular gases $^{12}CH_3F$, $^{13}CH_3F$, $^{12}CD_3F$, $^{13}CD_3F$, $^{12}CH_3Cl$, $^{12}CH_3Br$, $^{12}CH_3J$, $D_2^{16}O$, $D_2^{18}O$, $^{14}NH_3$, $^{14}ND_3$, $^{15}NH_3$ and $^{15}ND_3$. We observed FIR emission at more than 1000 laser transitions. In Table 1 the number of FIR emission ranges, the vibrational transition of the molecule, the region of FIR emission and the maximum quantum efficiency are listed.

Molecule	Lines	Vibration	Minimum Maximum frequency [cm^{-1}]		Efficiency [%]
$^{12}CH_3F$	56	ν_3	10	72	6
$^{13}CH_3F$	59	ν_3	15	71	1
$^{12}CD_3F$	27	ν_3	8	68	12
	137	ν_6	19	55	1
$^{13}CD_3F$	190	$\star$	11	61	12
$^{12}CH_3Cl$	84	ν_6	17	41	1
$^{12}CH_3Br$	9	ν_6	16	27	0.02
$^{12}CH_3J$	9	ν_6	12	20	0.04
	29	$2\nu_3$	13	27	0.1
$D_2^{16}O$	58	ν_2	26	236	5
$D_2^{18}O$	47	ν_2	25	240	5
$^{14}NH_3$	72	ν_2	16	200	12
$^{14}ND_3$	4	ν_4	67	186	0.3
$^{14}NH_2D$	50	ν_4	21	211	10
$^{14}NHD_2$	22	ν_4	36	168	0.4
$^{15}NH_3$	49	ν_2	30	180	2
$^{15}ND_3$	70	$\star\,(\nu_4)$	64	250	0.5

Table 1: Range of FIR emission at all investigated laser gases ($\star$: not yet assigned)

For the mechanism of generation of FIR radiation is due to stimulated Raman scattering, it is also possible to obtain FIR emission tunable in frequency. We observed maximum tuning ranges of more than 3 cm^{-1} in ammonia and 1.2 cm^{-1} in methylfluoride. Figure 2 shows a survey on all observed FIR-frequencies including the tuning ranges. Optical pumping the single-pass FIR-laser delivers high efficiencies up to quantum coversion coefficients of 0.1 (Figure 3). Using mode-locked pump pulses the corresponding FIR radiation consists of a train of ultrashort pulses with pulse durations of less than 100 ps (Fig. 4b).

Spectroscopic Applications

We applied both mid- and far-infrared laser radiation for the nonlinear spectroscopic investigations. Using the mid-infrared radiation of the high power CO_2-laser we studied the temperature dependence of the saturation of the local-vibrational-mode absorption of H$^-$ ions in CaF$_2$ crystals. Thereby we were able to determine the lifetime and the decay-mechanism of this localized mode [2]. In the far-infrared we observed nonlinear transmission resulting from resonant band modes in oxygen-doped silicon crystals. We showed that the strong optical excitation resulted in a nonthermal phonon distribution that decayed most likely by a process of spectral diffusion, caused by scattering at low-lying states [3].

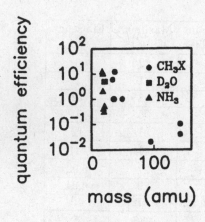

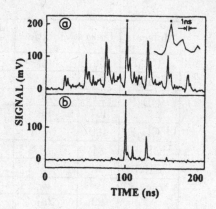

Fig.3: *Maximum photon conversion coefficients achieved for the gases under investigation*

Fig.4: *Modelocked CO_2-laser pulse (a) and resulting FIR-laser pulse (b)*

Another application of our laser system was to investigate the photoresponses spectral characteristics for different detector types. Thereby we measured the response time and detectivity of $YBa_2Cu_3O_{7-\delta}$-, metal resistor thin film-, germanium- and quantum well-detectors.

Summary We present a frequency tunable high power laser system, consisting of a continuously tunable high pressure CO_2-laser and an optically pumped single-pass FIR-laser. Investigating several isotopic molecular gases delivered more than 1000 frequency tuning ranges in the far-infrared region by stimulated Raman scattering. Thereby pulse durations of less than 100 ps and quantum efficiencies of up to 0.1 have been obtained. Applications of this radiation for nonlinear spectroscopic techniques in crystalline systems and the photoresponse measurement of several types of detectors have been demonstrated.

Financial support by the Bundesministerium für Forschung und Technologie, and the Bayerische Forschungsstiftung (FORSUPRA) are acknowledged.

References

[1] U. Werling, K.F. Renk, and Wan Chong-Yi, *Int. J. Infrared Millimeter Waves* **7**, 881 (1986).

[2] P.T. Lang, W.J. Knott, U. Werling, K.F. Renk, J.A. Campbell, and G.D. Jones, *Phys. Rev. B* **44**, 6780 (1991).

[3] U. Werling, and K.F. Renk, *Phys. Rev. B* **39**, 1286 (1989).

Zeitliches Absorptionsverhalten von amorphem Metall bei der Materialbearbeitung mittels Nd:YAG-Laser

L. Dorn, K.-S. Lee
Institutsbereich Fügetechnik/Schweißtechnik, TU Berlin
Straße des 17. Juni 135, D-10623 Berlin 12

1   Einleitung

Die genaue Kenntnis der Wechselwirkung zwischen Laserstrahl und Materie ist notwendig, um die wesentlichen physikalischen Prozesse bei der Laserbearbeitung zu verstehen. Eine wichtige Fragestellung ist dabei der Einkoppelmechanismus der Laserstrahlung in das Werkstück beim Schweißen. In dieser Untersuchung wurde der Einfluß der Strahlparameter auf das transiente Absorptionsverhalten eines amorphen Metalls beim Laserstrahlschweißen untersucht. Weiterhin wurde der Einfluß des Absorptionsverhaltens auf den Schweißvorgang und die Schweißergebnisse betrachtet.

2   Meßaufbau und Versuchsmaterial

Die Vorrichtung zur Messung von Reflexionssignal beim Schweißen besteht aus einer Ulbrichtkugel, zwei Fotodioden und der Verstärkungselektronik (Bild 1). Das Werkstück ist im unteren Bereich der Ulbrichtkugel eingespannt und der Laserstrahl wird auf die Werkstückoberfläche gerichtet. Der reflektierte Laserstrahl wird auf der Kugelinnenoberfläche diffus reflektiert und durch die Fotodiode detektiert. Parallel dazu wird der eingebrachte Laserstrahl vom Lasergerät durch eine Fotodiode gemessen.

Für die Schweißversuche wurde eine kommerziell hergestellte amorphe Metallfolie auf Kobaltbasis verwendet (Handelsbezeichnung Vitrovac 6025 F, Vacuumschmelze GmbH). Die Zusammensetzung wird mit Co69,3/Si7,9/Mo6,3/Fe3,5/Nb2,6/B10,4 (in Gewichtsprozent) angegeben. Die Folie ist 25 mm breit und 25 µm dick.

3   Versuchsergebnisse und Diskussion

3.1   Absorptionsverhalten in Abhängigkeit von der Leistungsdichte durch Variation der Fokuslage

Die Leistungsdichte läßt sich einerseits durch Variation der Strahlleistung und andererseits durch Variation der Fokuslage, d.h. durch den Abstand der Werkstückoberfläche zum Brennpunkt der Arbeitsoptik (Defokussierung), beeinflussen. Hier wurde die Leistungsdichte durch die Änderung der Fokuslage variiert. Die Leistungsdichteänderung durch Variation der Strahlleistung wird im Kap. 3.2 beschrieben.

Es wurden zwei unterschiedliche Meßreihen (mit einer Pulsleistung von 0,3 und 1,1 kW) durchgeführt. Um eine unterschiedliche Leistungsdichte zu erreichen, wurde die Fokuslage jeweils von -1

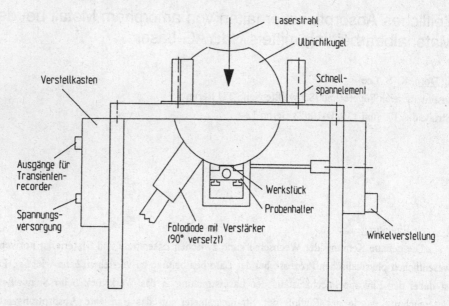

Bild 1: Reflexionsmeßeinrichtung

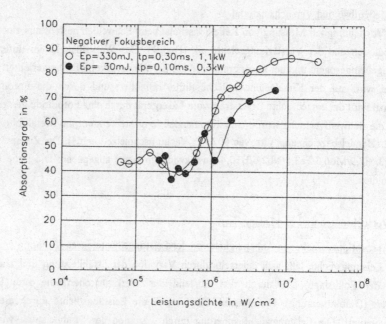

Bild 2: Absorptionsgrad in Abhängigkeit von der Leistungsdichte

bis +1 mm (bei 0,3 kW) und -4 bis +4 mm (bei 1,1 kW) variiert. Damit kein Bohrloch beim Schweißen entstand, wurde die Folienanzahl der Leistung angepaßt. Durch die Entstehung der Bohrung würde ein Teil des Laserstrahls nicht auf dem Werkstoff auftreffen, sondern durch das Werkstück hindurchtreten. Bei einer Pulsleistung von 0,3 kW wurden 10 Folien (Dicke von ca. 0,25 mm), und bei 1,1kW 30 Folien (Dicke von ca. 0,75 mm) überlappt angeordnet.

Für das Absorptionsverhalten wurden die Signale der eingebrachten und reflektierten Laserleistung gemessen, wobei die Differenz beider Signale dem Absorptionssignal entspricht. Zur Bestimmung der Laserenergie wurden die Signale über die Zeit integriert. Der Absorptionsgrad ist der prozentuale Anteil der eingekoppelten, d.h. nicht reflektierten Laserenergie.

Im Bild 2 werden die Absorptionsgrade im negativen Fokusbereich in Abhängigkeit von der Leistungsdichte dargestellt. Die Laserpulsleistung lag bei 0,3 und 1,1 kW. Die Fokuslage wurde jeweils von -1,0 bis 0 mm und von -4,0 bis 0 mm variiert. Bei einer Laserleistung von 0,3 kW liegt der Absorptionsgrad im Bereich der Leistungsdichte von $2*10^5$ bis $1*10^6$ W/cm^2 zwischen 35 und 55 %. Bei der niedrigen Pulsenergie schwankt der Absorptionsgrad stark. Ab einer Leistungsdichte von $1*10^6$ W/cm^2 steigt der Absorptionsgrad steil bis 73 % an.

Bei einer Leistung von 1,1 kW bleibt der Absorptionsgrad im Bereich der Leistungsdichte von $6*10^4$ bis $6*10^5$ W/cm^2 zwischen 40 und 50 %. Bei einer Leistungsdichte größer als $4*10^5$ W/cm^2 steigt der Absorptionsgrad sehr steil bis 87 % an. Ab einer Leistungsdichte von $1*10^7$ W/cm^2 fällt der Absorptionsgrad wieder leicht ab. Grund dafür ist Energieverlust durch Materialverdampfung.

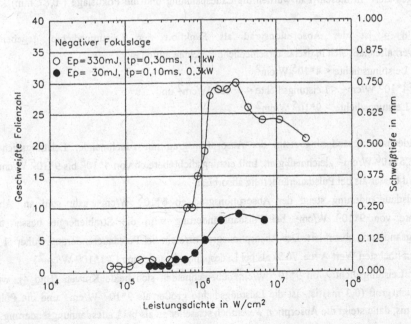

Bild 3: Geschmolzene Folienanzahl und Schweißtiefe in Abhängigkeit von der Leistungsdichte

Bei der höheren Laserpulsleistung (bei 1,1 kW) und ab einer Leistungsdichte von $1*10^6$ W/cm^2 wird mehr Laserenergie vom Werkstück absorbiert. Mit zunehmender Laserleistung beginnt die kritische Leistungsdichte früher, weil die zur Bildung der Metalldampfkapillare ausreichende Leistungsdichte schneller erreicht wird.

Bild 3 stellt die entsprechende Schweißtiefe (in mm) vom Bild 2 in Abhängigkeit von der Leistungsdichte dar.

Die erreichbare Schweißtiefe liegt bis zu einer Leistungsdichte von $4*10^5$ W/cm^2 bei ca. 50 μm. Ab einer Leistungsdichte von $4*10^5$ W/cm^2 nimmt die Schweißtiefe bis 750 μm rasch zu, wobei der Absorptionsgrad (Bild 2) steil ansteigt.

Ab einer Leistungsdichte von $3*10^6$ W/cm^2 verringert sich die Schweißtiefe für die beiden Pulsleistungen, obwohl der Absorptionsgrad weiter ansteigt. Grund dafür ist eine starke Spritzerbildung wegen der höheren Leistungsdichte von mehr als $3*10^6$ W/cm^2. Die Spritzer führen einen Teil der eingebrachten Laserenergie ab, was die erreichbare Schweißtiefe verringert.

3.2 Absorptionsverhalten in Abhängigkeit von der Leistungsdichte durch Variation der Pulsenergie

Die Pulsenergie läßt sich entweder durch Ladespannungs- oder Pulsdaueränderung variieren. Es wurden getrennt die Pulsdauer und die Ladespannung verändert, um die Einflüsse von der Pulsdauer und der Ladespannung auf das Absorptionsverhalten beim Schweißen zu untersuchen.

Bei der Ladespannungsänderung wurde die Pulsenergie von 86 bis 822 mJ erhöht, während die Pulsdauer von 0,30 ms und die Fokuslage von -0,87 mm konstant gehalten wurden. Bei der Pulsdaueränderung wurde die Pulsdauer von 0,1 bis 0,6 ms verlängert und dadurch die Pulsenergie von 33 bis 806 mJ geändert. In diesem Fall wurden die Ladespannung und die Fokuslage (-0,87 mm) konstant gehalten.

In Bild 4 ist der Absorptionsgrad als Funktion der Leistungsdichte gegeben. Das Absorptionsverhalten läßt sich in drei unterschiedliche Bereiche aufteilen:

- Leistungsdichte $< 4*10^5$ W/cm^2,
- $4*10^5$ W/cm^2 < Leistungsdichte $< 9*10^5$ W/cm^2 und
- Leistungsdichte $> 9*10^5$ W/cm^2 .

Bei Ladespannungsänderung steigt der Absorptionsgrad mit zunehmender Leistungsdichte von $2,5*10^5$ bis $2,5*10^6$ W/cm^2 gleichmäßig an. Im Leistungsdichtebereich von $4*10^5$ bis $9*10^5$ W/cm^2 wird der Laserstrahl besser als bei Pulsdaueränderung absorbiert.

Bei Pulsdaueränderung steigt der Absorptionsgrad ab $6*10^5$ W/cm^2 sehr steil an. Ab einer Leistungsdichte von $9*10^5$ W/cm^2 bei Pulsdaueränderung wird die Strahlenergie besser als bei Ladespannungsänderung absorbiert. Der Absorptionsgrad erreicht bei Pulsdaueränderung früher ($1,1*10^6$ W/cm^2) seinen höchsten Wert A=ca. 70 % als bei Ladespannungsänderung ($2,1*10^6$ W/cm^2).

Bei der Leistungsdichte von $9*10^5$ W/cm^2 überschneiden sich beide Kurven (Bild 4), weil die Pulsdauer gleich groß (0,3 ms) ist. Ist die Leistungsdichte größer als $9*10^5$ W/cm^2 und die Pulsdauer länger als 0,3 ms, dann steigt die Absorption wesentlich schneller an als bei Ladespannungsänderung.

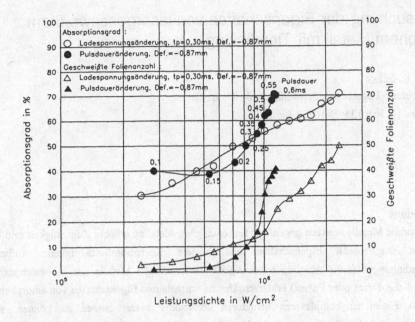

Bild 4: Absorptionsgrad und aufgeschmolzene Folienanzahl in Abhängigkeit von der Leistungsdichte (Variation der Ladespannung und Pulsdauer)

4 Zusammenfassung

Es wurden das Absorptionsverhalten mit der Variation von Fokuslage und Pulsenergie untersucht. Mit zunehmender Laserleistung stellt sich die kritische Leistungsdichte früher ein, weil eine für die Bildung der Metalldampfkapillare ausreichende Leistungsdichte schneller erreicht wird.

Der Absorptionsgrad erreicht bei Pulsdaueränderung früher ($1,1*10^6$ W/cm^2) seinen höchsten Wert A=ca. 70 % als bei Ladespannungsänderung ($2,1*10^6$ W/cm^2).

Die Autoren danken der Firma Vacuumschmelze GmbH für die Bereitstellung des Versuchsmaterials.

Untersuchung der Eigenschaften von lasergeschweißtem amorphem Metall mit Tiefkühlung

L. Dorn, K.-S. Lee

Institutsbereich Fügetechnik/Schweißtechnik, TU Berlin

Straße des 17. Juni 135, D-10623 Berlin 12

1 Einleitung

Amorphe Metalle besitzen gegenüber herkömmlichen Metallen erhöhte Zugfestigkeit und Duktilität, günstigere magnetische Eigenschaften und bessere Korrosionsbeständigkeit. Aufgrund der Herstellbedingungen (hohe Abkühlgeschwindigkeit) sind amorphe Metalle nur in bestimmten Formen (meist als Folie, Draht oder Pulver) erhältlich. Um die vorteilhaften Eigenschaften von amorphen Metallen durch das Fügen zu komplexeren Strukturen konstruktiv besser nutzen zu können, sollen die Möglichkeiten und Grenzen des Laserpulsschweißens untersucht werden.

Unter den normalen Raumbedingungen entsteht beim Schweißen ab einer Dicke von 100 μm unvermeidlich eine Kristallbildung. Um diese Kristallbildung zu vermeiden oder zu vermindern, wurde das Schweißen unter Tiefkühlumgebung durchgeführt.

2 Versuchseinrichtung und Versuchswerkstoff

Die Vorrichtung zum Schweißen amorpher Folien besteht aus einen Teflon-Behälter, der mit Stickstoff gekühlt werden kann, und aus einem Backenpaare, mit denen die Folien gespannt werden (Bild 1).

Für die Schweißversuche wurde eine kommerziell hergestellte amorphe Metallfolie auf Kobaltbasis verwendet (Vitrovac 6025 F, Vacuumschmelze GmbH). Die Zusammensetzung wird mit Co69,3/Si7,9/Mo6,3/Fe3,5/Nb2,6/B10,4 (in Gewichtsprozent) angegeben. Die Folie ist 25 mm breit und 25 μm dick.

3 Versuchsdurchführung

Es wurden 4 amorphe Folien mit und ohne Tiefkühlung durch flüssigen Stickstoff geschweißt. Die Folien wurden überlappt geschweißt und die Scherzugkraft in der obersten, untersten und mittleren Folien geprüft (Bild 2). Beim Schweißen wurde die Umgebungstemperatur bei ca. -130 °C konstant gehalten.

Für die Optimierung des Schweißverfahrens wurden die Strahlparameter möglichst so ausgesucht, daß sie die kürzeste Pulsdauer und die geringste Pulsenergie haben, damit das Werkstück so wenig wie möglich wärmebeeinflußt wird.

Auf die Tragfähigkeit lasergeschweißter Punktverbindungen wirken sich die Laserstrahlparameter Leistungsdichte und Impulsdauer wesentlich aus. Diese Einflußgrößen wurden an Überlappverbindungen der amorphen Metallfolien mit einer einzelnen Foliendicke von 25 μm experimentell untersucht. Die

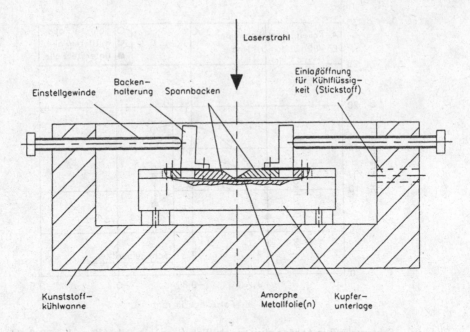

Bild 1: Spannvorrichtung zum Schweißen amorpher Folien

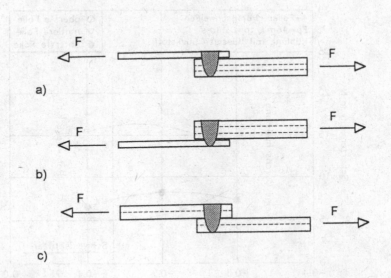

Bild 2: Drei Prüfarten beim Scherzugprüfung

884

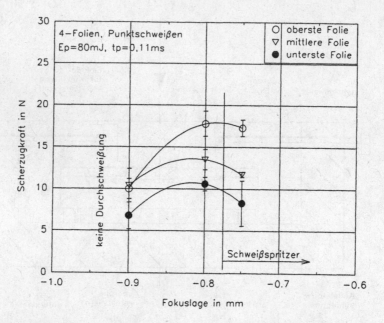

Bild 3: Scherzugkraft in Abhängigkeit von der Fokuslage beim Schweißen ohne Tiefkühlung

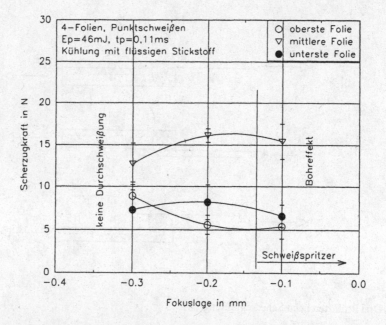

Bild 4: Scherzugkraft in Abhängigkeit von der Fokuslage beim Schweißen mit Tiefkühlung

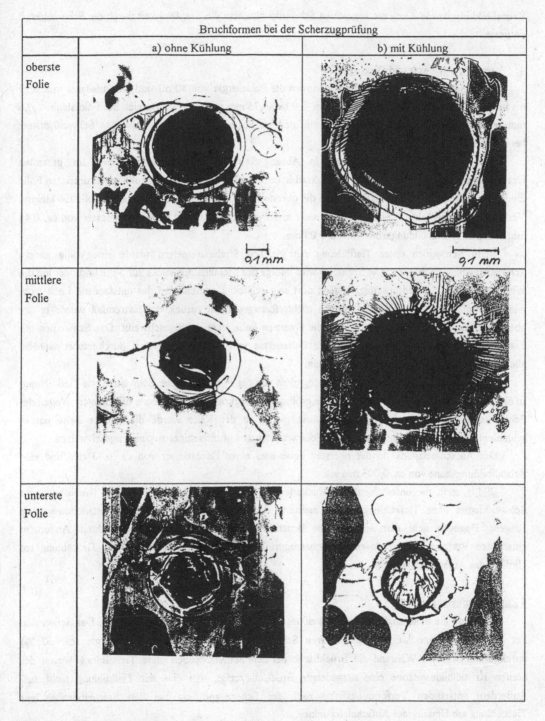

Bild 5: Bruchformen bei der Scherzugprüfung
a) Qp=80 mJ, E=7,1*10^5 W/cm^2, τp=0,11 ms, Def.=-0,8 mm
b) Qp=46 mJ, E=2,5*10^6 W/cm^2, τp=0,11 ms, Def.=-0,2 mm

886

Kristallbildung wurde durch die angeätzten Schweißpunkte und die Tragfähigkeit durch die Scherzugkraft dargestellt.

4   Versuchsergebnisse und Diskussion

Beim Schweißen ohne Tiefkühlung wurden die Pulsenergie von 80 mJ und die Pulsdauer von 0,11 ms konstant gehalten und die Fokuslage von -0,9 bis -0,75 mm geändert. Im Bereich der Fokuslage < -0,6 mm wurden die Folien nicht durchgeschweißt und bei Fokuslage > -0,75 mm wurden Schweißspritzer beobachtet.

Bild 3 stellt die Scherzugkraft in Abhängigkeit von der Fokuslage dar. Im gesamten Untersuchungsbereich erreicht die Scherzugkraft in der obersten Folie die größte und in der untersten Folie die kleinste Tragkraft. Grund dafür ist, daß die oberste Folie die größte und die unterste Folie die kleinste Verbindungsfläche haben. Der Aufschmelzpunkt in der oberste Folie zeigt einen Duchmesser von ca. 0,45 mm und eine Kristallbildungszone von ca. 0,03 mm.

Beim Schweißen unter Tiefkühlung mit gleichen Strahlparametern wurde eine völlig andere Schmelzdynamik beobachtet. Die Schmelze wird infolge des Metalldampfdrucks zur Seite und/oder auf die Werkstückoberfläche gedrückt und erstarrt dort sehr schnell. Dadurch wird das entstandene Loch nicht wieder geschlossen sondern bleibt als ein trichterförmiges Loch zurück. Im Extremfall wurde in der obersten Folie eine Bohrung erzeugt und die untersten Folie nicht durchgeschweißt. Das Schweißen mit Tiefkühlung konnte auch mit sehr kleiner Pulsenergie von 46 mJ erfolgreich durchgeführt werden, allerdings muß die Defokussierung gering sein.

Bild 4 zeigt die Scherzugkraft in Abhängigkeit von der Fokuslage beim Schweißen mit Tiefkühlung. In diesem Fall erreichte die mittlere Folie die größte Scherzugkraft und die oberste die kleinste. Wegen der Schweißspritzer und der besonderen Schmelzdynamik in der Kälte wurde die oberste Folie nur in schmalem Bereich verbunden. Die mittlere Folie war meistens großflächiger zusammengeschmolzen.

Der Aufschmelzpunkt in der obersten Folie wies einen Duchmesser von ca. 0,33 mm und eine Kristallbildungszone von ca. 0,005 mm auf.

Bild 5 stellt die unterschiedliche Brucharten bei der Scherzugprüfung dar. Die Bruchstelle von Schweißungen ohne Tiefkühlung zeigen meistens eine Bruchlinie, die von der Kristallbildungszone ausgeht. Dagegen sieht man keine solche Bruchlinie beim Schweißen mit Tiefkühlung. Außerdem entstanden Verformungslinien nach der Scherzugprüfung bei den Schweißproben mit Tiefkühlung am Umfang des Aufschmelzpunktes.

4 Zusammenfassung

Es wurde die Scherzugkraft der Schweißung mit und ohne Tiefkühlung untersucht. Das Schweißen mit Tiefkühlung kann wegen einer anderen Schmelzdynamik mit kleinerer Pulsenergie (ca. 60 %) durchgeführt werden. Während die Bruchfläche bei den Schweißproben ohne Tiefkühlung wegen der breiten Kristallbildungszone eine ausgedehnte Bruchlinie zeigt, tritt dies mit Tiefkühlung nicht auf. Außerdem entstanden Verformungslinien nach der Scherzugprüfung bei den Schweißproben mit Tiefkühlung am Umfang des Aufschmelzpunktes.

Die Autoren danken der Firma Vacuumschmelze GmbH für die Bereitstellung des Versuchsmaterials.

Optimierung der Oberflächenbehandlung bei Einwirkung pulsierender oszillierender Laserstrahlung

D. Lepski, L. Morgenthal, S. Völlmar
Fraunhofer-Einrichtung für Werkstoffphysik und Schichttechnologie (IWS)
Helmholtzstraße 20, D-O-8027 Dresden (ab 01.07.93: PLZ: 01171)

1. Aufgabenstellung

Hintergrund der Entwicklung war der Einsatz eines Strahlformungssystems mit zwei Schwingspiegeln (X-Y-Scanner), die den Laserstrahl rechtwinklig zueinander auslenken, an einem gepulst arbeitenden Nd-YAG-Laser. Für eine flexible Strahlformung zur Laseroberflächenveredlung (LOV), insbesondere zum Laserhärten, sind derartige Schwingspiegelsysteme besonders günstig, da sie bei einfachem Aufbau die Möglichkeit bieten, die mit dem Laserstrahl auf dem Werkstück zu belichtende Fläche in ihrer geometrischen Form sehr variabel zu gestalten. Mit derartigen Schwingspiegelsystemen kann in Verbindung mit der Variation der Spotgröße (Fokussierungsgrad) durch Parametervariation eine Vielzahl von nicht von vornherein übersehbaren Intensitätsverteilungen auf der Werkstückoberfläche in Form der entstehenden Lissajous-Figuren realisiert werden. Aus den eingestellten Parametern praktisch nicht mehr vorhersehbar sind jedoch die entstehenden Intensitätsverteilungen, wenn die Lissajous-Figuren an Lasern im Impulsbetrieb durch die Pulsdauer und die Tastfrequenz noch in willkürlich wählbare Teilstücke zerschnitten werden.

2. Lösungsweg

Die Berechnung der Verteilung der Intensität wird mit einem in FORTRAN77 geschriebenen Programm PULS realisiert.

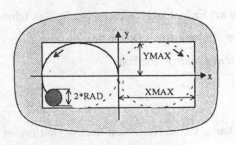

Bild 1: schematischer Strahlmittelpunktsweg bei zweidimensionaler Strahloszillation

Nach Bild 1 liegt für jeden Zeitpunkt innerhalb der vorgegebenen Bestrahlungszeit der Ort des Strahlmittelpunktes nach den vorgegebenen Scanneramplituden, -frequenzen und der Phasenverschiebung fest. Außerdem ist nach dem Zeitregime des Lasers (Laserimpuls und

Pulspause und der Phasenverschiebung zur Scannerbewegung) festzustellen, ob die Strahlung ein- oder ausgeschaltet ist. Das Strahlprofil des bewegten Laserstrahles wird im einfachsten Fall durch den Radius eines gleichmäßig ausgeleuchteten Kreises beschrieben. Für eine bessere Näherung kann das Strahlprofil aus Einzelteilen zusammengesetzt werden, für die jeweils die Relativintensität und ein äußerer und im Bedarfsfall auch ein innerer Radius vorzugeben sind. Die Normierung erfolgt nach der Gesamtleistung oder dem Spitzenwert im Strahlprofil.

Das Intensitätsprofil über der Oberfläche wird durch die Belegung eines Rastermaßes nach dem Schema des Bildes 2 beschrieben. Dazu wird über die Probenoberfläche ein Netz gelegt, dessen Ausdehnung und Intervalleinteilung vorzugeben sind, wobei das Programm an den Rändern noch jeweils 1 Zusatzintervall hinzufügt. Im aktuellen Programm liegt die Höchstgrenze der Intervalle in jeder Richtung bei 102.

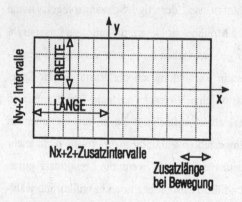

Bild 2: Rasterfeld der Intensitätsmatrix (schematisch)

Eine mögliche Bewegung in x-Richtung wird durch Hinzufügen von Zusatzintervallen berücksichtigt.

Aus der Lage des Mittelpunktes der Strahlung und den geometrischen Strahlgrößen (Radius bzw. Strahlprofil) wird ermittelt, welche Sektoren des Oberflächenrasters in diesem Zeitpunkt eine Bestrahlung erhalten. Dabei werden die vom Strahlprofil nur angeschnittenen Flächenanteile durch geschlossene Formeln beschrieben. Die Formeln berücksichtigen alle möglichen Fälle der Überdeckung von Kreisen und Rechtecken, wobei nur die Logik zur Auswahl der momentan zutreffenden Überlappung etwas Aufwand erfordert. Diese Berechnungen werden für alle Anteile des Strahlprofils durchgeführt, so daß am Ende dieser Teilrechnung die Zuwächse zu den verschiedenen Teilen des Oberflächenrasters bekannt sind. Dieser Prozeß wird in vorgebbaren Zeitschritten für eine gewünschte Integrationsdauer wiederholt.

3. Ergebnisse

Das vorliegende Grundprogramm wurde zur Nutzung auf verschiedenen Rechnerplattformen aufbereitet. Es kann unter den Betriebssystemen UNIX (für Rechner der workstation-Klasse) und MS-DOS (für PC), in Verbindung mit Standard-Graphikprogrammen zur anschaulichen 3D-Darstellung der berechneten Intensitätsverteilungen, genutzt werden.

Eine weiterentwickelte Programmvariante PROFIL bietet unter Einsatz des graphischen Systems PHIGS in Verknüpfung mit FORTRAN die Möglichkeit, das Modellierungsprogramm unter einer graphischen Oberfläche zu nutzen. Die Parametersätze (10 Scanner- und Laserparameter, bis zu 75 Parameter für das Strahlprofil und 7 Ablaufparameter) können als Datensätze im- und exportiert werden. Eine integrierte 2D-Grapik liefert für die schnelle Beurteilung von Parameterkombinationen eine erste Übersicht über den Weg des Strahlmittelpunktes im Rasterfeld der Intensitätsmatrix. Endergebnis ist auch hier wieder die Datenmatrix Intensitätsprofil.

Die berechnete Intensitätsmatrix kann als Eingangsdatenfeld für die weitere Prozeßsimulierung genutzt werden. Mit dem Personalcomputerprogramm GEOPT kann das infolge der modellierten Intensitätsverteilung im Werkstück induzierte Temperaturfeld sowie z.B. die daraus resultierende Härtungs- und Anlaßzonengeometrie beim Härten berechnet werden /LEPSKI/.

4. Anwendungsbeispiel

In Bild 3 ist zur vereinfachten Darstellung das bei eindimensionaler Strahloszillation jeweils ohne bzw. mit phasengesteuertem Impulsbetrieb entstehende Intensitätsprofil dargestellt. Um die Intensitätsüberhöhung am Spurrand für den cw-Laserbetrieb besonders deutlich zu machen, wurde das Verhältnis Schwingamplitude zu Laserspotradius mit 4:1 relativ groß gewählt.

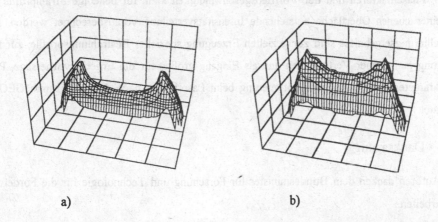

a)   b)

Bild 3: Berechnete Intensitätsprofile für eindimensionale Strahloszillation und
a) cw-Laserbetrieb
b) Impulslaserbetrieb mit definierter Phasenverschiebung zwischen dem Laserimpuls und der Scannerbewegung

890

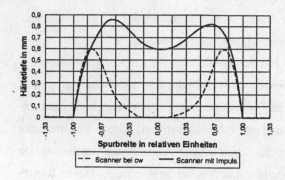

Bild 4: Mit GEOPT berechnete Härtespurprofile für die Intensitätsverteilungen aus Bild 3 (optimiert auf maximale Spurtiefe):

- o Laserleistung 3,6kW
- o Sollspurbreite 30mm
- o Oberflächentemperatur 1350°C
- o Abkühlzeit 1,5s

In Bild 4 sind nun die nach diesen Intensitätsprofilen mit dem Programm GEOPT berechneten Härtespurprofile dargestellt. Deutlich wird, daß beim cw-Betrieb aufgrund des großen A/R-Verhältnisses in der Spurmitte keine Härtung erfolgte. Beim Impulsbetrieb kann dagegen durch das Abschneiden der Randüberhöhung die Intensität in der Spurmitte angehoben werden, so daß auch in diesem Bereich für das Härten ausreichende Temperaturen erreicht werden.

5. Zusammenfassung

Es wurde ein Programmsystem zum Modellieren der unübersichtlichen Strahlenwege bei der Anwendung zweidimensionaler Schwingspiegelsysteme zur Strahlformung in Kombination mit pulsierenden Strahlquellen entwickelt. In Abhängigkeit von den Oszillationsfrequenzen, Amplituden, Phasenfaktoren und der Vortriebsgeschwindigkeit kann für beliebige Strahlprofile die auf einer ebenen Oberfläche entstehende Intensitätsverteilung vorausberechnet werden. Die Modellierungsergebnisse sind zur gezielten Erzeugung spezieller Bestrahlungsprofile, zur Beurteilung gemessener Verteilungen und als Eingangsgrößen für die Ergebnisvorhersage, Prozeßparameterbestimmung und -optimierung beim Laserhärten nach dem Programm GEOPT geeignet.

6. Danksagung

Die Autoren danken dem Bundesminister für Forschung und Technologie für die Förderung der Arbeiten.

7. Literatur

/LEPSKI/ D. Lepski, W. Reitzenstein: "Computergestützte Prozeßoptimierung bei der Laser-Umwandlungshärtung von Eisenwerkstoffen", Härterei-Techn. Mitteilungen 46(3), 178 (1991)

Time Resolved Photoelectron Spectroscopy After Laser Ionization

K.-M. Weitzel, M. Penno, and H. Baumgärtel
Institut für Physikalische und Theoretische Chemie, Freie Universität Berlin,
Takustraße 3, 1000 Berlin 33, FRG

In this contribution we will report on recent experiments aimed at studying the spectroscopy and the dynamics of state selected molecular ions. One of the major sources for kinetic and energetic information on unimolecular reactions of ions has been the one photon ionization of molecules with consecutive reaction. In this large field the most direct results emerged from investigating energy and/or state selected ions. In those experiments which do not inherently produce state selected ions the photoelectron photoion coincidence (PEPICO) technique [1] has to be applied. Typically these experiments have been performed at a resolution of about 30 meV (240 cm^{-1}) limited by the contribution from the photons and the electron energy. Applying this technique we have recently studied the J dependence of the dissociation energy Eo(J) [2,3]. It was at the same time the advantage and the limitation of this experiment that it did not require J selectivity but simply used the variation of the rotational temperature.

With the modern lasers becoming conveniently available it was clearly tempting to investigate the properties and the dynamics of ions at significantly higher resolution. As far as the spectroscopy of molecular ions is concerned a large number of rovibrationally resolved data has become available by applying the ZEKE technique [4]. The heart of the ZEKE technique lies at an extremely high electron energy resolution. The ultimate resolution is reached by combining this technique with the spectral resolution of a laser. However, this technique has mainly been restricted to the region of the ionization potential (IP) of molecules. Recently we have performed a ZEKE-ion coincidence experiment in the region of the dissociation threshold of the H loss reaction from acetylene ions [5]. This experiment has again been limited in photon resolution by the use of synchrotron radiation. As long as a laser system which allows to carry out the same experiment is not available we have to search for a different scheme for the ion preparation.

The next to ideal experiment would be to form v and J selected ions with one or two photons of one or two colors in a first step and then excite this ion with an additional photon to the reaction threshold. The different schemes are illustrated in fig. 1.

The most powerful but also most difficult approach is the one shown in fig. 1c. There the J selection is obtained in the first step and the v selection, i.e. v=0, in the second step. On the other hand scheme 1a in general provides only the v selectivity while scheme 1b provides only the J selectivity. However, it has been shown that also a high v selectivity can be observed in experiments following scheme 1b [6,7].

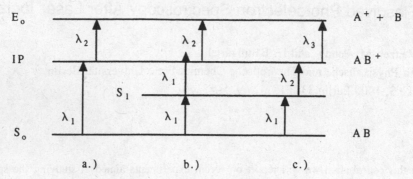

Fig. 1 Laser ionization schemes

We have recently set up an experiment also following scheme 1b, i.e. two photon ionization with consecutive further excitation of the ion by a third photon. As mentioned before, one has to check on the v selectivity of the ionization first. In this contribution we will present first results of this experiment. The formation of ions in different vibrational or rotational states will be reflected in the properties of the ejected photoelectron, i.e. its kinetic energy and its angular momentum. This is a classical task for photoelectron spectroscopy. Besides from the classical electrostatic analysers particularly magnetic bottle spectrometers [8] and time of flight spectrometers [9,10] have been applied recently in laser ionization studies.

Experimental Method

Molecules are ionized in the center of an electron ion spectrometer. The ionization region is either kept at a static homogeneous electric field or field free during the photon molecule interaction. In the latter case an electric field of about 10 V/cm is turned on after a delay time Δt. The electrons enter a field free drift region of 60 mm and are detected by a micro channel plate detector under a solid angle of 10 mrad. The photons for ionizing the molecules are provided by a tuneable dye laser system (Lasertechnik Berlin/Lambda Physik) which uniquely combines a spectral resolution of about 0.8 cm^{-1} (at 500 nm) with a temporal FWHM of the pulses of about 250 ps. The heart of this laser system is a short puls nitrogen laser which pumps a dye laser under grazing incidence conditions. There are several reasons for choosing the shortest possible laser puls corresponding to a desired spectral resolution. Short pulses bear less problems with competing processes like fragmentation or multiphoton absorption. They offer the additional advantage of making e$^-$ time of flight spectroscopy easily accessible. Here we apply a pulsed delayed extraction scheme to separate electrons with different kinetic energy and image them on to the detector. If the ionization occurs at a static homogeneous field all electrons will reach the MCP detector at the same time. If the ionization takes place field free the electrons will separate according to their kinetic energy. If the electric field is turned on after a delay time Δt only those electrons are detected which are seen by the detector under a certain solid angle. In general one will then see forward and backward scattered electrons separated by a dip caused by angular discrimination.

Results

We have applied the one color two photon ionization to the fluorobenzene (FB) molecule. This is a test molecule whose spectroscopy is quite well understood [11,12]. Fig. 2 shows the TPI spectrum in the region of the $S_1 < -- S_o$ (0-0) transition at room temperature. The major features of this spectrum are dominated by the absorption of the first photon. They agree with the literature [11,12] and the most prominent transitions are marked with arrows. Due to the small geometry changes the $\Delta v = 0$ transition should be prefered in the absorption of the second photon. Thus the ions should be formed in the v=0 state if we pump the 0-0 transition. Since the two photon energy corresponding to the 0-0 transition reaches about 199 meV (1602 cm^{-1}) above the ion ground state this excess energy then would have to be carried away by the electron. Therefore one may not expect to see low energy electrons. Fig. 3-5 show the e- TOF spectra recorded at 264.40 nm under various extraction field conditions.

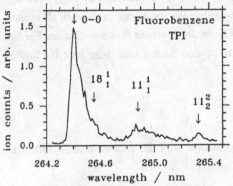

Fig. 2 TPI spectrum of FB at 300 K

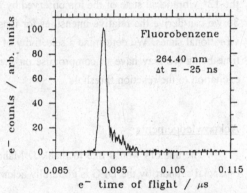

Fig. 3 e$^-$ TOF spectrum of FB,
λ = 264.4nm, Δt = -25ns

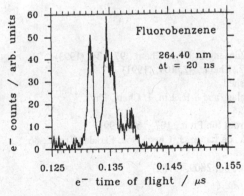

Fig. 4 e$^-$ TOF spectrum of FB,
λ = 264.4nm, Δt = 20ns

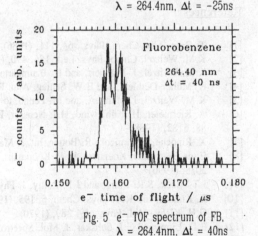

Fig. 5 e$^-$ TOF spectrum of FB,
λ = 264.4nm, Δt = 40ns

If the electric field is turned on at the time of the ionization or even earlier all electrons reach the detector at the same time. Thus the electron TOF spectrum (fig. 3) shows only one peak. If the electric field is delayed by 20 ns the e^- TOF spectrum (fig. 4) exhibits three distinct peaks. From experiments with monoenergetic electrons [5] we know that the two outer peaks correspond to forward and backward scattered electrons of the same kinetic energy. We assign these electrons to ions in the $v=0$ state. However, the spectrum also contains one peak in the center between forward and backward scattered electrons. This peak clearly corresponds to electrons with very low kinetic energy, thus to an ion in an excited vibrational state. Note that the FWHM of the peaks and their separation is on the order of a few ns. The experiment is thus only possible because of the short laser puls resulting in a total resolution of 500 ps. Fig. 5 shows the e^- TOF spectrum at a delay of 40 ns. There the electrons with 199 meV kinetic energy have been completely discriminated. The remaining electron peak is quite broad and shows a very shallow dip in the center. These electrons have kinetic energy of about 40 meV. By comparison with the photoelectron spectrum recorded by Boesl and coworkers [13] we assign these electrons to the $6a^2$ state of the ion. So far we have not been able to resolve the $6a^1$ and the 12^1 vibrational state of the ion observed by Boesl et al. in our e^- TOF spectra. However, from fig. 3b we can derive the relative intensities for ions being formed in the $v=0$ state to those in all excited vibrational states. We determine a selectivity of about 80 % for ions formed in the $v=0$ state. For the time being we may have to compromise on such a selectivity and simply take those ions for further excitation to the reaction threshold.

Acknowledgements

The authors would like to thank J. Mähnert for assistance with the data acquisition software. Financial support by the DFG is gratefully acknowledged.

References

[1]   T. Baer, Adv. Chem. Phys., 64, 111, (1986)
[2]   K.M. Weitzel, Chem. Phys. Lett., 186, 490, (1991)
[3]   K.M. Weitzel, J. Mähnert, and H. Baumgärtel, Ber. Bunsenges. Phys. Chem., 97, 134, (1993)
[4]   K. Müller-Dethlefs, and E.W. Schlag, Ann. Rev. Phys. Chem., 42, 109, (1991)
[5]   K.M. Weitzel, J. Mähnert, and M. Penno, to be published
[6]   A. Kiermeier, H. Kühlewind, H.J. Neusser, E.W. Schlag, and S.H. Lin, J. Chem. Phys., 88, 6182, (1988)
[7]   X. Ripoche, I. Dimicoli, R. Botter, Int. J. Mass Spectrom Ion Proc., 107, 165, (1991)
[8]   E. de Beer, B.G. Koenders, M.P. Koopmans, and C.A. de Lange, J. Chem. Soc. Faraday Trans., 86, 2035, (1990)
[9]   J.T. Meek, S.R. Long, and J.P. Reilly, J. Phys. Chem., 86, 2809, (1982)
[10]   K. Kimura, Int. Rev. Phys. Chem. 6, 195, (1987)
[11]   G.H. Kirby, Mol. Phys., 19, 289, (1970)
[12]   E.D. Lipp, and C.J. Seliskar, J. Mol. Spectrosc., 87, 255, (1981)
[13]   K. Walter, K. Scherm, U. Boesl, J. Phys. Chem., 95, 1188, (1991)

Post Deadline Papers

Congress-Chairman: H. Seidlitz

Dephasing-Induced Rabi Modulation in Inversionless Lasing

C. J. Hsu
Department of Physical Science, Univ. of New Brunswick, CANADA

ABSTRACT

Inversionless lasing can be achieved in a medium subject to a strong resonant pump which split the energy levels. In this work, by using a two-level system and a population modulation (pulsation), we demonstrate that the lasing can be understood as the amplification of the signal riding coherently on the modulated population. Though it is time average inversionless, a modulated excessively populated upper lever can give rise to amplification. In this mechanism, I find that the excessively populated level are three-fold split shifted by 0, $+\hbar\Omega$, and $-\hbar\Omega$ from the pump photon energy, where Ω may be called the induced Rabi frequency. This induced Rabi frequency is distinguished from the ordinary one by the fact that it is dependent on the relaxation rates. In a dephasing medium, the dipole dephasing can be a few orders faster than the decay rate and thus I find that the splitting behaviour is quite different from the splitting treated by others, and that the dephasing will enhance the lasing

I. Introduction

A two-level system driven by a strong resonant or near-resonant pump field is not stable, which can be disturbed by even a weak probe field. The probe-pump beating to modify the populations was introduced by Bleombergan and Shen,[1] but the significance was not clearly identified. With the same term Mollow[2] predicted stimulated emission 'without population inversion' which was verified experimentally in an atomic beam system.[3] 'Inversionless' lasing of this origin has been observed.[4],[5],[6] in gas system. In these studies, the lasing is either attributed to the transitions between two-fold split levels (or dressed states) [3] or considered as stimulated scattering without understanding the population property. [4-6] On the other hand, in the hole-burning studies for absorptions,[7] the population beating term is either ignored completely, or included in numerical treatments [8] in some physical conditions in which the physical meaning can not be readily obtained . Though the beating population has received more attention lately, the physical picture for the shifts and the the widths of the lasing component based on level transition is lacking and the explicit effect for split levels in comparison with saturation populations is not immediately clear. I have employed a three-component form with large pump detuning to analyze the anomalous width in stimulated Rayleigh scattering[9] and the holewidth narrowing in hole-burning.[10] In this work, I shall use the same method to demonstrate how the populations are redistributed and a three-fold split levels for the dressed system can be assigned.

This three-component form is valid for arbitrary pump intensity, frequencies for the pump and the probe fields, and as well as the relaxation rates. I shall demonstrate that the absorption frequency can be quite apart from that due to transitions between either bare levels or conventional dressed levels[11] when the pump is strong and detuned from resonance. In this study, I find that these stimulated emissions can also be considered as inversion lasing. The argument for being inversion transition is that the temporal phase can be dropped by choosing an appropriate rotating frame. Since this dressed levels are directly related to the pump frequency and relaxations rate of the medium, I feel that this approach has a potential to resolve the level splits in a multi-level system in which lasing may take place.

II. Three-fold split level dressed states

In a two-level system, the off-diagonal matrix element $\rho_{10}^{(1)}(\omega)$ governs the absorption (or amplification) of the probe at the frequency ω. By assuming the probe propagating at the (nearly) same direction as the pump, $\rho_{10}^{(1)}(\omega)$ can be solved from the Bloch equations. In order to understand how this term is related to the populations, I write this element in the form:

$$\rho_{10}^{(1)}(\omega) = \frac{\Omega_p n}{-i(\Delta\omega + \Delta\omega') + \gamma} + \frac{\Omega_p \Delta\rho^{(1)}(\Delta\omega')}{-i(\Delta\omega + \Delta\omega') + \gamma} \tag{1}$$

Here, $\Delta\omega$ is the pump detuned from the resonance of the medium, is the probe detuned from the pump, γ is the transverse relaxation rate, and $2|\Omega_p|$ and $2|\Omega_R|$ are the Rabi frequency of the probe and the pump respectively. There are two parts for the population difference: n is the usual saturation population difference given by $n_0 / [1 + 4 (\gamma/\gamma') |\Omega_R|^2/ (\Delta\omega^2 + \gamma)]$, where γ' is the longitudinal relaxation rate; and $\Delta\rho^{(1)}(\Delta\omega)$ is the amplitude for the beating population given by:

$$\rho_{00}^{(1)}(\Delta\omega') - \rho_{11}^{(1)}(\Delta\omega') = \Delta\rho^{(1)}(\Delta\omega')e^{-i\Delta\omega't} + \Delta\rho^{(1)*}(\Delta\omega')e^{i\Delta\omega't} \tag{2}$$

An explicit form for this amplitude has been given by Mollow. In order to understand the level structure, I like to write in a three-component form:

$$\Delta\rho^{(1)}(\Delta\omega') = -\pi\Omega p.(2\Omega_R^2) \left[\frac{X_0}{-i\Delta\omega' + \Gamma_0} + \frac{X_+}{-i(\Delta\omega' + \Omega) + \Gamma} + \frac{X_-}{-i(\Delta\omega' - \Omega) + \Gamma} \right] \tag{3a}$$

$$X_0 = (1 - \Gamma/z*) \frac{2\gamma - \Gamma_0}{\Omega^2 + (\Gamma - \Gamma_0)^2}, \qquad X_\pm = (\frac{\pm(\Omega \mp \Gamma)}{z*} + 1) \frac{\pm\Omega + 2\gamma - \Gamma}{[\pm\Omega - (\Gamma - \Gamma_0)](\pm2i\Omega)} \tag{3b}$$

$$n = n_0 / [1 + 4 (\gamma/\gamma') |\Omega_R|^2/ (\Delta\omega^2 + \gamma)] \tag{3c}$$

$$\Gamma_0 = \gamma' + (\gamma - \gamma') 4|\Omega_R|^2/ [(\Gamma_0 - \gamma)^2 + \Delta\omega^2 + 4| \Omega_R|^2] \tag{3d}$$

$$\Gamma = \tfrac{1}{2} (\gamma' + 2\gamma - \Gamma_0) \tag{3e}$$

$$\Omega = [(\gamma/\Gamma_0) (\Delta\omega^2 + \gamma) + (\gamma/\Gamma_0) 4| \Omega_R|^2 - \Gamma^2]^{\tfrac{1}{2}} \tag{3f}$$

In Equ. (3a) $z*$ is given by $i\Delta\omega + \gamma$. Γ_0 and Γ are induced relaxation rates, and may called induced Rabi frequency. We first calculate the induced relaxation rate Γ_0 by using iteration method initially with $\Gamma_o = \gamma'$. From this form the level split can be easily inspected, and it clearly shows that this system is dressed by the pump but the dressing is influenced by the relaxation rates as well as by the probe. This dressed system gives rise to a few features which are different from the usual dressed system: a) From the three-component form each of the two levels can be considered to split into three levels in contrast to two-fold split in the usual dressed system. b) All the levels generally have both a Lorentzian as well as a dispersive-like distribution in comparison to only a Lorentzian part in the usual dressed system. c) The level split with a reference to pump photon energy due to the fact that either absorption or emission arise from extra-resonances. d) In the earlier studies with only numerical display, the separation was believed equal to the usual Rabi frequency . In this study the separation can be accurately calculated with arbitrary pump intensity and detuning interacting with a medium with relaxation mechanism, and the former can be shown as limiting cases.

III. Inversion or inversionless lasing?

To understand the lasing is inversion or not, it is necessary to gain a clear knowledge for the beating population difference given in Equ. 2. In this equation, this population difference has a temporal phase and the phases for the pump and the probe in the product of ,which is purely the property of the fields. Though the this product can be very small due to the weak strength of in comparison to the saturation population difference, n, the effect can predominate over that of n,the saturation population difference, because the beating part is coupled to the pump as given in Equ. 1 when pump is strong. This fact can be seen from Fig. 1. The parameters used in all the graphs in this figure are : $\Delta\omega = -500$, $\gamma' = 2\gamma$, $2|\Omega_o| = 500\gamma$ and they are all in unit of γ . In Fig. 1 (a), it shows the absorption peak at $\Delta\omega' = -\Delta\omega$ (=500γ), or ω at ω_{1b} due to the first term in Equ.1 contributed by the saturation population difference, n , and in Fig. 1(b),due to the second term contributed by the beating population difference. By comparing these two graphs, one find surprisingly that the saturation population absorption peak equal to the gain dip from the beating population, and the net effect at $\Delta\omega' = 500\gamma$ is not observable in Fig. (c). In Fig.1 (c) and its inset are the usual three components. The two components in the inset are the lasing components at $\Delta\omega' = 0$ and $-\Omega$ in this case. In order to understand the lasing terms a better understanding of the beating term is essential. For this purpose, the Equ. 3 is written as:

$$\Delta\rho^{(1)}(\Delta\omega') = n\Omega\rho.(2\Omega_z^*)\,(\Delta n)$$
(4)

where Δn is the negative sum of the terms given in the bracket in Equ. 3 and this quantity represents the dressed medium property. The imaginary part of nonlinear susceptibility of this origin is proportional to real part of Δn , and the gain or absorption is independent of the phases of the pump and probe. In Fig. 2, the term $Real\,\Delta n$, which is proportional the inversion amplitude of the beating population, versus $\Delta\omega'$, probe detuning from the pump is plotted. In all the plots in this figure, $\gamma = 100\gamma'$,

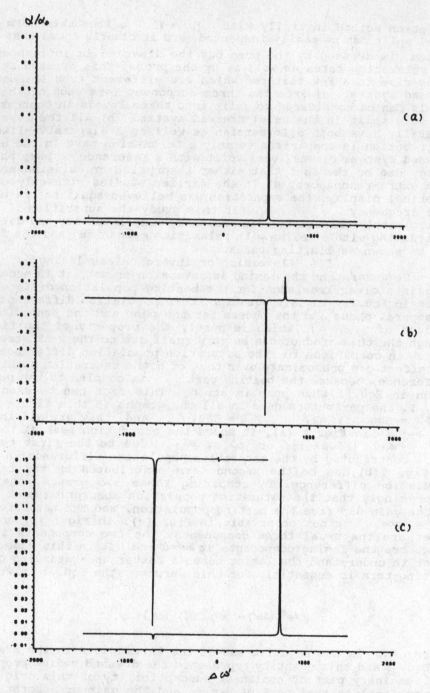

Fig. 1 Absorption as a function of the probe frequency detuned from the pump, $\Delta\omega$. It shows in (a) the contribution by the saturation population difference, in (b) by the beating population difference, and in (c) the overall contributions. The components in the inset of (c) are the lasing components. The parameters used are $\gamma = 2\gamma$, $2|\Omega_a|=500\gamma$, and pump detuning from resonance $\Delta\omega = -500\gamma$.

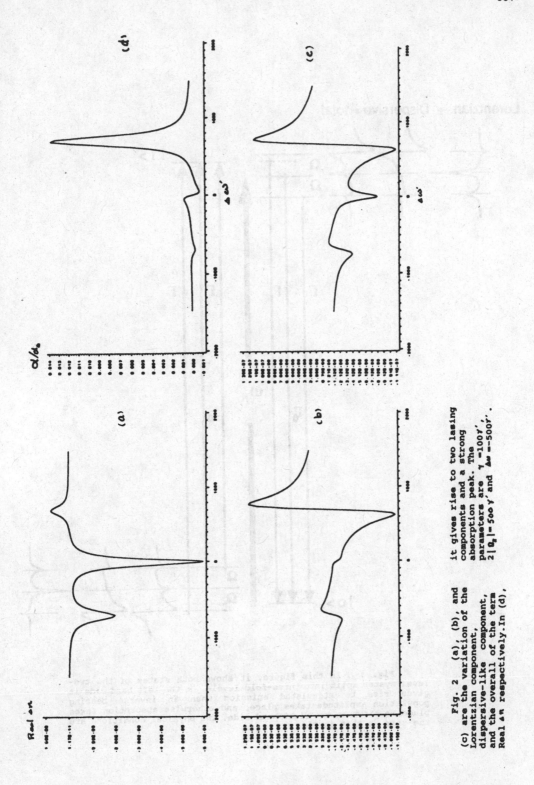

Fig. 2 (a), (b), and (c) are the variation of the Lorentzian component, dispersive-like component, and the overall of the term Real Δn respectively. In (d), it gives rise to two lasing components and a strong absorption peak. The parameters are $\gamma = 100\gamma'$, $2|g_x| = 500\gamma'$ and $\Delta\omega = -500\gamma'$.

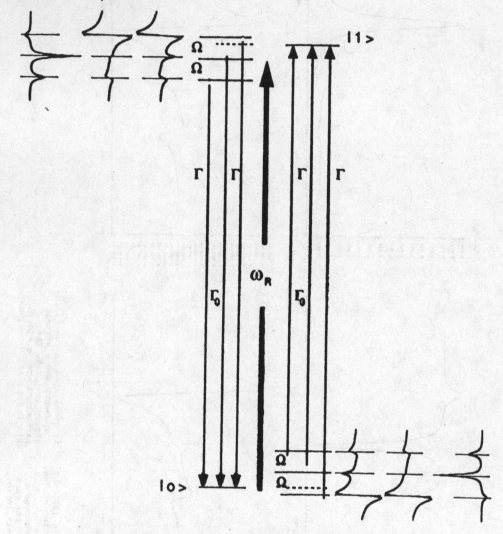

Lorentzian + Dispersive = total

Fig. 3   In this figure, it shows both states of the two-level system split into three-fold levels. On the left hand side it gives rise to stimulated emission when an inverse beating population amplitude takes place, and otherwise absorption takes place as shown on the right hand side. The physical conditions are the same as that in Fig. 2

$2|\Omega_R|=500\gamma$, $\Delta\omega = -500\gamma$. A large value γ is usually seen in a dephasing case. Because the X terms are generally complex, each of the three components comprise a Lorentzian as well as dispersive-like components, which are given respectively in Fig.2 (a) and (b), and the resultant is given in (c). The familiar corresponding absorption variation is plotted in Fig. 2(d). After resolving these shapes the energy levels for the dressed system can be determined and is displayed in Fig. 3. In this figure, I propose that: (1) the splitting is with reference to the pump photon energy since the lasing is due to extra-resonances; (2) the split of levels are three-fold because the transition rates Γ and Γ_0 are different and the magnitudes of X_+ and X_- are quite different when the pump is detuned from resonance; (3) the transition is allowed only between bare state←→dressed state and bare state←→bare state (the later has no effect due to the cancellation between two type populations as explained previously). In this figure, it shows this type transitions takes place only if there exists a probe. Stimulated emission occurs when the inversion for the amplitude of the beating population difference takes place and other-wise it will give absorption as shown on the left and right hand sides respectively in Fig.3. This fact indicates that, by choosing an appropriate rotating frame, the lasing can still be considered as an inverse transition. However, for the lasing being inversionless can still be justified based on the temporal factor which gives a zero average in the usual frame of reference.

VI. Conclusion

In this work, it shows that the beating population difference is more important than the saturation population in either absorption or stimulated emissions. In this mechanism with probe, a three-fold split dressed system is proposed, and lasing occurs when the inversion for the amplitude of the beating population difference occurs. The lasing can be considered either inversion or inversionless depending on the observer's frame of reference.

References:
1. N. Bloembergen and Y.R.Shen, Phys. Rev. A133, 37(1964).
2. R. B. Mollow, Phys.Rev. A5,2217(1972).
3. F. Y. Wu, S. Ezekiel, M. Ducloy, and B.R. Mollow, Phys.Rev. Lett. 38, 1077 (1977).
4. G. Khitrova, J. F. Valley, and H. M. Gibbs, Phys. Rev. Lett. 60, 1126 (1988).
5. D. Grandclement, G. Grynberg, and M. Pinard, Phys. Rev. Lett. 59, 40 (1987).
6. Mark T. Gruneisen, Kenneth R. MacDonald, and Robert W. Boyd, J. Opt.Soc. Am. B, 5,123 (1988).
7. W. R. Bennett, Phys. Rev. 126, 580 (1962), and Y.R. Shen, The principle of nonlinear Optics (John Wiley & Sons, 1984).
8. V. S. Letokhov and V. P. Chebotayev, Nonlinear Laser Spectroscopy (Springer-Verlag,Berlin, 1977), Chapter 11.
9. C. J. Hsu, J. Opt.Soc. Am. B, 7, 2155 (1990).
10. C. J. Hsu and C. H. Leung, submitted for publication.
11. C. Cohen-Tannoudji, in Frontiers in Laser Spectroscopy, Ed. R. Balian, S. Haroche and S. Liberman, (Amsterdam, North Holland).

Gain Coefficient of HF-Laser on two Quantum Transitions Under Optical Pumping

A.V. Michtchenko, F. Lara-Ochoa, A.D. Margolin*, V.M. Shmelyov*
Instituto de Quimica, U.N.A.M., Circuito Exterior, Cd. Universitaria, Del. Coyoacan,
C.P. 04510, Mexico, D.F., MEXICO.
*) Institute of Chemical Physics Russian Academy of Sciences,
Kosygin str., 4, 117977, Moscow, Russia.

The optical resonance pumping HF-laser has been studied theoretically and experimentally in many works[1-3]. In the case of optical resonance pumping laser's photon energy from high power optical source is resonantly absorbed by nonreacting molecules in the gas mixtures, and then an over population in the upper vibrational levels is created during vibrational-vibrational and vibrational-translational exchange between molecules in the gas mixture. Analyzing this non-Boltzman distribution of molecules on vibrational levels during these processes it can be seen that the gain coefficient on high vibrational levels (such as $v = 9 - 13$) is high enough and does not strongly depend on the number of vibrational level. Using the distributions of molecules on vibrational levels under optical pumping the calculations of the gain coefficient on two quantum transitions was made. It was found that in this case the gain coefficients on two quantum transitions on high vibrational levels ($v = 10 -13$) exceed the gain coefficient on one quantum transitions at the same vibrational levels. According to these results it is possible to use optical pumping to obtain laser radiation from these high vibrational levels. Using only two quantum transitions from upper vibrational levels it is possible to obtain laser radiation in other spectrum region with the increasing of frequency of laser radiation. For this purpose it is possible to use a selective resonator or to switch on the resonator at the proper moment of time.

1. Pummer H., Proch D., Schmailzl M., Kompa K.L. - *J. Phys. D, 1978, v. 11, No 2, p. 101*
2. Kwok M.A., Wilkins R.L. - *Appl.Optics, 1983, v. 22, No 17, p. 2721*
3. Sileo R.N., Cool T.A. - *J.Chem.Phys., 1976, v. 65, No 1, p. 117*

Short-Pulse High-Repetitive N$_2$-TE Lasers

S.V. Kukhlevsky, L. Kozma
Janus Pannonius University, Department of Physics, H-7624
Ifjusag u. 6, Pecs, Hungary

Abstract-We have studied the effects of shortening the excitation current on both the pulse-repetition rate (PRR) capabilities and the pulse duration of a low-pressure (P<200Torr) N$_2$-TE laser. It has been found that the laser pulsewidth decreases and the maximum PRR increases with the decreasing duration of the discharge current. A 1-ns laser pulse and PRR as high as 1000Hz have been obtained in nonflowing regime of the laser operation.

I. Introduction

Recently some attention has been devoted to the understanding and development of high PRR, transversely excited UV lasers without gas flow [1-3]. In these systems high PRR operation is achieved avoiding the common growth of arc channels by a resistively [1], inductively [2] stabilized discharge. Unfortunately, application of these method in N$_2$-TE lasers are not effective because of the insufficiency of their gain duration. Another method to avoid the growth of arc channels is to make the discharge duration as short as possible. In this case, short-pulse high-repetitive N$_2$-TE laser operation will be expected.

II. Experimental

The schematic diagram of the test N$_2$ laser is shown in Fig.1. This is the Polloni version [4] of a low-pressure N$_2$-TE laser with the C-C discharge excitation. The thyratron (Tr) TGE 500/16 switched the charge of the capacitor bank (C_1=1350pF) to the flat-plate capacitor (C=1350pF) through the inductance (L_1=5.5μH). The capacitor C discharged through inductance L uniformally distributed along the electrode (twenty inductances L_i in parallel). The way we placed the flat-plate capacitor in order to minimize the inductance L is shown in Fig.1.(b). The discharge pulsewidth was varied from 15 to 5.4ns by decreasing the value of L. We also studied the case when the flat-plate capacitor directly connected with the electrodes (the self-inductance of the discharge circuit and the discharge duration were calculated to be 1.7nH and 5ns respectively). A sealed-off glass cuveta having an area of 1x1cm^2 and a length of 30cm contained the gas. In the center of the cuveta 0.5mm thick and 30cm long copper electrodes were placed 3mm from one another. The

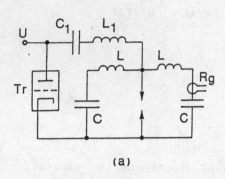

(a)

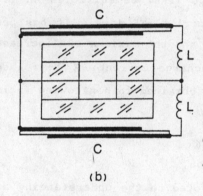

(b)

Fig.1

electrodes had a V-shape to induce preionisation. An aluminum flat mirror placed 1cm from the end of the plasma tube was used as a back reflector. During the present work the supply voltage had a constant value (9kV) and the gas pressure was varied from 75 to 200Torr. The discharge pulsewidth was measured by a Rogovsky-circuit placed around an inductance L_i and an oscilloscope (Tektronix 7104). The energy of the laser pulses was measured with a 14 NO type thermopile (Laser Instrumentation Ltd).

III. Results and Discussion

Fig.2 shows the laser energy at various PRR from 100 to 1000Hz for a gas pressure of 100Torr and for the different discharge durations

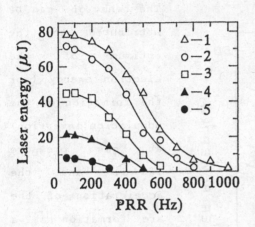

(1-5ns, 2-5.4ns, 3-8.5ns, 4-11ns and 5-15ns). One can see that the laser energy and PRR are increased with decreasing duration of the discharge pulse. As the pulse duration is decreased from 15 to 5ns, the maximum PRR is increased from 300 to 1000Hz. It has been found by observing the discharge visually that the small decrease in laser energy at a low PRR (up to about 200Hz) was caused by the onset of discharges in the glow plasma. As monotonic contraction of the plasma into the small-volume group of diffuse-arc discharges was observed. Consequently, it appears that the decrease of the active volume and the optical distortions are responsible for the dramatic degradation in the laser energy at the high repetition rates (PRR>200Hz).

Fig.2

the great number of diffuse-arc
PRR increased from 200 to 1000Hz a

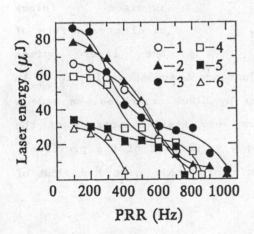

Fig.3

The high PRR operation can be better understood by comparing the repetitively-pulsed capabilities of the laser at different gas pressures. In Fig.3 the laser energy is displayed as a function of the PRR at the discharge pulsewidth of 5ns and for the various gas pressures from 75 to 200Torr (1-75Torr, 2-100Torr, 3-125Torr, 4-150Torr, 5-175Torr and 6-200Torr). This figure shows that the maximum PRR weakly depends on the gas pressure in the range of 75-150 Torr and quickly drops when pressure reaches 200Torr.

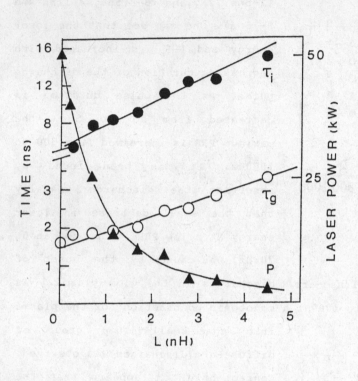

The behavior can be attributed to the optimum of the electron energy (for the formation of the population inversion) at gas pressure 125Torr and the acceleration of the arc formation at a higher gas pressure.

The dependence of the duration of the pumping current impulses (black circles) and FWHM of the laser pulses

Fig.4

(open circles) on the value of the inductance L are shown in Fig.4. In the figure we indicate the laser power by black triangles. We made a measurement when the inductance L was zero (self-inductance of the circuit was 1.7nH) and capacitor C had a value half of the previously used one, i.e. C=675pF. The laser pulse for this case had FWHM of 0.9ns.

IV. References

[1] Kan-Ich Fujit, A.G. Kearsley, A.J. Andrews, K.H. Errey, C.E. Webb: IEEE J. QE-17, 1315 (1981).
[2] R.C. Sze: J. Appl. Phys. 54, 1224 (1983).
[3] R. Buffa, L. Fini, M. Matera: Opt. Commun. 50, 397 (1984).
[4] R. Polloni: Opt. Quant. Electr. Lett. 8, 565 (1976).

Polarization Tests on TE/TM-Switching 1.3 μm Laser Diodes

M.Voß, A.Bärwolff, A.Klehr and R.Müller

Max-Born-Institut für Nichtlineare Optik und Kurzzeitspektroskopie

Rudower Chaussee 6, D-12489 Berlin, Germany

1. Introduction

Light emission from semiconductor lasers is predominantly TE polarized rather than TM polarized (with, respectively, the TE or TM field vector oscillating along the junction plane) due to a smaller reflectivity of the TM wave at the laser facets. However, biaxial tensile stress in the active region that is introduced by a considerable lattice mismatch between epitaxial layer and substrate, leads to laser operation in the TM mode /1,2/. This behavior is caused by a lower gain of the TE mode and a higher gain of the TM mode in comparison with a stress-free laser. Especially, strained InGaAsP lasers were found to exhibit dominant TM-mode emission as well as TE/TM-mode switching and polarization bistability, /3,4,5/. These phenomena are of interest both in view of potential device applications (optical switches and memories for optical computing and fibre-optic communications) and for improving understanding of nonlinear interaction between laser modes. Polarization measurements give insight into the physical mechanism of TE/TM switching.

This paper presents new results of polarization measurements of radiation emitted from ridge-waveguide (RW) InGaAsP/InP lasers at 1.3 μm wavelength. The degree of polarization, ρ, has been determined in dependence on injection current and temperature for below and above threshold operation of the laser. ρ is defined as $\rho = (P_{TE} - P_{TM}) / (P_{TE} + P_{TM})$ where P_{TE} and P_{TM} denote the power of the TE- and TM-polarized contribution, respectively, in the emitted radiation. Lasers of various emission characteristics have been investigated, in particular lasers showing polarization switching and bistability.

2. Laser structure and experimental setup

The laser structure shown in Fig.1 was grown on a (001) n-InP substrate (2) followed by a Sn-doped InP buffer layer (3), a 0.15 μm thick quaternary InGaAsP active layer (4) with a tensile stress in the range of $5 \times 10^8 - 1 \times 10^9$ dyn/cm^2, a 0.2 μm thick quaternary etch stopping layer (5) of composition $In_{1-x} Ga_x As_y P_{1-y}$ (x = 0.12; y = 0.27), a 1.6 μm thick InP cladding layer (6), and a Zn-doped (6....8 $\times 10^{18}$ /cm^3) InGaAsP cap layer (7). A CVD-SiO$_2$ isolation layer (8) is deposited over the entire wafer. At the top of the ridge a window is opened for the contact to the p-InGaAsP cap layer. The contacts (1) are made from gold. The cross section of the ridge exhibits a "waist" structure composed of a trapezoidal bottom part with a basis length of about 4.5 μm and a 3.5 μm wide rectangular upper part. Polarization bistability was found only in lasers with such a ridge structure while lasers with ridges of rectangular cross section commonly used did not show this effect.

The output beam of the laser was sent through a Foster polarization splitter which splits the light in the TE-polarized beam and the TM beam. Afterwards each of these two beams was fo-

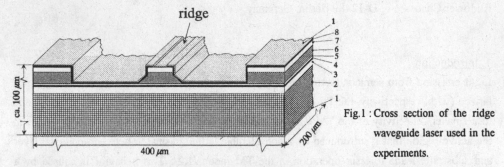

ridge

Fig.1 : Cross section of the ridge waveguide laser used in the experiments.

cused to a broad-area Ge-photodiode. Stable laser emission was achieved by means of a low-noise power supply and a temperature stabilization unit. A computer was used for the control of injection current and temperature as well as for a numerical analysis of the data.

3. Experimental results

In the following, we present data for RW-InGaAsP polarization bistable lasers.

Fig.2 shows the emitted power P and the polarization degree ρ as a function of injection

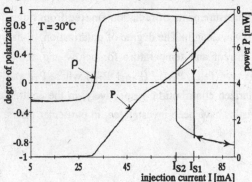

Fig.2: Measured light power, P, and degree of polarization, ρ, for a polarization-bistable RW-InGaAsP laser.

current for a polarization-bistable RW-laser. The bistable behavior of this laser is obvious from a hysteresis loop seen in the light power / current characteristics. Polarization-resolved measurements reveal that the lower hysteresis branch exhibits TE emission while the upper one shows TM emission. This is evident from the behavior of the ρ-curve: Lasing starts in the TE mode, $\rho > 0$, at about 30 mA and TE emission is sustained up to the switching current $I_{S1} \approx 73$ mA at which an abrupt transition from TE to TM emission occurs corresponding to a change in the sign of ρ. For $I > I_{S1}$ the TM mode is dominant ($\rho \approx -0.8$). When the current is reduced the laser switches back to TE emission at $I_{S2} \approx 66$ mA and ρ changes its sign again. The hysteresis width is given by the difference I_{S1}-$I_{S2} \approx 7$ mA. Well below threshold $\rho < 0$ was obtained corresponding to a larger power of the TM-polarized contribution in the spontaneous emission. Since the optical feedback provided by the laser facets becomes more efficient with increasing injection current TE-mode emission is favoured near the threshold, due to a higher single-pass gain in comparison with the TM mode, resulting in an enhancement of ρ. Finally, the degree of polarization increases rapidly and approaches the value ρ ≈ 1, i.e. $P_{TE} \gg P_{TM}$, just above the (TE-) threshold. Tensile stress in the RW-laser strongly depends on the lattice temperature T of the active region.

Fig.3 gives an example of the temperature dependence of the emission characteristics.

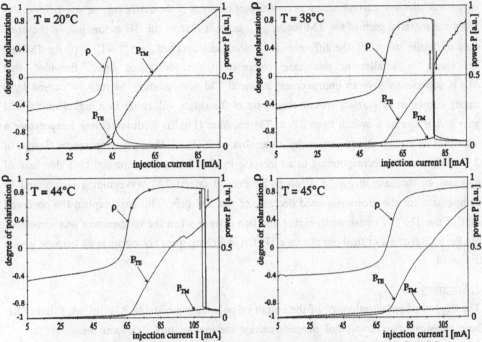

Fig.3:   Polarization- resolved power P_{TE} and P_{TM} and degree of polarization , ρ, of a RW-InGaAsP laser as functions of current measured at various temperatures.

Measured polarization-resolved light power/current characteristics and corresponding curves of the degree of polarization are depicted for temperature values between T=20°C and 45°C. At 20°C the laser operates essentially in the TM mode while at T=45°C the emitted light is TE polarized above threshold. There is a relatively small temperature interval between T=38°C and T=44°C where polarization bistability is found within small ranges of the injection current corresponding to dominant TE- or TM- mode emission on the lower or upper hysteresis branch, respectively. From Fig.3 it can be seen that the hysteresis loop is shifted to higher values of I and, simultaneously, the hysteresis width is enhanced as temperature increases. The same behavior was found for other polarization-bistable lasers. Beyond a critical value of temperature (lying between 44°C and 45°C in the case of Fig.3) hysteresis and polarization bistability disappear. The hysteresis width changes abruptly from a finite value to zero at the critical temperature. The behavior of ρ below the lasing threshold indicates a much stronger contribution of TM-polarized light in the spontaneous emission than the corresponding contribution following from Fig. 2.

4. Discussion

While the polarization of spontaneously emitted light is isotropic ($\rho = 0$) in a stress-free laser TE- or TM-polarized spontaneous light emission is favoured for compressive ($\rho > 0$) or tensile ($\rho < 0$) stress in case of well below threshold operation. Thus, the value of ρ as determined at very small injection current serves as an indicator of stress in the active region, see /6,7/.

Since the material gain of the TM mode, g_{TM}, exceeds that of the TE mode, g_{TE}, in the presence of tensile stress /2/, the difference in threshold currents $\left(\Delta I_{th} \equiv I_{th}^{TM} - I_{th}^{TE} > 0 \right)$ for TM and TE lasing is smaller in this case compared to a stress-free laser. Provided that ΔI_{th} is sufficiently low an improvement in lateral TM waveguiding, as may be caused by a strong depletion of carriers around the centre of the ridge, will result in a higher modal TM gain giving rise to a switch from TE to TM emission (Fig.2). With increasing temperature a reduction of tensile stress occurs (by an amount of 0.1% ... 1% at a temperature change of $\Delta T = 1°C$, see /8/) corresponding to an increase in ρ which is accompanied by a decrease of g_{TM} and an increase in g_{TE}. A stronger increase in lateral TM waveguiding is required to compensate for the aforementioned decrease of the TM gain. This may explain the observed shift of the TE/TM transition to higher injection current when the temperature was increased, see Fig.3. Beyond a critical temperature TE/TM switching does not occur at all because g_{TM} is too small.

5. Summary

The degree of polarization, ρ, of the output of polarization-bistable RW-InGaAsP lasers has been measured as a function of injection current and temperature. The main results are :

- large negative values (-0.45 ... -0.25) for ρ have been obtained in case of well below threshold operation which indicates a considerable tensile stress in the active region.
- as temperature increases the polarization state of the laser for above threshold operation may change from dominant TM ($\rho < 0$) to dominant TE ($\rho > 0$) emission via an intermediate state of TE/TM bistability ($\rho \gtrless 0$) at a small temperature interval.

Further work is required to obtain quantitative relations between ρ, for well below threshold operation, and features of polarization bistability as width of hysteresis and switching current.

The work was partially supported by the "Deutsche Forschungsgemeinschaft".
References
/1/   C.S. Adams and D.T. Cassidy, J. Appl. Phys., 1988, 64, 6631
/2/   T.C. Chong and C.G. Fonstad, IEEE J. Quantum Electron., 1989, QE-25, 171
/3/   Y.C. Chen and J.M. Liu, Opt. Quantum Electron., 1987, 19, S93
/4/   A. Klehr, A. Bärwolff, R. Müller, M. Voß, J.Sacher, W. Elsässer and G.O. Göbel,
       Electron. Lett., 1991, 27, 1680
/5/   A.Klehr, A.Bärwolff, G.Berger, R,Müller and M.Voß, Nonlinear Dynamics in Optical Systems,
       Technical digest, (Optical Society of America), 1992, Vol 16, 266
/6/   D.T. Cassidy and C.S. Adams, IEEE J. Quantum Electron., 1989, QE-25, 1156
/7/   M.J.B. Boermans, S. H. Hagen, A.Valster, M.N. Finke, J.M.M. van der Heyden,
       Electron. Lett., 1990, 26, 1439
/8/   Y. C. Chen and J. M. Liu, Appl. Phys. Lett., 1984, 45, 731

Phase and Structural Variations in the Surface Layers of Fe-Al-C System Monocrystals After Repeated Laser Treatment with Surface Fusion

V. Andryushchenko, K. Nikolaychuk. Intitute of metal Physics, Academy of Sciences of Ukraine, Vernadsky str. 36, Kiev, Ukraine.

1. Introduction.

Impulse laser treatment with fusion of metals surface is characterized by increasing in the dimensions of the treating zone, where thermohardening and nonhomogenious metal structure arise. This structure containes at least three layers. The external layer of metal surface has a structure simular one to the quenched from liquid state. The thermotreated zone and transitional layer are placed deeper.

As the result of the laser treatment to the steel surface, containing many carbides, disolving and increasing of remained austenite content is possible and other structural peculiarity is also possible.

In present investigation the task was to investigate the influance of repeated laser effect with fusion of a surface on the surface structure in monocrystals of the quanched steel specimens of Fe-Al-C system by X-ray method.

The solid solution of this alloys consisted of austenite with f.c.c. crystal lattice (γ-phase), $Fe_{4-y}Al_yC_x$ carbide (K-phase), tetrahonale α-martensite and aluminium ferrite with b.c.c. crystal lattice in the wide temperature and aluminium content range [1]. During the high temperature (1000-1250°C) quenching into a water of Fe-Al-C alloys the submicrobulks with highest content of alloying elements and atomic order degree are formed in γ-phase f.c.c. crystal lattice along [100] crystalographic directions, due to aluminium and carbon atoms inclination to the ordering. F.c.c. crystal lattice of this submicrobulks have higher parameter of the simple cell than that of γ-phase matrix f.c.c. lattice [2]. Both f.c.c. lattices are coherent and have the same orientation. On the increasing of aluminium content in the alloy and under thermotreatment also the ordered bulks are increased, coherence between both f.c.c. γ- and K-phase lattices are broken and the regions with ordered f.c.c. lattice are transformed into K-phase particles (3) without variation in crystalographic orientation. On the temperature decreasing K-phase does not vary its orientation and does not transform into martensite up to the liquid helium temperature. Therefore this phase is an additional hindrance to the completion of f.c.c.-b.c.c. crystal lattice transformation. Thus a coherence and mutual orientation of γ- and K-phase f.c.c. lattices and carbon atoms inculcated into γ-lattice will be determine the parameter and orientationa correlation (OC) of the α-martensite with b.c.c. lattice, formed from austenite.

1. α-martensite, formed from the austenite with f.c.c. lattice which was coherent with f.c.c. K-phase lattice, has anomaly high value of simple cell parameters correlation c/a and Gringer-Troyano twinning orientation (24-24 OC). The cause of anomaly tetrahonality and α-martensite twinning lies in a presence of coherence strains between f.c.c. and b.c.c. lattices, of K- and α-phase accordingly;

2. α-martensite, formed from the austenite with f.c.c. lattice, that is not coherent with K-phase f.c.c. lattice, has a tetrahonality degree corresponding to the content of carbon, disolved in the austenite, and orientated in accordanse with the Nishiyama orientation (12 OC).

In the work (4) it was showed, that during low temperature annealing at the first stages of a structure modification of α-martensite with anomaly high tetrahonality the cohrence brokening between f.c.c. austenite and b.c.c. martensite lattices occurs at first moment , then first stage of α-martensite disintegration followes. It was possible to fix these two processes by X-ray method separately due to the chemical composition variations, annealing temperature and time regimes.

2. Materials and experiments.

Monocrystal specimens of 0.8mm diameter and 10mm length have been cut by electrosparking along the dendrite stripes after quenching from $1150^{\circ}C$ into a water. The sampels for the investigation were selected to coincide with [100] axis of f.c.c. austenite lattice. These monocrystals have been subjected to five time impulse (8ms) laser treatment at power of 5 and 10 Jowls per specimen butt along [100] axis of f.c.c. austenite crystal lattice (fig.1). Series of X-ray filmings have been made in X-ray rotate camera (RKV-86) with a collimator opening to 0.2mm. Monocrystal specimens have been placed in the camera perpendicularly to the X-ray bunch and replaced up and down with step 0.25mm. Thus, monocrystal structure have been studied along the length from treated butt surface.

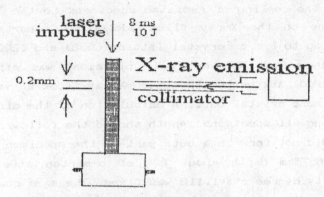

Fig.1 Experiment scheme.

3. Results and discussion.

The parent structure of the austenite specimens of Fe-4%wt. Al-2%wt. C alloy was f.c.c. crystal lattice (γ-phase) with submicro-bulks of K-phase formed along [100] direction. Their f.c.c. lattice was coherent with matrix γ-phase lattice. A disorientation of the bulks of coherent dissipation in the parent monocrystal was 2-3°C after quenching. X-ray films of rotation and swinging in range of 10-15° beside γ-phase (200) diffractional reflections have been calculated. The calculation showed, that f.c.c. monocrystal structure of specimens was preserved in generally during repeated laser treatment with butt surface fusion. However a bulks coherent dissipation disorientation was 8° beside the butt surface and decreased to 3° at the length of 2mm. Thereto a partial grain crushing (10%-15%) was observed at the surface layers. That was fixed on the X-ray films as the additional (200) γ-phase staines broaden along Debay rings. The austenite f.c.c. crystal lattice parameter after laser treatment to the specimen butt was equal to the parent parameter within experimental mistake (+0.0003nm) and did not vary along length of monocrystal. In order to obtain additional information about laser influence on γ-phase f.c.c. crystal lattice, the $\gamma \rightarrow \alpha$ transformation have been investigated. For Fe-4%Al-2%C alloy this transformation appeared during a cooling below -60°C and completed at the liquid nitrogen temperature. It was supposed that if α-martensite with anomaly tetrahonality would form during a cooling to liquid nitrogen temperature, then by this feature the conclusion about the coherence preservation between f.c.c. K-phase submicrobulks and γ-phase crystal lattice could be done.

Indeed, after the cooling of radiated specimens to the liquid nitrogen temperature on the X-ray films additional α-martensite staines corresponding to b.c.c. crystal lattice (002) and (200) reflactions have been arise. However their placement was differed considerably from (200) austenite staines on X-ray films at various distances from the mono crystal butt. A calculation of the diffraction al picture along all specimens length showed the following.

α-martensite did not form in a butt part of the specimen under 2mm. At the 0.25-0.75mm depth about 20% of α-marten site with anomaly tetrahonality degree c\a=1.112 was fixed. The same quantity of α-marten site was detected at the depth 0.75-1.0mm from the butt but with lower (c\a=1.091) tetrahonality degree. α-marten site had the lowest tetrahonality degree c\a=1.009 at 1.0-4.0mm. At the 4.0-4.5mm from the butt α-marten site with c\a=1.091 was appeared again and α-marten site with anomaly high tetrahonality degree c\a=1.113 was fixed deeper than 4.5mm.

The obtained data permited to make some conclusions about austenite f.c.c. crystal lattice structure of Fe-4%Al-2%C alloy after laser treatment.

At the first, surface layer (under 0.2mm) consisted of only γ-phase, which did not undergo γ→α transformation until liquid nitrogen temperature. Obviously it was caused by the fact that during the laser treatment a fusion layer was inriching by alloying elements, (the carbon generally), which remained in a solid solution of γ-phase during a high speed quenching from a liquid state, decreasing by thus the point of γ-α transformation below -196°C.

At the second, the formation of α-martensite with anomaly high tetrahonality degree in the surface layers 0.25-0.75 and deeper than 4.5mm indicates the presence of additional coherent straines between α-martensite b.c.c. lattice and K-phase submicrobulks f.c.c. lattice. Obviously, it is due to the presence of the coherence between γ- and K-phase f.c.c. crystal lattice prior the cooling.

At the third, the formation of α-martensite with c\a=1.091 at 0.75-1.0 and 4.0-4.5mm indicates about a carbon content only 2% and

the absence of additional coherent straines between b.c.c. and
f.c.c. α- and K-phase accordingly. Obviously it was caused by that,
that during laser treatment with surface fusion a coherence between
f.c.c. γ- and K-phase lattices was broken in these layers and so
after $\gamma \rightarrow \alpha$ transformation additional coherent straines were absent.

At the fourth, the formation of α-marten site with c\a=1.009
at 1.0-4.0mm from the butt is indicates about alloying element deficiency in this layer on the one hand or about the martensite disintegration completion during the impulse thermotreatment on the
other hand. Obviously, during the laser treatment the alloying elements (carbon generally) diffuse into the surface layer from the
depth (1.0-4.0mm), and during the following cooling $\gamma \rightarrow \alpha$ transformation happened in a poored by alloying element layer. In this layer
the coherence between f.c.c. γ- and K-phase lattice was broken during the alloying element diffusion.

Thus from above mentioned data one can make a conclusion about
the presence of 4.5mm zone with five structure layers, fixing by
X-ray, in Fe-4%AL-2%C monocrystal specimens, influenced by many time impulse laser along [100] crystalographic direction (Table 1.).
A five time impulse laser treatment did not lead to the austenite
monocrystal sructure variation deeper than 4.5mm.

Table 1. Structure and phase variatons along a specimens depth.

	Distance from the butt surface, mm	Austenite structure state after laser treatment	Phase composition after cooling to $-196°C$
1	0.2	γ-lattice f.c.c. crystal lattice inriched by C and Al	γ-phase
2	0.25-0.75	γ-phase and K-phase sub-microbulks f.c.c. lattices are coherent	$\gamma+\alpha_m$ with anomaly high tetrahonality
3	0.75-1.0	γ-phase and K-phase sub-microbulks f.c.c. lattices are not coherent	$\gamma+\alpha_m$

4	1.0-4.0	γ-phase f.c.c. lattice poored by C and Al	$\gamma+\alpha_m$ poored by carbon
5	4.0-4.5	γ-phase and K-phase sub-microbulks f.c.c. lattices are not coherent	$\gamma+\alpha_m$
6	4.5-10.0	γ-phase and K-phase sub-microbulks f.c.c. lattices are coherent	$\gamma+\alpha$ with anomaly high tetrahonality
7	0.25-1.0	After annealing 90 min at 270°C: γ-phase f.c.c. lattice poored by Al and C	$\gamma+\alpha_m$ poored by carbon
8	0.2	After repeated laser treatment: γ-phase f.c.c. lattice inriched by Al and C.	γ-phase
9	0.25-0.75	γ-phase and K-phase sub-microbulks f.c.c. lattices are coherent	$\gamma+\alpha_m$ with anomaly high tetrahonality

Literature references.

1. Nisida Keiso// Bull.Fac. Eng. -1968. -24. -p.71-108.

2. Tyapkin Y.D., Gulyayev A.A., Georgiyeva I.Y. // FMM. -1977. -v.43-6. - p.1297-1300.

3. Gulyayev A.A., Tyapkin Y.D. // MeTOM. -1982. -4. -p.9-11.

4. Lysak L.I., Drachinskaya A.G., Andryushchenko V.A. // Izvestiya AN SSSR, Metals. -1986. -N3. -p.135-136.

Superbroadband and Synchronized Multiline Oscillation of LiF:F$_2^-$ Color Center Laser

T.T.Basiev, S.B.Mirov, P.G.Zverev, V.V.Fedorfov, I.V.Kuznetsov*.

General Physics Institute, Russian Academy of Science,
Vavilov str., 38, Moscow, 117942 Russia

* "Alkor Technologies Inc.",
Stepan Rasin str., 8/50 St.Petersburg, 198035, Russia

Multifrequency and broadband lasers are often used nowdays in science and technology applications: in laser spectroscopy, investigation of different media, testing of solar cells and photomaterials, nanosecond color photography. Competition of amplification in laser active medium restrict the spectral range of simultaneous coexistence of different wavelengths of lasing. That's why the obtaining of laser oscillation in the whole amplification spectral region is an interesting physical problem. To solve this problem it is necessary to realize an independent oscillations of the certain parts of active media with the corresponding wavelengths. Authors [1] show the possibility to obtain superbroadband oscillation in the pulsed dye laser.

Recently, there have been the progress in the development of tunable LiF color center (CC) lasers [2]. Active color centers are intrinsic defects of crystal lattice. F$_2^-$ CC consists of two neighbouring anion vacancies that localize three electrons. F$_2^-$ CC are thermally stable at room temperature with multiyear stability. They can effectively absorb and emit photons with quantum yield about 60%. F$_2^-$ CC have wide absorption and luminescence bands with maximum at 1.12 μm, half-width 1400 cm^{-1}. They have high values of oscillation strength of electron-vibrational transitions, short radiation lifetime - 55 ns, sufficiently large Stocks shift of luminescence to obtain oscillations at 300 K according to four level scheme. The wavelength of radiation of widely spread neodimium lasers get into the absorption band of F$_2^-$ CC that allows to develop tunable in near IR from 1.1 to 1.25 μm high efficient lasers pumped by fundamental frequency of neodimium lasers.

In this paper we report about obtaining superbroadband laser oscillation in the whole amplification spectral region of F$_2^-$ CC in LiF crystals. Low density of active centers in CC crystals compared with the dye solution leads to a number of restrictions to the optical scheme of the broadband laser cavity. We use an original laser cavity that includs the grating (1200 lines per mm), focusing element and the mirror transparent for the pump

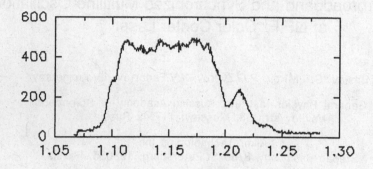

Fig.1. Superbroadband spectra of LiF:F_2^- color center laser with smooth spartial pump beam profile.

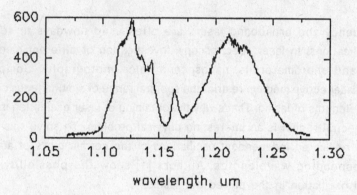

Fig.2. Superbroadband spectra of LiF:F_2^- color center laser with nonuniform distribution of pump laser.

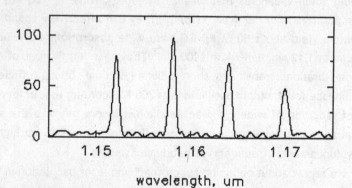

Fig.3. Spectral coding of Superbroadband spectra of LiF:F_2^- color center laser.

laser radiation and 100% reflectivity for oscillation wavelengths. The certain correlation of the distance between optical elements can realize the independent oscillation of the different parts of active medium with different wavelengths. Wide beam of the pump laser excites color centers in the large part of LiF active element. We use the first harmonic of nanosecond YAG:Nd laser with smooth spatial intensity profile. Pulse energy was as high as 25 mJ for 1.06 μm. Time duration in each pulse was about 8 ns. We obtain simultaneous oscillation of F_2^- CC at 1.1-1.24 μm with the efficiency 9-14%. We analyze laser spectrum of each pulse with polichromator and optical multichannel analyzer.

Fig.1 and 2 show two superbroadband spectra obtained in these experiments. The changes of laser energy versus wavelength is connected with the spatial distribution of the pump laser beam. Installation of the special mask or image controller into the pump beam results in synchronious multifrequency laser oscillation output with the special wavelength distribution or spectral coding (Fig.3). The linewidth of each line can be as narrow as 0.5 nm.

As there are a lot of interesting applications of such lasers in the visible range we tried to obtain second harmonic generation of broadband spectra in one nonlinear crystal. Authors [3,4] realized an optical scheme for doubling nonmonochromatic laser radiation with spectral width about 100 A by compensation of phase matching dispersion in nonlinear crystal with the grating. We develop the special focusing element for obtaining phase matching in one LiJO₃ nonlinear crystal for superbroadband radiation in near IR with spectral width more than 1000 A. We obtained simultaneous second harmonic generation of LiF:F_2^- superbroadband laser to the visible region from green to red, from 550 to 620 nm. Second harmonic generation conversion efficiency for the broadband laser radiation was about 10%.

As a result of this experimental study we design a superbroadband laser that can oscillate simultaneously in the near IR spectral region from 1.1 μm to 1.24 μm and in the visible from 550 nm to 620 nm with the possibility of simple and fast changing of the beam spectral profile according to user's requirements.

1. M.B.Danailov and I.P.Christov. Amplification of spatially dispersed ultrabroadband laser pulses. Optics Commun. vol.77, no.5,6, pp.394-401, 1990.

2. T.T.Basiev, S.B.Mirov, V.V.Osiko. Room temperature color center lasers. IEEE Journal of Quantum Electr., vol.24, no.6, pp.1052-1069, 1988.

3. V.D.Volosov and E.V.Goryachkina. Kvantovaya electronika, v.3, no.7, pp.1577-1583, 1976.

4. G.Szabo, Z.Bor. Appl. Phys. B, vol.50, p.51, 1990.

Narrowline High Efficient Tunable LiF:F$_2^+$ and LiF:F$_2^-$ Color Center Lasers for Near IR and Visible Spectral Regions

T.T.Basiev, V.V.Fedorov, S.B.Mirov, V.V.Ter-Mikirtichev, A.G.Papashvili, P.G.Zverev.

General Physics Institute, Russian Academy of Science,
Vavilov str., 38, Moscow, 117942 Russia

One of the main laser spectroscopy problems is creating of widely tunable powerful laser sources. Recently, there have been progress in the developing of color center lasers, that increase the possibility of laser spectroscopy in near IR spectral region [1]. One of them is LiF color center laser. Tunable range of LiF:F$_2^-$ laser is 1.09-1.25 μm and for LiF:F$_2^+$ laser is 0.84-1.1μm.

Investigation of the efficiency of LiF:F$_2^-$ color center laser was made in the nonselective cavity. An optimized neodimium pump laser based on YLiF$_4$:Nd^{3+} crystal worked at 1.047 μm wavelength. The pump beam was linearly polarized, 2 mm in diameter, time duration 30 ns and pulse energy 15 mJ. Optimization of color center crystal technology of preparation gives the increasing of the optical density and decreasing of the passive losses. The contrast of the crystals used in our experiments, that is Kactive/Kpassive was about 100. This results in the increase of the energetic efficiency from 30% up to 54%.

In the next experiment we investigate LiF:F$_2^-$ active element in the dispersive cavity. Laser cavity consists of the mirror with reflectivity more than 90% at 1.06 - 1.3 μm and the grating with 1200 lines per mm working in auto collimation scheme. Fig.1 shows the dependence of LiF:F$_2^-$ laser efficiency versus the laser wavelength. Curve (a) shows the dependence for the optimized pump source and active element. So after optimization of the pump source and laser crystal technology we can get increasing of the output efficiency from 20% to 38% and extending of the tuning spectral range from 1.09-1.25 μm up to 1.07-1.29 μm.

Narrow line tunable LiF:F$_2^-$ laser operation was investigated in the dispersive cavity that use a grazing incidence (Littman type) scheme with the second grating or a mirror as an end reflector for the first order diffraction from the first grating. Fig.1 shows the tuning curves for the pump pulse energy 45 mJ for this two schemes (b) and (c) respectively. The bandwidth of laser radiation was estimated with Fabry-Perot etalon and was less than 0,03 cm^{-1} in the first case (two gratings) and 0,05 cm^{-1} in the second (one grating and mirror).

Tunable Lasers based on LiF:F$_2^+$ CC crystals works in the nearby spectral region from 0.84 to 1.09 μm. F$_2^+$ CC have limited thermal stability at room temperature. Their half-decay time is about 12 hours. Earlier authors [2] proposed the technique of thermostabilization of thermally unstable F$_2^+$ CC by doping O$_2$ and OH$^-$ impurities and obtained oscillation using these impurity-vacancy centers at room temperature. However, besides the stabilizing effect there is some decline of the lasing characteristics and laser damage threshold of this crystals.

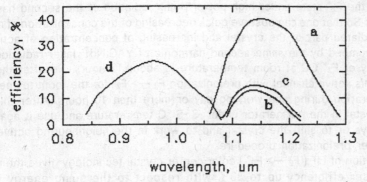

Fig.1. Tuning curves of LiF:F$_2^-$ (a,b,c) and LiF:F$_2^+$(d) lasers.

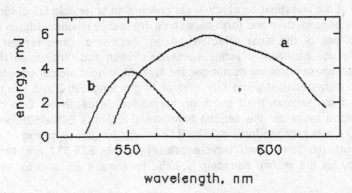

Fig.2. Second harmonic (a) and frequency mixing (b) of LiF:F$_2^-$ laser.

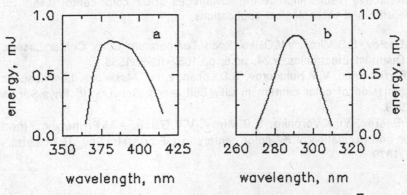

Fig.3. Third (a) and forth (b) harmonics of LiF:F$_2^-$ laser.

Authors [3] developed the original technique of photoionization of thermostable neutral F_2 CC to the F_2^+ state under high power pump radiation of the second harmonic of YAG:Nd^{3+} laser. So that one can observe quick decreasing of the concentration of neutral F_2 CC in the irradiated part of the crystal and increasing of concentration of active F_2^+ CC that will be pumped by the same second harmonic of YAG:Nd^{3+} laser radiation. The thermal unstability of F_2^+ CC at room temperature is about 12 hours. So after this it is possible to use this active channel with preionization $F_2 \longrightarrow F_2^+$ for the room temperature F_2^+ CC laser operation during one working day or more then 10 hours. After this it is possible keep crystal in the refrigerator under -3 -5 °C temperature and use it again and again for next days or to shift the crystal and to work in the neighbouring active zone of the crystal after preionization procedure.

The optimization of LiF:($F_2 \longrightarrow F_2^+$) color center crystal technology gives the record values of the laser efficiency up to 38% with respect to the pump energy in the nondispersive cavity and 24% pump efficiency in the dispersive cavity (Fig.1, d). The linewidth of tunable laser radiation in MALSAN dispersive cavity was about 1.5 cm^{-1} and divergency 0.5 mRad. The use of intracavity etalon reduce the linewidth less than 0.3cm^{-1}.

We also report about investigation of nonlinear conversion of tunable LiF color center laser radiation to the second, third and forth harmonics, frequency mixing. Lithium niobate crystal is known as one of the most effective second harmonic converter for LiF:F_2^- spectral region. We use nonlinear crystals specially chosen and cutted, so that our nonlinear element has phase matching conditions for 1.14 μm at right incident angle. Third and forth harmonics were obtained using KDP crystal 40 mm long. Fig.2 and 3 (a,b) show the tuning curves for the second, third and forth harmonics, respectively. One can see that the tuning spectral range for the second harmonic (Fig.2,a) is 545 - 625 nm, third harmonic - 360 - 420 nm and for the forth is 273 - 312 nm, for frequency mixing of tunable near IR radiation with the pump laser wavelength (Fig.2,b) is 538-577 nm. Maximum conversion efficiency for the second harmonic is 30%, for third is 8% and for the forth - 14%.

Lithium Iodate crystal is the most effective second harmonic doubler for LiF:F_2^+ color center laser region. We obtain second harmonic conversion to the spectral range from 420 to 540 nm with efficiency up to 30 % in the maximum of the tuning curve.

Thus the above results improve the advantages of LiF color center lasers for many modern scientific and technological applications.

1. T.T.Basiev, S.B.Mirov, V.V.Osiko. Room-Temperature Color Center Lasers, IEEE Journal of Quantum Electronics, v.24, no.6, pp.1052-1069, 1988.

2. I.A.Parfianovich, V.M.Hulugurov, B.D.Lobanov, N.T.Maximova. Luminescence and stimulated emission of color centers in LiF, Bull. Acad. Sci. USSR, Phys.Ser., vol.43, pp.20-27, 1979.

3. T.T.Basiev, Yu.K.Voronko, S.B.Mirov, V.V.Osiko, A.M.Prohorov. Kinetics of accumulation and oscillation of F_2^+ color centers in LiF crystals. Sov. JETF Lett., vol.30, pp.626-629, 1979.

New Generation of 2-12 μm Laser Radiation Photodetectors

J. Piotrowski, VIGO SENSOR SA, Warsaw, Poland.
H. Schmidt, DoroTek, Berlin, Germany.
Z. Drozdowicz, Oriel Corp. Stratford, CT. USA.
Z. Puzewicz, IEK-WAT, Warsaw, Poland.

Introduction

The ultimate signal-to-noise performance of the semiconductor photodetector is limited by the statistical fluctuations of the thermal generation and recombination rates in the volume of photodetector material [1-2]. High generation and recombination rates make room temperature IR photodetectors very noisy. Cooling is a very effective but impractical way of suppression of the thermal processes which makes cryogenically cooled detectors expensive and inconvenient in use.

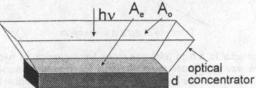

Schematic of photodetector with optical concentrator

To analyze the limitations of IR detectors performance and possible ways to overcome them without cryogenic cooling let us consider a simplified model of a photodetector as a slab of homogeneous semiconductor with thickness d. The device optical area A_0 is usually equal to its actual "electrical" area A_e, but may be different, for example by the use of an optical concentrator.

The generation-recombination limited detectivity is [3]

$$D^* = \frac{\lambda}{hc} \cdot (\frac{A_o}{A_e})^{1/2} \cdot \frac{\eta}{d^{1/2}} \cdot \frac{1}{[2(G+R)]^{1/2}}$$

where λ is the wavelength, h is the Planck constant, c is the light velocity, q is the electron charge, η is the quantum efficiency, G and R are the generation and recombination rates.

Problem: how to achieve perfect detection without cryocooling?
Minimizing G+R

The thermal generation/ recombination rates can be minimized by a careful selection of the semiconductor bandgap and doping profiles. Such a selection was achieved by using a newly developed variable gap Hg-Cd-Zn-Te epilayers with optimized composition and doping level profiles. The use of a lightly doped p-type material is preferable, since the Auger generation rate is smaller compared to that in n-type material.

Another possibility is to suppress thermal generation using stationary depletion of semiconductor [4]. One of possible ways was to apply the magnetoconcentration effect [5].

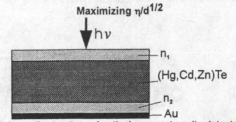

Schematic structure of optical resonant cavity detector

Since the detectivity is proportional to the factor $\eta/d^{1/2}$, a high quantum efficiency must be achieved in thin devices. This is usually difficult to realize, since the absorption of longwave radiation in uncooled narrow gap semiconductor is weak. The quantum efficiency can be increased in devices with non-reflective frontside and highly reflective backside surfaces of the sensitive element. The effective absorption can be increased further using interference effect to set up the resonant cavity [3], This can be achieved in a multilayer structures, shown in which the semiconductor flake is sandwiched between two dielectric layers and a back metal reflecting mirror. The optimized resonant cavity structure with carefully selected thicknesses and reflection coefficients of all layers exhibit detectivity enhanced by a factor of ~2.5 at peak wavelength compared to optimized conventional device.

The interference effects strongly modify spectral response of the device, however, and the optical cavity gain can be achieved only in narrow spectral regions.

Increasing A_o/A_e ratio

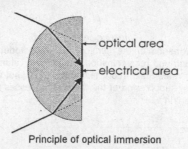

Principle of optical immersion

The optical immersion of detector to a high refraction index hemispherical or hyperhemispherical lens is an example of a very effective way to increase the A_o/A_e ratio [6]. Due to immersion, the linear size area of detector increases by a factor of n or n^2 for hemispherical and hyperhemispherical immersion, respectively. As a result, the detectivity can be increased by the same factor.

The problems arising from difficulties in mechanical matching of detector and lens material, transmission and reflection losses were solved by the use of monolithic technology developed at Vigo [7]. In this technology the immersion lens is formed directly in transparent CdZnTe substrate (n=2.7) of epitaxial HgCdTe layers used for the sensitive elements.

The optical immersion offers additional advantage of reduction of bias power dissipation by a large factors of n^2 and n^4 in case of hemi- and hyperhemispherical immersion, respectively. In the case of monolithic devices the factors are 7 and 49. The monolithic approach permits simple and economical manufacturing, the devices are rugged, mechanically stable and they can operate in a very broad spectral band with minimized reflection and absorption loses.

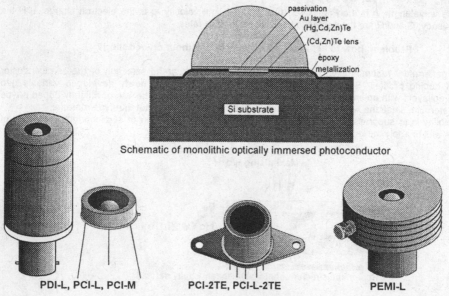

Schematic of monolithic optically immersed photoconductor

PDI-L, PCI-L, PCI-M   PCI-2TE, PCI-L-2TE   PEMI-L

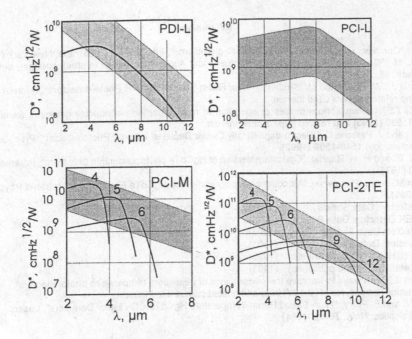

Spectral detectivities of immersed photodetectors [8]

Optically immersed photodetectors: uncooled Dember detectors PDI-L, photoconductors PCI-L, PCI-M, photoelectromagnetic detectors PEMI-L and 2-stage TE-cooled photoconductors PCI-L-TE, PCI-L-2TE.

The performance of the immersed devices have been highly improved [9] compared to that of conventional non-immersed devices [10] operated at the same conditions. It should be noted that the gain in performance due to optical immersion is achieved without sacrifice in response time. For example, the response time of the 10.6 μm photoconductors are shorter than 1nsec and 10 nsec for the uncooled and two-stage TE-cooled devices, respectively. Near BLIP performance has been achieved in the 3-5 μm regions using two-stage Peltier coolers. Four- or five stage TE coolers would be necessary for near BLIP 8-12 μm photoconductors. Optically immersed photovoltaic detectors have been also demonstrated [9].

Some gains due to optical immersion are specific for a particular type of photodetector [7-9]. For example, the reduction of bias power dissipation is important in case of photoconductors or for some types of photodiodes which require strong bias for the best performance. The immersion also reduces highly the heat load of heat sinks and cooling devices, making possible to reduce size, weight, price and power consumption of cooling systems.

In case of photodiodes and Dember effect detectors, which have a very low resistance-area product at near room temperature, immersion increases the device resistance by a large factor ($\approx$50). This is often necessary condition to achieve the device potential performance in practice. The reduction of capacitance by the same factor is of utmost importance for high frequency photodiodes which make possible to improve dramatically the high frequency performance of RC limited devices.

In the case of photoelectromagnetic detectors, the reduced size of active elements make possible to achieve a strong magnetic field with miniature permanent magnets.

SUMMARY
The performance of 2-14 μm near room temperature photodetectors have been highly improved by the use of variable gap semiconductor structures with optimized gap/doping profiles, -high quantum efficiency design, monolithic optical immersion. We believe, that further efforts would eventually make possible to achieve near-BLIP performance and fast response of 2-14 μm photodetectors without cryogenic cooling.

928

References

1. Rose A., "Concepts in photoconductivity and allied problems", Interscience Publishers, New York (1963).
2. White A. M., "Generation-Recombination Processes and Auger suppression in small bandgap detectors", J. Cryst. Growth, **86**, 840 (1988).
3. Piotrowski J., W. Galus and M. Grudzien, "Near Room-Temperature IR Photo-detectors", Infrared Phys., **31**, 1 (1991) and related papers cited therein.
4. Elliott C. T., "Non-equilibrium modes of operation of narrow gap semiconductor devices", Semicond. Sci. Technol., **5**, 530 (1990) and related papers cited therein.
5. Djuric Z., and J. Piotrowski, "Electromagnetically Carrier Depleted Infrared Photodetector". Proc. SPIE **1540** Infrared Technology, 1540-1569 (1992).
6. Slavek J. E. and H. H. Randal, "Optical immersion of HgCdTe photoconductive detectors", Infrared Phys., **15** 339-340 (1975).
7. Grudzien M. and J. Piotrowski, "Monolithic optically immersed HgCdTe IR detectors", Infrared Phys., **29**, 251-253 (1989).
8. VIGO Detectors Data Sheets   (1993).
   DOROTEK Detectors Data Sheets   (1993).
   BSA Detectors Data Sheets   (1993).
   BEC Detectors Data Sheets   (1993).
   EDINBURGH INSTR. Data Sheets   (1993).
   ORIEL Catalogue, (to be published). (1993).
9. Piotrowski J., "New ways to improve the performance of near-room temperature photodetectors", Opto-Electronics Review, No. 1, 9-12 (1992) and related papers cited therein.
10. Galus W. and F. Perry, "High-Speed Room Temperature HgCdTe CO_2 Laser Detectors", Laser Focus/Electrooptics, Nov., 76-82)1984).

Printed in the United States
By Bookmasters